Handbook of Coriander (*Coriandrum sativum*)

Coriander (*Coriandrum sativum* L., family Umbelliferae/Apiaceae) is one of the most popular spices globally. Different parts of *Coriandrum sativum* are edible and widely used as a seasoning due to their unique flavor and aroma. *Coriandrum sativum* medical uses have been recognized since ancient times. Coriander leaves (cilantro) and coriander fruit (seed) are used in different ethnic foodstuffs, meat and poultry dishes, soup, pudding, bread, and seafood dishes. *Coriandrum sativum* is rich in linalool, vitamin A, vitamin B12, vitamin C, folate, and phenolics. *Coriandrum sativum* fixed oil is rich in sterols, tocols, and bioactive phytochemicals. Petroselinic acid is the major fatty acid in *Coriandrum sativum* fixed oil and exhibits health-promoting traits.

Coriandrum sativum is recommended as a food preservative to replace synthetic antioxidants because of its antioxidant and antibacterial traits. Furthermore, *Coriandrum sativum* cilantro and seeds are rich in water-soluble and lipid-soluble phytochemicals with unique anticancer, anxiolytic, neuroprotective, migraine-relieving, hypoglycemic, hypolipidemic, anticonvulsant, analgesic, and anti-inflammatory traits. Those medical benefits and their integration into daily life render *Coriandrum sativum* an excellent functional food. Regarding the cosmetic industry, *Coriandrum sativum* is used as an ingredient in conventional Ayurvedic cosmetic formulations to normalize skin color. In addition, *Coriandrum sativum* volatile oil finds use as an ingredient in perfumes.

*Handbook of Coriander (***Coriandrum sativum***): Chemistry, Functionality, and Applications* is a valuable resource for pharmaceutical and nutraceutical developers, as well as novel food developers and R&D researchers in a variety of fields that use herbs, spices, and medicinal plants.

Key Features:

- Explores the chemistry of *Coriandrum sativum* phytochemicals, oils, and extracts
- Discusses *Coriandrum sativum* active constituents and their health-enhancing traits
- Presents the applications of *Coriandrum sativum* phytochemicals, oils, and extracts
- Addresses the growing application areas, including horticulture, functional food, clinical nutrition, pharmaceuticals, and cosmetics

Authored by international scientists and industry experts, this book is a great resource for food chemistry, clinical nutrition, biochemistry, pharmacology, and horticulture researchers and students, as well as developers of novel food, cosmetics, and pharmaceuticals, in addition to R&D researchers in different sectors that utilize herbs, spices, and medical plants.

Handbook of Coriander (*Coriandrum sativum*)

Chemistry, Functionality, and Applications

Edited by
Mohamed Fawzy Ramadan

CRC Press
Taylor & Francis Group
Boca Raton London New York

CRC Press is an imprint of the
Taylor & Francis Group, an **informa** business

MATLAB® and Simulink® are trademarks of the MathWorks, Inc. and are used with permission. The MathWorks does not warrant the accuracy of the text or exercises in this book. This book's use or discussion of MATLAB® and Simulink® software or related products does not constitute endorsement or sponsorship by the MathWorks of a particular pedagogical approach or particular use of the MATLAB® and Simulink® software.

First edition published 2023
by CRC Press
6000 Broken Sound Parkway NW, Suite 300, Boca Raton, FL 33487-2742

and by CRC Press
4 Park Square, Milton Park, Abingdon, Oxon, OX14 4RN
CRC Press is an imprint of Taylor & Francis Group, LLC

© 2023 Mohamed Fawzy Ramadan; individual chapters, the contributors

Library of Congress Cataloging-in-Publication Data

Names: Ramadan, Mohamed Fawzy, editor.
Title: Handbook of coriander (Coriandrum sativum) : chemistry, functionality, and applications / edited by Mohamed Fawzy Ramadan.
Description: First edition. | Boca Raton : CRC Press, 2023. | Includes bibliographical references and index.
Identifiers: LCCN 2022029000 (print) | LCCN 2022029001 (ebook) | ISBN 9781032068527 (hardback) | ISBN 9781032069333 (paperback) | ISBN 9781003204626 (ebook)
Subjects: LCSH: Coriander. | Coriander--Seeds. | Coriander--Composition. | Coriander--Therapeutic use. | Coriander--Industrial applications.
Classification: LCC SB351.C69 H36 2023 (print) | LCC SB351.C69 (ebook) | DDC 583/.9882--dc23/eng/20220822
LC record available at https://lccn.loc.gov/2022029000
LC ebook record available at https://lccn.loc.gov/2022029001

ISBN: 9781032068527 (hbk)
ISBN: 9781032069333 (pbk)
ISBN: 9781003204626 (ebk)

DOI: 10.1201/9781003204626

Typeset in Times
by Deanta Global Publishing Services, Chennai, India

Dedicated to the soul of my father, Professor Fawzy Ramadan Hassanien, and my beloved family.

Contents

SECTION I   Coriander: Cultivation, Composition, and Applications

SECTION II Coriander Leaves: Chemistry, Technology, Functionality, and Applications

SECTION III Coriander Fixed Oil: Chemistry, Technology, Functionality, and Applications

SECTION IV   Coriander Essential Oil: Chemistry, Technology, Functionality, and Applications

SECTION V   Coriander Extracts: Chemistry, Technology, Functionality, and Applications

Preface

Spices, herbs, and medicinal plants that have been shown to have health-promoting properties are fascinating. *Coriandrum sativum* leaves, seeds, and roots are rich in volatile oils, fixed oils, fatty acids, tocols, sterols, and carotenoids. Their yields and compositions are influenced by ecotype, variety, genotype, planting season and condition, plant part, growth stage, harvesting time, and extracting process. The health-promoting effects of *C. sativum* include its protective traits against cancer, neurodegenerative diseases, and metabolic syndrome.

Regarding the several published contributions on the functional, nutritional, and pharmacological effects of *Coriandrum sativum*, this handbook tries to create multidisciplinary discussions of the chemical profile and biological potential, as well as food and non-food uses of *C. sativum*, *C. sativum* oils, *C. sativum* bioactive compounds, and *C. sativum* extracts. The book also explores the proper uses of *Coriandrum sativum* in developing nutraceuticals, pharmaceuticals, novel food, and drugs.

The book contains chapters within various sections:

Section I. Coriander: Cultivation, Composition, and Applications
Section II. Coriander Leaves: Chemistry, Technology, Functionality, and Applications
Section III. Coriander Fixed Oil: Chemistry, Technology, Functionality, and Applications
Section IV. Coriander Essential Oil: Chemistry, Technology, Functionality, and Applications
Section V. Coriander Extracts: Chemistry, Technology, Functionality, and Applications

Intending to provide a comprehensive contribution to the scientific community involved in food science, horticulture, clinical nutrition, health, and pharmacology, this book comprehensively reviews the aspects that led to the recent advances in *C. sativum* biochemistry, production, and functionality. The editor hopes that the handbook will be a rich source for researchers and developers in related disciplines.

The editor sincerely thanks all contributors for their valuable contributions and their cooperation. In addition, the help and support of the Taylor & Francis staff, especially Stephen Zollo and Laura Piedrahita, were essential for completing the project and are highly appreciated.

Prof. Mohamed Fawzy Ramadan
Makkah, Saudi Arabia

MATLAB® is a registered trademark of The MathWorks, Inc. For product information, please contact:

The MathWorks, Inc.
3 Apple Hill Drive
Natick, MA 01760-2098 USA
Tel: 508 647 7000
Fax: 508-647-7001
E-mail: info@mathworks.com
Web: www.mathworks.com

About the Editor

Mohamed Fawzy Ramadan is a Professor of Food Chemistry and Biochemistry at the Department of Clinical Nutrition, Faculty of Applied Medical Sciences Umm Al-Qura University, Makkah, Saudi Arabia. Since 2014, Prof. Ramadan has been a Professor in the Agricultural Biochemistry Department, Faculty of Agriculture at Zagazig University, Egypt.

Prof. Ramadan earned his PhD (*Dr. rer. Nat.*) in Food Chemistry from the Berlin University of Technology (Germany, 2004). He conducted post-doctoral research at ranked universities such the University of Helsinki (Finland), the Max-Rubner Institute (Germany), Berlin University of Technology (Germany), and the University of Maryland (USA). In 2010, he was appointed Visiting Professor (100% research) at King Saud University. In 2012, he was appointed Visiting Professor (100% teaching) at the School of Biomedicine, Far Eastern Federal University, in Vladivostok, Russian Federation.

Prof. Ramadan has published more than 300 research papers and reviews in international peer-reviewed journals. He has also edited and published several books and book chapters (with a Scopus *h*-index of 44 and more than 6000 citations). In addition, he has been an invited speaker at several international conferences. Since 2003, Prof. Ramadan has been a reviewer and editor of several highly cited international journals such as the *Journal of Medicinal Food* and *Journal of Advanced Research*.

Prof. Ramadan has received several prizes, including the Abdul Hamid Shoman Prize for Arab Researcher in Agricultural Sciences (2006), Egyptian State Prize for Encouragement in Agricultural Sciences (2009), European Young Lipid Scientist Award (2009), AU-TWAS Young Scientist National Awards (Egypt) in Basic Sciences, Technology and Innovation (2012), TWAS-ARO Young Arab Scientist (YAS) Prize in Scientific and Technological Achievement (2013), and Atta-ur-Rahman Prize in Chemistry (2014).

List of Contributors

Khalid Abbas
Government College University
Faisalabad, Punjab, Pakistan

Hamza Ben Abdallah
Institut National de recherché et d'Analyse
    Physico-chimique (INRAP)
Ariana, Tunisia

Muhammad Faizan Afzal
Government College University
Faisalabad, Punjab, Pakistan

Muhammad Haseeb Ahmad
Government College University
Faisalabad, Punjab, Pakistan

Mushtaq Ahmad
Quaid-i-Azam University
Islamabad, Pakistan

Rabia Shabir Ahmad
Government College University
Faisalabad, Punjab, Pakistan

Sidra Nisar Ahmed
The Women University Multan
Pakistan
and
Quaid-i-Azam University
Islamabad, Pakistan

Sara Aimen
The Women University Multan
Pakistan

Haseeb Anwar
Government College University
Faisalabad, Punjab, Pakistan

Muhammad Sajid Arshad
Government College University
Faisalabad, Pakistan

Hamide Filiz Ayyildiz
Selcuk University
Konya, Türkiye

Stéphane Ballas
Ovalie Innovation
Auch, France

Aamna Balouch
University of Sindh
Jamshoro, Pakistan

Aiman K. H. Bashir
Cape Peninsula University of Technology
Cape Town, South Africa

Erman Beyzi
Erciyes University
Kayseri, Turkey

Ayesha Bibi
Government College University
Faisalabad, Pakistan

Stevan Blagojević
Institute of General and Physical Chemistry
Belgrade, Serbia

Mohammad Hossein Boskabady
Mashhad University of Medical Sciences
Mashhad, Iran

Ahmad Cheikhyoussef
University of Namibia
Windhoek, Namibia

Natascha Cheikhyoussef
Ministry of Higher Education, Technology, and
    Innovation
Windhoek, Namibia

Monika Choudhary
Punjab Agricultural University
Ludhiana, Punjab, India

Saša Đurović
Institute of General and Physical Chemistry
Belgrade, Serbia
and
Peter the Great Saint-Petersburg Polytechnic
  University
Graduate School of Biotechnology and Food
  Industries
Saint-Petersburg, Russia

Mariam I. Gamal El-Din
Ain Shams University
Abbassia, Cairo, Egypt

Mohamed Helmy El-Morsy
Umm Al-Qura University
Makkah Al Mukarramah, Saudi Arabia
and
Desert Research Center
Mataryia, Cairo, Egypt

Philippe Evon
Université de Toulouse
Toulouse Cedex 4, France

Felicitas M. Fwanyanga
University of Namibia
Windhoek, Namibia

Zahra Ghorbanzadeh
Agricultural Biotechnology Research Institute
  of Iran (ABRII)
Karaj, Iran

Kok Ming Goh
UCSI University
Kuala Lumpur, Malaysia

Saba Ghufran
Xinjiang University
Urumqi, China

Adem Gunes
Erciyes University
Kayseri, Turkey

Zinar Pinar Gumus
Ege University
Izmir, Turkey

Rasmieh Hamid
Cotton Breeding Department
Cotton Research Institute of Iran (CRII)
Gorgan, Iran

Imam Hasan
Bangladesh Agricultural University
Mymensingh, Bangladesh

Bazla Naseer Hashmi
Government College University
Faisalabad, Pakistan

Qamar Ul Hassan
University of Agriculture
Faisalabad, Pakistan

Lydia Ndinelao Horn
University of Namibia
Windhoek, Namibia

Karim Hosni
Institut National de recherché et d'Analyse
  Physico-chimique (INRAP)
Ariana, Tunisia

Mahmoud Hosseini
Mashhad University of Medical Sciences
Mashhad, Iran

Ghulam Hussain
Government College University
Faisalabad, Punjab, Pakistan

Ahmed A. Hussein
Cape Peninsula University of
  Technology
Bellville, ZA

Ali Imran
Government College University
Faisalabad, Punjab, Pakistan

Muhammad Imran
Government College University
Faisalabad, Punjab, Pakistan

Mohammad Rafiqul Islam
Bangladesh Agricultural University
Mymensingh, Bangladesh

Umaima Ismail
University of Sindh
Jamshoro, Pakistan

Feba Jacob
Kerala Agricultural University
Thrissur, India

Muhammad Saqaf Jagirnai
University of Sindh
Jamshoro, Pakistan

Nazish Jahan
University of Agriculture
Faisalabad, Pakistan

Md. Anwar Jahid
Patuakhali Science and Technology University
Barishal, Bangladesh

Zubair Javed
Government College University
Faisalabad, Punjab, Pakistan

Miroslava Kačániová
Slovak University of Agriculture
Nitra, Slovakia

Allah Nawaz Khan
Chinese Academy of Sciences
Xiangshan, Beijing, China

Muhammad Kamran Khan
Government College University
Faisalabad, Punjab, Pakistan

Huseyin Kara
Selcuk University
Konya, Türkiye

Kevser Karaman
Erciyes University
Kayseri, Turkey

Amarjeet Kaur
Punjab Agricultural University
Ludhiana, Punjab

Onur Ketenoglu
Eskisehir Osmangazi University
Eskisehir, Turkey

Abdul Rauf Khaskheli
Shaheed Mohtarma Benazir Bhutto Medical
    University
Larkana, Pakistan

Mohammad Reza Khazdair
Mashhad University of Medical Sciences
Mashhad, Iran
and
Birjand University of Medical Sciences
Birjand, Iran

Mustafa Kiralan
Balikesir University
Balikesir, Turkey

Abdul Hameed Kori
University of Sindh
Jamshoro, Pakistan

Anant Kumar
Vellore Institute of Technology
Tamil Nadu, India

Laurent Labonne
Université de Toulouse
Toulouse Cedex 4, France

Zahid Hussain Laghari
University of Sindh
Jamshoro, Pakistan

Li Ann Lew
UCSI University
Kuala Lumpur, Malaysia

Jacek Łyczko
Wrocław University of Environmental and Life
    Sciences
Wrocław, Poland

Gayathri Mahalingam
Vellore Institute of Technology
Tamil Nadu, India

Sarfaraz Ahmed Mahesar
University of Sindh
Jamshoro, Pakistan

Liza Malik
Government College University
Faisalabad, Punjab, Pakistan

Othmane Merah
Université de Toulouse
Toulouse Cedex 4, France

Yassine M'Rabet
Institut National de recherché et d'Analyse
    Physico-chimique (INRAP)
Ariana, Tunisia

Muhammad Mohsin
University of Agriculture
Faisalabad, Pakistan

Asma Mukhtar
Government College University
Faisalabad, Pakistan

Eduardo P. Mulima
University of Púnguè
Manica, Mozambique

Humaira Muzaffar
Government College University
Faisalabad, Punjab, Pakistan

Muhammad Nadeem
University of Veterinary and Animal Sciences
Lahore, Pakistan

Soha Navaid
Government College University
Faisalabad, Punjab, Pakistan

Laaraib Nawaz
Government College University
Faisalabad, Punjab, Pakistan

Muhammad Noman
Government College University
Faisalabad, Punjab, Pakistan

Kar Lin Nyam
UCSI University
Kuala Lumpur, Malaysia

Hanan El-Sayed Osman
Umm Al-Qura University
Makkah Al Mukarramah, Saudi Arabia
and
Botany and Microbiology Department
Faculty of Science
Al-Azhar University
Cairo, Egypt

Aijaz Ahmed Otho
University of Sindh
Jamshoro, Pakistan

Muhammad Abdul Rahim
Government College University
Faisalabad, Punjab, Pakistan

Mohamed Fawzy Ramadan
Umm Al-Qura University
Makkah, Saudi Arabia

Neelum Rashid
Mirpur University of Science and Technology
Mirpur, Azad Kashmir

Muhammad Rashid
Government College University
Faisalabad, Punjab, Pakistan

Sofia Rashid
Quaid-i-Azam University
Islamabad, Pakistan
and
Comsats University
Islamabad, Pakistan

Nakul Ravishankar
Vellore Institute of Technology
Tamil Nadu, India

Vijayasarathy S.
Vellore Institute of Technology
Vellore, Tamil Nadu, India

Idowu Jonas Sagbo
Cape Peninsula University of Technology
Cape Town, South Africa

Swati Sahoo
Epigeneres Biotech Pvt. Ltd.,
Lower Parel, Mumbai, India

Sana Saleem
Government College University
Faisalabad, Punjab, Pakistan

Abdul Sami
Government College University
Faisalabad, Punjab, Pakistan

Saira Sattar
University of Okara
Okara, Pakistan

Syed Tufail Hussain Sherazi
University of Sindh
Jamshoro, Pakistan

Hadia Shoaib
University of Sindh
Jamshoro, Pakistan

Mairton Gomes da Silva
Federal University of Recôncavo of Bahia
Bahia, Brazil

Valérie Simon
Université de Toulouse
Toulouse Cedex 4, France

Arashdeep Singh
Punjab Agricultural University
Ludhiana, Punjab, India

Sibel Turan Sirke
Erciyes University
Kayseri, Turkey

Brijesh Sukumaran
Sunandan Divatia School of Science
NMIMS (Deemed-to-be) University
Mumbai, India

Thierry Talou
Université de Toulouse
Toulouse Cedex 4, France

Chin Xuan Tan
Universiti Tunku Abdul Rahman
Jalan Universiti Bandar Barat
Kampar Perak, Malaysia

Seok Shin Tan
Monash University Malaysia
Bandar Sunway
Selangor, Malaysia

Seok Tyug Tan
Management and Science University
Shah Alam
Selangor, Malaysia

Rukam Singh Tomar
Junagadh Agricultural University
Junagadh Gujarat, India

Abner Tomas
Ndaka Mushrooms and Processing
Oniipa, Namibia

Zeliha Ustun-Argon
Necmettin Erbakan University
Konya, Turkey

Thierry Véronèse
Ovalie Innovation
Auch, France

Ghulam Yaseen
Quaid-i-Azam University
Islamabad, Pakistan

Cigdem Yengin
Ege University
Izmir, Turkey

Fadia S. Youssef
Ain Shams University
Abbassia, Cairo, Egypt

Mohammad Zafar
Quaid-i-Azam University
Islamabad, Pakistan

Mariem Ziedi
Institut National de recherché et d'Analyse
    Physico-chimique (INRAP)
Ariana, Tunisia

1 Introduction to Handbook of Coriander (*Coriandrum sativum*) *Chemistry, Functionality, and Applications*

Mohamed Fawzy Ramadan

CONTENTS

1.1 UNITED NATIONS SUSTAINABLE DEVELOPMENT GOALS AND HEALTH-ENHANCING PLANTS

The United Nations Sustainable Development Goals (UNSDGs, https://sustainabledevelopment.un.org) comprise a vision of a peaceful, fairer, and sustainable world. "*Good Health and Well-Being*" is the third UNSDG (https://sdgs.un.org/goals/goal3), which aims to enhance health using health-enhancing plants and environmentally friendly methodologies in the food industry (Ramadan, 2021).

New plant-based products with functional traits could be designed to enhance human health. Current innovations will affect the way we eat in the future (McClements, 2019). Novel nutraceuticals and pharmaceuticals have been developed from spices, herbs, and medicinal plants. Globally, the World Health Organization (WHO) mentioned that approximately 80% of the global population depends upon traditional medicine. WHO highlighted the importance of exploring medicinal plants for healthcare benefits (i.e., safety, efficacy, quality control, quality assurance, dosage, clinical trials, toxicity, drug interaction, and therapeutic uses). With the developments in clinical nutrition, there is tremendous interest in aromatic plants as phytoconstituent-rich sources for nutraceuticals, novel foods, and drugs. Interest in plant-based active constituents and oils has increased due to their health-promoting roles (Kiralan et al., 2014; Ramadan, 2021; Elimam et al., 2022).

1.2 CORIANDER (*CORIANDRUM SATIVUM*): CHEMISTRY, FUNCTIONALITY, AND APPLICATIONS

The consumption of spices, medicinal plants, and herbs rich in health-enhancing phytoconstituents might expand consumers' life span. The abundant manifestation of bioactivity of phytoconstituents

DOI: 10.1201/9781003204626-1

1

in medicinal plants, herbs, and spices makes them natural active compounds (Ramadan and Moersel 2003; Singh et al., 2006; Gantait et al., 2022).

Coriandrum sativum Linn. (coriander, family Umbelliferae or Apiaceae) is one of the earliest used spices (Yousuf et al. 2014; Moniruzzaman et al. 2014; Meena et al. 2014; Gantait et al. 2022) having beneficial medicinal impacts (Rajeshwari and Andallu 2015; Gantait et al. 2022; Sobhani et al. 2022). *Coriandrum sativum* originated in the eastern Mediterranean and is grown in Europe, Africa, and Asia. Different parts of *C. sativum* are edible and utilized as a seasoning due to their unique flavor. The medical uses of *Coriandrum sativum* have been recognized since ancient times. Coriander leaves (cilantro) and coriander fruit (seed) are used in curry meat dishes, puddings, bread, soups, poultry and seafood dishes, and various ethnic foodstuffs. *C. sativum* contains high levels of vitamin B12, folate, vitamin C, vitamin A, and phenolics. Besides, *C. sativum* is considered an alternative food preservative due to its antioxidant and antimicrobial potential (Moniruzzaman et al., 2014; Meena et al., 2014; Gantait et al., 2022).

C. sativum cilantro, roots, and seeds contain bioactive phytochemicals (i.e., gallic acid, thymol, and bornyl acetate) that exhibit unique neuroprotective, anticancer, anxiolytic, migraine-relieving, hypolipidemic, anticonvulsant, analgesic, hypoglycemic, and anti-inflammatory effects. Linalool is the main bioactive constituent responsible for several coriander therapeutic traits. Other *C. sativum* active constituents are volatile oil, fatty acids, tocols, sterols, and carotenoids, wherein their yields and compositions are affected by variety, genotype, ecotype, planting season and condition, plant part, growth stage, harvesting time, and extracting process (Ramadan and Moersel, 2002; 2003; 2004; Ramadan, Kroh, Moersel, 2003; Ramadan, Amer, and Awad, 2008). Meanwhile, *C. sativum* essential oils, fixed oils, extracts, water-soluble compounds, and phenolics exist in aerial parts and seeds (Meena et al., 2014; Gantait et al., 2022). Besides, *C. sativum* volatile oil ranks second highest in the global annual production.

According to WHO (2019), cardiovascular diseases are the leading cause of mortality globally. *C. sativum* phytochemicals have high potential in cardiovascular health and have exhibited cardio-protective, antihyperlipidemic, cardiometabolic disorder-inhibiting traits, and angiotensin-converting enzyme-inhibiting effects. On the other hand, due to the COVID-19 pandemic, the conventional Ayurvedics system showed an impact compared to modern medicine, and they have the advantage of being cost-effective with lesser side effects (Gidwani et al., 2022).

The antioxidant potentials of *C. sativum* provide a key mechanism in its health-promoting effects against cancer, neurodegenerative diseases, and metabolic syndrome. These therapeutic effects and their integration into daily life render *C. sativum* a promising novel food. *C. sativum* has been conventionally utilized as digestive and appetite stimulants, and diuretic, lipid-lowering, glucose-lowering, and antimicrobial agents. It has also been used for treating digestive disorders, central nervous system diseases, and airway disorders (Gantait et al., 2022; Sobhani et al., 2022).

C. sativum is utilized as an ingredient in conventional Ayurvedic cosmetic formulations to normalize skin color in the cosmetic industry. *C. sativum* volatile oil as an ingredient in cosmetics and perfume was spotlighted in 2000 bce. Moreover, *C. sativum* fixed oil is rich in sterols, tocols, and other bioactive phytochemicals. Petroselinic acid is the main fatty acid in coriander crude oil and exhibits several biological and health-promoting traits (Mahleyuddin et al., 2022).

1.3 CORIANDER (*CORIANDRUM SATIVUM*) MARKET

The popularity of healthy cuisines is linked to the increased demand for *Coriandrum sativum* worldwide. *C. sativum* has an outstanding international market because of the increase in the world population with its consumption requirements. FAOSTAT (https://www.fao.org/faostat) reported that, in 2020, the total yield of anise, badian, fennel, and coriander reached 11,362 hectograms per hectare (hg/ha). Arizio and Curioni (2011) reported that the *Coriandrum sativum* importing countries (*ca.* 63% of world imports) are led by Malaysia, Sri Lanka, the United Kingdom, the USA, and Japan.

Europe accounts for *ca.* 15.0% of global *Coriandrum sativum* seed imports. In Europe, the UK is the highest *Coriandrum sativum* seed importer. However, opportunities for other suppliers could exist in other growing or large markets, including the Netherlands, Poland, Germany, and France.

Coriander seeds are the round-shaped, brown-colored, dried fruit of *C. sativum*. *Coriandrum sativum* farmers remove the fruit seeds from the stem with threshing machines or combine harvesters, or beat them with sticks. '*Microcarpum*' and '*Macrocarpum*' are the major *Coriandrum sativum* varieties utilized for seed production. European *C. sativum* seed imports are likely to increase at a *ca.* 2% annual growth rate. Consumption growth and imports are anticipated to be driven by healthy eating trends, high usage of *C. sativum* seeds as an ingredient in food products, the volatile oil industry, and the increasing interest in healthy cuisines. Between 2015 and 2018, European *C. sativum* seed imports increased to *ca.* 24,000 tonnes worth €26 million. Bulgaria is the leading EU producer of *C. sativum*, where high volumes are used in the volatile oil industry. The second-highest European producing country is Spain, wherein the production targets spice production more. After Bulgaria, Italy is considered the second-highest European exporter and producer of *C. sativum* seeds; Italy focuses on producing *C. sativum* seeds for sowing (https://www.cbi.eu/market-information/spices-herbs/coriander-seeds/market-potential).

1.4 CORIANDER (*CORIANDRUM SATIVUM*) IN THE INTERNATIONAL SCIENTIFIC LITERATURE

Coriandrum sativum is highly attractive for international research. Hundreds of contributions were published on *Coriandrum sativum*. A search with the keyword '*Coriandrum sativum*' in PubMed (March 2022) resulted in 593 published contributions belonging to *Coriandrum sativum* production, cultivation, and the bioactivity of phyto-extracts, amino acids, seed oil, fatty acids, active compounds, and industrial uses.

A careful search for *Coriandrum sativum* in Scopus (www.scopus.com) showed that the number of documents published on *Coriandrum sativum* is exceptionally high (*approx.* 2200 till March 2022). Of the published contributions, *ca.* 1970 were research contributions, 70 conference articles, 100 review contributions, and 17 book chapters. The contributions counts on *Coriandrum sativum* from 2000 to 2020 are presented in Figure 1.1. The contributions annually published on *Coriandrum sativum* have increased from 23 articles in 2000 to 145 articles in 2020. In the scientific community, these indicators reflect the interest and importance of *Coriandrum sativum* as a research topic. Between 2000 and 2020, Figure 1.2 represents the distribution of document types

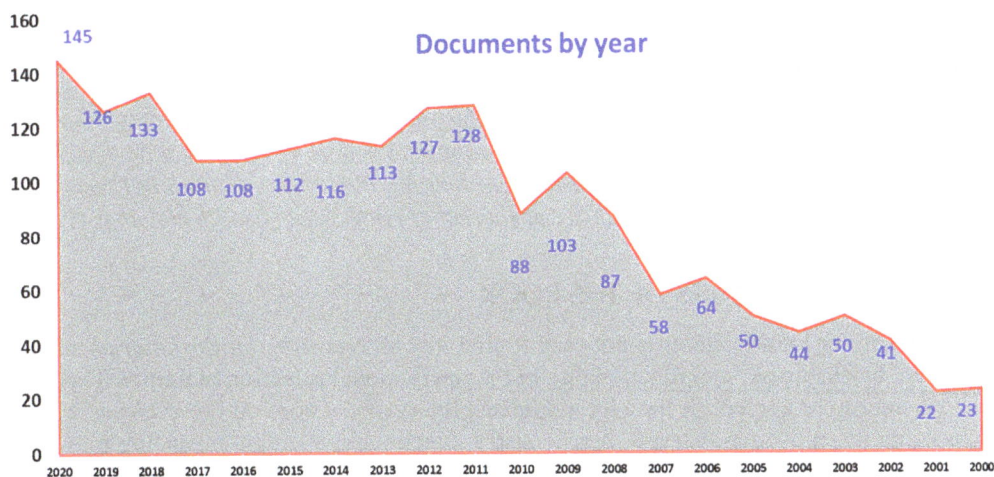

FIGURE 1.1 Scholarly output on *Coriandrum sativum* from 2000 to 2020 (www.scopus.com).

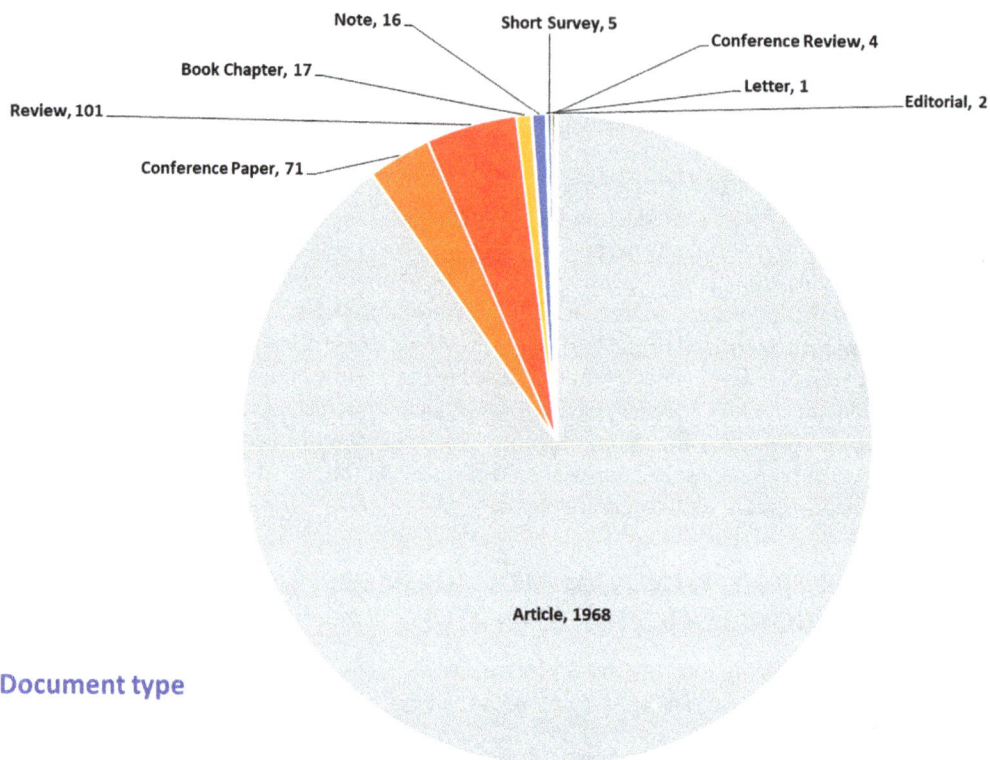

FIGURE 1.2   Distribution by types of document on *Coriandrum sativum* from 2000 to 2020 (www.scopus .com).

on *Coriandrum sativum*, which include research contributions (1968), review articles (101), and conference papers (71). The contributions are related to the subject fields (Figure 1.3) of agricultural and biological sciences (1227 contributions), biochemistry, genetics, and molecular biology (432 contributions), pharmacology, toxicology, and pharmaceutics (406 contributions), chemistry (321 contributions), medicine (308 contributions), environmental science (264 contributions), immunology and microbiology (128 contributions), and engineering (97 contributions).

Scientists from India (604), the USA (189), Brazil (152), Iran (142), Pakistan (106), Turkey (91), Egypt (80), China (70), Italy (63), Germany (59), Japan (57), and Saudi Arabia (51) emerged as major authors (Figure 1.4). The scientific journals with the highest numbers of contributions were *Acta Horticulturae* (40), *Food Chemistry* (37), *Industrial Crops and Products* (34), *Journal of Ethnopharmacology* (34), *Indian Journal of Agricultural Sciences* (26), *Journal of Agricultural and Food Chemistry* (24), *Journal of Essential Oil Bearing Plants* (24), *Journal of Essential Oil Research* (23), *Horticultura* Brasileira (21), and *Journal of Food Science and Technology* (19).

1.5   AIMS AND FEATURES OF THE BOOK

To the best of our knowledge, it is not easy to find a book reporting on *Coriandrum sativum* cultivation, biochemistry, and functionality in the international scientific literature. Therefore, this book aims to be a scientific base for multidisciplinary discussions on *Coriandrum sativum*, emphasizing its cultivation, harvest, biochemistry, functionality, health-enhancing traits, processing, and technology. The impact of conventional and innovative processing on the recovery of bioactive constituents from *Coriandrum sativum* bio-wastes and agro-byproducts is discussed. In addition, this handbook reports the potential uses of *Coriandrum sativum* in foodstuffs, cosmetics, and pharmaceuticals.

Computer Science, 15, 0% Arts and Humanities, 10, 0% Health Professions, 13, 0%

Economics, Econometrics and Finance, 5, 0%

Neuroscience, 8, 0%

Multidisciplinary, 32, 1%

Mathematics, 12, 0%

Energy, 23, 1%

Business, Management and Accounting, 6, 0%

Social Sciences, 24, 1%
Physics and Astronomy, 36, 1%

Decision Sciences, 1, 0%

Veterinary, 49, 1%
Earth and Planetary Sciences, 23, 1%

Agricultural and Biological Sciences, 1227, 34%

Nursing, 52, 1%
Immunology and Microbiology, 128, 4%

Materials Science, 40, 1%
Engineering, 97, 3%
Chemical Engineering, 95, 3%

Pharmacology, Toxicology and Pharmaceutics, 406, 11%

Medicine, 308, 8%

Chemistry, 321, 9%

Biochemistry, Genetics and Molecular Biology, 432, 12%

Environmental Science, 264, 7%

Documents by subject area

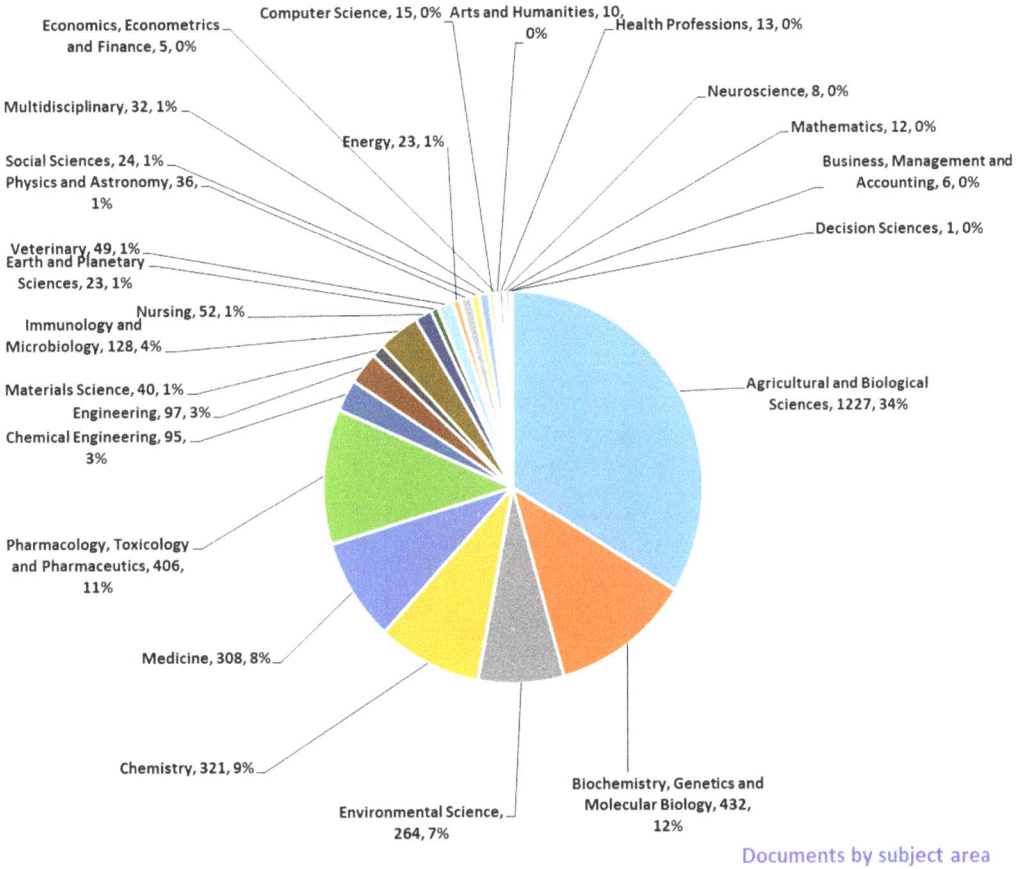

FIGURE 1.3 Distribution by subject area of documents on *Coriandrum sativum* from 2000 to 2020 (www.scopus.com).

Document by Country

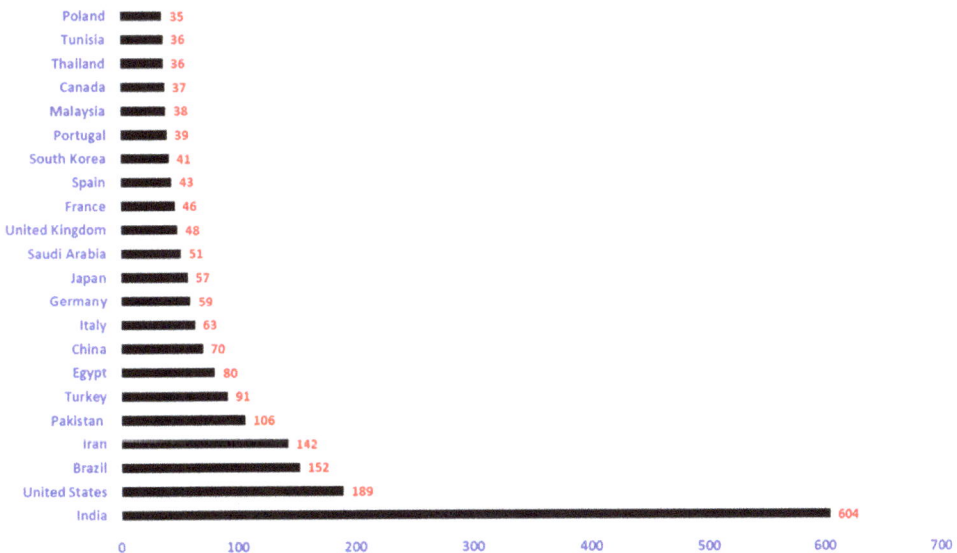

Country	Documents
Poland	35
Tunisia	36
Thailand	36
Canada	37
Malaysia	38
Portugal	39
South Korea	41
Spain	43
France	46
United Kingdom	48
Saudi Arabia	51
Japan	57
Germany	59
Italy	63
China	70
Egypt	80
Turkey	91
Pakistan	106
Iran	142
Brazil	152
United States	189
India	604

FIGURE 1.4 Distribution by country of documents on *Coriandrum sativum* from 2000 to 2020 (www.scopus.com).

The book chapters discuss recent advances in phytochemistry, cultivation, and food sciences research. The book contains comprehensive chapters under main sections, namely:

Section I: Coriander: Cultivation, Composition, and Applications
Section II: Coriander Leaves: Chemistry, Technology, Functionality, and Applications
Section III: Coriander Fixed Oil: Chemistry, Technology, Functionality, and Applications
Section IV: Coriander Essential Oil: Chemistry, Technology, Functionality, and Applications
Section V: Coriander Extracts: Chemistry, Technology, Functionality, and Applications

REFERENCES

Arizio O., Curioni A. (2011) The global and regional coriander market (*Coriandrum sativum* L.). *REVISTA COLOMBIANA DE CIENCIAS HORTÍCOLAS* 5: 263–278.

Elimam D.M., Ramadan M.F., Elshazly A.M., Farag M.A. (2022) Introduction to Mediterranean fruits bio-wastes: Chemistry, functionality and techno-applications. In: Ramadan, M. F., Farag, M. A. (eds) *Mediterranean Fruits Bio-wastes*. Springer, Cham. https://doi.org/10.1007/978-3-030-84436-3_1

Gantait S., Sharangi A.B., Mahanta M., Meena N.K. (2022) Agri-biotechnology of coriander (*Coriandrum sativum* L.): An inclusive appraisal. *Applied Microbiology and Biotechnology* 106(3): 951–969. https://doi.org/10.1007/s00253-022-11787-4

Gidwani B., Bhattacharya R., Shukla S.S., Pandey R.K. (2022) Indian spices: Past, present and future challenges as the engine for bio-enhancement of drugs: Impact of COVID-19. *Journal of the Science of Food and Agriculture*. https://doi.org/10.1002/jsfa.11771

Kiralan M., Özkanb G., Bayrak A., Ramadan M.F. (2014) Physicochemical properties and stability of black cumin (*Nigella sativa*) seed oil as affected by different extraction methods. *Industrial Crops and Products* 57: 52–58.

Mahleyuddin N.N., Moshawih S., Ming L.C., Zulkifly H.H., Kifli N., Loy M.J., Sarker M.M.R., Al-Worafi Y.M., Goh B.H., Thuraisingam S., Goh H.P. (2022) *Coriandrum sativum* L.: A review on ethnopharmacology, phytochemistry, and cardiovascular benefits. *Molecules* 27(1): 209. https://doi.org/10.3390/molecules27010209

McClements D.J. (2019) The science of foods: Designing our edible future. In: McClements, D. J. (ed.) *Future Foods: How Modern Science Is Transforming the Way We Eat*. Springer International Publishing, Cham.

Meena S.K., Jat N.L., Sharma B., Meena V.S. (2014) Effect of plant growth regulators and sulphur on productivity of coriander (*Coriandrum sativum* L.) in Rajasthan. *Ecoscan* 6: 69–73.

Moniruzzaman M., Rahman M.M., Hossain M.M., Karim A.J.M.S., Khaliq Q.A. (2014) Response of coriander (*Coriandrum sativum* L.) foliage to different rates and methods of nitrogen application. *Bangladesh Journal of Agricultural Research* 39(2): 359–371.

Rajeshwari C.U., Andallu B. (2015) Bioactive phytoconstituents in methanolic extract and ethyl acetate fraction of methanolic extract of coriander (*Coriandrum sativum* L.) seeds. *Annals of Phytomedicine* 4: 71–76.

Ramadan M.F. (2021) Introduction to black cumin (*Nigella sativa*): Chemistry, technology, functionality and applications. In: Ramadan, M. F. (ed.) *Black Cumin (Nigella sativa) Seeds: Chemistry, Technology, Functionality, and Applications. Food Bioactive Ingredients*. Springer, Cham. https://doi.org/10.1007/978-3-030-48798-0_1

Ramadan M.F., Amer M.M.A., Awad A. (2008) Coriander (*Coriandrum sativum* L.) seed oil improves plasma lipid profile in rats fed diet containing cholesterol. *European Food Research and Technology* 227: 1173–1182. https://doi.org/10.1007/s00217-008-0833-y

Ramadan M.F., Kroh L.W., Moersel J.-T. (2003) Radical scavenging activity of black cumin (*Nigella sativa* L.), coriander (*Coriandrum sativum* L.) and niger (*Guizotia abyssinica* Cass.) crude seed oils and oil fractions. *Journal of Agricultural and Food Chemistry* 51(24): 6961–6969. https://doi.org/10.1021/jf0346713

Ramadan M.F., Moersel J.-T. (2002) Oil composition of coriander (*Coriandrum sativum* L.) fruit-seeds. *European Food Research and Technology* 215: 204–209. https://doi.org/10.1007/s00217-002-0537-7

Ramadan M.F., Moersel J.-T. (2003) Analysis of glycolipids from black cumin (*Nigella sative* L.), coriander (*Coriandrum sativum* L.) and niger (*Guizotia abyssinica* Cass.) oilseeds. *Food Chemistry* 80: 197–204. https://doi.org/10.1016/S0308-8146(02)00254-6

Ramadan M.F., Moersel J.-T. (2004) Oxidative stability of black cumin (*Nigella sativa* L.), coriander (*Coriandrum sativum* L.) and niger (*Guizotia abyssinica* Cass.) upon stripping. *European Journal of Lipid Science and Technology* 106(1): 35–43. https://doi.org/10.1002/ejlt.200300895

Singh G., Maurya S., De Lampasona M.P., Catalan C.A.N. (2006) Studies on the essential oils, part 41. Chemical composition, antifungal, antioxidant and sprout suppressant activities of coriander (*Coriandrum sativum*) essential oil and its oleoresin. *Flavour and Fragrance Journal* 21(3): 472–479.

Sobhani Z., Mohtashami L., Amiri M.S., Ramezani M., Emami S.A., Simal-Gandara J. (2022) Ethnobotanical and phytochemical aspects of the edible herb *Coriandrum sativum* L. *Journal of Food Science* 87(4): 1386–1422. https://doi.org/10.1111/1750-3841.16085

World Health Organization. Cardiovascular diseases. Available online: https://www.who.int/health-topics/cardiovasculardiseases#tab=tab_1 (accessed on 25 March 2022)

Yousuf M.N., Brahma S., Kamal M.M., Akter S., Chowdhury M.E.K. (2014) Effect of nitrogen, phosphorus, potassium, and sulphur on the growth and seed yield of coriander (*Coriandrum sativum* L.). *Bangladesh Journal of Agricultural Research* 39: 303–309.

Section I

Coriander

Cultivation, Composition, and Applications

2 Coriander Cultivation and Agricultural Practices

Lydia N. Horn, Eduardo P. Mulima
and Felicitas M. Fwanyanga

CONTENTS

2.1   INTRODUCTION

Coriander (*Coriandrum sativum* L.) is an important spice plant that has been known to humankind since 5000 BC (Arora *et al.*, 2021; Al-Snafi, 2016). This annual spice crop belongs to the Apiaceae family and has been cultivated for many years worldwide; it contains 3700 species, including carrots, celery, and parsley. It is known to be one of the first spices to be used by humankind as a flavoring substance. It is indigenous to the Mediterranean region and famous in India, Morocco, Russia, Hungary, Poland, Romania, Guatemala, Mexico, Turkey, and Argentina (Coskuner and Karababa, 2007). The belief in the Mediterranean region as its origin was based on discovering its desiccated fruits in various archaeological sites in that region (Arora *et al.*, 2021). In addition, it has a wide range of diversity that could be distinguished in different groups (Diederichsen, 1996). The most commonly used parts of the plant are the fresh leaves and dried seeds during cooking. It is a good source of antioxidants, and it may help people use less salt and reduce their sodium intake (Rao and Garg, 2020). In addition, coriander fruits have an aromatic odor and pleasant aromatic taste (Banerjee, 2020). The odor and taste are due to the essential oil content that varies from 0.1 to 1% in the dry seeds (Arora *et al.*, 2021; Sharmeen *et al.*, 2021; Shivanand, 2010). The main components of the essential oil include linalool, linalyl acetate, geranial, camphor, limonene, geranyl acetate, and

DOI: 10.1201/9781003204626-3

γ-terpinene (Shivanand, 2010; Sharmeen *et al.*, 2021). Linalool is the extract's main bioactive compound (70%). It is used as an additive for processed food, beverages, and fragrance ingredients in cosmetics and household detergent (Sarkic and Stappen, 2018). It is the main compound responsible for cilantro seeds' antimicrobial and anti-diabetic effects. As a medicinal plant, coriander has been used as an antifungal (Basilico and Basilico, 1999), antioxidant (Chithra and Leelamma, 1999), hypolipidemic, antimicrobial (Silva and Domingues, 2015 and Sourmahi *et al.*, 2015), hypocholesterolemic, and anticonvulsant agent (Raza and Choudhary, 2000). In the traditional medicinal system, the leaves and fruits of the plant have been used to treat skin disease, and its paste is applied by mixing in water to wash the face and forehead. Also, it has been used as a carminative, stimulating, diuretic, tonic, stomachic agent to get rid of bad breath and mouth diseases (Aissaoui *et al.*, 2008). Coriander is a warm-climate crop and could be well cultivated on almost all soils even though it prefers fertile and well-drained soil for maximum performance (Kassu *et al.*, 2018).

2.2   HISTORY AND ORIGIN

Coriandrum sativum originated in the Italian parts of the Mediterranean region, and it spread as a spice plant to India, China, Russia, Central Europe, and Morocco, wherein the plant has been cultivated since human antiquity (Arora *et al.*, 2021; Foudah *et al.*, 2021; Al-Snafi, 2016). Pathak *et al.* (2011) reported that coriander is indigenous to Italy. However, it is widely cultivated in the Netherlands, Central and Eastern Europe, the Mediterranean (Morocco, Malta, and Egypt), China, India, and Bangladesh. Coriander is mentioned in the papyrus of Ebers and the writings of Cato and Pliny and was also well known in England before the Norman Conquest (Shivanand, 2010). Ukraine is the major producer of coriander oil and controls the world price on a supply and demand basis. In India it is chiefly found in Madhya Pradesh, Maharashtra, Rajasthan, Andhra Pradesh, Tamil Nadu, Karnataka, and Bihar (Shivanand, 2010).

2.3   PRODUCTION PRACTICES

Even though coriander production can be done at any time of the year, the crop is highly sensitive to dry and warm climates, and its production is mainly for leaf purposes and a higher grain yield (Banerjee, 2020). The best season for grain production is when it is grown in dry and cold weather, free from frost, especially during the flowering and fruit setting stage (Banerjee, 2020). Cloudy weather is not favorable, especially during the flowering and fruiting stages, because this weather attracts pests and diseases which attack the crop. In addition, heavy rain is detrimental to the crop as it causes much damage; therefore, irrigation while undercover in a shaded place is much recommended for coriander (Singh *et al.*, 2020). The crop can be cultivated on almost all types of soils, supplied by sufficient organic matter (Banerjee, 2020). Coriander yield is determined by the interaction of various agronomic factors, with the main factors being soil nutrition and pest control. The coriander's nutrition is crucial, of which nitrogen is the most important external nutrient for seed yield and plant growth. Coriander yield is determined by the interaction of various agronomic factors, wherein the main factors are soil nutrition and pest control. According to Singh *et al.* (2020), plant nutrition is crucial to coriander, with nitrogen being the most important external nutrient factor that regulates seed yield. Well-maintained nutrients and proper irrigation could enhance coriander seed yield by 10–70% (Singh *et al.*, 2020).

2.4   MORPHOLOGY

Coriander is an annual herb up to 90 cm tall or 20 to 140 cm depending on agro-climatic conditions (Yasir *et al.*, 2019). The plant has many branches and sub-branches with small leaves used as an herb. The new leaves are oval, but aerial leaves are elongated, while the flowers are white, having slightly brinjal-like shades, while fruits are round in shape (Pathak *et al.*, 2011). The whole plant, especially the unripe fruit, is characterized by a strong disagreeable odor (Shivanand, 2010). Different varieties of *Coriandrum sativum* that differ in fruit size have been reported (Al-Snafi, 2016). In addition, coriander plant is characterized by its aromatic smell and flavor (Burdock and Carabin, 2009). The lower leaves are broad with crenate-lobed margins, and the upper one is finely cut with linear lobes. The flowers can be white or pinkish arranged in compound terminal umbels, while the fruits are nearly globular in shape and yellow-brown schizocarp. The plant requires a warm, dry summer with short, rainy winters and is cultivated as a cold-weather crop (Foudah *et al.*, 2021).

2.5 PHYTOCHEMISTRY

Coriander has been known and used as a flavor plant for many dishes, and it is also used to treat some diseases and could be used for its anticancer and antifungal properties, to reduce pain and inflammation, to improve skin health, and as a natural preservative (Mandal and Mandal, 2015). According to Shivanand (2010), green coriander seeds contain up to 1.8% volatile oil. Depending on the source, the distilled oil contains 65–70% of (+)-linalool (coriandrol) and smaller amounts of α-pinene, γ-terpinene, limonene, and ρ -cymene together with various non-linalool alcohols and esters. Other constituents isolated from the fruits include flavonoids, coumarin, *iso*-coumarines, phthalides, and phenolic acids. Al-Snafi (2016) reported that the phytochemical screening of coriander displayed the presence of essential oil, tannins, terpenoids, reducing sugars, alkaloids, phenolics, flavonoids, fatty acids, sterols, and glycosides. Equally important, the essential oil from coriander leaves was reported to be dominated by aldehydes and alcohols. The major constituents were 2E-decenal (15.9%), decanal (14.3%), 2E-decen-1-ol (14.2%), and n-decanol (13.6%). Other constituents present in fairly good amounts are 2E-tridecen-1-al (6.75%), 2E-dodecenal (6.23%), dodecanal (4.36%), undecanol (3.37%), and undecanal (3.23%) (Matasyoh *et al.*, 2009). The nutritional composition is important to determine its contribution to people's health. According to USDA (2013), the nutrient composition of coriander is based on the leaves and seeds (Table 2.1).

2.6 PHARMACOLOGICAL USES

2.6.1 ANTIOXIDANT PROPERTIES

The properties of coriander cannot be underestimated due to its nutritional benefits and its health or medicinal benefits. One of the most well-known and characterized functions of coriander is the antioxidant activity. Antioxidants are compounds that inhibit oxidation, a chemical reaction that can produce free radicals and chain reactions that may damage the cells of organisms. Marangoni and Moura (2011) reported that the use of coriander essential oil presented a stronger synthetic antioxidant effect than that of butylhydroxytoluene (BHT) on the retardation of lipid oxidation. Coriander seeds were also reported to possess antioxidative traits (Deepa and Anuradha, 2011).

Jia *et al.* (2012) suggested that coriander can be safely used as a natural antioxidant and an alternative to synthetic additives for long-term storage in the feed or food industry. A study on the antioxidant properties of coriander root suggested another potential antioxidant property of coriander oil in preventing oxidative stress-related diseases (Tang *et al.*, 2013). Furthermore, Crespo *et al.* (2019) evaluated antioxidant activity in mixtures of three essential oils: *Apium graveolens*, *Thymus vulgaris*, and *Coriandrum sativum*, using the Simplex Lattice Mixture Design, and the mixture of the three oils showed the best antioxidant activity and also had the highest synergistic effect. Another study investigating the anti-inflammatory and antioxidant effects of coriander extract on liver ischemia reperfusion injury at light microscopic and biochemical levels revealed decreased apoptotic cell death and liver enzymes in liver ischemia/reperfusion injury (Kükner *et al.*, 2021). The essential oil extracted from the coriander chemotype grown in the Al-Kharj region of Saudi Arabia was noted to contain low antioxidant potential (Foudah *et al.*, 2021).

TABLE 2.1

Nutrient Composition of Coriander Found in Leaf and Seed (USDA, 2013)

Nutrient	Amount (per 100g)	
	Coriander Leaf	Coriander Seed
Water	7.30 g	8.86 g
Energy	279 kcal	298 kcal
Protein	21.9 g	12.3 g
Total lipid (fat)	4.78 g	17.7 g
Carbohydrate, by difference	52.1 g	54.9 g
Fiber, total dietary	10.4 g	41.9 g
Calcium, Ca	1246 mg	709 mg
Iron, Fe	42.4 mg	16.3 mg
Magnesium, Mg	694 mg	330 mg
Phosphorus, P	481 mg	409 mg
Potassium, K	4466 mg	1267 mg
Sodium, Na	211 mg	35.0 mg
Zinc, Zn	4.72 mg	4.70 mg
Vitamin C, total ascorbic acid	566 mg	21.0 mg
Thiamin	1.25 mg	0.24 mg
Riboflavin	1.50 mg	0.29 mg
Niacin	10.7 mg	2.13 mg
Vitamin B-12	0.00 sg	0.00 sg
Vitamin A, RAE	293 sg	0.00 sg
Vitamin A, IU	5850 IU	0 IU
Vitamin D (D2 + D3)	0.00 sg	0.0 sg
Vitamin D	0 IU	0 IU
Fatty acids, total saturated	0.115 g	0.990 g
Fatty acids, total monounsaturated	2.23 g	13.5 g
Fatty acids, total polyunsaturated	0.32 g	1.75 g
Cholesterol	0.0 g	0.0 g

2.6.2 ANTIMICROBIAL EFFECTS

The antibacterial activities of different medicinal plants have been known for a long time, and several research activities on the antibacterial effect of spices, essential oils, and their derivatives have been reported. Matasyoh *et al.* (2009) screened coriander oil for antimicrobial activity against both Gram-positive and Gram-negative bacteria and a pathogenic fungus. The oil showed pronounced antibacterial and antifungal activity against all microbes tested (Matasyoh *et al.*, 2009). The study by Oudah and Ali (2010) on cold aqueous extract of seeds for 48 h revealed antibacterial activity for all tested bacteria. Another study on ethanol extracts of seeds, leaves, and stems for 48 h showed antibacterial activity (Soares *et al.*, 2012). According to Duarte *et al.* (2012), apart from being widely used as an essential oil, coriander oil can act as a potential improver agent of antibiotics against *Acinetobacter baumannii*. Sambasivaraju and Za's (2016) study on five bacteria (*Staphylococcus aureus*, *Escherichia coli*, *Klebseilla*, *Pseudomonas*, and *Salmonella*) using antibiotic coriander oil revealed a highest antibacterial action against *E. coli* which was better than Gentamicin. Recently, Foudah *et al.* (2021) explored the antimicrobial effects of coriander leaf essential oil cultivated in Saudi Arabia, and they confirmed that the essential oil extracted from coriander possesses superior antimicrobial effect.

2.7 TRADITIONAL/MEDICINAL BENEFITS

Coriander is one of the well-known and favored traditional spices (Choudhary and Punjabi, 2012). Green coriander contains about 169 mg/100 g of vitamin C and 12 mg/100 g of vitamin A. Coriander has a high nutrition component, being rich in vitamins, minerals, and iron, basically due to its green leaves and dried fruits (Bhat *et al.*, 2014). It is very low in saturated lipids and cholesterol, and it has abundant thiamine, zinc, and dietary fiber. Apart from aromatic and culinary value, the main component of the essential oil, which is made up of hydrocarbon and oxygenated compounds, produces several fragrant substances with various smells. The plant parts are used to treat different disorders in traditional medicine (Foudah *et al.*, 2021).

2.8 CULTIVATION

2.8.1 Varieties

Coriander is an annual herb that prefers cool weather, and it is cultivated either as a summer or winter annual crop (Banerjee, 2020; Al-Snafi, 2016). The genus *Coriandrum* contains *C. sativum* and *C. tordylium* (Fenzl) Bornm, of which *C. sativum* is widely cultivated, while *C. tordylium* grows wild (Turin *et al.*, 2021). Being a tropical plant, coriander prefer a frost-free tropical climate that is cool and comparatively dry during flowering and seed formation (Kassu *et al.*, 2018). Coriander cultivation is strongly related to its leaves or seeds (Bhat *et al.*, 2014; Yasir *et al.*, 2019). There are varieties bred for each purpose, where the varieties for seed will produce seed quicker than a leaf variety. The leaf varieties will also develop flowers but not faster than the seed varieties to ensure the plants focus their energy on growing new leaves as required.

2.8.2 Field Preparation

Coriander is grown well in the winter season; however, the timing of sowing depends on the purpose. For example, it can be grown throughout the year for leaf purposes (Kassu *et al.*, 2018). For good germination and growing, it is recommended that coriander seeds be planted in light, and well-drained soil (Singh and Singh, 2014). It is recommended that planting be done immediately after ploughing to avoid soil moisture loss and break the clods. Where the soil moisture is insufficient, watering before planting is recommended to help germinate seeds. Where cultivation depends on the rain, especially in communal areas or farms where dry land cultivation is practiced, ploughing should be done immediately after the rain for soil moisture conservation.

2.8.3 Sowing

Coriander herb likes cold-weather, well-cultivated soils, and can tolerate poor soil fertility and withstand full sun, with the best month for leaf production being late spring and autumn (Kassu *et al.*, 2018). The plant can also grow well in pots or trays filled with good multipurpose compost. Planting can be done with five seeds per hill at the spacing of 20 cm between rows and 20 cm between plants (Singh and Singh, 2014). This is done to allow the plants to grow to their full size. The seed rate is 10–15 kg/ha, and seeds should be soaked in plain water overnight and split in two halves in the morning before sowing for better germination (Singh and Singh, 2014). When planting, the best and recommended practice is to plant directly into the soil to avoid transplanting, which distorts the crop and negatively affects it. Germination takes up to three weeks; therefore, it is better for the seeds to be sown at three-week intervals (Turin *et al.*, 2021). The plants need to be watered in dry periods; ensure that the soil never dries out.

2.8.4 Pest and Diseases

Coriander is susceptible to various disease-causing pathogens such as vascular wilt, stem root, root rot, charcoal rot, seedling blight, stem gall, leaf blight, anthracnose, powdery mildew, seed rot, grain mold, bacterial blight, soft rot, seedling root rot, reniform and root-knot nematodes, and phyllody as well as virus diseases (Khare *et al.*, 2017). Several known resistant varieties are the best option to control the diseases and nutrition effects. In order to achieve year-round cultivation, production can be done under protected structures such as shade nets and greenhouses (Turin *et al.*, 2021). In a field investigation on an insect-pests scenario of coriander in Rabi, it was reported that the coriander crop received an infestation of 15 pest species from different orders and families (Meena *et al.*, 2018). Three species of aphids (*Hyadaphis coriandri*, *Aphis gossypii*, and *Myzus persicae*) and thrips (*Thrips tabaci*, *Scirtothrips dorsalis*, and *Frankliniella schultzei*), and one species each of seed wasp (*Systole albipennis*), seed bug (*Nysius* sp.), painted bug (*Bagrada hilaris*), and tobacco caterpillar (*Spodoptera litura*) were reported as pests of major status (Meena *et al.*, 2018).

2.9 HARVESTING

Coriander reaches maturity 100–150 days after sowing, and this depends on the variety used and climatic conditions in an area. According to Kassu *et al.* (2018) and Singh and Singh (2014), straw yield (q/ha) of coriander under irrigation is obtained from 10 to 15 q/ha. Under unirrigated (barani) cultivation, the average yield is low, being 4 to 5 q/ha, and in a mixed crop, it is 1.5 to 3.0 q/ha. Under managed conditions, yields up to 20 q/ha in unirrigated, and up to 25 q/ha in irrigated conditions can be obtained. The crop grown exclusively for leaf purposes yielded green leaves of about 50–80 q/ha with 3–4 cuttings depending upon cultivars, climate, and management practices. For fresh leaves, 1 or 2 cuttings are taken 45 and 60 days after sowing (Kassu *et al.*, 2018). Depending on the purpose of the plant, for leaves harvest when the plants are big and robust enough to cope, while for seeds, wait until the flowers have died off. It has to be done when the grain changes its color to straw or light brown. If harvesting is delayed, seed shattering and splitting may occur. Harvesting leaves is done by plucking the leaves, cutting each leaf off the stem, or sniping the whole stem. For seeds, the stem is cut and placed in a paper bag, with stems sticking out and hanging upside down in a cool, dry place to allow drying. After three weeks, shake the bag, and seeds will fall out from the flowers; keep them dry. The seeds can be used for spices or re-sowing in the following cropping season.

2.10 POSTHARVEST TECHNOLOGY

Coriander leaves are used as fresh green vegetables or salad leaves, while the seeds are processed into oil and used for a range of medical applications (Yasir *et al.*, 2019). The oil has antibacterial properties and treats colic, neuralgia, and rheumatism (Ali, 2008). The correct stage for harvesting the coriander seed is 90 days from planting, ripe with a full aroma. Since ripening is progressive on the plant, harvesting is ideal when between half and two-thirds of the seeds are ripe. The coriander is mainly consumed in an unprocessed form, but it can also be processed to simplify the international commercialization, viability, and palatability (Singh *et al.*, 2020). The plant is commonly used for making sauces and salsas or blended into powder for flavoring products like meat, fish, sodas, pickles, bakery, and curry recipes. To process this herb, it is essential to know the optimum harvest period to ensure maximum plant material production and maintain the best property of the spice product (Douglas *et al.*, 2005). The leaves and seed are the parts of the plant that are commonly processed, either in green or ripe fruits.

The seed is pale yellow with a characteristic odor and taste (Diederichsen, 1996). The harvested fresh leaves are properly washed and kept in shape for drying to be powdered in a disintegrator and micro-pulverized to avoid losing flavor before packing (Bhat *et al.*, 2014). The drying process is carried out in different ways: sun drying, microwave drying, freeze-drying, etc. Microwave drying showed higher chlorophyll content than those drying methods, and the green color was preserved better than the air-dried and freeze-dried samples (Bhat *et al.*, 2014). The other methods to process the leaves are converted into various products like purees and pastes used in fast food trades. The mature brown seeds can be used and ground to form a powder. Also, seeds can be dried and stored as whole or infused to make delicious vinegar. The ripe seeds may be sometimes processed to obtain the essential oil. Coriander seeds are an excellent source of volatile and crude oil, tannins, saponins, cellulose, pentosans, and pigments. The oil is mainly obtained by steam distillation and supercritical CO_2 extraction. It is characterized by its odor of linalool and warm aromatic flavor (Bhat *et al.*, 2014).

2.10.1 Drying

There are various ways of drying coriander; this varies with the purpose of utilization. Seeds are dried in paper bags facing down with the stem sticking out, while the whole plant is spread out in the sun to dry and left to wither for two days until the moisture content is about 18% (Ali, 2008). After drying, threshing can remove the seeds, and dry them in the shade to prevent overheating until the moisture content is about 9%. This process could be done using driers at a temperature below 100°C to avoid reducing volatile oils.

2.10.2 Oil Extraction

The seed is ground immediately prior to distillation to increase oil yield and reduce the distillation time (Bhat *et al.*, 2014). The essential oil content varies from 0.1 to 1.5% and contains a range of different compounds (Foudah *et al.*, 2021). Steam distillation is a traditional method for the extraction of essential oil from coriander seeds. However, the alteration of chemical constituents of essential oils is a significant problem associated with hydro-distillation methods; this leads to the destruction of the heat-sensitive compounds (Geed *et al.*, 2012). Solvent extraction is an alternative method to steam distillation methods, but the method also has the problem of destroying thermally liable constituents due to the application of high temperatures (Stahl *et al.*, 1987). This problem could be addressed by using green processing technology, i.e., supercritical CO_2 extraction (Said *et al.*, 2014).

ACKNOWLEDGMENTS

The authors are grateful for the opportunity to share our information on the coriander crop and appreciate the University of Namibia and the University of Púnguè Mozambique for providing a conducive environment for us to write this book chapter.

REFERENCES

Aissaoui, A., El-Hilaly, J., Israili, Z. H., & Lyoussi, B. (2008). Acute diuretic effect of continuous intravenous infusion of an aqueous extract of *Coriandrum sativum* L. in anesthetized rats. *Journal of Ethnopharmacology*, *115*(1), 89–95. https://doi.org/10.1016/j.jep.2007.09.007

Ali, A. (2008). *Processing of Coriander. Practical Action: Technology Challenging Poverty. Practical Brief.* The Schumacher Centre for Technology & Development.

Al-Snafi, A. E. (2016). A review on chemical constituents and pharmacological activities of *Coriandrum sativum*. *IOSR Journal of Pharmacy (IOSRPHR)*, *06*(07), 17–42. https://doi.org/10.9790/3013-067031742

Arora, V., Adler, C., Tepikin, A., Ziv, G., Kahane, T., Abu-Nassar, J., Golan, S., Mayzlish-Gati, E., & Gonda, I. (2021). Wild coriander: An untapped genetic resource for future coriander breeding. *Euphytica*, *217*, 1–11.

Banerjee, A. (2020). Complete guide to coriander farming: Varieties, climate requirement, harvesting, yield and economics. *AGRIPEDIA*, September 2020.

Basilico, M. Z., & Basilico, J. C. (1999). Inhibitory effects of some spice essential oils on *Aspergillus ochraceus* NRRL 3174 growth and ochratoxin A production. *Letters in Applied Microbiology*, *29*, 238–241.

Bhat, S., Kaushai, P., Kaur, M., & Sharma, H. K. (2014). Coriander (*Coriandrum sativum* L.): Processing, nutrition and functional aspects. *African Journal of Plant Science, Academic Journal*, *8*(1), 25–33.

Burdock, G. A., & Carabin, I. G. (2009). Safety assessment of coriander (*Coriandreum sativum* L.) essential oil as a food ingredient. *Food and Chemical Toxicology*, *47*(1), 22–34.

Chithra, V., & Leelamma, S. (1999). *Coriandrum sativum* changes the levels of lipid peroxides and activity of antioxidant enzymes in experimental animals. *Indian Journal of Biochemistry & Biophysics*, *36*, 59–61.

Choudhary, R., & Punjabi, N. (2012). Knowledge of farmers about coriander production technology. *Rajasthan Journal of Extension Education*, *20*, 233–237.

Coskuner, Y., & Karababa, E. (2007). Physical properties of coriander seeds (*Coriandrum sativum* L.). *Journal of Food Engineering*, *80*, 408–416.

Crespo, Y. A., Bravo Sánchez, L. R., Quintana, Y. G., Cabrera, A. S. T., Bermúdez del Sol, A., & Mayancha, D. M. G. (2019). Evaluation of the synergistic effects of antioxidant activity on mixtures of the essential oil from *Apium graveolens* L., *Thymus vulgaris* L. and *Coriandrum sativum* L. using simplex-lattice design. *Heliyon*, *5*(6). https://doi.org/10.1016/j.heliyon.2019.e01942

Deepa, B., & Anuradha, C. V. (2011). Antioxidant potential of *Coriandrum sativum* L. seed extract. *Indian Journal of Experimental Biology*, *49*(1), 30–38.

Diederichsen, A. (1996). *Coriander (Coriandrum sativum L.): Promoting the Conservation and Use of Underutilized and Neglected Crops 3*. Rome: Institute of Plant Genetics and Crop Plant Research, Gatersleben/International Plant Genetic Resources Institute.

Douglas, M., Heyes, J., & Small Field, B. (2005). *Herbs, Spices and Essential Oilspost Harvest Operation in Developing Countries*. New Zealand: NZ Institute for Crop and Food Research Ltd.

Duarte, A., Ferreira, S., Silva, F., & Domingues, F. C. (2012). Synergistic activity of coriander oil and conventional antibiotics against *Acinetobacter baumannii*. *Phytomedicine*, *19*(3–4), 236–238. https://doi.org/10.1016/j.phymed.2011.11.010

Foudah, A. I., Alqarni, M. H., Alam, A., Ayman Salkini, M., Ibnouf Ahmed, E. O., & Yusufoglu, H. S. (2021). Evaluation of the composition and *in vitro* antimicrobial, antioxidant, and anti-inflammatory activities of Cilantro (*Coriandrum sativum* L. leaves) cultivated in Saudi Arabia (Al-Kharj). *Saudi Journal of Biological Sciences*, *28*(6), 3461–3468. https://doi.org/10.1016/j.sjbs.2021.03.011

Geed, S. R., Said, P. P., Pradhan, R. C., & Rai, B. N. (2012). Extraction of essential oil from coriander seed. *International Journal of Food and Nutritional Sciences*, *3*(3), 7–9.

Jia, H., Ren, H., Deng, C., Kato, H., & Endo, H. (2012). Effects of Chinese parsley (*Coriandrum sativum*) on oxidative stabilities of diet during storage as compared with a synthetic antioxidant. *International Journal of Food Properties*, *15*(6), 1394–1407. https://doi.org/10.1080/10942912.2010.526277

Kassu, K. T., Dawit, H. H., Wubengeda, A. Y., Almaz, A. T., & Asrat, M. T. (2018). Yield and yield components of coriander under different sowing dates and seed rates in tropical environment. *Advances in Horticultural Science*, *32*(2), 193–203.

Khare, M., Tiwari, S., & Sharma, Y. (2017). Disease problems in the cultivation of coriander (*Coriandrum sativum* L.) and their management leading to production of high quality pathogen free seed. *International Journal of Seed Spices*, *7*, 1–7.

Kükner, A., Söyler, G., Toros, P., Dede, G., Meriçli, F., Işık, S., Edebal, O., & Özoğul, C. (2021). Protective effect of *Coriandrum sativum* extract against inflammation and apoptosis in liver ischaemia/reperfusion injury. *Folia Morphologica (Poland)*, *80*(2), 363–371. https://doi.org/10.5603/FM.A2020.0060

Mandal, S., & Mandal, M. (2015). Coriander (*Coriandrum sativum* L.) essential oil: Chemistry and biological activity. *Asian Pacific Journal of Tropical Biomedicine*, *5*(6), 421–428.

Marangoni, C., & Moura, N. F. De. (2011). Antioxidant activity of essential oil from *Coriandrum sativum* L. in Italian salami. *Ciência e Tecnologia de Alimentos*, *31*(1), 124–128. https://doi.org/10.1590/s0101-20612011000100017

Matasyoh, J. C., Maiyo, Z. C., Ngure, R. M., & Chepkorir, R. (2009). Chemical composition and antimicrobial activity of the essential oil of *Coriandrum sativum*. *Food Chemistry*, *113*(2), 526–529. https://doi.org/10.1016/j.foodchem.2008.07.097

Meena, N. K., Lal, G., Kant, K., Meena, R. S., & Meena, S. R. (2018). Pest scenario of cumin (*Cuminum cyminum* L.) and population dynamics in semi-arid region of Rajasthan. *International Journal Seed Spices*, *8*(1), 780–783.

Oudah, I. M., & Ali, Y. H. (2010). Evaluation of aqueous and ethanolic extraction for coriander seeds, leaves and stems and studying their antibacterial activity. *Iraqi National Journal of Nursing Science*, *23*(2), 1–7.

Pathak, N. L., Kasture, S. B., Bhatt, N. M., & Rathod, J. D. (2011). Phytopharmacological properties of *Randia dumetorum* as a potential medicinal tree: An overview. *Journal of Applied Pharmaceutical Science*, *1*(10), 24–26.

Rao, S., & Garg, G. (2020). Importance of the pest and pathogen control system with special emphasis on coriander crop on the Indian subcontinent. In *Ethnopharmacological Investigation of Indian Species*, 11.

Raza, M., & Choudhary, M. I. (2000). Mediacal plants with anticonvulsant activities. *Studies in Natural Products Chemistry*, 22, 507–553.

Said, P. P., Pradhan, R. C., & Rai, B. N. (2014). A green separation of *Lagenaria siceraria* seed oil. *Industrial Crops and Products*, 52, 796–800.

Sambasivaraju, D., & Za, F. (2016). Evaluation of antibacterial activity of *Coriandrum sativum* (L) against gram-positive and gram-negative bacteria. *International Journal of Basic and Clinical Pharmacology*, 5(6), 2653–2656.

Sarkic, A., & Stappen, I. (2018). Essential oil and their single compounds in cosmetics- A critical review. *Cosmetics*, 5(1), 11.

Sharmeen, J. B., Mahomoodally, F. M., Zengin, G., & Maggi, F. (2021). Essential oil as natural sources of fragrance compounds for cosmetics and cosmeceuticals. *Molecules*, 26(3), 666.

Shivanand, P. (2010). *Coriandrum sativum*: A biological description and its uses in the treatment of various diseases. *International Journal of Pharmacy and Life Sciences*, 1(3), 119–126.

Silva, F., & Domingues, F. C. (2015). Antimicrobial activity of coriander oil and its effectiveness as food preservative. *Critical Reviews in Food Science and Nutrition*, 57, 35–47.

Singh, H., Singh, S., & Panghal, V. P. S. (2020). Effect of irrigation and fertilization on growth and yield in coriander: A review. *Agricultural Reviews*. doi: 10.18805/ag.R-1816

Singh, S., & Singh, D. R. (2014). *Coriander (Coriandrum sativum L.). Package of Practices (PoP) for Vegetable Cultivation in Andaman and Nicobar Islands*. Port Blair: ICAR-Central Island Agricultural Research Institute.

Soares, B. V., Morais, S. M., Dos Santos Fontenelle, R. O., Queiroz, V. A., Vila-Nova, N. S., Pereira, C. M. C., Brito, E. S., Neto, M. A. S., Brito, E. H. S., Cavalcante, C. S. P., Castelo-Branco, D. S. C. M., & Rocha, M. F. G. (2012). Antifungal activity, toxicity and chemical composition of the essential oil of *Coriandrum sativum* L. fruits. *Molecules*, 17(7), 8439–8448. https://doi.org/10.3390/molecules17078439

Sourmaghi, M. H. S., Kiaee, G., Golfakhrabadi, F., Jamalifar, H., & Khanavi, M. (2015). Comparison of essential oil composition and antimicrobial activity of *Coriandrum sativum* L. extracted by hydrodistillation and microwave-assisted hydrodistillation. *Journal of Food Science and Technology*, 52, 2452–2457.

Stahl, E., Quirin, K. W., & Gerard, P. (1987). *Verdichtete Gaze zur Extraktion und Raffination*. Heidelberg: Springer.

Tang, E. L. H., Rajarajeswaran, J., Fung, S. Y., & Kanthimathi, M. S. (2013). Antioxidant activity of *Coriandrum sativum* and protection against DNA damage and cancer cell migration. *BMC Complementary and Alternative Medicine*, 13(347), 1–13. https://doi.org/10.1186/1472-6882-13-347

Turin, E., Shestopalov, M., Radchenko, A., Ganockaya, T., Rostova, E., Karaeva, N., Svyatuk, Y. V., & Susskiy, A. (2021). Optimization of agricultural practices in winter crops *Coriandrum sativum* L. In *IOP Conference Series: Earth and Environmental Science*, Vol. 624, pp. 012103. IOP Publishing.

USDA. (2013). *National Nutrient Database for Standard Reference 26 Full Report (All Nutrients)*. Nutrient data for 2013, spices, coriander seed.

Yasir, M., Khanam, T., & Hafeel, M. H. M. (2019). A review on Kishneez (*Coriander sativum* Linn): A potential herb. *The Pharma Innovation Journal*, 8(9), 503–506.

3 Greenhouse Production of Coriander

*Mairton Gomes da Silva, Hans Raj Gheyi
and Tales Miler Soares*

CONTENTS

3.1 INTRODUCTION

Coriander (*Coriandrum sativum* L.) is one of the most important spices and has been used in folk medicine since ancient times (Wei et al., 2019). It is cultivated in several countries (Rashed and Darwesh, 2015; Fikadu-Lebeta et al., 2019; Fattahi et al., 2021), driven by its use in cooking, for giving flavor and aroma to different dishes (Guha et al., 2014; Sharangi and Roychowdhury, 2014a; Gastón et al., 2016; Pacheco-López et al., 2016; Abbassi et al., 2018).

Fresh coriander leaves are widely used in salads (Sagarika et al., 2014; Mandal and Mandal, 2015; Izgi et al., 2017; Almeida et al., 2019; Afshari et al., 2021; Arora et al., 2021), being rich in vitamins, proteins and other minerals (Nadeem et al., 2013; Katiyar et al., 2014; Kassu et al., 2018; Hosseini et al., 2021), with insignificant contents of cholesterol and saturated fats (Anjukrishna et al., 2021).

Dried leaves and seeds are also used as spices and food ingredients (Guha et al., 2016; Gholizadeh et al., 2018). In addition to its culinary use (leaves and seeds), its essential oil (as discussed in other chapters) is used in the food, cosmetic and pharmaceutical industries due to its medicinal properties (Khalid, 2012; Ramezani et al., 2009; Rahimi et al., 2013; Hassan and Ali, 2014; Katar et al., 2016; Unlukara et al., 2016; Izgi, 2020; Khodadadi et al., 2021).

To produce fresh biomass, coriander is cultivated throughout the year. According to Table 3.1, in the study by Gowtham and Mohanalakshmi (2018) in Coimbatore, India, the production of fresh coriander biomass varied according to the time of year, as recorded in other studies in Northeastern Brazil, such as in Mossoró, RN (Lima et al., 2007), and Itabaiana, SE (Tavares, 2016).

Due to climate change, under open field conditions, the trend is for the production of coriander, as well as other leafy vegetables, to be increasingly affected by extreme events, whether by water restriction (i.e., coriander irrigation is compromised by lack of water) or by excess. As several authors point out (Ghamarnia and Daichin, 2013; Saxena et al., 2013; Meneses et al., 2014; Sharangi and Roychowdhury, 2014b; Harisha et al. 2019), the lack of water supply, depending on the growth stage, can harm the yield and quality of coriander.

DOI: 10.1201/9781003204626-4

TABLE 3.1

Leaf Fresh Biomass per Area (kg m⁻²)* of Coriander Grown in Different Months of Sowing and under Cultivation Conditions in Protected Environment and Open Field, in Coimbatore, India

	Cultivation Conditions	
Time of Sowing	Protected Environment*	Open Field*
September	2.34	1.16
October	2.46	1.26
November	2.17	1.13
December	2.04	1.10
January	1.83	1.05
February	1.64	0.83
March	1.83	
April	1.49	
May	2.14	

Source: Adapted from Gowtham and Mohanalakshmi (2018).

* Values originally in the unit t ha⁻¹.

Excess water causes leaf burning by leaf blight (*Alternaria dauci*), which is the main disease that affects coriander in the hot and rainy season of the year, which can cause product shortage and increase its market price (Reis and Lopes, 2016).

To mitigate adverse effects of the climate, such as excess and/or lack of rain, high temperatures, etc., among other abiotic and biotic factors, cultivation in a protected environment (a greenhouse) has been used and has promoted gains in production when compared to the open field, as shown in Table 3.1 in the study by Gowtham and Mohanalakshmi (2018).

This chapter is focused on the production of fresh coriander biomass in a protected environment in different countries under different growing conditions.

3.2 IMPORTANCE OF CORIANDER IN WORLDWIDE AND IN BRAZIL

The importance of coriander globally is very high, given the large number of studies involving this crop under different growing conditions, as shown in Tables 3.1, 3.2 and 3.3. Coriander is cultivated in practically every country in the world (Elhindi et al., 2016), and its production is mostly to supply local markets (Reis and Lopes, 2016). With the cultivation of coriander throughout its territory (Godara et al., 2014; Lal et al., 2017; Chethan et al., 2019), India is the country that stands out as the largest producer, consumer and exporter in the world (Singh, 2014; Lal et al., 2014a; Verma et al., 2014; Crespo et al., 2017; Shashidhar et al., 2017; Kumawat et al., 2017; Hongal et al., 2018), followed by Mexico, Russia and Iran (Fatemi et al., 2020). Although India has a larger area cultivated with coriander, production per unit area is rather low compared to other countries (Singh et al., 2018).

In Brazil, mainly in the North and Northeast regions (Pereira et al., 2011; Almeida et al., 2014; Santana et al., 2019; Freitas et al., 2020), as well as in other countries (Meyering et al., 2020; McAusland et al., 2020; Serri et al., 2021), fresh biomass in the form of bunches is one of the main ways of marketing coriander for use in cooking and in fresh salads (Cardoso et al., 2017).

For the production of fresh biomass, coriander is cultivated throughout the year in Brazil (Souza et al., 2017; Silva et al., 2018a), in general, by small farmers, in domestic and community gardens (Ferreira Neto et al., 2014; Pinto et al., 2018; Menezes et al., 2020a), thus playing a prominent role

TABLE 3.2

Shoot Fresh Biomass (SFB) per Plant of Coriander Grown under Different Cultivation Conditions in Protected Environment

Location	Cultivar	Cultivation Conditions	Growth Days/Season	SFB (g plant^{-1})	Reference
Mossoró – RN, Brazil	'Português' 'Super Verdão' 'Tabocas' 'Verdão'	In pots with soil mixed with cotton compost at 180 g pot^{-1}	26 DAE/winter	3.33c 12.60b 16.60b 20.40a	Pereira et al. (2011)[#I]
Recife – PE, Brazil	'Tabocas' 'Verdão'	In pots with sand	45 DAS	14.42aA[1] 5.11bA[2] 12.61aB[1] 5.54bA[2]	Bonifacio et al. (2014)[##]
Cáceres – MT, Brazil	'Português' 'Verdão' 'Português' 'Verdão'	In seedbeds in the soil	43 DAS/winter 50 DAS/winter	13.65a 15.28a 17.31b 24.76a	Marsaro et al. (2014)[###]
Recife – PE, Brazil	'Verdão'	In pots with substrates	60 DAS/spring – summer	0.93[3] 1.75[4]	Vasconcelos et al. (2014)
Recife – PE, Brazil	'Verdão'	In pots with soil	22 DAS + 22 DAT	6.82[5] 4.12[6]	Lira et al. (2015)[II]
Tehran, Iran		In pots with soil	14 DAS + 45 DAAT	~ 4.70a ~ 2.80b	Ahmadi and Souri (2018)[SIII]
Cruz das Almas – BA, Brazil	'Verdão SF 177'	Plastic bags with substrates using tap water for irrigation	35 DAAT	6.54	Passos et al. (2018)
Islamabad, Pakistan		In pots with soil		3.26[7]	Ullah et al. (2018)[I]
Jeddah, Saudi Arabia		In pots with soil using fertilizer treatment with plant growth-promoting bacteria (KB-10) and silicon	30 DAS/autumn	~ 7.90	Al-Garni et al. (2019)[SS]
Chapadinha – MA, Brazil	'King' 'Tabocas'	In pots with soil mixed with fermented organic substrate and using only water for irrigation	30 DAS/autumn–winter	2.16a 1.86a	Araújo et al. (2019)[SSSIV]
Tehran, Iran		In pots with sand	14 DAE + 35 DAAT/spring	~ 3.30[8]	Mohammadipour and Souri (2019)
Arapiraca – AL. Brazil		In pots with soil	30 DAS/winter	0.69[9]	Silva et al. (2019b)[I]

(Continued)

TABLE 3.2 (CONTINUED)
Shoot Fresh Biomass (SFB) per Plant of Coriander Grown under Different Cultivation Conditions in Protected Environment

Location	Cultivar	Cultivation Conditions	Growth Days/Season	SFB (g plant⁻¹)	Reference
Cruz das Almas – BA, Brazil	'Verdão'	In pots with soil without cattle manure	40 DAS/winter	3.66[10]	Costa (2020)
		In pots with soil mixed with cattle manure		6.68[11]	
Sari, Iran		In pots with soil	45 DAS + 30 DAAT	~ 0.87a[12]	Rabiei et al. (2020)[&]
				~ 0.88a[12]	
				~ 0.82A[13]	
				~ 0.88B[13]	

[#] Means followed by the same letter do not differ significantly at $p = 0.05$ by Tukey test.

[##] Lowercase letters compare the means of the salinity levels ([1] 0 mM NaCl – control and [2] 100 mM NaCl) for each coriander cultivar, and uppercase letters compare the means of the cultivars at each salinity level at $p = 0.05$ by Tukey test.

[###] In each cultivation period, means followed by the same letter do not differ significantly at $p = 0.05$ by Tukey test.

[3] and [4] Values obtained when the coriander plants were grown in the nutrient solutions of Castellane and Araújo (1994) and Furlani et al. (1999) under electrical conductivities (ECsol) of 2.07 dS m⁻¹ ($SFB = -0.176**ECsol^2 + 0.729**ECsol + 0.173$; ** significant at $p \leq 0.01$ by t-test) and 1.63 dS m⁻¹ ($SFB = -0.664**ECsol^2 + 2.169**ECsol - 0.021$; ** significant at $p \leq 0.01$ by t-test), respectively.

[5] and [6] Values obtained when the coriander plants were grown under electrical conductivity of the soil saturation extract (ECse) of 1.86 dS m⁻¹ under soil moisture at 100% ($SFB = -2.2841**ECse + 24.697**$; ** significant at $p \leq 0.01$ by t-test) and 75% ($SFB = -1.3957**ECse + 14.966$; ** significant at $p \leq 0.01$ by t-test) of container capacity, respectively.

[s] Means followed by the same letter do not differ significantly at $p = 0.05$ by LSD test, obtained without salt stress and under irrigation water salinity of 2.0 dS m⁻¹ (from mixed K_2SO_4 + $MgNO_3$), respectively.

[7] Value obtained under ECse of 4.0 dS m⁻¹.

[ss] Value obtained without salt stress.

[sss] Means as a function of the coriander cultivars followed by the same letter do not differ significantly at $p = 0.05$ by Tukey test.

[8] Value obtained when the coriander plants were grown under glycine concentration of 10 mg L⁻¹.

[9] Value obtained when the coriander plants were grown under irrigation water depth (IWD) of 171.13% of the crop evapotranspiration ($SFB = -0.0004*IWD^2 + 0.1369*IWD - 4.844$; * significant at $p \leq 0.05$ by t-test).

[10] and [11] Values obtained when the coriander plants were grown under doses of cassava wastewater (CW) of 60.00 m³ ha⁻¹ ($SFB = 0.0521CW + 0.536$) and 47.71 m³ ha⁻¹ ($SFB = -0.0014CW^2 + 0.1336CW + 3.4949$), respectively.

[12] Means followed by the same lowercase letter are not significantly different at $p \leq 0.05$ by LSD test, obtained under salinity levels of 0 mM NaCl – control and 80 mM NaCl, respectively.

[13] Means followed by the same uppercase letter do not differ significantly at $p \leq 0.05$ by LSD test, obtained without plant growth-promoting rhizobacteria (PGPR) and with PGPR, respectively.

[&] Value originally in the unit mg plant⁻¹.

[I], [II], [III] and [IV] Initially obtained yields of 10, 3, 20 and 30 plants pot⁻¹, respectively.

DAE – days after emergence; DAS – days after sowing; DAT – days after transplanting; DAAT – days after application of treatments.

TABLE 3.3
Shoot Fresh Biomass per Area (SFBA) of Coriander Grown under Different Cultivation Conditions in a Protected Environment

Location	Cultivar	Cultivation Conditions	Growth Days/ Season	SFBA (kg m^{-2})	Reference
Brasília – DF, Brazil	'Verdão'	In wooden seedbeds with substrates	33 DAS/autumn[#]	3.05[1]&[&]	Kaneko (2006)[I]
			58 DAS/autumn[##]	1.35[2]	
Recife – PE, Brazil	'Tabocas'	In seedbeds in the soil under irrigation water depth of 120% of the reference evapotranspiration	46 DAS/summer	2.26[3]	Albuquerque Filho et al. (2009)[II]
			52 DAS/autumn	2.41[4]	
Cáceres – MT, Brazil	'Português' 'Verdão'	In seedbeds in the soil	43 DAS/winter	0.73a 0.81a	Marsaro et al. (2014)[S][II]
	'Português' 'Verdão'		50 DAS/winter	0.92b 1.32a	
Viçosa – MG, Brazil	'Vedete'	In seedbeds in the soil	17 DAS + 29 DAT/spring	2.90[&&]	Angeli et al. (2016)[SS]
Fortaleza – CE, Brazil	'Verdão'	In trays with substrate with volume of 0.031 dm^3 cell^{-1} with 12 plants	30 DAS/spring	1.56	Olimpio (2017)[II]
Capitão Poço – PA, Brazil	'Verdão'	In wooden seedbeds with soil mixed with aquatic macrophytes as substrate at 50%	37 DAS/winter	7.09a	Esteves Junior and Oliveira Junior (2019)[SSS][II]
		In wooden seedbeds with soil		2.09b	
Recife – PE, Brazil	'Verdão'	In masonry seedbeds with soil	35 DAS/winter	1.51[5] 1.62[6]	Zamora et al. (2021)[I]

[#] and [##] Experiments carried out during the months from March to April and from April to June, respectively.

[1] and [2] Values obtained when the coriander plants were grown in 'Cedrinho' and 'Samaúma' substrates with addition of 50.0 and 16.7% of cattle manure, respectively.

[3] and [4] Values obtained when the coriander plants were grown under doses of hydrogel polymer (HP) of 6.30 dag kg^{-1} (SFBA = -0.054*HP^2 + 0.68*HP + 20.46; * significant at $p \leq 0.05$ by t-test) and 7.91 dag kg^{-1} (SFBA = -0.11***HP^2 + 1.74***HP + 17.20; *** significant at $p \leq 0.001$ by t-test), respectively.

[I] and [II] Values originally in the units g m^{-2} and t ha^{-1}, respectively.

[S] In each cultivation period, means followed by the same letter do not differ significantly at $p = 0.05$ by Tukey test.

[&] and [&&] Productions of 96 and 29 bunches m^{-2}, respectively.

[SS] Value obtained under irrigation depth (ID) of 63 mm and nitrogen dose (ND) of 94 kg ha^{-1} (SFBA = -0.00077**ID^2 – 0.00017**ND^2 + 0.09638***ID + 0.03184***ND – 1.60642; *** and ** significant at $p \leq 0.001$ and $p \leq 0.01$ by t-test, respectively).

[SSS]Means followed by the same letter do not differ significantly at $p = 0.05$ by Tukey test.

[5] and [6] Values obtained when the coriander plants were grown under ID of 120% (SFBA = 14.175**ID – 194.83; ** significant at $p \leq 0.01$ by t-test) and 88.42% (SFBA = -0.4397**ID^2 + 77.753**ID – 1815.5; ** significant at $p \leq 0.01$ by t-test) of the crop evapotranspiration by continuous and pulse irrigations, respectively.

DAS – days after sowing; DAT – days after transplanting.

in economic and social development (Grangeiro et al., 2011; Sousa et al., 2011; Silva et al., 2013; Menezes et al., 2020b). Due to the fragility of fresh coriander, its leaves are susceptible to rapid loss of water after harvest (Freitas et al., 2019), so its circulation is restricted to the local market and its commercialization is daily among producers, intermediaries and final consumers in open markets and/or door-to-door sales (Bomfim, 2017).

Given this form of informal marketing of fresh coriander in Brazil (Reis and Lopes, 2016), as well as in other countries (Lal et al., 2014b), there is a great lack of statistical data on its production (Rocha, 2017), because only a small portion of what is produced reaches the fruit and vegetable distribution centers, and is thus computed. Under protected environments, coriander production data are even more scarce.

Therefore, estimates of coriander fresh biomass production per unit area are important, as shown in Table 3.3, where the result of the survey of several studies carried out under different cultivation conditions is summarized.

3.3 CULTIVATION OF CORIANDER IN A PROTECTED ENVIRONMENT

3.3.1 ADVANTAGES AND DISADVANTAGES

As an advantage, due to environmental control, coriander can be cultivated at any time of the year (Dixit, 2007; Mohanalakshmi et al., 2019), which decreases the seasonality of the product on offer (Guha et al., 2013; Rajasekar et al., 2013). As shown in Table 3.1, the cultivation of coriander in a protected environment has promoted higher yields than those obtained in outdoor cultivation (Gowtham and Mohanalakshmi, 2018), depending on the cultivation system and the degree of environmental control (Dixit, 2007; Guha et al., 2016; Desai et al., 2017).

Also according to Table 3.1, Gowtham and Mohanalakshmi (2018) evaluated different sowing dates for coriander. According to the results, they found that, even in months with lower fresh biomass yields under cultivation in a protected environment, such production was, in general, higher than those obtained in the open field. The production in a protected environment was approximately two-fold higher than those obtained in the field, for the sowing months of September, October, November and February. Similarly, Desai et al. (2017) recorded a higher yield of fresh coriander biomass (0.58 kg m^{-2})[*] in a protected environment (using a red screen with 50% shading) compared to the open field (0.01 kg m^{-2}).[*]

The main disadvantage of cultivation in a protected environment concerns the high initial financial investment for its construction (Dixit, 2007). However, as Anderson and Jia (1996) point out, the investment in a structure (cover) to protect from the adverse environment for the cultivation of coriander is justified because it is a short cycle crop, varying between 50 and 60 days for the production of fresh biomass. Therefore, this short production cycle leads to high space utilization.

3.3.2 CULTIVATION TECHNIQUES

Within the protected environment, coriander cultivation has been carried out in wooden beds filled with different substrates (Kaneko, 2006; Esteves Junior and Oliveira Junior, 2019; Figure 3.1A), in beds in the soil itself (Marsaro et al., 2014; Angeli et al., 2016; Figure 3.1B), in masonry beds filled with soil (Silva et al., 2019a; Zamora et al., 2019; Pamplona et al., 2021; Figure 3.1C), in tiles filled with the substrate (Cardoso et al., 2019; Figure 3.1D), in pots (Donega et al., 2013; Marichali et al., 2014; Okkaoğlu et al., 2015; Shams et al., 2016; Oliveira et al., 2017; Sá et al., 2017; Mehrabani et al., 2018; França et al., 2020; Amiripour et al., 2021; Fatemi et al., 2021; Figure 3.1E) and trays (Oliveira et al., 2010; Sales et al., 2015; Machado and Marreiros, 2016; Sá et al., 2016; Olimpio, 2017; Figures 3.1F)

---

[*] Values originally in the unit kg per 100 m^2; experiment carried out during summer season, in Navsari, India.

FIGURE 3.1 Coriander grown in wooden seedbeds filled with different substrates (A), seedbeds in the soil (B), masonry seedbeds filled with soil (C), in tiles filled with the substrate (D), in pots with soil (E), in trays with the substrate (F), and under hydroponic conditions (G). A – source: Esteves Junior and Oliveira Junior (2019), B – source: Barros et al. (2018), C and E – source: Menezes (2018), F – source: Olimpio (2017), and D and G – (photos by the authors).

with soil and/or combined with different substrates and in hydroponics – soilless cultivation[*] (Silva et al., 2016a; 2018b; 2020a; 2020b; Figure 3.1G).

As already mentioned, the commercialization of fresh coriander biomass is in the form of bunches for use in cooking. The composition of coriander bunches depends on the form of cultivation, for example, the sowing of seeds in large quantities in furrows/rows at certain spacing is more common (Oliveira et al., 2004; Agasimani, 2014; Özyazici, 2021; Figure 3.1B). Cultivation has also been carried out in holes opened at certain distances, either by direct sowing (Cavalcante et al., 2017; Figure 3.1C) or by seedlings produced in trays (Angeli et al., 2016; Santos et al., 2021; Figure 3.1F) and later taken to the field. In other words, in this last type of cultivation, the bunch of plants comes from the hole itself (Linhares et al., 2015; Zamora, 2018).

Therefore, coriander yield is directly influenced by plant density. There is no standard in terms of the mass of the coriander bunch, being variable between locations even within the same region. This is well exemplified in Table 3.3 in the studies conducted by Kaneko (2006) in Brasília and Angeli et al. (2016) in Viçosa, both in Brazil, in which the authors recorded the numbers of 96 and 29 bunches m^{-2} of coriander plants, corresponding to productions of 3.05 and 2.90 kg m^{-2}, respectively. While in the first study the bunch was approximately 30 g, in the second study it was 100 g, which explains this difference in the number of bunches per unit area. Bunches with 100 g standard

---

[*] Chapter 4 'Production of Coriander Using the Hydroponic Technique' written only on this theme.

have been adopted in the municipality of Mossoró in the Northeast region of Brazil (Barros Júnior et al., 2004).

3.4 PRODUCTION OF CORIANDER UNDER DIFFERENT GROWING CONDITIONS IN A PROTECTED ENVIRONMENT

Table 3.2 shows the compendium of several studies for the variable shoot fresh biomass (SFB) of coriander per plant, in which different aspects were addressed, such as the performance of coriander cultivars (Pereira et al., 2011; Bonifacio et al., 2014; Marsaro et al., 2014; Araújo et al., 2019), salt stress (Bonifacio et al., 2014; Lira et al., 2015; Ahmadi and Souri, 2018; Ullah et al., 2018; Al-Garni et al., 2019; Rabiei et al., 2020), irrigation depths (Silva et al., 2019b), the application of attenuators to mitigate salt stress (Mohammadipour and Souri, 2019; Rabiei et al., 2020) and the use of cassava wastewater in combination with cattle manure (Costa, 2020).

There is a wide variation in coriander SFB values (Table 3.2). This can be due to different factors, such as cultivar, planting time and duration of the crop, in addition to the cultivation conditions used (imposed treatments). In the studies carried out in Brazil, in general, there are higher productions of SFB per plant of 'Verdão' coriander.

Therefore, in Brazil, the realization of a large number of studies with the 'Verdão' cultivar is explained, as it tolerates higher temperatures compared to other cultivars. For example, Pereira et al. (2011) evaluated different cultivars of coriander and recorded SFB production of 'Verdão' coriander six-fold greater than that of 'Portuguese' coriander. In the study by Marsaro et al. (2014), when the plants were harvested at 43 days after sowing (DAS), there was no significant difference between the means of the 'Portuguese' and 'Verdão' coriander cultivars. At 50 DAS, the SFB production of 'Verdão' coriander was higher; however, this superiority did not exceed 1.5-fold compared to that of 'Portuguese' coriander. In these studies, the SFB of 'Verdão' coriander was greater than 20 g plant^{-1} (Table 3.2). Therefore, these results show that the cultivation of 'Portuguese' coriander is more suitable for the climatic conditions of the second study site.

As shown in Table 3.2, according to the study by Pereira et al. (2011), the SFB production of 'Tabocas' coriander was only lower than that of 'Verdão' coriander. Different results were observed by Bonifacio et al. (2014), who recorded 14% higher SFB production of 'Tabocas' coriander compared to 'Verdão' coriander under cultivation without salt stress (0 mM NaCl), while under salinity (100 mM NaCl) there was no significant difference between cultivars. However, when evaluating the relative yield (RY),[*] expressed as RY = (with salinity/without salinity – control) × 100, 'Verdão' coriander was superior (RY ~ 44%) compared to 'Tabocas' coriander (RY ~35 %). These results show that the effect of salt stress varied according to the coriander cultivar, as well as the concentration of salts to which the plants were exposed, as in the study by Rabiei et al. (2020), who found no significant difference between the means of SFB obtained without (0 mM NaCl) and with salt stress (80 mM NaCl).

The use of one cultivar to the detriment of others is often done for cultural reasons, that is, without considering the production capacity and edaphoclimatic conditions of those places. This type of behavior is characteristic of small producers who live in places far from big centers. Therefore, under contrasting climate conditions, studies involving the evaluation of different coriander cultivars are of great importance (Pereira et al., 2011; Bonifacio et al., 2014; Marsaro et al., 2014; Araújo et al., 2019).

Also according to Table 3.2, the SFB of 'Verdão' coriander did not exceed 1.80 g plant^{-1} in the study by Vasconcelos et al. (2014), while Lira et al. (2015) recorded values above 4.00 g plant^{-1}. Such studies were carried out in the same place but at different times. This partially explains this discrepancy in the results; another complementary explanation concerns the density of plants since 16 and 3 plants pot^{-1} were cultivated in these studies, respectively. This was reinforced in the study by Silva

---

[*] Variable established by the authors of this chapter.

et al. (2019b), also conducted in Northeastern Brazil; they recorded coriander SFB less than 0.70 g plant^{-1} under cultivation with 10 plants pot^{-1}. That is, the greater the number of coriander plants per unit area, the lower the production per plant, as reported by other authors (Silva et al., 2016b; Martins et al., 2018), reaching the point that the increase in density attains a point of maximum production per unit area (Almeida et al., 2019), and from this point onwards there is a decrease in production due to the competition that is established (Olimpio, 2017).

As verified for the variable SFB per plant (Table 3.2), for SFB per area (Table 3.3) the results are also quite variable depending on the different conditions of coriander cultivation. For example, Kaneko (2006) recorded higher production (3.05 kg m^{-2})[*] in the first cycle compared to the second cropping cycle (1.35 kg m^{-2}),[*] both conducted in the same year, although this last cycle was carried out for a longer period. In the study by Albuquerque Filho et al. (2009), SFB[†] productions were close in two consecutive cycles in the same year. However, the values were lower than those recorded by Kaneko (2006) in the first cultivation cycle. This difference can be explained by the number of plants used per unit of area. Additionally, another explanation concerns the differences between cultivars, as mentioned above (as shown in Table 3.2); 'Verdão' coriander has shown higher yields of fresh biomass compared to other cultivars.

Clearly, in the study conducted by Marsaro et al. (2014), the smallest number of plants per unit area influenced the yield of coriander, for example, SFB did not exceed 1.40 kg m^{-2} (as shown in Table 3.3). In this study, only one plant was cultivated per hole, adopting a spacing of 0.05 m × 0.25 m between holes and rows, respectively. With this spacing, totaling 80[‡] plants m^{-2}, this value is ten- and three-fold lower than that used in the studies of Kaneko (2006) and Albuquerque Filho et al. (2009), respectively. In the study by Angeli et al. (2016) with a total of 300[‡] plants m^{-2} (3 plants hole^{-1}, 0.05 m × 0.20 m between holes and rows, respectively), a production of SFB (2.90 kg m^{-2}) approaches the value recorded by Kaneko (2006) in the first cultivation cycle.

Also according to Table 3.3, the highest SFB of coriander was recorded in the study of Esteves Junior and Oliveira Junior (2019), of the order of 7.09 kg m^{-2} under cultivation with the incorporation of organic compost into the soil, this value being more than three-fold the production obtained when the cultivation was only in soil.

3.5   WATER PRODUCTIVITY OF CORIANDER IN A PROTECTED ENVIRONMENT

Water productivity, expressed by the production of fresh coriander biomass divided by the volume of water applied, ranged between 7.52 and 49.81 kg m^{-3} (Table 3.4). Therefore, there is a large variation in water productivity values, reinforcing what has already been mentioned for SFB production per plant (Table 3.2) and per area (Table 3.3).

This higher water productivity (49.81 kg m^{-3}) in the study by Angeli et al. (2016) is due to the smaller volume of water used to produce 1 kg of fresh biomass. For example, the production of SFB was approximately 28 and 20% higher than the SFB obtained by Albuquerque Filho et al. (2009) in the first and second cycles, respectively; while water productivity was approximately 100 and 51% higher in the study by Angeli et al. (2016).

3.6   CONCLUSIONS

Due to climate change, crops have been increasingly affected, especially leafy vegetables such as coriander. Therefore, maintaining regularity in terms of production and, above all, quality is a great challenge for producers. In this context, to mitigate the adverse effects of unfavorable weather, the

---

[*] 800 plants m^{-2} were cultivated, estimated from plot size of 0.25 m × 0.20 m (40 plants plot^{-1}) by the authors of this chapter.

[†] 240 plants m^{-2} were cultivated, estimated from plot size (lysimeter with 1.38 m in diameter, with a surface area of 1.50 m^2 – 360 plants lysimeter^{-1}) by the authors of this chapter.

[‡] Values calculated by the authors of this chapter.

TABLE 3.4
Water Productivity (WP) of the Shoot Fresh Biomass of Coriander Grown under Different Cultivation Conditions in Protected Environment

Location	Cultivar	Cultivation Conditions	WP (kg m^{-3})	Reference
Recife – PE, Brazil	'Tabocas'	In seedbeds in the soil	23.58[1]	Albuquerque Filho et al. (2009)[I]
			32.89[2]	
Viçosa – MG, Brazil	'Vedete'	In seedbeds in the soil	49.81[3]	Angeli et al. (2016)[I]
Viçosa – MG, Brazil	'Vedete'	In seedbeds in the soil	7.52[4]	Delazari et al. (2019)[II]
Recife – PE, Brazil	'Verdão'	In masonry seedbeds with soil	20.40[5]	Zamora et al. (2019)[I]
			12.98[6]	

[1] and [2] Values obtained when the coriander plants were grown under doses of hydrogel polymer (HP) of 16.00 dag kg^{-1} (WP = 0.39*HP + 17.34; * significant at $p \leq 0.05$ by t-test) and 8.18 dag kg^{-1} (WP = -0.11***HP^2 + 1.80**HP + 25.53; *** and ** significant at $p \leq 0.001$ and $p \leq 0.01$ by t-test, respectively), and irrigation water depths (ID) of 60 and 120% of the reference evapotranspiration, respectively.

[3] Value obtained under ID of 63 mm and nitrogen dose (ND) of 105 kg ha^{-1} (WP = -0.5844***ID + 0.2036***ND + 65.2479; *** significant at $p \leq 0.001$ by t-test);

[4] value obtained under ID of 25% of the crop evapotranspiration (ETc) and ND of 70 kg ha-1 (WP = -0.037***ID + 8.4452; *** significant at $p \leq 0.001$ by t-test).

[5] and [6] Values obtained (originally in the unit g L^{-1}) when the coriander plants were grown under ID of 71.60% ETc (WP = -0.0045**ID^2 + 0.6444**ID − 2.6691; ** significant at $p \leq 0.01$ by t-test) and 97.42% ETc (WP = -0.0013**ID^2 + 0.2533**ID + 0.6378; ** significant at $p \leq 0.01$ by t-test) by pulse and continuous irrigation, respectively.

[I]   Same growth periods and cultivation conditions shown in Table 4.3.

[II]   Experiment carried out during the spring season for 15 days after sowing + 29 days after transplanting.

cultivation of coriander in a protected environment has been increasingly recurrent, not only in research but also in commercial production. Therefore, based on the various aspects covered in this chapter, it is concluded that there is great potential for growing coriander in a protected environment, thus ensuring regularity in the production of fresh biomass throughout the year. Additionally, a database regarding the production of fresh coriander biomass per unit area was generated from the compendium of several studies conducted under different climatic conditions, which may serve as support in future consultations for implementing coriander cultivation in a protected environment.

REFERENCES

Abbassi, A., Mahmoudi, H., Zaouali, W., M'Rabet, Y., Casabianca, H., & Hosni, K. (2018). Enzyme-aided release of bioactive compounds from coriander (*Coriandrum sativum* L.) seeds and their residue by-products and evaluation of their antioxidant activity. *Journal of Food Science and Technology*, 55(8), 3065–3076. https://doi.org/10.1007/s13197-018-3229-4.

Afshari, M., Pazoki, A., & Sadeghipour, O. (2021). Foliar-applied silicon and its nanoparticles stimulate physio-chemical changes to improve growth, yield and active constituents of coriander (*Coriandrum sativum* L.) essential oil under different irrigation regimes. *Silicon*. https://doi.org/10.1007/s12633-021-01101-8.

Agasimani, A. (2014). Evaluation of coriander (*Coriandrum sativum* L.) genotypes in hill zone of Karnataka. *The Asian Journal of Horticulture*, 9(1), 174–177.

Ahmadi, M., & Souri, M. K. (2018). Growth and mineral content of coriander (*Coriandrum sativum* L.) plants under mild salinity with different salts. *Acta Physiologiae Plantarum*, 40(11), 194. https://doi.org/10.1007/s11738-018-2773-x.

Albuquerque Filho, J. A. C., Lima, V. L. A., Menezes, D., Azevedo, C. A. V., Dantas Neto, J., & Silva Júnior, J. G. (2009). Características vegetativas do coentro submetido a doses do polímero hidroabsorvente e lâminas de irrigação. *Revista Brasileira de Engenharia Agrícola e Ambiental*, 13(6), 671–679. https://doi.org/10.1590/S1415-43662009000600002.

Al-Garni, S. M. S., Khan, M. M. A., & Bahieldin, A. (2019). Plant growth-promoting bacteria and silicon fertil-izer enhance plant growth and salinity tolerance in *Coriandrum sativum*. *Journal of Plant Interactions*, 14(1), 386–396. https://doi.org/10.1080/17429145.2019.1641635.

Almeida, B. C., Lemos Neto, H. S., Guimarães, M. A., Sampaio, I. M. G., & Silva, L. S. (2019). Desempenho agroeconômico do coentro em diferentes densidades de semeadura. *Revista de Ciências Agrárias*, 62, 1–7. https://doi.org/10.22491/rca.2019.2973.

Almeida, B. O., Prins, C. L., Carvalho, A. J. C., Freitas, M. S. M., Stida, W. F., & Lima, T. C. (2014). Crescimento de coentro submetido a diferentes tensões de água no substrato. *Horticultura Brasileira*, 31(Supplement 2), 1665–1672.

Amiripour, A., Jahromi, M. G., Soori, M. K., & Torkashvand, A. M. (2021). Changes in essential oil composi-tion and fatty acid profile of coriander (*Coriandrum sativum* L.) leaves under salinity and foliar-applied silicon. *Industrial Crops and Products*, 168, 113599. https://doi.org/10.1016/j.indcrop.2021.113599.

Anderson, R. G., & Jia, W. (1996). Greenhouse production of garlic chives and cilantro. In Janick, J. (Ed.), *Progress in new crops* (pp. 594–597), ASHS Press, Alexandria.

Angeli, K. P., Delazari, F. T., Nick, C., Ferreria, M. G., & Silva, D. J. H. (2016). Yield components and water use efficiency in coriander under irrigation and nitrogen fertilization. *Revista Brasileira de Engenharia Agrícola e Ambiental*, 20(5), 415–420. https://doi.org/10.1590/1807-1929/agriambi .v20n5p415-420.

Anjukrishna, V. U., Raj, N. M., Anitha, P., & Aneesha, A. K. (2021). Effect of seed treatments, spacing and season of sowing on yield and quality of coriander (*Coriandrum sativum* L.) under rain shelter. *Journal of Spices and Aromatic Crops*, 30(1), 90–99. https://doi.org/10.25081/josac.2021.v30.i1.6424.

Araújo, J. S., Brito Filho, A. L., Meneses, K. C., Coutinho, R. S., Cabral, I. S., Fortes, E. M., Rodrigues, R. C., Parente, M. O. M., Parente, H. N., Jesus, A. P. R., & Santos, J. O. (2019). Production of coriander (*Coriandrum sativum* L.) in organic substrates and fertigation with biodigester effluents. *International Journal of Advanced Engineering Research and Science*, 6(8), 167–170. https://doi.org/10.22161/ijaers .68.21.

Arora, V., Adler, C., Tepikin, A., Ziv, G., Kahane, T., Abu-Nassar, J., Golan, S., Mayzlish-Gati, E., & Gonda, I. (2021). Wild coriander: An untapped genetic resource for future coriander breeding. *Euphytica*, 217, 138. https://doi.org/10.1007/s10681-021-02870-4.

Barros, K. F. G., Santana, A. C., Martins, C. M., Campos, P. S. S. (2018). Valor da água virtual de hortaliças comercializada em Benevides - PA. *Nucleus*, 15(1), 9–24. https://doi.org/10.3738/1982.2278.1739.

Barros Júnior, A. P., Bezerra Neto, F., Negreiros, M. Z., Oliveira, E. Q., Silveira, L. M., & Câmara, M. J. T. (2004). Desempenho agronômico de cultivares comerciais de coentro em cultivo solteiro sob condições de temperatura elevada e ampla luminosidade. *Revista Caatinga*, 17(2), 82–86.

Bomfim, J. M. F. (2017). *Pegada hídrica e desempenho econômico da cultura do coentro (Coriandrum sati-vum L.) no agreste sergipano* [Master Thesis, Universidade Federal de Sergipe]. https://ri.ufs.br/bit-stream/riufs/4251/1/JOSE_MURILHO_FARIAS_BOMFIM.pdf.

Bonifacio, A., Silva Júnior, G. S., Silva, L. E., Rodrigues, A. C., Willadino, L. G., & Camara, T. J. R. (2014). Respostas fisiológicas e bioquímicas de cultivares de coentro submetidas à salinidade. *Proceedings of the Brazilian Symposium on Salinity and Brazilian Meeting on Irrigation Engineering*, Fortaleza, Brazil. https://doi.org/10.12702/ii.inovagri.2014-a738.

Cardoso, M. O., Berni, R. F., Antonio, I. C., & Kano, C. (2017). Growth, production and nutrients in corian-der cultivated with biofertilizer. *Horticultura Brasileira*, 35(4), 583–590. https://doi.org/10.1590/S0102 -053620170417.

Cardoso, M. O., Berni, R. F., Chaves, F. C. M., & Pinheiro, J. O. C. (2019). *Índices agroeconômicos do coentro cultivo em substratos de fibra de coco com fertirrigação*. 26 pp. Embrapa Amazônia Ocidental, Manaus, Brazil. http://ainfo.cnptia.embrapa.br/digital/bitstream/item/208251/1/BP-29.pdf.

Castellane, P. D., & Araújo, J. A. C. (1994). *Cultivo sem solo: Hidroponia*. 43 p. FUNEP, Jaboticabal, Brazil.

Cavalcante, L. S., Santos, F. F., Pereira, M. T. S., Oliveira, T. A., Silva, T. R. G., & Santos, M. A. L. (2017). Desempenho da cultura do coentro em função de diferentes lâminas de água. *Proceedings of the Inovagri International Meeting*, Fortaleza, Brazil. https://doi.org/10.7127/IV-INOVAGRI-MEETING -2017-RES1170226.

Chethan, T., Vishnuvardhana, Ramachandra, R. K., & Anjanappa, M. (2019). Evaluation of coriander (*Coriandrum sativum* L) genotypes for growth and herbage yield in the northern transitional zone of Karnataka. *International Journal of Chemical Studies*, 7(2), 182–185.

Costa, A. G. (2020). *Caracterização da manipueira e sua utilização para a produção do coentro 'Verdão' e da pimenta 'Biquinho'* [Doctoral Thesis, Universidade Federal do Recôncavo da Bahia]. https://www .ufrb.edu.br/pgea/images/Teses/ANDREZZA_GRASIELLY_COSTA.pdf.

Crespo, E. C., Chulim, Á. C., Rosales, L. J. L., Benítez, G. A., Pineda, J. P., & Montoya, R. B. (2017). Extracción de N-P-K en *Coriandrum sativum* 'Pakistan' en hidroponia. *Revista Mexicana de Ciencias Agrícolas*, 8(2), 355–367. https://doi.org/10.29312/remexca.v8i2.56.

Delazari, F. T., Copati, M. G. F., Silva, G. H., Gomes, R. S., Silva, D. J. H., & Nick, C. (2019). Morphological and ecophysiological indicators for coriander under irrigation depths and nitrogen levels. *Journal of Agricultural Science*, 11(3), 549–557. https://doi.org/10.5539/jas.v11n3p549.

Desai, C. D., Desai, G. B., Desai, C. S., Patel, S. D., & Mehta, V. S. (2017). Effect of different colour shade nets on biomass yield and quality of fenugreek, coriander and garlic. *Multilogic in Science*, 7(23), 185–187.

Dixit, A. (2007). Performance of leafy vegetables under protected environment and open field condition. *The Asian Journal of Horticulture*, 2(1), 197–200.

Donega, M. A., Mello, S. C., Moraes, R. M., & Cantrell, C. L. (2013). Nutrient uptake, biomass yield and quantitative analysis of aliphatic aldehydes in cilantro plants. *Industrial Crops and Products*, 44, 127–131. https://doi.org/10.1016/j.indcrop.2012.11.004.

Esteves Junior, F. C., & Oliveira Junior, M. V. R. (2019). *Cultivo de coentro em substrato contendo macrófita aquática ou cama de aviário* [Monograph, Universidade Federal Rural da Amazônia]. bdta.ufra.edu.br /jspui//handle/123456789/1339.

Elhindi, K. M., El-Hendawy, S., Abdel-Salam, E., Schmidhalter, U., Rehman, S. U., & Hassan, A.-A. (2016). Foliar application of potassium nitrate affects the growth and photosynthesis in coriander (*Coriander sativum* L.) plants under salinity. *Progress in Nutrition*, 18(1), 63–73.

Fatemi, H., Pour, B. E., & Rizwan, M. (2020). Isolation and characterization of lead (Pb) resistant microbes and their combined use with silicon nanoparticles improved the growth, photosynthesis and antioxidant capacity of coriander (*Coriandrum sativum* L.) under Pb stress. *Environmental Pollution*, 266(Part 3), 114982. https://doi.org/10.1016/j.envpol.2020.114982.

Fatemi, H., Pour, B. E., & Rizwan, M. (2021). Foliar application of silicon nanoparticles affected the growth, vitamin C, flavonoid, and antioxidant enzyme activities of coriander (*Coriandrum sativum* L.) plants grown in lead (Pb)-spiked soil. *Environmental Science and Pollution Research*, 28, 1417–1425. https:// doi.org/10.1007/s11356-020-10549-x.

Fattahi, B., Arzani, K., Souri, M. K., & Barzegar, M. (2021). Morphophysiological and phytochemical responses to cadmium and lead stress in coriander (*Coriandrum sativum* L.). *Industrial Crops and Products*, 171, 113979. https://doi.org/10.1016/j.indcrop.2021.113979.

Ferreira Neto, M., Miranda, R. S., Prisco, J. T., & Gomes-Filho, E. (2014). Changes in growth parameters and biochemical mechanisms of coriander plants irrigated with saline water. *Proceedings of the Brazilian Symposium on Salinity and Brazilian Meeting on Irrigation Engineering*, Fortaleza, Brazil. https://doi .org/10.12702/ii.inovagri.2014-a514.

Fikadu-Lebeta, W., Diriba-Shiferaw, G., & Mulualem-Azene, M. (2019). The need of integrated nutrient management for coriander (*Coriandrum sativum* L.) production. *International Journal of Food & Nutrition*, 4(1), 1–13.

França, T. S., Paz, K. M. V., Souza, A. F., Santos, L. J. S., & Santos, M. A. L. (2020). Resposta do coentro irrigado com água salina. *Revista Ambientale*, 12(1), 29–33. https://doi.org/10.48180/ambientale .v12i1.192.

Freitas, B. L. L. C., Almeida, R. H., & Simplício, J. B. (2019). Coriander storage (*Coriandrum sativum* L) as alternative to minimize postharvest damages. *Amazonian Journal of Plant Research*, 3(3), 383–390.

Freitas, M. S. M., Gonçalves, Y. S., Lima, T. C., Santos, P. C., Peçanha, D. A., Vieira, M. E., Carvalho, A. J. C., & Vieira, I. J. C. (2020). Potassium sources and doses in coriander fruit production and essential oil content. *Horticultura Brasileira*, 38(3), 268–273. https://doi.org/10.1590/S0102-053620200305.

Furlani, P. R., Bolonhezi, D., Silveira, L. C. P., & Faquin, V. (1999). Nutrição mineral de hortaliças, preparo e manejo de soluções nutritivas. *Informe Agropecuário*, 20(200/201), 90–98.

Gastón, M. S., Cid, M. P., Vázquez, A. M., Decarlini, M. F., Demmel, G. I., Rossi, L. I., Aimar, M. L., & Salvatierra, N. A. (2016). Sedative effect of central administration of *Coriandrum sativum* essential oil and its major component linalool in neonatal chicks. *Pharmaceutical Biology*, 54(10), 1954–1961. https://doi.org/10.3109/13880209.2015.1137602.

Ghamarnia, H., & Daichin, S. (2013). Effect of different water stress regimes on different coriander (*Coriandrum sativum* L.) parameters in a semi-arid climate. *International Journal of Agronomy and Plant Production*, 4(4), 822–832.

Gholizadeh, A., Dehghani, H., Khodadadi, M., & Gulick, P. J. (2018). Genetic combining ability of coriander genotypes for agronomic and phytochemical traits in response to contrasting irrigation regimes. *PLoS One*, 13(6), e0199630. https://doi.org/10.1371/journal.pone.0199630.

Godara, A. S., Gupta, U. S., Lal, G., & Singh, R. (2014). Influence of organic and inorganic source of fertilizers on growth, yield and economics of coriander (*Coriandrum sativum* L.). *International Journal of Seed Spices*, 4(2), 77–80.

Gowtham, T., & Mohanalakshmi, M. (2018). Influence of growing environment on growth and yield parameters of coriander under shade net and open field condition. *Madras Agricultural Journal*, 105(7/9), 332–335. https://doi.org/10.29321/MAJ.2018.000155.

Grangeiro, L. C., Freitas, F. C. L., Negreiros, M. Z., Marrocos, S. T. P., Lucena, R. R. M., & Oliveira, R. A. (2011). Crescimento e acúmulo de nutrientes em coentro e rúcula. *Revista Brasileira de Ciências Agrárias*, 6(1), 11–16. https://doi.org/10.5039/agraria.v6i1a634.

Guha, S., Debnath, S., & Sharangi, A. B. (2016). Influence of growing conditions on yield and essential oil of coriander during year-round cultivation. *International Journal of Agriculture Sciences*, 8(5), 1021–1026.

Guha, S., Sharangi, A. B., & Debnath, S. (2013). Effect of different sowing times and cutting management on phenology and yield of off season coriander under protected cultivation. *Trends in Horticultural Research*, 3(1), 27–32. https://doi.org/10.3923/thr.2013.27.32.

Guha, S., Sharangi, A. B., & Debnath, S. (2014). Phenology and green leaf yield of coriander at different sowing dates and harvesting times. *Journal of Food, Agriculture & Environment*, 12(3/4), 251–254. https://doi.org/10.1234/4.2014.5393.

Harisha, C. B., Asangi, H. A., & Singh, R. (2019). Growth, yield, water use efficiency of coriander (*Coriandrum sativum*) affected by irrigation levels and fertigation. *Indian Journal of Agricultural Sciences*, 89(7), 1167–1172.

Hassan, F. A. S., & Ali, E. F. (2014). Impact of different water regimes based on class-A pan on growth, yield and oil content of *Coriandrum sativum* L. plant. *Journal of the Saudi Society of Agricultural Sciences*, 13(2), 155–161. https://doi.org/10.1016/j.jssas.2013.05.001.

Hongal, A., Basavaraja, N., Hongal, S., Hegde, N. K., & Kulkarni, S. (2018). Evaluation of coriander (*Coriandrum sativum* L.) genotypes for yield and quality under hill zone (Zone-9) of Karnataka, India. *International Journal of Current Microbiology and Applied Sciences*, 7(5), 2494–2502. https://doi.org/10.20546/ijcmas.2018.705.287.

Hosseini, M., Boskabady, M. H., & Khazdair, M. R. (2021). Neuroprotective effects of *Coriandrum sativum* and its constituent, linalool: A review. *Avicenna Journal of Phytomedicine*, 11(5), 436–450. https://doi.org/10.22038/AJP.2021.55681.2786.

Izgi, M. N. (2020). Effects of nitrogen fertilization on coriander (*Coriandrum sativum* L.): Yield and quality characteristics. *Applied Ecology and Environmental Research*, 18(5), 7323–7336. https://doi.org/10.15666/aeer/1805_73237336.

Izgi, M. N., Telci, I., & Elmastaş, M. (2017). Variation in essential oil composition of coriander (*Coriandrum sativum* L.) varieties cultivated in two different ecologies. *Journal of Essential Oil Research*, 29(6), 494–498. https://doi.org/10.1080/10412905.2017.1363090.

Kaneko, M. G. (2006). *Produção de coentro e cebolinha em substratos regionais da Amazônia à base de madeira em decomposição (paú)* [Master Thesis, Universidade de Brasília]. https://repositorio.unb.br/handle/10482/5071.

Kassu, K. T., Dawit, H. H., Wubengeda, A. Y., Almaz, A. T., & Asrat, M. T. (2018). Yield and yield components of coriander under different sowing dates and seed rates in tropical environment. *Advances in Horticultural Sciences*, 32(2), 193–203. https://doi.org/10.13128/ahs-21304.

Katar, D., Kara, N., & Katar, N. (2016). Yields and quality performances of coriander (*Coriandrum sativum* L.) genotypes under different ecological conditions. *Turkish Journal of Field Crops*, 21(1), 79–87. https://doi.org/10.17557/tjfc.77478.

Katiyar, R. S., Nainwal, R. C., Singh, D., Chaturvedi, V., & Katiyar, S. K. (2014). Effect of spacing and varieties on growth and yield of coriander (*Coriandrum sativum* L.) on reclaimed sodic waste soil. *Progressive Research*, 9(Special), 811–814.

Khalid, K. A. (2012). Effect of phosphorous fertilization on anise, coriander and sweet fennel plants growing under arid region conditions. *Medicinal and Aromatic Plant Science and Biotechnology*, 6(Special 1), 127–131.

Khodadadi, M., Fotokian, M. H., & Kareem, S. H. S. (2021). Genetics of water deficit stress resistance through phenotypic plasticity in coriander. *Scientia Horticulturae*, 286, 110233. https://doi.org/10.1016/j.scienta.2021.110233.

Kumawat, R., Singh, D., Kumawat, K. R., Choudhary, M., & Kumawat, S. (2017). Effect of moisture stress on yield of coriander (*Coriandrum sativum* L.) genotypes by moisture stress indices. *Journal of Pharmacognosy and Phytochemistry*, 6(4), 1493–1498.

Lal, G., Harisha, C. B., Meena, N. K., Meena, R. D., & Choudhary, M. K. (2017). Performance of coriander varieties (*Coriandrum sativum* L.) under organic management system. *International Journal of Seed Spices*, 7(1), 8–11.

Lal, G., Mehta, R. S., Maheria, S. P., & Sharma, Y. (2014a). Influence of sulphur and zinc on growth and yield of coriander (*Coriandrum sativum* L.). *International Journal of Seed Spices*, 4(2), 32–35.

Lal, G., Saran, P. L., Devi, G., Deepak., & Raj, R. (2014b). Seed production technology of coriander (*Coriandrum sativum*). In Chaudhary, A. K., Rana, K. S., Dass, A., & Srivastav, M (Eds.), *Advances in vegetable agronomy* (1st ed., pp. 214–222), PGS, IARI and DARE, ICAR, New Delhi.

Lima, J. S. S., Bezerra Neto, F., Negreiros, M. Z., Freitas, K. K. C., & Barros Júnior, A. P. (2007). Desempenho agroeconômico de coentro em função de espaçamentos e em dois cultivos. *Revista Ciência Agronômica*, 38(4), 407–413.

Linhares, P. C. F., Oliveira, J. D., Almeida, A. M. B., Neves, A. P. M., Cunha, L. M. M., Paiva, A. C. C., Pereira, B. B. M., & Medeiros, A. P. (2015). Espaçamento e densidades de plantas no surgimento de doenças e pragas e no estiolamento do coentro. *Informativo Técnico do Semiárido*, 9(1), 35–38.

Lira, R. M., Santos, A. N., Silva, E. F. F., Silva, J. S., Barros, M. S., & Gordin, L. C. (2015). Cultivo de coentro em diferentes níveis de salinidade e umidade do solo. *Revista Geama*, 1(3), 293–303.

Machado, F. R., & Marreiros, E. O. (2016). Avaliação de substratos e seu enriquecimento na emergência e desenvolvimento do coentro (*Coriandrum sativum*). *Edição Especial*, 110–121.

Mandal, S., & Mandal, M. (2015). Coriander (*Coriandrum sativum* L.) essential oil: Chemistry and biological activity. *Asian Pacific Journal of Tropical Biomedicine*, 5(6), 421–428. https://doi.org/10.1016/j.apjtb.2015.04.001.

Marichali, A., Dallali, S., Ouerghemmi, S., Sebei, H., & Hosni, K. (2014). Germination, morpho-physiological and biochemical responses of coriander (*Coriandrum sativum* L.) to zinc excess. *Industrial Crops and Products*, 55, 248–257. https://doi.org/10.1016/j.indcrop.2014.02.033.

Marsaro, R., Melo, K. D. A., Seabra Junior, S., & Borges, L. S. (2014). Produção de cultivares de coentro em diferentes telados e campo aberto. *Cultivando o Saber*, 7(4), 362–373.

Martins, B. N. M., Candian, J. S., Tavares, A. E. B., Jorge, L. G., & Cardoso, A. I. I. (2018). Densidade de plantio na produção de coentro. *Revista Mirante*, 11(7), 18–27.

McAusland, L., Lim, M.-T., Morris, D. E., Smith-Herman, H. L., Mohammed, U., Hayes-Gill, B. R., Crowe, J. A., Fisk, I. D., & Murchie, E. H. (2020). Growth spectrum complexity dictates aromatic intensity in coriander (*Coriandrum sativum* L.). *Frontiers in Plant Science*, 11, 462. https://doi.org/10.3389/fpls.2020.00462.

Mehrabani, L. V., Kamran, R. V., Khurizadeh, S., & Nezami, S. S. (2018). Response of coriander to salinity stress. *Journal of Plant Physiology and Breeding*, 8(2), 89–98. https://doi.org/10.22034/JPPB.2018.9804.

Meneses, A. T., Silva, A. R., Meireles, D. A., Silva, A. R., & Oliveira, A. P. (2014). Desempenho vegetativo e agroeconômico do coentro em função de diferentes espaçamentos. *Horticultura Brasileira*, 31(Supplement 2), 828–833.

Menezes, S. M. (2018). *Estado nutricional e acúmulo de nutrientes em coentro fertirrigado por gotejamento contínuo e pulsado* [Master Thesis, Universidade Federal Rural de Pernambuco]. http://www.tede2.ufrpe.br:8080/tede2/handle/tede2/7695.

Menezes, S. M., Silva, G. F., Silva, M. M., Morais, J. E. F., Santos Júnior, J. A., Menezes, D., & Rolim, M. M. (2020b). Continuous and pulse fertigation on dry matter production and nutrient accumulation in coriander. *Dyna*, 87(212), 18–25. http://doi.org/10.15446/dyna.v87n212.78569.

Menezes, S. M., Silva, G. F., Zamora, V. R. Ó., Silva, M. M., Silva, A. C. R. A., & Silva, E. F. F. (2020a). Nutritional status of coriander under fertigation depths and pulse and continuous drip irrigation. *Revista Brasileira de Engenharia Agrícola e Ambiental*, 24(6), 364–371. https://doi.org/10.1590/1807-1929/agriambi.v24n6p364-371.

Meyering, B., Hoeffner, A., & Albrecht, U. (2020). Reducing preharvest bolting in open-field-grown cilantro (*Coriandrum sativum* L. cv. Santo) through use of growth regulators. *HortScience*, 55(1), 63–70. https://doi.org/10.21273/HORTSCI14614-19.

Mohammadipour, N., & Souri, M. K. (2019). Effects of different levels of glycine in the nutrient solution on the growth, nutrient composition, and antioxidant activity of coriander (*Coriandrum sativum* L.). *Acta Agrobotanica*, 72(1), 1759. https://doi.org/10.5586/aa.1759.

Mohanalakshmi, M., Boomiga, M., & Gowtham, T. (2019). Effect of season and growing condition on yield and quality parameters of coriander (*Coriandrum sativum* L.). *International Journal of Chemical Studies*, 7(3), 2989–2993.

Nadeem, M., Anjum, F. M., Khan, M. I., Tehseen, S., El-Ghorab, A., & Sultan, J. I. (2013). Nutritional and medicinal aspects of coriander (*Coriandrum sativum* L.). A review. *British Food Journal*, 115(5), 743–755. https://doi.org/10.1108/0007070131133152.

Okkaoğlu, H., Sönmez, Ç., Şimşek, A. Ö., & Bayram, E. (2015). Effect of salt stress on some agronomical characteristics and essential oil content of coriander (*Coriandrum sativum* L.) cultivars. *Journal of Applied Biological Sciences*, 9(3), 21–24.

Olimpio, L. S. (2017). *Recipientes e densidades de cultivo na produção de coentro em ambiente protegido* [Monograph, Universidade Federal do Ceará]. http://www.repositorio.ufc.br/bitstream/riufc/31097/1/2017_tcc_lsolimpio.pdf.

Oliveira, A. P., Araújo, L. R., Mendes, J. E. M. F., Dantas Júnior, O. R., & Silva, M. S. (2004). Resposta do coentro à adubação fosfatada em solo com baixo nível de fósforo. *Horticultura Brasileira*, 22(1), 87–89. https://doi.org/10.1590/S0102-05362004000100017.

Oliveira, K. P., Freitas, R. M. O., Nogueira, N. W., Praxedes, S. C., & Oliveira, F. N. (2010). Efeito da irrigação com água salina na emergência e crescimento inicial de plântulas de coentro cv. Verdão. *Revista Verde de Agroecologia e Desenvolvimento Sustentável*, 5(2), 201–208.

Oliveira, T. A., Santos, F. F., Pereira, M. T. S., Cavalcante, L. S., Barbosa Júnior, M. R., & Santos, M. A. L. (2017). Avaliação da cultura do coentro sob diferentes níveis de salinidade. *Proceedings of the Inovagri International Meeting*, Fortaleza, Brazil. https://doi.org/10.7127/iv-inovagri-meeting-2017-res1800219.

Özyazici, G. (2021). Influence of organic and inorganic fertilizers on coriander (*Coriandrum sativum* L.) agronomic traits, essential oil and components under semi-arid climate. *Agronomy*, 11, 1427. https://doi.org/10.3390/agronomy11071427.

Pacheco-López, N. A., Cano-Sosa, J., Poblano, F., Ingrid, M., Rodríguez-Buenfil, I. M., & Ramos-Díaz, A. (2016). Different responses of the quality parameters of *Coriandrum sativum* to organic substrate mixtures and fertilization. *Agronomy*, 6, 21. https://doi.org/10.3390/agronomy6020021.

Pamplona, L. J. C., Ferreira, L. L., Silva, F. B. M. D., Morais, C. D. O., Alencar, R. D., Porto, V. C. N., & Alves, C. S. (2021). Cobertura de solo modifica a performance de coentro. *Research, Society and Development*, 10(4), e4661048963. https://doi.org/10.33448/rsd-v10i4.8963.

Passos, D. R. C., Silva, R. F., & Castro, D. M. (2018). Crescimento de plantas de coentro submetidas à água tratada por "toque terapêutico". *Cadernos de Agroecologia*, 13(1), 1–6.

Pereira, M. F. S., Linhares, P. C. F., Maracajá, P. B., Moreira, J. C., & Guimarães, M. C. D. (2011). Desempenho agronômico de cultivares de coentro (*Coriandrum sativum* L.) fertilizado com composto. *Revista Verde de Agroecologia e Desenvolvimento Sustentável*, 6(3), 235–239.

Pinto, A. A., Camara, F. T., Pinto, L. A., Tavares, M. S., & Lima, A. I. S. (2018). Desenvolvimento e produtividade do coentro em função da adubação nitrogenada. *Agrarian Academy*, 5(9), 160–168. https://doi.org/10.18677/Agrarian_Academy_2018a16.

Rabiei, Z., Hosseini, S. J., Pirdashti, H., & Hazrati, S. (2020). Physiological and biochemical traits in coriander affected by plant growth-promoting rhizobacteria under salt stress. *Heliyon*, 6, e05321. https://doi.org/10.1016/j.heliyon.2020.e05321.

Rahimi, A. R., Babaei, S., Mashayekhi, K., Rokhzadi, A., & Amini, S. (2013). Anthocyanin content of coriander (*Coriandrum sativum* L.) leaves as affected by salicylic acid and nutrients application. *International Journal of Bioscience*, 3(2), 141–145. https://doi.org/10.12692/ijb/3.2.141-145.

Rajasekar, M., Arumugam, T., & Kumar, S. R. (2013). Influence of weather and growing environment on vegetable growth and yield. *Journal of Horticulture and Forestry*, 5(10), 160–167. https://doi.org/10.5897/JHF2013.0317.

Ramezani, S., Rahmanian, M., Jahanbin, R., Mohajeri, F., Rezaei, M. R., & Solaimani, B. (2009). Diurnal changes in essential oil content of coriander (*Coriandrum sativum* L.) aerial parts from Iran. *Research Journal of Biological Sciences*, 4(3), 277–281.

Rashed, N. M., & Darwesh, R. K. (2015). A comparative study on the effect of microclimate on planting date and water requirements under different nitrogen sources on coriander (*Coriandrum sativum*, L.). *Annals of Agricultural Science*, 60(2), 227–243. https://doi.org/10.1016/j.aoas.2015.10.009.

Reis, A., & Lopes, C. A. (2016). *Doenças do coentro no Brasil*. 6 pp. Embrapa Hortaliças, Brasília, Brazil. https://www.infoteca.cnptia.embrapa.br/infoteca/bitstream/doc/1066501/1/CT157.pdf.

Rocha, A. O. (2017). *Manejo da podridão de raiz e colo em coentro (Coriandrum sativum L.)* [Master Thesis, Universidade Federal de Alagoas]. http://www.repositorio.ufal.br/jspui/handle/riufal/2984.

Sá, F. V. S., Souto, L. S., Paiva, E. P., Ferreira Neto, M., Silva, R. A., Silva, M. K. N., Mesquita, E. F., Almeida, F. A., & Alves Neto, A. (2016). Tolerance of coriander cultivars under saline stress. *African Journal of Agricultural Research*, 11(39), 3728–3732. https://doi.org/10.5897/AJAR2016.11390.

Sá, M. B., Santos, W., Santos, R. S. S., Silva, J. J. G., Silva, B. L., & Santos, M. A. L. (2017). Análise da produtividade da cultura do coentro em ambiente protegido em função da aplicação de diferentes lâminas de irrigação no agreste alagoano. *Proceedings of the Inovagri International Meeting*, Fortaleza, Brazil. https://doi.org/10.7127/IV-INOVAGRI-MEETING-2017-RES2890380.

Sagarika, G., Sharangi, A. B., & Debnath, S. (2014). Phenology and green leaf yield of coriander at different sowing dates and harvesting times. *Journal of Food, Agriculture & Environment*, 12(3/4), 251–254.

Sales, M. A. L., Moreira, F. J. C., Eloi, W. M., Ribeiro, A. A., Sales, F. A. L., & Monteiro, R. N. F. (2015). Germinação e crescimento inicial do coentro em substrato irrigado com água salina. *Revista Brasileira de Engenharia de Biossistemas*, 9(3), 221–227. https://doi.org/10.18011/bioeng2015v9n3p221-227.

Santana, L. D., Pinto, A. A., Galdino, A. G. S., Camara, F. T., & Lopes, N. S. (2019). Desempenho agronômico do coentro verdão em função do arranjo populacional na região do Cariri cearense. *Applied Research & Agrotechnology*, 12(2), 37–45. https://doi.org/10.5935/PAeT.V12.N2.03.

Santos, M. R., Moreira, C. F., Leonardo, F. A. P., Moreira, A. L., Silva, T. B. M., Carreiro, L. G., Sousa, R. A., Coelho, W. A. A., & Faria, T. S. F. (2021). Produção de coentro em função do tipo de plantio e densidade de semeadura. In Oliveira, R. J (Org.), *Extensão rural práticas e pesquisas para o fortalecimento da agricultura familiar* (1st ed., pp. 563–577), Científica Digital, Guarujá. https://doi.org/10.37885/201202460.

Saxena, S. N., Rathore, S. S., Kakani, R. K., Lal, G., Meena, R. S., Singh, H., & Singh, B. (2013). Effect of water stress on morpho-physiological parameters of coriander (*Coriandrum sativum* L.). *Annals of Arid Zone*, 52(1), 17–21.

Serri, F., Souri, M. K., & Rezapanah, M. (2021). Growth, biochemical quality and antioxidant capacity of coriander leaves under organic and inorganic fertilization programs. *Chemical and Biological Technologies in Agriculture*, 8, 33. https://doi.org/10.1186/s40538-021-00232-9.

Shams, M., Ramezani, M., Esfahan, S. Z., Esfahan, E. Z., Dursun, A., & Yildirim, E. (2016). Effects of climatic factors on the quantity of essential oil and dry matter yield of coriander (*Coriandrum sativum* L.). *Indian Journal of Science and Technology*, 9(6), 1–4. https://doi.org/10.17485/ijst/2016/v9i6/61301.

Sharangi, A. B., & Roychowdhury, A. (2014a). Phenology and yield of coriander (*Coriandrum sativum* L.) at different sowing dates. *Journal of Plant Sciences*, 9(2), 32–42. https://doi.org/10.3923/jps.2014.32.42.

Sharangi, A. B., & Roychowdhury, A. (2014b). Phenology and yield of coriander as influenced by sowing dates and irrigation. *The Biascan*, 9(4), 1513–1520.

Shashidhar, M. D., Pujari, R., Sharatbabu, A. G., Geeta, B. P., Arif, A., & Dharamatti, V. (2017). Cultivation of coriander (*Coriandrum sativum* L.): A review article. *International Journal of Pure & Applied Bioscience*, 5(3), 796–802. https://doi.org/10.18782/2320-7051.4040.

Silva, A. H. S., Silva, M. M., Zamora, V. R. O., Silva, G. F., Santos Júnior, J. A., & Freire, M. M. (2019a). Viabilidade econômica da cultura do coentro sob gotejamento por pulsos e lâminas crescentes de irrigação. *Proceedings of the Inovagri International Meeting*, Fortaleza, Brazil.

Silva, J. C., Costa, L. F. F., Oliveira, J. A., Farias, A. V. A., Santos, L. E. A. S., & Santos, M. A. L. (2019b). Consumo hídrico do coentro sob estratégia de irrigação com água salina no agreste alagoano. *Revista Ambientale*, 11(1), 70–79. https://doi.org/10.34032/ambientale.v11i1.223.

Silva, M. G., Oliveira, I. S., Soares, T. M., Gheyi, H. R., Santana, G. O., & Pinho, J. S. (2018b). Growth, production and water consumption of coriander in hydroponic system using brackish waters. *Revista Brasileira de Engenharia Agrícola e Ambiental*, 22(8), 547–552. https://doi.org/10.1590/1807-1929/agriambi.v22n8p547-552.

Silva, M. G., Soares, T. M., Gheyi, H. R., Costa, I. P., & Vasconcelos, R. S. (2020a). Growth, production and water consumption of coriander grown under different recirculation intervals and nutrient solution depths in hydroponic channels. *Emirates Journal of Food and Agriculture*, 32(4), 281–294. https://doi.org/10.9755/ejfa.2020.v32.i4.2094.

Silva, M. G., Soares, T. M., Gheyi, H. R., Oliveira, I. S., & Silva Filho, J. A. (2016b). Crescimento e produção de coentro hidropônico sob diferentes densidades de semeadura e diâmetros dos canais de cultivo. *Irriga*, 21(2), 312–326. https://doi.org/10.15809/irriga.2016v21n2p312-326.

Silva, M. G., Soares, T. M., Gheyi, H. R., Oliveira, I. S., Silva Filho, J. A., & Carmo, F. F. (2016a). Frequency of recirculation of the nutrient solution in the hydroponic cultivation of coriander with brackish water. *Revista Brasileira de Engenharia Agrícola e Ambiental*, 20(5), 447–454. https://doi.org/10.1590/1807-1929/agriambi.v20n5p447-454.

Silva, M. G., Soares, T. M., Gheyi, H. R., Oliveira, M. G. B., & Santos, C. C. (2020b). Hydroponic cultivation of coriander using fresh and brackish waters with different temperatures of the nutrient solution. *Engenharia Agrícola*, 40(6), 674–683. https://doi.org/10.1590/1809-4430-Eng.Agric.v40n6p674-683/2020.

Silva, V. P. R., Sousa, I. F., Tavares, A. L., Silva, T. G. F., Silva, B. B., Holanda, R. M., Brito, J. I. B., Braga, C. C., Souza, E. P., & Silva, M. T. (2018a). Evapotranspiration, crop coefficient and water use efficiency of coriander grown in tropical environment. *Horticultura Brasileira*, 36(4), 446–452. https://doi.org/10 .1590/S0102-053620180404.

Silva, V. P. R., Tavares, A. L., & Sousa, I. F. (2013). Evapotranspiração e coeficientes de cultivo simples e dual do coentro. *Horticultura Brasileira*, 31(2), 255–259. https://doi.org/10.1590/S0102-05362013000200013.

Singh, H., Panghal, V. P. S., & Duhan, D. S. (2018). Irrigation and nutritional effect on growth and seed yield of coriander (*Coriandrum sativum* L.). *Journal of Plant Nutrition*, 41(13), 1705–1710. https://doi.org/10 .1080/01904167.2018.1459693.

Singh, S. P. (2014). Effect of bio-fertilizer azospirillum on growth and yield parameters of coriander (*Coriandrum sativum* L.) cv. Pant Haritima. *International Journal of Seed Spices*, 4(2), 73–76.

Sousa, V. L. B., Lopes, K. P., Costa, C. C., Pôrto, D. R. Q., & Silva, D. S. O. (2011). Tratamento pré germinativo e densidade de semeadura de coentro. *Revista Verde de Agroecologia e Desenvolvimento Sustentável*, 6(2), 21–26.

Souza, Ê. G. F., Souza, A. R. E., Soares, E. B., Barros Júnior, A. P., Silveira, L. M., & Bezerra Neto, F. (2017). Green manuring with *Calotropis procera* for the production of coriander in two growing seasons. *Ciência e Agrotecnologia*, 41(5), 533–542. https://doi.org/10.1590/1413-70542017415013417.

Tavares, A. L. (2016). *Fenometria, produtividade e necessidades hídricas das culturas da alface e do coentro em clima tropical* [Doctoral Thesis, Universidade Federal de Campina Grande]. http://dspace.sti.ufcg .edu.br:8080/jspui/handle/riufcg/1441.

Ullah, M. A., Rasheed, M., & Mahmood, I. A. (2018). Salt tolerance of coriander (*Coriandum sativum*) as medicinal plant under integrated salinity and sodicity conditions. *Academia Journal of Medicinal Plants*, 6(10), 342–346. https://doi.org/10.15413/ajmp.2018.0168.

Unlukara, A., Beyzi, E., Ipek, A., & Gurbuz, B. (2016). Effects of different water applications on yield and oil contents of autumn sown coriander (*Coriandrum sativum* L.). *Turkish Journal of Field Crops*, 21(2), 200–209. https://doi.org/10.17557/tjfc.46160.

Vasconcelos, L. S. B., Bezerra Neto, E., Nascimento, C. W. A., & Barreto, L. P. (2014). Desenvolvimento de plantas de coentro em função da força iônica da solução nutritiva. *Pesquisa Agropecuária Pernambucana*, 19(1), 11–19. https://doi.org/10.12661/pap.2014.003.

Verma, P., Doshi, V., & Solanki, R. K. (2014). Genetic variability assessed in coriander (*Coriandrum sativum* L.) over years under environmental conditions of South Eastern Rajasthan (Hadoti Region). *International Journal of Seed Spices*, 4(2), 94–95.

Wei, J.-N., Liu, Z.-H., Zhao, Y.-P., Zhao, L.-L., Xue, T.-K., & Lan, Q.-K. (2019). Phytochemical and bioactive profile of *Coriandrum sativum* L. *Food Chemistry*, 286, 260–267. https://doi.org/10.1016/j.foodchem .2019.01.171.

Zamora, V. R. O. (2018). *Gotejamento por pulsos sob cinco lâminas de fertirrigação na produtividade da cultura do coentro* [Master Thesis, Universidade Federal Rural de Pernambuco]. http://www.tede2.ufrpe .br:8080/tede2/handle/tede2/7638.

Zamora, V. R. O., Silva, M. M., Santos Júnior, J. A., Silva, G. F., Menezes, D., & Almeida, C. D. G. C. (2021). Assessing the productivity of coriander under different irrigation depths and fertilizers applied with continuous and pulsed drip systems. *Water Supply*, 21(5), 2099–2108. https://doi.org/10.2166/ws.2021 .008.

Zamora, V. R. O., Silva, M. M., Silva, G. F., Santos Júnior, J. A., Menezes, D., & Menezes, S. M. (2019). Pulse drip irrigation and fertigation water depths in the water relations of coriander. *Horticultura Brasileira*, 37(1), 22–28. https://doi.org/10.1590/S0102-053620190103.

4 Production of Coriander Using the Hydroponic Technique

*Mairton Gomes da Silva, Tales Miler Soares
and Hans Raj Gheyi*

CONTENTS

4.1 INTRODUCTION

Coriander (*Coriandrum sativum* L.) is one of the spices that occupy a prominent position world-wide, adding flavor and aroma to various foods (Rashed and Darwesh, 2015; López et al., 2016). In addition to its use in cooking, coriander is used in the food, cosmetic and pharmaceutical industries due to its medicinal properties (Uitterhaegen et al., 2018; Khodadadi et al., 2021).

In Brazil, especially in the North and Northeast regions (Santos Júnior et al., 2015; Silva et al., 2016a; 2020a; 2020b; 2022), as well as in other parts of the world (Divya et al., 2014; Jamila et al., 2019; Mohammadipour and Souri, 2019; Wei et al., 2019; Nguyen et al., 2020; Amiripour et al., 2021; Özyazici, 2021), fresh biomass is the main form of coriander consumption.

Although coriander is a crop more tolerant to high temperatures than other leafy vegetables (Silva et al., 2016b), its production has been affected by abiotic stresses and climatic factors (Khodadadi et al., 2021). As a result of soil degradation and increasingly scarce water resources, maintaining the production standards and quality of coriander and other leafy vegetables has become increasingly difficult.

In this context, soilless cultivation (hydroponics) has expanded rapidly since this cultivation technique can be implemented in small areas, and there is no need for crop rotation as practiced in soil. The main advantage of this type of cultivation concerns the higher water use efficiency than conventional planting, as seen in Silva et al. (2021), a desirable characteristic for places with a limited supply of good quality water. Furthermore, in hydroponics, other advantages can be obtained, such as producing better quality all year round, higher yield and early harvest. Additionally, the responses of coriander to salinity have been promising in hydroponics (Cazuza Neto et al., 2014a; Silva et al., 2015a; 2015b; 2018; 2020c; 2022).

Many studies have been conducted with coriander under hydroponic conditions because, in this type of cultivation, the nutrients are more accessible to plants, ensuring earlier results compared to

DOI: 10.1201/9781003204626-5

cultivation in soil (Daflon et al., 2014; Kulkarni et al., 2016; AlQuraidi et al., 2019; Hu et al., 2019). Furthermore, many studies envision hydroponics not only in experimentation but also as a commercial enterprise, in which typical structures of commercial cultivation are used (Lennard and Ward, 2019; Silva et al., 2020a; 2020b; 2020c; 2022). In this case, the results found in these studies could be achieved in commercial hydroponic enterprises.

This chapter addresses different aspects related to the hydroponic production of fresh biomass of coriander, such as the main hydroponic systems used, sowing density, hydroponic channels, depths and circulation frequencies of the nutrient solution in the cultivation channels and cultivation under the combined and/or individual stresses of salinity and temperature in the root-zone, among others.

4.2 HYDROPONIC SYSTEMS USED IN CORIANDER CULTIVATION

In Brazil, coriander is a rising crop under hydroponic cultivation, mainly for fresh biomass production. Like other leafy vegetables, such as lettuce, both on a commercial scale and in research, the most used system for the cultivation of this species is the nutrient film technique (NFT) (Donegá, 2009; Luz et al., 2012; Cazuza Neto al., 2014a; Silva et al., 2015a; Pessoa, 2020; Figures 4.1A and 4.1B). In this system, the channels are installed with a certain slope (Figure 4.1B). With this, a thin film of the solution runs through the channel when the system is in operation; when the system is turned off, the solution stops flowing.

Other systems often used in studies with coriander are deep flow technique (DFT) or conventional floating (Silva et al., 2020a; Figures 4.2A and 4.2B), DFT adapted in circular section pipes (Santos Júnior et al., 2015; Cavalcante et al., 2016; Silva et al., 2018; 2019; 2020b; Cruz et al., 2020; Oliveira et al., 2020; Figures 4.2C and 4.2D) and cultivation in channels filled with substrate, irrigating with nutrient solution (Rebouças et al., 2013; Oliveira et al., 2016; Maia, 2017; Diniz et al., 2019; Figure 4.2E).

4.2.1 HYDROPONIC SYSTEMS COMPONENTS

In the conventional DFT system, plants require support for accommodation above the nutrient solution, such as the use of polystyrene sheets (Figures 4.2A and 4.2B). This system (in buckets, tanks, pools, pots, etc.) can be of the closed type without renewal of the solution, requiring continuous

FIGURE 4.1 Coriander plants grown in polypropylene hydroponic channels of rectangular section (A) and with flattened bottom (B) using a nutrient film technique (NFT) system. A – source: Donegá (2009) and B – (photo by the authors).

FIGURE 4.2 Coriander plants grown in the conventional deep flow technique (DFT) system in benches (A and B), in the adapted DFT system with PVC tubes of circular section (C and D) and in channels with substrate (E). A–D (photo by the authors) and E – source: Maia (2017).

aeration as done by Neffati and Marzouk (2009) and Hazeri et al. (2012), or with the renewal of the solution at each frequency of circulation as done by Silva et al. (2020b)[*] in studies with coriander.

In the NFT system, cultivation channels in rectangular sections of polypropylene (Figure 4.1A) and channels with a flattened bottom (Figure 4.1B), developed specifically for hydroponics to accommodate the plants, have been used. Despite the benefits of these hydroponic channels with such shapes, in many locations far from production/distribution centers, the access has been restricted, mainly due to shipping costs. In this context, sewage-type (white color, as shown in Figure 4.3A) and irrigation-type (blue color, as shown in Figures 4.3B and 4.3C) PVC pipes have been used. These channels have been used in the DFT system adapted in pipes (Figures 4.2C, 4.2D, 4.3A, and 4.3B) and the NFT system (Figure 4.3C).

---

[*] The circulation of nutrient solutions was based on alternating intervals of 0.25 h; the system was turned on for 0.25 h and turned off for 0.25 h, from 06:00 to 18:00 h; from 18:00 to 06:00 h, the system was turned on for 0.25 h, every 2 h. Similar programming was also used in the NFT system.

FIGURE 4.3   Coriander plants grown in the adapted DFT system with PVC pipes (sewage-type) of a circular section of 100 mm in diameter (A) and irrigation-type pipes of 50 and 75 mm in diameter (B), and in the NFT system with irrigation-type pipes of 50 mm in diameter (C). A – source: Cavalcante et al. (2016), B – (photo by the authors) and C – (photo is courtesy of the hydroponic producer, São Felipe – BA, Brazil).

Regarding the length of the cultivation channels, it is recommended not to use very long channels. For example, currently, companies sell channels with lengths shorter than 6.0 m (the standard length of sewage- and irrigation-type pipes). The choice of the hydroponic channels as to their length must first consider their arrangement within the protected environment. For example, there should be easy access between the benches, and there should not be a great height difference between the beginning and end of the bench, as the ends can be very low and/or high in the case of level ground.

Long cultivation channels, when employed, may result in a gradient of production because of changes in the ionic proportions of the nutrient solution; that is, plants located at the end of the channel receive a lower amount of nutrients, including dissolved oxygen (Silva et al., 2016b). In this context, studies were conducted in the NFT hydroponic system with 4.5-m-long channels (Luz et al., 2012) and in the DFT hydroponic system in 6.0-m-long channels (Silva et al., 2016b; 2017a), evaluating the coriander 'Verdão' cultivated in different positions in the cultivation channel (initial, intermediate and final). In the study of Luz et al. (2012), despite the use of short channels, shoot fresh biomass (SFB) production was approximately 27% lower in plants at the end of the channels compared to those harvested at the beginning of the channels.

In the study conducted by Silva et al. (2016b), the production gradient was less pronounced; when plants were cultivated in the intermediate and final positions of the channels, the mean SFB was approximately 20% lower than the mean obtained in the initial position of the channels, in cultivation with nutrient solution circulation at 0.25-h intervals. However, when the solutions were circulated with longer intervals, for example, every 4 or 8 h, there was no significant gradient in the production of coriander harvested in the different positions of the cultivation channel. These results show that the effect of plant position in the cultivation channel was minimized under stress conditions with low aeration of the solution. This assertion is supported by the study conducted by Silva et al. (2017a), who recorded that there was no significant difference in the SFB of coriander cultivated in the different channel positions when subjected to water salinity levels of 4.91 and 7.00 dS m⁻¹.

In the first studies with coriander in the adapted DFT system, the following cultivation conditions were used: 0.040 m solution depth in a 100-mm-diameter cultivation channel and solution circulation frequency of twice a day (Santos Júnior et al., 2015; Cavalcante et al., 2016) and a 0.045 m depth in a 75-mm-diameter channel and solution circulation frequency of every two hours (Silva et al., 2016a).

In a later study conducted by Silva et al. (2016b), as shown in Figure 4.4A, the differences in coriander SFB production were minimal when plants were cultivated under different intervals of solution circulation in the channels with a 0.045 m depth. In the study carried out by Silva et al. (2020b), as shown in Figure 4.4B, when the intervals between recirculations were increased, the means of SFB were approximately 37 and 45% lower under intervals of 12 and 24 h (twice and once a day, respectively) compared to the control (0.25 h intervals), in an experiment carried out in the autumn

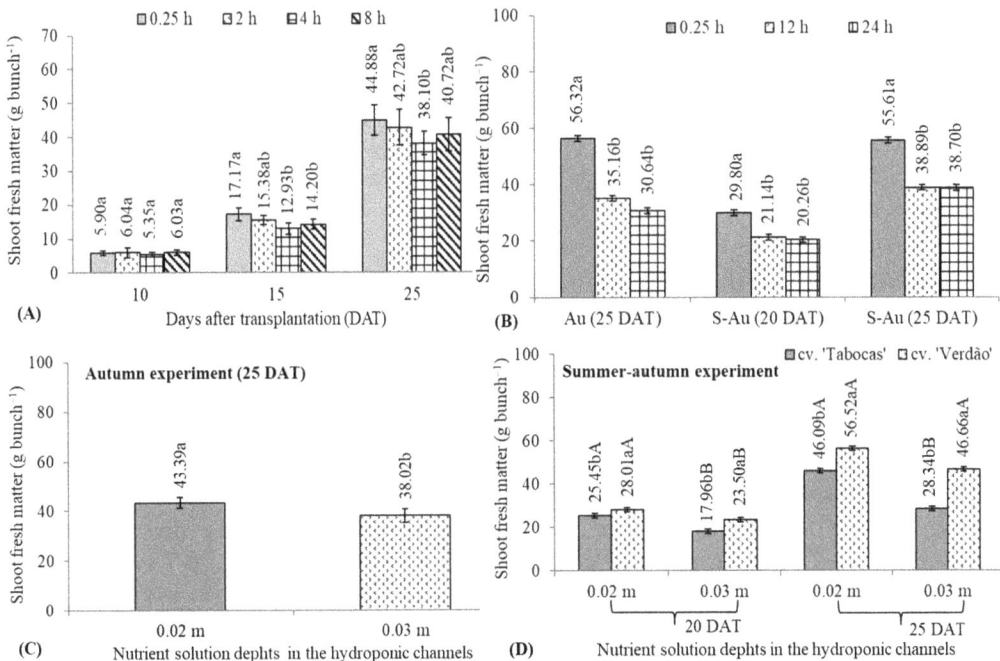

FIGURE 4.4 Shoot fresh biomass (bunch with 12 plants) of coriander grown under different recirculation intervals (A and B) and depths of nutrient solution (C and D) in the hydroponic channels. A – source: adapted from Silva et al. (2016b) and B, C and D – source: Silva et al. (2020b). Notes: A and B (in each cultivation period) and C – means followed by the same letter are not significantly different at $p = 0.05$ by Tukey test; D – in each cultivation period, lowercase letters compare the means of the solution depths in each coriander cultivar, and uppercase letters compare the means of the coriander cultivars in respective solution depth at $p = 0.05$ by Tukey test; Au – autumn; S-Au – summer-autumn.

for 25 days. When the experiment was carried out in the summer-autumn for the same period, the losses under the 12 and 24 h intervals were lower, approximately 30%.

Silva et al. (2020b) also evaluated solution depths in the hydroponic channels (Figures 4.4C and 4.4D). In general, the highest coriander yields were obtained with the lowest depth (0.02 m), and the highest depth of solution (0.03 m) affected the production of the coriander cultivar 'Tabocas' more drastically in comparison to coriander cultivar 'Verdão'. As conclusions, the authors indicate that the greater volume of solution in the channel did not benefit coriander cultivation; that is, it may have impaired the aeration of the solution. Therefore, in this case, it is preferable to use a smaller depth of solution (0.02 m), resulting in the saving of water and fertilizers to prepare the nutrient solution.

Concerning the diameter of the hydroponic channels, Silva et al. (2016a) evaluated the cultivation of coriander in PVC irrigation-type channels with 50 and 75 mm diameters in the adapted DFT system and Soares et al. (2017) in polypropylene-type channels with 58 and 90 mm diameters in the NFT system. In the first study, the authors recorded 21% higher coriander yield (SFB of the bunch of plants) in channels with 75 mm diameter than in those with 50 mm diameter (there was an overflow of the solution at the end of the cultivation cycle, proving to be unfeasible when the channels were level). In commercial production, as shown in Figure 4.3C, irrigation-type channels with 50 mm diameter have been used to cultivate coriander in the NFT system.

In the study carried out by Soares et al. (2017), as shown in Table 4.1 on the topic 'Coriander Sowing Density', when coriander was cultivated in channels with the largest diameter of 90 mm, there was higher SFB production per plant compared to the production obtained in channels with a diameter of 58 mm. However, when evaluating SFB per area (kg m^{-2}), the opposite behavior occurred; that is, there was higher production under cultivation in the 58-mm-diameter channels. This is due to the spacing used; for example, in the channels with 58 mm diameter, the spacings were 0.10 m × 0.10 m between bunches of plants and channels, whereas, in the channels with 90 mm diameter, the spacings were 0.25 m × 0.30 m, respectively. Therefore, the number of bunches of plants per linear meter was variable between the cultivation channels with different diameters. Therefore, in terms of production, this study shows that it is feasible to use hydroponic channels with smaller diameters in the NFT system, as has been done in commercial production.

4.2.2 Oxygenation versus Temperature and Circulation Frequency of the Nutrient Solution

In hydroponic cultivation, it is of great importance to maintain adequate conditions of oxygenation of the circulating solution. For example, in the NFT system, the fall of the solution that is not used by plants when returning to the reservoir by drainage favors its oxygenation.

According to Silva et al. (2020c), under high-temperature conditions, there is a reduction in dissolved oxygen (DO) availability in the solution; consequently, there may be an increase in the incidence of *Pythium* sp., impairing plant growth and development. In the study carried out by Silva et al. (2020a), the temperatures and DO concentrations of the solution in the DFT system adapted in channels (Figures 4.5A and 4.5B) and the conventional DFT system (Figures 4.5C and 4.5D) under solution depths were measured, in the morning and afternoon at different stages of coriander growth. In general, changes in DO concentrations were more pronounced in the DFT system in channels (Figure 4.5A), with mean values in the first two evaluations (at 6 and 11 days after transplanting – DAT) ranging from 5.9–6.8 mg L^{-1} (morning) to 5.3–5.5 mg L^{-1} (afternoon). In the other evaluations, DO concentrations were on average 5.0 mg L^{-1} under cultivation with a 0.02 m depth in the morning. In the afternoon, DO concentrations ranged from 4.2 to 3.5 mg L^{-1}.

In the DFT system adapted in pipes, temperatures on the DO are evident; with the increase in the temperatures throughout the day, there were reductions in the DO concentrations. In the morning, the temperatures of the solutions did not exceed 30°C, while in the afternoon, they were always

TABLE 4.1

Shoot Fresh Biomass per Plant (SFBP) and per Area (SFBA) of Coriander Grown During 32 Days after Sowing (Winter Season) in NFT Hydroponic System under Different Cultivation Conditions, in Lagoa Seca – PB, Brazil

Cultivar	Cultivation Conditions	SFBP (g plant^{-1})	SFBA (kg m^{-2})[#]
'Tabocas'	**Cultivation of 5 plants per cell (phenolic foam):**		
	Spacing of 0.10 m × 0.10 m between bunches of plants and channels of 58 mm in diameter	3.11[*]	2.46
	Spacing of 0.25 m × 0.30 m between bunches of plants and channels of 90 mm in diameter, respectively	5.02[**]	0.46
	Cultivation of 8 plants per cell:		
	Spacing of 0.10 m × 0.10 m between bunches of plants and channels of 58 mm in diameter	3.11[*]	2.05
	Spacing of 0.25 m × 0.30 m between bunches of plants and channels of 90 mm in diameter, respectively	5.02[**]	0.70
'Verdão'	**Cultivation of 5 plants per cell:**		
	Spacing of 0.10 m × 0.10 m between bunches of plants and channels of 58 mm in diameter	2.96	2.60
	Spacing of 0.25 m × 0.30 m between bunches of plants and channels of 90 mm in diameter, respectively	3.98	0.46
	Cultivation of 8 plants per cell:		
	Spacing of 0.10 m × 0.10 m between bunches of plants and channels of 58 mm in diameter	2.96	3.49
	Spacing of 0.25 m × 0.30 m between bunches of plants and channels of 90 mm in diameter, respectively	3.98	0.86

Source: Adapted from Soares et al. (2017).

[#] Values converted from the unit kg ha^{-1}.

[*] and [**] Values are the means from two quantities of plants per phenolic foam cell (there was no significant difference between the means as a function of quantities of plants); similar behavior was observed for cultivar 'Verdão'.

higher than 30°C (Figure 4.5B). In the conventional DFT system, there was little oscillation of the DO concentrations along the cultivation period, regardless of the solution depth, ranging from 5.9–7.0 to 5.4–6.3 mg L^{-1} in the morning and afternoon, respectively (Figure 4.5C). According to data from the same study, as seen in Silva et al. (2017b), the temperatures of the solutions remained relatively constant throughout the day, so the temperatures taken simultaneously with the DO concentrations also changed little as a function of the measurement times, regardless of the solution depths in the benches (Silva et al., 2020a; as shown in Figure 4.5D).

Therefore, it is evident that the materials used in these systems were determinants for temperature attenuation. For example, due to the low thermal transmittance of the polystyrene sheets in the conventional DFT system (Figures 4.2A and 4.2B), the solution's heating was lower than the DFT system in pipes. These higher temperatures of the solution (consequently lower DO concentrations) in the DFT in pipes are due to the incidence of radiation on the cultivation channel, which favored the exchange of heat with the solution.

Despite the lower DO concentrations in the DFT system in pipes and the conventional DFT system (Figure 4.6A), the quality of coriander plants and even their SFB production were not affected (Figure 4.6B).

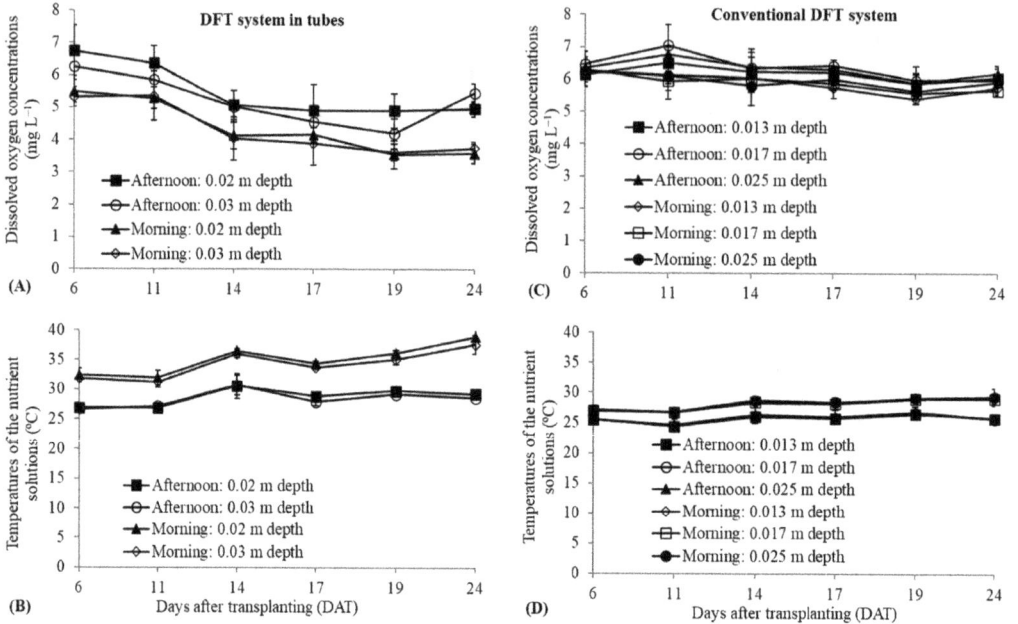

FIGURE 4.5 Temperatures and dissolved oxygen concentrations of the nutrient solution in the hydroponic channels of the adapted DFT system (A and B) and the cultivation benches of the conventional DFT system (C and D) for coriander grown under different solution depths. Adapted from Silva et al. (2020a).

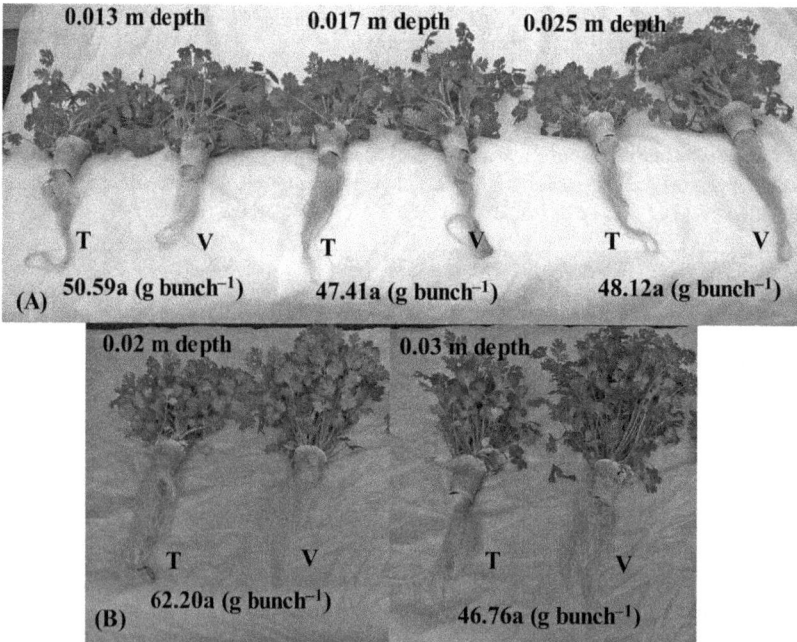

FIGURE 4.6 Visual aspect and shoot fresh biomass (bunch with 12 plants) of coriander cultivar 'Tabocas' (T) and cultivar 'Verdão' (V) grown in the conventional DFT system (A) and in the adapted DFT system (B) under different nutrient solution depths. Adapted from Silva et al. (2020a). In each cultivation system, means followed by the same letter are not significantly different at $p = 0.05$ by Tukey test; the means were obtained as a function of the two coriander cultivars.

4.3 CORIANDER SOWING DENSITY

As already mentioned, bunches of fresh biomass are the main way of commercializing coriander, mainly for cooking. Therefore, sowing is performed with a certain number of seeds per unit of cultivation (cup/container). Different growth media (substrates) for the production of coriander seedlings can be used, for example, as shown in Figures 4.2B and 4.2D, plastic cups filled with coconut fiber (with holes at the bottom for roots to pass through). Similar procedures for seedling production have been adopted in commercial production. Upon reaching the harvest point, the whole cup with the plants (Figure 4.7A) is taken to the consumer market. Also, polystyrene and/or polyethylene trays filled with different substrates have been used for the production of seedlings (Donegá, 2009; Figures 4.7B. Seedlings have also been produced in phenolic foams (Figure 4.7D).

Tables 4.1 and 4.2 show the compendium of studies with coriander under different hydroponic systems and cultivation conditions, considering the variable of greatest economic interest (SFB). These studies were carried out in different seasons and different stages of growth (harvest point). This partially explains the different results for SFB production. In addition to the peculiarities of each hydroponic system and differences between cultivars, another explanation concerns the sowing density, that is, the number of plants per unit of cultivation.

FIGURE 4.7 Bunch of coriander plants for commercialization (A), production of seedlings in trays containing substrate – seedling stage (B), plants with commercial size (C) and seedlings production in phenolic foam (D). A – (photo courtesy of the hydroponic producer, Serra Preta – BA, Brazil), B – source: Olimpio (2017), C – source: Donegá (2009) and D – (photo by the authors).

As shown in Table 4.2, in the study carried out by Silva et al. (2016a), the SFB per plant decreased as the number of plants per unit of cultivation increased, ranging from 1.92 to 1.24 g plant⁻¹ at the densities of 12 and 60 plants cup⁻¹, respectively. However, when the SFB of the bunch of plants was evaluated, regardless of density, the SFB was the same, with an average of 44.42 g bunch⁻¹. This means that individual losses were offset with more plants per cup.

Also, in Table 4.2, in the study carried out by Silva, M. G. (personal communication), with cultivation in the NFT system, the author observed high SFB production per plant, ranging from 9.26 and 5.31 g plant⁻¹ at the densities of 6 and 30 plants cup⁻¹, respectively.

Notably, the lower number of 6 plants cup⁻¹ in the study of Silva, M. G. (personal communication), led to higher SFB per plant (9.26 g plant⁻¹). Higher SFB values were recorded, equal to 11.5 g plant⁻¹ (Cazuza Neto et al., 2014a),[*] 34.6 g plant⁻¹ (Currey et al., 2019)[†] and 19.2 g plant⁻¹ (Nguyen et al., 2019).[‡]

When evaluating SFB production per unit of area (kg m⁻²) (Tables 4.1 and 4.2), the results are quite different; that is, studies with the highest SFB per plant did not necessarily find higher SFB per area or vice versa. These results are related to the spacings used between bunches of plants and/or individual plants in the cultivation rows (hydroponic channels and/or channels with substrate). For example, as shown in Table 4.2, despite the lower individual SFB (1.10 g plant⁻¹), there was a higher SFB per area (2.92 kg m⁻²) in the study carried out by Rebouças et al. (2013) compared to the values of 1.56 kg m⁻² (3.12 g plant⁻¹) in the study carried out by Oliveira et al. (2016), and 1.18 kg m⁻² (5.93 g plant⁻¹) in the study carried out by Maia (2017). In these studies, coriander was cultivated in channels with the substrate (fertigation with nutrient solutions), where the sowing was similar to that used in soil, that is, seeds in a large amount (high competition) were arranged in furrows (holes) opened in the substrate. In part, this explains such results.

On the other hand, the studies carried out by Santos Júnior et al. (2015) and Silva et al. (2016a) with the DFT system in pipes, as a consequence of the highly dense spacing between bunches of plants (every 0.07 m), reported the highest SFB yields per area, of the order of 5.90 and 3.19 kg m⁻², respectively.

4.4   CORIANDER CULTIVATION SUBJECTED TO ISOLATED AND/OR COMBINED STRESSES OF SALINITY AND ROOT-ZONE TEMPERATURE

Salinity is the principal abiotic stress that can dramatically affect crop yield. Salinity has increased rapidly in different arid and semi-arid regions due to inadequate agricultural management. This is because, in traditional cultivation in soil, due to the scarcity of low-salinity water, brackish waters have been used for the irrigation of crops (Silva et al., 2018; 2020c). In this context, the technique of soilless cultivation (hydroponics) has been increasingly recommended, mainly for the production of leafy vegetables.

In hydroponic cultivation, the responses of plants to salinity can be better than in the soil for the same level of water salinity, considering the following hypothesis: while in traditional soil-based cultivation, the plant responds to two components of water stress, which are drought stress and osmotic stress, in hydroponics, as there is no water/air interface between solid particles, the medium has no matric potential, effectively resulting in the effect of only one of these stresses: osmotic stress (Silva et al., 2018; 2021; 2022).

---

[*] Value obtained under control treatment (without salt stress) – cultivation in the NFT system with bunch of 8 plants of coriander 'Verdão' (11 DAS + 26 DAT).

[†] Value obtained under high photosynthetic daily light integrals (~ 18 mol m⁻² d⁻¹) and electrical conductivity levels of the solution (0.5, 1.0, 2.0, 3.0 or 4.0 dS m⁻¹) – cultivation in the NFT system with 1 plant per phenolic foam cube of coriander 'Santo' (21 DAS + 28 DAT).

[‡] Value obtained by the sum of the fresh biomass of leaves and stems under root-zone temperature at 25°C and photosynthetic photon flux density of 300 μmol m⁻² s⁻¹ – cultivation in the conventional DFT system with 1 plant of coriander per sponge cube (18 DAS + 19 DAT).

TABLE 4.2
Shoot Fresh Biomass per Plant (SFBP) and per Area (SFBA) of Coriander Grown under Different Cultivation Conditions and Hydroponic Systems

Location	Cultivar	System/Cultivation Conditions	Growth Days/ Season	SFBP (g plant^{-1})	SFBA (kg m^{-2})	Reference
Mossoró – RN, Brazil	'Verdão'	Channels with substrate	28 DAE/ spring	1.10[1]I	2.92I	Rebouças et al. (2013)
Mossoró – RN, Brazil	'Verdão'	Channels with substrate/ cultivation conditions with 50 plants m^{-2} and channels spaced at 0.10 m	35 DAS/ winter	3.12[2]	1.56II	Oliveira et al. (2016)
Mossoró – RN, Brazil	'Verdão'	Channels with substrate/ cultivation conditions with 20 plants m^{-2} and channels spaced at 0.10 m	42 DAS	5.93[3]	1.18II	Maia (2017)
Campina Grande – PB, Brazil	'Tabocas'	DFT in PVC tubes/1 g of seeds per cup and spacing of 0.07 m used between bunches of plants (cups)	28 DAS/ winter		5.90	Santos Júnior et al. (2015)
Cruz das Almas – BA, Brazil	'Verdão'	DFT in PVC tubes/ cultivation of 12, 24, 36, 48 and 60 plants cup^{-1} and spacing of 0.07 m × 0.20 m between bunches of plants and channels, respectively	13 DAS + 25 DAT/spring	1.92, 1.75, 1.58, 1.41 and 1.24[4]	3.19[5]II	Silva et al. (2016a)
Cruz das Almas – BA, Brazil	'Tabocas' 'Verdão'	Conventional DFT/spacing of 0.25 m × 0.25 m between bunches of plants	10 DAS + 25 DAT/ autumn	3.59 4.53	0.69II 0.87II	Silva et al. (2020a)[6]
	'Tabocas' 'Verdão'	DFT in PVC tubes/spacing of 0.25 m × 0.30 m between bunches of plants and channels, respectively		3.83 5.25	0.61II 0.84II	
Cruz das Almas – BA, Brazil	'Tabocas' 'Verdão'	NFT/spacing of 0.25 m × 0.25 m between bunches of plants and channels, respectively	13 DAS + 25 DAT/ summer	3.44 6.33	0.66II 1.22II	Silva et al. (2020c)[6]
Cruz das Almas – BA, Brazil	'Verdão'	NFT/spacing of 0.25 m × 0.25 m between bunches of plants and channels, respectively	10 DAS + 25 DAT/ autumn-winter	7.22	1.38II	Silva et al. (2022)[6]
Cruz das Almas – BA, Brazil	'Verdão'	NFT/cultivation of 6, 12, 18, 24 and 30 plants cup^{-1} and spacing of 0.25 m × 0.25 m between bunches of plants and channels, respectively	10 DAS + 25 DAT/ summer	9.26, 8.28, 7.29, 6.30 and 5.31[7]	0.84, 0.95, 1.06, 1.17 and 1.28[8]II	Silva, M. G. (personal communication)[9]

[1] Sum of the fresh biomass of leaves and stems.

I Results obtained under control treatment.

[2] Result obtained with a concentration of nutrient solution at 100%.

II Values calculated by the authors of this chapter.

[3] Value obtained when the coriander plants were grown without salt stress and ascorbic acid concentration (AAC) of 0 mol L^{-1} (SFBP = -4.257^{**}AAC^2 + 4.611^{**}AAC + 5.926; ** significant at $p \leq 0.01$ by t-test).

TABLE 4.2 (CONTINUED)

Shoot Fresh Biomass per Plant (SFBP) and per Area (SFBA) of Coriander Grown under Different Cultivation Conditions and Hydroponic Systems

[4] Values estimated based on model SFBP = –0.014**SD + 2.0836** (SD is sowing densities – number of plants cup^{-1}; ** significant at $p \le 0.01$ by t-test).

[5] There was no significant effect on the shoot fresh biomass of the bunch of plants (SFBB), with a mean of 44.42 g $bunch^{-1}$.

[6] Density of 12 plants cup^{-1} and the results were obtained under control treatment.

[7] Values estimated based on model SFBP = –0.1649**SD + 10.2548** (** significant at $p \le 0.01$ by t-test).

[8] Values estimated based on SFBB using the model SFBB = 1.1301**SD + 46.0758** (** significant at $p \le 0.01$ by t-test).

[9] Experiment carried out during the months of January to February 2021, and results obtained under control treatment (without salt stress).

DAE – days after emergence; DAS – days after sowing; DAT – days after transplanting.

Under hydroponic conditions, there is a large number of studies with coriander involving different strategies for the use of brackish waters, as shown in Table 4.3, for example, the use of brackish waters only to replenish water consumption by plants, in this case, the initial salinity of the nutrient solution is only caused by fertilizers used in the preparation of the nutrient solution, and the electrical conductivity of the solution (ECsol) is gradually increased with the addition of salts present in the brackish water (Figure 4.8A); the exclusive use of brackish waters to prepare the solution and replenish water consumption (Silva et al., 2016b; 2018; 2020c; 2022), with increasing ECsol during the cultivation cycle (Figure 4.8B); and the use of brackish waters only to prepare the solution (Silva et al., 2022), with ECsol remaining relatively constant.

As shown in Table 4.3, different yields of coriander are obtained using brackish waters according to the strategy employed. For example, in the study carried out by Silva et al. (2015a), when brackish waters were used only to replenish the water consumption of coriander, regardless of the salinity level, SFB yields were the same (there was no significant difference, with an average of 49.79 g $bunch^{-1}$). With a similar strategy, under electrical conductivity of the water (ECw) of 6.50 dS m^{-1}, Silva et al. (2022) recorded a mean SFB of 76.53 g $bunch^{-1}$, a value close to that obtained under the condition without salt stress (ECw 0.30 dS m^{-1}, 86.59 g $bunch^{-1}$), although they are statistically different.

In addition to the strategy employed for brackish waters, the different results presented in Table 4.3 are also explained by the salinity levels adopted at exposure to salts and the cultivation period. In addition, the response of coriander to salt stress depends on the type of salts to which it has been subjected. In these studies presented in Table 4.3, the brackish waters were artificially produced by the addition of NaCl; however, natural brackish waters contain in their composition several other salts. For example, in the study carried out by Ahmadi and Souri (2018) with coriander irrigated with artificially produced waters (with the same ECw level of 4.0 dS m^{-1}) with different mixtures of salts, these authors recorded lower (by approximately 35%) mean SFB using NaCl compared to the mixture of $KCl + NaCl + CaCl_2$. Furthermore, in the studies of Silva, M. G. (personal communication, as shown in Table 4.4), and Oliveira et al. (2020), as shown in Table 4.6, brackish waters with different cationic natures were also produced.

It is expected that the responses of plants to salinity are better in hydroponic cultivation; however, under natural conditions, plants are also exposed to complex interactions with other abiotic stresses, such as high temperatures when cultivation is carried out in warmer climate regions. Furthermore, as reported by other authors (Silva et al., 2017c; Cocetta et al., 2018), the heating of the hydroponic nutrient solution further increases salinity in the cultivation medium due to the increase of the root-zone temperature, as can be seen in Figure 4.9. In this context, some studies were conducted with coriander to evaluate the

TABLE 4.3

Shoot Fresh Biomass of the Bunch (SFBB) of Plants of the Coriander Grown under Isolated and/or Combined Stresses of Salinity and Root-Zone Temperature (RZT) with Different Strategies Using Fresh and Brackish Waters

Cultivar	System/Growth Days/Season	Strategies Using Brackish Waters	ECw (dS m⁻¹)	RZT (°C)	SFBB (g bunch⁻¹)	Reference
'Verdão'	NFT/11 DAS + 26 DAT/winter	EUWS	0.30–7.73	Ambient temperature	92.66–30.13#	Cazuza Neto et al. (2014a)[1]
'Verdão'	NFT/10 DAS + 24 DAT/spring	WSRW	0.43–8.53	(29.04 ± 3.08)[I(&)]	49.79##	Silva et al. (2015a)
'Verdão'	DFT in pipes/11 DAS + 25 DAT/summer	EUWS	0.32 and 4.91	Ambient temperature	50.33a and 32.87b	Silva et al. (2016b)[2]Y
'Verdão'	DFT in pipes/8 DAS + 25 DAT/winter	EUWS	0.26–7.00	Ambient temperature	44.05–25.72##	Silva et al. (2018)[2]
'Tabocas'	NFT/13 DAS + 25 DAT/summer	EUWS	0.30 and 6.50	(28.26 ± 0.81 and 28.37 ± 0.77)[I]	41.31abB and 37.05bB	Silva et al. (2020c)[2]W
			6.50	(30.38 ± 0.27)[II]	27.01cB	
			0.30 and 6.50	(32.28 ± 0.26 and 32.25 ± 0.16)[II]	43.00aB and 17.99dB	
'Verdão'		EUWS	0.30 and 6.50	(28.26 ± 0.81 and 28.37 ± 0.77)[I]	75.98aA and 42.86cA	
			6.50	(30.38 ± 0.27)[II]	34.16dA	
			0.30 and 6.50	(32.28 ± 0.26 and 32.25 ± 0.16)[II]	48.02bA and 25.57eA	
'Verdão'	NFT/10 DAS + 25 DAT/autumn–winter	EUWS	0.30 and 6.50	(23.90 ± 0.36 and 24.00 ± 0.55)[I]	86.59a and 45.28c	Silva et al. (2022)[2]Z
			0.30 and 6.50	(30.17 ± 0.32 and 30.35 ± 0.24)[II]	72.86b and 36.13d	
		WSRW	6.50	(23.51 ± 0.33)[I]	76.53b	
		WSPS	6.50	(24.26 ± 1.03)[I]	41.59c	

EUWS – Exclusive use of brackish waters.

WSRW – Brackish waters used only to replenish the water consumed by plants.

WSPS – Brackish waters used only to prepare the nutrient solution.

\# Values estimated based on model SFBB = −8.4152**ECw + 95.1832** (ECw – electrical conductivity of the water; ** significant at $p \leq 0.01$ by t-test).

[1] and [2] Bunches formed by 8 and 12 plants, respectively.

[I] RZT under ambient conditions.

(&) Values obtained only in the control treatment (without salt stress) – extracted from Silva et al. (2017c).

\## There was no significant difference between the means as a function of ECw, and the bunch was formed by 24 plants.

Y Means followed by same letters indicate no significant differences at $p = 0.05$ by Tukey test.

\### Values estimated based on model SFBB = −2.7191**ECw + 44.7593** (** significant at $p \leq 0.01$ by t-test).

W The coriander cultivars were grown in the same hydroponic channel; means followed by same letters indicate no significant differences at $p = 0.05$ by Tukey test – lowercase letters compare the means of the combinations (temperatures of the nutrient solutions using fresh and brackish waters) in each cultivar, and uppercase letters compare the means of the cultivars in each combination.

[II] Constant RZT during the experiment.

Z Means followed by same letters indicate no significant differences at $p = 0.05$ by Scott-Knott test.

DAS – days after sowing; DAT – days after transplanting.

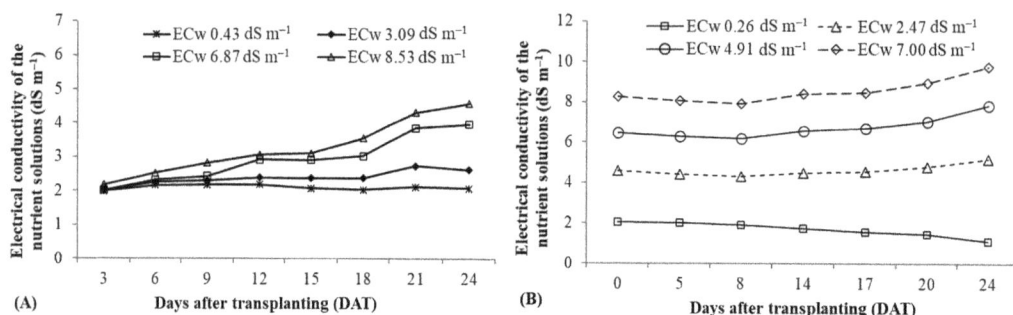

FIGURE 4.8 Electrical conductivity of the nutrient solutions using brackish waters only to replenish the water consumption of coriander plants (A) and exclusive use (B). A – Adapted from Silva et al. (2015a) and B – (data by the authors).

dynamics of the interaction between salinity and temperature in the root-zone (Santos et al., 2019; Silva et al., 2020c; 2022). Thus, a better understanding of coriander responses to salt stress and temperature stress in the root-zone can help develop strategies to mitigate the effect of such stresses.

By standardization, the ECsol is established at 25°C. As observed in Figures 4.9A and 4.9B, under this condition of ECsol measurement, the values remain relatively constant throughout the day. It is worth pointing out that the ECsol measurements at 25°C are of methodological importance, serving to compare salinity under different temperature conditions for the EC of either water and/or nutrient solution. However, the crop responds to the reality of the rhizosphere; that is, it is conditioned to the actual ECsol and not to the ECsol parameterized at 25°C. Therefore, for actual ECsol, daily amplitudes of 1.12 and 1.50 dS m⁻¹ were observed in cultivation under ambient temperature conditions in the root-zone (Figure 4.9A) and at 32°C (Figure 4.9B), that is, further increasing the EC of the saline solution.

In the studies carried out by Silva et al. (2020c; 2022), coriander was subjected only to the salinity level of 6.5 dS m⁻¹ combined with constant temperatures in the root-zone of 30 and/or 32°C (Table 4.3). As expected, the lowest yields of coriander were obtained when these stresses were combined. This effect becomes even more pronounced according to the growing season; for example, considering the condition of cultivation under salinity and temperature of 30°C for 'Verdão' coriander, the SFB means were approximately 55% (Silva et al., 2020c) and 50% (Silva et al., 2022) lower compared to the control, in experiments conducted in summer and autumn-winter, respectively.

Also, in the study conducted by Silva et al. (2020c), as shown in Table 4.3, carried out in summer in NFT hydroponics, under salt stress and temperature of 32°C, production of 'Verdão' coriander was approximately 66% lower compared to control (without salt stress and ambient temperature). It can also be verified that this interaction of stresses was less pronounced on the cultivar 'Tabocas', as the coriander production was approximately 56% lower under the same combination of stresses. However, the higher production of 'Verdão' coriander compared to 'Tabocas' coriander is highlighted.

In this case, some strategies can be adopted to minimize possible losses of coriander production when subjected to the combined and/or individual stresses of salinity and temperature in the root-zone, for example, to increase the number of plants per unit of cultivation, as done in the studies carried out by Silva, M. G. (personal communication), and Santos Júnior, J. A. (personal communication), as shown in Table 4.4.

In the study carried out by Silva, M. G. (personal communication, as shown in Table 4.4), in addition to the density of coriander plants, salts of different cationic natures in a mixed and/or individual way to produce water with a salinity of 6.50 dS m⁻¹ were evaluated. At densities between 6 and 24 plants, the individual SFB yields varied according to the type of salt, whereas under higher density (30 plants), the means of production under salinity with different salts did not differ significantly.

TABLE 4.4

Shoot Fresh Biomass per Plant (SFBP) and per Bunch (SFBB) of Coriander Grown Hydroponically Using Variable Sowing Densities and Brackish Waters of Different Cationic Natures to Prepare Nutrient Solution

Cationic Natures[1]#	Sowing Densities (SD) (plants cup^{-1})##					Equation
	6	12	18	24	30	
	Shoot Fresh Biomass per Plant (g)					
Control (ECw 0.25 dS m^{-1})	9.56a	7.89a	7.19a	6.49a	5.31a	SFBP = –0.1649**SD + 10.2548**
NaCl	3.25c	2.82c	2.73c	2.64c	2.52b	SFBP = –0.0271**SD + 3.2798**
KCl	5.31b	4.31b	2.59c	3.54b	2.83b	SFBP = –0.0954**SD + 5.4336**
MgCl$_2$	3.69c	2.34c	2.72c	2.19c	2.17b	SFBP = –0.0532**SD + 3.5774**
CaCl$_2$	3.44c	3.92b	3.83b	2.54c	2.31b	SFBP = –0.0607**SD + 4.3004**
NaCl + CaCl$_2$ + MgCl$_2$ (7:2:1 equivalent ratio)	4.22c	4.14b	3.58b	2.31c	2.98b	SFBP = –0.0717**SD + 4.7362**
NaCl + CaCl$_2$ + KCl (7:2:1 equivalent ratio)	3.91c	3.26c	3.16b	2.55c	2.55b	SFBP = –0.0572**SD + 4.1152**
NaCl + CaCl$_2$ + MgCl$_2$ + KCl (7:2:0.5:0.5 equivalent ratio)	4.97b	3.98b	3.52b	3.04b	2.86b	SFBP = –0.0860**SD + 5.2202**
	Shoot Fresh Biomass per Bunch (g)					
Control (ECw 0.25 dS m^{-1})	47.03a	64.90a	70.56a	72.44a	77.16a	SFBB = 1.1301**SD + 46.0758**
NaCl	15.81b	26.08b	24.73c	24.73c	30.66c	SFBB = 0.4725**SD + 15.8976**
KCl	26.78b	33.02b	34.14b	36.99b	37.86c	SFBB = 0.4355**SD + 25.9162**
MgCl$_2$	19.17b	16.06c	25.60c	24.20c	18.51d	ns (SFBB = mean = 20.71)
CaCl$_2$	21.18b	27.48b	32.25b	25.17c	31.99c	ns (SFBB = mean = 27.61)
NaCl + CaCl$_2$ + MgCl$_2$ (7:2:1 equivalent ratio)	26.68b	34.78b	32.05b	30.01c	32.24c	ns (SFBB = mean = 31.15)
NaCl + CaCl$_2$ + KCl (7:2:1 equivalent ratio)	22.35b	29.69b	26.93c	28.55c	46.42b	SFBB = 0.7832**SD + 16.6920**
NaCl + CaCl$_2$ + MgCl$_2$ + KCl (7:2:0.5:0.5 equivalent ratio)	24.20b	32.61b	35.81b	35.92b	37.76c	SFBB = 0.5072**SD + 24.1286**

Sowing Densities (g cup^{-1})[2]	ECsol (dS m^{-1})			Equation
	1.49	2.77	6.44	
	Shoot Fresh Biomass per Bunch (g)			
1.0	49.14[1]		30.00	SFBB = –3.8657**ECsol + 54.8980
1.5		57.23[1]	21.53	SFBB = –2.6578**ECsol^2 + 14.75**ECsol + 36.768
2.0	mean = 47.00		ns	

[1] Silva, M. G. (personal communication) – experiment carried out during the months of January to February 2021, corresponding to 35 days (10 days after sowing (DAS) + 25 days after transplanting).

For each individual or combined cationic nature, water was produced with an electrical conductivity of 6.5 dS m^{-1}.

Means followed by same letters in the columns indicate no significant differences at $p = 0.05$ by Scott-Knott test.

** Significant at $p \leq 0.01$ by t-test.

[2] Santos Júnior, J. A. (personal communication) – four levels of electrical conductivity of the nutrient solution (ECsol) in the cultivation of coriander 'Tabocas' were used, one without salt stress (ECsol 1.49 dS m^{-1}) and the other three (3.14, 4.87 and 6.44 dS m^{-1}) were prepared by mixing brackish water and rainwater; the experiment was carried out during the months of July to August 2013, corresponding to 35 DAS.

[1] Highest means of SFBB using the respective equations.

ns – Not significant.

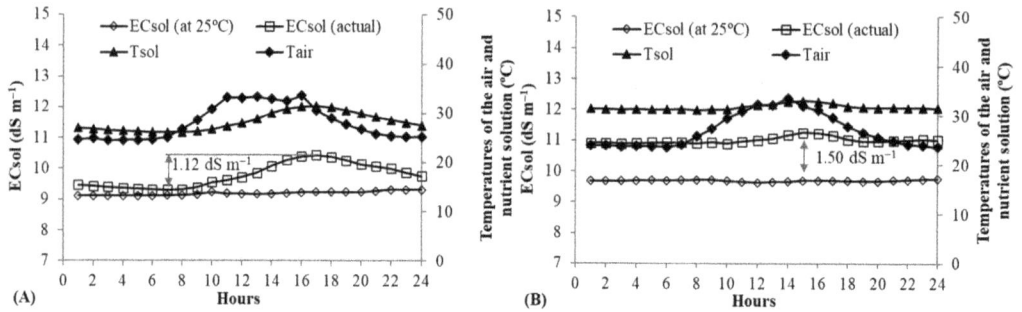

FIGURE 4.9 Values of electrical conductivity of the nutrient solution with temperature compensation at 25°C – ECsol (at 25°C) and without compensation – ECsol (actual), and temperatures of the air (Tair) and nutrient solution (Tsol) during the day, in coriander grown under salt stress (6.5 dS m^{-1}) and ambient RZT (A) and at 32°C RZT (B), using an NFT hydroponic system in the summer season. Adapted from Silva (2019).

In general, as the number of plants per unit of cultivation increases, individual production of SFB is lower, a behavior similar to that recorded by Silva et al. (2016a), as shown in Table 4.2.

When evaluating SFB production per bunch of plants in the study carried out by Silva, M. G. (personal communication, as shown in Table 4.4), the results were quite varied according to the types of salts for each plant density. As expected, the SFB of the bunch increased with the increase in the number of plants per unit of cultivation, despite the lower SFB production per plant. The greater number of plants per unit of cultivation compensated for these losses and promoted higher production in the bunch, a behavior similar to that recorded by Silva et al. (2016a). Under control treatment without salt stress (ECw of 0.25 dS m^{-1}), the SFB yields estimated based on the model presented in Table 4.4 ranged between 52.86 and 79.98 g bunch^{-1} under cultivation with densities of 6 and 30 plants cup^{-1}, respectively.

Under the condition of cultivation with brackish water (ECw of 6.5 dS m^{-1}) produced from the salts MgCl$_2$, CaCl$_2$ and NaCl + CaCl$_2$ + MgCl$_2$ (7:2:1, equivalent ratio) in the study carried out by Silva, M. G. (personal communication, as shown in Table 4.4), the increase in the number of plants per unit of cultivation did not result in significant gains in bunch production, with averages of 20.71, 27.61 and 31.15 g bunch^{-1}, respectively. The highest densities could not sustain the compensation of the individual reduction of fresh biomass per plant. The largest increase in coriander production between 6 and 30 plants cup^{-1} occurred under NaCl + CaCl$_2$ + KCl (7:2:1, equivalent ratio), approximately 19 g bunch^{-1}. Under salinity (ECw of 6.5 dS m^{-1}) produced with the salts KCl, NaCl and NaCl + CaCl$_2$ + MgCl$_2$ + KCl (7:2:0.5:0.5, equivalent ratio), the increments were approximately 10, 11 and 12 g bunch^{-1}, respectively.

Also, in the study carried out by Silva, M. G. (personal communication, as shown in Table 4.4), even under high salinity (ECw of 6.5 dS m^{-1}), the increase in the number of plants per unit of cultivation compensated for the reduction under this condition of salt stress, with SFB production of approximately 40 g for a bunch composed of 30 plants (under KCl, CaCl$_2$ + MgCl$_2$ + KCl and NaCl + CaCl$_2$ + MgCl$_2$ + KCl) compared to the production of approximately 53 g for a bunch composed of 6 plants in cultivation without salt stress (ECw 0.25 dS m^{-1}). Similarly, in the study carried out by Santos Júnior, J. A. (personal communication, as shown in Table 4.4), the SFB yields under the ECsol levels of 1.49 (control) and 4.5 dS m^{-1} were approximately 49 g bunch^{-1} at sowing densities of 1.0 and 1.5 g of seeds per unit of cultivation, respectively.

In summary, under hydroponic conditions, regardless of the combined and/or individual stresses of salinity and temperature in the root-zone, the coriander yields were satisfactory (quality and production aspects). Furthermore, the visual quality of the plants produced is emphasized, a characteristic that is important for acquiring the product. In the studies presented in Tables 4.3 and 4.4, in general, there were no toxic symptoms in coriander plants (Figure 4.10). Deleterious effects of salinity were recorded only under the highest ECw level of

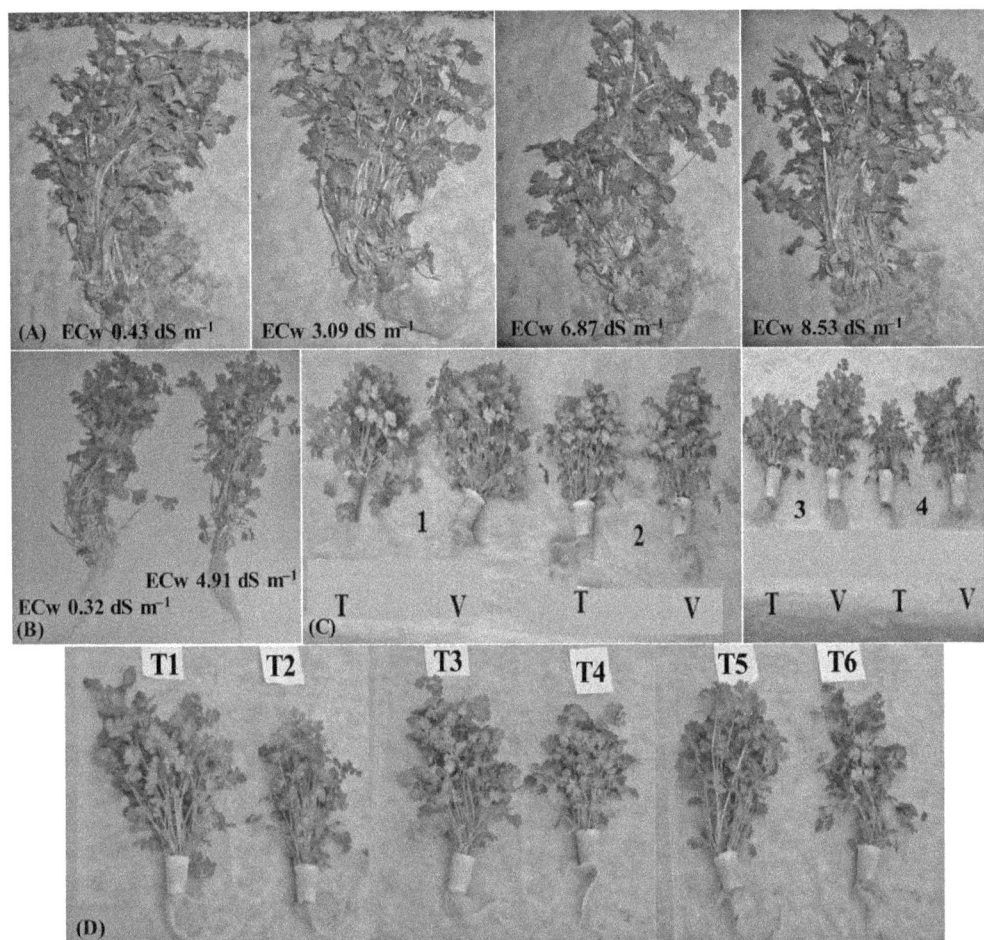

FIGURE 4.10 The visual aspect of coriander grown in hydroponic conditions. ECw – electrical conductivity of water; T – cultivar 'Tabocas'; V – cultivar 'Verdão'; 1 and 2 – plants grown with root-zone temperature (RZT) in ambient conditions under ECw of 0.30 and 6.50 dS m^{-1}, respectively; 3 and 4 – plants grown under ECw of 6.50 dS m^{-1} with RZT at 30 and 32°C, respectively; T1 and T2 – plants grown under ambient RZT with exclusive use of waters with ECw of 0.30 and 6.50 dS m^{-1}, respectively; T3 and T4 – plants grown under RZT at 30°C with exclusive use of waters with ECw of 0.30 and 6.50 dS m^{-1}, respectively; T5 and T6 – plants grown under ambient RZT with brackish water (ECw of 6.50 dS m^{-1}) used only to replenish the water consumed by plants and only to prepare the nutrient solution, respectively. A – (photo by the authors, concerning the publication from Silva et al., 2015a), B – (photo of the authors, concerning the publication from Silva et al., 2016b), C – source: Silva et al. (2020c) and D – source: Silva et al. (2022).

7.73 dS m^{-1} in the study carried out by Cazuza Neto et al. (2014a), with burning on the edges of coriander leaves. However, in the studies of Silva et al. (2020c) (Figure 4.10C) and Silva, M. G. (personal communication, as shown in Figure 4.11), conducted in the summer, both using water with ECw of 6.5 dS m^{-1}, necrosis in the older leaves was recorded. In the first study, the number of symptomatic leaves per bunch of plants did not exceed 10 leaves, while in the second study, the symptomatic leaves varied according to the type of salt used to prepare the brackish water, as well as according to the density of plants per unit of cultivation.

Therefore, based on these last-mentioned results, it may be strategic to avoid the exclusive use of brackish waters with high salinity levels in warmer seasons, so its use is suggested only to replenish the water consumption of plants, as shown in Table 4.4.

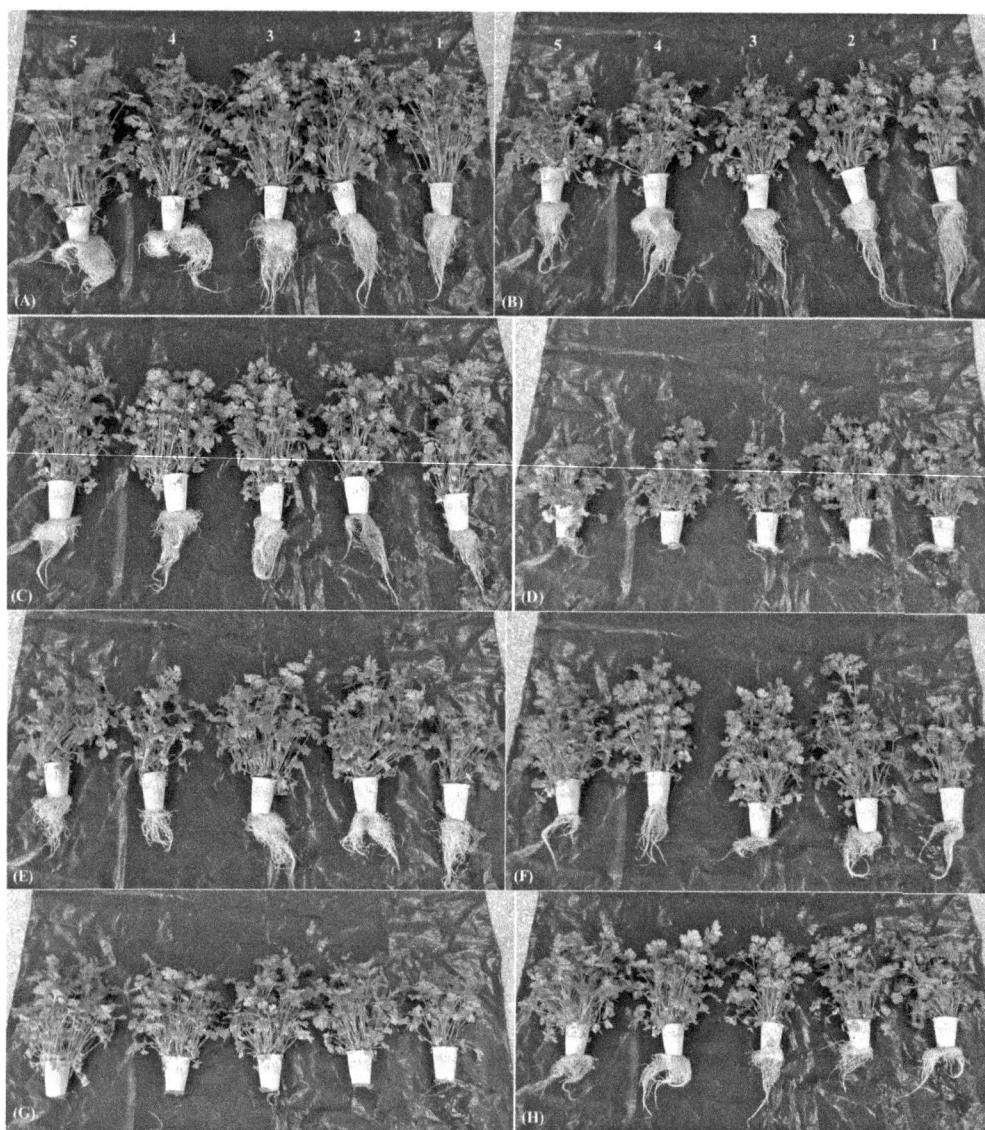

FIGURE 4.11 Visual aspect of coriander grown in hydroponic conditions without salt stress – ECw of 0.25 dS m^{-1} (A) and under salt stress – ECw of 6.50 dS m^{-1} with NaCl (B), KCl (C), MgCl$_2$ (D), CaCl$_2$ (E), NaCl + CaCl$_2$ + MgCl$_2$ (F), NaCl + CaCl$_2$ + KCl (G) and NaCl + CaCl$_2$ + MgCl$_2$ + KCl (H). 1, 2, 3, 4 and 5 – bunches formed by 6, 12, 18, 24 and 30 plants, respectively. A–H (photos by the authors).

4.5 SALINITY TOLERANCE OF CORIANDER UNDER HYDROPONIC CONDITIONS

The response of coriander to salinity can be better understood based on the relative yield. The acceptable percentage of relative yield depends mainly on the salinity level of the water available for cultivation. For example, in the case of coriander, relative yields of approximately 90% were obtained by Cazuza Neto et al. (2014a) and Silva et al. (2018) under the water salinity levels of 1.40 and 1.85 dS m^{-1} (values estimated based on the equations presented in Table 4.3), using brackish water throughout the production process. On the other hand, in studies with only one level of ECw,

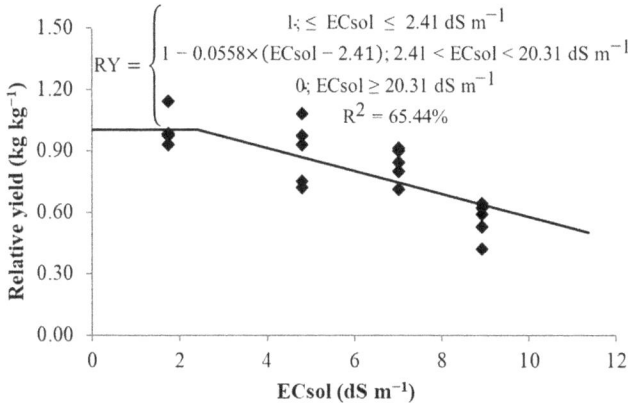

$$RY = \begin{cases} 1; \le ECsol \le 2.41 \ dS \ m^{-1} \\ 1 - 0.0558\times(ECsol - 2.41); \ 2.41 < ECsol < 20.31 \ dS \ m^{-1} \\ 0; \ ECsol \ge 20.31 \ dS \ m^{-1} \end{cases}$$

$R^2 = 65.44\%$

FIGURE 4.12 Relative yield (RY) of coriander as a function of mean time-weighted electrical conductivity of the nutrient solution (ECsol) in a DFT system in tubes. Source: (data of the authors, extracted from Silva et al., 2018).

that is, 4.91 dS m⁻¹ (Silva et al., 2016b) and 6.50 dS m⁻¹ (Silva et al., 2020c; 2022),[*] the relative yields of coriander were approximately 65, 56 and 52%, respectively. These studies were conducted with the cultivar 'Verdão'.

Also, in the study carried out by Silva et al. (2020c), it was verified that for the cultivar 'Tabocas', there was no significant difference between the ECw means of 0.30 and 6.50 dS m⁻¹, as well as in the study of Santos Júnior, J. A. (personal communication, as shown in Table 4.4), with this same cultivar subjected to ECsol levels of 1.49, 3.14, 4.87 and 6.44 dS m⁻¹ with sowing density of 2.0 g cup⁻¹. For a seeding density of 1.0 g cup⁻¹, the relative yield was 61% under ECsol of 6.44 dS m⁻¹.

Based on information extracted from Silva et al. (2018), the salinity tolerance of coriander was evaluated based on the relative production of fresh shoot biomass, using the plateau model with linear reduction proposed by Mass and Hoffman (1977). With this, the weighted means of ECsol as a function of the measurements performed during coriander cultivation were used. As shown in Figure 4.12, a salinity threshold (ST) was estimated for coriander under hydroponic conditions of 2.41 dS m⁻¹ based on the ECsol, with a percentage reduction of 5.58% per unit increase of ECsol above the threshold. According to the classification proposed by Ayers and Westcot (1985) with salinity ranges for crops, hydroponic coriander was classified as moderately sensitive to salinity, with ST ranging from 1.3 to 3.0 dS m⁻¹. However, since this classification is based on the electrical conductivity of the soil saturation extract, there is a specific difficulty in comparing it with ECsol.

4.6 WATER USE EFFICIENCY OF THE HYDROPONIC CORIANDER

As shown in Tables 4.5 and 4.6, different water consumption values are recorded to produce one bunch of coriander composed of different numbers of plants, which partially explains this variation. Another explanation concerns the growing season (influences in the crop evapotranspirometric demand) and crop cycle duration, as well as the hydroponic system employed. Silva (2019) recorded the maximum water consumption, equal to 3.86 L bunch⁻¹ (Table 4.5), in an experiment conducted in the summer. Values lower than 2.0 L bunch⁻¹ (Santos Júnior et al., 2015; Silva et al., 2016b; 2020a; 2020b; as shown in Table 4.5) and lower than 1.0 L bunch⁻¹ (Oliveira et al., 2020; as shown in Table 4.6) have been recorded.

[*] Root-zone temperature under ambient conditions.

TABLE 4.5
Water Consumption (WC) and Water Use Efficiency (WUE) of the Shoot Fresh Biomass of the Bunch of Plants of the Coriander Grown Hydroponically under Different Cultivation Conditions

Cultivation Conditions

	ECw (dS m⁻¹)	WC (L bunch⁻¹)	WUE (kg m⁻³)[I]	Reference
	0.30–7.73	2.57–1.10[II]	38.68–25.41[III]	Cazuza Neto et al. (2014b)[$]
Spacing of 0.07 and 0.15 m used between bunches of plants under ECsol of 2.50 dS m⁻¹		(1.13 and 1.58)[Y]	~ 82 and ~ 66	Santos Júnior et al. (2015)[$$]
	0.32 and 4.91	(1.45a–0.97b)[W]	34.56[#]	Silva et al. (2016b)[&]
	0.43–8.53	2.21[##]	27.05–18.91[###]	Silva et al. (2017c)
[1]Ambient and 30°C RZT	0.30	(3.86 and 3.80		Silva (2019)[&]
[1]Ambient, 30 and 32°C RZT	6.50	3.54, 3.29 and 2.75		
[2]Ambient RZT	0.30 and 6.50	1.72 and 1.44	[3]56.96a and 42.02b	
[2]30°C RZT	0.30 and 6.50	2.02 and 1.09[Y]	[3]39.34c and 34.23c[W]	
Nutrient solution depth of 0.013, 0.017 and 0.025 m	0.41	1.51[4]	33.67[4]	Silva et al. (2020a)[&]
[5]Nutrient solution recirculation at 0.25 and 24 h	0.30	(1.93a and 1.45b)[W]	(29.43a and 21.44b)[W]	Silva et al. (2020b)[&]
[6]Nutrient solution recirculation at 0.25 and 24 h	0.41	[7](3.03a and 2.61a)[W]	(20.05a and 17.48a)[W]	
Sowing densities of 1.0, 1.5 and 2.0 g cup⁻¹ under ECsol of 1.49 dS m⁻¹		(2.95, 2.89 and 2.99)[Y]		Santos Júnior, J. A. (personal communication)
Sowing densities of 1.0, 1.5 and 2.0 g cup⁻¹ under ECsol of 6.44 dS m⁻¹		(1.31, 1.47 and 2.01)[Y]		

[I]   Values of WUE originally in the unit g L⁻¹, and calculated based on the fresh shoot biomass divided by the water consumption per bunch of plants, considering the same growth periods and cultivation conditions shown in Table 4.3.

[II]   Values estimated based on model WC = -0.1815**ECw $+ 2.646$** (ECw – electrical conductivity of the water; ** significant at $p \leq 0.01$ by t-test).

[III]   Values estimated based on model WUE = -1.6502**ECw $+ 38.172$** (** significant at $p \leq 0.01$ by t-test).

[Y]   Only mean values.

[W]   Means followed by same letters indicate no significant differences at $p = 0.05$ by Tukey test.

[#]   There was no significant difference between the means as a function of ECw (extracted from Silva, 2014).

[##]   There was no significant difference between the means as a function of ECw.

[###]   Values estimated based on model WUE = -1.008**ECw $+ 27.509$** (** significant at $p \leq 0.01$ by t-test).

[1] and [2] Experiments carried out in the summer, autumn and winter, respectively.

[$]   and [&] Bunches formed by 8 and 12 plants, respectively.

[$$]   Sowing densities of 1.0, 1.5 and 2.0 g cup⁻¹.

[3]   Extracted from Silva (2019).

[4]   There was no significant difference between the means as a solution depth in the hydroponic channels of the conventional DFT system.

[5] and [6] Experiments carried out in the autumn and summer-autumn, respectively.

[7]   Values estimated when the coriander plants were grown in channels with a nutrient solution depth of 0.02 m.

TABLE 4.6

Water Consumption (WC) and Water Use Efficiency (WUE) of the Shoot Fresh Biomass of the Bunch of Plants of the Coriander Grown Hydroponically under Different Cultivation Conditions, in Recife – PE, Brazil[1]

Cationic Natures	WC		WUE[##]	
	Equation	(L bunch^{-1})[#]	Equation	(kg m^{-3})[#]
NaCl	$WC = -0.0358^{**}ECsol + 0.79$	0.73	$WUE = -1.2778^{**}ECsol^2 + 5.4829^{ns}ECsol + 48.1462$	53.65
CaCl$_2$	$WC = 0.0049^{**}ECsol^2 - 0.0778^{**}ECsol + 0.83$	0.72	$WUE = -4.7595^{**}ECsol + 59.7825$	52.17
MgCl$_2$	$WC = -0.0228^{**}ECsol + 0.7262$	0.69	$WUE = -3.2634^{**}ECsol^2 + 21.3863^{**}ECsol + 20.1231$	45.99

Source: Adapted from Oliveira et al. (2020).

[1] Four levels of electrical conductivity of the nutrient solution (ECsol) in the cultivation of coriander 'Verdão' (bunch with 15 plants) were used, one without salt stress (ECsol 1.6 dS m^{-1}) and the other three (3.2, 4.8 and 6.4 dS m^{-1}) obtained with salts of different cationic nature.

[#] Values obtained under control treatment (ECsol 1.6 dS m^{-1}).

[##] Values of WUE originally in the unit g L^{-1}.

[**] Significant at $p \leq 0.01$ by t-test; ns – not significant.

The lower the water consumption, the higher the water use efficiency (WUE). In the study of Santos Júnior et al. (2015), the WUE was approximately 82 kg m^{-3}, while in the study carried out by Silva et al. (2020b), the WUE did not exceed 20 kg m^{-3}.

4.7 CONCLUSIONS

The demand for food to meet the population's needs is progressively increasing, as with the leafy vegetables. However, in the aggravating climate change scenario, water resources, especially with low salinity, are increasingly scarce for the irrigation of crops. Therefore, it is necessary to use systems that require a lower volume of water but keep yields within competitiveness standards. In this context, soilless cultivation (hydroponics) has several advantages over conventional planting, mainly due to the higher water use efficiency. Furthermore, in hydroponics, it has been possible to use low-quality water sources, such as brackish waters, to cultivate coriander. Therefore, based on the various aspects addressed in this chapter, it is concluded that the potential for coriander cultivation under hydroponic conditions to produce fresh biomass for use in cooking in different regions is great.

REFERENCES

Ahmadi, M., & Souri, M. K. (2018). Growth and mineral content of coriander (*Coriandrum sativum* L.) plants under mild salinity with different salts. *Acta Physiologiae Plantarum*, 40(11), 194. https://doi.org/10.1007/s11738-018-2773-x.

AlQuraidi, A. O., Mosa, K. A., & Ramamoorthy, K. (2019). Phytotoxic and genotoxic effects of copper nanoparticles in coriander (*Coriandrum sativum*-Apiaceae). *Plants*, 8(1), 19. https://doi.org/10.3390/plants8010019.

Amiripour, A., Jahromi, M. G., Soori, M. K., & Torkashvand, A. M. (2021). Changes in essential oil composition and fatty acid profile of coriander (*Coriandrum sativum* L.) leaves under salinity and foliar-applied silicon. *Industrial Crops and Products*, 168, 113599. https://doi.org/10.1016/j.indcrop.2021.113599.

Ayers, R. S., & Westcot, D. W. (1985). *Water quality for agriculture* (Irrigation and Drainage, Paper 29), FAO, Rome.

Caracciolo, F., El-Nakhel, C., Raimondo, M., Kyriacou, M. C., Cembalo, L., De Pascale, S., & Rouphael, Y. (2020). Sensory attributes and consumer acceptability of 12 microgreens species. *Agronomy*, 10(7), 1043. https://doi.org/10.3390/agronomy10071043.

Cavalcante, A. R., Santos Júnior, J. A., Gheyi, H. R., Dias, N. S., & Paz, V. P. S. (2016). Produção e composição mineral do coentro em sistema hidropônico de baixo custo. *Irriga*, 21(4), 685–696. https://doi.org/10.15809/irriga.2016v21n4p685-696.

Cazuza Neto, A., Soares, T. M., Bione, M. A., Freitas, F. T. O., Melo, D. M., & Silva Filho, J. A. (2014a). Qualidade do molho de coentro produzido em água salobra em sistema hidropônico NFT. *Proceedings of the Brazilian Symposium on Salinity and Brazilian Meeting on Irrigation Engineering*, Fortaleza, Brazil. https://doi.org/10.12702/ii.inovagri.2014-a493.

Cazuza Neto, A., Soares, T. M., Bione, M. A., Freitas, F. T. O., Melo, D. M., & Silva Filho, J. A. (2014b). Efeito da salinidade no consumo hídrico do molho de coentro produzido em sistema hidropônico NFT. *Proceedings of the Brazilian Symposium on Salinity and Brazilian Meeting on Irrigation Engineering*, Fortaleza, Brazil. https://doi.org/10.12702/ii.inovagri.2014-a494.

Cocetta, G., Mishra, S., Raffaelli, A., & Ferrante, A. (2018). Effect of heat root stress and high salinity on glucosinolates metabolism in wild rocket. *Journal of Plant Physiology*, 231, 261–270. https://doi.org/10.1016/j.jplph.2018.10.003.

Cruz, R. I. F., Navarro, F. E. C., Pereira, P. F. A., Bognola, A. F., Oliveira, T. F., & Santos Júnior, J. A. (2020). Aspectos fotossintéticos e produção do coentro sob soluções nutritivas preparadas em diferentes águas salobras. *Proceedings of the Inovagri Meeting Virtual*, Fortaleza, Brazil.

Currey, C. J., Walters, K. J., & Flax, N. J. (2019). Nutrient solution strength does not interact with the daily light integral to affect hydroponic cilantro, dill, and parsley growth and tissue mineral nutrient concentrations. *Agronomy*, 9(7), 389. https://doi.org/10.3390/agronomy9070389.

Daflon, D. S. G., Freitas, M. S. M., Carvalho, A. J. C., Monnerat, P. H., & Prins, C. L. (2014). Sintomas visuais de deficiência de macronutrientes e boro em coentro. *Horticultura Brasileira*, 32(1), 28–34. https://doi.org/10.1590/S0102-05362014000100005.

Diniz, A. A., Dias, N. S., Souza, F. I., Sá, F. V. S., Araújo, N. O., & Fernandes, A. L. M. (2019). Produção e qualidade do coentro cultivado com solução nutritiva em fibra de coco. *Revista Brasileira de Agricultura Irrigada*, 13(2), 3306–3313. https://doi.org/10.7127/rbai.v13n200994.

Divya, P., Puthusseri, B., & Neelwarne, B. (2014). The effect of plant regulators on the concentration of carotenoids and phenolic compounds in foliage of coriander. *LWT- Food Science and Technology*, 56(1), 101–110. https://doi.org/10.1016/j.lwt.2013.11.012.

Donegá, M. A. (2009). *Relação K:Ca e aplicação de silício na solução nutritiva para o cultivo hidropônico de coentro* [Master Thesis, Escola Superior de Agricultura "Luiz de Queiroz"/Universidade de São Paulo]. https://teses.usp.br/teses/disponiveis/11/11136/tde-25022010-080813/publico/Mateus_Donega.pdf.

Hazeri, N., Valizadeh, J., Shakeri, A., & Rajabpour, M. (2012). Evaluation of essential oil and mineral composition of coriander (*Coriandrum sativum* L.) among growth conditions of hydroponic, field and greenhouse. *Journal of Essential Oil Bearing Plants*, 15(6), 949–954. https://doi.org/doi:10.1080/0972060x.2012.10662598.

Hu, J., Wu, X., Wu, F., Chen, W., White, J. C., Yang, Y., Wang, B., Xing, B., Tao, S., & Wang, X. (2019). Potential application of titanium dioxide nanoparticles to improve the nutritional quality of coriander (*Coriandrum sativum* L.). *Journal of Hazardous Materials*, 389, 121837. https://doi.org/10.1016/j.jhazmat.2019.121837.

Jamila, F., Fouad, A., Zakia, B., Rachid, B., Nabil, S., & Rachid, M. (2019). Effect of silicon application on *Coriandrum sativum* (L) under salt stress. *International Journal of Scientific & Engineering Research*, 10(11), 26–33.

Khodadadi, M., Fotokian, M. H., & Kareem, S. H. S. (2021). Genetics of water deficit stress resistance through phenotypic plasticity in coriander. *Scientia Horticulturae*, 286, 110233. https://doi.org/10.1016/j.scienta.2021.110233.

Kulkarni, S., Abraham, P. S., Mohanty, N., Kadam, N. N., & Thakur, M. (2016). Sustainable raft based hydroponic system for growing spinach and coriander. In Pawar, P. M., Ronge, B. P., Balasubramaniam, R., & Seshabhattar, S (Eds.), *Proceedings of the International Conference on Advanced Technologies for Societal Applications* (pp. 117–125), Springer International Publishing, New York.

Lennard, W., & Ward, J. (2019). A comparison of plant growth rates between an NFT hydroponic system and an NFT aquaponic system. *Horticulturae*, 5(2), 27. https://doi.org/10.3390/horticulturae5020027.

López, N. A. P., Cano-Sosa, J., Cantón, F. P., Rodríguez-Buenfil, I. M., & Ramos-Díaz, A. (2016). Different responses of the quality parameters of *Coriandrum sativum* to organic substrate mixtures and fertilization. *Agronomy*, 6(2), 21. https://doi.org/10.3390/agronomy6020021.

Luz, J. M. Q., Andrade, L. V., Dias, F. F., Silva, M. A. D., Haber, L. L., & Oliveira, R. C. (2012). Produção hidropônica de coentro e salsa crespa sob concentrações de solução nutritiva e posições das plantas nos perfis hidropônicos. *Bioscience Journal*, 28(4), 589–597.

Maia, P. M. E. (2017). *Cultivo de hortaliças em substrato de fibra de coco sob estresse salino e aplicação foliar de ácido ascórbico* [Doctoral Thesis, Universidade Federal de Lavras]. http://repositorio.ufla.br/jspui/handle/1/13228.

Mohammadipour, N., & Souri, M. K. (2019). Effects of different levels of glycine in the nutrient solution on the growth, nutrient composition, and antioxidant activity of coriander (*Coriandrum sativum* L.). *Acta Agrobotanica*, 72(1), 1759. https://doi.org/10.5586/aa.1759.

Neffati, M., & Marzouk, B. (2009). Roots volatiles and fatty acids of coriander (*Coriandrum sativum* L.) grown in saline medium. *Acta Physiologiae Plantarum*, 31(3), 455–461. https://doi.org/10.1007/s11738-008-0253-4.

Nguyen, D. T. P., Lu, N., Kagawa, N., Kitayama, M., & Takagaki. M. (2020). Short-term root-zone temperature treatment enhanced the accumulation of secondary metabolites of hydroponic coriander (*Coriandrum sativum* L.) grown in a plant factory. *Agronomy*, 10(3), 413. https://doi.org/10.3390/agronomy10030413.

Nguyen, D. T. P., Lu, N., Kagawa, N., & Takagaki, M. (2019). Optimization of photosynthetic photon flux density and root-zone temperature for enhancing secondary metabolite accumulation and production of coriander in plant factory. *Agronomy*, 9(5), 224. https://doi.org/10.3390/agronomy9050224.

Olimpio, L. S. (2017). *Recipientes e densidades de cultivo na produção de coentro em ambiente protegido* [*Course Completion Work, Universidade Federal do Ceará*]. http://www.repositorio.ufc.br/bitstream/riufc/31097/1/2017_tcc_lsolimpio.pdf.

Oliveira, F. A., Souza Neta, M. L., Oliveira, M. K. T., Silva, R. T., Martins, D. C., & Costa, J. P. B. M. (2016). Production of coriander in substrate fertigated with increasing nutrient concentration. *Revista de Ciências Agrárias*, 59(3), 275–279. https://doi.org/10.4322/rca.2241.

Oliveira, T. F., Navarro, F. E. C., Silva, I. A. C., Pereira, P. F. A., Cruz, R. I. F., & Santos Júnior, J. A. (2020). Consumo e eficiência hídrica do coentro sob soluções nutritivas preparadas em diferentes águas salobras. *Proceedings of the Inovagri Meeting Virtual*, Fortaleza, Brazil.

Özyazici, G. (2021). Influence of organic and inorganic fertilizers on coriander (*Coriandrum sativum* L.) agronomic traits, essential oil and components under semi-arid climate. *Agronomy*, 11(7), 1427. https://doi.org/10.3390/agronomy11071427.

Pessoa, U. C. M. (2020). *Uso de águas subterrâneas salobras do semiárido pernambucano para produção de coentro em sistema hidropônico NFT* [Master Thesis, Universidade Federal Rural de Pernambuco]. http://www.pgea.ufrpe.br/sites/ww3.pgea.ufrpe.br/files/documentos/dissertacao_uriel_calisto_moura_pessoa.pdf.pdf.

Rashed, N. M., & Darwesh, R. K. (2015). A comparative study on the effect of microclimate on planting date and water requirements under different nitrogen sources on coriander (*Coriandrum sativum*, L.). *Annals of Agricultural Sciences*, 60(2), 227–243. https://doi.org/10.1016/j.aoas.2015.10.009.

Rebouças, J. R. L., Ferreira Neto, M., Dias, N. S., Souza Neto, O. N., Diniz, A. A., & Lira, R. B. (2013). Cultivo hidropônico de coentro com uso de rejeito salino. *Irriga*, 18(4), 624–634. https://doi.org/10.15809/irriga.2013v18n4p624.

Santos, C. C., Soares, T. M., Silva, M. G., Oliveira, M. G. B., & Gheyi, H. R. (2019). Uso de água salobra no cultivo hidropônico do coentro com temperatura da solução nutritiva controlada. *Proceedings of the Seminário de Pesquisa em Engenharia de Água e Solo*, Cruz das Almas, Brazil.

Santos Júnior, J. A., Gheyi, H. R., Cavalcante, A. R., Medeiros, S. S., Dias, N. S., & Santos, D. B. (2015). Water use efficiency of coriander produced in a low-cost hydroponic system. *Revista Brasileira de Engenharia Agrícola e Ambiental*, 19(12), 1152–1158. https://doi.org/10.1590/1807-1929/agriambi.v19n12p1152-1158.

Silva, M. G. (2014). *Uso de água salobra e frequência de recirculação de solução nutritiva para produção do coentro hidropônico* [Master Thesis, Universidade Federal do Recôncavo da Bahia]. https://www1.ufrb.edu.br/pgea/images/Teses/MAIRTON-GOMES-DA-SILVA.pdf.

Silva, M. G. (2019). *Coentro hidropônico sob diferentes condições de cultivo relacionadas à solução nutritiva: temperatura, salinidade e recirculação* [Doctoral Thesis, Universidade Federal do Recôncavo da Bahia]. https://www1.ufrb.edu.br/pgea/images/Teses/MAIRTON_GOMES_DA_SILVA.pdf.

Silva, M. G., Costa, I. P., Alves, L. S., Soares, T. M., & Gheyi, H. R. (2020a). Cultivo de coentro nos sistemas hidropônicos DFT convencional *vs.* adaptado em tubos de PVC sob diferentes lâminas de solução nutritiva. *Proceedings of the Inovagri Meeting Virtual*, Fortaleza, Brazil.

Silva, M. G., Oliveira, I. S., Soares, T. M., Gheyi, H. R., Santana, G. O., & Pinho, J. S. (2017a). Uso de águas salobras no cultivo hidropônico de coentro sob intervalos de recirculação da solução nutritiva. *Proceedings of the Inovagri International Meeting*, Fortaleza, Brazil. https://doi.org/10.12702/iii .inovagri.2015-a007.

Silva, M. G., Oliveira, I. S., Soares, T. M., Gheyi, H. R., Santana, G. O., & Pinho, J. S. (2018). Growth, production and water consumption of coriander in hydroponic system using brackish waters. *Revista Brasileira de Engenharia Agrícola e Ambiental*, 22(8), 547–552. https://doi.org/10.1590/1807-1929/agri-ambi.v22n8p547-552.

Silva, M. G., Silva, P. C. C., Cova, A. M. W., Gheyi, H. R., & Soares, T. M. (2022). Experiências com o uso de águas salobras em hidroponia no Nordeste brasileiro. In Cerqueira, P. R. S., Lacerda, C. F., Araujo, G. G. L., Gheyi, H. R., & Simões, W. L. (Eds.), *Agricultura irrigada em ambientes salinos* (pp. 290–321), Codevasf, Brasília. https://www.codevasf.gov.br/acesso-a-informacao/institucional/biblioteca-geraldo -rocha/publicacoes/outras-publicacoes/agricultura-irrigada-em-ambientes-salinos.pdf.

Silva, M. G., Soares, T. M., Costa, I. P., Gheyi, H. R., & Alves, L. S. (2017b). Monitoramento de elementos meteorológicos e temperatura da solução nutritiva hidropônica em ambiente protegido com uso do Arduino. *Proceedings of the Inovagri International Meeting*, Fortaleza, Brazil. https://doi.org/10.7127/ iv-inovagri-meeting-2017-res0740728.

Silva, M. G., Soares, T. M., & Gheyi, H. R. (2019). Épocas de colheita de duas cultivares de coentro (*Coriandrum sativum* L.) em condições hidropônicas. *Proceedings of the Seminário de Pesquisa em Engenharia de Água e Solo*, Cruz das Almas, Brazil.

Silva, M. G., Soares, T. M., Gheyi, H. R., Costa, I. P., & Vasconcelos, R. S. (2020b). Growth, production and water consumption of coriander grown under different recirculation intervals and nutrient solution depths in hydroponic channels. *Emirates Journal of Food and Agriculture*, 32(4), 281–294. https://doi .org/10.9755/ejfa.2020.v32.i4.2094.

Silva, M. G., Soares, T. M., Gheyi, H. R., Oliveira, I. S., Freitas, F. T. O., & Rafael, M. R. S. (2017c). Consumo hídrico do coentro em hidroponia NFT com o uso de águas salobras para reposição do consumo evapo-transpirado. *Proceedings of the Inovagri International Meeting*, Fortaleza, Brazil. https://doi.org/10 .7127/iv-inovagri-meeting-2017-res0740720.

Silva, M. G., Soares, T. M., Gheyi, H. R., Oliveira, I. S., Rafael, M. R. S., & Souza, T. C. N. (2015b). Intervalos de recirculação de solução nutritiva no cultivo hidropônico de coentro com o uso de água doce e salo-bra. *Proceedings of the Inovagri International Meeting*, Fortaleza, Brazil. https://doi.org/10.12702/iii .inovagri.2015-a007.

Silva, M. G., Soares, T. M., Gheyi, H. R., Oliveira, I. S., & Silva Filho, J. A. (2016a). Crescimento e produção de coentro hidropônico sob diferentes densidades de semeadura e diâmetros dos canais de cultivo. *Irriga*, 21(2), 312–326. https://doi.org/10.15809/irriga.2016v21n2p312-326.

Silva, M. G., Soares, T. M., Gheyi, H. R., Oliveira, I. S., Silva Filho, J. A., & Carmo, F. F. (2016b). Frequency of recirculation of the nutrient solution in the hydroponic cultivation of coriander with brackish water. *Revista Brasileira de Engenharia Agrícola e Ambiental*, 20(5), 447–454. https://doi.org/10.1590/1807 -1929/agriambi.v20n5p447-454.

Silva, M. G., Soares, T. M., Gheyi, H. R., Oliveira, M. G. B., & Santos, C. C. (2020c). Hydroponic cultivation of coriander using fresh and brackish waters with different temperatures of the nutrient solution. *Engenharia Agrícola*, 40(6), 674–683. https://doi.org/10.1590/1809-4430-Eng.Agric.v40n6p674-683/2020.

Silva, M. G., Soares, T. M., Gheyi, H. R., Santos, C. C., & Oliveira, M. G. B. (2022). Hydroponic cultivation of coriander intercropped with rocket subjected to saline and thermal stresses in the root-zone. *Revista Ceres*, 69(2), 148–157. https://doi.org/10.1590/0034-737X202269020004.

Silva, M. G., Soares, T. M., Oliveira, I. S., Santos, J. C. S., Pinho, J. S., & Freitas, F. T. O. (2015a). Produção de coentro em hidroponia NFT com o uso de águas salobras para reposição do consumo evapotranspirado. *Revista Brasileira de Agricultura Irrigada*, 9(4), 246–258. https://doi.org/10.7127/RBAI.V9N400319.

Soares, C. S., Silva, J. A., & Silva, G. N. (2017). Produção de coentro em diferentes espaçamentos dos canais hidropônicos. *Pesquisa Agropecuária Pernambucana*, 22, e201701. https://doi.org/10.12661/pap.2017.001.

Uitterhaegen, E., Burianová, K., Ballas, S., Véronèse, T., Merah, O., Talou, T., Stevens, C. V., Evon, P., & Simon, V. (2018). Characterization of volatile organic compound emissions from self-bonded boards resulting from a coriander biorefinery. *Industrial Crops and Products*, 122, 57–65. https://doi.org/10 .1016/j.indcrop.2018.05.050.

Wei, J. N., Liu, Z. H., Zhao, Y. P., Zhao, L. L., Xue, T. K., & Lan, Q. K. (2019). Phytochemical and bioactive profile of *Coriandrum sativum* L. *Food Chemistry*, 286, 260–267. https://doi.org/10.1016/j.foodchem .2019.01.171.

5 Coriander Transcriptome
Trends, Scope, and Utilization for Coriander Improvement

*Zahra Ghorbanzadeh, Rasmieh Hamid,
Feba Jacob and Rukam Singh Tomar*

CONTENTS

5.1 *CORIANDRUM SATIVUM*

Coriandrum sativum (2n=2X=22) is a hardy annual plant that belongs to the Apiaceae family, natively grown in Mediterranean Europe and Western Asia. The Apiaceae family consists of more than 434 genera which are further split into 3700 species (https://en.wikipedia.org/wiki/Apiaceae), comprising numerous eminent crop plants like celery (*Apium graveolens*) and carrot (*Daucus carota*). Coriander is a vegetable produced with universal importance, and its production tripled from 1994 to 2016. Asia is considered the top production source of this plant worldwide, accounting for 71.4% of its total production (http://faostat3.fao.org). Coriander boasts of high nutrients and is rich in vitamin C and carotene (Anitha and Hore 2018). *C. sativum* grows to *ca.* 25–60 cm or 6–24 inches in height. Its umbel flowers are pale pink to white, Figure 5.1D–F, and emit a strong fragrance, a characteristic of cilantro and coriander. This plant's natural flowering season is from June to July, producing round fruits known as coriander (Carrubba and Lombardo 2020). *C. sativum* has two major varieties, namely *vulgare*, and *microcarpum*. The former produces larger fruits with essential oil (EO) yields of 0.1–0.35% (v/w), whereas the latter yields smaller fruits with 0.8–1.8% (v/w) harvest of EO (Silva, Domeño et al. 2020). The coriander seed is a globular dry schizocarp, about 3–5 mm in size, with two mericarps, Figure 5.1G. One to two true seeds can be spotted within each schizocarp. This seed consists of three tissues: outer seed coat, endosperm, and the embryo. The embryo is very small compared to the endosperm, providing nutrients to the developing embryo at the time of germination (Diederichsen, Banniza et al. 2020). The term cilantro refers to the immature *C. sativum* plant, or simply the leaves. This name was originally derived from the Spanish synonym for the plant, and since "cilantro" is widely used in Spanish-speaking regions, cilantro has turned into the commonly accepted name for leafy portions of this plant that are yet immature (Mohammed and Qoronfleh 2020).

DOI: 10.1201/9781003204626-6

FIGURE 5.1   *Coriandrum sativum* (A) seedling, (B) young foliage, (C) late flower blossoms, (D) characteristic umbel organization of flowers, (E) fresh seeds, and (F) mature seeds and oil.

Notwithstanding coriander's agricultural, medical, and economic value, there is limited information about the molecular markers, molecular genetics, and specific functional genes. In recent years, genetic information has been more accurate and quickly accessible due to the advent of next-generation sequencing (NGS) technologies. However, the enormous genome size and ploidy levels make the study of coriander challenging. Nevertheless, transcriptome analysis is an efficient method of characterization, for the identification of various molecular mechanisms and proteins (Wang, Gerstein et al. 2009). Several studies have used transcriptome for gene discovery and studied various plant pathways' molecular mechanisms. For example, Zhou et al. (2015) used the Illumina platform to execute the transcriptome analysis of Chinese chive (*Alliumtuberosum Rottler* ex Spr.) to discover genes and design SSR markers. Fu et al. (2013) created large-scale sequence information of celery through Illumina paired-end sequencing and successfully developed 2997 EST-SSR markers from the plant.

In this chapter, we would like to provide an "on the fly" portrait of the use of either microarray or RNA-seq-based data sets in coriander research studies. Researchers have gleaned much information from transcriptome data sets by assuming guilt-by-association, given the relative simplicity of doing so.

5.2   CORIANDER IMPORTANCE

Coriander has a long history of use which dates back to *ca*. 400 BC when it was used as a Greek traditional medicine by Hippocrates. Greeks and Romans also used it to add flavor to their wine (Brian 2018). This herb was entitled the "spice of happiness" by the Egyptians, who believed it to be an aphrodisiac. They also used it as a cooking ingredient that treated digestive illnesses (Prachayasittikul, Prachayasittikul et al. 2018). Finally, when it was brought to Great Britain in the 13th century, it was used in midwifery to accelerate the process of childbirth, and thus the seeds of *C. sativum* have been widely used in a large variety of ways for over 7000 years (Kassahun 2020). Spices, i.e., plants with strong aromas and flavors, have absorbed the attention of man for many years. They were probably first used for fragrances and perfumes, flavoring and condiments, food preservatives, and even curatives and aphrodisiac stimulants (Alamgir 2018). Coriander is mostly raised for its fruits and leaves (Figure 5.1B and 5.1E) with a distinctive fragrance and pleasant mild sweet yet slightly spicy taste (Jansen 1981). Furthermore, coriander seeds and leaves make the diet nutritious and add aroma and taste to it (El-Sayed and Youssef 2019). It was reported that coriander seeds and leaves provide the diet with dietary fiber, iron, magnesium, and manganese (Kumar, Kumar et al. 2020). Moreover, the fresh leaves, dried leaves, and seeds of coriander have enormous amounts of protein, fat, carbohydrate, calcium, phosphorous, sodium, zinc, carotene, thiamine, riboflavin, niacin, tryptophan, vitamin B6, folate, vitamin A, vitamin D, vitamin B12, vitamin C, and vitamin E (Haytowitz and Bhagwat 2010, Bhat, Kaushal et al. 2014). Large seeded kinds of coriander are mainly grown for seed spice and vegetable leaf production (Kassahun 2020). The finely ground coriander fruit is the main ingredient in curry powder (Diederichsen, Banniza et al. 2020), mixed spice, flavored bread and cakes, and flavored spirits (Diwan, Bisen and Maida 2018). The coriander leaves, cilantro, are the key ingredients in daily meal preparations. The leaves give a pleasant flavor to soups and meals prepared from meat and fish. In addition, the young leaves are used in Mexican salsa, far eastern dishes, and oriental foods (Tulsani, Hamid et al. 2020). A newer opening position for this crop is the organic market, in which it is commanding high quality and price. There has been a sharp increase in demand for various coriander purees and pastes because they are a crucial part of the fast-food industry (YALDIZ and ÇAMLICA). Coriander seeds and leaves are used for medical purposes worldwide, and it is on the top of the list of healing spices (Singh, Agrawal and Mishra 2020). Coriander seeds are customarily used to treat indigestion, rheumatism, and joint pain and are an anti-parasite (Singh, Tanwar and Agarwal 2015). Recent studies have proved that when used directly with dishes, the seed affects carbohydrate metabolism (Prachayasittikul et al. 2018). Coriander works well as a home remedy for biliousness (Kačániová et al. 2020) stomachache, and it has anti-diabetic and anti-inflammatory properties, for which either fruits are boiled in water and then drunk or the leaves are chewed (Mandal and Mandal 2015). Fresh coriander leaves contain an antibacterial compound dodecenal, which is a safe and natural method to fight against *Salmonella*, which is a common cause of deadly foodborne illness. It is also realized that dodecenal is twice as effective as the usual antibiotic medicine gentamicin at eradicating *Salmonella* under laboratory tests (Rajeshwari and Andallu 2011, Mandal and Mandal 2015). Coriander essential oil, which exists in the seeds and leaves, plays an important role and is used in different industrial applications (Silva et al. 2011). It is also responsible for presenting qualities such as the spicy aroma and taste (Prachayasittikul et al. 2018). Coriander essential oil has a much better odor than many other commercial essential oils and can be used as the primary material for manufacturing many other products (Nadeem et al. 2013). The essential oil distilled from the fruits is employed for use in perfume manufacturing, soaps, chocolate, cocoa, candy, tobacco, meat products, canned soups, baked foods, and alcoholic beverages like gin and in order to conceal the unpleasant odors in pharmaceutical preparations (Uitterhaegen et al. 2016, Msaada et al. 2017). Volatile components in essential oil, obtained from both seeds and leaves, have proved to stop growth of a range of microorganisms (Sahib, Anwar et al. 2013). Coriander oil has lipid peroxidation and antioxidant properties, and anti-inflammatory activities, sedative-hypnotic activities, diuretic activities, anti-helminthic activities, hypoglycemic activity, and metal detoxification, anti-anxiety, anti-proliferative, and carminative activities of the essential oil are also reported (Bhat et al. 2014, Msaada et al. 2017, Silva and Domingues 2017). Oleoresin from coriander is utilized as a flavoring substance, as an ingredient in pharmaceutical formulation and perfumery (Sharma and Sharma 2012, Pujari et al. 2019). It is also stated that coriander essential oil is essential for growth and proper brain function. Another important factor in coriander is fatty acids; the main kinds detected in coriander are petroselenic, linoleic, oleic, and palmitic acids (Nguyen et al. 2020). The fatty acids of coriander are essential as they include high levels of petroselenic acid. The high content of dietary petroselinic acid in coriander oil has proven to be quite effective in diminishing the arachidonic acid level in the heart (Kassahun 2020) and liver (Nguyen et al. 2015, Nguyen et al. 2020). Petroselenic acid has a promising non-food application in oleo chemistry in which it is applied as a plastic lubricant during nylon manufacture. Petroselenic acid is likely to be used in the production of medium chain acids because it can be split into lauric and adipic acids by oxidative cleavage (Uitterhaegen 2014). Lauric acid is believed to contain a certain amount of surfactants and edible products, and adipic acid for nylon synthesis (Krist 2020). At present, adipic acid is taken from mineral oil through a process which releases gases such as N_2O destroying the ozone layer and contributing to global warming. Adipic acid, derived from coriander seed, is a product which is more environmentally friendly (Fayyad et al. 2017). Oleic acid is a major element used in the food industries, in particular for salad cream dressing and mayonnaise preparation. Residues from distillation can be used as livestock feed, and the fatty acids have potential usage as lubricants. Another advantage of coriander, which originates from its reproductive biology, is that it produces a large amount of nectar, which in turn promotes pollination (Mandal and Mandal 2015) (Figure 5.1D). It was reported that, in Russia, one hectare of coriander gives 500 kg of honey collected by honeybees (Elhaj 2016). Other studies conducted by Rajeshwari and Andallu reveal many other health benefits of coriander for controlling mouth ulcers, swellings, conjunctivitis, anemia, diarrhea, menstrual disorders, small pox, and skin disorders.

5.3 TRANSCRIPTOMES

The transcriptome constitutes the collection of all RNA molecules like mRNA, rRNA, tRNA, and other non-coding RNA molecules in an organism. The genome stores the information about an organism's inheritance while the transcriptome echoes its performance in diverse physiological adaptations. Therefore, transcriptome analysis, especially the quantification and analysis of mRNA population, might aid in exploring the current condition as well as foretelling the impending activities of an organism (Wang et al. 2019). Formerly, microarrays were used to understand the differential gene expression pertaining to specific conditions but provided less information. NGS methods are being used to overcome this drawback (Tam et al. 2014). RNA-sequencing (RNA-seq) is a whole transcriptome sequencing method that scrutinizes gene expression at the transcriptional level, thus providing abundant information about non-coding regions, identifying genes, and determining the structure of transcripts (Stark et al. 2019). There are two approaches to this transcriptome analysis: align-then-assemble and assemble-then-align. Align-then-assemble depends entirely on the reference genome, while assemble-then-align is used when the reference genome is unavailable and performs a de novo assembly (Deng et al. 2012). Transcript abundance can be quantified by several methods, including serial gene expression analysis, microarray, digital gene expression (DGE), and RNA-seq. DGE and RNA-seq have helped in pinpointing the molecular information of plant transcriptome and variation in gene expression among highly similar samples (Saeidian et al. 2020). DGE can quantify transcript abundance counts directly and is more cost-efficient than RNA-seq. However, RNA-seq is a more flexible method and can perceive full-transcript sequence, transcript abundance, exon boundaries, and alternative splicing. Moreover, each transcript can be mapped several times in RNA-seq. The sequencing depth of RNA-seq is correlated with the transcript abundance but is not equal to it (Wyman 2020).

Gene expression profiling has been extensively utilized to decipher the association between the cellular expression patterns and the resultant ecologically influenced disease phenotypes. Detection technologies based on PCR employ species-specific primers and provide improved information on the biology of plant/microbe communications, especially regarding the ecosystem, etiology, and epidemiology of plant pathogenic microorganisms (Shivaraj, Deshmukh et al. 2017). However, although RT PCR is a powerful technique for absolute comparison of all transcripts of the inspected tissue, it has certain drawbacks, i.e., it is critically influenced by the use of calibration and reference materials. Furthermore, the lack of quality template (owing to the ineffective RNA extraction methods) and the occurrence of the high content of PCR inhibitory compounds, probably plant polysaccharides and polyphenolics, hinder the practical application of PCR for diagnosing plant/microbe consortium tissues (Deepak et al. 2007).

Genome-wide transcription studies have improved our knowledge about the functions of the gene and the complicated mechanisms that regulate gene expression (Soutourina 2018). Most genome-wide information is obtained through gene expression level (i.e., variations in mRNA quantity) (Oikonomopoulos et al. 2020). It is generally taken for granted that identical RNA molecules are transcribed from each gene; however, several different isoforms may be produced from a single gene using alternative promoters, exons, and terminators (Raabe and Brosius 2015). Thus, alternative RNA molecules are produced during transcription that vary in length, function, localization, and expression pattern (Fernie and Usadel 2013, Ingolia 2014, Stark et al. 2019). Multiple transcript isoforms that are alternatively spliced are produced in eukaryotes by most of the genes, dramatically increasing the genome's protein-coding potential (Thatcher et al. 2014). Large-scale transcriptome analysis, with the aid of NGS technologies, has turned out to be quite helpful to gain candidates for such unidentified genes (Gao and Shi 2019, Rathod et al. 2020). The NGS technologies are likely to generate high-throughput reads at a reasonably low price and are now being used in several research areas, including genome sequencing, marker discovery, and, especially, transcriptome analysis (Reuter et al. 2015, Goodwin et al. 2016). Transcriptome analysis, owing to the great depth of sequencing provided by the use of NGS technologies, can analyze almost all of the expressed sequences (Kchouk et al. 2017), including rare transcripts in a precise tissue at a particular developmental stage (van Dijk et al. 2018). As a result, this approach is highly advantageous, specifically for transcriptome analysis in non-model organisms such as coriander (Figure 5.2). Transcriptome analysis is a powerful achievement for identifying different molecular mechanisms. Planting large genomes and ploidy levels turn transcriptomes into a powerful tool for characterizing and identifying proteins (Wang et al. 2009). There are several reports about the benefits of transcriptome analysis in gene discovery and learning molecular mechanisms of different plant pathways. Zhou et al. (2015) used the Illumina platform for transcriptome analysis in Chinese chive (*Allium tuberosum Rottler* ex Spr.) for both gene discovery and SSR designing. Zhang et al. (2014) carried out in-depth transcriptome sequencing of coriander flower buds and created 3059 genic SSRs. Fu et al. (2013), by Illumina paired-end sequencing, generated a large amount of sequence information on celery and also developed 2997 EST-SSR markers. Tulsani et al. (2020) used three various tissues of coriander (flower, leaves, and seed) and Ion torrent S5 sequencing technology to identify the tissue transcriptome of *Coriandrum sativum* and to develop a set of EST-SSRs. In this work, transcriptome landscaping helped in characterizing the genes and developing EST-SSRs,

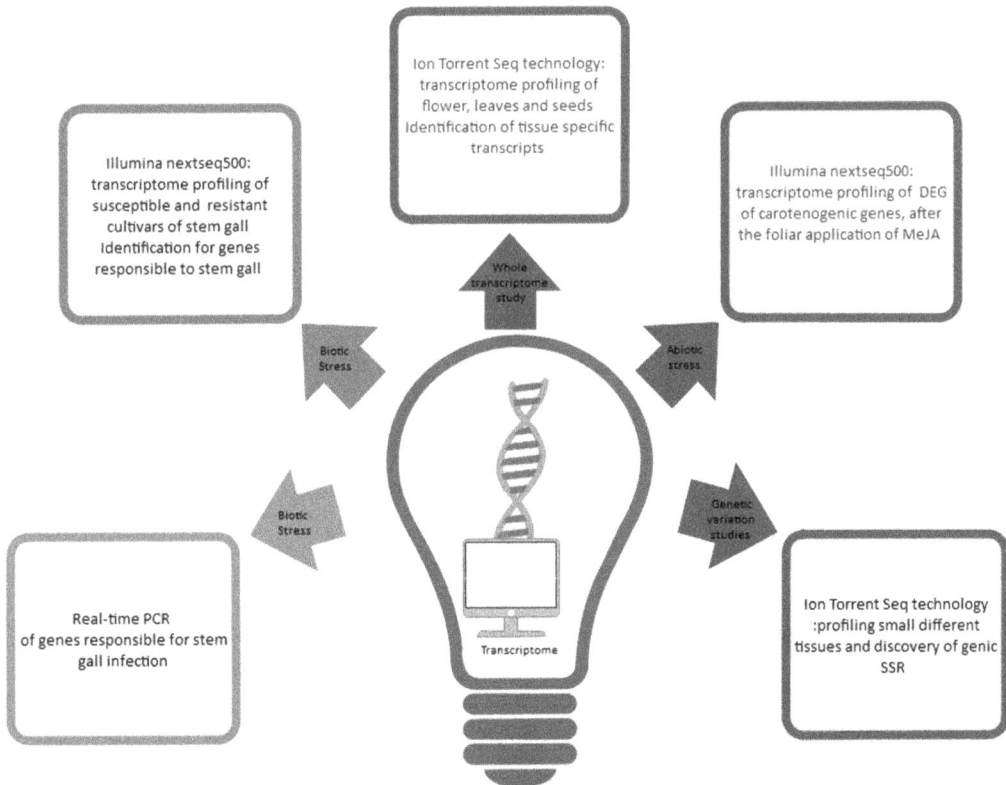

FIGURE 5.2 Representation of the transcriptome studies conducted in coriander.

which are noteworthy add-ons to the prevalent genomic resources of coriander, and would further permit coriander genome annotation.

5.4 TRANSCRIPTOME IN BIOTIC AND ABIOTIC STRESS

Coriander suffers from various biotic and abiotic stresses; among biotic stresses, diseases are the most restricting elements in its production. An aggregate of 20 fungal pathogens was recorded in coriander, and some common fungal diseases are blight (*Alternaria* spp.), powdery mildew (*Erysiphe polygoni*), wilt (*Fusarium oxysporium* F. sp. *corianderii*), stem gall (*Protomyces macrospores* Unger), and *Batticaloa* (Chattopadhyay et al., 2018; Khare et al. 2017). Stem gall disease, a destructive disease caused by *P. macrospores* Unger, reduces the quality of the seeds and causes a reduction in the yield (Kumar et al. 2014).

Coriander genetic improvement for stem gall disease resistance or tolerance is crucial in breeding programs (Figure 5.3) (Leharwan and Gupta 2019). Disease resistance or responsive genes are great

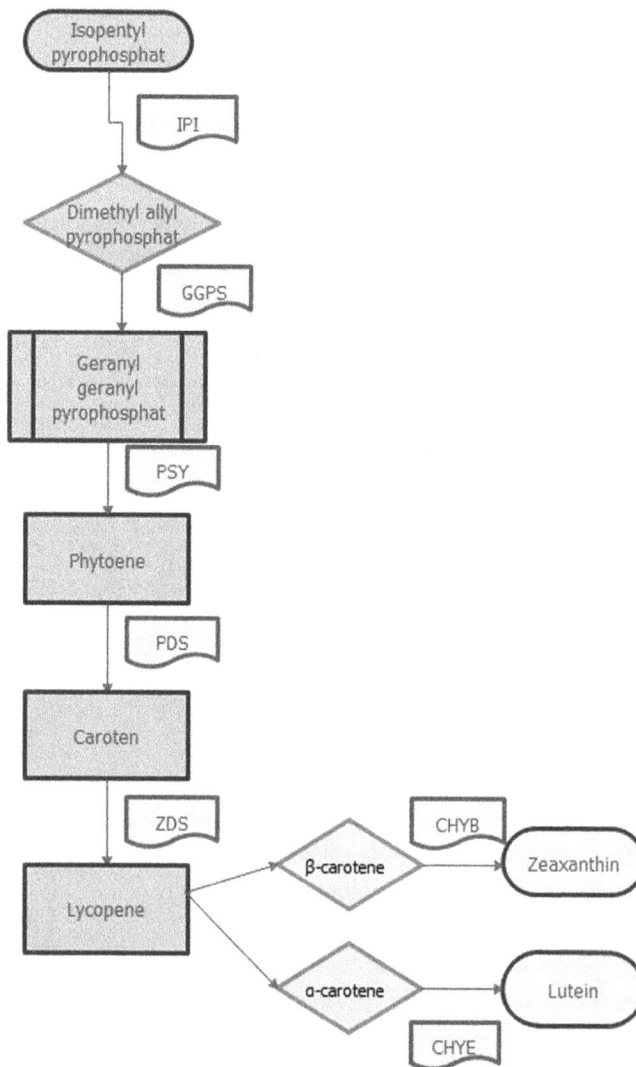

FIGURE 5.3 Carotenoid biosynthetic pathway in coriander. GGPP, geranylgeranyl pyrophosphate; PSY, phytoene synthase; PDS, phytoene desaturase; ZDS, ζ-carotene desaturase.

targets to develop resistant cultivars. Transcriptome studies, by profiling genes in different tissues, various developmental stages, and responses to biotic and abiotic stresses, could facilitate gene identification research (Tariq et al. 2018). Choudhary et al. utilized leaf samples from resistant (ACr-1) and susceptible (CS-6) coriander cultivars to perform transcriptome sequencing using the Illumina NextSeq500 platform. After the inferior reads and adapter sequences were trimmed, 49,163,108 and 43,746,120 high-quality reads were retained in total, which on further assembly resulted in validated transcripts of 59,933 and 56,861. They predicted a total of 52,506 and 48,858 coding sequences, from which 50,506 and 46,945 were annotated using the NCBI nr database. Through gene ontology analysis, 19,099 and 17,625 terms were annotated; 24 different functional pathway categories were acquired through pathway analysis. Among these, signal transduction, transport, catabolism, translation and carbohydrate metabolism pathways, etc., dominated. Differentially expressed genes analysis predicted 13,123 commonly expressed coding sequences, out of which 431 and 400 genes were significantly up- and down-regulated in which R genes, stress-inducible transcription factors like ERF, NAC, bZIP, MYB, DREB, and WRKY, and antifungal-related genes were predicted (Choudhary et al.).

Choudhary et al. (2020) assessed the expression level of genes related to disease resistance like LRR, USP, GDSL, ANK, and PDR using real-time PCR in coriander cultivars like AgCr-1, ACr-1, ACr-2, CO-2, and CS-6 that are popular in India. The highest expression was observed in cultivar AgCr-1 followed by Pant Haritma, Hisar Sugandh, and ACr-1, with the lowest expression in Hisar Anand, ACr-2, CO-2, Rajendra Swathi, and CS-6. The relevance of the conserved domain of the genes was known through domain analysis. The genes identified may have possible importance in coriander and can be further utilized in programs for crop improvement. Contrasting cultivars are often an honest source for the evaluation of candidate genes and may be further used as potential markers. They also could be utilized in a hybridization program for incorporating new genes and making the adapted cultivars of the region disease-resistant. Biotic and abiotic stresses cause variations in the expression levels of carotenogenic genes and also the number of carotenoids, especially during developmental processes (Yungyuen et al. 2018). This suggests a well-coordinated regulatory mechanism in carotenoid biosynthesis and accumulation. Plant hormones, chiefly, MeJA and salicylic acid, play vital roles in coordinating stress responses in plants.

Divya et al. (2018) examined the effect of the foliar application of MeJA and gene-specific inhibitors of carotenogenesis on the regulation and differential expression of carotenogenic genes, mainly CsPDS, CsZDS, CsCCD, and CsCHYE, and the accumulation of major carotenoids and chlorophylls. Furthermore, they analyzed the association of transcript levels of crucial carotenogenic genes and the buildup of major carotenoids. Moreover, they also explored the vital mechanisms that enhance carotenoids, in the presence of MeJA and various gene-specific inhibitors, by studying the transcriptional expression profile of ten carotenogenic genes. Foliar application of MeJA (10 μM) up-regulated CsPDS (phytoene desaturase), CsZDS (ς carotene desaturase), CsCHYE (carotene ε- hydroxylase), and CsLCYE (lycopene β-cyclase) genes, and there was a strong correlation between their transcript levels and carotenoid content. There was an increase of 3.9- and 6.1-fold three days after the treatment for β-carotene and lutein, respectively. Gene-specific inhibitors fosmidomycin, norflurazon, and amitrol were used to confirm the regulatory effect of essential genes like CsPDS, CsZDS, CsLCYE, and LCYB. Norflurazon, the phytoene desaturation inhibitor, reduces the β-carotene and lutein content, correlated to CsPDS and CsZDS gene induction. These results exhibit that carotenogenic genes are strongly induced by the signaling networks controlled by MeJA, leading to the buildup of carotenoids. These data may help develop accurate approaches for reworking the carotenoid pathway so that the required level of a selected carotenoid in leafy vegetables is attainable.

5.5 OTHER TRANSCRIPTOME STUDIES

Coriander mericarp oil consists of many fatty acids, other than essential oil (EO), while the leaves have very little EO. Therefore, coriander mericarp oil has high lipid content; approx. 28.4% of the total mericarp weight is highly beneficial in the food industry (Nguyen 2015). Petroselinic acid ($18:1\Delta 6$), a monounsaturated cis Δ-6 fatty acid, has many potential applications in nutraceutical, functional foods, and pharmaceutical industries. About 80% of petroselinic acid has been obtained from coriander fruit's oil (*Coriandrum sativum* L.), making it a rich source for petroselinic acid biosynthesis (Janković, Govedarica et al. 2020). Through transcriptome analysis, Yang et al. investigated the candidate genes responsible for the biosynthesis of petroselinic acid in fruits of *Coriandrum sativum* L. In this study, samples of fruit from three progressive stages with the quick build-up of petroselinic acid were utilized for sequencing RNA with the Illumina Hiseq4000 platform. Characterized acyl-ACP desaturase was confirmed after thoroughly examining these accumulated transcriptome data and discriminatory expressed genes (DEGs). Additionally, other candidate genes pertaining to the biosynthesis of petroselinic acid were analyzed in-depth by quantitative real-time PCR (q-PCR) and bioinformatics tools. The all-embracing analysis of transcriptome can reveal a model for the biosynthesis of petroselinic acid in coriander fruits, which may help identify the key genes in the pathway for petroselinic metabolism (Yang et al. 2020).

Plant terpenes are a vast and diverse class of naturally derived mixtures, significant in the medical, perfume, and culinary industries. *Coriandrum sativum* (coriander) seeds produce EO abundant in monoterpenes and volatile C10 terpenes. To study EO metabolism, the seeds were sequenced via Illumina technology at three developmental stages (early, mid, and late) (Galata et al. 2014). Analysis of the differential transcript abundance among the different stages in the selected terpene biosynthetic genes showed two terpene production pathways, active in the seeds constitutively, having slight up-regulation in the mid-developmental stage. All the involved genes, with active photosynthesis and fatty acid biosynthesis and metabolism, were also identified from the coriander transcript library. In addition to validating the reliability of the transcriptome sequence data, a terpene synthase candidate gene, CsγTRPS, encoding a 611 amino acid protein, was expressed in bacteria, and Ni-NTA affinity chromatography was used to purify the recombinant protein obtained. Enzymatic assays with the precursor to monoterpenes, geranyl diphosphate (GPP), revealed that this 65.86 kDa recombinant protein catalyzed the conversion of GPP to γ-terpinene, with apparent Vmax, Km, and kcat values of 2.2 ± 0.2 pkat/mg, 66 ± 13 μM and $1.476 \times 10^{-4\ \mathrm{s}^{-1}}$, respectively. This knowledge will encourage future studies regarding the production of essential and fatty acid oil in coriander and assist in improving the EO of coriander through plant breeding or metabolic engineering.

5.6 NON-CODING RNA

MicroRNAs (miRNAs) are small and non-coding RNAs of 18–24 nucleotides that control the target gene expression at the transcriptional and post-transcriptional levels in plants (Hamid et al. 2020). They play a vital role in numerous processes like growth and development, cell proliferation, and response to stresses in plants (Megha et al. 2018). Up to now, there are rare reports of miRNAs identification in coriander. Mir Drikvand et al., using the seed and leaf transcriptome, identified the conserved miRNAs and their target genes in coriander. These miRNAs showed differential expression between seed and leaf tissues, and their target genes' roles in various biological processes were confirmed. In general, by the identification of the regulatory roles of these miRNAs on a broad spectrum of gene networks and biological processes of cilantro in this study, they can be used as candidate genes to boost qualitative and quantitative yield and resistance to several stresses in this plant (Mir Drikvand et al. 2019).

5.7 MOLECULAR MARKERS

Nowadays, molecular marker-assisted selection (MAS) has become a crucial step for breeding in its conventional form. Having enough knowledge about germplasm is of great value in crop breeding as it produces important information for the parental line selection in the breeding program (Pareek et al. 2017). Most studies used random molecular markers and a few polymorphic co-dominant markers (Furan and Geboloğlu 2017). Lately, simple sequence repeats (SSRs) have been broadly utilized in extensive research (Abou El-Nasr et al. 2013). SSR markers (co-dominant) are being used extensively for high-throughput genotyping and map construction due to their abundance, random distribution within the genome, high polymorphism information content (PIC), and stable co-dominance. However, the relatively low polymorphism and high chances of no gene-rich regions in the genome restrict their use. In sharp contrast, genomic SSRs are highly polymorphic and are likely to be extensively distributed across the genome, leading to better genetic diversity analysis. On the basis of the original sequences, there are two types of SSRs, i.e., containing genomic SSRs and EST-SSRs. EST-SSRs are derived from the expressed regions of the genome, that were more conserved in the process of evolution than non-coding sequences. Moreover, EST-SSR markers are more transferable when compared with genomic SSRs (Varshney et al. 2005). EST-SSR markers are the best tools to recognize candidate functional genes. When weighed against traditional methods, RNA-seq technology is an economical and quicker approach to develop EST-SSRs (Hamid et al. 2018). Expressed sequence tags (ESTs) and genomic sequences information can clarify the breeding study in coriander. Tulsani et al. (2020) used transcriptome data from seeds, leaves, and flowers, identifying a total of 9746 SSR-containing loci. The SSR identified as most abundant were the dinucleotide repeat motifs (45.5%), followed by tri- (34.6%), tetra- (4.5%), penta- (1.5%), and hexanucleotide repeats (1%). They designed 3795 primers, out of which 120 randomly selected primers were validated in 14 accessions of coriander grown in India. This is a useful study for preliminary information regarding the transcriptome and genic markers that can be useful in coriander breeding and genetic diversity estimation.

5.8 PLASTID SEQUENCES

The dynamic arrangement of the plant mitochondrial genome provides an effective model for studying genome structure and evolution. Furthermore, the growing availability of plant organelle and nuclear genome sequence data allows for a better understanding of the processes that drive plant genome evolution. Moreover, the plastid, mitochondrial, and nuclear genomes interact structurally and functionally (Zoschke and Bock 2018, Lee and Kang 2020). Transfer of DNA across these three compartments has been observed in higher plants, except for transfer into the plastid genome (Forsythe et al. 2019).

Wang et al. (2021) revealed the first full mitochondrial genome of the valuable medicinal and culinary plant *Coriandrum sativum* using Oxford Nanopore and Illumina sequencing methods. The whole mitogenome was organized into two circular-mapping forms, of 82,926 bp (cir1) and 224,590 bp (cir2). The cir1 mitogenome had 28 genes, including 14 protein-coding genes, 2 rRNA genes, and 12 tRNA genes. The cir2 mitogenome had 62 genes, including 41 protein-coding genes, 5 rRNA genes, and 16 tRNA genes. Based on phylogenetic research, *Coriandrum sativum* was shown to be the most closely related to *Daucus carota*. The genome sequence data that support the study's conclusions are freely accessible in NCBI GenBank at (https://www.ncbi.nlm.nih.gov/) with the accession numbers MW477237–MW477238. In addition, PRJNA729412, SUB3752698, and SAMN19115529 are the related BioProject, SRA, and Bio-Sample numbers.

5.9 CORIANDER GENOMICS DATABASE (CGDB)

Accompanying the rapid progress of sequencing technologies, many genomes are being sequenced and released (Chen et al. 2019, McCombie et al. 2019). Therefore, tons of plant genome databases allow the researchers to inspect and download omics data sets. Despite these statistics' being incomplete, even for a single plant or families such as *Arabidopsis*, there are more than 40 available genome databases (http://www.arabidopsis.org/index.jsp), cotton (https://cottonfgd.org/), rice (http://rice.plantbiology.msu.edu), peanut (https://peanutbase.org/), maize (https://www.maizegdb

.org), carrot (https://world-crops.com/carrot/), strawberry (Hirakawa et al. 2014), Chinese cabbage (Cheng et al. 2011), and pineapple (Xu et al. 2018).

The Apiaceae family has only a few draft genome sequences that have been released, and only the carrot genome has been completely sequenced so far (Simon et al. 2019). Nevertheless, there is no such database available for coriander, which restricts the ability of researchers to obtain genomic and other types of omics data for coriander. Furthermore, although a carrot genome database has been built, the genome size, gene number, and several repeat sequences are significantly greater in coriander when compared to carrots (Song et al. 2020). As a result, it is essential to create a platform for coriander genome analysis utilizing the information from carrot genome data as a reference.

The coriander genome and the carrot genome can be efficiently used to review the evolution of the Apiaceae gene family. The boom in coriander research will provide large quantities of transcriptome, metabolome, proteome, and phenomics data in the future. Song et al. established the Coriander Genomics Database (CGDB), a large repertoire of all the omics data sets, to explore the coriander genome, transcriptome, and metabolome. BLAST, a well-known search tool, was used, which assisted the researchers in finding the genes of interest in coriander and other related species in this database. The GB can be effortlessly applied to find details about each gene. This website is intended to be the chief coriander research database. The intuitive CGDB browser allows us to search, visualize, download, and observe the cross-species collinearity while it provides unlimited access to the latest omics data sets. This database will enable a dynamic global community of coriander researchers and thus simplify comparative genomic evolution studies in Apiaceae (Song et al. 2020).

5.10   CONCLUDING REMARKS AND FUTURE PERSPECTIVES

Modern breakthroughs of technology in high-throughput sequencing and genotyping have enormously transformed the genomics landscape in plant research. The cost of NGS has been lowered, and hence it is being used by researchers globally to respond to various biological questions accurately. Coriander is a widely accepted seed flavor and vegetable; nevertheless, the transcriptome and genomic information limitations have impeded further investigations into it. Transcriptome studies pave the way for the better realization of processes such as adaptation and speciation in coriander. Thus, any pure biosynthetic pathway study, such as the coriander terpene synthase characterization and transcript sequencing produced, will lend a helping hand to future work concerning metabolic engineering.

REFERENCES

Abou El-Nasr, T., M. Ibrahim, K. Aboud and A. El-Enany Magda (2013). "Assessment of genetic variability for three coriander (*Coriandrum sativum* L.) cultivars grown in Egypt, using morphological characters, essential oil composition and ISSR Markers." *World Applied Sciences Journal* 25(6): 839–849.

Alamgir, A. (2018). *Therapeutic Use of Medicinal Plants and Their Extracts: Volume 1*, Springer.

Anitha, M. and J. Hore (2018). "Production technology of some major and minor spice crops." In *Indian Spices*, Editors: Amit Baran Sharangi. Springer: 95–175.

Bhat, S., P. Kaushal, M. Kaur and H. Sharma (2014). "Coriander (*Coriandrum sativum* L.): Processing, nutritional and functional aspects." *African Journal of Plant Science* 8(1): 25–33.

Brian, K. (2018). *Spices and Herbs: An Essential Guide to Spice Up Your Health and Flavor Your Diet*, EWJ Publishing via PublishDrive.

Carrubba, A. and A. Lombardo (2020). "Plant structure as a determinant of coriander (*Coriandrum sativum* L.) seed and straw yield." *European Journal of Agronomy* 113: 125969.

Chattopadhyay, A., J. Purohit, D. Patel and M. Tetarwal (2018). "Management of diseases of spice crops through microbes." In *Biological Control of Crop Diseases: Recent Advances & Perspectives*, Editors: Dinesh Singh, B.N. Chakraborty, P. Sharma & R. N. Pandey. 249–270.

Chen, F., Y. Song, X. Li, J. Chen, L. Mo, X. Zhang, Z. Lin and L. Zhang (2019). "Genome sequences of horticultural plants: Past, present, and future." *Horticulture Research* 6(1): 1–23.

Cheng, F., S. Liu, J. Wu, L. Fang, S. Sun, B. Liu, P. Li, W. Hua and X. Wang (2011). "BRAD, the genetics and genomics database for Brassica plants." *BMC Plant Biology* 11(1): 1–6.

Choudhary, S., M. B. Naika and R. Meena (2020). "Identification and expression analysis of candidate genes associated with stem gall disease in Coriander (*Coriandrum sativum* L.) cultivars." *Molecular Biology Reports* 47(7): 5403–5409.

Choudhary, S., M. B. Naika, R. Sharma, R. Meena, R. Singh and G. Lal. (2019). "Transcriptome analysis of coriander: A dual purpose crop unravels stem gall resistance genes." *Journal of Genetics* 98: 19.

Deepak, S., K. Kottapalli, R. Rakwal, G. Oros, K. Rangappa, H. Iwahashi, Y. Masuo and G. Agrawal (2007). "Real-time PCR: Revolutionizing detection and expression analysis of genes." *Current Genomics* 8(4): 234–251.

Deng, X., Z. Li and W. Zhang (2012). "Transcriptome sequencing of *Salmonella enterica* serovar *Enteritidis* under desiccation and starvation stress in peanut oil." *Food Microbiology* 30(1): 311–315.

Diederichsen, A., S. Banniza, C. Armstrong-Cho and T. Sander (2020). "*Coriandrum sativum* L.-Coriander." In *Medicinal, Aromatic and Stimulant Plants*, Editors: Johannes Novak & Wolf-Dieter Blüthner. Springer: 265–281.

Divya, P., B. Puthusseri, M. A. Savanur, V. Lokesh and B. Neelwarne (2018). "Effects of methyl jasmonate and carotenogenic inhibitors on gene expression and carotenoid accumulation in coriander (*Coriandrum sativum* L.) foliage." *Food Research International* 111: 11–19.

Diwan, G., B. Bisen and P. Maida (2018). "Effect of nitrogen doses and row spacing on growth and seed yield of coriander (*Coriandrum sativum* L.)." *International Journal of Chemical Studies* 6(4): 2768–2772.

El-Sayed, S. M. and A. M. Youssef (2019). "Potential application of herbs and spices and their effects in functional dairy products." *Heliyon* 5(6): e01989.

Elhaj, S. O. (2016). *Evaluation of the Efficacy of Coriander Extracts against Wilt Disease Caused by Fusarium solani in Potato Plant*, Sudan University of Science and Technology.

Fayyad, G., N. Ibrahim, A. Yaakob and P. P. B. dan Bioteknologi (2017). "Evaluation of biological activities of seeds of *Coriandrum sativum*." *International Journal of Scientific and Engineering Research* 8(7): 1058–1063.

Fernie, A. and B. Usadel (2013). "The plant transcriptome-from integrating observations to models." *Frontiers in Plant Science* 4: 48.

Forsythe, E. S., J. Sharbrough, J. C. Havird, J. M. Warren and D. B. Sloan (2019). "CyMIRA: The cytonuclear molecular interactions reference for Arabidopsis." *Genome Biology and Evolution* 11(8): 2194–2202.

Fu, N., Q. Wang and H.-L. Shen (2013). "De novo assembly, gene annotation and marker development using Illumina paired-end transcriptome sequences in celery (*Apium graveolens* L.)." *PloS One* 8(2): e57686.

Furan, M. and M. Geboloğlu (2017). "Assessment of genetic variation on some cultivated Turkish coriander (*Coriandrum sativum* L.) varieties based on ISSR and SRAP markers." *Yüzüncü Yil Üniversitesi Journal of Agricultural Sciences* 27(2): 245–251.

Galata, M., L. S. Sarker and S. S. Mahmoud (2014). "Transcriptome profiling, and cloning and characterization of the main monoterpene synthases of *Coriandrum sativum* L." *Phytochemistry* 102: 64–73.

Gao, Z. and T. Shi (2019). "*Prunus mume* transcriptomics." In *The Prunus mume Genome*, Springer: 93–99.

Goodwin, S., J. D. McPherson and W. R. McCombie (2016). "Coming of age: Ten years of next-generation sequencing technologies." *Nature Reviews Genetics* 17(6): 333.

Hamid, R., F. Jacob, H. Marashi, V. Rathod and R. S. Tomar (2020). "Uncloaking lncRNA-mediated gene expression as a potential regulator of CMS in cotton (*Gossypium hirsutum* L.)." *Genomics* 112(5): 3354–3364.

Hamid, R., R. S. Tomar, H. Marashi, S. M. Shafaroudi, B. A. Golakiya and M. Mohsenpour (2018). "Transcriptome profiling and cataloging differential gene expression in floral buds of fertile and sterile lines of cotton (*Gossypium hirsutum* L.)." *Gene* 660: 80–91.

Haytowitz, D. B. and S. Bhagwat (2010). "USDA database for the oxygen radical absorbance capacity (ORAC) of selected foods, release 2." *US Department of Agriculture* 3(1): 10–48.

Hirakawa, H., K. Shirasawa, S. Kosugi, K. Tashiro, S. Nakayama, M. Yamada, M. Kohara, A. Watanabe, Y. Kishida and T. Fujishiro (2014). "Dissection of the octoploid strawberry genome by deep sequencing of the genomes of Fragaria species." *DNA Research* 21(2): 169–181.

Ingolia, N. T. (2014). "Ribosome profiling: New views of translation, from single codons to genome scale." *Nature Reviews Genetics* 15(3): 205–213.

Janković, M. R., O. M. Govedarica and S. V. Sinadinović-Fišer (2020). "The epoxidation of linseed oil with in situ formed peracetic acid: A model with included influence of the oil fatty acid composition." *Industrial Crops and Products* 143: 111881.

Jansen, P. C. M. (1981). *Spices, Condiments and Medicinal Plants in Ethiopia, Their Taxonomy and Agricultural Significance*, Pudoc.

Kačániová, M., L. Galovičová, E. Ivanišová, N. L. Vukovic, J. Štefániková, V. Valková, P. Borotová, J. Žiarovská, M. Terentjeva and S. Felšöciová (2020). "Antioxidant, antimicrobial and antibiofilm activity of coriander (*Coriandrum sativum* L.) essential oil for its application in foods." *Foods* 9(3): 282.

Kassahun, B. M. (2020). "Unleashing the exploitation of coriander (*Coriander sativum* L.) for biological, industrial and pharmaceutical applications." *Academic Research Journal of Agricultural Science and Research* 8(6): 552–564.

Kchouk, M., J.-F. Gibrat and M. Elloumi (2017). "Generations of sequencing technologies: From first to next generation." *Biology and Medicine* 9(3).

Khare, M., S. Tiwari and Y. Sharma (2017). "Disease problems in the cultivation of coriander (*Coriandrum sativum* L.) and their management leading to production of high quality pathogen free seed." *International Journal of Seed Spices* 7(1): 1–7.

Krist, S. (2020). "Coriander seed oil." In *Vegetable Fats and Oils*, Springer: 265–271.

Kumar, D., S. Kumar and C. Shekhar (2020). "Nutritional components in green leafy vegetables: A review." *Journal of Pharmacognosy and Phytochemistry* 9(5): 2498–2502.

Kumar, G., S. Yadav, J. Patel, A. Sarkar and L. Awasthi (2014). "Management of stem gall disease in coriander using Pseudomonas and Trichoderma (bioagents) and fungicides." *Journal of Pure and Applied Microbiology* 8(6): 4975–4978.

Lee, K. and H. Kang (2020). "Roles of organellar RNA-binding proteins in plant growth, development, and abiotic stress responses." *International Journal of Molecular Sciences* 21(12): 4548.

Leharwan, M. and M. Gupta (2019). "Stem gall of coriander: A review." *Agricultural Reviews* 40(2): 121–128.

Mandal, S. and M. Mandal (2015). "Coriander (*Coriandrum sativum* L.) essential oil: Chemistry and biological activity." *Asian Pacific Journal of Tropical Biomedicine* 5(6): 421–428.

McCombie, W. R., J. D. McPherson and E. R. Mardis (2019). "Next-generation sequencing technologies." *Cold Spring Harbor Perspectives in Medicine* 9(11): a036798.

Megha, S., U. Basu and N. N. Kav (2018). "Regulation of low temperature stress in plants by microRNAs." *Plant, Cell & Environment* 41(1): 1–15.

Mir Drikvand, R., S. S. Sohrabi, S. M. ʼSohrabi and K. Samiei (2019). "Identification and characterization of conserved miRNAs of *Coriandrum sativum* L. using next-generation sequencing data." *Crop Biotechnology* 9(25): 59–74.

Mohammed, S. G. and M. W. Qoronfleh (2020). "Vegetables." In *Personalized Food Intervention and Therapy for Autism Spectrum Disorder Management*, Springer: 225–277.

Msaada, K., M. B. Jemia, N. Salem, O. Bachrouch, J. Sriti, S. Tammar, I. Bettaieb, I. Jabri, S. Kefi and F. Limam (2017). "Antioxidant activity of methanolic extracts from three coriander (*Coriandrum sativum* L.) fruit varieties." *Arabian Journal of Chemistry* 10: S3176–S3183.

Nadeem, M., F. M. Anjum, M. I. Khan, S. Tehseen, A. El-Ghorab and J. I. Sultan (2013). "Nutritional and medicinal aspects of coriander (*Coriandrum sativum* L.): A review." *British Food Journal* 115(5): 743–755.

Nath, P. (2015). "Coriander A potential medicinal herb." *Indian Food Industry Mag*, 34(2).

Nguyen, Q. H. (2015). *Study on Bioaccumulation and Integrated Biorefinery of Vegetable Oil and Essential Oil of Coriander (Coriandrum sativum L.)*. Institut National Polytechnique Toulouse, France.

Nguyen, Q.-H., T. Talou, M. Cerny, P. Evon and O. Merah (2015). "Oil and fatty acid accumulation during coriander (*Coriandrum sativum* L.) fruit ripening under organic cultivation." *The Crop Journal* 3(4): 366–369.

Nguyen, Q.-H., T. Talou, P. Evon, M. Cerny and O. Merah (2020). "Fatty acid composition and oil content during coriander fruit development." *Food Chemistry* 326: 127034.

Oikonomopoulos, S., A. Bayega, S. Fahiminiya, H. Djambazian, P. Berube and J. Ragoussis (2020). "Methodologies for transcript profiling using long-read technologies." *Frontiers in Genetics* 11: 606.

Pareek, N., M. Jakhar and C. Malik (2017). "Analysis of genetic diversity in coriander (*Coriandrum sativum* L.) varieties using random amplified polymorphic DNA (RAPD) markers." *Journal of Microbiology and Biotechnology Research* 1(4): 206–215.

Prachayasittikul, V., S. Prachayasittikul, S. Ruchirawat and V. Prachayasittikul (2018). "Coriander (*Coriandrum sativum*): A promising functional food toward the well-being." *Food Research International* 105: 305–323.

Pujari, R., B. Sunanda, A. Kurubar, J. Narayan, T. Chetan and S. Kale (2019). "Collection and evaluation of coriander varieties for growth and seed purpose in UKP command area." *International Journal of Current Microbiology and Applied Sciences* 8(6): 3125–3130.

Raabe, C. A. and J. Brosius (2015). "Does every transcript originate from a gene." *Annals of the New York Academy of Sciences* 1341: 136–148.

Rajeshwari, U. and B. Andallu (2011). "Medicinal benefits of coriander (*Coriandrum sativum* L.)." *Spatula DD* 1(1): 51–58.

Rathod, V., R. Hamid, R. S. Tomar, R. Patel, S. Padhiyar, J. Kheni, P. Thirumalaisamy and N. S. Munshi (2020). "Comparative RNA-Seq profiling of a resistant and susceptible peanut (*Arachis hypogaea*) genotypes in response to leaf rust infection caused by *Puccinia arachidis*." *3 Biotech* 10: 1–15.

Reuter, J. A., D. V. Spacek and M. P. Snyder (2015). "High-throughput sequencing technologies." *Molecular Cell* 58(4): 586–597.

Saeidian, A. H., L. Youssefian, H. Vahidnezhad and J. Uitto (2020). "Research techniques made simple: Whole-transcriptome sequencing by RNA-Seq for diagnosis of monogenic disorders." *Journal of Investigative Dermatology* 140(6): 1117–1126. e1111.

Sahib, N. G., F. Anwar, A. H. Gilani, A. A. Hamid, N. Saari and K. M. Alkharfy (2013). "Coriander (*Coriandrum sativum* L.): A potential source of high-value components for functional foods and nutraceuticals-A review." *Phytotherapy Research* 27(10): 1439–1456.

Sharma, M. and R. Sharma (2012). *Coriander. Handbook of Herbs and Spices*, Elsevier: 216–249.

Shivaraj, S., R. K. Deshmukh, R. Rai, R. Bélanger, P. K. Agrawal and P. K. Dash (2017). "Genome-wide identification, characterization, and expression profile of aquaporin gene family in flax (*Linum usitatissimum*)." *Scientific Reports* 7(1): 1–17.

Silva, F., C. Domeño and F. C. Domingues (2020). *Coriandrum sativum L.: Characterization, Biological Activities, and Applications. Nuts and Seeds in Health and Disease Prevention*, Elsevier: 497–519.

Silva, F. and F. C. Domingues (2017). "Antimicrobial activity of coriander oil and its effectiveness as food preservative." *Critical Reviews in Food Science and Nutrition* 57(1): 35–47.

Silva, F., S. Ferreira, J. A. Queiroz and F. C. Domingues (2011). "Coriander (*Coriandrum sativum* L.) essential oil: Its antibacterial activity and mode of action evaluated by flow cytometry." *Journal of Medical Microbiology* 60(10): 1479–1486.

Simon, P., M. Iorizzo, D. Grzebelus and R. Baranski (2019). *The Carrot Genome*, Springer.

Singh, D., A. Tanwar and P. Agrawal (2015). "An overview on coriander." *Journal of Biomedical and Pharmaceutical Research* 4(2).

Singh, S., N. Agrawal and I. Mishra (2020). *Pharmacology and Phytochemistry of Coriander. Ethnopharmacological Investigation of Indian Spices*, IGI Global: 173–196.

Song, X., F. Nie, W. Chen, X. Ma, K. Gong, Q. Yang, J. Wang, N. Li, P. Sun and Q. Pei (2020). "Coriander genomics database: A genomic, transcriptomic, and metabolic database for coriander." *Horticulture Research* 7(1): 1–10.

Soutourina, J. (2018). "Transcription regulation by the Mediator complex." *Nature Reviews Molecular Cell Biology* 19(4): 262.

Stark, R., M. Grzelak and J. Hadfield (2019). "RNA sequencing: The teenage years." *Nature Reviews Genetics* 20(11): 631–656.

Tam, S., R. De Borja, M.-S. Tsao and J. D. McPherson (2014). "Robust global microRNA expression profiling using next-generation sequencing technologies." *Laboratory Investigation* 94(3): 350–358.

Tariq, R., C. Wang, T. Qin, F. Xu, Y. Tang, Y. Gao, Z. Ji and K. Zhao (2018). "Comparative transcriptome profiling of rice near-isogenic line carrying Xa23 under infection of *Xanthomonas oryzae* pv. oryzae." *International Journal of Molecular Sciences* 19(3): 717.

Thatcher, S. R., W. Zhou, A. Leonard, B.-B. Wang, M. Beatty, G. Zastrow-Hayes, X. Zhao, A. Baumgarten and B. Li (2014). "Genome-wide analysis of alternative splicing in *Zea mays*: Landscape and genetic regulation." *The Plant Cell* 26(9): 3472–3487.

Tulsani, N. J., R. Hamid, F. Jacob, N. G. Umretiya, A. K. Nandha, R. S. Tomar and B. A. Golakiya (2020). "Transcriptome landscaping for gene mining and SSR marker development in coriander (*Coriandrum sativum* L.)." *Genomics* 112(2): 1545–1553.

Uitterhaegen, E. (2014). *Coriander Oil–Extraction, Applications and Biologically Active Molecules*, Master's dissertation, Faculty of Bioscience Engineering Universiteit Gent.

Uitterhaegen, E., K. A. Sampaio, E. I. Delbeke, W. De Greyt, M. Cerny, P. Evon, O. Merah, T. Talou and C. V. Stevens (2016). "Characterization of French coriander oil as source of petroselinic acid." *Molecules* 21(9): 1202.

van Dijk, E. L., Y. Jaszczyszyn, D. Naquin and C. Thermes (2018). "The third revolution in sequencing technology." *Trends in Genetics* 34(9): 666–681.

Varshney, R. K., A. Graner and M. E. Sorrells (2005). "Genic microsatellite markers in plants: Features and applications." *TRENDS in Biotechnology* 23(1): 48–55.

Wang, B., V. Kumar, A. Olson and D. Ware (2019). "Reviving the transcriptome studies: An insight into the emergence of single-molecule transcriptome sequencing." *Frontiers in Genetics* 10: 384.

Wang, Y., Q. Lan, X. Zhao, L. Wang, W. Yu, B. Wang and Y. Wang (2021). "The complete mitochondrial genome of *Coriandrum sativum*." *Mitochondrial DNA Part B* 6(8): 2391–2392.

Wang, Z., M. Gerstein and M. Snyder (2009). "RNA-Seq: A revolutionary tool for transcriptomics." *Nature Reviews Genetics* 10(1): 57.

Wyman, D. E. (2020). *Full Characterization of Transcriptomes Using Long Read Sequencing*, UC Irvine.

Xu, H., Q. Yu, Y. Shi, X. Hua, H. Tang, L. Yang, R. Ming and J. Zhang (2018). "PGD: Pineapple genomics database." *Horticulture Research* 5(1): 1–9.

Yaldiz, G. and M. Çamlica "Evaluation of yield and quality of coriander (*Coriandrum sativum*) in Turkey and world." *Full Text Proceedings Book*.

Yang, Z., C. Li, Q. Jia, C. Zhao, D. C. Taylor, D. Li and M. Zhang (2020). "Transcriptome analysis reveals candidate genes for petroselinic acid biosynthesis in fruits of *Coriandrum sativum* L." *Journal of Agricultural and Food Chemistry* 68(19): 5507–5520.

Yungyuen, W., G. Ma, L. Zhang, M. Futamura, M. Tabuchi, K. Yamawaki, M. Yahata, S. Ohta, T. Yoshioka and M. Kato (2018). "Regulation of carotenoid metabolism in response to different temperatures in citrus juice sacs *in vitro*." *Scientia Horticulturae* 238: 384–390.

Zhang, W., D. Tian, X. Huang, Y. Xu, H. Mo, Y. Liu, J. Meng and D. Zhang (2014). "Characterization of flower-bud transcriptome and development of genic SSR markers in Asian lotus (*Nelumbo nucifera* Gaertn.)." *PloS One* 9(11): e112223.

Zhou, S.-M., L.-M. Chen, S.-Q. Liu, X.-F. Wang and X.-D. Sun (2015). "De novo assembly and annotation of the Chinese chive (*Allium tuberosum* Rottler ex Spr.) transcriptome using the Illumina platform." *PLoS One* 10(7): e0133312.

Zoschke, R. and R. Bock (2018). "Chloroplast translation: Structural and functional organization, operational control, and regulation." *The Plant Cell* 30(4): 745–770.

6 Morphological Characteristics of Coriander (*Coriandrum sativum* L.)

Hanan El-Sayed Osman and Mohamed Helmy El-Morsy

CONTENTS

6.1 INTRODUCTION

Coriander is considered a multipurpose herb grown mainly for its foliage and seeds, with numerous food-related biological activities and multiple functional uses (Burdock and Carabin, 2009). Besides their culinary uses as a flavoring agent for different dishes, including meat sauce, candy, pickles, and alcoholic beverages, among others, the fruits are extensively used for pharmacological purposes due to their stomachic, spasmolytic, carminative, hypoglycemic, lypolytic, antioxidant, antimicrobial, antifungal, anticancerous, and antimutagenic activities (Burt, 2004).

Traditional herbal medicine is gaining more popularity and is still widely practiced. This has attracted the attention of many researchers and encouraged them to screen plants of medicinal interest to study their bioactive compounds (Bochra et al., 2015). Accordingly, studies have been conducted on some medicinal plants and have focused on their bioactive compounds' biological activities (Bakkali et al., 2008, Christaki et al., 2012, Properzi et al., 2012). Coriander (*Coriandrum sativum* L.) is considered an example of an essential and exciting medicinal plant. Hippocrates used coriander in time-honored Greek medicines (460–377 BC). The Egyptians called this herb a "spice of happiness," perhaps because it was widely thought to be an aphrodisiac. Indeed, coriander seeds have been used to treat numerous digestive complaints such as indigestion, nausea, and dysentery, while coriander leaves stimulate appetite and help in easing digestion (Wangensteen et al., 2004; Maroufi et al., 2010).

It is an annual herbaceous plant that originated from the Mediterranean area but is extensively cultivated in Central Europe, Asia, and North Africa as a culinary and medicinal herb. Furthermore,

DOI: 10.1201/9781003204626-7

it is successfully grown in various conditions (Seidemann, 2005; Uhl, 2000). The Romans and Greeks also used coriander to flavor wine and medication. Afterward, the Romans introduced it into Great Britain (Livarda and van der Veen, 2008).

The whole dried seeds are ground and widely used as a spice in the Mediterranean region and as a significant ingredient of the curry powder in Indian food (Bochra et al., 2015). Additionally, the fresh leaves commonly known as cilantro or Chinese parsley are extensively utilized in eastern cooking and Indian cuisine as food flavoring or to mask the unpleasant odors of certain foods. They are also an essential ingredient in Thai and Vietnamese cuisine (Gil et al., 2002). Finally, it is noteworthy that all parts of this plant have been utilized as traditional remedies to treat various disorders in traditional medicine (Gil et al., 2002).

6.1.1 Scientific Classification of the Plant

Kingdom: Plantae
Division: Magnoliophyta
Class: Magnoliopsida
Order: Apiales
Family: Apiaceae
Genus: *Coriandrum*
Binomial name: *Coriandrum sativum* L.

6.2 COMMON NAMES

Table 6.1 summarizes the common names of coriander species.

6.3 DISTRIBUTION

Coriander is cultivated worldwide for its seeds with a good smell as a spice and its fresh leaves and roots as a condiment. This crop originated in the Mediterranean areas. After domestication, coriander moved westward and northward to Europe and Russia and eastward to the Indian Subcontinent (Nawata et al., 1995). In all South-East Asian countries, coriander is grown as a culinary herb and vegetable. Cropping for its fruits is restricted to higher altitudes. In South-East Asia, as in many other parts of the world, coriander is usually grown as a small-scale horticultural crop. Large-scale production exists in southern Russia, Ukraine, and other East European countries (Nadeem et al., 2013).

6.4 MORPHOLOGICAL FEATURES

Coriander is an erect annual herb and, depending on the climatic conditions of its cultivation region, is grown as a summer or winter annual crop. When flowering, the plant can reach a height of 0.20–1.40 meters.

6.4.1 Root

Cilantro roots have a pure white central tap root covered in tiny hair-like rootlets, typically a darker shade of tan (Figure 6.1). They have an aromatic, somewhat peppery flavor that is more pungent than the commonly used leaves (Khan et al., 2014).

6.4.2 Stem

The stem is more or less erect and sympodial, monochasial-branched, sometimes with several side branches at the basal node (Figure 6.2). Each branch finishes with an inflorescence. The color of

TABLE 6.1
Common Names of the Coriander Species

Language	Name
Arabic	Kuzbara, Kuzbura
Armenian	Chamem
Chinese	Yuan sui, hu sui
Czech	Koriandr
Danish	Coriander
Dutch	Coriander
English	Coriander, collender, Chinese parsley
Ethiopian	(Amharic) Dembilal
French	Coriandre, persilarabe
Georgian	(Caucasus) kinza, kindza, kindz
German	Koriander, Wanzendill, Schwindelkorn
Greek	Koriannon, korion
Hindi	Dhania, dhanya
Hungarian	Coriander
Italian	Coriandolo
Japanese	Koendoro
Malay	Ketumbar
Persian	Geshnes
Polish	Kolendra
Rumanian	Coriándru
Portuguese	Coentro
Russian	Koriandr, koljandra, kinec, kinza, vonjueezel'e, klopovnik
Sanskrit	Dhanayaka, kusthumbari
Spanish	Coriandro, cilantro, cilandrio, culantro
Turkish	Kisnis
Swiss	Chrapfechörnli, Böbberli, Rügelikümmi

Source: Goetsch et al., 1984.

FIGURE 6.1 Coriander tap root covered with tiny root hair.

FIGURE 6.2 Coriander stem is finely grooved and richly branched.

the more or less ribbed stem is green and sometimes turns red or violet during the flowering period. The stem of the mature plant is hollow, and its basal parts can reach a diameter of up to 2 cm (Diederichen, 1996).

6.4.3 LEAVES

Leaves alternate, somewhat variable in shape, size, and number, with a yellow-green, margined sheath surrounding the supporting stem for up to three-quarters of its circumference (Khan et al., 2014). Petiole and rachis subterete, sulcate and light green, whereas blade white waxy, shiny green often with darker green veins (Figure 6.3). Basal leaves are usually simple, withering early, often in a rosette, blade ovate in outline, deeply cleft, or parted into three incised-dentate lobes. The petiole is 0–15 cm long, blade ovate or elliptical in outline, up to 30 cm × 15 cm,

FIGURE 6.3 Coriander leaves alternate on the stem, and most of them are pinnate or bipinnate.

FIGURE 6.4 Umbellet inflorescence.

usually pinnately divided into 3–11 leaflets. Each blade of the simple lower leaves is divided into 3–7 simple leaf-like lobes (Diederichen, 1996). All higher leaves are compound; petiole restricted to the sheath, blade divided into three leaflets of which the central one is largest, each often variously divided into ultimately sublinear, entire, acute lobes. During the flowering period, the leaves sometimes turn red or violet.

6.4.4 INFLORESCENCE

A compound umbel. The primary branches of a compound umbel are called rays, the members of the involucres are bracts, and the members of the involucel subtending the umbellet sarebractlets (Khan et al., 2014). The umbel has two to eight primary rays of different lengths so that the umbellets are located at the same level (Figure 6.4).

6.4.5 FLOWER

Six different kinds of flowers are present; outermost sterile flower: pedicellate, bracteates, incomplete, zygomorphic. Outer female flower: pedicellate, bracteate, unisexual, zygomorphic, epigynous. Outer male flower: pedicellate, bracteate, unisexual, zygomorphic. Outer bisexual flower: pedicellate, bracteate, bisexual, zygomorphic, epigynous. Inner male flower: unisexual, actinomorphic. Inner bisexual flower: bisexual, complete, actinomorphic, epigynous (Figure 6.5).

6.4.5.1 Calyx

Five sepals, gamosepalous, in zygomorphic flowers. Two anterior sepals are larger, valvate, green, persistent, epigynous; all sepals are equal-sized in actinomorphic flowers.

6.4.5.2 Corella

The flowers have five petals; the peripheral flowers of every umbellet are asymmetric, as the petals toward the outside of the umbellets are lengthened. The central flowers are circular, with small inflexed petals (Figure 6.5). The color of the petals is pale pink or sometimes white.

6.4.5.3 Androecium

Five stamens, incurved in bud condition, free epigynous; filaments long, anther dorsifixed and extrose. After the flower opens, the white filaments are visible between the petals because they are bent, and the pollen sacs at their top are hidden in the center of the flower. This stage is the best for artificial emasculation of the flowers because the filaments are easy to distinguish, and they have not yet spread any pollen grains (Bochra et al., 2015).

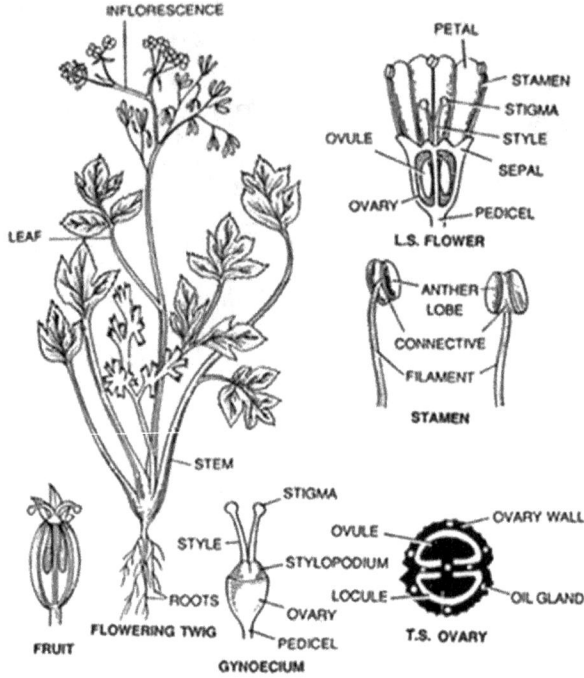

FIGURE 6.5   Coriander flowers.

6.4.5.4   Gyencium

Coriander has an inferior ovary, bicarpellary, syncarpous, bilocular, axile placentation, one ovule in each loculus, two flat stigmas, two long styles which flatten at the base into a bilobedepigynous disc called stylopodium.

6.5   FRUITS

The fruits are spherical or ovate, with a diameter of up to 6 mm. Usually, the schizocarp does not spontaneously split into two mericarps. Instead, the two mericarps have a sclerotified pericarp at the convex outside, while the pericarp on the concave inside is pellicular (Seidemann, 2005). The tiny carpophore is visible in the center of the hollow fruits. Every mericarp has six longitudinal, straight side ribs on the convex outside, alternating with five waved, often hardly visible main ribs

FIGURE 6.6 Coriander fruit, the small oval and aromatic fruit is a cremocarp, yellowish-brown, 3 to 5 mm in diameter, with wavy longitudinal ridges.

(Figure 6.6). On the convex inside, two longitudinal vittae contain the essential oil of the ripe fruit (Figure 6.6). Starting from the root, there are schizogenic channels in all parts of the plant which contain essential oils.

REFERENCES

Bakkali, F., Averbeck, S., Averbeck, D., Idaomar, M. 2008. Biological effects of essential oils - A review. *Food Chemistry and Toxicology* 46, 446–475.

Burdock, G., Carabin, I. G. 2009. Safety assessment of coriander (*Coriandrum sativum* L.) essential oil as a food ingredient. *Food and Chemical Toxicology* 47(1), 22–34. https://doi.org/10.1016/j.fct.2008.11.006.

Burt, S. 2004. Essential oils: Their antibacterial properties and potential applications in foods - A review. *International Journal of Food Microbiology* 94, 223–253.

Christaki, E., Bonos, E., Giannenas, I., Florou-Paneri, P. 2012. Aromatic plants as a source of bioactive compounds. *Agriculture* 2, 228–243.

Diederichen, A. 1996. *Coriander: Coriandrum sativum L.* International Plant Genetic Resources Institute, Promoting the conversion and use of underutilized and neglected crops.

Gil, A., De La Fuente, E. B., Lenardis, A. E., López Pereira, M., Suárez, S. A., Bandoni, A. 2002. Coriander essential oil composition from two genotypes grown in different environmental conditions. *Journal of Agriculture and Food Chemistry* 50, 2870–2877.

Khan, I. U., Dubey, W., Gupta, V. 2014. Taxonomical aspect of coriander (*Coriandrum sativum* L.). *International Journal of Current Research* 6(11), 9926–9930.

Laribi, B., Kouki, K., M'Hamdi, M., Bettaieb, T. 2015. Coriander (*Coriandrum sativum* L.) and its bioactive constituents. *Fitoterapia*. https://doi.org/10.1016/j.fitote.2015.03.012

Livarda, A., van der Veen, M. 2008. Social access and dispersal of condiments in North-West Europe from the Roman to the medevil period. *Vegetation History and Archeobotany* 5, 1–9.

Maroufi, K. Farahani, H. A., Darvishi, H. H. 2010. Importance of coriander (*Coriandrum sativum* L.) between the medicinal and aromatic plants. *Advances in Environmental Biology* 4(3), 433–436.

Nadeem, M., Anjum, F. M., Khan, M. I., Tehseen, S., El-Ghorab, A., Sultan, J. I. 2013. Nutritional and medicinal aspects of coriander (*Coriandrum sativum* L.): A review. *British Food Journal* 115(5), 743–755. https://doi.org/10.1108/00070701311331526

Nawata, E., Itanai, J., Masanaga, Y.I. 1995. The distribution and dissemination pathway of coriander in Asia. *Acta Horticulturae* 390, 167–176. https://doi.org/10.17660/ActaHortic.1995.390.24

Properzi, A., Angelini, P., Bertuzzi, G., Venanzoni, R. 2012. Some biological activities of essential oils. *Medicinal & Aromatic Plants* 2(5), 1–4.

Seidemann, J. 2005. *World Spice Plants: Economic, Usage, Botany, Taxonomy*, Springer-Verlag Berlin Heidelberg, p. 591.

Uhl, S.R. 2000. Coriander. In *Handbook of Spices, Seasonings, and Flavorings*, Technomic Publishing Co., Inc., pp. 94–97.

Wangensteen, H. Samuelsen, A.B. Malterud, K.E. 2004. Antioxidant activity in extracts from coriander. *Food Chemistry* 88, 293–297.

7 Morphological and Physiological Characterization of *Coriandrum sativum* L. under Different Concentrations of Lead

Sidra Nisar Ahmed, Mushtaq Ahmad, Mohammad Zafar,
Sara Aimen, Ghulam Yaseen, Neelum Rashid, and Sofia Rashid

CONTENTS

7.1 INTRODUCTION

Coriandrum sativum L. (coriander), commonly called Dhaniya, belongs to the genus *Coriandrum* and the Apiaceae family (Evans, 2002; Hedberg, Edwards, and Nemomissa, 2003; Shirkhodaei, Darzi, and HAJ, 2014). It is an annual herbaceous plant growing up to two meters, cultivated as a spring herb in most countries and in Mediterranean and Southeast Asian countries grown as a winter crop (Kadhim, 2021;

DOI: 10.1201/9781003204626-8

Randall, 2012; Rondon et al., 2011; Shirkhodaei et al., 2014). Coriander is cultivated globally as a naturalized and commercial crop (Wiersema and León, 2013). Its stem is slender, branched, and glabrous, and the leaves are pale yellow in appearance; it is distributed throughout the world and the majority of coriander species are grown all over Pakistan, especially in hilly areas and plains under irrigated conditions (Aftab, Haider, Ali, Malik, & Journal, 2021; Al-Snafi, 2016; Mir & Persian, 1992; Ramadan & Mörsel, 2002; Randall, 2012). Coriander, being broadly distributed, has 3 subspecies and 10 varieties classified on the basis of taxonomy describing morphological features like flower color, leaf shape and size of fruit, and the weight of 1000 grains (Silva, Domeño, & Domingues, 2020).

C. sativum is an herb and a spice and possesses nutritional and medicinal properties (Delaquis, Stanich, Girard, & Mazza, 2002; Diederichsen, 1996). The coriander seeds and vegetative plant parts are essential for numerous biological functions and health-related activities (Begnami, Duarte, Furletti, & Rehder, 2010; Guerra et al.; Matasyoh, Maiyo, Ngure, & Chepkorir, 2009; Saeed & Tariq, 2007; YALDIZ & ÇAMLICA). For example, its seeds are rich in essential oil, secondary metabolites, flavonoids, and phenolic compounds; due to these properties, the oil is mainly used in pharmaceutical industries (Carrubba, la Torre, Prima, Saiano, & Alonzo, 2002; Diederichsen, 1996; Ghobadi, Ghobadi, & Engineering, 2010; Msaada, Taarit, Hosni, Hammami, & Marzouk, 2009; Tylewicz, Nowacka, Martín-García, Wiktor, & Caravaca, 2018). The ground seed, also called the fruit, is used in pickling spices; it gives aroma or flavor to soups, dishes, bread, cakes, pastries, candies, puddings, alcoholic beverages, and frozen dairy desserts. The fruit oil is commonly used in detergents, creams, lotions, surfactants, emulsifiers, and perfumes (Coşkuner & Karababa, 2007). Coriander can also be used internally as tonics. It is used to treat syncope and memory loss. The gargling of fresh leafy juice is effective in stomatitis and sore throat. A leaf paste is applied locally for swellings and boils and over the forehead and temples for headaches (Pathak Nimish, Kasture Sanjay, Bhatt Nayna, & Rathod Jaimik, 2011; Randall, 2012).

The essential oil of coriander is used in the flavoring industry (cocoa, chocolate, and liquor) (Devi & Sharangi, 2019) as well as the pharmaceutical industry due to its joint pain, carminative, rheumatism, anticancer, antimutagenic (Pathak Nimish et al., 2011; Rajeshwari & Andallu, 2011), antitumor (Chithra & Leelamma, 2000), antifungal (Basilico & Basilico, 1999), antioxidant (Chithra & Leelamma, 1999; Pathak Nimish et al., 2011), antiulcer and antidiabetic (Gray & Flatt, 1999), hypolipidemic (Chithra & Leelamma, 2000), analgesic (Pathak Nimish et al., 2011; Rajeshwari & Andallu, 2011), anticonvulsant (Hosseinzadeh & Madanifard, 2000), diuretic (Pathak Nimish et al., 2011; Rajeshwari & Andallu, 2011), gastrointestinal, hepatoprotective, antidiarrheal (Nithya, 2015), and anti-inflammatory (Al-Snafi, 2016; Neha Mohan, Suganthi, & Gowri, 2013; Wichtl, 1994) characteristics.

Around 60–80% of the essential oil is a yellowish or colorless liquid obtained from the plant body. Camphor, geranyl acetate, pinene, linalool, *p*-cymene, limonene, and geraniol are important constituents (Ayodeji Ahmed, 2020; Pathak Nimish et al., 2011). Its seeds contain minerals, vitamins, and dietary fiber (Ensminger, 1986; Holland, McCance, Widdowson, Unwin, & Buss, 1991; Aumatell, 2012; Dorman & Deans, 2000).

Nowadays, heavy metals (HM) pollution is considered an important issue globally because its concentration increases day by day and cannot be degraded (Al Naggar, Khalil, & Ghorab, 2018). As a result, HM undergo bioaccumulation and biomagnification and become a part of the food chain and have adverse impacts on human health (Ahmad, Akhtar, Zahir, & Jamil, 2012; Liu, Yang, Xie, Xia, & Fan, 2012; Saher & Siddiqui, 2019; Tao, Guo, & Ren, 2015; Vandecasteele, Samyn, Quataert, Muys, & Tack, 2004). Industrialization is the leading cause of HM pollution in agricultural soil (Ahmad et al., 2012; Kuriakose & Prasad, 2008; Q. Li et al., 2013; W. Li, Khan, Yamaguchi, & Kamiya, 2005). HM are added to the soil by the application of fertilizers and pesticides (Ahmad et al., 2012; Ćurguz, Raičević, Veselinović, Tabakovic-Tošić, & Vilotić, 2012; Ghobadi et al., 2010; Hasan, Fariduddin, Ali, Hayat, & Ahmad, 2009; Hussain et al., 2013; Jain, 2013; Msaada et al., 2009). Pakistan is an agricultural country; it is facing extreme water shortages. To overcome this problem, wastewater from different industries is an attractive choice for farmers.

These wastewaters contain HM like mercury, arsenic, cadmium, chromium, iron, copper, nickel, etc. These HM are accumulating in agri-land and are harmful and toxic for organisms; therefore, they are considered "serious threats" for plants, animals, and humans (Chibuike, Obiora, & science, 2014; Jamal et al., 2013; Nicholson, Smith, Alloway, Carlton-Smith, & Chambers, 2003; Wong, Li, Zhang, Qi, & Min, 2002). Lead (Pb) is one of these HM accumulating in soil and plants. Therefore, Pb is a potential pollutant that continuously deteriorates soil health (Huang et al., 2019; Xia et al., 2020). Lead in solution form can readily be absorbed by plants and accumulate in roots (Wierzbicka et al., 2007). Inside the plant, lead interrupts physiological, biochemical, and metabolic processes such as water status, nitrate assimilation, plant growth and development, and seed germination resulting in poor growth rates and development (Hockmann, Tandy, Studer, Evangelou, & Schulin, 2018; Lamhamdi, Bakrim, Aarab, Lafont, & Sayah, 2011; Uslu, Babur, Alma, & Solaiman, 2020; Wierzbicka et al., 2007). The significant sources of increasing lead toxicity in the environment are mining, tanning, the automobile industry, and the paint and lead battery industries (Alkorta et al., 2004; Babel & Kurniawan, 2003; Das, Vimala, & Karthika, 2008). Plants take up Pb through their root system and leaves as well (Sharma & Dubey, 2005). Lead above optimum concentration decreases the photosynthetic ability, chlorophyll content, carotenoid content, and carbon dioxide assimilation rate of plants; thus, it damages cell structures, causes the reduction of physical and biochemical activities, and exerts negative impacts on plant growth (Seregin & Kozhevnikova, 2008).

It is essential to understand the phytotoxic effect of lead due to excessive lead pollution and lack of knowledge of lead toxicity, the bioaccumulation problems in plants, and the ways that lead interacts with plants and the environment. Coriander's economic and nutritional conditions and its significance as a medicinal herb make it necessary for us to study and examine the effects of Pb on coriander herbs. This mainly includes possible damages through a significant increase in Pb contamination caused by automobile exhausts, excessive use of fertilizers and pesticides, and crop irrigation with sewage water. With the importance of this work, there is a dire need to experiment to investigate the effects of Pb on coriander.

7.2 EXPERIMENTAL DESIGN

Lead (Pb) uptake in coriander irrigated with sewage water is significant worldwide. A pot experiment was performed to study the impact of lead nitrate $Pb(NO_3)_2$ on coriander in the wire net house at the Department of Botany, Women University Multan, Pakistan. The experimental coriander plant belongs to the family Apiaceae (Tylewicz, Nowacka, Martín- García, Wiktor, & Caravaca, 2018). The study's main objective is to examine the morphological and physiological responses to Pb. Certified coriander seeds were obtained from the Ayub Agriculture Research Institute, Faisalabad, Pakistan. The clay pots of 30 cm height and 28 cm diameter contain 8 kg soil per pot used to raise coriander seedlings (Figure 7.1). Seeds of coriander were exposed to different treatments of $Pb(NO_3)_2$ per kg of soil. The treatments used for experiments were 0, 25 ppm, 50 ppm, 100 ppm, and 200 ppm of $Pb(NO_3)_2$. Each treatment contained eight replicates. The 40 pots were prepared for experimental trial were aligned in randomized block design for analysis. The data was recorded for three replicates per treatment during the flowering and after harvest, wherein morphological and physiological parameters were measured.

7.2.1 ANALYSIS OF SOIL TEXTURE

Before starting the experiment, soil texture analysis was done by "X-ray spectrophotometer" (PW 1660-Philips). The data in Table 7.1 reveals that the humified soil exhibits different characteristics. By examining the soil samples displayed, the soil is sandy-loam soil with percentages as follows: clay 20.5% and silica sand 68.8% with a pII of 7.36. The soil C, K, N, and P were analyzed. The amount of carbon (C) in the soil was 2.70%, potassium (K) content in the soil was 2.13%, nitrogen (N) content in the soil was 1.52%, and phosphorus (P) content in the soil was 17.69%. Also, 4.16% soil organic matter (OM) along with 1.78% electrical conductivity (EC) were detected as given in Table 7.1.

FIGURE 7.1   Pot filling, seed sowing, and arrangement of clay pots.

TABLE 7.1
Examination of Physicochemical Characteristics of Soil Samples

S	Texture	Silica Sand	Clay	EC	OM	pH	C	K	N	P
1	SL	63.3	19.67	1.67	4.84	7.21	2.67	213.25	1.72	17.67
2	SL	69.5	20.60	1.71	5.16	7.30	2.81	215.07	1.45	17.53
3	SL	70.2	21.4	1.76	3.62	7.54	2.78	211.60	1.50	17.50
4	SL	68.4	20.52	1.88	3.57	7.40	2.57	212.40	1.30	17.48
5	SL	67.6	20.44	1.92	3.63	7.35	2.68	214.52	1.67	18.31
Mean	SL	67.80	20.53	1.78	4.16	7.36	2.70	213.36	1.52	17.69

S:	Sample	SL:	Sandy loam
OM:	Organic Matter	EC:	Electrical conductivity
C:	Carbon	P:	Phosphorus
N:	Nitrogen	K:	Potassium

7.2.2   Morphological and Plant Biomass Attributes

7.2.2.1   Total Height of Plants = Root Length + Shoot Length (cm plant^{-1})

The coriander's total height under lead treatments and control was measured after harvesting the plants by measuring tape from the tip of the shoot to root endings. Mean values were calculated. $Pb(NO_3)_2$ caused a reduction in total plant height compared to control. According to Figure 7.2 and

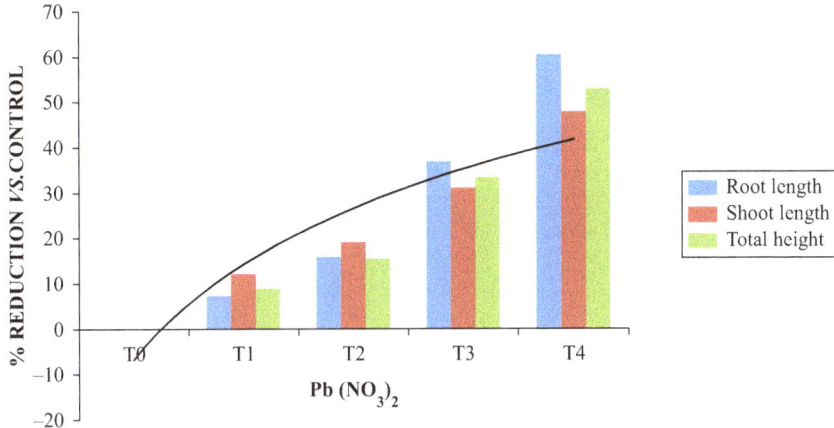

FIGURE 7.2 Effect of Pb on root length, shoot length, and the total height of coriander.

Table 7.2(a), for treatments T1, T2, T3, and T4, the decline in plant height was 9%, 15%, 33.3%, and 52.9%, respectively. The root length, shoot length, and total height of plants with treatments have a strong negative correlation. The value of $r = -0.993**$ for root length, $r = -0.980**$ for shoot length, and $r = -0.990**$ which shows there is a significant strong negative correlation. However, a higher level of $Pb(NO_3)_2$ has more effect on the total height of the plant than lower lead treatments. Similarly, the reductions in root length with increasing lead doses at all treatments were 7.23%, 15.7%, 36.7%, and 60.3% from T1 to T4, respectively. Some researchers explored the effects of Pb on plant height, shoot length, and root length. Pb also decreases germination index and germination percentage (Aslam, Bhat, Choudhary, & Ansari, 2017; Sorrentino et al., 2018).

7.2.2.2 Number of Branches per Plant and Height of First Branch

The number of lead-treated plants and control branches was counted, and their mean values were observed. The decline in the total number of branches under Pb stress was 8%, 15%, 25%, and 45.5% for T1, T2, T3, and T4, respectively. At the same time, the reduction in the height of the first branch was 7.43%, 7.61%, 5.78%, and 4.82% with all treatments (T1–T4). Normal soil (control) does not affect the number of branches and height of the first branch. A high Pb concentration causes a gradual reduction in height and number of branches, as shown in Figure 7.3. According to Table 7.2 (a), the results were statistically significant, and a strong negative correlation between the concentration of $Pb(NO_3)_2$ and number of branches $r = -0.961**$ and first branch height $r = -0.980**$ was observed (Figure 7.4). The height of primary and secondary branches was reported by the application of ZnSO on coriander (Marichali, Dallali, Ouerghemmi, Sebei, & Hosni, 2014).

7.2.2.3 Stem Diameter (mm)

Results showed that different lead doses have a toxic impact on stem diameter, illustrated in Figure 7.5 and Table 7.2 (a). Lead nitrate stem diameter concentrations decrease, i.e., 4.5% at T1, 7.4% at T2, and 11.5% and 36.2% at T3 and T4, respectively. The coefficient of correlation value between different doses of lead and stem diameter is $r = -0.0980**$, indicating that the results were statistically significant with a strong negative correlation between lead and stem diameter treatments. The researcher reported that the application of zinc on coriander reduces the diameter of the stem, its primary branches, and the diameter of secondary branches (Marichali et al., 2014).

7.2.2.4 Analysis of Chlorophyll/SPAD Contents

The leaf chlorophyll content was measured with a SPAD meter. Chlorophyll was measured from various positions in each plant, and its content was examined. The decline in SPAD content in all

TABLE 7.2

(a) Correlation Coefficient Matrix between Morphological Traits of *Coriandrum sativum* under Different Pb(NO$_3$)$_2$ Treatments

Trait	Pb(NO$_3$)$_2$	RL (cm)	SL (cm)	TPL (cm)	No. Br	1st brH	SD (mm)	SPAD	Um/P	Fr/P
Pb(NO$_3$)$_2$										
RL (cm)	−0.993**									
SL (cm)	−0.980**	0.989**								
TPL (cm)	−0.990**	0.999**	0.993**							
No. Br	−0.961**	0.975**	0.996**	0.980**						
1st brH	−0.980**	0.993**	0.990**	0.992**	0.986**					
SD (mm)	−0.980**	0.950*	0.931**	0.944**	0.901*	0.925*				
SPAD	−0.982**	0.992**	0.998**	0.993**	0.994**	0.997**	0.931*			
Um/P	−0.995**	1.000**	0.990**	0.998**	0.975**	0.993**	0.957**	0.993**		
Fr/P	−0.975**	0.992**	0.996**	0.994**	0.993**	0.997**	0.914*	0.998**	0.991**	

* Correlation is significant at the 0.05 level (2-tailed).

** Correlation is significant at the 0.01 level (2-tailed). Pb(NO$_3$)$_2$ = Lead nitrate

RL (cm) = Root length in centimeters

SL (cm) = Shoot length in centimeters

TPL (cm) = Total plant length in centimeters

No. Br = Number of branches

1st brH = Height of first branch

S.D (mm) = Stem diameter measured in millimeters

SPAD = Soil plant analysis determination

Um/P = Numbers of umbels per plant

Fr/P = Numbers of fruit per plant

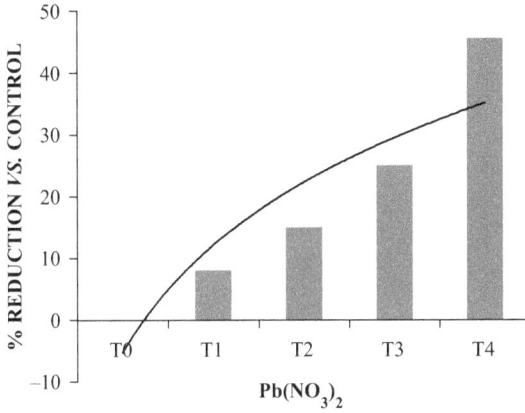

FIGURE 7.3   Phytotoxic effect of Pb(NO$_3$)$_2$ on the number of branches.

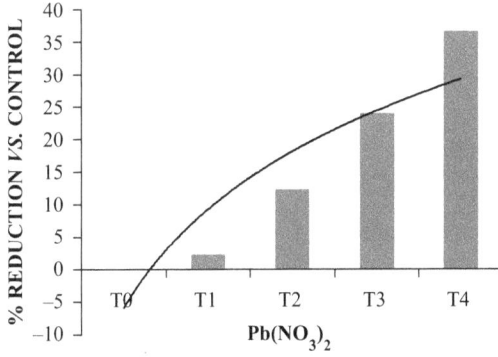

FIGURE 7.4   Phytotoxic effect of Pb(NO$_3$)$_2$ on height of the first branch in coriander.

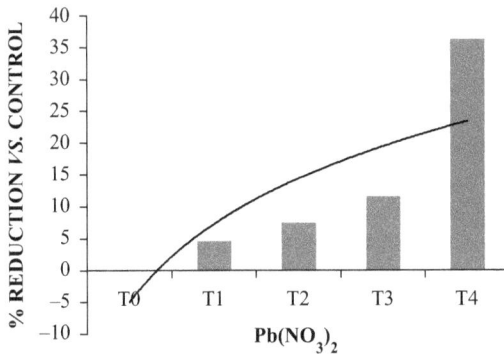

FIGURE 7.5   Phytotoxic effect of Pb on stem diameter in coriander plants.

Pb-treated plants was 4.22%, 12.07%, 20.3%, and 31.28% at T1, T2, T3, and T4, respectively. The results showed that Pb has a significant effect on the chlorophyll content of coriander at the highest doses of T3 and T4 (100 and 200 ppm), as demonstrated in Figure 7.6. Similarly, mercury at high concentration has a strong phytotoxic influence on cells and reduces the chlorophyll content (Liu, Wang, Chen, Xu, & Wang, 2010).

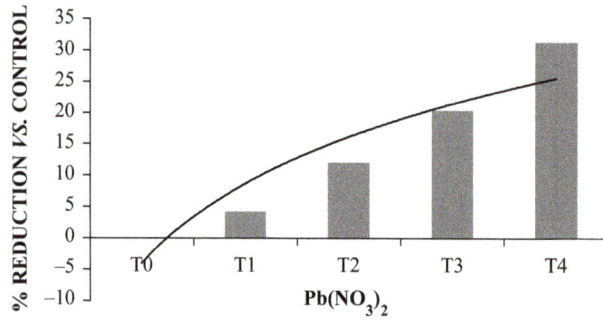

FIGURE 7.6   Effect of Pb on leaf SPAD or chlorophyll value.

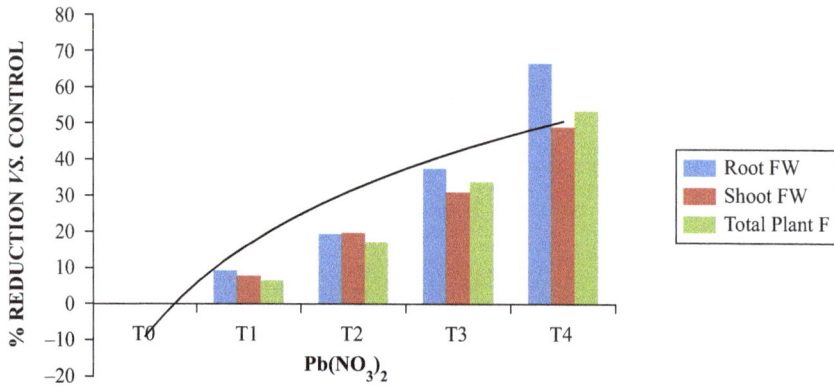

FIGURE 7.7   Effect of Pb on shoot, root, and total plant fresh weight.

7.2.2.5   Biomass Plant^{-1} (Root + Shoot Fresh Weight and Dry Weight = Total Plant Fresh and Dry Weight)

Figure 7.7 and Table 7.2(a) illustrate that Pb stress reduces the biomass content of plants under the various concentrations of lead nitrate. The decline in biomass value increases with increasing concentrations/ treatments, which suggests that the correlation value is strongly negative between various concentrations of Pb and biomass content, indicating significant results. Shoot fresh weight reduction was 8%, 20%, 31%, and 49% and root reduction was 8%, 19%, 36%, and 67% successively at T1, T2, T3, and T4. Coefficient of correlation for total plant fresh weight $r = -0.989$** and for total plant dry weight $r = -0.963$**. The total plant fresh weight reduction percentages were observed at 6%–53% at T1–T4, respectively. The dry weight was significantly affected by the highest concentration T3-100 ppm 54% and T4-200 ppm of Pb(NO$_3$)$_2$ 71% reduction. Significant reduction in coriander biomass content was observed at 200 ppm and showed that coriander biomass was strongly influenced by lead phytotoxicity. Pb decreases root and shoot length, plant tolerance index and biomass, and the dry weight of root and shoot (Aslam et al., 2017; Farooqi, Iqbal, Kabir, & Shafiq, 2009; Kabir, Iqbal, Shafiq, & Farooqi, 2008; Seema, Shabnam, Malik, & Ali, 2013). The influence of zinc, lead, copper, and cadmium on vegetable crops reduced the biomass content up to 9%–32% (Singh & Aggarwal, 2006).

7.3   YIELD ATTRIBUTES

7.3.1   Number of Umbels Plant^{-1}

According to Table 7.2 (a), the total number of umbels per plant per treatment was influenced by lead toxicity. The umbel numbers declined when high doses were applied gradually, showing a

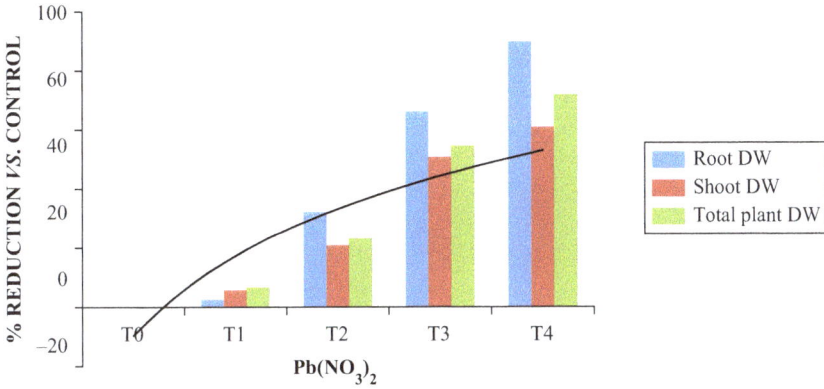

FIGURE 7.8 Effect of Pb on the shoot, root, and total plant dry weight.

significant negative strong coefficient $r = -0.995**$. Figure 7.9 presents a graph of the reduction of coriander's umbel production under lead doses due to lead-contaminated soil. Umbel number reduction percentages at treatments T1, T2, T3, and T4 were 7%, 16%, 35%, and 60%, respectively. The effect of zinc sulfate on coriander was reported, and the outcomes were also the same as the case with lead nitrate, a reduction in umbels per plant with increasing zinc doses (Marichali et al., 2014).

7.3.2 Number of Umbellets Umbel^{-1}

The total number of umbellets per umbel was decreased compared to control at all treatments of lead nitrate presented by Figure 7.9 and Table 7.2 (b). The significant reduction in umbellet number per umbel in each Pb-treated plant was observed at high doses of $Pb(NO_3)_2$; at doses of 25 ppm and 50 ppm, the decline in percentages was 14% and 29%, respectively.

7.3.3 Number of Flowers Umbel^{-1}

Pb $(NO_3)_2$ negatively affects flower production. Flower formation decreased as Pb doses went on increasing compared with control as given in Table 7.2 (b), which showed statistically less significant results and a strongly negative correlation value between flower numbers and Pb doses, $r =$

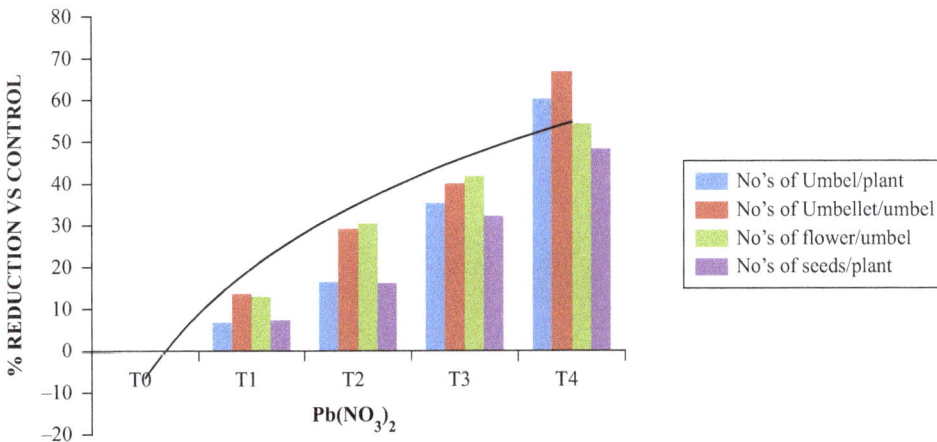

FIGURE 7.9 Effect of $Pb(NO_3)_2$ on coriander yield parameters.

TABLE 7.2

(b) Correlation Coefficient Matrix between Biomass and Yield Traits of *Coriandrum sativum* under Different Heavy Metal Pb(NO3)2 Treatments

Traits	$Pb(No_3)_2$	RFW	SFW	TPFW	RDW	SDW	TPDW	um/U	F/U	S/P
$Pb(No_3)_2$										
RFW	-0.997**									
SFW	-0.983**	0.993**								
TPFW	-0.989**	0.997**	0.996**							
RDW	-0.966**	0.981**	0.990**	0.993**						
SDW	-0.919*	0.941*	0.954*	0.962**	0.985**					
TPDW	-0.963**	0.978**	0.984**	0.991**	0.999**	0.985**				
um/U	-0.981**	0.989**	0.998**	0.989**	0.978**	0.932*	0.970**			
FU	-0.932*	0.953*	0.982**	0.968**	0.977**	0.947**	0.969**	0.982**		
S/P	-0.984**	0.994**	0.996**	0.999**	0.995**	0.964**	0.994**	0.989**	0.973**	

*   Correlation is significant at the 0.05 level (2-tailed).

** Correlation is significant at the 0.01 level (2-tailed).

RFW = Root fresh weight

SFW = Shoot fresh weight

TPFW = Total plant fresh weight

RDW = Root dry weight

SDW = Shoot dry weight

TPDW = Total plant fresh weight

um/U = Numbers of umbellet per umbel

F/U = Numbers of flowers per umbel

S/P = Numbers of seeds per plant

–0.932*. The decline in flower number at all treatments (T1, T2, T3, T4) of Pb was 13%, 30%, 42%, and 54%, as shown in Figure 7.9.

7.3.4 Number of Seeds Umbel^{-1}

According to data in Figure 7.9 and Table 7.2 (b), seed numbers are influenced by the phytotoxic effect of lead stress. The total number of seeds in each umbel per Pb-treated plant was counted and compared with control plants (without Pb doses). The correlation coefficient value between Pb and number of seeds per umbel is $r = -0.984$**. Seed yield was reduced by 14% at different doses of cadmium at 5.75 mg/kg soil (Wani, Khan, & Zaidi, 2007).

7.3.5 Fruit Numbers

According to data in Table 7.2 (a), lead stress impacts the number of fruits per plant. Reductions in fruit numbers at lead treatments of 25–200 ppm (T1–T4) were 6%, 17%, 30%, and 44%, respectively. This decline in yield was due to lead toxicity. The influence of zinc, lead, copper, and cadmium on all vegetable crops reduced the yield parameters by about 9–67% (Singh & Aggarwal, 2006).

7.4 BIOCHEMICAL ANALYSIS

7.4.1 Essential Oil Content

Lead treatments strongly influence the essential oil in contrast with control, as represented in Figure 7.10. The essential oil decreases when the doses of lead increase.

7.4.2 Determination of Essential Oil Components

The coriander treated with different doses of lead showed the following percentages of essential oil components. The percentages of components at four treatments were linalool 88.6–80.2%, geraniol 2.6–0.9%, geranyl acetate 2.52–1.45%, camphor 1.65-0.85%, and γ-terpinene 1.37–0.25% at T0–T4 as represented in Table 7.3. Zn causes a reduction in the yield of essential oil compounds found in the fruits and negatively impacts the content of oxygenated monoterpenes including camphor, linalool, and geraniol (Marichali et al., 2014).

7.4.3 Assessment of Photosynthetic Pigments

Photosynthetic pigments are affected by lead toxicity. The reduction in photosynthetic pigments such as chlorophyll-a, -b, total chlorophyll, and carotenoid contents against $Pb(NO_3)_2$ concentrations was recorded as shown in Figure 7.11. At the highest dose of lead, the maximum reduction

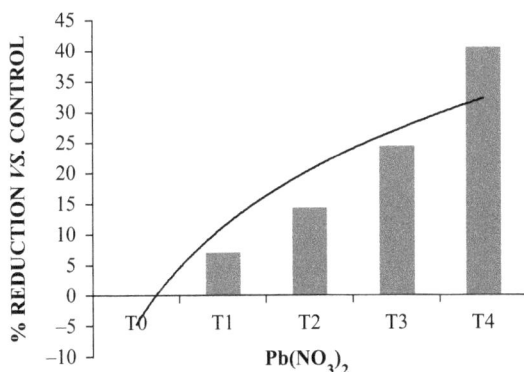

FIGURE 7.10 Effect of lead nitrate on essential oil content in coriander plants.

TABLE 7.3

Components of Essential Oil in Coriander under Treatments of Lead Nitrate

Lead Nitrate	Linalool	Geraniol	Geranyl Acetate	Camphor	γ-Terpinene
T0	88.62%	2.61%	2.52%	1.65%	1.37%
T1	86.56%	2.53%	2.42%	1.46%	1.25%
T2	85.40%	2.42%	2.35%	1.32%	1.05%
T3	83.52%	1.84%	2.05%	1.13%	0.81%
T4	80.23%	0.91%	1.45%	0.85%	0.25%

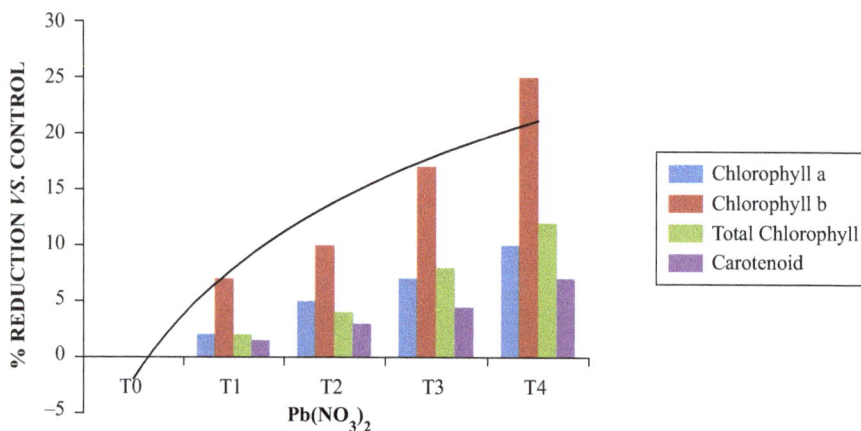

FIGURE 7.11   Impact of lead nitrate on the decline in photosynthetic pigments.

was observed in chlorophyll-a (10%), followed by 25% in chlorophyll-b, 12% in the total chlorophyll content, and 7% in carotenoid contents. Researchers have reported similar results on the effects of HM on photosynthetic pigments of the plant (Dobroviczká et al., 2013; Hattab, Dridi, Chouba, Kheder, & Bousetta, 2009; Myśliwa-Kurdziel & Strzałka, 2002; Pietrini et al., 2010; Piršelová, Kuna, Lukáč, & Havrlentová, 2016; Sorrentino et al., 2018).

7.5   CONCLUSION

The current study presents that lead is toxic for coriander plants. The application of $Pb(NO_3)_2$ in waste-water for irrigation purposes is toxic for coriander plants. The effects of 25–200 ppm lead on coriander's growth, morphology, and physiology are significant. $Pb(NO_3)_2$ causes a reduction in growth, biomass, and physiological and morphological attributes at different treatments. Furthermore, attributes of the affected plant have shown "stressed physiological activity of plants" that affected plant growth-biomass and yield parameters. Hence, $Pb(NO_3)_2$ affects plant quantity and decreases the quality of plants.

7.5.1   RECOMMENDATIONS

- Industrial wastewater is unfit for the irrigation of crops because it contains HM.
- Industrial wastewater should be discharged into separate reservoirs rather than rivers or agriculture soils.
- To avoid environmental pollution, there should be proper wastewater treatment to control HM.
- Using some techniques like biochar and chelators effectively could improve plant growth and yield.

REFERENCES

Aftab, A., Haider, M., Ali, Q., & Malik, A. (2021). Genetic evaluation for morphological traits of *Coriandrum sativum* grown under salt stress. *Biological and Clinical Sciences Research Journal*, 2021(1), e006–e006. http://bcsrj.com/ojs/index.php/bcsrj/article/view/52.

Ahmad, I., Akhtar, M. J., Zahir, Z. A., & Jamil, A. (2012). Effect of cadmium on seed germination and seedling growth of four wheat (*Triticum aestivum* L.) cultivars. *Pakistan Journal of Botany*, 44(5), 1569–1574.

Al Naggar, Y., Khalil, M. S., & Ghorab, M. (2018). Environmental pollution by heavy metals in the aquatic ecosystems of Egypt. *Open Access to Journal of Toxicology*, 3(1), 555603. https://doi.org/10.19080/OAJT.2018.03.555603.

Al-Snafi, A. E. (2016). A review on chemical constituents and pharmacological activities of *Coriandrum sativum*. *Journal of Pharmacy*, 6(7), 17–42. https://doi.org/10.9790/3013-067031742.

Alkorta, I., Hernández-Allica, J., Becerril, J., Amezaga, I., Albizu, I., & Garbisu, C. (2004). Recent findings on the phytoremediation of soils contaminated with environmentally toxic heavy metals and metalloids such as zinc, cadmium, lead, and arsenic. *Reviews in Environmental Science and Biotechnology*, 3(1), 71–90. https://doi.org/10.1023/B:RESB.0000040059.70899.3d.

Aslam, R., Bhat, T. M., Choudhary, S., & Ansari, M. (2017). An overview on genotoxicity of heavy metal in a spice crop (*Capsicum annuum* L.) in respect to cyto-morphological behaviour. *Caryologia*, 70(1), 42–47. https://doi.org/10.1080/00087114.2016.1258884.

Aumatell, M. R. (2012). Gin: Production and sensory properties. In *Alcoholic Beverages* (pp. 267–280): Elsevier.

Ayodeji Ahmed, A. (2020). GLIMPSE OF FISH AS PERISHABLE STAPLE. *Al-Qadisiyah Journal for Agriculture Sciences*, 10(2), 349–375. https://doi.org/10.33794/qjas.2020.167497.

Babel, S., & Kurniawan, T. (2003). Low-cost adsorbents for heavy metals uptake from contaminated water: A review. *Journal of Hazardous Materials*, 97(1–3), 219–243. https://doi.org/10.1016/S0304-3894 (02)00263-7.

Basilico, M., & Basilico, J. (1999). Inhibitory effects of some spice essential oils on Aspergillus ochraceus NRRL 3174 growth and ochratoxin A production. *Letters in Applied Microbiology*, 29(4), 238–241. https://doi.org/10.1046/j.1365-2672.1999.00621.

Begnami, A., Duarte, M., Furletti, V., & Rehder, V. (2010). Antimicrobial potential of *Coriandrum sativum* L. against different Candida species in vitro. *Food Chemistry*, 118(1), 74–77. https://doi.org/10.1016/j.foodchem.2009.04.089

Carrubba, A., la Torre, R., Prima, A. D., Saiano, F., & Alonzo, G. (2002). Statistical analyses on the essential oil of Italian coriander (*Coriandrum sativum* L.) fruits of different ages and origins. *Journal of Essential Oil Research*, 14(6), 389–396. https://doi.org/10.1080/10412905.2002.9699899

Chibuike, G. U., & Obiora, S. C. (2014). Heavy metal polluted soils: Effect on plants and bioremediation methods. *Applied and Environmental Soil Science*. https://doi.org/10.1155/2014/752708.

Chithra, V., & Leelamma, S. (1999). *Coriandrum sativum* changes the levels of lipid peroxides and activity of antioxidant enzymes in experimental animals. *Indian Journal of Biochemistry and Biophysics*, 59–61. http://hdl.handle.net/123456789/15417.

Chithra, V., & Leelamma, S. (2000). *Coriandrum sativum*-effect on lipid metabolism in 1, 2- dimethyl hydrazine induced colon cancer. *Journal of Ethnopharmacology*, 71(3), 457–463.

Coşkuner, Y., & Karababa, E. (2007). Physical properties of coriander seeds (*Coriandrum sativum* L.). *Journal of Food Engineering*, 80(2), 408–416.

Ćurguz, V. G., Raičević, V., Veselinović, M., Tabakovic-Tošić, M., & Vilotić, D. (2012). Influence of heavy metals on seed germination and growth of *Picea abies* L. Karst. *Polish Journal of Environmental Studies*, 21(2).

Das, N., Vimala, R., & Karthika, P. (2008). Biosorption of heavy metals: An overview. *Indian Journal of Biotechnology*, 7, 159–169.

Dobroviczká, T., Piršelová, B., Meszaros, P., Blehová, A., Libantova, J., Moravčiková, J., & Matušíková, I. (2013). Effects of cadmium and arsenic ions on content of photosynthetic pigments in the leaves of *Glycine max* (L.) Merrill. *Pakistan Journal of Botany*, 45(1), 105–110. https://www.researchgate.net/publication/278683605.

Delaquis, P. J., Stanich, K., Girard, B., & Mazza, G. (2002). Antimicrobial activity of individual and mixed fractions of dill, cilantro, coriander and eucalyptus essential oils. *International Journal of Food Microbiology*, 74(1–2), 101–109.

Devi, A. R., & Sharangi, A. (2019). Morphological charater and seed yield potential of coriander genotypes under gangetic Alluvial Region of West Bengal. *International Journal of Current Microbiological Application*, 8(4), 775–782.

Diederichsen, A. (1996). *Coriander: Coriandrum sativum* L. (Vol. 3): Bioversity International.

Dorman, H. D., & Deans, S. G. (2000). Antimicrobial agents from plants: Antibacterial activity of plant volatile oils. *Journal of Applied Microbiology*, 88(2), 308–316.

Ensminger, A. (1986). *Food for Health: A Nutrition Encyclopedia*.

Evans, W. (2002). *Trease and Evans: Pharmacognocy*: WB Saunders. Fifteenth International Edition.

Farooqi, Z., Iqbal, M. Z., Kabir, M., & Shafiq, M. (2009). Toxic effects of lead and cadmium on germination and seedling growth of *Albizia lebbeck* (L.) Benth. *Pakistan Journal of Botany*, 41(1), 27–33. http://www.pakbs.org/pjbot/PDFs/41(1)/PJB41(1)027.pdf.

Ghobadi, M., Ghobadi, M. J. (2010). The effects of sowing dates and densities on yield and yield components of coriander (*Coriandrum sativum* L.). *International Journal of Agricultural and Biosystems Engineering*, 4(10), 725–728. https://doi.org/10.5281/zenodo.1079068.

Gray, A. M., & Flatt, P. R. (1999). Insulin-releasing and insulin-like activity of the traditional anti-diabetic plant *Coriandrum sativum* (coriander). *British Journal of Nutrition*, 81(3), 203–209. https://doi.org/10.1017/S0007114599000392.

Hasan, S. A., Fariduddin, Q., Ali, B., Hayat, S., & Ahmad, A. (2009). Cadmium: Toxicity and tolerance in plants. *Journal of Environmental Biology*, 30(2), 165–174. https://doi.org/10.1007/978-3-319-03777-6_4.

Hattab, S., Dridi, B., Chouba, L., Kheder, M. B., & Bousetta, H. (2009). Photosynthesis and growth responses of pea *Pisum sativum* L. under heavy metals stress. *Journal of Environmental Sciences*, 21(11), 1552–1556. https://doi.org/10.1016/S1001-0742(08)62454-7.

Hedberg, I., Edwards, S., Nemomissa, S., & Uppsala University, Uppsala, & the National Herbarium, A. A. U., Addis Ababa. (2003). *Flora of Ethiopia and Eritrea. Apiaceae to Dipsacaceae*; Department of Systematic Botany, Uppsala University, Uppsala and the National Herbarium, Addis Ababa University, Addis Ababa. 4(1). https://www.jstor.org/stable/23655176.

Hockmann, K., Tandy, S., Studer, B., Evangelou, M. W., & Schulin, R. (2018). Plant uptake and availability of antimony, lead, copper and zinc in oxic and reduced shooting range soil. *Journal of Environmental Pollution*, 238, 255–262. https://doi.org/10.1016/j.envpol.2018.03.014.

Holland, B., McCance, R. A., Widdowson, E. M., Unwin, I., & Buss, D. (1991*). Vegetables, Herbs and Spices: Fifth Supplement to McCance and Widdowson's The Composition of Foods* (Vol. 5): Royal Society of Chemistry. https://doi.org/10.1002/food.19920360456.

Hosseinzadeh, H., & Madanifard, M. (2000). Anticonvulsant effects of *Coriandrum sativum* L. seed extracts in mice. *Archives of Iranian Medicine*, 3, 1–4.

Huang, H., Rizwan, M., Li, M., Song, F., Zhou, S., He, X., … Cao, M. (2019). Comparative efficacy of organic and inorganic silicon fertilizers on antioxidant response, Cd/Pb accumulation and health risk assessment in wheat (*Triticum aestivum* L.). *Journal of Environmental Pollution*, 255, 113146. https://doi.org/10.1016/j.envpol.2019.113146.

Hussain, A., Abbas, N., Arshad, F., Akram, M., Khan, Z. I., Ahmad, K., … Mirzaei, F. (2013). Effects of diverse doses of Lead (Pb) on different growth attributes of *Zea mays* L. *Agricultural Sciences*, 4(5). https://doi.org/10.4236/as.2013.45037.

Jain, R. K. J. (2013). Study of heavy metals effect in response to linum seed germination. *African Journal of Plant Science*, 7(3), 93–109. https://doi.org/10.5897/AJPS11.169.

Jamal, Q., Durani, P., Khan, K., Munir, S., Hussain, S., Munir, K., & Anees, M. (2013). Heavy metals accumulation and their toxic effects. *Journal of Bio-Molecular Sciences*, 1(1), 27–36.

Kabir, M., Iqbal, M. Z., Shafiq, M., & Farooqi, Z. (2008). Reduction in germination and seedling growth of *Thespesia populnea* L., caused by lead and cadmium treatments. *Pakistan Journal of Botany*, 40(6), 2419–2426.

Kadhim, A. J. (2021). Effect of biofertilizers and animal manure on morphophysiological characteristics and amount of coriander (*Coriandrum sativum* L.) essential oil under drought stress conditions. Paper presented at the IOP Conference Series: *Journal of Earth and Environmental Science*. https://doi.org/10.1088/1755-1315/735/1/012047.

Kirtikar, K. R., & Basu, B. D. (1999). *Indian Medicinal Plants*. Pharmacognosy, Nirali Prakashan. 6.19–16.28. Dehra Dun International Book Distributors, India.

Kuriakose, S. V., & Prasad, M. (2008). Cadmium stress affects seed germination and seedling growth in *Sorghum bicolor* (L.) Moench by changing the activities of hydrolyzing enzymes. *Journal of Plant Growth Regulation*, 54(2), 143–156. https://doi.org/10.1007/s10725-007-9237-4.

Lamhamdi, M., Bakrim, A., Aarab, A., Lafont, R., & Sayah, F. (2011). Lead phytotoxicity on wheat (*Triticum aestivum* L.) seed germination and seedlings growth. *Journal of Comptes Rendus Biologies*, 334(2), 118–126. https://doi.org/10.1016/j.crvi.2010.12.006.

Li, Q., Lu, Y., Shi, Y., Wang, T., Ni, K., Xu, L., ... Giesy, J. P. (2013). Combined effects of cadmium and fluoranthene on germination, growth and photosynthesis of soybean seedlings. *Journal of Environmental Sciences*, 25(9), 1936–1946. https://doi.org/10.1016/s1001-0742(12)60264-2.

Li, W., Khan, M. A., Yamaguchi, S., & Kamiya, Y. (2005). Effects of heavy metals on seed germination and early seedling growth of *Arabidopsis thaliana*. *Journal of Plant Growth Regulation*, 46(1), 45–50. https://doi.org/10.1007/s10725-005-6324-2.

Liu, D., Wang, X., Chen, Z., Xu, H., & Wang, Y. (2010). Influence of mercury on chlorophyll content in winter wheat and mercury bioaccumulation. *Plant, Soil and Environment*, 56(3), 139–143. https://doi.org/10.17221/210/2009-PSE.

Liu, S., Yang, C., Xie, W., Xia, C., & Fan, P. (2012). The effects of cadmium on germination and seedling growth of *Suaeda salsa*. *Journal of Procedia Environmental Sciences*, 16, 293–298. https://doi.org/10.1016/j.proenv.2012.10.041.

Marichali, A., Dallali, S., Ouerghemmi, S., Sebei, H., & Hosni, K. (2014). Germination, morpho-physiological and biochemical responses of coriander (*Coriandrum sativum* L.) to zinc excess. *Industrial Crops and Products*, 55, 248–257. https://doi.org/10.1016/j.indcrop.2014.02.033.

Matasyoh, J., Maiyo, Z., Ngure, R., & Chepkorir, R. (2009). Chemical composition and antimicrobial activity of the essential oil of *Coriandrum sativum*. *Journal of Food Chemistry*, 113(2), 526–529. https://doi.org/10.1016/j.foodchem.2008.07.097.

Msaada, K., Taarit, M. B., Hosni, K., Hammami, M., & Marzouk, B. (2009). Regional and maturational effects on essential oils yields and composition of coriander (*Coriandrum sativum* L.) fruits. *Journal of Scientia Horticulturae*, 122(1), 116–124. https://doi.org/10.1016/j.scienta.2009.04.008.

Myśliwa-Kurdziel, B., & Strzałka, K. (2002). Influence of metals on biosynthesis of photosynthetic pigments. In *Physiology and Biochemistry of Metal Toxicity and Tolerance in Plants* (pp. 201–227): Springer. https://doi.org/10.1007/978-94-017-2660-3_8.

Neha Mohan, P., Suganthi, V., & Gowri, S. (2013). Evaluation of anti-inflammatory activity in ethanolic extract of *Coriandrum sativum* L. using carrageenan induced paw oedema in albino rats. Journal of Der Pharma Chemica, 5(2), 139–143.

Nicholson, F. A., Smith, S. R., Alloway, B., Carlton-Smith, C., & Chambers, B. (2003). An inventory of heavy metals inputs to agricultural soils in England and Wales. *Journal of Science of the Total Environment*, 311(1–3), 205–219. https://doi.org/10.1016/S0048-9697(03)00139-6.

Nithya, V. J. W. (2015). Evaluation of antidiarrheal activity on *Coriandrum sativum* Linn. In Wistar albino rats. *Journal of Peace Research*, 4, 638–643. http://hdl.handle.net/123456789/31972.

Pathak Nimish, L., Kasture Sanjay, B., Bhatt Nayna, M., & Rathod Jaimik, D. J. (2011). Phytopharmacological properties of *Coriander sativum* as a potential medicinal tree: An overview. *Journal of Applied Pharmaceutical Science*, 1(4), 20–25.

Pietrini, F., Zacchini, M., Iori, V., Pietrosanti, L., Ferretti, M., & Massacci, A. (2010). Spatial distribution of cadmium in leaves and its impact on photosynthesis: Examples of different strategies in willow and poplar clones. *Plant Biology*, 12(2), 355–363. https://doi.org/10.1111/j.1438-8677.2009.00258.x.

Piršelová, B., Kuna, R., Lukáč, P., & Havrlentová, M. (2016). Effect of cadmium on growth, photosynthetic pigments, iron and cadmium accumulation of Faba Bean (*Vicia faba* cv. Aštar). *Agriculture*, 62(2), 72–79. https://doi.org/10.1515/agri-2016-0008.

Rajeshwari, U., & Andallu, B. (2011). Medicinal benefits of coriander (*Coriandrum sativum* L). *Spatula DD*, 1(1), 51–58.

Ramadan, M., & Mörsel, J.-T. J. (2002). Oil composition of coriander (*Coriandrum sativum* L.) fruit-seeds. *Journal of European Food Research and Technology*, 215(3), 204–209. https://doi.org/10.5455/SPATULA.20110106123153.

Randall, R. (2012). *A Global Compendium of Weeds*: Department of Agriculture and Food Western Australia, 1124 pp. http://www.cabi.org/isc/FullTextPDF/2013/20133109119.pdf.

Rondon, F. C., Bevilaqua, C. M., Accioly, M. P., Morais, S. M., Andrade-Junior, H. F., Machado, L. K., ... Rodrigues, A. C. M. (2011). In vitro effect of *Aloe vera*, *Coriandrum sativum* and *Ricinus communis* fractions on *Leishmania infantum* and on murine monocytic cells. *Veterinary Parasitology*, 178(3–4), 235–240. https://doi.org/10.1016/j.vetpar.2011.01.007.

Saeed, S., & Tariq, P. J. (2007). Antimicrobial activities of *Emblica officinalis* and *Coriandrum sativum* against gram positive bacteria and *Candida albicans*. *Pakistan Journal of Botany*, 39(3), 913–917. https://pubmed.ncbi.nlm.nih.gov/17337425.

Saher, N. U., & Siddiqui, A. S. (2019). Occurrence of heavy metals in sediment and their bioaccumulation in sentinel crab (*Macrophthalmus depressus*) from highly impacted coastal zone. *Chemosphere*, 221, 89–98. https://doi.org/10.1016/j.chemosphere.2019.01.008.

Seema, M., Shabnam, I., Malik, M., & Ali, A. (2013). Differential growth and photosynthetic responses and pattern of metal accumulation in sunflower (*Helianthus annuus* L.) cultivars at elevated levels of lead and mercury. *Pakistan Journal of Botany*, 45(Suppl. 1), 367–374.

Seregin, I., & Kozhevnikova, A. D. (2008). Roles of root and shoot tissues in transport and accumulation of cadmium, lead, nickel, and strontium. *Russian Journal of Plant Physiology*, 55(1), 1–22. https://doi.org/10.1134/S1021443708010019.

Sharma, P., & Dubey, R. S. (2005). Lead toxicity in plants. *Brazilian Journal of Plant Physiology*, 17, 35–52. https://doi.org/10.1590/S1677-04202005000100004.

Shirkhodaei, M., Darzi, M. T., & Haj, S. H. M. R. (2014). Influence of vermicompost and biostimulant on the growth and biomass of coriander (*Coriandrum sativum* L.). *International Journal of Advanced Biological And Biomedical Research*, 706–714. https://doi.org/10.33945/Sami/Ijabbr.

Silva, F., Domeño, C., & Domingues, F. C. (2020). *Coriandrum sativum* L.: Characterization, biological activities, and applications. In *Nuts and Seeds in Health and Disease Prevention* (pp. 497–519): Elsevier.

Singh, S., & Aggarwal, P. (2006). Effect of heavy metals on biomass and yield of different crop species. *Indian Journal of Agricultural Sciences*, 76(11), 688.

Sorrentino, M., Capozzi, F., Amitrano, C., Giordano, S., Arena, C., & Spagnuolo, V. (2018). Performance of three cardoon cultivars in an industrial heavy metal-contaminated soil: Effects on morphology, cytology and photosynthesis. *Journal of Hazardous Materials*, 351, 131–137. https://doi.org/10.1016/j.jhazmat.2018.02.044.

Tao, L., Guo, M., & Ren, J. (2015). Effects of cadmium on seed germination, coleoptile growth, and root elongation of six pulses. *Polish Journal of Environmental Studies*, 24(1), 255–299. https://doi.org/10.15244/pjoes/29942.

Tylewicz, U., Nowacka, M., Martín-García, B., Wiktor, A., & Caravaca, A. M. G. (2018). Target sources of polyphenols in different food products and their processing by-products. In *Polyphenols: Properties, Recovery, and Applications* (pp. 135–175): Elsevier, https://doi.org/10.1016/B978-0-12-813572-3.00005-1.

Uslu, O. S., Babur, E., Alma, M. H., & Solaiman, Z. M. (2020). Walnut shell biochar increases seed germination and early growth of seedlings of fodder crops. *Agriculture*, 10(10), 427. https://doi.org/10.3390/agriculture10100427.

Vandecasteele, B., Samyn, J., Quataert, P., Muys, B., & Tack, F. M. (2004). Earthworm biomass as additional information for risk assessment of heavy metal biomagnification: A case study for dredged sediment-derived soils and polluted flood plain soils. *Environmental Pollution*, 129(3), 363–375. https://doi.org/10.1016/j.envpol.2003.12.007.

Wani, P., Khan, M., & Zaidi, A. (2007). Impact of heavy metal toxicity on plant growth, symbiosis, seed yield and nitrogen and metal uptake in chickpea. *Australian Journal of Experimental Agriculture*, 47(6), 712–720. https://doi.org/10.1071/EA05369.

Wichtl, M. (1994). *Coriandri fructus*. In *Herbal Drugs and Phytopharmaceuticals*: CRC Press.

Wierzbicka, M., Przedpełska, E., Ruzik, R., Ouerdane, L., Połeć-Pawlak, K., Jarosz, M., … Szakiel, A. (2007). Comparison of the toxicity and distribution of cadmium and lead in plant cells. *Protoplasma*, 231(1), 99–111. https://doi.org/10.1007/s00709-006-0227-6.

Wong, S., Li, X., Zhang, G., Qi, S., & Min, Y. (2002). Heavy metals in agricultural soils of the Pearl River Delta, South China. *Environmental Pollution*, 119(1), 33–44. https://doi.org/10.1016/S0269-7491(01)00325-6.

Xia, Y., Luo, H., Li, D., Chen, Z., Yang, S., Liu, Z., … Gai, C. (2020). Efficient immobilization of toxic heavy metals in multi-contaminated agricultural soils by amino-functionalized hydrochar: Performance, plant responses and immobilization mechanisms. *Environmental Pollution*, 261, 114217. https://doi.org/10.1016/j.envpol.2020.114217.

Yaldiz, G., & Çamlica, M. J. F. T. P. B. Evaluation of yield and quality of coriander (*Coriandrum sativum*) in Turkey and world. *Full Text Proceedings Book*. Department of Field Crops, Faculty of Agriculture and Natural Sciences, Abant İzzet Baysal University, 14280 Bolu, Turkey.

8 Structural Composition of Coriander Seed

Abner Tomas, Natascha Cheikhyoussef, Ahmad Cheikhyoussef, Idowu Jonas Sagbo, and Ahmed A. Hussein

CONTENTS

8.1 INTRODUCTION

Coriander (*Coriandrum sativum* L.) is an annual plant with unique aromatic characteristics and a long history in ethno-food uses as a culinary herb (Mandal & Mandal, 2015). It has been previously reported as one of the oldest herbal plants belonging to the Apiaceae (Umbelliferae) family (Al-Marzoqi et al., 2015). The Apiaceae family is mostly made up of aromatic plants with hollow stems and consists of approximately 455 genus and 3600 species, of which more than 150 plants have been generally recognized as safe (GRAS) by the Food and Drug Administration (FDA) with several condiments or vegetables having important nutritional and medicinal properties (Costa et al., 2015; Demir & Korukluoglu, 2020).

Leaves, stems, and coriander seeds are commonly used in traditional culinary settings as spices to provide pleasant color and flavor (Iqbal et al., 2018). Coriander seeds have been useful in food preparation as a flavoring agent, adjuvant, and preservation, and in preventing food-borne diseases and food spoilage (Kajal & Singh, 2018). For example, studies have reported using coriander leaves and seeds in the preparation of curry and traditional cuisines and garnishing dishes before serving (Coşkuner & Karababa, 2007). Coriander seeds are also added to dishes as an aromatic spice and act as digestive agents to facilitate the digestion process (Mandal & Mandal, 2015). Essential oils extracted from coriander leaves and seeds have been reported to be useful in beverages, fish and meat products, food applications, pickles, and sweets due to their pleasant aroma and free radical scavenging activity (Ashraf et al., 2020).

Coriander has been reported as a rich source of compounds possessing several biological activities (Eidi & Eidi, 2011; Mandal and Mandal, 2015). Interestingly, studies have reported significant pharmacological effects of coriander leaves and seeds, including anticancer, antimicrobial, antidiabetic, anti-inflammatory, antioncogenic, antioxidant, hepatoprotection, hypoglycemic, hypocholesterolemic, and neuroprotection effects (Wangensteen et al., 2004; Deepa & Anuradha, 2011; Shahwar et al., 2012; Al-Marzoqi et al., 2015; Iqbal et al., 2018). In addition, coriander seed oils have been reported to be used in folk medicine to treat anxiety, depression, and Alzheimer's disease

DOI: 10.1201/9781003204626-9

(Ashraf et al., 2020). These bioactive functions are due to multiple bioactive components like polyphenols, alkaloids, flavonoids, terpenoids, fatty acids, tocols, and vitamins (Yildiz, 2016).

8.2 CORIANDER SEEDS' PROXIMATE COMPOSITION

Coriander seeds are globular and aromatic with a spicy and somewhat bittersweet taste (Ceska et al., 1988). The seeds are the most commonly used part of the plant, with volatile oil (1%) and fixed oil (25%) being the important constituents of the seeds (Sahib et al. 2013).

The properties of coriander seeds have been evaluated, such as moisture content, varying from 7.10% to 18.94% (Coşkuner & Karababa, 2007) and 8% to 16% (Sharanagat & Goswami, 2014). In addition, coriander seeds contain several compounds such as fatty acids, sterols, tocols, minerals, essential oils, and phenolics (Laribi et al., 2015).

The dehulling of coriander seeds yielded a 97% purity and increased oil content (100%), crude protein (70%), and reduced crude fiber and other carbohydrates by 28% and 65%, respectively (Evangelista et al., 2015). Evangelista et al. (2015) recommended that pre-pressing and solvent extraction are more suitable for pure dehulled seeds because of their high oil content. Coriander seeds are considered an important source of lipids and minerals (Table 8.1) (Bhat et al., 2014) such as potassium (1267 mg/100 g), calcium (709 mg/100 g), phosphorus (409 mg/100 g), magnesium (330 mg/100 g), sodium (35 mg/100 g), and zinc (4.70 mg/100 g) (USDA, 2013). The content of vitamins such as folate is 200 µg/100 g, vitamin C (21 mg/100 g), thiamin (0.23 mg), riboflavin (0.29 mg), and niacin (2.13 mg) in coriander fresh seeds were also reported (Iwatani, Arcot & Shreshtha, 2003; USDA, 2013). The proximate composition (%) of coriander seeds is shown in Table 8.1.

Sriti et al. (2013) reported the presence of eight fatty acids in the oil extracted from the coriander seed, with petroselinic acid being the major fatty acid with 75–77%, followed by 12–13% linoleic acid, 5% oleic acid, and 3% palmitic acid, in addition to β-sitosterol being the major sterol with 33–35% of total sterols (Sriti et al., 2013). Petroselinic acid (Figure 8.1) has been reported later on to be the main component in the oil (Uitterhaegen et al., 216). Coriander seeds contain 13–29% of vegetable oil (Ramadan & Mörsel, 2002) and 0.35% essential oil (Msaada et al., 2007). Using ultrahigh-resolution total correlation NMR spectroscopy, Verma, Parihar, and Baishya (2019) identified

TABLE 8.1
Proximate Composition (%) of *Coriandrum sativum* L. Seeds

Proximate Composition (per 100 g)	Diederichsen, 1996	Peter, 2004	Hussain et al., 2009	Shahwar et al., 2012	USDA, 2013
Moisture	11.3	NR	6.65	6.2	8.86
Starch	10.5	11	NR	NR	NR
Lipid/crude fat	19.1	20	9.83	9.12	17.7
Protein/crude protein	11.4	11	11.7	12.5	12.3
Crude fiber	28.4	30	5.53	37.14	41.9
Ash	4.98	NR	8.03	8.59	8.59
Carbohydrate	1.92	NR	63.7	26.3	54.9
Energy	NR	NR	390.39	NR	298
Cu (ppm)	NR	NR	41.4	NR	NR
Zn (ppm)	NR	NR	59.4	NR	4.70
Fe (ppm)	NR	NR	150.4	NR	16.32
Co	NR	NR	4.8	NR	NR
Cd	NR	NR	2.4	NR	NR

Note: NR = not reported

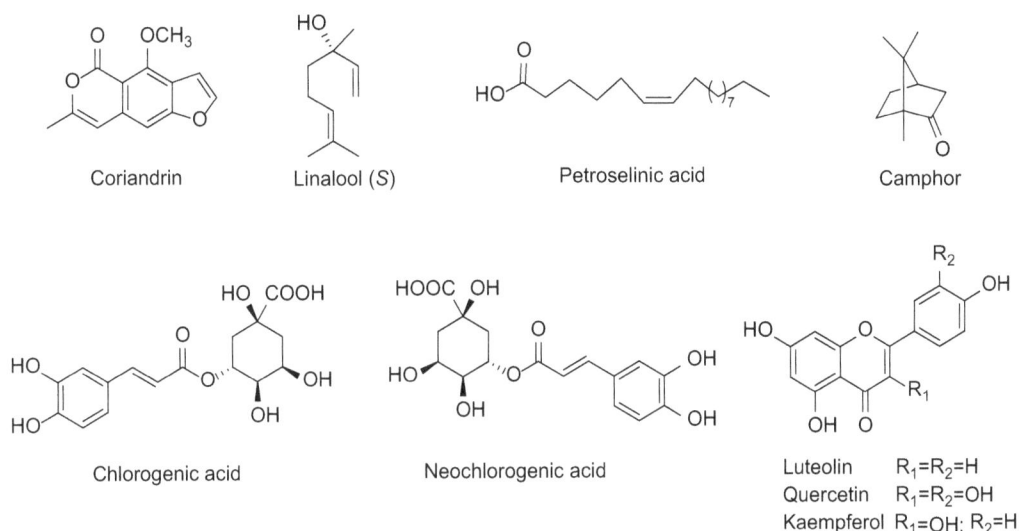

Coriandrin Linalool (*S*) Petroselinic acid Camphor

Chlorogenic acid Neochlorogenic acid Luteolin $R_1=R_2=H$
Quercetin $R_1=R_2=OH$
Kaempferol $R_1=OH; R_2=H$

FIGURE 8.1 Chemical structure for the common and most active constituents in coriander seed extracts.

36 common metabolites in *C. sativum* L. seed extract belonging to organic acids, amino acids, and carbohydrates.

8.3 CORIANDER SEEDS' BIOACTIVE CONSTITUENTS

Several bioactive compounds have been extracted from coriander seeds and are shown in Table 8.2. One effective way to produce coriander plant with an increased level of bioactive constituents is via *in vitro* culture (Matkowski, 2008; Dias et al., 2011; Ogita, 2015; Steingroewer et al., 2018).

8.3.1 PHYTOCHEMICAL ANALYSIS

Six phenolic acids derivatives, including chlorogenic acid (5-*O*-caffeoylquinic acid), caffeic acid, *p*-coumaric acid, and ferulic acid, have been identified using HPLC-DAD-ESI/MS by comparing their UV and mass characteristics and retention time with standard commercial databases (Barros et al., 2012). A wide variety of monoterpenoids, aromatic glycosides, and monoterpene glycoside sulfates, and a wide range of essential oils, monoterpenes, and fatty acids were reported in the chemical composition of coriander seeds (Pacholczyk-Sienicka, Ciepielowski & Albrecht, 2021). The water-soluble constituents of coriander seeds contain 4 new monoterpenoid glycosides, 2 new monoterpenoid glucoside sulfates, and 2 new aromatic glycosides amongst the 33 compounds that have been isolated from the water-soluble portion of the methanolic extract of coriander seeds (Ishikawa, Kondo & Kitajima, 2003). In addition, two other glycosides of 2-*C*-methyl-D-erythritol were also isolated from the coriander seeds, as reported by Kitajima et al. (2003).

The total phenolic contents (TPC) and total flavonoid contents (TFC) of coriander seeds were reported to be 15.5–64.9 mg gallic acid equivalent (GAE)/g extract and 4.61–13.8 mg quercetin acid equivalent (QE)/g respectively (Abbassi et al., 2018). Demir and Korukluoglu (2020) used methanol and ethanol to extract phenolics and reported 4.2, 2.1 mg GAE/g of dry weight (DW), respectively. An earlier study by Shahwar et al. (2012) indicated that methanol extract of coriander seeds had a TPC of 29.2 mg/g and the *n*-hexane extract of seed had 11.4 mg/g. Another study by Gallo et al. (2010) used two different methods of microwave-assisted extraction (MAE) and ultrasound-assisted extraction (UAE) for the extraction of phenolics from coriander seed and reported it to contain 41.8 and 82.0 mg GAE/100g of DW, respectively. Deepa and Anuradha (2011) reported a TPC of the coriander seeds of 12.2 GAE/g and a TFC of 12.6 QE/g.

TABLE 8.2

Bioactive Compounds Contained in *Coriandrum sativum* L. Seeds

No.	Compound	References
1.	Quercetin 3-glucuronide, isoquercitin, and rutin	Kunzemann & Herrmann (1977)
2.	Coriandrin and dihydrocoriandrin	Ceska et al. (1988)
3.	Isocoumarine	Taniguchi et al. (1996)
4.	Fatty acids: petroselinic acid and linoleic acid	Ramadan & Mörsel, (2002); Sriti et al. (2009, 2010)
5.	Glycosides of 2-*C*- methyl-D- erythritol	Kitajima et al. (2003)
6.	Monoterpenoid glycosides	Ishikawa, Kondo & Kitajima, (2003)
7.	Camphor, linalool, geranyl acetate	Raal et al. (2004)
8.	Caffeic acid and kaempferol	Shan et al. (2005)
9.	β-carotene, β-cryptoxanthin epoxide, lutein-5,6-epoxide, violaxanthin, and neoxanthin	Guerra et al. (2005)
10.	Monoterpenes, linalyl acetate	Msaada et al. (2007, 2009)
11.	Tocols: tocopherol and tocotrienol Sterols: stigmasterol and β-sitosterol	Sriti et al. (2010)
12.	Flavonoids (apigenin, quercetin, and kaempferol)	Sulaiman et al. (2011)
13.	Anthocyanins	Barros et al. (2012); Msaada et al. (2017)
14.	Phenolic acids (caffeoyl *N*-tryptophan hexoside)	Barros et al. (2012)
15.	Chlorogenic acid, ferulic acid, *p*-coumaric, and caffeic acids	Barros et al. (2012)
16.	Tannins	Msaada et al. (2017)
17.	Quercetin; quercetin-3-*O*-rutinoside; kaempferol; 3,3',4',5,7-flavanpentol; luteolin; 7-*O*-ethylquercetin; apigenin	Kajal & Singh (2018)
18.	Organic acids, amino acids, and carbohydrates	Verma et al. (2019)
19.	Linalool, ascorbyl palmitate, petroselinic acid, docosahexanoic acid, linoleic acid, α-linolenic acid, oleic acid	Kajal & Singh (2019)
20.	Quercetin, kaempferol	Demir & Korukluoglu (2020)
21.	Caffeic acid	Demir & Korukluoglu (2020)

Msaada et al. (2017) evaluated TFC and total condensed tannin contents (TCTC) in the methanol extract of three varieties of *C. sativum* seeds originating from Tunisia, Syria, and Egypt. TFC ranging from 2.03 to 2.51 mg EC/g DW and a TCTC from 0.09 to 0.17 mg EC/g DW were reported with the TFC and TCTC values obtained in methanol extract of the Syrian variety (Msaada et al., 2017). The Tunisian coriander variety has high phenolic acids (81.4%); meanwhile, the Syrian coriander variety is rich in flavonoid (61.3%) due to the high presence of luteolin (18.1%) and rutin trihydrate (13.0%). For the Egyptian coriander variety, the methanol extract has equal content of phenolic acids (49.1%) and flavonoids (50.8%) (Msaada et al., 2017). They further reported 21 compounds in the 3 coriander varieties including 11 phenolic acids including caffeic, chlorogenic, *o*-coumaric, *p*-coumaric, ferulic, gallic, rosmarinic, salicylic, *trans*-cinnamic acids, *trans*-hydroxycinnamic acid, and vanillic, and 10 flavonoids including apigenin, coumarine, kaempferol, luteolin, naringin, quercetin, quercetin-3-*O*-rhamnoside, resorcinol, and rutin (Msaada et al., 2017). However, another study confirmed the presence of phenolic acids and derivatives only but not flavonoids, with caffeoyl N-tryptophan hexoside being the highest phenolic derivative (45.3 mg/kg) meanwhile, *N*-caffeoyltryptophan being the lowest in the methanolic extract (Barros et al., 2012). They further reported the composition of polyphenols content of the coriander seeds to include caffeic acid, 3-*O*-caffeoylquinic acid, *p*-coumaric acid, 3-*O*-caffeoylquinic acid, ferulic acid, ferulic acid derivative, *N*-feruloyltryptophan hexoside, and *N*-feruloyltryptophan. β-Carotene and total carotenoid content of 160 and 1010 μg/100 g, respectively, were also determined in coriander seeds (Kandlakunta, Rajendran, & Thingnganing, 2008).

8.3.2 Volatile Constituents

The most important constituents of coriander seeds are fixed oils (9.3–37.6%) and volatile oils (0.3–1.8%), and these have been reported by several studies with linalool being the main constituent (60–70%) of coriander seeds' volatile oil irrespective of the isolation procedure and the operational conditions such as temperature (Coşkuner & Karababa, 2007; Sriti et al., 2010a; Mhemdi et al., 2011; Zeković et al., 2017; Hanifei et al., 2021).

Two common methods of steam and hydro-distillation are the major reported methods to extract volatile oils from coriander seeds, with the yield ranging from 0.1 to 0.39% (Msaada et al., 2007; Nejad et al., 2010) being affected by several factors such as cultivar origin, climatic conditions, and the geographic position of the growth region (Laribi et al., 2015). The major volatile compounds in coriander seed essential oil are linalool (37.6–87.5%) (Figure 8.1), γ-terpinene (4.17–14.4%), α-pinene (0.02–4.09%), β-pinene (0.05–1.82%), camphor (0.14–6.2%), decanal (0.16–4.69%), *p*-cymene (0.45–3.6%), geranyl acetate (0.83–17.5%), limonene (0.02–3.10%), geraniol (0.81–2.23%), α-cedrene (3.87), camphene (1.78%), and D-limonene (1.36%) (de Figueiredo et al., 2004; Msaada et al., 2007; Nejad et al., 2010; Shahwar et al., 2012; Laribi et al., 2015; Mandal and Mandal, 2015).

It was reported that coriander seeds from Russia have a higher concentration of camphor (69.7%) (Figure 8.1) and lower concentration of linalool (2.96%); meanwhile, the seeds from New Zealand have camphor (5.1%) and linalool (65.8%) (Figure 8.1) (Misharina, 2001; Smallfield et al., 2001). The solid-phase micro-extraction-gas chromatography-mass spectrometry analysis of the volatile fraction of the extract of coriander seeds from Italy identified several terpenes such as α-pinene, 2-carene, *p*-mentha-1,3,3-triene, eucalyptol, β-terpinyl-acetate, β-myrcene, β-ocymene, γ-terpinene, *trans*-linalool oxide, terpinolene, borneol, linalool, camphor, terpinen-4-ol, α-terpineol, *cis*-geraniol, lavandulyl acetate, and geranyl acetate (Palmieri et al., 2020). The major components in the essential oil obtained from India's coriander seed have been identified as myristic acid, petroselinic acid, linalool *cis*-isoapiole 4-bromo-2-adamantanol, and (-)-carvone (Shrirame et al., 2018).

8.3.3 Non-Volatile Constituents

The non-volatiles from coriander seeds were extracted *via* solvent extraction using methanol and *n*-hexane solvents, of which the extraction with methanol had a higher yield (5.43%) than *n*-hexane extraction of 0.9% (Shahwar et al., 2012). A quantitative analysis of coriander seeds extracts (methanol and ethanol) using LC-QTOF-MS was reported by Demir and Korukluoglu (2020) and revealed the presence of vanillic acid, caffeic acid, syringic acid, 4-coumaric acid, neochlorogenic acid, chlorogenic acid, ferulic acid, 3-hydroxycinnamic acid, 2-hydroxycinnamic acid, epigallocatechin, catechin, epigallocatechin-3-gallate, epicatechin, naringin, *trans*-resveratrol, myricetin, *cis*-resveratrol, quercetin, luteolin, and kaempferol. They further emphasized that chlorogenic acid and neochlorogenic acid were the most abundant phenolic compounds, followed by luteolin and kaempferol (Figure 8.1).

The oil obtained from the seeds of coriander has a light yellow color and has recently been proposed as a novel food ingredient in the European Union (Nguyen et al., 2020). The oil yield ranges from 19.24 to 28.4% of dry weight (Ramadan & Mörsel, 2003; Msaada et al., 2009; Sriti et al., 2009). It has a distinctive smell and is unique in its high content of petroselinic acid (C18:1n-12), ranging from 65.7% to 76.6%, followed by linoleic acid ranging from 13.0 to 16.7% and some other fatty acids such as oleic and palmitic acids (Sahib et al., 2013). Acylated steryl glucoside, steryl glucoside, and glucocerebroside are the major glycolipids in the seed oil, and triacylglycerols are the main neutral lipid ranging from 93.0 to 95.6% of total lipids (Ramadan & Mörsel, 2003; Sahib et al., 2013).

Coriander oil is a good source of tocols (327.47 µg/g), whereby tocotrienol is much higher (275.8 µg/g) than tocopherol (51.6 µg/g). In terms of total sterol content, it ranges from 36.9 to 51.8 mg/g oil, with stigmasterol and β-sitosterol being the most predominant sterols (21.7–29.8% and 24.8–36.8%) markers in coriander seeds respectively, in addition to cholesterol (1.02–2.18%) which was also detected in Tunisian coriander seeds (Sriti et al., 2009; Sriti et al., 2010b), but not in the coriander seeds from Germany; instead, ergosterol and lanosterol were detected (Ramadan & Mörsel, 2003). Different ripening stages and rainfall amounts significantly affect coriander fruit's oil content and composition, as investigated by Nguyen et al. (2020) for fruits grown in France. The oil yield was highest in plants receiving more rain, while the fatty acid, petroselinic acid, was the highest in fruits collected before fully ripened (Nguyen et al., 2020).

8.4 FOOD USES

Coriander seeds are ground into powder and used as a flavoring agent in the preparation of meat/poultry products and fish products, bakery, curry, stews and sauce recipes, sodas, and pickle formulations (Ravi et al., 2007; Sahib et al., 2013; Singletary, 2016; Prachayasittikul et al., 2018). In addition, Shahwar et al. (2012) advised including coriander seeds in cuisine and food preparations as it will increase the content of antioxidants and prevent oxidative deterioration of food.

8.5 BIOLOGICAL ACTIVITY OF CORIANDER SEEDS

Several studies reported that coriander seed is a rich source of several bioactive compounds, including flavonoids (such as quercetin, isoquercetin, and rutin), terpenes, phenolic acids (ferulic, gallic, and caffeic acids), carotenoids, fatty acids rich in petroselinic acid, coumarins, tannins, sterols, antheole, borneol, vitamins, and tocopherols which could be responsible for its therapeutic action (Nadeem et al., 2013; Laribi et al., 2015; Singletary, 2016). In addition, *C. sativum* seeds possess carminative, cooling, and hypotensive properties (Sairam, 1998; Taniguchi et al., 1996).

Linalool (2,6-dimethyl-2,7-octadien-6-ol), an abundant compound in coriander seeds, is believed to be responsible for most disease pathway modulation and protective effects against cancer metabolic syndrome and neurodegenerative diseases (Prachayasittikul et al., 2018; An et al., 2021). Coriandrin (Figure 8.1) exhibits UVA-dependent antiviral activity against various enveloped viruses with notable *in vitro* activity against HIV-L2 (Kraus & Ridgeway, 1994). The cholesterol-lowering property of coriander seeds is attributed to the increasing activity of plasma lecithin-cholesterol acyltransferase, leading to enhanced cholesterol degradation to fecal bile acids and neutral sterols that might be responsible for its hypocholesterolemic effect (Chithra & Leelamma, 1997; Dhanapakiam et al., 2008). Zeb et al. (2018) combined garlic and coriander seed powder and administered it to hyperlipidemic patients at a dose rate of 2 g/day to study its impact on body mass index (BMI), lipid profile, and blood pressure. They confirmed that the supplements influenced and improved the patients' BMI, HDL, total cholesterol, triglycerides, LDL, and systolic blood pressure.

Coriander seed is known in folk medicine for its wide range of healing properties for intestinal parasites as it is larvacidal, bactericidal, and fungicidal. In gastrointestinal complaints, it has been used in the treatment of abdominal problems such as stomach ulcers, anorexia, dyspepsia, flatulence, griping pain, indigestion, vomiting, sub-acid gastritis, and diarrhea (Taniguchi et al., 1996; Jabeen et al., 2009; Chahal et al., 2017). In addition, coriander seed is also used as an anti-edemic, anthelmintic, antihypertensive, anti-inflammatory, antiseptic, emmenagogue, and myorelaxant, and for diabetes and rheumatism (Aissaoui et al., 2008; Eidi et al., 2009; Eidi & Eidi, 2011; Deepa & Anuradha, 2011; Hosseinzadeh et al., 2016).

In terms of coriander seeds' antioxidant activity, methanol extracts showed a significant radical scavenging activity of 64.4% compared to *n*-hexane extract of 52.6% at a concentration of 500 µg/mL

(Shahwar et al., 2012). The addition of *C. sativum* seed powder to the diet of streptozotocin-induced diabetic rats led to a noticeable lowering of blood glucose. In addition, it increased insulin levels *via* inhibiting the peroxidative process, damage reactivating the antioxidant enzymes and antioxidant levels in diabetic rats (Deepa & Anuradha, 2011). They further recommended that the increased dietary intake of coriander seeds might help decrease the oxidative stress in diabetes mellitus. More recent studies are reporting the *in vitro* modulation of aldose reductase, sorbitol accumulation, and advanced glycation end products by a flavonoid-rich extract from the *C. sativum* seeds (Kajal, & Singh, 2018) and the impact of petroleum ether extract of *C. sativum* seeds on the mitigation of diabetic nephropathy in streptozotocin nicotinamide (NAD) induced type 2 diabetes model (Kajal, & Singh, 2019) to elucidate these extracts' possible mechanism of action for the potential antidiabetic therapy from *C. sativum*.

The antimicrobial and antifungal activity of crude extract from coriander seeds has also been reported against pathogenic bacteria such as *Salmonella choleraesuis* (Kubo et al., 2004), *Staphylococcus aureus*, *Klebsiella pneumoniae*, and *Pseudomonas aeruginosa*, as well as antifungal activity against *Penicillium lilacinum* and *Aspergillus niger* (Zardini et al., 2012).

The ethanol extract of *C. sativum* seeds had a protective role against tacrine-induced orofacial dyskinesia whereby the administration of 100, 200 mg/kg for 15 days *in vivo* resulted in decreased tacrine-induced vacuous chewing movements, tongue protrusions, and orofacial bursts; at the same time, the locomotion and cognition compared to the tacrine-treated group were significantly increased (Mohan, Yarlagadda, & Chintala, 2015).

8.6 CONCLUSIONS

Coriander seeds are beneficial to human health due to many bioactive compounds such as fatty acids, sterols, tocols, and volatile compounds. In addition, they are a rich source of phenolic acids and a potential source of food flavoring and antioxidants. Coriander seeds possess several biological activities that can be explored to develop functional food for the health and nutraceutical industry. The seeds contain unique fatty acids, particularly petroselinic acid, at a notable concentration. The essential oil rich in linalool can also be prepared from coriander seeds.

REFERENCES

Abbassi, A., Mahmoudi, H., Zaouali, W., M'Rabet, Y., Casabianca, H., & Hosni, K. (2018). Enzyme-aided release of bioactive compounds from coriander (*Coriandrum sativum* L.) seeds and their residue by-products and evaluation of their antioxidant activity. *Journal of Food Science and Technology*, *55*(8), 3065–3076. https://doi.org/10.1007/s13197-018-3229-4

Aissaoui, A., El-Hilaly, J., Israili, Z. H., & Lyoussi, B. (2008). Acute diuretic effect of continuous intravenous infusion of an aqueous extract of *Coriandrum sativum* L. in anesthetized rats. *Journal of Ethnopharmacology*, *115*(1), 89–95. https://doi.org/10.1016/j.jep.2007.09.007

Al-Marzoqi, A. H., Hameed, I. H., & Idan, S. A. (2015). Analysis of bioactive chemical components of two medicinal plants (*Coriandrum sativum* and *Melia azedarach*) leaves using gas chromatography-mass spectrometry (GC-MS). *African Journal of Biotechnology*, *14*(40), 2812–2830. https://doi.org/10.5897/AJB2015.14956

An, Q., Ren, J. N., Li, X., Fan, G., Qu, S. S., Song, Y., Li, Y., & Pan, S. Y. (2021). Recent updates on bioactive properties of linalool. *Food & Function*, *12*(21), 10370–10389. https://doi.org/10.1039/d1fo02120f

Ashraf, R., Ghufran, S., & Sultana, B. (2020). Chapter 31: Cold pressed coriander (*Coriandrum sativum* L.) seed oil. In: M. F. Ramadan, ed. *Cold Pressed Oils: Green Technology, Bioactive Compounds, Functionality, and Applications*. s.l.: Elsevier, pp. 345–356. https://doi.org/10.1016/B978-0-12-818188-1.00031-1

Barros, L., Duenas, M., Dias, M. I., Sousa, M. J., & Santos-Buelga, C. (2012). Phenolic profiles of in vivo and in vitro grown *C. sativum* L. *Food Chemistry*, *132*(2), 841–848. https://doi.org/10.1016/j.foodchem.2011.11.048

Bhat, S., Kaushal, P., Kaur, M., & Sharma, H. K. (2014). Coriander (*Coriandrum sativum* L.): Processing, nutritional and functional aspects. *African Journal of Plant Science*, *8*(1), 25–33. https://doi.org/10.5897/AJPS2013.1118

Ceska, O., Chaudary, S. K., Warrington, P., Ashwood-Smith, M. J., Bushnell, G. W., & Poulton, G. A. (1988). Coriandrin, a novel, highly photoactive compound isolated from *Coriandrum sativum*. *Phytochemistry*, *27*(7), 2083–2087. https://doi.org/10.1016/0031-9422(88)80101-8

Chahal, K. K., Singh, R., Kumar, A., & Bhardwaj, U. (2017). Chemical composition and biological activity of *Coriandrum sativum* L.: A review. *Indian Journal of Natural Products and Resources*, *8*(3), 193–203. http://nopr.niscair.res.in/handle/123456789/43357

Chithra, V., & Leelamma, S. (1997). Hypolipidemic effect of coriander seeds (*Coriandrum sativum*): Mechanism of action. *Plant Foods for Human Nutrition (Dordrecht, Netherlands)*, *51*(2), 167–172. https://doi.org/10.1023/a:1007975430328

Coşkuner, Y., & Karababa, E. (2007). Physical properties of coriander seeds (*Coriandrum sativum* L.). *Journal of Food Engineering*, *80*(2), 408–416. https://doi.org/10.1016/j.jfoodeng.2006.02.042

Costa, D. C., Costa, H. S., Albuquerque, T. G., Ramos, F., & Castilho, M. C. (2015). Advances in phenolic compounds analysis of aromatic plants and their potential applications. *Trends in Food Science & Technology*, *45*(2), 336–354. https://doi.org/10.1016/j.tifs.2015.06.009

de Figueiredo, R. O., Marques, M. O. M., Nakagawa, J., & Ming, L. C. (2004). Composition of coriander essential oil from Brazil. *Acta Horticulturae*, *629*, 135–137. https://doi.org/10.17660/ActaHortic.2004.629.18

Deepa, B., & Anuradha, C. V. (2011). Antioxidant potential of *Coriandrum sativum* L. seed extract. *Indian Journal of Experimental Biology*, *49*(1), 30–38.

Demir, S., & Korukluoglu, M. (2020). A comparative study about antioxidant activity and phenolic composition of cumin (*Cuminum cyminum* L.) and coriander (*Coriandrum sativum* L.). *Indian Journal of Traditional Knowledge*, *19*(2), 383–393.

Dhanapakiam, P., Joseph, J. M., Ramaswamy, V. K., Moorthi, M., & Kumar, A. S. (2008). The cholesterol lowering property of coriander seeds (*Coriandrum sativum*): Mechanism of action. *Journal of Environmental Biology*, *29*(1), 53–56.

Dias, M. I., Barros, L., Sousa, M. J., & Ferreira, I. C. (2011). Comparative study of lipophilic and hydrophilic antioxidants from in vivo and in vitro grown *Coriandrum sativum*. *Plant Foods for Human Nutrition*, *66*(2), 181–186. https://doi.org/10.1007/s11130-011-0227-3

Diederichsen, A. (1996). Coriander. Promoting the conservation and use of underutilized and neglected crops. 3. In: Purseglove, J. W., Brown, E. G., Green, C. L., & Robbins, S. R. J., eds. *Spices*, Vol. 2. New York: Longman, pp. 736–788.

Eidi, M., & Eidi, A. (2011). Chapter 47: Effect of coriander (*Coriandrum sativum* L.) seed ethanol extract in experimental diabetes. In: V. R. Preedy, R. R. Watson & V. B. Patel, eds. *Nuts and Seeds in Health and Disease Prevention*. s.l.: Elsevier, pp. 395–400. https://doi.org/10.1016/B978-0-12-375688-6.10047-7

Eidi, M., Eidi, A., Saeidi, A., Molanaei, S., Sadeghipour, A., Bahar, M., & Bahar, K. (2009). Effect of coriander seed (*Coriandrum sativum* L.) ethanol extract on insulin release from pancreatic beta cells in streptozotocin-induced diabetic rats. *Phytotherapy Research: PTR*, *23*(3), 404–406. https://doi.org/10.1002/ptr.2642

Evangelista, R. L., Evangelista, M. H., Cermak, S. C., & Isbell, T. A. (2015). Dehulling of coriander fruit before oil extraction. *Industrial Crops and Products*, *69*, 378–384. https://doi.org/10.1016/j.indcrop.2015.02.057

Gallo, M., Ferracane, R., Graziani, G., Ritieni, A., & Fogliano, V. (2010). Microwave assisted extraction of phenolic compounds from four different spices. *Molecules*, *15*(9), 6365–6374. https://doi.org/10.3390/molecules15096365

Guerra, N. B., de Almeida Melo, E., & Filho, J. M. (2005). Antioxidant compounds from coriander (*Coriandrum sativum* L.) etheric extract. *Journal of Food Composition and Analysis*, *18*(2–3), 193–199. https://doi.org/10.1016/j.jfca.2003.12.006

Hanifei, M., Mehravi, S., Khodadadi, M., Severn-Ellis, A. A., Edwards, D., & Batley, J. (2021). Detection of epistasis for seed and some phytochemical traits in coriander under different irrigation regimes. *Agronomy*, *11*(9), 1891. https://doi.org/10.3390/agronomy11091891

Hosseinzadeh, S., Ghalesefidi, M. J., Azami, M., Mohaghegh, M. A., Hejazi, S. H., & Ghomashlooyan, M. (2016). In vitro and in vivo anthelmintic activity of seed extract of *Coriandrum sativum* compared to niclosamid against *Hymenolepis nana* infection. *Journal of Parasitic Diseases*, *40*(4), 1307–1310. https://doi.org/10.1007/s12639-015-0676-y

Hussain, J., Khan, A. L., Rehman, N., Zainullah, Khan, F., Hussain, S. T., & Shinwari, Z. K. (2009). Proximate and nutrient investigations of selected medicinal plants species of Pakistan. *Pakistan Journal of Nutrition*, *8*(5), 620–624. https://doi.org/10.3923/pjn.2009.620.624

Iqbal, M. J., Butt, M. S., & Suleria, H. A. R. (2018). Coriander (*Coriandrum sativum* L.): Bioactive molecules and health effects. In: Mérillon, J. M., & Ramawat, K., eds. *Bioactive Molecules in Food. Reference Series in Phytochemistry*. Cham: Springer. https://doi.org/10.1007/978-3-319-54528-8_44-1

Ishikawa, T., Kondo, K., & Kitajima, J. (2003). Water-soluble constituents of coriander. *Chemical & Pharmaceutical Bulletin*, 51(1), 32–39. https://doi.org/10.1248/cpb.51.32

Iwatani, Y., Arcot, J., & Shreshtha, A. K. (2003). Determination of folate contents in some Australian vegetables. *Journal of Food Composition and Analysis*, 16(1), 37–48. https://doi.org/10.1016/S0889 -1575(02)00159-X

Jabeen, Q., Bashir, S., Lyoussi, B., & Gilani, A. H. (2009). Coriander fruit exhibits gut modulatory, blood pressure lowering and diuretic activities. *Journal of Ethnopharmacology*, 122(1), 123–130. https://doi .org/10.1016/j.jep.2008.12.016

Kajal, A., & Singh, R. (2018). An allied approach for in vitro modulation of aldose reductase, sorbitol accumulation and advanced glycation end products by flavonoid rich extract of *Coriandrum sativum* L. seeds. *Toxicology Reports*, 5, 800–807. https://doi.org/10.1016/j.toxrep.2018.08.001

Kajal, A., & Singh, R. (2019). *Coriandrum sativum* seeds extract mitigate progression of diabetic nephropathy in experimental rats via AGEs inhibition. *PloS One*, 14(3), e0213147. https://doi.org/10.1371/journal .pone.0213147

Kandlakunta, B., Rajendran, A., & Thingnganing, L. (2008). Carotene content of some common (cereals, pulses, vegetables, spices and condiments) and unconventional sources of plant origin. *Food Chemistry*, 106(1), 85–89. https://doi.org/10.1016/j.foodchem.2007.05.071

Kitajima, J., Ishikawa, T., Fujimatu, E., Kondho, K., & Takayanagi, T. (2003). Glycosides of 2-C-methyl-D-erythritol from the fruits of anise, coriander and cumin. *Phytochemistry*, 62(1), 115–120. https://doi.org /10.1016/s0031-9422(02)00438-7

Kraus, G. A., & Ridgeway, J. (1994). Total Synthesis of Coriandrin. *The Journal of Organic Chemistry*, 59(17), 4735–4737. https://doi.org/10.1021/jo00096a013

Kubo, I., Fujita, K., Kubo, A., Nihei, K., & Ogura, T. (2004). Antibacterial activity of coriander volatile compounds against *Salmonella choleraesuis*. *Journal of Agricultural and Food Chemistry*, 52(11), 3329–3332. https://doi.org/10.1021/jf0354186

Kunzemann, J., & Herrmann, K. (1977). Isolation and identification of flavon(ol)-O-glycosides in caraway (*Carum carvi* L.), fennel (*Foeniculum vulgare* Mill.), anise (*Pimpinella anisum* L.), and coriander (*Coriandrum sativum* L.), and of flavon-C-glycosides in anise. I. Phenolics of spices. *Zeitschrift fur Lebensmittel-Untersuchung und -Forschung*, 164(3), 194–200. https://doi.org/10 .1007/BF01263030

Laribi, B., Kouki, K., M'Hamdi, M., & Bettaieb, T. (2015). Coriander (*Coriandrum sativum* L.) and its bioactive constituents. *Fitoterapia*, 103, 9–26. https://doi.org/10.1016/j.fitote.2015.03.012

Mandal, S., & Mandal, M. (2015). Coriander (*Coriandrum sativum* L.) essential oil: Chemistry and biological activity. *Asian Pacific Journal of Tropical Biomedicine*, 5, 421–428. https://doi.org/10.1016/j.apjtb.2015 .04.001

Matkowski, A. (2008). Plant in vitro culture for the production of antioxidants--A review. *Biotechnology Advances*, 26(6), 548–560. https://doi.org/10.1016/j.biotechadv.2008.07.001

Misharina, T. A. (2001). Effect of conditions and duration of storage on composition of essential oil from coriander seeds. *Prikladnaia Biokhimiia i Mikrobiologiia*, 37, 726–732.

Mohan, M., Yarlagadda, S., & Chintala, S. (2015). Effect of ethanolic extract of *Coriandrum sativum* L. on tacrine induced orofacial dyskinesia. *Indian Journal of Experimental Biology*, 53(5), 292–296.

Msaada, K., Hosni, K., Taarit, M. B., Chahed, T., Kchouk, M. E., & Marzouk, B. (2007). Changes on essential oil composition of coriander (*Coriandrum sativum* L.) fruits during three stages of maturity. *Food Chemistry*, 102(4), 1131–1134. https://doi.org/10.1016/j.foodchem.2006.06.046

Msaada, K., Hosni, K., Taarit, M. B., Hammami, M., & Marzouk, B. (2009). Effects of growing region and maturity stages on oil yield and fatty acid composition of coriander (*Coriandrum sativum* L.) fruit. *Scientia Horticulturae*, 120(4), 525–531. https://doi.org/10.1016/j.scienta.2008.11.033

Msaada, K., Jemia, M. B., Salem, N., Bachrouch, O., Sriti, J., Tammar, S., Bettaieb, I., Jabri, I., Kefi, S., & Limam, F. (2017). Antioxidant activity of methanolic extracts from three coriander (*Coriandrum sativum* L.) *Arabian Journal of Chemistry*, 10(2), CS3176–CS3183. https://doi.org/10.1016/j.arabjc.2013 .12.011

Nadeem, M., Muhammad Anjum, F., Issa Khan, M., Tehseen, S., El-Ghorab, A., & Iqbal Sultan, J. (2013). Nutritional and medicinal aspects of coriander (*Coriandrum sativum* L.): A review. *British Food Journal*, 115(5), 743–755. https://doi.org/10.1108/00070701311331526

Nejad Ebrahimi, S., Hadian, J., & Ranjbar, H. (2010). Essential oil compositions of different accessions of *Coriandrum sativum* L. from Iran. *Natural Product Research*, 24(14), 1287–1294. https://doi.org/10 .1080/14786410903132316

Nguyen, Q. H., Talou, T., Evon, P., Cerny, M., & Merah, O. (2020). Fatty acid composition and oil content during coriander fruit development. *Food Chemistry*, *326*, 127034. https://doi.org/10.1016/j.foodchem .2020.127034

Ogita, S. (2015). Plant cell, tissue and organ culture: The most flexible foundations for plant metabolic engineering applications. *Natural Product Communications*, 10(5), 815–820. https://doi.org/10.1177 /1934578X1501000527

Pacholczyk-Sienicka, B., Ciepielowski, G., & Albrecht, Ł. (2021). The application of NMR spectroscopy and chemometrics in authentication of spices. *Molecules*, 26(2), 382. https://doi.org/10.3390/ molecules26020382

Palmieri, S., Pellegrini, M., Ricci, A., Compagnone, D., & Lo Sterzo, C. (2020). Chemical composition and antioxidant activity of thyme, hemp and coriander extracts: A comparison study of maceration, soxhlet, UAE and RSLDE techniques. *Foods (Basel, Switzerland)*, 9(9), 1221. https://doi.org/10.3390/ foods9091221

Peter, K. V. (2004). *Handbook of Herbs and Spices, Vol. 2*. Abnigton Hall, England: Woodhead Publishing Ltd., pp. 158–174.

Prachayasittikul, V., Prachayasittikul, S., Ruchirawat, S., & Prachayasittikul, V. (2018). Coriander (*Coriandrum sativum*): A promising functional food toward the well-being. *Food Research International*, *105*, 305– 323. https://doi.org/10.1016/j.foodres.2017.11.019

Raal, A., Arak, E., & Orav, A. (2004). Chemical composition of coriander seed essential oil and their conformity with EP standards. *AGRAARTEADUS: Journal of Agricultural Science*, *15*, 234–239.

Ramadan, M., & Mörsel, J. T. (2002). Oil composition of coriander (*Coriandrum sativum* L.) fruit-seeds. *European Food Research and Technology*, 215(3), 204–209. https://doi.org/10.1007/s00217-002-0537-7

Ramadan, M., & Mörsel, J. T. (2003). Analysis of glycolipids from black cumin (*Nigella sativa* L.), coriander (*Coriandrum sativum* L.) and niger (*Guitoza abyssinica* Cass.) oil seeds. *Food Chemistry*, 80(2), 197–204. https://doi.org/10.1016/S0308-8146(02)00254-6

Ravi, R., Prakash, M., & Bhat, K. (2007). Aroma characterization of coriander (*Coriandrum sativum* L.) oil samples. *European Food Research and Technology*, *225*, 367–374. https://doi.org/10.1007/s00217-006 -0425-7

Sahib, N. G., Anwar, F., Gilani, A. H., Hamid, A. A., Saari, N., & Alkharfy, K. M. (2013). Coriander (*Coriandrum sativum* L.): A potential source of high-value components for functional foods and nutraceuticals-a review. *Phytotherapy Research: PTR*, 27(10), 1439–1456. https://doi.org/10.1002/ptr.4897

Sairam, T. V. (1998). *Home Remedies: A Handbook of Herbal Cures for Common Ailments*. New Delhi: Penguin Books India, p. 75.

Shan, B., Cai, Y. Z., Sun, M., & Corke, H. (2005). Antioxidant capacity of 26 spice extracts and characterization of their phenolic constituents. *Journal of Agricultural and Food Chemistry*, 53(20), 7749–7759. https://doi.org/10.1021/jf051513y

Shahwar, M. K., El-Ghorab, A. H., Anjum, F. M., Butt, M. S., Hussain, S., & Nadeem, M. (2012). Characterization of coriander (*Coriandrum sativum* L.) seeds and leaves: Volatile and non volatile extracts. *International Journal of Food Properties*, 15(4), 736–747. https://doi.org/10.1080/10942912.2010.500068

Sharanagat, V. S., & Goswami, T. K. (2014). Effect of moisture content on physio-mechanical properties of coriander seeds (*Coriandrum sativum*). *Agricultural Engineering International: CIGR Journal*, 16(3), 166–172.

Singletary, K. (2016). Coriander: Overview of potential health benefits. *Nutrition Today*, 51(3), 151–161. https://doi.org/10.1097/NT.0000000000000159

Smallfield, B. M., van Klink, J. W., Perry, N. B., & Dodds, K. G. (2001). Coriander spice oil: Effects of fruit crushing and distillation time on yield and composition. *Journal of Agricultural and Food Chemistry*, 49(1), 118–123. https://doi.org/10.1021/jf001024s

Sriti, J., Talou, T., Wannes, W. A., Cerny, M., & Marzouk, B. (2009). Essential oil, fatty acid and sterol composition of Tunisian coriander fruit different parts. *Journal of the Science of Food and Agriculture*, 89(10), 1659–1664. https://doi.org/10.1002/jsfa.3637

Sriti, J., Wannes, W. A., Talou, T., Mhamdi, B., Cerny, M., & Marzouk, B. (2010b). Lipid profiles of Tunisian coriander (*Coriandrum sativum*) seed. *Journal of the American Oil Chemists' Society*, 87(4), 395–400. https://doi.org/10.1007/s11746-009-1505-1

Sriti, J., Wannes, W. A., Talou, T., Mhamdi, B., Hamdaoui, G., & Marzouk, B. (2010a). Lipid, fatty acid and tocol distribution of coriander fruit's different parts. *Industrial Crops and Products*, *31*(2), 294–300. https://doi.org/10.1016/j.indcrop.2009.11.006

Steingroewer, J., et al. (2018). Monitoring of plant cells and tissues in bioprocesses. In: Pavlov, A., & Bley, T., eds. *Bioprocessing of Plant in Vitro Systems. Reference Series in Phytochemistry*. Cham: Springer. https://doi.org/10.1007/978-3-319-54600-1_7

Sulaiman, S. F., Sajak, A. A. B., Ooi, K. L., & Seow, E. M. (2011). Effect of solvents in extracting polyphenols and antioxidants of selected raw vegetables. *Journal of Food Composition and Analysis*, *24*(4–5), 506–515. https://doi.org/10.1016/j.jfca.2011.01.020

Taniguchi, M., Yanai, M., Xiao, Y. Q., Kido, T., & Baba, K. (1996). Three isocoumarines from *Coriandrum sativum*. *Phytochemistry*, *42*(3), 843–846. https://doi.org/10.1016/0031-9422(95)00930-2

Uitterhaegen, E., Sampaio, K. A., Delbeke, E. I., De Greyt, W., Cerny, M., Evon, P., Merah, O., Talou, T., & Stevens, C. V. (2016). Characterization of French coriander oil as source of petroselinic acid. *Molecules (Basel, Switzerland)*, *21*(9), 1202. https://doi.org/10.3390/molecules21091202

USDA National Nutrient Database for Standard Reference Release 26 Full Report (All Nutrients) Nutrient data for 2013, Spices, coriander seed.

Verma, A., Parihar, R., & Baishya, B. (2019). Identification of metabolites in coriander seeds (*Coriandrum Sativum* L.) aided by ultrahigh resolution total correlation NMR spectroscopy. *Magnetic Resonance in Chemistry: MRC*, *57*(6), 304–316. https://doi.org/10.1002/mrc.4850

Wangensteen, H., Samuelsen, A. B., & Malterud, K. E. (2004). Antioxidant activity in extracts from coriander. *Food Chemistry*, *88*(2), 293–297. https://doi.org/10.1016/j.foodchem.2004.01.047

Yildiz, H. (2016). Chemical composition, antimicrobial, and antioxidant activities of essential oil and ethanol extract of *Coriandrum sativum* L. leaves from Turkey. *International Journal of Food Properties*, *19*(7), 1593–1603. https://doi.org/10.1080/10942912.2015.1092161

Zardini, H. Z., Tolueinia, B., Momeni, Z., Hasani, Z., & Hasani, M. (2012). Analysis of antibacterial and antifungal activity of crude extracts from seeds of *Coriandrum sativum*. *Gomal Journal of Medical Sciences*, *10*(2), 167–171.

Zeb, F., Safdar, M., Fatima, S., Khan, S., Alam, S., Muhammad, M., Syed, A., Habib, F., & Shakoor, H. (2018). Supplementation of garlic and coriander seed powder: Impact on body mass index, lipid profile and blood pressure of hyperlipidemic patients. *Pakistan Journal of Pharmaceutical Sciences*, *31*(5), 1935–1941.

Zeković, Z., Bera, O., Đurović, S., & Pavlić, B. (2017). Supercritical fluid extraction of coriander seeds: Kinetics modelling and ANN optimization. *The Journal of Supercritical Fluids*, *125*, 88–95. https://doi.org/10.1016/j.supflu.2017.02.006

9 Phytochemicals and Secondary Metabolites of Coriander

Jacek Łyczko

CONTENTS

9.1 INTRODUCTION

The phytochemical composition and secondary metabolites of *Coriandrum sativum* L. (coriander), a widely known and utilized plant, have been analyzed numerous times. In addition, research has been conducted regarding its pharmacological and overall bioactive usefulness and in light of its excellent seasoning properties. This perspective makes coriander a fascinating medicinal and aromatic plant, wherein phytochemicals are considered antibacterial, antioxidant, anticancer, anti-inflammatory, and generally health-promoting agents. Furthermore, the distinct coriander aroma, both in the case of the seeds (dried fruits) and green parts (cilantro), has impacted various cuisines from Mexican, through African, up to Indian ones.

In coriander, the features responsible are high essential oil content (seeds), rich phenolics profile (cilantro), and lipids content (seeds). Additionally, the distribution of particular phytochemicals and secondary metabolites means that the discussion on coriander constituents should always be divided into two objects: seeds and cilantro. Moreover, coriander is an excellent example of a medicinal and aromatic plant, which presents a different chemical composition in this classification. Finally, before the detailed discussion about coriander chemical constituents, it has to be underlined that this composition is strongly dependent on the plant varieties, region and meteorological conditions of cultivation, time of harvest, and post-harvest treatment. In terms of varieties, the two most popular ones are *C. sativum* L. var. *vulgare alef* and *C. sativum* L. var. *microcarpum* DC. Other mentioned features, which strongly influence the plant phytochemical composition, are more unstable, and their influence on composition should be evaluated in each case.

DOI: 10.1201/9781003204626-10

9.2   CORIANDER SEEDS

9.2.1   ESSENTIAL OIL

Coriander seeds are the main source of coriander essential oil. Depending on the origin, the essential oil yield varies from approximately 0.2%–1.5% v/w (Anwar et al. 2011; Mandal and Mandal 2015) up to even 5.2% v/w (Orav, Arak, and Raal 2011). Additionally, the *microcarpum* variety has been recognized as having slightly higher essential oil content than the *vulgare* variety (Wei et al. 2019; Laribi et al. 2015).

The chemical components of coriander seed essential oil represent multiple chemical groups: alcohols, aldehydes, ketones, esters, and hydrocarbons (terpenes). The most characteristic for coriander seeds group are alcohols, with the most characteristic representant, linalool, followed by geraniol and terpinen-4-ol (Figure 9.1). Also, terpenes should be pointed out as a group with characteristic representatives: γ-terpinene, *p*-cymene, limonene, camphene, myrcene, and α-pinene (Mandal and Mandal 2015).

As mentioned before, it is hard to specify the content of particular constituents of essential oils regarding the unstable cultivation conditions. Nevertheless, the European Pharmacopeia indicates the limits of the most required phytochemicals (Table 9.1). Furthermore, the broad review of scientific reports regarding coriander seeds essential oils shows that usually, linalool content varies between 37.6% and 87.5% (again with significantly higher content for the *microcarpum* variety than *vulgare*, of 63.5–71.0% and 42.1–52.7%, respectively), γ-terpinene 0.1–13.6%, geraniol 1.2–4.6%, and geranyl acetate 0.1–4.7% (Laribi et al. 2015; Mandal and Mandal 2015). To underline the volatility of the share of phytochemicals, other studies may be mentioned where geranyl acetate content reached 15.9% and as major constituents were given β-caryophyllene (3.26%), camphor (3.02–6.33%), and α-pinene (2.16%) (Khani and Rahdari 2012; Gębarowska

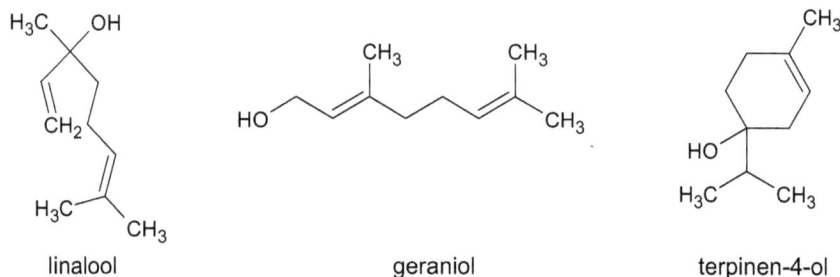

FIGURE 9.1   Structures of characteristics for coriander seeds essential oil alcohols.

TABLE 9.1
Limits of Characteristic Constituents of Coriander Seed Essential Oil

Constituent	European Pharmacopeia (%)
α-Pinene	3.0–7.0
Limonene	1.5–5.0
γ-Terpinene	1.5–8.0
p-Cymene	0.5–4.0
Camphor	3.0–6.0
Linalool	65.0–78.0
α-Terpineol	0.1–1.5
Geranyl acetate	0.5–4.0
Geraniol	0.5–3.0

TABLE 9.2

The Percentage of Major Constituents of Coriander Seed Essential Oil Regarding the Origin of Coriander

Constituent	Algeria (Zoubiri and Baaliouamer 2010)	India (Padalia et al. 2011)	Iran (Khani and Rahdari 2012)	Pakistan (Anwar et al. 2011)	Poland (Gębarowska et al. 2019)
	(%)				
α-Pinene	3.41	2.87	0.12	1.63	2.16
Limonene	1.23	2.55	0.62	0.26	1.18
γ-Terpinene	-	0.95	0.11	4.17	3.72
p-Cymene	1.76	0.83	2.52	1.12	0.54
Camphor	1.85	0.79	3.02	0.38	6.33
Linalool	73.1	55.4	57.5	69.6	77.5
α-Terpineol	0.20	0.12	0.18	-	0.60
Geranyl acetate	-	2.18	15.0	4.99	0.69
Geraniol	-	1.11	0.24	-	4.43

et al. 2019). An overview of major coriander seeds' essential oil constituents in various scientific sources is presented in Table 9.2.

Due to the dominant linalool content in coriander seed essential oil, its case should be more deeply discussed. Overall, naturally, linalool occurs as a mixture of two enantiomers: S-(+)-isomer and R-(-)-isomer (Figure 9.2). S-(+)-linalool is even commonly named coriandrol. Nevertheless, this should not suggest that linalool present in coriander seeds essential oil is pure S-(+)-isomer, although it is in the majority. The enantioselective gas chromatography analyses show that the S-(+)-isomer and R-(-)-isomer ratio is approximately 4:1 or higher (Gaydou and Randriamiharisoa 1987; Oliver 2003; Gębarowska et al. 2019).

9.2.1.1 Odor-Active Compounds

As for other medicinal and aromatic plants, just a part of volatile constituents is responsible for the characteristic fragrance of the plant. Furthermore, no percentage of particular compounds can be given in this subsection and the corresponding subsection regarding cilantro odor-active compounds. The reason for that is a well-known fact that for the characteristic fragrance of medicinal and aromatic plants, it is not necessarily the constituent with the highest abundance that is responsible. Commonly, compounds with a very low share may have the most significant impact on the distinct odor of the sample (Eyres et al. 2005; Cerutti-Delasalle et al. 2016).

The coriander seed essential oil is characterized as floral, herbal, green, spicy, sweet, rose-like, pleasant, cooling, earthy. Amongst the most abundant coriander seed essential oil constituents, more than half have a significant impact on its fragrance, namely α-pinene, linalool, α-terpineol, geraniol, and geranyl

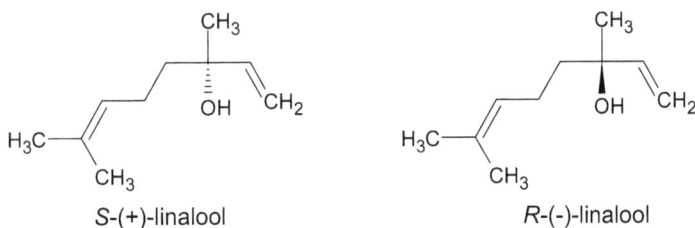

S-(+)-linalool   R-(-)-linalool

FIGURE 9.2 Isomers of linalool.

TABLE 9.3

Odor Descriptors of Odor-Active Compounds Present in Coriander Seed Essential Oil

Odor-Active Compound	Input in Coriander Seed Essential Oil Fragrance (Ravi, Prakash, and Bhat 2007)	Aroma Description Given in *The Good Scent Company* Website
Citronellol	Strong floral, rose, sweet like	Floral, leather, waxy, rose, bud, citrus
Cuminal	Spicy, harsh	Spicy, cumin, green, herbal
Geraniol	Fresh, sweet, rose-like	Sweet, floral, fruity, rose, waxy, citrus
Geranyl acetate	Pleasant, floral rose, herbal	Floral, rose, lavender, green, waxy
Linalool	Floral, grassy, pleasant, citrussy	Citrus, floral, sweet, bois de rose, woody, green, blueberry
α-Pinene	Woody, spicy, oily	Woody, pine, terpenic, cooling, camphoreous, fresh, herbal
α-Terpineol	Sweet, lilac odor	Pine, woody, resinous, cooling, lemon, lime, citrus, floral

acetate. Additionally, cuminal and citronellol are listed (Ravi, Prakash, and Bhat 2007). The aroma descriptors of coriander seeds' essential oil odor-active compounds are given in Table 9.3.

9.2.2 LIPIDS

The content of lipids in coriander is strongly dependent on the maturation stage (defined as the day after flowering) of coriander seeds. If the extraction of lipids is carried out with immature seeds, the lipid yield may be up to five times lower than the yields obtained from fully maturated seeds. Lipid content measured on the 5th, 7th, 9th, 16th, and 16th–20th days after flowering was determined as 2.70, 14.4, 15.4, 24.7, and 25.8%, respectively. What needs to be underlined is that the time to reach the fully maturated stage strongly depends on the geographical origin and variety of coriander (Msaada et al., 2009).

Coriander seed lipids are composed of various phytochemicals, namely neutral lipids (mono-acylglycerol – 0.57%, diacylglycerol – 1.88%, triacylglycerol – 95.5%) and free fatty acids, 2.05%. A second important group of which coriander seeds oil consists are sterols, with approximately 0.37–0.51% (w/w). Finally, the presence of phospholipids should be highlighted with major representatives, phosphatidylcholine and phosphatidylethanolamine, with an abundance of 35.9% and 33.8% of phospholipids, respectively. Further representatives of phospholipids are phosphatidic acid (8.11%) and phosphatidylglycerol (6.68%) (Laribi et al. 2015; Sriti et al. 2010).

The high health-promoting value of coriander seed oil is created particularly by fatty acids (both free and bounded into esters), especially unsaturated ones, and phytosterols. The most characteristic representatives of fatty acids are petroselinic acid (C18:1n-12), linoleic acid (C18:2n-6), oleic acid (C18:1n-9), palmitic acid (C16:0), and stearic acid (C18:0) (Table 9.4). Regarding the sterols content, the most abundant ones, unsurprisingly, are stigmasterol and β-sitosterol, with shares in total sterols content of 21.7–29.8% and 24.8–36.8%, respectively. Furthermore, there is some content of cholesterol, Δ^7-avenasterol, Δ^5-avenasterol, Δ^7-stugmasterol, Δ^5, 24-stigmastadienol, campestrol, ergosterol, and lanosterol (Ramadan and Mörsel 2002; Sriti et al. 2010; 2009; Laribi et al. 2015).

9.2.3 POLYPHENOLS

The total content of polyphenols and flavonoids in coriander seeds is extremely hard to define since the applied extraction methods (including mainly organic solvents extraction) may cause significant

TABLE 9.4
Profile of Fatty Acids Present in the Lipid Fraction of Different Coriander Seeds

Fatty Acid	(%)	
	(Ramadan and Mörsel 2002)	(Sriti et al. 2010)
C14:0	-	0.08
C16:0	3.96	3.50
C16:1n-7	0.41	0.23
C18:0	2.91	0.78
C18:1n-12	65.7	76.6
C18:1n-9	7.85	5.47
C18:2n-6	16.7	13.0
C18:3n-6	1.22	-
C20:0	0.25	0.10
C18:3n-3	0.20	0.15
C20:1n-9	0.30	-
C22:1n-9	0.16	-
C22:6n-3	0.34	-
Saturated fatty acids	-	4.46
Monounsaturated fatty acids	-	82.3
Polyunsaturated fatty acids	-	13.2

differences in the resulting extraction. Furthermore, the indistinct definition of polyphenols and inconsistencies in qualifying, for instance, flavonoids as overall polyphenols or separate groups also brings many misunderstandings. Nevertheless, for coriander, polyphenols and flavonoids are commonly given separately (Laribi et al. 2015; Msaada et al. 2017; Wei et al. 2019).

Most references provide only the overall content of polyphenols and flavonoids without identifying particular secondary metabolites. However, as pointed out above, the spectrum of data regarding polyphenols and flavonoids content is broad and approximately suggests that total polyphenols content (expressed as gallic acid equivalents (GAE) per 1 g) ranges between 1 mg GAE/g and 998 mg GAE/g (Zeković et al. 2016; Laribi et al. 2015). Furthermore, the polyphenols profile is also reported in a broad spectrum from a few up to more than 20 compounds. Commonly identified compounds are catechin, chlorogenic acid, ferulic acid, synaptic acid, and vanillic acid. At the same time, the highest abundance is related to catechin (12.6–92.0 mg/100g), coumaric acid (9.67–43.9 mg/100 g), daidzein (29.3–47.7 mg/100 g), sinapic acid (5.50–22.3 mg/100 g), chlorogenic acid (15.6 mg/100 g), *trans*-ferulic acid (2.93–57.6 mg/100 g), and 3,4-dimethoxycinnamic acid (117–942 mg/100 g) (Demir and Korukluoglu 2020; Zeković et al. 2016).

9.3 CILANTRO

9.3.1 Essential Oil

Even though cilantro is widely utilized in numerous cuisines as a seasoning with a strong, characteristic aroma, which may dominate the smell of food, its essential oil content is extremely low – approximately 0.12–0.15% v/w (Laribi et al. 2015; Padalia et al. 2011). Also, the constituents of cilantro essential oil, compared to coriander seed essential oil, are different. Aliphatic alcohols and aldehydes dominate the profile of volatile phytochemicals of cilantro (Łyczko et al., 2021; de Melo et al., 2019). The broadest group of constituents are (*E*)-alk-2-enals, namely (*E*)-dec-2-enal,

TABLE 9.5

The Percentage of Major Constituents of Cilantro Essential Oil Regarding the Origin of Coriander

Constituent	Bangladesh (Bhuiyan, Begum, and Sultana 2009)	India (Padalia et al. 2011)	Kenya (Laribi et al. 2015)	Pakistan (Shahwar et al. 2012)	Poland (Łyczko et al. 2021)
(*E*)-dec-2-enal	-	18.2	15.9	32.2	0.46
(*E*)-dodec-2-enal	1.3	8.72	-	7.51	4.18
(*E*)-tetradec-2-enal	-	-	-	6.56	0.30
(*E*)-tridec-2-enal	1.0	-	-	3.0	0.36
Dec-9-en-1-ol	-	11.6	-	-	0.33
Decanal	-	-	14.3	1.73	4.53
Decanol	-	5.77	13.6	2.18	4.76
Dodecanal	1.3	5.81	-	4.07	1.76
n-tetradecanol	-	6.09	-	-	-

(*E*)-dodec-2-enal, (*E*)-tridec-2-enal, and (*E*)-tetradec-2-enal. An overview of major cilantro essential oil constituents in various sources is presented in Table 9.5.

Regarding the most characteristic volatile constituent of coriander, linalool, it is not identified in most studies in the cilantro essential oils. Nevertheless, in some cases, the scientific reports point out that linalool is present in cilantro essential oil, even in considerable numbers (1.44–13.9%) (Shahwar et al. 2012; Łyczko et al. 2021). Additionally, some less common compounds like limonene, (Z)-hex-3-en-1-ol acetate, nonane, and 3-carene were reported in significant amounts – 2.43%, 34.5%, 9.28%, and 1.57%, respectively (Łyczko et al. 2021).

9.3.1.1 Odor-Active Compounds

For cilantro, the most characteristic phytochemical constituents, (*E*)-alk-2-enals and aliphatic alcohols, are responsible for the herb's characteristic aroma (Cadwallader et al. 1999; Eyres et al. 2005). Generally, the aroma of cilantro is determined as herbal, spicy, earthy, fresh, floral/fruity, and pleasant (Priyadarshi et al., 2016). In addition, citrusy and soapy notes may be pointed out (Cadwallader et al. 1999).

In light of the substantial similarity of odor-active compounds of coriander, it is unsurprising that those cluster in distinct odor groups. Therefore, the two most important groups are (Z)-dec-2-enal, which is overlooked in the majority of earlier cited studies, and a group consisting of (*E*)-dodec-2-enal, (*E*)-dodec-2-en-1-ol, and *n*-dodecanol. The aroma influences of those groups are coriander ID-aldehydic-pungent-spicy and coriander ID-floral-pungent fragrances, respectively. Furthermore, the next volatile phytochemical is often not detected in most studies – β-ionone (floral, rose/violet scents). Finally, (*E*)-dec-2-enal, with its aldehydic-fatty-pungent fragrance, has significant input into the cilantro aroma (Eyres et al. 2005). The second part of cilantro aroma is green-cut grass notes which is due to the presence of hex-3-en-1-ol, both (*E*)- and (*Z*)-isomers (Cadwallader et al. 1999; Eyres et al. 2005).

9.3.2 LIPIDS

Compared to coriander seeds, cilantro is a much less attractive source of lipid fraction. Nevertheless, the content of lipids, mainly fatty acids, is given in a few sources. The overall fatty acids content is determined as 4.18–6.12%, depending on the part of the plant. As the predominant constituent, polyunsaturated fatty acid α-linolenic (C18:3n-3) is highlighted with the percentage of the total lipid

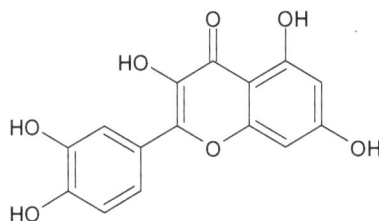

FIGURE 9.3 Quercetin structure.

fraction ranging between 39.4 and 41.1%. In terms of the abundance of fatty acids, were linoleic acid, heptadecanoic acid, and palmitic acid, while ones like oleic acid or stearic acid were found just in trace amounts (Prachayasittikul et al. 2018; Laribi et al. 2015).

9.3.3 POLYPHENOLS

As in previous cases, the determined content of total and particular polyphenols may differ depending on the applied extraction solvent, plant origin, or coriander variety. In terms of use for organic extraction solvent, the measured total phenolic content varies from 18.67 mg GAE/g (*n*-hexane extract) to 30.25 mg GAE/g (methanol extract) up to 449 mg GAE/g (ethanol extract) (Priyadarshi and Naidu 2019; Barros et al. 2012).

The profile of polyphenolic constituents of cilantro extract is more diversified than the profile of coriander seeds and consists of almost 60 compounds. Amongst polyphenols, the most numerous and the most abundant group is quercetin (Figure 9.3) and its derivatives (quercetin-3-rhamnoside, quercetin-3-*O*-rutinoside, quercetin-3-*O*-glucuronide, quercetin-3-*O*-glucoside), which may represent even 78.70% of cilantro's polyphenols (Barros et al. 2012). This group is followed by dimethoxycinnamoyl hexoside, kaempferol-3-*O*-rutinoside, and *p*-coumaroylquinic acid. Furthermore, the presence of vanillic acid, *p*-coumaric acid, ferulic acid (*cis*- and *trans*-), luteolin derivatives, apigenin derivatives, and some more has been reported (Barros et al. 2012; Nambiar, Daniel, and Guin 2010).

9.4 CONCLUSION

Coriandrum sativum, in terms of phytochemicals and secondary metabolites, is a highly complex object. Both coriander fruits and cilantro represent different chemical profiles regarding all major chemical groups: lipids, essential oils, and phenolics. The research conducted on this valuable medicinal and aromatic plant has delivered a large amount of data regarding its phytochemistry; nevertheless, interest in this topic is continuously growing. This chapter's brief review of the coriander phytochemicals and secondary metabolites, besides the major constituents of particular chemical groups, presents how the analytical approach used for the analysis may influence the obtained results. Therefore, a consistent and unambiguous report on the profiles and quantities of coriander phytochemicals is challenging. Nevertheless, the chapter presents the generally recognized major constituents of coriander divided into two objects: coriander seeds (fruits) and cilantro (green, aerial parts).

REFERENCES

Anwar, Farooq, Muhammad Sulman, Abdullah Ijaz Hussain, Nazamid Saari, Shahid Iqbal, and Umer Rashid. 2011. "Physicochemical Composition of Hydro-Distilled Essential Oil from Coriander (*Coriandrum sativum* L.) Seeds Cultivated in Pakistan." *Journal of Medicinal Plants Research* 5 (15): 3537–44. https://doi.org/10.5897/JMPR.9000978.

Barros, Lillian, Montserrat Dueñas, Maria Inês Dias, Maria João Sousa, Celestino Santos-Buelga, and Isabel C.F.R. Ferreira. 2012. "Phenolic Profiles of *in vivo* and *in vitro* Grown *Coriandrum sativum* L." *Food Chemistry* 132 (2): 841–48. https://doi.org/10.1016/j.foodchem.2011.11.048.

Bhuiyan, Nazrul Islam, Jaripa Begum, and Mahbuba Sultana. 2009. "Chemical Composition of Leaf and Seed Essential Oil of *Coriandrum sativum* L. from Bangladesh." *Bangladesh Journal of Pharmacology* 4: 150–53. https://doi.org/10.3329/bjp.v4i2.2800.

Cadwallader, Keith R., Ranee Surakarnkul, S.-P. Yang, and Thomas E. Webb. 1999. "Character-Impact Aroma Components of Coriander (*Coriandrum sativum* L.) Herb." In *Flavor Chemistry of Ethnic Foods*, edited by F. Shahidi, and C.-T. Ho, Boston: Springer: 77–84. https://doi.org/10.1007/978-1-4615-4783-9_7.

Cerutti-Delasalle, Céline, Mohamed Mehiri, Cecilia Cagliero, Patrizia Rubiolo, Carlo Bicchi, Uwe J. Meierhenrich, and Nicolas Baldovini. 2016. "The (+)-*cis*- and (+)-*trans*-Olibanic Acids: Key Odorants of Frankincense." *Angewandte Chemie* 128 (44): 13923–27. https://doi.org/10.1002/ange.201605242.

Demir, Sema, and Mihriban Korukluoglu. 2020. "A Comparative Study about Antioxidant Activity and Phenolic Composition of Cumin (*Cuminum cyminum* L.) and Coriander (*Coriandrum sativum* L.)." *Indian Journal of Traditional Knowledge* 19 (2): 383–93.

Eyres, Graham, Jean-Pierre Dufour, Gabrielle Hallifax, Subramaniam Sotheeswaran, and Philip J. Marriott. 2005. "Identification of Character-Impact Odorants in Coriander and Wild Coriander Leaves Using Gas Chromatography-Olfactometry (GCO) and Comprehensive Two-Dimensional Gas Chromatography-Time-of-Flight Mass Spectrometry (GCxGC-TOFMS)." *Journal of Separation Science* 28: 1061–74. https://doi.org/10.1002/jssc.200500012.

Gaydou, Emile M., and Robert P. Randriamiharisoa. 1987. "Gas Chromatography Diastereomer Separation of Linalool Derivatives: Application to the Determination of the Enantiomeric Purity of Linalool in Essential Oils." *Journal of Chromatography A* 396:378–381. https://doi.org/10.1016/S0021-9673(01)94078-6.

Gębarowska, Elżbieta, Maria Pytlarz-Kozicka, Joanna Nöfer, Jacek Łyczko, Maciej Adamski, and Antoni Szumny. 2019. "The Effect of *Trichoderma* spp. On the Composition of Volatile Secondary Metabolites and Biometric Parameters of Coriander (*Coriandrum sativum* L.)." *Journal of Food Quality* 2019: 1–7. https://doi.org/10.1155/2019/5687032

Khani, Abbas, and Tahere Rahdari. 2012. "Chemical Composition and Insecticidal Activity of Essential Oil from *Coriandrum sativum* Seeds against *Tribolium confusum* and *Callosobruchus Maculatus*." *ISRN Pharmaceutics* 2012: 1–5. https://doi.org/10.5402/2012/263517.

Laribi, Bochra, Karima Kouki, Mahmoud M'Hamdi, and Taoufik Bettaieb. 2015. "Coriander (*Coriandrum sativum* L.) and Its Bioactive Constituents." *Fitoterapia* 103: 9–26. https://doi.org/10.1016/j.fitote.2015.03.012.

Łyczko, Jacek, Klaudia Masztalerz, Leontina Lipan, Hubert Iwiński, Krzysztof Lech, Ángel A. Carbonell-Barrachina, and Antoni Szumny. 2021. "*Coriandrum sativum* L.-Effect of Multiple Drying Techniques on Volatile and Sensory Profile." *Foods* 10 (2): 403. https://doi.org/10.3390/foods10020403.

Mandal, Shyamapada, and Manisha Mandal. 2015. "Coriander (*Coriandrum sativum* L.) Essential Oil: Chemistry and Biological Activity." *Asian Pacific Journal of Tropical Biomedicine* 5 (6): 421–428. https://doi.org/10.1016/j.apjtb.2015.04.001.

Melo, Ana Cristina G.R. de, Maria Daniele V. dos Santos, Moisés Felix de Carvalho Neto, Jacqueline A. Takarashi, Vany P. Ferraz, Edvan A. Chagas, Pollyana C. Chagas, and Antonio A. de Melo Filho. 2019. "Phytochemical Trial and Bioactivity of the Essential Oil from Coriander Leaves (*Coriandrum sativum*) on Pathogenic Microorganisms." *Chemical Engineering Transactions* 75: 403–8. https://doi.org/10.3303/CET1975068.

Msaada, Kamel, Karim Hosni, Mouna Ben Taarit, Mohamed Hammami, and Brahim Marzouk. 2009. "Effects of Growing Region and Maturity Stages on Oil Yield and Fatty Acid Composition of Coriander (*Coriandrum sativum* L.) Fruit." *Scientia Horticulturae* 120 (4): 525–31. https://doi.org/10.1016/j.scienta.2008.11.033.

Msaada, Kamel, Mariem Ben Jemia, Nidhal Salem, Olfa Bachrouch, Jazia Sriti, Sonia Tammar, Iness Bettaieb, et al. 2017. "Antioxidant Activity of Methanolic Extracts from Three Coriander (*Coriandrum sativum* L.) Fruit Varieties." *Arabian Journal of Chemistry* 10: 3176–83. https://doi.org/10.1016/j.arabjc.2013.12.011.

Nambiar, Vanisha S, Mammen Daniel, and Parul Guin. 2010. "Characterization of Polyphenols from Coriander Leaves (*Coriandrum sativum*), Red Amaranthus (*A. paniculatus*) and Green Amaranthus (*A. frumentaceus*) Using Paper Chromatography: And Their Health Implications." *Journal of Herbal Medicine and Toxicology* 4 (1): 173–77.

Oliver, James E. 2003. "(S)(+)-Linalool from Oil of Coriander." *Journal of Essential Oil Research* 15 (1): 31–33. https://doi.org/10.1080/10412905.2003.9712256.

Orav, Anne, Elmar Arak, and Ain Raal. 2011. "Essential Oil Composition of *Coriandrum sativum* L. Fruits from Different Countries." *Journal of Essential Oil-Bearing Plants* 14 (1): 118–23. https://doi.org/10.1080/0972060X.2011.10643910.

Padalia, Rajendra Chandra, Neha Karki, Archana Negi Sah, and Ram Swaroop Verma. 2011. "Constituents of Leaf and Seed Essential Oil of *Coriandrum sativum* L." *Journal of Essential Oil Bearing Plants* 14 (5): 610–16. https://doi.org/10.1080/0972060X.2011.10643979.

Prachayasittikul, Veda, Supaluk Prachayasittikul, Somsak Ruchirawat, and Virapong Prachayasittikul. 2018. "Coriander (*Coriandrum sativum*): A Promising Functional Food toward the Well-Being." *Food Research International* 105: 305–23. https://doi.org/10.1016/j.foodres.2017.11.019.

Priyadarshi, Siddharth, Hafeeza Khanum, Ramasamy Ravi, Babasaheb Baskarrao Borse, and Madeneni Madhava Naidu. 2016. "Flavour Characterisation and Free Radical Scavenging Activity of Coriander (*Coriandrum sativum* L.) Foliage." *Journal of Food Science and Technology* 53 (3): 1670–78. https://doi .org/10.1007/s13197-015-2071-1.

Priyadarshi, Siddharth, and Madeneni Madhava Naidu. 2019. "A Comparative Study on Nutritional, Fatty Acids, Carotenoids, Aroma and Antioxidative Characteristics of *Microcarpum DC* and *Vulgare alef* Varieties of Coriander Foliage." *Indian Journal of Traditional Knowledge* 18 (3): 458–67.

Ramadan, Mohamed, and Jörg-Thomas Mörsel. 2002. "Oil Composition of Coriander (*Coriandrum sativum* L.) Fruit-Seeds." *European Food Research and Technology* 215: 204–9. https://doi.org/10.1007/s00217 -002-0537-7.

Ravi, Ramasamy, Maya Prakash, and K. Keshava Bhat. 2007. "Aroma Characterization of Coriander (*Coriandrum sativum* L.) Oil Samples." *European Food Research and Technology* 225: 367–74. https:// doi.org/10.1007/s00217-006-0425-7.

Shahwar, Muhammad Khuram, Ahmed Hassan El-Ghorab, Faqir Muhammad Anjum, Masood Sadiq Butt, Shahzad Hussain, and Muhammad Nadeem. 2012. "Characterization of Coriander (*Coriandrum sativum* L.) Seeds and Leaves: Volatile and Non Volatile Extracts." *International Journal of Food Properties* 15 (4): 736–47. https://doi.org/10.1080/10942912.2010.500068.

Sriti, Jazia, Thierry Talou, Wissem Aidi Wannes, Muriel Cerny, and Brahim Marzouk. 2009. "Essential Oil, Fatty Acid and Sterol Composition of Tunisian Coriander Fruit Different Parts." *Journal of the Science of Food and Agriculture* 89 (10): 1659–64. https://doi.org/10.1002/jsfa.3637.

Sriti, Jazia, Wissem Aidi Wannes, Thierry Talou, Baya Mhamdi, Muriel Cerny, and Brahim Marzouk. 2010. "Lipid Profiles of Tunisian Coriander (*Coriandrum sativum*) Seed." *Journal of the American Oil Chemists' Society* 87 (4): 395–400. https://doi.org/10.1007/s11746-009-1505-1.

Wei, Jing-Na, Zheng-Hui Liu, Yun-Ping Zhao, Lin-Lin Zhao, Tian-Kai Xue, and Qing-Kuo Lan. 2019. "Phytochemical and Bioactive Profile of *Coriandrum sativum* L." *Food Chemistry* 286: 260–67. https:// doi.org/10.1016/j.foodchem.2019.01.171.

Zeković, Zoran, Muammer Kaplan, Branimir Pavlić, Elmas Oktem Olgun, Jelena Vladić, Oltan Canli, and Senka Vidović. 2016. "Chemical Characterization of Polyphenols and Volatile Fraction of Coriander (*Coriandrum sativum* L.) Extracts Obtained by Subcritical Water Extraction." *Industrial Crops and Products* 87: 54–63. https://doi.org/10.1016/j.indcrop.2016.04.024.

Zoubiri, Safia, and Aoumeur Baaliouamer. 2010. "Essential Oil Composition of *Coriandrum sativum* Seed Cultivated in Algeria as Food Grains Protectant." *Food Chemistry* 122 (4): 1226–28. https://doi.org/10 .1016/j.foodchem.2010.03.119.

10 Antimicrobial Activity of Coriander

Hadia Shoaib, Hamide Filiz Ayyildiz,
Umaima Ismail, and Huseyin Kara

CONTENTS

10.1 INTRODUCTION

Microorganisms, including bacteria and fungi, are the leading causative agent in various human ailments [1]. In this regard, research on antimicrobial agents is a vast field in medical sciences. Furthermore, increasing antibiotic resistance raises the need for and significance of discovering new antimicrobial compounds. Plants have been employed as a traditional source of medicine for many years. Nowadays, the search for naturally occurring antioxidants and antimicrobial compounds is gaining importance because of their therapeutic potential; such compounds are associated with reduced side effects [2,3]. History reveals that plants have served as a great source of effective antimicrobial agents [4]. It is estimated that between 250,000 and 500,000 species exist on earth; some have contributed significantly to the field of medicine [5].

Researchers have been attracted to medicinal plants in recent years because most herbs and spices possess antioxidant activity, making them an essential constituent in nutraceuticals, pharmaceuticals, and food processing [2]. Coriander, an annual herb, belongs to Umbelliferae or Apiaceae, also known as cilantro, Arab parsley, or Chinese parsley. This plant is native to the European Mediterranean area. The Chinese initiated its cultivation in the 1st century BC, and now it is cultivated throughout the world [6] but expanded in central Europe and South Asian countries. Pakistan, India, Iran, Turkey, Canada, Romania, Poland, Morocco, Mexico, Argentina, Hungary, former members of the Soviet Union, Guatemala, the Czech Republic, and Slovakia are commercial coriander producers [7]. Coriander is reported as one of the oldest spices, used for more than 5000 years. Egyptian, Sanskrit, and Roman literature mentioned the use of coriander. Egyptians called this herb the Spice of Happiness [8]. This highly recognized medical tree is generally known as Dhanya.

Fatty acids, flavonoids, sterols, and EOs have been obtained from different parts of this herb. The leaves, seeds, flowers, and fruits of the plant are widely used and possess substantial activities including antimicrobial, antioxidant, antimutagenic, antidiabetic [9], hypolipidemic, diuretic, anticonvulsant, soporific-hypnotic [10], antifeeding, antiulcer, antiprotozoal, hepatoprotective, and anxiolytic activity. It also plays a defensive role in lead poisoning, heavy metal detoxification, and post-coital infections [11,12].

This herb is manipulated as a traditional medicine in many countries to treat digestive, urinary, and respiratory systems disorders; it is also employed in diabetes, inflammation, insomnia, anxiety, and convulsion [13,14]. Chinese medical theory considered coriander a warm-natured, acrid tasting herb used to treat measles with no eruption and cold without sweats, while its fruit is used in dyspepsia and lack of appetite. Various folk recipes have reported using Chinese parsley as a therapeutic agent in eczema, rectocele, and insufficient postpartum milk [15].

10.2 BOTANICAL DESCRIPTION OF CORIANDER

Only two known genera of this plant, including *C. sativum* L. and its wild relative *C. tordylium*, are cultivated worldwide. Leaves and seeds are the root cause of the cultivation of this plant. Coriander is a slender, branched, glabrous, soft, annual, and perennial herb. The height of this plant depends on the agro-climatic conditions ranging between 20 and 140 cm; following sowing, roots emerge upon the crop's maturation after 2–3 months. Fruits are threshed out after drying, winnowed, and stored in bags. The seeds of this plant have a citrus flavor because of the terpenes, linalool, and pinene in them. The color of the stem is hollow green and may turn red or violet during the flowering season. It is more or less erect, sympodial, monochasial-branched, and sometimes has several branches at the basal node. The flowers are small, shortly stalked umbels, with a pink and white color; the leaves are green with a shiny, waxy underside, of two types: the upper leaves are reduced to a small leaf sheath, and the lower ones are stalked [16]. The roots of this plant are spindle-shaped. The fruits are globular or ovate, straw yellow-colored, consisting of two pericarps, a crown-like part with five needle ridges at the top, and a diameter of 6 mm (Figure 10.1).

EOs of this plant lies in the longitudinal vitae responsible for its characteristic "bug" smell. The aromatic profile of different parts of this plant varies significantly upon maturation. Consequently,

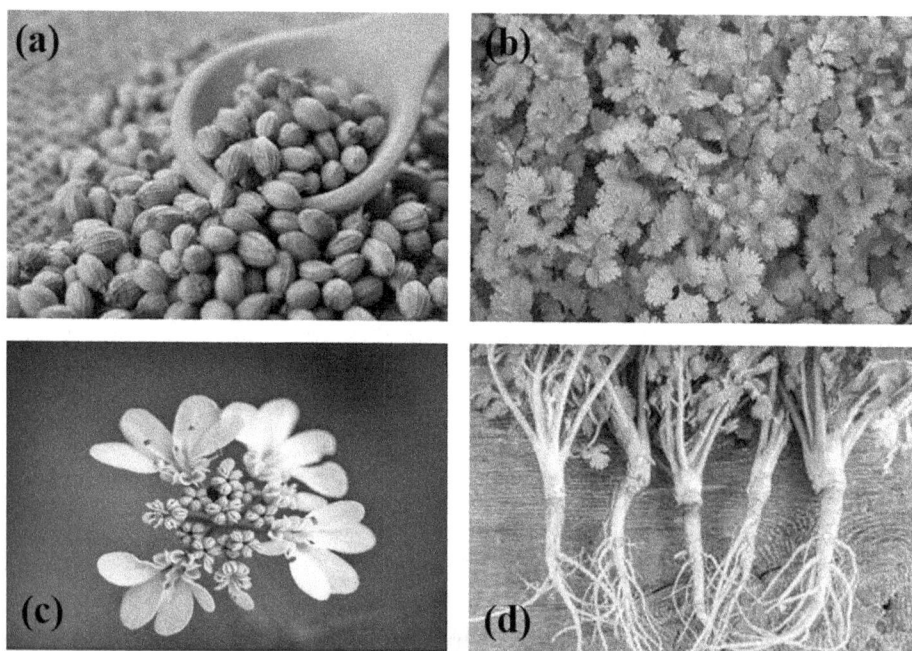

FIGURE 10.1   Optical photograph of different parts of the coriander plants: (a) fully grown seeds, (b) leaves, (c) flowers, and (d) roots [17].

the odor of ripe fruits is different from unripe seeds and green leaves. The name of this plant is derived from the Greek word "Korion" which means bug [18–20].

There are two varieties of this plant: *vulgare* and *microcarpum*. The leaves and fruits of *vulgare* are generally larger than *microcarpum*, and the yield of EOs of *vulgare* fruit (0.1–0.5%) is lower than the fruits of *microcarpum* (0.8–2.1%) [21]. Table 10.1 shows the classification and common names of *Coriander*.

This herbaceous plant is best grown between February and October. The early stages of growth require a cold climate, and warm weather in the later stages. Moist to moderately heavy soils with minimal or no irrigation is best for its growth. The coriander is a cross-pollinated crop. Grades of plant cross-pollination are reported as 50% [22] to 60% [23]. India is the main producer of coriander

TABLE 10.1
The Classification and Common Names of Coriander [25,26]

Kingdom	Plantae
Subkingdom	Trachiobionta
Superdivision	Spermatophyta
Division	Magnoliophyta
Class	Magnoliopsida
Subclass	Rosidae
Order	Apiales
Family	Apiaceae
Genus	*Coriandrum* L.
Species	*Coriandrum sativum* L.
Language	**Name**
Arabic	Kuzbura
Armenian	Chamem
Chinese	Yuan sui, hu sui
Czech	Koriandr
Danish	Coriander
Dutch	Coriander
English	Coriander, collender, Chinese parsley
Ethiopian (Amharic)	Dembilal
French	Coriandre, persilarabe
Georgian (Caucasus)	Kinza, kindza, kindz
German	Koriander, Wanzendill, Schwindelkorn
Greek	Koriannon, korion
Hindi	Dhania, dhanya
Hungarian	Coriander
Italian	Coriandolo
Japanese	Koendoro
Malay	Ketumbar
Persian	Geshnes
Polish	Kolendra
Rumanian	Coriándru
Portuguese	Coentro
Russian	Koriandr, koljandra, kinec, kinza, vonjueezel'e, klopovnik
Sanskrit	Dhanayaka, kusthumbari
Spanish	Coriandro, cilantro, cilandrio, cilantro
Turkish	Kisnis
Swiss	Chrapfechörnli, Böbberli, Rügelikümmi

seeds with the cultivation of 5.25×10^5 hectares, having an annual yield of 3.10×10^5 tonnes. The coriander oil market is controlled by Ukraine and India [24].

10.3 USES OF CORIANDER

The coriander plant is considered a significant and widely used medicinal plant. It is employed in treating various disorders in the folk medicine systems of different civilizations. Furthermore, this herb is widely employed in the food, health, cosmetics, soft drink, and chocolate industries [15].

All parts of this plant are edible and widely used as culinary ingredients. Fresh leaves and dried seeds are considered important contents of food. Its green foliage is enriched with proteins, carbohydrates, fiber, carotene, vitamin B, vitamin C, and mineral elements (including calcium, phosphorous, and iron). Coriander is employed as a vegetable in salads. When incorporated into food, the leaves and seeds act as a preservative and provide typical flavor because of the presence of EOs [27,28]. Its chemical constituents include linalool, pinene, cymene, geraniol, borneol, and phellandrene. However, the chemical constituents of EOs found in the leaves and seeds of this plant are very different from one another [29]. The seeds contain EOs, mainly linalool, while the leaves contain decanal, decanol, cyclodecane, and dodecane [30]. The seeds are the richest source of lipids as they are 28% of the total seed weight [31]. Ground coriander seeds are a major ingredient of curry powder in Indian food. In Thai and Vietnamese cuisine, fresh green leaves are widely used because of their unique aromatic odor [32].

This plant's traditional uses are galactagogue, antiseptic, carminative, and flavoring agent [33]. Ayurvedic literature reported a decoction of coriander seeds to reduce blood lipid levels. Regarding this context, research was conducted on triton-induced hyperlipidemic rats to evaluate coriander's therapeutic and prophylactic activity in hyperlipidemia. The reference used in this study was the commercially available herbal drug "liponil". Consequently, it has been found that coriander at a dose of 1g/kg of body weight reduced triglycerides and cholesterol levels in rats.

Coriander seeds are used to cure indigestion, cough, bronchitis, vomiting, diarrhea, dysentery, rheumatism, and pain in joints. It is also used as an antiedemic, anti-inflammatory, antiseptic, emmenagogue, antidiabetic, antihypertensive, lipolytic, and myorelaxant and also possesses nerve soothing properties [15] (Figure 10.2). In Moroccan and Palestinian pharmacopeia, coriander as a traditional diuretic and for the treatment of urinary tract infections has been mentioned [34–36].

The leaves of coriander showed more potent antioxidant activity than the seeds [37]. Another study reported that the aqueous extracts of seeds exhibited antioxidant activity both *in vitro* and *in vivo* [38]. Various methods evaluated fresh coriander juice's time- and dose-dependent *in vivo* antioxidant activity [39]. This spice reduced lipid peroxidation by 300–600%, increased the antioxidant enzyme activities (catalase by 57–75%, superoxide dismutase by 57–62%, and glutathione peroxidase by 80–83%), and reduced liver damage [8]. Naveen and Farhath [2] observed that coriander seed extract minimized drug-induced oxidative stress and protected the system against toxicity. The antioxidant property of coriander seed was related to the large amounts of tocopherols, carotenoids, and phospholipids, which acted through different mechanisms.

10.4 ANTIMICROBIAL ACTIVITIES OF CORIANDER

The quality and safety of prepared and processed food are crucial in the food industry. Deterioration of food quality is caused by microorganisms present in it, and the ingestion of such food leads to infections; that is why food manufacturers try their best to prepare food products free from microorganisms. About one-third of the world's food production is lost annually due to microbial spoilage of food [40].

The antimicrobial activity of medicinal plants and their extracts has been identified since antiquity. However, due to the increased resistance to pathogens and the need for new food preservatives, the potential use of essential oils as antimicrobial agents has been the subject of new investigations. For example, EOs obtained from the extracts of coriander seeds and leaves exhibit varying degrees of germicidal effects toward a range of pathogenic microorganisms [41]. Table 10.2 and Table 10.3

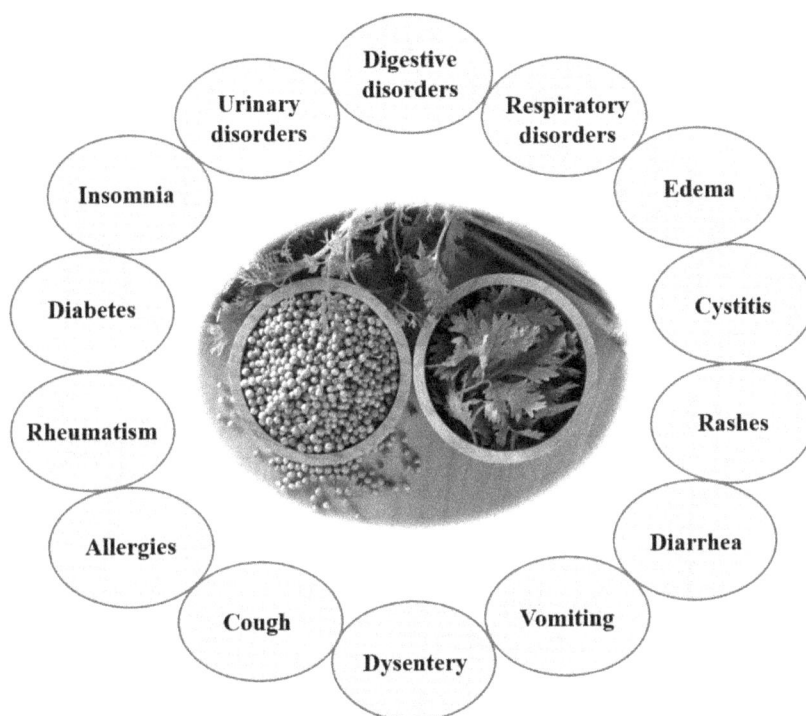

FIGURE 10.2 Therapeutic impact of coriander on human health.

represent a broad spectrum of antimicrobial activity of coriander against yeasts, molds, and Gram-positive and Gram-negative microorganisms.

Methods to evaluate antimicrobial susceptibility include diffusion, dilution, and bioautographic methods because such methods are semi-quantitative (diffusion) or quantitative (dilution) [42]. Disc or agar diffusion assay is used to evaluate the antimicrobial activity of coriander extracts in which the disc is directly impregnated with the oil and is kept on the agar surface; this method can also be performed by placing it in a vapor phase in which the disc is placed on the lid of the *Petri* dish. The second method proposed some advantages and it allows the diffusion mechanism to be inferred because of the volatility of EOs [43]. In order to determine the quantitative minimal inhibitory concentration (MIC) and minimum bactericidal/fungicidal (MBC/MFC), a broth or agar dilution method is employed. Researchers usually adopt the clinical and laboratory standards institute guidelines to determine antibacterial and antifungal MIC values. Various factors such as the type of EO/extract, the volume of inoculum, growth phase, pH of the media, incubation time, and incubation temperature can affect the observation of these tests; that is why there is no standard to determine the antimicrobial activity of coriander extracts [42].

Antibacterial and antifungal activity is correlated with the destructive effects of EO active constituents (aldehydes and alcohols) on the microbial biofilm and interference with the quorum-sensing (QS) system; this makes coriander EOs a synergistic drug or substitute for antibiotics [44,45]. Alcohol extracts of coriander, iron-chelating activity, and the hydrophobic character of phenolic compounds are responsible for bacterial cell damage [33].

10.4.1 ANTIBACTERIAL ACTIVITY OF CORIANDER

Various studies have reported the antibacterial activity of different parts of the coriander plant. EO of coriander, fruit possesses potent antibacterial activity towards several bacterial strains tested by

TABLE 10.2

Antibacterial Activity of Different Parts of Coriander with Minimal Inhibitory Concentrations (MIC), Minimal Bactericidal Concentrations (MBC) against Gram-Positive and Gram-Negative Bacteria

Region	Parts of Plant	Type of Bacteria	Microorganisms/ Species	Strain	Method	Incubation Temperature	Incubation Time	MIC	MBC	Reference
Germany	Coriander fruits EO	Gram-positive bacteria	*S. pyogenes*	Clinical isolate	Agar dilution method	37°C	24 h	0.3 mg/mL	1.1 mg/mL	[46]
Turkey	Coriander leaves			ATCC 176	Discdiffusion method	37°C	24 h	ND	ND	[49]
Germany	Coriander fruits EO		*S. viridans*	Clinical isolate	Agar dilution method	37°C	24 h	0.7 mg/mL	0.7 mg/mL	[46]
Germany	Coriander fruits EO		MSSA	Clinical isolate	Agar dilution method	37°C	24 h	2.2 mg/mL	9.0 mg/mL	[46]
Germany	Coriander fruits EO		MRSA	Clinical isolate	Agar dilution method	37°C	24 h	2.2 mg/mL	6.5 mg/mL	[46]
Germany	Coriander fruits EO		*E. faecalis*	Clinical isolate	Agar dilution method	37°C	24 h	4.4 mg/mL	27.9 mg/mL	[46]
Turkey	Coriander seeds EO			ATCC 29212	Broth microdilution method	37°C	24 h	1.562 µg/mL	ND	[50]
Germany	Coriander fruits EO		*E. faecium*	Clinical isolate	Agar dilution method	37°C	24 h	2.3 mg/mL	101.3 mg/mL	[46]
Turkey	Coriander seeds EO			Food isolate	Broth microdilution method	37°C	24 h	<0.195 µg/mL	-	[50]
Germany	Coriander fruits EO		VRE	Clinical isolate	Agar dilution method	37°C	24 h	2.2 mg/mL	73.0 mg/mL	[46]

(*Continued*)

TABLE 10.2 (CONTINUED)

Antibacterial Activity of Different Parts of Coriander with Minimal Inhibitory Concentrations (MIC), Minimal Bactericidal Concentrations (MBC) against Gram-Positive and Gram-Negative Bacteria

Region	Parts of Plant	Type of Bacteria	Microorganisms/ Species	Strain	Method	Incubation Temperature	Incubation Time	MIC	MBC	Reference
Kenya	Coriander leaves		S. aureus	ATCC 25923	Broth dilution method	37°C	24 h	108 mg/ mL	-	[51]
India	Coriander seed EO			Food isolate	Agar well diffusion method	30°C/37°C	24 h	0.16 mg/ mL	-	[52]
Portugal	Coriander EO			ATCC 25923	Broth microdilution method	37°C	24 h	0.2 %, v/v	1.6 %, v/v	[48]
Serbia	Commercial EO of coriander			ATCC 25923	Microdilution method	37°C	24 h	2.8 μl/ mL	11.4 μl/ mL	[53]
Turkey	Coriander seeds EO			ATCC 25923	Broth microdilution method	37°C	24 h	12.5 μg/ mL	-	[50]
India	Methanolic coriander seeds extract			MTCC 96	Agar well diffusion method	37°C	24 h	4.16 μg/ mL	-	[54]
Pakistan	Coriander seeds EO			-	-	-	-	0.3 %	-	[55]
Turkey	Coriander leaves			ATCC 29213	Discdiffusion method	37°C	24 h	-	-	[49]
Saudi Arabia	Coriander leaves EO			ATCC 26923	Agar well diffusion method	37°C	24 h	0.05 %	-	[56]

(Continued)

TABLE 10.2 (CONTINUED)

Antibacterial Activity of Different Parts of Coriander with Minimal Inhibitory Concentrations (MIC), Minimal Bactericidal Concentrations (MBC) against Gram-Positive and Gram-Negative Bacteria

Region	Parts of Plant	Type of Bacteria	Microorganisms/ Species	Strain	Method	Incubation Temperature	Incubation Time	MIC	MBC	Reference
Iran	Coriander leaves EO			ATCC 25913	Agar diffusion method	37°C	24 h	5 µg/mL	5 µg/ mL	[57]
Iran	Coriander seeds EO			ATCC 25913	Agar diffusion method	37°C	24 h	20 µg/ mL	40 µg/ mL	[57]
Turkey	Coriander seeds EO		*S. epidermidis*	DSMZ 20044	Broth microdilution method	37°C	24 h	3.125 µg/mL	-	[50]
Turkey	Coriander seeds EO		*E. aerogenes*	ATCC 13048	Broth microdilution method	37°C	24 h	3.125 µg/mL	-	[50]
Turkey	Coriander seeds EO		*E. durans*	Food isolate	Broth microdilution method	37°C	24 h	100 µg/ mL	-	[50]
Portugal	Coriander EO			ATCC 11778	Broth microdilution method	37 C	24 h	0.1 %, v/v	>3.2 %, v/v	[48]
Portugal	Coriander EO		*E. faecalis*	ATCC 29212	Broth microdilution method	37 C	24 h	0.8 %, v/v	>3.2 %, v/v	[48]
Turkey	Coriander leaves			ATCC 29122	Discdiffusion method	37°C	24 h	-	-	[49]
India	Coriander seeds EO		*L. monocytogenes*	Food isolate	Agar well diffusion method	30°C/37°C	24 h	0.20 mg/ mL	-	[52]
Turkey	Coriander leaves			BC 8353	Discdiffusion method	37°C	24 h	-	-	[49]

(Continued)

TABLE 10.2 (CONTINUED)
Antibacterial Activity of Different Parts of Coriander with Minimal Inhibitory Concentrations (MIC), Minimal Bactericidal Concentrations (MBC) against Gram-Positive and Gram-Negative Bacteria

Region	Parts of Plant	Type of Bacteria	Microorganisms/Species	Strain	Method	Incubation Temperature	Incubation Time	MIC	MBC	Reference
Turkey	Coriander seeds EO			Food isolate	Broth microdilution method	37°C	24 h	6.25	-	[50]
Turkey	Coriander seeds EO		L. innocua	Food isolate	Broth microdilution method	37°C	24 h	0.390 µg/mL	-	[50]
India	Coriander seeds EO		M. luteus	Food isolate	Agar well diffusion method	30°C/37°C	24 h	0.33 mg/mL	-	[52]
Kenya	Coriander leaves		B. subtilis	Clinical isolate	Broth dilution method	37°C	24 h	108 mg/mL	-	[51]
Turkey	Coriander seeds EO			DSMZ 1971	Broth microdilution method	37°C	24 h	<0.195 µg/mL	-	[50]
Saudi Arabia	Coriander leaves EO			ATCC 11774	Agar well diffusion method	37°C	24 h	-	-	[56]
Turkey	Coriander leaves			ATCC 6633	Discdiffusion method	37°C	24 h		-	[49]
India	Methanolic coriander seeds extract		B. pumilus	MTCC 7411	Agar well diffusion method	37°C	24 h	4.16 µg/mL	-	[54]
Pakistan	Coriander seeds EO		B. cerius			-	-	0.3 %	-	[55]
Turkey	Coriander leaves			BC 6830	Discdiffusion method	37°C	24 h		-	[49]
Kenya	Coriander leaves	Gram-negative bacteria	E. coli	ATCC 25922	Broth dilution method	37°C	24 h	163 mg/mL	-	[51]

(Continued)

TABLE 10.2 (CONTINUED)

Antibacterial Activity of Different Parts of Coriander with Minimal Inhibitory Concentrations (MIC), Minimal Bactericidal Concentrations (MBC) against Gram-Positive and Gram-Negative Bacteria

Region	Parts of Plant	Type of Bacteria	Microorganisms/Species	Strain	Method	Incubation Temperature	Incubation Time	MIC	MBC	Reference
Germany	Coriander fruits EO			Clinical isolate	Agar dilution method	37°C	24 h	2.3 mg/mL	2.3 mg/mL	[46]
India	Coriander seeds EO			Food isolate	Agar well diffusion method	30°C/37°C	24 h	0.14 mg/mL	-	[52]
Portugal	Coriander EO			ATCC 25922	Broth microdilution method	37°C	24 h	0.2 %, v/v	0.2 %, v/v	[48]
Turkey	Coriander seeds EO			ATCC 25922	Broth microdilution method	37°C	24 h	50 µg/mL	-	[50]
Serbia	Commercial EO of coriander			ATCC 25922	Microdilution method	37°C	24 h	2.8 µl/mL	2.8 µl/mL	[53]
India	Methanolic coriander seeds extract			MTCC 119	Agar well diffusion method	37°C	24 h	4.16 µg/mL	-	[54]
Pakistan	Coriander seeds EO			-	-	-	-	0.3 %	-	[55]
Turkey	Coriander leaves			1402	Discdiffusion method	37°C	24 h	-	-	[49]
Turkey	Coriander leaves			O157:H7	Discdiffusion method	37°C	24 h	-	-	[49]
Saudi Arabia	Coriander leaves EO			ATCC 11229	Agar well diffusion method	37°C	24 h	-	-	[56]

(Continued)

TABLE 10.2 (CONTINUED)

Antibacterial Activity of Different Parts of Coriander with Minimal Inhibitory Concentrations (MIC), Minimal Bactericidal Concentrations (MBC) against Gram-Positive and Gram-Negative Bacteria

Region	Parts of Plant	Type of Bacteria	Microorganisms/Species	Strain	Method	Incubation Temperature	Incubation Time	MIC	MBC	Reference
Iran	Coriander leaves EO		K. pneumoniae	ATCC 8739	Agar diffusion method	37°C	24 h	40 µg/mL	80 µg/mL	[57]
Iran	Coriander seeds EO			ATCC 8739	Agar diffusion method	37°C	24 h	160 µg/mL	160 µg/mL	[57]
Germany	Coriander fruits EO			Clinical isolate	Agar dilution method	37°C	24 h	2.3 mg/mL	6.3 mg/mL	[46]
Kenya	Coriander leaves			Clinical isolate	Broth dilution method	37°C	24 h	163 mg/mL	-	[51]
Saudi Arabia	Coriander leaves EO			NCTC 9633	Agar well diffusion method	37°C	24 h	0.0125 %	-	[56]
Turkey	Coriander leaves		K. pneumoniae ozaenae	BC 34	Discdiffusion method	37°C	24 h	-	-	[49]
Turkey	Coriander leaves			BC 30	Discdiffusion method	37°C	24 h	-	-	[49]
Turkey	Coriander leaves			BC 32	Discdiffusion method	37°C	24 h	-	-	[49]
Portugal	Coriander EO			ATCC 13883	Broth microdilution method	37 C	24 h	0.2 %, v/v	0.2 %, v/v	[48]
Turkey	Coriander seed EO			Food isolate	Broth microdilution method	37°C	24 h	0.390 µg/mL	-	[50]
India	Coriander seeds EO		S. typhimurium	MTCC 3224	Agar well diffusion method	30°C/37°C	24 h	0.19 mg/mL	-	[52]

(Continued)

TABLE 10.2 (CONTINUED)

Antibacterial Activity of Different Parts of Coriander with Minimal Inhibitory Concentrations (MIC), Minimal Bactericidal Concentrations (MBC) against Gram-Positive and Gram-Negative Bacteria

Region	Parts of Plant	Type of Bacteria	Microorganisms/ Species	Strain	Method	Incubation Temperature	Incubation Time	MIC	MBC	Reference
Kenya	Coriander leaves			Clinical isolate	Broth dilution method	37°C	24 h	130 mg/mL	-	[51]
Turkey	Coriander leaves			RSSK 95091	Discdiffusion method	37°C	24 h	-	-	2020
Portugal	Coriander EO			ATCC 13311	Broth microdilution method	37 C	24 h	0.4 %, v/v	0.8 %, v/v	[48]
Turkey	Coriander seeds EO			SL 1344	Broth microdilution method	37°C	24 h	<0.195 µg/mL	-	[50]
Turkey	Coriander seeds EO		*S. enteritidis*	ATCC 13075	Broth microdilution method	37°C	24 h	<0.195 µg/mL	-	[50]
Turkey	Coriander seeds EO		*S. infantis*	Food isolate	Broth microdilution method	37°C	24 h	<0.195 µg/mL	-	[50]
Turkey	Coriander seeds EO		*S. kentucky*	Food isolate	Broth microdilution method	37°C	24 h	<0.195 µg/mL	-	[50]
Iran	Coriander leaves EO		*S. enterica*	PTCC 1709	Agar diffusion method	37°C	24 h	80 µg/mL	80 µg/mL	[57]
Iran	Coriander seeds EO			PTCC 1709	Agar diffusion method	37°C	24 h	160 µg/mL	320 µg/mL	[57]
Portugal	Coriander EO		*P. aeruginosa*	ATCC 27853	Broth microdilution method	37 C	24 h	1.6 %, v/v	1.6 %, v/v	[48]

(Continued)

TABLE 10.2 (CONTINUED)
Antibacterial Activity of Different Parts of Coriander with Minimal Inhibitory Concentrations (MIC), Minimal Bactericidal Concentrations (MBC) against Gram-Positive and Gram-Negative Bacteria

Region	Parts of Plant	Type of Bacteria	Microorganisms/Species	Strain	Method	Incubation Temperature	Incubation Time	MIC	MBC	Reference
Kenya	Coriander leaves			ATCC 27853	Broth dilution method	37°C	24 h	0 mg/mL	-	[51]
Turkey	Coriander leaves			BC 4372	Discdiffusion method	37°C	24 h	-	-	[49]
India	Methanolic coriander seed extract			MTCC 741	Agar well diffusion method	37°C	24 h	4.16 µg/mL	-	[54]
Turkey	Coriander seeds EO			DSMZ 50071	Broth microdilution method	37°C	24 h	0.390 µg/mL	-	[50]
Turkey	Coriander seeds EO		P. fluorescence	Food isolate	Broth microdilution method	37°C	24 h	3.125 µg/mL	-	[50]
Turkey	Coriander leaves			BC 7324	Discdiffusion method	37°C	24 h		-	[49]
Turkey	Coriander leaves		P. pseudoalcaligenes	BC 3445	Discdiffusion method	37°C	24 h		-	[49]
Serbia	Commercial EO of coriander		P. mirabilis	Clinical isolate	Microdilution method	37°C	24 h	11.4 µl/mL	45.4 µl/mL	[53]
Kenya	Coriander leaves			Clinical isolate	Broth dilution method	37°C	24 h	217 mg/mL	-	[51]
Turkey	Coriander leaves			BC 2644	Discdiffusion method	37°C	24 h	-	-	[49]
Turkey	Coriander leaves		A. lwoffi	BC 2819	Discdiffusion method	37°C	24 h	-	-	[49]

(Continued)

TABLE 10.2 (CONTINUED)

Antibacterial Activity of Different Parts of Coriander with Minimal Inhibitory Concentrations (MIC), Minimal Bactericidal Concentrations (MBC) against Gram-Positive and Gram-Negative Bacteria

Region	Parts of Plant	Type of Bacteria	Microorganisms/ Species	Strain	Method	Incubation Temperature	Incubation Time	MIC	MBC	Reference
Turkey	Coriander leaves		*A. piechaudii*	BC 0236	Discdiffusion method	37°C	24 h	-	-	[49]
Turkey	Coriander leaves		*F. indologenes*	BC 1520	Discdiffusion method	37°C	24 h	-	-	[49]
Turkey	Coriander leaves		*Y. enterocolitica*	Clinical isolate	Discdiffusion method	37°C	24 h	-	-	[49]
Iran	Coriander leaves EO			PTCC 1477	Agar diffusion method	37°C	24 h	2.5 µg/ mL	10 µg/ mL	[57]
Iran	Coriander seed EO			PTCC 1477	Agar diffusion method	37°C	24 h	80 µg/ mL	320 µg/ mL	[57]
Iran	Coriander leaves EO		*V. cholerae*	PTCC 1611	Agar diffusion method	37°C	24 h	5 µg/mL	10 µg/ mL	[57]
Iran	Coriander seed EO			PTCC 1611	Agar diffusion method	37°C	24 h	20 µg/ mL	80 µg/ mL	[57]

*(-) = not mentioned

TABLE 10.3

Antifungal Activity of Different Parts of Coriander with Minimal Inhibitory Concentrations (MIC), Minimal Fungal Concentrations (MFC) against Molds and Yeasts

Region	Parts of Plant	Type of Fungus	Microorganism/ Species	Strain	Method	Incubation Temperature	Incubation Time	MIC	MFC	Reference
Pakistan	Coriander seeds EO		Aspergillus niger	-	Disc diffusion method	37°C	-	0.3 %	-	[55]
Turkey	Coriander leaves	Molds		BC 102	-		24 h	ND	-	[49]
Tunisia	Coriander seeds EO		Aspergillus flavus	Food isolate	-	-	-	0.7%	-	[59]
Pakistan	Coriander seeds EO			-	-		-	0.3 %	-	[55]
Pakistan	Coriander seeds EO		Aspergillus parasiticus	-	-		-	0.3 %	-	[55]
Pakistan	Coriander seeds EO		Aspergillus fumigatus	-	-		-	0.3 %	-	[55]
Pakistan	Coriander seeds EO		Penicillium digiatum	-	-		-	0.3 %	-	[55]
Turkey	Coriander leaves		Paecilomyces varioti	108	Disc diffusion method	37°C	24 h	-	-	[49]
Turkey	Coriander leaves		Penicillium brevicompactum	BC 109	Disc diffusion method	37°C	24 h	-	-	[49]
Turkey	Coriander leaves		Penicillium roquefortii	BC 111	Disc diffusion method	37°C	24 h	-	-	[49]
Turkey	Coriander leaves		Trichothecium roseum	BC 116	Disc diffusion method	37°C	24 h	-	-	[49]
Turkey	Coriander leaves		Cladosporium herbarum	BC 106	Disc diffusion method	37°C	24 h	-	-	[49]
Kenya	Coriander leaves	Yeasts	Candida albicans	Clinical isolate	Broth dilution method	37°C	24 h	163 mg/ mL	-	[51]
Turkey	Coriander seeds EO			DSMZ 1386	Broth microdilution method	37°C	24 h	<0.195 µg/mL	-	[50]
Saudi Arabia	Coriander leaves EO			ATCC 10231	Agar well diffusion method	37°C	24 h	0.0125 %	-	[56]
Serbia	Commercial EO of coriander			ATCC 10231	Microdilution method	37°C	24 h	>0.45 µl/ mL	>0.45 µl/ mL	[53]

(Continued)

TABLE 10.3 (CONTINUED)
Antifungal Activity of Different Parts of Coriander with Minimal Inhibitory Concentrations (MIC), Minimal Fungal Concentrations (MFC) against Molds and Yeasts

Region	Parts of Plant	Type of Fungus	Microorganism/ Species	Strain	Method	Incubation Temperature	Incubation Time	MIC	MFC	Reference
Turkey	Coriander leaves			ATCC 1223	Disc diffusion method	37°C	24 h	-	-	[49]
Turkey	Coriander leaves			ATCC 90029	Discdiffusion method	37°C	24 h	-	-	[49]
Brazil	Coriander EO			CBS 562	Agar diffusion method	35°C	24–48 h	15.6 µg/mL	31.2 µg/mL	[60]
Brazil	Coriander EO	Candida tropicalis		CBS 94	Agar diffusion method	35°C	24–48 h	31.2 µg/mL	62.5 µg/mL	[60]
Brazil	Coriander EO	Candida krusei		CBS 573	Agar diffusion method	35°C	24–48 h	15.6 µg/mL	31.2 µg/mL	[60]
Brazil	Coriander EO	Candida dubliniensis		CBS 7987	Agar diffusion method	35°C	24–48 h	31.2 µg/mL	62.5 µg/mL	[60]
Brazil	Coriander EO	Candida rugosa		CBS 12	Agar diffusion method	35°C	24–48 h	15.6 µg/mL	31.2 µg/mL	[60]
Pakistan	Coriander seeds EO	Candidalipolytica		-	-	-	-	0.3 %	-	[55]
Turkey	Coriander leaves	Candida crusei		ATCC 14243	Disc diffusion method	37°C	24 h	-	-	[49]
Pakistan	Coriander seeds EO	Saccharomyces cerevisiae		-	-	-	-	0.3 %	-	[55]

*(-) = not mentioned

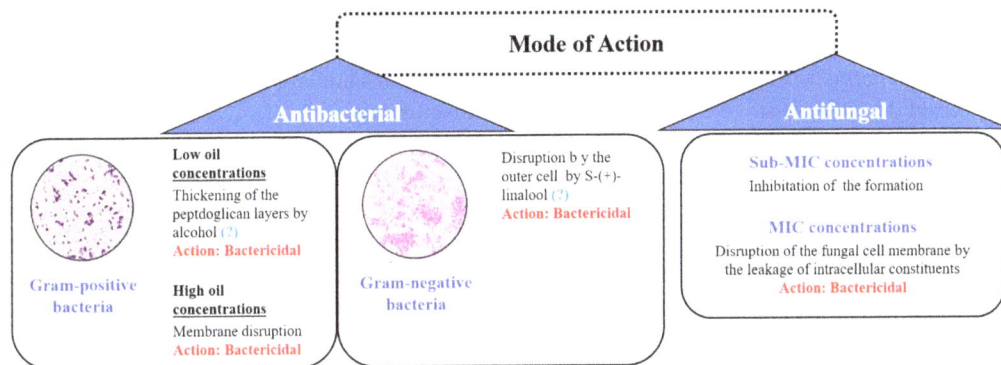

FIGURE 10.3 Coriander oil mechanism of action on bacterial and fungal cells [58].

Casetti et al. [46], including *S. aureus*, MSSA and MRSA, *S. viridans*, *S. pyogenes*, *E. faecalis*, *E. faecium*, VRE, *E. coli*, and *K. pneumoniae*, by an agar dilution method which is due to the presence of phenolics. Polyphenols contain antioxidant and antimicrobial activity against oxidizing and potential free radicals. Therefore, coriander shows good antimicrobial potential due to the presence of phenolics. Phenolics have evolved as antioxidant and antimicrobial agents against environmental stress due to various oxidizing and potentially harmful free radicals [47].

Silva and collaborators have represented the mechanism of antimicrobial activity of coriander oil. Coriander oil is expected to have similar activity against Gram-positive and Gram-negative microorganisms at a cellular level. The bactericidal action of coriander oil results from membrane permeabilization leading to the subsequent loss of cellular functions such as membrane polarization, efflux activity, and metabolic activity. Later on, it was suggested that a distinct action of coriander extract on Gram-positive and Gram-negative bacteria is because of the variation in cell wall composition [48] (Figure 10.3). Antibacterial studies of different parts of coriander (i.e., leaves, stems, and fruits) against pathogenic bacterial strains are shown in Table 10.2.

10.4.2 ANTIFUNGAL ACTIVITY OF CORIANDER

EOs derived from coriander plants (i.e., leaves and seeds) have exhibited perfect antifungal activity. The antifungal activity of different parts of coriander has been studied against various fungal strains, including *Aspergillus niger* (*A. niger*), *Aspergillus flavus* (*A. flavus*), *Aspergillus parasiticus* (*A. parasiticus*), *Aspergillus fumigatus* (*A. fumigatus*), *Penicillium digitatum* (*P. digitatum*), *Paecilomyces variotii* (*P. variotii*), *Penicillium brevicompactum* (*P. brevicompactum*), *Penicillium roquefortii* (*P. roquefortii*), *Trichothecium roseum* (*T. roseum*), *Cladosporium herbarum* (*C. herbarum*), *Candida albicans* (*C. albicans*), *Streptococcus pyogenes* (*S. pyogenes*), Methicillin-resistant *Staphylococcus aureus* (MRSA), *Candida tropicalis* (*C. tropicalis*), *Candida krusei* (*C. krusei*), *Candida dubliniensis* (*C. dubliniensis*), *Candida rugosa* (*C. rugosa*), *Candida lipolytica* (*C. lipolytica*), *Candida crusei* (*C. crusei*), and *Saccharomyces cerevisiae* (*S. cerevisiae*) as shown in Table 10.3.

EO from coriander leaves demonstrated high antifungal activities, especially against *Candida albicans*, with a MIC value of 163 mg/mL [51]. Furthermore, research conducted by Silva et al. [58] demonstrated the mechanism of action of coriander on fungal cells. It was found that EO disrupts the cytoplasmic membrane leading to leakage of intracellular components (DNA) increase in ionic permeability, which causes a cell to turn from turgid to withered appearance least to the alteration of morphology (Figure 10.3).

10.4.3 OTHER ANTIMICROBIAL ACTIVITIES OF CORIANDER

An increase in the number of antimicrobial agents and antibiotic-resistant microorganisms has created a scenario where researchers focus on identifying a compound that possesses potent antimicrobial activity. It has been found that the concomitant use of antimicrobial agents with natural compounds augments the antibiotic activity or can reverse the resistance [58]. Research has shown the synergistic effect of coriander with some antibiotics and antifungal drugs, while some have stated the antagonistic effect of coriander oil with antibiotics.

Toroglu et al. stated that coriander oil had an antagonistic effect with some antibiotics against Gram-positive and Gram-negative bacteria, while against *S. aureus*, coriander oil synergizes with gentamicin [61]. It shows additive effects for *A. baumannii* with chloramphenicol ciprofloxacin, gentamicin, and tetracycline [62]. The colonization of microorganisms has excellent resistance to antimicrobials and disinfectants due to biofilms. Studies on the antimicrobial activity of coriander oil have focused on the planktonic cells, the ability of microorganisms to form biofilms, and the antibiofilm activity of the oil. Research has been carried out to evaluate the action of coriander oil on the formation of biofilms by *Candida albicans*, characterized by an increased lag phase and a decrease in the growth of biofilm at a concentration of 0.125 mg/mL [63].

Another study evaluated the antibiofilm activity of coriander oil against *Stenotropomonas maltophilia* (*S. maltophilia*) and *Bacillus subtilis* (*B. subtilis*). The authors concluded that the oil shows the strongest antibiofilm activity against *S. maltophilia* compared to *B. subtilis*. The minimum biofilm inhibitory concentrations, $MBIC_{50}$, and $MBIC_{90}$, of *S. maltophilia* were found at 7.49 µL/mL and 7.96 µL/mL, while for *B. subtilis* they were 7.42 µL/mL and 6.95 µL/mL, respectively. These results encourage the use of coriander oil as an antibiofilm compound [64]. Another study evaluated coriander oil's ability to inhibit the formation or eradication of *Acinetobacter baumannii* biofilms. It has been found that a concentration more than two to four times the MIC value possesses a greater ability to inhibit biofilm formation, causing an 85% decrease in metabolic activity and biofilm mass reduction. Preformed biofilms show a 75–90% reduction in total biomass and metabolic activity after 24 h of incubation [62].

Research conducted by Lopez and co-workers revealed that essential oil constituents are less effective in a vapor phase. Camphor, 1,8-cineole, *p*-cymene, and linalool showed no activity against foodborne pathogens, while linalool indicates slight inhibitory activity against *S. choleraesuis* and *C. albicans*. In addition, the micelle formation of the lipophilic molecules in the aqueous phase suppresses the adjunction of essential oil components to the microorganisms [65,66].

10.5   CONCLUSION

The chapter summarizes several studies on the antimicrobial properties of coriander and its extracts from various parts, indicating that coriander has the potential to be used as a medicinal plant. Coriander has a long history of use as traditional medicine, and it has been used to treat respiratory conditions like asthma and bronchiolitis. This chapter gives a summary of antibacterial and antifungal activity. The primary antibacterial and antifungal activities components are EOs, fatty acids, polyphenols, tocopherols, sterols, and carotenoids. Coriander EOs and extracts are effective antimicrobial agents because they damage microbial cell membranes. Because of coriander's protective effects on body cells, it is possible to create healthy foods. Hopefully, the important information reported on coriander and its extracts from various parts for antimicrobial activities will help generate more interest in coriander by defining novel pharmacological and clinical applications. Thus, it may help develop new drug formulations.

REFERENCES

1. Chopra, I. (2007). The increasing use of silver-based products as antimicrobial agents: A useful development or a cause for concern? *Journal of Antimicrobial Chemotherapy*, 59(4), 587–590.
2. Naveen, S., Siddalinga, S., & Khanum, F. (2011). Antioxidant potential of some common plant sources. *International Journal of Pharmaceutical Research and Development*, 3(1), 154–174.

3. Jin, L. L., Song, S. S., Li, Q., Chen, Y. H., Wang, Q. Y., & Hou, S. T. (2009). Identification and characterisation of a novel antimicrobial polypeptide from the skin secretion of a Chinese frog (*Rana chensinensis*). *International Journal of Antimicrobial Agents*, *33*(6), 538–542.

4. Satish, S., Mohana, D., Ranhavendra, M., & Raveesha, K. (2007). Antifungal activity of some plant extracts against important seed borne pathogens of *Aspergillus* sp. *Journal of Agricultural Technology*, *3*(1), 109–119.

5. Borris, R. P. (1996). Natural products research: Perspectives from a major pharmaceutical company. *Journal of Ethnopharmacology*, *51*(1–3), 29–38.

6. Laribi, B., Kouki, K., M'Hamdi, M., & Bettaieb, T. (2015). Coriander (*Coriandrum sativum* L.) and its bioactive constituents. *Fitoterapia*, *103*, 9–26.

7. López, P. A., Widrlechner, M. P., Simon, P. W., Rai, S., Boylston, T. D., Isbell, T. A., et al. (2008). Assessing phenotypic, biochemical, and molecular diversity in coriander (*Coriandrum sativum* L.) germplasm. *Genetic Resources and Crop Evolution*, *55*(2), 247–275.

8. Verma, A., Pandeya, S., Yadav, S. K., Singh, S., & Soni, P. (2011). A review on *Coriandrum sativum* (Linn.): An ayurvedic medicinal herb of happiness. *Journal of Advances in Pharmacy and Healthcare Research*, *1*(3), 28–48.

9. Singh, G., Kapoor, I., Pandey, S., Singh, U., & Singh, R. (2002). Studies on essential oils: Part 10; antibacterial activity of volatile oils of some spices. *Phytotherapy Research*, *16*(7), 680–682.

10. Hosseinzadeh, H., & Madanifard, M. (2000). Anticonvulsant effects of *Coriandrum sativum* L. seed extracts in mice. *Archives of Iranian Medicine*, *3*(4), 1–4.

11. Nadeem, M., Anjum, F. M., Khan, M. I., Tehseen, S., El-Ghorab, A., & Sultan, J. I. (2013). Nutritional and medicinal aspects of coriander (*Coriandrum sativum* L.): A review. *British Food Journal*, *115*(5), 743–755.

12. Zare-Shehneh, M., Askarfarashah, M., Ebrahimi, L., Kor, N. M., Zare-Zardini, H., Soltaninejad, H., et al. (2014). Biological activities of a new antimicrobial peptide from *Coriandrum sativum*. *International Journal of Biosciences*, *4*(6), 89–99.

13. Taherian, A. A., Vafaei, A. A., & Ameri, J. (2012). Opiate system mediate the antinociceptive effects of *Coriandrum sativum* in mice. *Iranian Journal of Pharmaceutical Research: IJPR*, *11*(2), 679.

14. Gastón, M. S., Cid, M. P., Vázquez, A. M., Decarlini, M. F., Demmel, G. I., Rossi, L. I., et al. (2016). Sedative effect of central administration of *Coriandrum sativum* essential oil and its major component linalool in neonatal chicks. *Pharmaceutical Biology*, *54*(10), 1954–1961.

15. Beyzi, E., Karaman, K., Gunes, A., & Beyzi, S. B. (2017). Change in some biochemical and bioactive properties and essential oil composition of coriander seed (*Coriandrum sativum* L.) varieties from Turkey. *Industrial Crops and Products*, *109*, 74–78.

16. Pathak Nimish, L., Kasture Sanjay, B., Bhatt Nayna, M., & Rathod Jaimik, D. (2011). Phytopharmacological properties of *Coriander sativum* as a potential medicinal tree: An overview. *Journal of Applied Pharmaceutical Science*, *1*(4), 20–25.

17. Chakravarty, D., Erande, M. B., & Late, D. J. (2015). Graphene quantum dots as enhanced plant growth regulators: Effects on coriander and garlic plants. *Journal of the Science of Food and Agriculture*, *95*(13), 2772–2778.

18. Ceska, O., Chaudhary, S., Warrington, P., Ashwood-Smith, M., Bushnell, G., & Poultont, G. (1988). Coriandrin, a novel highly photoactive compound isolated from *Coriandrum sativum*. *Phytochemistry*, *27*(7), 2083–2087.

19. Diederichsen, A. (1996). *Coriander. Promoting the conservation and use of underutilized and neglected crops*. 3. Italy: International Plant Genetic Resources Institute.

20. Small, E. (1997). *Coriander, culinary herbs*. Ottawa: NRC Research.

21. Wei, J.-N., Liu, Z.-H., Zhao, Y.-P., Zhao, L.-L., Xue, T.-K., & Lan, Q.-K. (2019). Phytochemical and bioactive profile of *Coriandrum sativum* L. *Food Chemistry*, *286*, 260–267.

22. Dash, B. (2011). Antibacterial activities of methanol and acetone extracts of fenugreek (*Trigonella foenum*) and coriander (*Coriandrum sativum*). *Life Sciences and Medicine Research*, *27*, 1–8.

23. Khalid, S., Ali, Q., Hafeez, M., & Malik, A. (2021). Perception regarding self-medication of antibiotics in general public sector university of Southern Punjab: A comparison between medical and non-medical students. *Biological and Clinical Sciences Research Journal*, *2021*(1).

24. Sahib, N. G., Anwar, F., Gilani, A. H., Hamid, A. A., Saari, N., & Alkharfy, K. M. (2013). Coriander (*Coriandrum sativum* L.): A potential source of high-value components for functional foods and nutraceuticals-A review. *Phytotherapy Research*, *27*(10), 1439–1456.

25. Pimenov, M. G., & Leonov, M. V. e. (1993). *The genera of the Umbelliferae: A nomenclator*. Kew: Royal Botanic Gardens.

26. Goettsch, E., Engels, J., & Demissie, A. (1984). Crop diversity in Konso agriculture. *Germplasm Newsletter, 7*, 18–26.

27. Kalemba, D., & Kunicka, A. (2003). Antibacterial and antifungal properties of essential oils. *Current Medicinal Chemistry, 10*(10), 813–829.

28. Bhat, S., Kaushal, P., Kaur, M., & Sharma, H. (2014). Coriander (*Coriandrum sativum* L.): Processing, nutritional and functional aspects. *African Journal of Plant Science, 8*(1), 25–33.

29. Önder, A. (2018). Coriander and its phytoconstituents for the beneficial effects. In H. A. El-Shemy (Ed.), *Potential of Essential Oils* (pp. 165–185). London: Intechopen.

30. Mandal, S., & Mandal, M. (2015). Coriander (*Coriandrum sativum* L.) essential oil: Chemistry and biological activity. *Asian Pacific Journal of Tropical Biomedicine, 5*(6), 421–428.

31. Yeung, E. C., & Bowra, S. (2011). Embryo and endosperm development in coriander (Coriandrum sativum). *Botany, 89*(4), 263–273.

32. Gil, A., De La Fuente, E. B., Lenardis, A. E., López Pereira, M., Suárez, S. A., Bandoni, A., et al. (2002). Coriander essential oil composition from two genotypes grown in different environmental conditions. *Journal of Agricultural and Food Chemistry, 50*(10), 2870–2877.

33. Wong, P. Y., & Kitts, D. D. (2006). Studies on the dual antioxidant and antibacterial properties of parsley (*Petroselinum crispum*) and cilantro (*Coriandrum sativum*) extracts. *Food Chemistry, 97*(3), 505–515.

34. Lal, A., Kumar, T., Murthy, P. B., & Pillai, K. S. (2004). Hypolipidemic effect of *Coriandrum sativum* L. in triton-induced hyperlipidemic rats, *Indian Journal of Experimental Biology, 42*, 909–912.

35. Eddouks, M., Maghrani, M., Lemhadri, A., Ouahidi, M.-L., & Jouad, H. (2002). Ethnopharmacological survey of medicinal plants used for the treatment of diabetes mellitus, hypertension and cardiac diseases in the south-east region of Morocco (Tafilalet). *Journal of Ethnopharmacology, 82*(2–3), 97–103.

36. Abu-Rabia, A. (2005). Herbs as a food and medicine source in Palestine. *Asian Pacific Journal of Cancer Prevention, 6*(3), 404.

37. Wangensteen, H., Samuelsen, A. B., & Malterud, K. E. (2004). Antioxidant activity in extracts from coriander. *Food Chemistry, 88*(2), 293–297.

38. Satyanarayana, S., Sushruta, K., Sarma, G., Srinivas, N., & Raju, G. S. (2004). Antioxidant activity of the aqueous extracts of spicy food additives-evaluation and comparison with ascorbic acid in in vitro systems. *Journal of Herbal Pharmacotherapy, 4*(2), 1–10.

39. Panjwani, D., Mishra, B., & Banji, D. (2010). Time dependent antioxidant activity of fresh juice of leaves of *Coriandrum sativum. International Journal of Pharmaceutical Sciences and Drug Research, 2*(1), 63–66.

40. Alboofetileh, M., Rezaei, M., Hosseini, H., & Abdollahi, M. (2014). Antimicrobial activity of alginate/clay nanocomposite films enriched with essential oils against three common foodborne pathogens. *Food Control, 36*(1), 1–7.

41. Chao, S. C., Young, D. G., & Oberg, C. J. (2000). Screening for inhibitory activity of essential oils on selected bacteria, fungi and viruses. *Journal of Essential Oil Research, 12*(5), 639–649.

42. Burt, S. (2004). Essential oils: Their antibacterial properties and potential applications in foods-a review. *International Journal of Food Microbiology, 94*(3), 223–253.

43. Clemente, I., Aznar, M., Silva, F., & Nerín, C. (2016). Antimicrobial properties and mode of action of mustard and cinnamon essential oils and their combination against foodborne bacteria. *Innovative Food Science & Emerging Technologies, 36*, 26–33.

44. Alves, S., Duarte, A., Sousa, S., & Domingues, F. C. (2016). Study of the major essential oil compounds of *Coriandrum sativum* against *Acinetobacter baumannii* and the effect of linalool on adhesion, biofilms and quorum sensing. *Biofouling, 32*(2), 155–165.

45. Duarte, A., Ferreira, S., Silva, F., & Domingues, F. (2012). Synergistic activity of coriander oil and conventional antibiotics against *Acinetobacter baumannii. Phytomedicine, 19*(3–4), 236–238.

46. Casetti, F., Bartelke, S., Biehler, K., Augustin, M., Schempp, C., & Frank, U. (2012). Antimicrobial activity against bacteria with dermatological relevance and skin tolerance of the essential oil from *Coriandrum sativum* L. fruits. *Phytotherapy Research, 26*(3), 420–424.

47. Mechchate, H., Es-Safi, I., Amaghnouje, A., Boukhira, S., A Alotaibi, Al-Zharani, M., et al. (2021). Antioxidant, anti-inflammatory and antidiabetic proprieties of LC-MS/MS identified polyphenols from coriander seeds. *Molecules, 26*(2), 487.

48. Silva, F., Ferreira, S., Queiroz, J. A., & Domingues, F. C. (2011). Coriander (*Coriandrum sativum* L.) essential oil: Its antibacterial activity and mode of action evaluated by flow cytometry. *Journal of Medical Microbiology, 60*(10), 1479–1486.

49. Gundogdu, E., Tanrıverdi, E., & Yildiz, H. (2020). Antimicrobial activity of *Coriandrum sativum* L. and its effect on microbiological properties of yoghurt. *Journal of Agricultural Science and Technology, 22*(5), 1247–1256.

50. Özkinali, S., Şener, N., Gür, M., Güney, K., & Olgun, Ç. (2017). Antimicrobial activity and chemical composition of coriander & galangal essential oil. *Indian Journal Pharmaceutical Education, 51,* 221–223.

51. Matasyoh, J., Maiyo, Z., Ngure, R., & Chepkorir, R. (2009). Chemical composition and antimicrobial activity of the essential oil of *Coriandrum sativum. Food Chemistry, 113*(2), 526–529.

52. Bag, A., & Chattopadhyay, R. R. (2015). Evaluation of synergistic antibacterial and antioxidant efficacy of essential oils of spices and herbs in combination. *PloS One, 10*(7), e0131321.

53. Bogavac, M., Karaman, M., Janjušević, L., Sudji, J., Radovanović, B., Novaković, Z., et al. (2015). Alternative treatment of vaginal infections–in vitro antimicrobial and toxic effects of *Coriandrum sativum* L. and *Thymus vulgaris* L. essential oils. *Journal of Applied Microbiology, 119*(3), 697–710.

54. Dua, A., Garg, G., Kumar, D., & Mahajan, R. (2014). Polyphenolic composition and antimicrobial potential of methanolic coriander (Coriandrum sativum) seed extract. *International Journal of Pharmaceutical Sciences and Research (IJPSR), 5*(6), 2302–2308.

55. Ali, S. A. Q., & Malik, A. Antimicrobial activity of *Coriander sativum, Journal of Pharmaceutical Research International, 32*(48), 74–81.

56. Foudah, A. I., Alqarni, M. H., Alam, A., Salkini, M. A., Ahmed, E. O. I., & Yusufoglu, H. S. (2021). Evaluation of the composition and in vitro antimicrobial, antioxidant, and anti-inflammatory activities of Cilantro (*Coriandrum sativum* L. leaves) cultivated in Saudi Arabia (Al-Kharj). *Saudi Journal of Biological Sciences, 28*(6), 3461–3468.

57. Rivers, K., Hanna-Mahase, C., Frankson, M., Smith, F., & Peter, S. (2013). Association between obesity and impaired glucose tolerance in new providence adolescents as demonstrated by the HbA 1c Test. *West Indian Medical Journal, 62*(8).

58. Silva, F., & Domingues, F. C. (2017). Antimicrobial activity of coriander oil and its effectiveness as food preservative. *Critical Reviews in Food Science and Nutrition, 57*(1), 35–47.

59. Lasram, S., Zemni, H., Hamdi, Z., Chenenaoui, S., Houissa, H., Tounsi, M. S., et al. (2019). Antifungal and antiaflatoxinogenic activities of *Carum carvi* L., *Coriandrum sativum* L. seed essential oils and their major terpene component against *Aspergillus flavus. Industrial Crops and Products, 134,* 11–18.

60. Freires, I. d. A., Murata, R. M., Furletti, V. F., Sartoratto, A., Alencar, S. M. d., Figueira, G. M., et al. (2014). *Coriandrum sativum* L. (coriander) essential oil: Antifungal activity and mode of action on *Candida* spp., and molecular targets affected in human whole-genome expression. *PLoS One, 9*(6), e99086.

61. Toroglu, S. (2011). In-vitro antimicrobial activity and synergistic/antagonistic effect of interactions between antibiotics and some spice essential oils. *Journal of Environmental Biology, 32*(1), 23–29.

62. Duarte, A. F., Ferreira, S., Oliveira, R., & Domingues, F. C. (2013). Effect of coriander oil (*Coriandrum sativum*) on planktonic and biofilm cells of *Acinetobacter baumannii. Natural Product Communications, 8*(5), 1934578X1300800532.

63. Furletti, V., Teixeira, I., Obando-Pereda, G., Mardegan, R., Sartoratto, A., Figueira, G., et al. (2011). Action of *Coriandrum sativum* L. essential oil upon oral *Candida albicans* biofilm formation. *Evidence-Based Complementary and Alternative Medicine, 2011.*

64. Kačániová, M., Galovičová, L., Ivanišová, E., Vukovic, N. L., Štefániková, J., Valková, V., et al. (2020). Antioxidant, antimicrobial and antibiofilm activity of coriander (*Coriandrum sativum* L.) essential oil for its application in foods. *Foods, 9*(3), 282.

65. Lopez, P., Sanchez, C., Batlle, R., & Nerin, C. (2005). Solid-and vapor-phase antimicrobial activities of six essential oils: Susceptibility of selected foodborne bacterial and fungal strains. *Journal of Agricultural and Food Chemistry, 53*(17), 6939–6946.

66. López, P., Sanchez, C., Batlle, R., & Nerín, C. (2007). Vapor-phase activities of cinnamon, thyme, and oregano essential oils and key constituents against foodborne microorganisms. *Journal of Agricultural and Food Chemistry, 55*(11), 4348–4356.

11 Coriander as Antioxidant in Biological Animal Models

Haseeb Anwar, Ghulam Hussain, Muhammad Imran, Laaraib Nawaz, Soha Navaid, and Sana Saleem

CONTENTS

DOI: 10.1201/9781003204626-12

11.1 INTRODUCTION

With the advancement in knowledge, attention is now being given at the molecular level to the micronutrients like flavonoids, minerals, carotenoids, anthocyanins, and vitamins associated with natural plant sources for their therapeutic interventions. As they possess several beneficial phytonutrients, several plant species like herbs are important. Among various herbaceous plants, coriander, scientifically known as *Coriandrum sativum* L., is a wonderful annual small plant belonging to the Apiaceae family, which is a sort of carrot family. The name coriander originated from the Greek word, i.e., "Koris," which means bedbug, due to its pungent, putrid odor. The fresh leaves of coriander used as flavoring agents are also referred to as cilantro, as well as Mexican, Japanese, or Chinese parsley (Mahendra and Bisht 2011b).

Coriander (*Coriandrum sativum* L.) comprises two species, i.e., one is a cultivated plant species (*Coriandrum sativum*), while the other is a wild species (*Coriandrum tordylium*). Green leafy vegetables are of considerable importance as they are enriched with numerous health benefits. The health benefits of coriander can also be seen from the fact that almost 40% of drugs that are prescribed contain components of coriander as an integral part. Coriander is referred to as a functional food that has promising health benefits for the well-being of humankind. Coriander is a vast reservoir of essential bioactive compounds that have created new avenues in drug discovery (Kačániová et al. 2020).

11.1.1 Historical Background and Occurrence

Coriander has been known for its health benefits and usages since 1550 BC in Greek Eber Papyrus, and it is also referred to in the literature of Sanskrit as long ago as 5000 BC. In order to give wine a pungent taste and odor, Greeks widely used coriander as a flavoring agent. Hippocrates in *ca.* 460–377 BC first used coriander as Greek medicine. One of the ancient findings from the Egyptian tomb of Ramses was seeds of coriander. Egyptians have a strong affiliation with the small herb, i.e., coriander is called the "Spice of Pleasure," due to its excitatory nature as a sexual stimulant. In the Indo-Pak subcontinent, coriander is known for its anti-diarrheal potential and for treating upset stomach by boiling the fresh coriander leaves in water with some other spices. The Romans cultivated coriander in Great Britain as they used to import coriander from far-off places to benefit from this valuable herb. Many citations have been gathered from early manuscripts since the 13th century about coriander as a beneficial herb in accelerating child-birth. Furthermore, dried seeds of coriander have been in use for 7000 years approximately, and since the 1900s, coriander oil has been used as a fragrant ingredient (Burdock and Carabin 2009).

It is widely present in the Western Mediterranean region and the regions of Southern Europe. On the whole, this wondrous herb is cultivated globally (Weiss 2002). It is mostly cultivated in the Indo-Pak subcontinent and Asia, Central Europe, South and Western Australia, and Morocco. Globally, with 300,000 tons of production per annum, India is the biggest consumer, producer, and exporter of coriander. The seeds of coriander, termed its fruit, are widely used as a flavoring agent in beverages, candies, cookery, tobacco, and the perfumery industry (Wangensteen, Samuelsen, and Malterud 2004). Coriander also contains 0.03 to 2.6% essential oils, with numerous health and medicinal purposes (Nadeem et al., 2013).

11.1.2 Classification

The classification of coriander is described in Table 11.1.

11.1.3 Constituents

There are two main constituents of coriander, i.e., coriander leaves and coriander seeds.

TABLE 11.1

Classification of *Coriandrum sativum*

Kingdom	Plantae	References
Subkingdom	Tracheobionta (vascular plants)	(Mahendra and Bisht 2011b)
Super-division	Spermatophyta (seed plants)	
Division	Magnoliophyta (flowering plants)	
Class	Magnoliopsida	
Sub-class	Rosidae	
Order	Apiales	
Family	Apiaceae (carrot family)	
Genus	*Coriandrum* L.	
Species	*Coriandrum sativum* L.	

11.1.3.1 Fresh Coriander Leaves

Fresh leaves of coriander contain 87.9% moisture content, 6.5% carbohydrates, and 3.3% protein, 1.7% total ash content, 0.14% calcium, 0.06% phosphorus, 0.01% iron, 60 mg/100 g vitamin, 135 mg/100 g vitamin C and vitamin B-2 complex, and 0.8 mg/100 g niacin. Almost 1 cup or 100 g of fresh coriander leaves (cilantro) contains approximately 30 g of crude fiber, 20 g of lipids, 11 g of starch, and nearly 11 g of proteins (Devi et al. 2020) (Table 11.2).

11.1.3.2 Coriander Seeds

The seeds of coriander, which are considered its fruit part, contain 28.4% crude fiber, 19.1% lipids, 11.4% crude protein, 11.3% water, 10.5% starch, 4.98% mineral contents, 0.84% essential oils, and 1.92% of sugar (Yashin et al. 2017). In addition, coriander seeds also have anti-microbial attributes against some microorganisms that are saprophytic (El-Sayed and Ahmed 2017) (Table 11.3).

11.2 ANTIOXIDANT BEHAVIOR OF CORIANDER

Our body is equipped with a complex defensive system of antioxidants with a versatile ability to counteract the noxious effects of oxidative damage caused by free radicals (Alam, Bristi, & Rafiquzzaman, 2013). We may benefit from antioxidants in three ways, either naturally (produced inside the body or endogenously) or from outside exogenous sources *via* natural sources or artificial sources. Plant-based foods are thought to be rich in beneficial antioxidants. Antioxidants procured from various plant-based natural products are termed "phytochemical" or "phytonutrient", or simply "plant-based nutrient". Recent studies have shown that antioxidants procured from various plant-based natural products have

TABLE 11.2

Nutritional Value of Coriander Present in Its Leaves

Nutrients and Calories Present in Leaves of Coriander			
Calories	3.68 calories	Vitamin C	45%
Lipid	0.083 g	Vitamin A	225%
Carbs	0.587 g	Vitamin K	258%
Protein	0.341 g	Vitamin B-6	11%
		Folates	15%
		Manganese	18%
		Iron	22%

Source: Momin et al., 2012.

TABLE 11.3

Constituents of Coriander Present inside Coriander Seeds

Components and Constituents of Coriander	Percentage (%)	Components and Constituents of Coriander	Percentage (%)
Protein	3.3	**Mineral contents**	4.98
Carbohydrates	6.5	**Essential oil**	0.84
Calcium	0.14	**Linalool**	49
Ash	1.7	**Terpinolene**	7
Iron	0.01	**α-Pinene**	6-8
Phosphorus	0.06	**Oxygenated monoterpenes**	89.9
Water	11.3	**Total phenolic content**	38.0
Crude protein	11.4	**λ-terpinene**	8.8
Lipid	19.1	**Camphor**	3.7
Crude fiber	28.4	**Limonene**	2.3
Starch	10.5	**p-cymene**	1.5
Sugar	1.92	**Neryl acetate**	
Genaryl acetate	17.6		

Source: Nadeem et al. 2013.

various pharmacological effects as they may have anti-mutagenic, anti-microbial, anti-allergic, anti-carcinogenic, and anti-diabetic effects (Shori, 2015). Maximum possible efforts are now going on to procure antioxidants from natural sources to avoid severe and chronic side effects of synthetic antioxidants. As a result of metabolic and physiologic processes occurring in the body, several reactive oxygen species in free radicals are produced. Due to their unstable nature, they react with biomolecules (DNA, proteins, and lipids) by snatching their electrons to stabilize themselves (Rhee, Park, and Cho 2009). In doing so, these reactive oxygen species, which are termed oxidants, cause oxidative damage and result in the occurrence of oxidative stress, which is the imbalance between reactive oxygen species and the body's ability to cope with their harmful effects by producing specific other molecules having the capacity to neutralize the damaging effect of free radicals/oxidants. Antioxidants have the potential to scavenge free radicals in the form of oxidants and stop them from damaging our cells (Rhee, Park, and Cho 2009). Natural plant sources, including various herbs, possess antioxidant properties to prevent the phenomenon of lipid peroxidation and microbial growth (Friedman 2017).

Various diseases associated with cell death and tissue damage may occur due to oxidative damage, and these pathophysiological conditions include cardiovascular diseases, diabetes, cancer, cardiovascular diseases, cancer, atherosclerosis, nerve injury, neural disorders, obesity, gastritis, arthritis, atherosclerosis, aging, etc. That is why treating such diseases naturally *via* consuming antioxidants is of great interest to medical doctors and nutritionists (Atefeh, Azamia, and Abd0lham1'dAngaj 2010). Studies have proved the beneficial aspects of using natural products with anti-inflammatory, anti-cancer, anti-aging, anti-oxidative, and hypolipidemic potential (Cho et al. 2006). However, now, the anti-oxidative potential of various parts of coriander has been evaluated, confirming that the seeds of coriander have less antioxidant potential than the leaves as they possess the maximum potential of anti-oxidative activity (Wangensteen, Samuelsen, and Malterud 2004).

Sage, thyme, and rosemary, which are culinary herbs, have the maximum anti-oxidative potential (Jaswir, Man, and Kitts 2000). Besides having anti-oxidative potential, culinary herbs are also widely used to prevent the spoilage of food, and also they are supposed to prevent the growth of food-borne pathogens. Furthermore, herbs like cilantro (*Coriandrum sativum*) and parsley (*Petroselinum crispum*) act as flavoring agents to give flavors to beverages and food cuisines (Wong and Kitts 2006). Another essential attribute associated with culinary herbs is that they are also used as preservative agents because they have abundant phytonutrients. On comparing cilantro and parsley, the leaves of parsley are enriched with more phenolic contents than cilantro. However, the phenolic contents of both parsley and cilantro possess anti-bacterial and anti-oxidative characteristics. The radical scavenging activities of methanol extract of both herbs were evaluated using an iron-induced linoleic acid model system which showed a positive correlation with reducing the oxidative damage. Furthermore, an *in vivo* trial on rat model estimated that the inclusion of fresh parsley leaf in diets of rats significantly increased the antioxidant potential after the plasma was screened for antioxidants; it was further confirmed after human trials that parsley leaves significantly decreased the level of oxidative damage (Wong and Kitts 2006).

11.3   MECHANISM OF ACTION OF CORIANDER AS ANTIOXIDANT

Phytonutrients like flavonoids, carotenoids, and polyphenolics are considered potential therapeutic agents to treat various health ailments due to their chelating activities and antioxidant potentials. Coriander acts as a cytoprotective agent as the leaves of coriander can inhibit or slow down the oxidative damage induced by some radiation exposure. So, the use of coriander ameliorated the altered levels of biochemical variables by reducing oxidative stress markers like glutathione and lipid peroxidation along with an increase in antioxidant markers like catalase, malondialdehyde, and total antioxidant activity statuses (Fahmy, Shreif, & Gharib, 2014). It was reported that the antioxidant potential of coriander seed oil was far better than crude seed oil from niger and black cumin seeds because coriander seeds contain a higher level of bioactive compounds like un-saponifiables posing maximum radical scavenging activity (Ramadan et al., 2003).

One study suggested that compounds enriched with polyphenolic contents are considered to cause suppression of oxidative stress induced by hydrogen peroxide (Hashim et al., 2005). In addition, *in vitro* studies have shown that bioactive components from fresh leaves of coriander have the potential to scavenge superoxide anion free radicals (Campanella et al. 2003), while in studies in which oxidative damage was induced by acetic acid, methanol extracts of coriander and parsley have shown remarkable results in scavenging hydroxyl free radicals which are generated as a result of oxidative reaction induced by ascorbic acid (Fejes et al. 2000).

The antioxidant potential related to coriander's polyphenolic contents helps reduce oxidative damage in lymphocytes induced by hydrogen-peroxide induction. Oxidative damage caused by hydrogen peroxide significantly decreased the amount of antioxidant relative enzymes, like glutathione peroxidase, catalase, superoxide dismutase, glutathione-s-transferase, and glutathione reductase, along with the rise in thiobarbituric acid-reacting substances (TBARS). Treatment with fractions of polyphenols obtained from extracts of coriander and parsley (50 μg/mL) increased the potency of antioxidant enzymes and ultimately caused a significant reduction in the concentration of TBARS (Hashim et al. 2005) (Table 11.4).

TABLE 11.4
Biological Activities of Coriander

Biological Activities of Coriander	
Antioxidant	Cardio-protective
Anti-microbial	Anti-diabetic
Anti-inflammatory	Anti-cancer
Neuro-protective	Nephro-protective
Gastro-protective	Hepato-protective
Hypoglycemic	Hypo-cholesterolemic
Anxiolytic	Gastrointestinal
Reproductive	Cytotoxic

Source: Author.

11.4 CORIANDER IN BIOLOGICAL ANIMAL MODELS

Linalool and acyl linalyl are the main ingredients of coriander known to have several biological properties, e.g., antioxidant, anti-microbial, hypoglycemic, anxiolytic, hypolipidemic, analgesic, and anti-inflammatory (Laribi et al. 2015).

11.4.1 CORIANDER'S ANTIOXIDANT EFFECT IN LIVER MODEL

11.4.1.1 Liver Model of Albino Mice

Lead acetic acid derivation was found to cause nodular injury and blockage within the central vein in the liver. The liver was found ordinary within the group treated with coriander extract. There was no nodular injury and a blockage within the central vein of the liver. The net and microscopic design of the liver was found typical. *Coriandrum sativum* can inhibit or reduce the oxidative harm caused by lead in mice. The nodular injury was not found (Jahid, Khan, and Islam 2020).

11.4.1.2 Liver Model of Rabbits

Coriandrum sativum can create a great critical variety in most of the assessed tests (a few biochemical parameters (ALT, AST, high mountain, urea, and creatinine, oxidative stress marker (MDA), antioxidants enzymes, and histological variations), and moderate down the oxidative harm actuated by lead harmfulness in rabbits (Donia 2019).

11.4.1.3 Liver Model of Wistar Rats

Leaf extract of coriander can diminish the degree of minuscule harm to Wistar rodent liver due to mercury administration. The leaf extract of coriander treats damaged liver by decreasing the concentration of overwhelming metals that gather within the liver and minimizing levels of liver chemical markers of damaged liver. Damage to the liver membrane caused by necrosis can discharge liver chemical markers such as SGOT, ALP, and SGPT, which can be measured through blood serum. These chemicals are quantitative markers valuable for deciding the degree and sort of damaged liver cell. Leaf extract of coriander diminished the degree of liver harm due to mercury administration. The main compound of the leaf extract of coriander with potential as a hepatoprotective is flavonoids that can minimize the release of reactive oxygen species through concealment of singlet oxygen, suppressing chemicals that release reactive oxygen species like xanthine oxidase, cyclooxygenase, monooxygenase, and lipoxygenase, and decreasing free radical responses in lipid peroxidation. From this behavior, the leaf extract of coriander can anticipate harm to the liver caused by free radicals from the digestion system and the collection of poisonous substances, i.e., mercury (Ardiani, Sudaryanto, Istiadi, & Hadi, 2020). Coriander has a hepatoprotective role against CCl_4 toxicity *in vivo* in rats. There was a decline in the weight of the liver and other biomarkers like AST, ALT, bilirubin, and basic phosphatase in rats treated with the extract of coriander. The defensive impact might be due to phenolics, most particularly quercetin and *iso*-quercetin (Janbaz, Saeed, and Gilani 2004).

11.4.1.4 Liver and Kidney Model of Rabbits

Carbon tetrachloride administration can harm both hepatic and renal tissues as investigated by an unexpected rise of their biomarkers. Moreover, the harmfulness brought down the endogenous antioxidant chemicals, while treatment with coriander significantly brought down the destructive impacts of CCl4 by saving cancer prevention agents and scavenging free radicals. Coriander in feed decreased the levels of ALT, ALP, and AST. Coriander treatment appears to be a successful apparatus to counter lipid peroxidation. The rebuilding of TBARS may be owing to the movement of antioxidants including glutathione reductase, superoxide dismutase, glutathione peroxidase, computerized axial tomography, and glutathione-s-transferase (Iqbal et al. 2018).

11.4.1.5 Liver and Kidney Model of Mice

The antioxidant impact of coriander, in contrast, aganist lead nitrate actuated harmfulness in mice was reported. The oxidative push actuated in mice by measurements of lead nitrate (40 mg/kg bw orally) for a week. Medication using coriander essentially diminished antagonistic impacts associated with most biochemical parameters modified in creatures medicated with lead, related to the oxidative push of the liver and kidneys. The histological investigation uncovered that at maximum dosages, extract of *Coriandrum sativum* was able to reestablish the hepatic and renal tissues to their typical design in contrast to the tissues harmed examined within the lead nitrate deprived of supplementation of extract. From all the above, it can be concluded that *Coriandrum sativum* protects against lead harmfulness and warrants the distinguishing proof and separation of dynamic compounds dependable for its antioxidant impacts (Kansal et al., 2011).

11.4.2 CORIANDER'S ANTIOXIDANT EFFECT IN GROWTH

The growth model of chicken: Supplementation of chicken feed with 1% coriander powder led to an increase. In expansion, at the hepatic level, dietary supplementation with coriander powder driven to vascular hypertrophy, the nearness of perivascular leukocyte infiltrates, and the appearance of cytoplasmic vacuolation of hepatocytes within the peri-lobular territories. Histomorphometric examination of the information gotten in this considers and histological changes within the liver give unused data on the utilization of coriander seed powder in broiler feed as a potential promoter of their development (Bodea et al. 2020).

    Fish model of growth: Coriander powder supplements the diet, promoting growth, feed conversion ratio, protein retention, and protein productive value. The proportion of fat in the edible part of the fish fed was slightly lower. The protein efficiency ratio and protein efficient value of coriander powder were significantly superior. Using cilantro powder in the edible portion of experimental fishes, the dry matter and raw protein increased, while the percentage of fat decreased and the ash content remained unchanged (Al-Shakarchi and Mohammad 2021).

11.4.3 CORIANDER'S ANTIOXIDANT EFFECT IN KIDNEY MODEL

Renal model of rats: Coriander showed a significant dose-dependent beneficial impact on lead actuated nephrotoxicity in rats in a few parameters like reduction of serum and renal lead accumulation, standardization of serum creatinine and uric acid with the direction of electrolytes levels in serum and urine rebuilding within the δ ALAD chemical action, raising enzymatic and non-enzymatic levels, and diminishing the oxidants present in plasma and kidney (El-Masry et al. 2016).

    Renal model of male mice: The consumption of coriander leaf extracts gives a capable resistance to oxidative push within the kidney, likely due to diminished concentrations in heavy metals such as cadmium, arsenic, and iron. It is expected that reduction in the concentration of arsenic to the detection constraints could be a major element of the resistance (Nishio, Tamano, Morioka, Takeuchi, & Takeda, 2019).

    Renal model of gentamicin-treated rats: Coriander could be a source of phytochemical nephroprotective potential, with flavonoids and polyphenols as the foremost vital constituents. Coriander extract treatment prevents the increase of nitrogen, creatinine, and urea in blood serum. The rat liver treated with gentamicin showed the existence of inflammation and necrosis. At the same time, the treatment groups showed no cell death and the existence of no inflammation together with ordinary renal design, showing the chemoprotective impact against gentamicin-initiated nephron harmfulness. The general histopathological review reported on the coriander's defensive impact on the kidney's basic life structures (Lakhera et al., 2015).

    Renal model of cadmium chloride administered rats: The current investigation has been attempted to assess the defensive impact of coriander in cadmium chloride actuated harmfulness quality in Wistar rats. Cadmium shows a critical rise within the enzymes (ALT, AST, acidic phosphatase, basic phosphatase) and cholesterol and LPO in serum and hepatic tissues. Treatment with aqueous coriander extract negatively impacts these parameters in cadmium-intoxicated rats. Furthermore, the treatment with cadmium caused a concomitant decrease within the enzymatic and non-enzymatic cancer prevention agents like superoxide dismutase, catalase, decreased glutathione, and serum protein individually. In contrast, treatment with watery extracts of *Coriandrum sativum* resulted in the checked enhancement of the cancer prevention agents and protein compared to control rats. The study concluded that the treatment with fluid extracts of *Coriandrum sativum* created a special defensive effect against cadmium-induced toxicity in albino rats (John, Shobana, and Keerthana 2014).

11.4.3.1 Renal and Hypertension

Rat model of hypertension: The administration of diuretic action of plant extract proceeds in Wistar rats of both sexes. The outcomes showed that the laxative impact of *Coriandrum sativum* was affirmed due to a critical increase in diuresis in rats, compared with the standard dose. Subsequently, a laxative is suggested as an excellent option for the action and administration of hypertension (Jabeen et al., 2009).

11.4.4 Coriander's Antioxidant Effect in the Nervous System Model

Due to high membrane lipid components, the central nervous system (CNS) is highly susceptible to oxidative injuries. Furthermore, oxidative stress has been linked to the progress of various CNS disorders and neurobehavioral disorders. (Sharma et al. 2009). Antioxidant, anti-dyslipidemic, anti-microbial, anti-diabetic, anti-epileptic, anxiolytic, anti-stress amphetamine, anti-mutagenic, neuroprotective, and anti-hypertension properties are only a few of the culinary and therapeutic properties of this plant (Singh, Tanwar, and Agrawal 2015).

11.4.4.1 Coriander's Antioxidant Effect against Alzheimer's

Mouse models of Alzheimer's disease: Researchers looked at the anxiolytic, antidepressant, and antioxidant properties of inhaled cilantro volatile oil extracted from *Coriandrum sativum* against a beta-amyloid mouse model of Alzheimer's disease. Swimming and other parameters significantly improved after exposure to coriander volatile oil, suggesting an anxiolytic and antidepressant-like impact. Furthermore, cilantro volatile oil decreased catalase activity while raising glutathione levels in the hippocampus. Several contacts to coriander volatile oil, according to our findings, may help patients with Alzheimer's disease relieve stress, anxiety, and oxidative stress (Cioanca et al., 2014).

11.4.4.2 Coriander's Antioxidant Effect against Neurodegeneration

Male Wistar rats model of neurodegeneration: By impairing cognition and memory, lead exposure is recognized to cause apoptotic neurobehavioral and neurodegeneration disorders. *Coriandrum sativum* has a shielding effect against lead poisonousness. Cilantro seed was extracted against lead-induced oxidative stress in male Wistar rats. Delta-amino levulinic acid dehydratase activity was measured in the blood as an indicator enzyme for lead toxicity. The hydroalcoholic seed extract of cilantro alleviated oxidative stress caused by lead in tissue-specific ways (Mahendra and Bisht 2011a).

11.4.4.3 Coriander's Antioxidant Effect against Epilepsy

Rat model of epilepsy: It has been suggested that in epileptic seizures, oxidative stress plays a major role. It has been shown that *Coriandrum sativum* has antioxidant properties. On the other hand, *Coriandrum* has also been shown to have depressant effects on the CNS. The effects of a hydroalcoholic extract of the aerial portion the plant on oxidative damage in brain tissues following seizures caused by pentylenetetrazole were examined in the rat model (Velaga et al. 2014).

11.4.4.4   Coriander's Antioxidant Effect against Anxiety

Anxiety model of mice: Using diazepam as a control, researchers looked at the anti-anxiety efficacy of a coriander hydroalcoholic extract in mice. Its aqueous extract has anxiolytic properties, which were tested in male albino mice using an elevated plus-maze as an animal model of anxiety. Compared to the control group, aqueous extract at different dosages 50, 100, and 500 mg/kg decreased spontaneous action and neuromuscular coordination (Mahendra and Bisht 2011a). All of these findings led researchers to believe that cilantro extract could be used as a neuroleptic and muscle relaxant (Emamghoreishi, Khasaki, and Aazam 2005).

11.4.5   Coriander's Antioxidant Effect in Cardiac and Lipolytic Model

Coriander seed powder has a cardiovascular shielding effect in insulin-resistant diabetic patients, as it reduces several contents of metabolic syndrome, decreases atherogenesis, and increases cardioprotective indices (Parsaeyan 2012). The herb has also been used pharmaceutically in respiratory disorders such as cough and bronchitis, although no study is available to give evidence for this (Sahib et al., 2013).

11.4.5.1   Coriander's Antioxidant Effect against Myocardial Infarction Model

Heart failure model of rats: Coriander has high lipolytic activity in addition to anti-microbial properties, suggesting that it can play a role in preventing myocardial infarction. High levels of lipids or cholesterol cause arterial stenosis, which causes a reduction in blood flow and, as a result, low oxygen to cardiac tissues, leading to cardiac necrosis or myocardial infarction (Dhyani et al. 2020). Coriander serves as a reducing agent or antioxidant, which helps to prevent oxidation during myocardial infarction. Ischemic heart disease can be prevented by consuming antioxidants from natural sources. Plants are abundant in antioxidant-active chemical constituents derived from natural sources. Coriander leaves have been shown to have potent antioxidant activity, which explains why they help prevent myocardial infarction (Zeb et al., 2018).

11.4.5.2   Coriander's Antioxidant Effect against a Hyperlipidemic Model

Hyperlipidemic rats model: Coriander's hypolipidemic impact was investigated (*Coriandrum sativum*). Coriander was offered to triton-induced hyperlipidemic rats. Coriander was discovered to reduce lipid uptake while increasing lipid breakdown. Based on these findings, coriander was thought to be a preventive and therapeutic herb for hyperlipidemia (Lal et al., 2004). Cilantro seeds were introduced into the food, and the influence of cilantro seed administration on lipid metabolism was reported. The seeds had an intense hypolipidemic impact. Its hypocholesterolemic influence appeared to be due to increased plasma LCAT activity, as well as increased cholesterol depletion to fecal bile acids and neutral sterols (Dhanapakiam et al., 2007).

11.4.6   Coriander's Antioxidant Effect in the Reproductive Model

11.4.6.1   In Testicular Injury of Mice Model

Lead acetate was found to cause a reduction in testis weight, isolation of primary spermatocyte from spermatids, histopathological characteristics in seminiferous tubules, and an abnormal structure of spermatogenic cells in testis seminiferous tubules. Coriander extracts have been reported to be helpful in the treatment of lead-intoxicated mice with normal histoarchitecture of the testis. In mice, *Coriandrum sativum* may prevent or delay the oxidative damage caused by lead. They also found that cilantro inhibited elevated AST and ALT activities and cholesterol levels, suggesting that the extract could defend against renal, hepatic, and testicular injury (Islam et al. 2019).

11.4.6.2 Post-Coital Antifertility Activity in Female Rats Model

Female fertility in rats was tested using an aqueous extract of fresh coriander seeds. On days 5, 12, and 20 of the pregnancy, researchers looked at the effects on the estrus cycle, fertilization, fetal loss, abortion, teratogenicity, and serum progesterone levels. The weight and length of the fetuses born by treated rats with the extract did not differ significantly, and there were no defects in the organs of the offspring. However, on day 5 of pregnancy, the extracts caused a substantial drop in serum progesterone levels, explaining the anti-implantation effect shown in this research (Momin, Acharya, and Gajjar 2012).

11.4.7 Coriander's Antioxidant Effect for Microflora

Broiler chicks model for microflora: Coriander improved efficiency indices and ileum microflora in broiler chicks. It also caused an immune response in the people who were examined. In addition, the birds' gastrointestinal health and general quality of life will improve if harmful bacteria in the digestive tract are reduced. According to the findings, coriander powder or extract can be used as an antibiotic substitute in poultry processing (Hosseinzadeh et al., 2014). In addition, coriander powder in the feed and cilantro extract in the water may be used as natural feed additives and growth regulators in poultry diets, replacing synthetic antibiotics (Naeemasa et al., 2015).

11.4.8 Coriander's Antioxidant Effect in Diabetes

Coriander seed supplementation improved glycemic control indexes, lipid profile, and oxidative status. In addition, coriander seed powder supplement reduced serum glucose, insulin resistance, insulin, TC, LDL-C, TG, and MDA levels while significantly raising serum TAC levels (Zamany et al., 2021).

Diabetic mice model: Because of its anti-inflammatory and antioxidant properties, the polyphenol fraction of coriander seeds may be used as a treatment complement for type 1 diabetes and type 2 diabetes. Diabetic mice were given coriander seeds that displayed overall remarkable antidiabetic and anti-hyperglycemic activity, suggesting that it could play a key role in managing diabetes and its complications (Mechchate et al. 2021).

Diabetic rats model: Coriander seeds can treat hyperglycemia and diabetes complications caused by oxidative stress. Coriander seeds effectively reduce blood glucose levels and reduce oxidative damage in organs such as the kidney due to diabetes mellitus. As a result, it can be inferred that it has hypoglycemic and antioxidant properties (Mishra et al.). Its seed extract is traditionally used as a diabetic treatment. In diabetic rats, incorporating ground cilantro seed extract into their diet resulted in a significant decrease in blood glucose and increased insulin levels (Deepa and Anuradha 2011).

11.4.9 Coriander Antioxidant Effect in Inflammatory Bowel Disease

Coriander has excellent anti-inflammatory, hypoglycemic, and hypocholesterolemic properties. In addition, it is also effective in reducing intestinal weight. Coriander seeds and herbs also have highly potent hepatoprotective and antioncogenic effects (Iqbal, Butt, and Suleria 2017).

Ulcerative colitis model of mice: The protective effects of *Coriandrum sativum* are also seen in acetic acid-induced colitis in mice. The many medicinal effects of coriander include inflammation, analgesic, antioxidant, and antispasmodic effects, and its various uses, especially as edible vegetables or spices (Heidari, Sajjadi, and Minaiyan 2016).

11.4.10 Coriander's Antioxidant Effect in Gastric Mucosal Protection

Rat model of gastric injury: Rat models investigated the effect of coriander pretreatment on gastrointestinal injuries. Analysis of gastric mucus and ulcers induced by indomethacin showed that defensive actions of coriander did not fluctuate due to mucus or endogenously stimulated pros-taglandins. The protective effect against ethanol-induced damage of the gastric tissue might be related to the free-radical scavenging property of different antioxidant constituents, e.g., coumarins, terpenes, flavonoids, linalool, polyphenolic, and catechins are found in coriander. In addition, the suppression of ulcerative colitis may be due to the arrangement pattern of a protected coating of hydrophobic interactions (Al-Mofleh et al., 2006).

11.5 CONCLUSION

Coriander is a commonly found herb; it has many essential micronutrients and macronutrients present in its leaves and seeds. It is used in different animal models as a therapeutic approach. These models have proven their medicinal properties, including healing potential in myocardial dysfunction or metabolic disorders (diabetes, hypertension, and atherosclerosis). It is also used in specific neurodegenerative and psychotic diseased models showing its modulating potential. Furthermore, the reproductive potential was also modulated after using coriander. These characteristics mark its potentiating abilities in healing, modulating, and fighting against various agents.

REFERENCES

Alam, M. N., Bristi, N. J., & Rafiquzzaman, M. (2013). Review on in vivo and in vitro methods evaluation of antioxidant activity. *Saudi Pharmaceutical Journal*, 21(2), 143–152.

Al-Mofleh, I. A., A. A. Alhaider, J. S. Mossa, M. O. Al-Sohaibani, S. Rafatullah, and S. Qureshi. 2006. "Protection of Gastric Mucosal Damage by *Coriandrum sativum* L. Pretreatment in Wistar Albino Rats." *Environmental Toxicology & Pharmacology* 22, no. 1: 64–69.

Al-Shakarchi, Hani Hashim, and Mahmoud A. Mohammad. 2021. "Effects of Different Levels of Cinnamon and Coriander Powders on Growth Performance, Feed Utilization, Survival Rate and Body Chemical Composition of Common Carp *Cyprinus carpio* L." *Plant Archives* 21, no. 1: 46–54.

Ardiani, Z. R., Sudaryanto, S., Istiadi, H., & Hadi, H. (2020). The effect of coriander leaves extract on the degree of wistar rats liver microscopic damage was given mercuric chloride orally. *Diponegoro Medical Journal (Jurnal Kedokteran Diponegoro)*, 9(6), 436–441.

Atefeh, H., Mahnaz Azamia, and S. Abdolhamid Angaji. 2010. "Medicinal Effects of *Heracleum persicum* (Golpar)." *Middle-East Journal of Scientific Research* 5, no. 3: 174–176.

Bodea, Alexandra, Dorel Dronca, Liliana Petculescu Ciochină, Eliza Simiz, Ioan Peţ, Lavinia Ştef, Tiberiu Iancu, Nicolae Păcală, Mirela Ahmadi, and Roxana Popescu. 2020. "The Effect of Coriander on the Histological Structure of the Intestine and Liver in Broilers." *Scientific Papers: Animal Science & Biotechnologies. Lucrari Stiintifice: Zootehnie si Biotehnologii* 53, no. 2.

Burdock, George A., and Ioana G. Carabin. 2009. "Safety Assessment of Coriander (*Coriandrum sativum* L.) Essential Oil as a Food Ingredient." *Food & Chemical Toxicology* 47, no. 1): 22–34.

Campanella, L., A. Bonanni, G. Favero, and M. Tomassetti. 2003. "Determination of Antioxidant Properties of Aromatic Herbs, Olives and Fresh Fruit Using an Enzymatic Sensor." *Analytical & Bioanalytical Chemistry* 375, no. 8: 1011–1016.

Cho, Jae Youl, S.-C. Park, T.-W. Kim, K.-S. Kim, J.-C. Song, H.-M. Lee, H.-J. Sung, M.-H. Rhee, S.-K. Kim, H.-J. Park, E. S. Yoo, C. H. Lee, and M. H. Rhee. 2006. "Radical Scavenging and Anti-Inflammatory Activity of Extracts from Opuntia humifusa Raf." *Journal of Pharmacy & Pharmacology* 58, no. 1: 113–119.

Cioanca, Oana, Lucian Hritcu, Marius Mihasan, Adriana Trifan, and Monica Hancianu. 2014. "Inhalation of Coriander Volatile Oil Increased Anxiolytic-Antidepressant-Like Behaviors and Decreased Oxidative Status in Beta-Amyloid (1–42) Rat Model of Alzheimer's Disease." *Physiology & Behavior* 131: 68–74.

Deepa, B., and C. V. Anuradha. 2011. "Antioxidant Potential of *Coriandrum sativum* L. Seed Extract." *Indian Journal of Experimental Biology* 49: 30–38.

Devi, Suman, Ena Gupta, Mamta Sahu, and Pragya Mishra. 2020. "Proven Health Benefits and Uses of Coriander (*Coriandrum sativum* L.)." In *Ethnopharmacological Investigation of Indian Spices*. IGI Global.

Dhanapakiam, P., J. Mini Joseph, V. K. Ramaswamy, M. Moorthi, and A. Senthil Kumar. 2007. "The Cholesterol Lowering Property of Coriander Seeds (*Coriandrum sativum*): Mechanism of Action." *Journal of Environmental Biology* 29, no. 1: 53.

Dhyani, Neha, Adila Parveen, Aisha Siddiqi, M. Ejaz Hussain, and Mohammad Fahim. 2020. "Cardioprotective Efficacy of *Coriandrum sativum* (L.) Seed Extract in Heart Failure Rats Through Modulation of Endothelin Receptors and Antioxidant Potential." *Journal of Dietary Supplements* 17, no. 1: 13–26.

Donia, Gehad Rashad. 2019. "Protective Effect of Coriander (*Corindrum sativum*) against Lead Toxicity in Rabbits." *European Journal of Biomedical* 6, no. 13: 520–532.

El-Masry, Shimaa, H. A. Ali, Nora M. El-Sheikh, and Safaa Mostafa Awad. 2016. "Dose-Dependent Effect of Coriander (*Coriandrum sativum* L.) and Fennel (*Foeniculum vulgare* M.) on Lead Nephrotoxicity in Rats." *International Journal of Research Studies in Biosciences* 4: 36–45.

El-Sayed, S. A. E., and Sarah Yousef Ahmed. 2017. "Effects of Coriander Seeds Powder (*Coriandrum sativum*) as Feed Supplements on Growth Performance Parameters and Immune Response in Albino Rats." *International Journal of Livestock Research* 7, no. 2: 54–63.

Emamghoreishi, Masoumeh, Mohammad Khasaki, and Maryam Fath Aazam. 2005. "*Coriandrum sativum*: Evaluation of Its Anxiolytic Effect in the Elevated Plus-Maze." *Journal of Ethnopharmacology* 96, no. 3: 365–370.

Fahmy, H. A., Shreif, N. H., & Gharib, O. A. (2014). The protective effect of *Coriandium sativum* extract on hepato-renal toxicity induced in irradiated rats. *European Journal of Medicinal Plants*, 4(2), 196.

Fejes, S. Z., A. Blazovics, E. Lemberkovics, G. Petri, E. Szöke, and A. Kery. 2000. "Free Radical Scavenging and Membrane Protective Effects of Methanol Extracts from *Anthriscus cerefolium* L. (Hoffm.) and *Petroselinum crispum* (Mill.) Nym. ex A W Hill." *Phytotherapy Research* 14, no. 5: 362–365.

Friedman, Mendel. 2017. "Antimicrobial Activities of Plant Essential Oils and Their Components against Antibiotic-Susceptible and Antibiotic-Resistant Foodborne Pathogens." In *Essential Oils & Nanotechnology for Treatment of Microbial Diseases*, 14–38. CRC Press.

Hashim, M. S., S. Lincy, V. Remya, M. Teena, and L. Anila. 2005. "Effect of Polyphenolic Compounds from *Coriandrum sativum* on H2O2-Induced Oxidative Stress in Human Lymphocytes." *Food Chemistry* 92, no. 4: 653–660.

Heidari, Bahareh, Seyed Ebrahim Sajjadi, and Mohsen Minaiyan. 2016. "Effect of *Coriandrum sativum* Hydroalcoholic Extract and Its Essential Oil on Acetic Acid-Induced Acute Colitis in Rats." *Avicenna Journal of Phytomedicine* 6, no. 2: 205.

Hosseinzadeh, Hesam, Ali Ahmad Alaw Qotbi, Alireza Seidavi, David Norris, and David Brown. 2014. "Effects of Different Levels of Coriander (*Coriandrum sativum*) Seed Powder and Extract on Serum Biochemical Parameters, Microbiota, and Immunity in Broiler Chicks." *Scientific World Journal* 2014.

Iqbal, Muhammad Jawad, Masood Sadiq Butt, Aamir Shehzad, and Muhammad Asghar. 2018. "Evaluating Therapeutic Potential of Coriander Seeds and Leaves (*Coriandrum sativum* L.) to Mitigate Carbon Tetrachloride-Induced Hepatotoxicity in Rabbits." *Asian Pacific Journal of Tropical Medicine* 11, no. 3: 209.

Iqbal, Muhammad Jawad, Masood Sadiq Butt, and Hafiz Ansar Rasul Suleria. 2017. "Coriander (*Coriandrum sativum* L.): Bioactive Molecules and Health Effects." *Bioactive Molecules in Food*: 1–37.

Islam, Mohammad Rafiqul, Md. Anwar Jahid, Md. Zahirul Islam Khan, and Md. Nabiul Islam. 2019. "Prophylactic Effects of Vitamin E and Coriander (*Coriandrum sativum*) Extract on Lead-Induced Testicular Damage in Swiss Albino Mice." *Journal of Advanced Biotechnology and Experimental Therapeutics* 2, no. 1: 31–35.

Jabeen, Qaiser, Samra Bashir, Badiaa Lyoussi, and Anwar H. Gilani. 2009. "Coriander Fruit Exhibits Gut Modulatory, Blood Pressure Lowering and Diuretic Activities." *Journal of Ethnopharmacology* 122, no. 1: 123–130.

Jahid, M. A., Mohammad Zahirul Islam Khan, and Mohammad Rafiqul Islam. 2020. "Prophylactic Effect of Vitamin E and Coriander (*Coriandrum sativum*) Seed Extract against Lead Toxicity in Liver of Swiss Albino Mice." *Journal of Advanced Biotechnology and Experimental Therapeutics* 3, no. 3: 263–267.

Janbaz, Khalid H., S. A. Saeed, and A. H. Gilani. 2004. "Studies on the Protective Effects of Caffeic Acid and Quercetin on Chemical-Induced Hepatotoxicity in Rodents." *Phytomedicine* 11, no. 5: 424–430.

Jaswir, Irwandi, Yaakob B. Che Man, and David D. Kitts. 2000. "Synergistic Effects of Rosemary, Sage, and Citric Acid on Fatty Acid Retention of Palm Olein during Deep-Fat Frying." *Journal of the American Oil Chemists' Society* 77, no. 5: 527–533.

John, N., A. Arul, G. Shobana, and K. Keerthana. 2014. "Protective Effect of *Coriander sativum* L. on Cadmium Induced Toxicity in Albino Rats." *World Journal of Pharmacy & Pharmaceutical Sciences (WJPPS)* 3, no. 8: 525–534.

Kačániová, Miroslava, Lucia Galovičová, Eva Ivanišová, Nenad L. Vukovic, Jana Štefániková, Veronika Valková, Petra Borotová, Jana Žiarovská, Margarita Terentjeva, and Soňa Felšöciová. 2020. "Antioxidant, Antimicrobial and Antibiofilm Activity of Coriander (*Coriandrum sativum* L.) Essential Oil for its Application in Foods." *Foods* 9, no. 3: 282.

Kansal, Leena, Veena Sharma, Arti Sharma, Shweta Lodi, and S. H. Sharma. 2011. "Protective Role of *Coriandrum sativum* (Coriander) Extracts against Lead Nitrate Induced Oxidative Stress and Tissue Damage in the Liver and Kidney in Male Mice." *Index Medicus for the South-East Asia Region (IMSEAR).*

Lakhera, Abhijeet, Aditya Ganeshpurkar, Divya Bansal, and Nazneen Dubey. 2015. "Chemopreventive Role of *Coriandrum sativum* against Gentamicin-Induced Renal Histopathological Damage in Rats." *Interdisciplinary Toxicology* 8, no. 2: 99.

Lal, A., T. Kumar, P. Balakrishna Murthy, and K. Sadasivan Pillai. 2004. "Hypolipidemic Effect of *Coriandrum sativum* L. in Triton-Induced Hyperlipidemic Rats." *Indian Journal of Experimental Biology* 42: 909–912.

Laribi, Bochra, Karima Kouki, Mahmoud M'Hamdi, and Taoufik Bettaieb. 2015. "Coriander (*Coriandrum sativum* L.) and its Bioactive Constituents." *Fitoterapia* 103: 9–26.

Mahendra, Poonam, and Shradha Bisht. 2011a. "Anti-Anxiety Activity of *Coriandrum sativum* Assessed Using Different Experimental Anxiety Models." *Indian Journal of Pharmacology* 43, no. 5: 574.

Mahendra, Poonam, and Shradha Bisht. 2011b. "*Coriandrum sativum*: A Daily Use Spice with Great Medicinal Effect." *Pharmacognosy Journal* 3, no. 21: 84–88.

Mechchate, Hamza, Imane Es-Safi, Amal Amaghnouje, Smahane Boukhira, Amal A. Alotaibi, Mohammed Al-Zharani, Fahd A. Nasr, Omar M. Noman, Raffaele Conte, and El Hamsas El Youbi Amal. 2021. "Antioxidant, Anti-Inflammatory and Antidiabetic Proprieties of LC-MS/MS Identified Polyphenols from Coriander Seeds." *Molecules* 26, no. 2: 487.

Mishra, Bipin B., Shree R. Padmadeo, Kumud R. Thakur, Deepak K. Jha, Kumar Pranay VyomeshVibhaw, and Pankaj Kumar. "Hypoglycemic and Antioxidative Potential of *Coriandrum sativum* Seed Extract in Alloxan Induced Diabetic Rats." *Bioscience Biotechnology Research Communications* 14, no. 1: 275–281.

Momin, Abidhusen H., Sawapnil S. Acharya, and Amit V. Gajjar. 2012. "*Coriandrum sativum*-Review of Advances in Phytopharmacology." *International Journal of Pharmaceutical Sciences & Research* 3, no. 5: 1233.

Nadeem, Muhammad, Faqir Muhammad Anjum, Muhammad Issa Khan, Saima Tehseen, Ahmed El-Ghorab, and Javed Iqbal Sultan. 2013. "Nutritional and Medicinal Aspects of Coriander (*Coriandrum sativum* L.): A Review." *British Food Journal* 115, no. 5: 743.

Naeemasa, M., A. A. Alaw Qotbi, A. Seidavi, D. Norris, D. Brown, and M. Ginindza. 2015. "Effects of Coriander (*Coriandrum sativum* L.) Seed Powder and Extract on Performance of Broiler Chickens." *South African Journal of Animal Science* 45, no. 4: 371–378.

Nishio, R., Tamano, H., Morioka, H., Takeuchi, A., & Takeda, A. (2019). Intake of heated leaf extract of *Coriandrum sativum* contributes to resistance to oxidative stress via decreases in heavy metal concentrations in the kidney. *Plant Foods for Human Nutrition*, 74(2), 204–209.

Ramadan, M. F., Kroh, L. W., & Mörsel, J. T. (2003). Radical scavenging activity of black cumin (*Nigella sativa* L.), coriander (*Coriandrum sativum* L.), and niger (*Guizotia abyssinica* Cass.) crude seed oils and oil fractions. *Journal of agricultural and Food Chemistry*, 51(24), 6961–6969.

Parsaeyan, Nayereh. 2012. "The Effect of Coriander Seed Powder Consumption on Atherosclerotic and Cardioprotective Indices of Type 2 Diabetic Patients." *Iranian Journal of Diabetes and Obesity* 4(2).

Rhee, Man Hee, Hwa-Jin Park, and Jae Youl Cho. 2009. "*Salicornia herbacea*: Botanical, Chemical and Pharmacological Review of Halophyte Marsh Plant." *Journal of Medicinal Plants Research* 3, no. 8: 548–555.

Sahib, Najla Gooda, Farooq Anwar, Anwarul-Hassan Gilani, Azizah Abdul Hamid, Nazamid Saari, and Khalid M Alkharfy. 2013. "Coriander (*Coriandrum sativum* L.): A Potential Source of High-Value Components for Functional Foods and Nutraceuticals-A Review." *Phytotherapy Research* 27, no. 10: 1439–1456.

Sharma, Deep Raj, Aditya Sunkaria, Amanjit Bal, Yangchen D. Bhutia, R. Vijayaraghavan, S. J. S. Flora, and Kiran Dip Gill. 2009. "Neurobehavioral Impairments, Generation of Oxidative Stress and Release of Pro-apoptotic Factors after Chronic Exposure to Sulphur Mustard in Mouse Brain." *Toxicology & Applied Pharmacology* 240, no. 2: 208–218.

Shori, A. B. (2015). Screening of antidiabetic and antioxidant activities of medicinal plants. *Journal of Integrative Medicine*, 13(5), 297–305.

Singh, Deeksha, Asmita Tanwar, and Parul Agrawal. 2015. "An Overview on Coriander." *Journal of Biomedical and Pharmaceutical Research* 4, no. 2.

Velaga, Manoj Kumar, Prabhakara Rao Yallapragada, Dale Williams, Sharada Rajanna, and Rajanna Bettaiya. 2014. "Hydroalcoholic Seed Extract of *Coriandrum sativum* (Coriander) Alleviates Lead-Induced Oxidative Stress in Different Regions of Rat Brain." *Biological Trace Element Research* 159, no. 1: 351–363.

Wangensteen, Helle, Anne Berit Samuelsen, and Karl Egil Malterud. 2004. "Antioxidant Activity in Extracts from Coriander." *Food Chemistry* 88, no. 2: 293–297.

Weiss, Edward A. 2002. *Spice Crops*. CABI Publishing.

Wong, Peter Y. Y., and David D. Kitts. 2006. "Studies on the Dual Antioxidant and Antibacterial Properties of Parsley (*Petroselinum crispum*) and Cilantro (*Coriandrum sativum*) Extracts." *Food Chemistry* 97, no. 3: 505–515.

Yashin, Alexander, Yakov Yashin, Xiaoyan Xia, and Boris Nemzer. 2017. "Antioxidant Activity of Spices and Their Impact on Human Health: A Review." *Antioxidants* 6, no. 3: 70.

Zamany, Sanaz, Aida Malek Mahdavi, Saeed Pirouzpanah, and Ali Barzegar. 2021. "The Effects of Coriander Seed Supplementation on Serum Glycemic Indices, Lipid Profile and Parameters of Oxidative Stress in Patients with Type 2 Diabetes Mellitus: A Randomized Double-Blind Placebo-Controlled Clinical Trial."

Zeb, Falak, Mahpara Safdar, Sadia Fatima, Saleem Khan, Sahib Alam, Mumtaz Muhammad, Ayesha Syed, Fozia Habib, and Hira Shakoor. 2018. "Supplementation of Garlic and Coriander Seed Powder: Impact on Body Mass Index, Lipid Profile and Blood Pressure of Hyperlipidemic Patients." *Pakistan Journal of Pharmaceutical Sciences* 31, no. 5.

12 Neuroprotective Potency of Coriander

Mahmoud Hosseini, Mohammad Hossein Boskabady, and Mohammad Reza Khazdair

CONTENTS

12.1 INTRODUCTION

Coriander (*Coriandrum sativum* L.) is an annual aromatic plant originating from Mediterranean areas, belonging to the family Apiaceae. The leaves are small branches and sub-branches, the flowers are white and slightly, and the fruits are almost ovate globular, with multiple longitudinal ridges on the surface (Yeung and Bowra 2011). The essential oils and extracts of aromatic plants have been used in food preservation, pharmaceuticals, and alternative medicine. All segments of coriander are edible and traditionally used to treat different disorders (Sahib et al. 2013). In addition, the fresh leaves could be used for garnishing and are used in many foods like chutneys, salads, and soups (Bhat et al. 2014). Fresh coriander contains 84% water, a low level of saturated lipids, cholesterol, but a high level of thiamine, zinc, and dietary fibers (Kandlakunta, Rajendran, and Thingnganing 2008). Coriander seeds are also considered a source of vitamins, lipids, and minerals, including calcium, potassium, magnesium, phosphorus, sodium, and zinc (Iwatani, Arcot, and Shrestha 2003, Bhat et al. 2014). The main coriander oil (EO) components are linoleic and linolenic acids (Sahib et al. 2013). Therefore, nutritional supplementation with coriander seeds is suggested to decrease saturated but increase unsaturated fatty acids (Ertas et al. 2005).

In Iranian traditional medicine, coriander attenuates digestive problems, flatulence, diarrhea and colic, and other gastrointestinal diseases (Zargari 1991). In addition, the hepatoprotective

DOI: 10.1201/9781003204626-13

properties of coriander due to the antioxidant potential of its phenolic compounds (Pandey et al. 2011) and the cardiovascular protective effect of the plant were demonstrated (Dhanapakiam et al. 2007). Furthermore, the plant also showed antioxidant activities in different organs (Wangensteen, Samuelsen, and Malterud 2004, Kansal et al. 2011), including preventive effects on brain histology changes such as gliosis, lymphocytic infiltration, and cellular edema (Vekaria et al. 2012).

Coriander is also potentially considered for the growth and functioning of the brain. The leaves of coriander (5, 10, and 15%, w/w) improved memory scores, prevented the process of aging in mice dose-dependently, reversed the memory deficits induced by scopolamine and diazepam, and significantly reduced brain cholinesterase activity and serum total cholesterol levels (Mani et al. 2011). The preventive effect of the hydroalcoholic extract of coriander (50, 100, and 200 mg/kg) on neuronal damages induced by pentylenetetrazole (PTZ) in a rat model of seizure was also demonstrated (Pourzaki et al. 2017). A study on the effects of ethanol extract of coriander seeds on learning in second-generation mice showed that the plant extract did not improve learning in a short period after training, while it improved learning in the long term (Zargar-Nattaj et al. 2011).

There is increasing interest in discovering effective drugs focusing on the folk medicinal and medicinal plant. However, it should be noted that many modern drugs have plant origins, through observation of therapeutic practices of native people (Gilani 2005, Balick and Cox 1996). Therefore, in the present chapter, the neuropharmacological properties of coriander and its components, including their sedative, anti-anxiety, anti-depressive, anticonvulsant and anti-seizures, memory improving, anti-Parkinson's, and protective effects on neurotoxicity are summarized.

12.2 CHEMICAL COMPOSITIONAL OF CORIANDER

Coriander is cultivated in different soil, climatic, seasonal, and other ecological conditions and, therefore, has a broad range of compositions. Different EO components and different chemical compounds have been extracted from different coriander parts, which seems to depend on the variety of the plant (Diederichsen and Hammer 2003). EO of premature fruits contain mainly compounds such as geranyl acetate (46.2%) and linalool (10.9%), while in mature fruits, linalool (87.5%), followed by *cis*-dihydrocarvone (2.36%), were detected as the main compounds (Msaada et al. 2007). The chemical composition of coriander seed essential oil from Sistan and Baluchestan province, Iran, is linalool (57.5%), geranyl acetate (15.9%), β-caryophyllene (3.26%), camphor (3.02%), and *p*-cymene (2.5%). Chemical analysis indicated clearly that linalool was the main component of coriander EO (Sriti Eljazi et al. 2017).

The coriander seeds EO from different geographical origins also showed variation in their constituents. For example, the major components in plants of European include linalool (58.0–80.3%), γ-terpinene (0.3–11.2%), and α-pinene (0.2–10.9%) (Orav, Arak, and Raal 2011), while the main oil components of the plants from Brazil were linalool (77.4%), γ-terpinene (4.64%), and α-pinene (5.50%) (de Figueiredo et al. 2004). Additionally, the seed oil of plants grown in Iran contains linalool (40.9–79.9%), neryl acetate (2.3–14.2%), and γ-terpinene (0.1–13.6%) (Nejad Ebrahimi, Hadian, and Ranjbar 2010).

The oils extracted from the plant's leaves have been shown to contain 44 compounds, most of which are aromatic acids, including 2-decenoic (30.8%), *E*-11-tetradecenoic acids (13.4%), capric acid (12.7%), and undecyl alcohol (6.4%). Other compounds, including 2-undecenal (3.87%), dodecanoic acid (2.63%), cyclododecane (2.45%), decanal (1.35%), and decamethylene glycol (1.15%), have also reported as present in the leaves of the plant (Nurzyńska-Wierdak 2013). Monoterpenoids such as limonene and β-myrcene are also reported as the constituents of coriander EO (Tashinen and Nykänen 1975). The chemical composition of EO from the seed of coriander is given in Table 12.1.

TABLE 12.1

Main Chemical Composition of the Essential Oil from Coriander

Compound	Composition %	Origin	References
Linalool, *cis*-dihydrocarvone	(87.5%), (2.36%)	Seed oil	(Msaada et al. 2007)
Linalool, geranyl acetate, β-caryophyllene	(57.5%), (15.9%), (3.26%)	Seed oil	(Sriti Eljazi et al. 2017)
Linalool, γ-terpinene, α-pinene	(58.0--80.3%), (0.3–11.2%), (0.2–10.9%)	Seed oil	(Orav, Arak, and Raal 2011)
Linalool, γ-terpinene, α-pinene	(77.4%), (4.64%), (5.5%)	Seed oil	(de Figueiredo et al. 2004)
Linalool, neryl acetate, γ-terpinene	(40.9–79.9%), (2.3–14.2%), (0.1–13.6%)	Seed oil	(Nejad Ebrahimi, Hadian, and Ranjbar 2010)
2-decenoic	(30.8%)	Leaves	(Nurzyńska-Wierdak 2013)
E-11-tetradecenoic acids	(13.4%)		
Capric acid	(12.7%)		
undecyl alcohol	(6.4%)		
2-undecenal	(3.87%)		
Dodecanoic acid	(2.63%)		
Cyclododecane	(2.45%)		
Decanal	(1.35%)		
Decamethylene glycol	(1.15%)		

12.3 SEDATIVE, ANTI-ANXIETY, AND ANTI-DEPRESSIVE PROPERTIES OF CORIANDER AND LINALOOL

12.3.1 CORIANDER

Fresh leaves and coriander seeds have been recommended for relieving insomnia in Iranian folk medicine (Mir 1992, Zargari 1991). Pre-sleep eating of a combination of chopped fresh leaves (30 g) or seeds of the plant with tea, as a single dose, has been suggested to relieve anxiety and insomnia (Mir 1992). Similar uses have also been advised in other folk medicines (Duke 2002). These effects were also confirmed using animal models. Intraperitoneal (i.p.) administration of aqueous extracts (200, 400, and 600 mg/kg, i.p.), hydroalcoholic extracts (400 and 600 mg/kg), and 600 mg/kg coriander EO in male albino mice increased sleep duration to 160–220, 130–180, and 210 min, respectively. These results showed that EO and extracts of coriander exhibit sedative and hypnotic activities in a pentobarbital-induced hypnotic model (Emamghoreishi and Heidari-Hamedani 2015). The sleep-prolonging effects of the fractions of aerial parts of the plant, including water (WF), ethyl acetate (EAF), and *n*-butanol (NBF) fractions, were also confirmed in mice. The hydroalcoholic extract (HAE), EAF, and NBF significantly prolonged sleep duration, but only the NBF decreased sleep latency significantly (Rakhshandeh, Sadeghnia, and Ghorbani 2012). The anxiolytic effects of 10, 25, 50, and 100 mg/kg (i.p.) of seed aqueous extract and their effects on motor activity and neuromuscular function in an animal model were evaluated. In elevated plus-maze, the seeds' aqueous extract (100 mg/kg, i.p.) showed an anxiolytic effect by increasing the time spent on and the entries to the open arms. Furthermore, coriander extract (50, 100, and 500 mg/kg) was accompanied by motor activity impairments and neuromuscular dysfunction (Emamghoreishi, Khasaki, and Aazam 2005). The hydroalcoholic extract of the seeds (100 and 200 mg/kg) showed anti-anxiety effects comparable to the effects of 0.5 mg/kg diazepam; however, 50 mg/kg of the extract did not show anti-anxiety activity (Mahendra and Bisht 2011). The anti-anxiety activity of coriander leaf aqueous extract (50, 100, and 200 mg/kg, i.p.) has also been evaluated in mice and compared with the effects of diazepam (5 mg/kg, i.p.) as a standard drug. Treatment with 200 mg/kg of extract increased the time spent and the number of entries into the open arms of the elevated plus-maze, which was assumed to be due to the anxiolytic effect of the extract (Pathan, Kothawade, and Logade 2011). Interestingly, the seed extract (100 and 200 mg/kg) also had an anxiolytic effect similar to the effects of the leaf extract. Locomotion activity and frequency of rearing in the groups treated by 200 mg/kg of the extract was also lower than that of the control group. Furthermore, the mentioned extracts significantly increased social interactions of treated animals (Mahendra and Bisht 2011). The fresh leaf extract obtained (10 ml/kg, i.p.) attenuated immobility time and was more effective than the extract of dried plant. It was concluded that the extract of green but not dried plant has anti-depressive effect (Kishore and Siddiqui 2003).

Aqueous extract and fixed oil of the seeds showed anti-depressant activity at the end of the treatment of 14 consecutive days using tail suspension and forced swimming tests. Furthermore, the aqueous extract (200 and 400 mg/kg, p.o.) and also an extract prepared by diethyl ether (2 and 4 ml/kg, p.o.) showed a significant anti-depressant-like activity, which was similar to the effects of commonly used anti-depressant drugs fluoxetine 20 mg/kg (p.o.) and imipramine 15 mg/kg (p.o.). Furthermore, aqueous seed extract also inhibited monoamine oxidase-B (MAO-B) activity, which was suggested as a mechanism of anti-depressant activity (Kharade et al. 2011). Furthermore, administration of coriander seed essential oil and its major component linalool intracerebroventricularly (i.c.v) (0.86, 8.6, and 86 µg/chick, i.c.v. for both agents) on locomotor activity in chicks significantly decreased escapes, defecations, and crossing numbers in an open field, while increasing the sleeping posture (Gastón et al. 2016).

The inhalation of coriander oil (1% and 3%) significantly increased time spent in the open arms, open-arm entries, and the number of crossings (exploratory locomotor activity) in β-amyloid (1-42) rat model of Alzheimer's disease (Cioanca et al. 2014). The coriander oil also significantly reduced immobility but increased swimming time and glutathione content. Furthermore, amyloid deposits are scarce in rats treated with *C. sativum* oil, which are apparent and abundant in the hippocampus of rats treated with i.c.v. β-amyloid (1-42), (Cioanca et al. 2014).

The hydroalcoholic extract of coriander (50, 100, and 200 mg/kg) reduced the burst discharges' duration, frequency, and amplitude while prolonging the latency of the seizure attacks. In addition, the extract also prevented the production of dark neurons and apoptotic cells (Table 12.2) in different areas of the hippocampus (Pourzaki et al. 2017). Neuro-pharmacological effects of coriander and linalool are shown in Figure 12.1.

12.3.2 LINALOOL

The inhalation of linalool (1% and 3%) increased the sleeping time and was similar to the effects of 1 mg/kg diazepam in anxiety-induced mice. It also reduced body temperature but could not affect the bar descent latency in the rotarod performance test. Linalool (3%) also significantly impaired exploratory and motor performances in spontaneous locomotor activity evaluation (de Moura Linck et al. 2009).

Similar to the administration of diazepam (1 mg/kg, i.p.), inhalation of linalool (3%) increased the time spent in the light compartment of the light/dark test box and also increased the latency in the first crossing between the light and dark compartments, but had no effect on the crossing number in anxiety-induced mice (Linck et al. 2010). Furthermore, both diazepam and linalool (1%) improved the social interaction behaviors. Both linalool (3%) and diazepam postponed the first attack and significantly reduced the attacks' number and duration in the aggressive behavior test (Linck et al. 2010).

In the elevated plus-maze test, inhalation of mass fraction (w/w) of linalool oxide (0.6%, 2.5%, and 5.0%) and injection of diazepam (0.5 mg/kg, i.p.) were able to increase the open arm entries and the time spent in open arms. Additionally, both linalool oxide and diazepam prolonged the time spent in the light compartment and the crossing numbers between the light and dark compartments of the light-dark box (Souto-Maior et al. 2011). Administration of linalool

TABLE 12.2

Sedative, Anti-Anxiety, and Anti-Depressive Effects of Coriander and Linalool

Extract of Coriander/ Linalool	Type of Study	Doses	Effects
Aqueous fresh leaves	Traditional medicine	30 g	Fresh leaves (30 g) or seeds of the plant with tea, as a single dose, has been suggested to relieve anxiety and insomnia (Mir 1992).
Aqueous seed extract	Pentobarbital-induced hypnosis in mice	100, 200, 400, and 600 mg/ kg, i.p.	Exhibits sedative and hypnotic activities in pentobarbital-induced hypnotic model (Emamghoreishi and Heidari-Hamedani 2015).
Aqueous seed extract	Plus-maze in mice	10, 25, 50, and 100 mg/kg, i.p.	Increased the time spent on and the entries to the open arms. Furthermore, some extracts were accompanied by motor activity impairments and neuromuscular dysfunction (Emamghoreishi, Khasaki, and Aazam 2005).
Hydroalcoholic seed extract	Induced anxiety in mice	50, 100, and 200 mg/kg, i.p.	Showed anti-anxiety effects comparable to the effects of 0.5 mg/kg diazepam (Mahendra and Bisht 2011).
Leaves aqueous extract	Induced anxiety in mice	50, 100, and 200 mg/kg, i.p.	Increased the time spent in and the number of entries into the open arms of the elevated plus-maze, which was assumed to be due to the anxiolytic effect of the extract (Pathan, Kothawade, and Logade 2011).
Hydroalcoholic seed extract	Induced anxiety in mice	50, 100, and 200 mg/kg, i.p.	The extract also had an anxiolytic effect similar to the effects of the leaf extract. Locomotion activity and frequency of rearing in the groups treated with 200 mg/ kg of the extract were also lower than that of the control group (Mahendra and Bisht 2011).
Fresh and dried plant extract	Induced depression in mice	10 ml/kg, i.p.	The green but not dried plant extract has an anti-depressive effect (Kishore and Siddiqui 2003).
Aqueous extracts—extract obtained by diethyl ether	Induced depression in mice	200 and 400 mg/kg-2 and 4 ml/kg, p.o.	Inhibited monoamine oxidase-B (MAO-B) activity and showed anti-depressant like activity comparable to the effects of fluoxetine and imipramine (Kharade et al. 2011).
Coriander oil and linalool	Induced anxiety in chick	0.86, 8.6, and 86 µg/chick, i.c.v.	Decreased escapes, defecations, and crossing numbers in the open field, while increasing the sleeping posture (Gastón et al. 2016).
Inhalation coriander oil	Induced Alzheimer's disease in rat	1% and 3%	Significantly increased time spent in the open arms, open-arm entries, and the number of crossings (exploratory locomotor activity) in β-amyloid (1-42) rat model of Alzheimer's disease (Cioanca et al. 2014).
C. sativum hydroalcoholic extract	Induced seizure in rat	50, 100, and 200 mg/kg	Reduced duration, frequency, and amplitude of the burst discharges while prolonging the latency of the seizure attacks. The extract also prevented cells in different areas of the hippocampus (Pourzaki et al. 2017).
Inhaled linalool	Induced anxiety in mice	1% and 3%	Linalool 3% decreased exploring and motor activities. Linalool (1% and 3%) increased sleeping time and reduced body temperature (de Moura Linck et al. 2009).
Inhaled linalool	Induced anxiety in mice	1% and 3%	Linalool (3%) increased exploring behavior comparable to the effects of diazepam (1.0 mg/kg, i.p.). Linalool (1%) increased social interactions (Linck et al. 2010).

(Continued)

TABLE 12.2 (CONTINUED)
Sedative, Anti-Anxiety, and Anti-Depressive Effects of Coriander and Linalool

Extract of Coriander/ Linalool	Type of Study	Doses	Effects
Inhaled linalool oxide	Induced anxiety in mice	0.65%, 2.5%, and 5.0% (w/w)	Increased open entries and open arm duration and increased the time spent on the light compartment (Souto-Maior et al. 2011).
Linalool	Induced depression in rat	30 mg/kg, p.o.	Showed antidepressant activity by increased latency time and self-cleaning time while decreasing immobility time in rats (dos Santos et al. 2018).
Linalool	Induced anxiety in mice	100 mg/kg, i.p.	Improved cognitive performance and decreased the prolonged time in the dark chamber of model mice. Linalool also reduced the pathological lesions induced by AB to the normal range. It also increased SOD and GPX activities in the hippocampus and cortex compared to the induced animals while decreasing the cortex's MDA level close to the sham group (Xu et al. 2017).
Linalool	Induced Alzheimer's disease in mice	25 mg/kg, p.o.	Improved learning and spatial memory and significantly reduced extracellular β-amyloidosis, astrogliosis, and microgliosis as well as reducing the levels of the pro-inflammatory markers including p38 MAPK, COX2, and IL-1β in hippocampi and amygdalae (Sabogal-Guáqueta, Osorio, and Cardona-Gómez 2016).

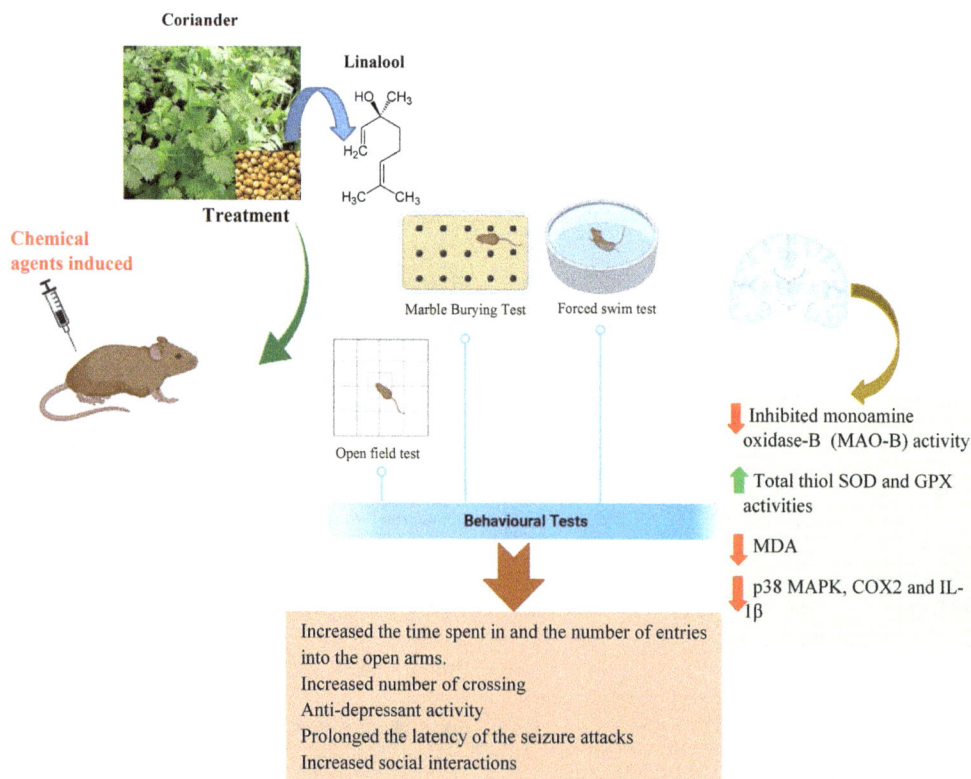

FIGURE 12.1   Neuro-pharmacological effects of coriander and linalool.

(30 mg/kg, i.p.) showed significant anti-depressant activity by increased latency time and self-cleaning time, while decreasing immobility time in rats (dos Santos et al. 2018). Linalool (100 mg/kg, i.p.) significantly improved the cognitive performance of model mice in the Morris water maze (MWM) test in amyloid-beta (Aβ_{1-40})-induced cognitive deficits of mice. Treatment with linalool significantly decreased the prolonged time in the dark chamber of model mice. Treatment with linalool also reduced the pathological lesions induced by Aβ to normal range. In addition, linalool (100 mg/kg) significantly increased SOD and GP$_X$ activities in the hippocampus and cortex compared to the induced animals while decreasing the MDA level in the cortex close to the sham group (Xu et al. 2017). Administration of linalool (25 mg/kg, p.o.), every 48 h for 3 months, improved learning and spatial memory and behavior during the elevated plus-maze on mice model of Alzheimer's disease. Linalool also significantly reduced extracellular β-amyloidosis, astrogliosis, and microgliosis, as well as reducing the levels of the pro-inflammatory markers including p38 MAPK, COX2, and IL-1β in hippocampi and amygdalae (Sabogal-Guáqueta, Osorio, and Cardona-Gómez 2016) (Table 12.2).

12.4 ANTICONVULSANT AND ANTI-SEIZURE EFFECTS OF CORIANDER AND LINALOOL

12.4.1 CORIANDER

The anticonvulsant activity of aqueous and ethanol extracts of coriander seeds was studied using electroshock and pentylenetetrazole (PTZ)-induced seizure models. Administration of aqueous (0.5 g/kg, i.p.) and ethanol extracts (3.5 and 5 g/kg, i.p.) shortened tonic seizure period and inhibited maximal electroshock-induced seizures. Moreover, both of the extracts postponed the onsets of clonic convulsions. The ethanol extract effect was similar to phenobarbital (20 mg/kg) effects in the PTZ test (Hosseinzadeh and Madanifard 2000).

Intraperitoneal administration of aqueous extracts (5, 10, and 15 mg/kg, i.p.), as well as its essential oils, 30 min before PTZ injection (85 mg/kg, i.p.) was adequate to postpone myoclonic and clonic seizure onsets in mice in a dose-dependent manner (EMAM and HEYDARI 2008). Similarly, the aqueous and ethanol extracts (200, 400, 600, and 800 mg/kg) and essential oil of the seeds were administered 30 min before PTZ injection (90 mg/kg, i.p.) to evaluate the anticonvulsant effects. The results showed that 600 and 800 mg/kg of all extracts increased myoclonic and clonic seizures latencies (Emamghoreishi and Heidari-Hamedani 2010).

The effect of an hydroalcoholic extract of coriander leaves on oxidative damage in the brain tissues of PTZ-induced seizure was investigated (Karami et al. 2015). Administration of the extract (100, 500, and 1000 mg/kg, i.p.) postponed the onsets of the minimal clonic (MCS) and generalized tonic-clonic (GTCS) seizures. Furthermore, cortical and hippocampal tissue MDA levels in the animals pre-treated with the extract decreased significantly. Pre-treatment with 500 mg/kg of the extract also significantly improved total thiol contents in the cortical tissues (Karami et al. 2015). Similarly, the effects of water fraction (WF), *n*-butanol fraction (NBF), and ethyl acetate fraction (EAF) of coriander (25 and 100 mg/kg, i.p.) on the brain tissue oxidative damage due to seizures induced by PTZ were investigated previously. The GTCS latency in WF and EAF treated rats (100 mg/kg, i.p) was longer than that of the PTZ group. Interestingly, WF, NBF, and EAF attenuated hippocampal MDA levels. Additionally, both WF and NBF increased cortical and hippocampal thiol contents (Anaeigoudari et al. 2016) (Table 12.3).

12.4.2 LINALOOL

The anticonvulsant effects of linalool on experimental seizure models related to glutamate, including N-methyl-D-aspartate (NMDA), quinolinic acid-induced convulsions, and the behavioral and

TABLE 12.3

Anticonvulsant and Anti-Seizure Activity of Coriander and Linalool

Extract/Component	Type of Study	Doses	Results
Aqueous seed extract	Induced seizure in mice	0.5 g/kg, i.p.	Shortened tonic seizure period and inhibited maximal electroshock-induced seizures. Moreover, it postponed the onsets of clonic convulsions (Hosseinzadeh and Madanifard 2000).
Ethanolic seed extract		3.5 and 5 g/kg, i.p.	Tonic seizure duration was decreased, clonic convulsion onsets were postponed. Moreover, the extract effect was similar to phenobarbital (20 mg/kg) effects in the PTZ test (Hosseinzadeh and Madanifard 2000).
Aqueous and ethanolic extracts and essential oil		200, 400, 600, and 800 mg/kg, i.p.	Increased myoclonic and clonic seizure latencies (Emamghoreishi and Heidari-Hamedani 2010).
Hydroalcoholic seed extract and essential oil		5, 10, and 15 mg/kg, i.p.	Postponed myoclonic and clonic seizure onsets in mice in a dose-dependent manner (Emam and Heydari 2008).
Hydroalcoholic leaf extract	PTZ- induced seizure in rat	100, 500, and 1000 mg/kg, i.p.	Postponed the onsets of the minimal clonic (MCS) and generalized tonic-clonic (GTCS) seizures. Furthermore, cortical and hippocampal tissue MDA levels were decreased significantly. Also significantly improved total thiol contents in the cortical tissues (Karami et al. 2015).
Aerial parts fractions		25 and 100 mg/kg, i.p.	GTCS latency was significantly improved. In addition, hippocampal MDA concentrations were decreased cortical and hippocampal tissues thiol contents were improved (Anaeigoudari et al. 2016).
Linalool	*In vitro*	0.3 mM or 1.0 mM	Modulated glutamate activation expression in the rat cortex membrane cells (Elisabetsky, Brum, and Souza 1999).
Linalool	Quinolinic acid-induced seizure in rat	350 mg/kg, i.p.	Delayed NMDA (270 mg/kg, i.p.) induced seizure onsets comparable to the effects of diazepam. Also, significantly inhibited quinolinic acid (9.2 mM, i.c.v.) induced seizure and significantly inhibited and delayed the behavioral expression in PTZ (60 mg/kg, s.c.) kindling model (Elisabetsky, Brum, and Souza 1999).
Linalool	Quinolinic acid-induced seizure in mice	1.0 or 3.0 mM	Reduced potassium-stimulated glutamate release and glutamate uptake (90%) (Brum et al. 2001).

neurochemical factors related to PTZ-kindling, were investigated. Treatment with linalool (0.3 mM or 1.0 mM) modulated glutamate activation expression in the rat cortical membrane cells (*in vitro*). Linalool at a dose (350 mg/kg, i.p.) and diazepam also significantly prolonged the delay time to the onset of NMDA (270 mg/kg, i.p.) induced seizures. Administration of linalool (15, 30 and 45mM, i.c.v.) significantly inhibits quinolinic acid (9.2 mM, i.c.v.) induced seizure attack in a dose-dependent manner. In addition, linalool (2.2 and 2.5 g/kg, p.o.) significantly inhibited and delayed the behavioral expression in PTZ (60 mg/kg, s.c.) kindling model (Elisabetsky, Brum, and Souza 1999). The anticonvulsant effects of linalool in mice cortical synaptosomes showed linalool (1.0 or 3.0 mM) reduced potassium-stimulated glutamate release as well as glutamate uptake (90%) (Brum et al. 2001) (Table 12.3).

12.5 MEMORY IMPROVING EFFECTS OF CORIANDER AND LINALOOL

12.5.1 CORIANDER

Rats assessed daily inhalation of coriander volatile oil (1% and 3%) for 21 days on spatial memory performance in the $A\beta_{1-40}$ induced Alzheimer's disease model. Volatile oils (both 1% and 3%) improved and reversed deleterious effects of ($A\beta_{1-40}$, i.c.v.) injection and increased the percentage of spontaneous alternations and number of arm entries, and reduced working and improved errors in the radial arm maze task. In addition, the volatile oil also improved hippocampal tissue oxidative stress markers in $A\beta_{1-40}$ treated rats by increasing GP_X levels and reducing SOD, lactate dehydrogenase (LDH), and MDA levels in a dose-response manner. In addition, the amyloid deposits were scarce in volatile oil-treated animals. Also, cleavages of DNA were absent in the coriander oils groups (Cioanca et al. 2013).

Similarly, both 1% and 3% of volatile oil increased the time spent in and the entering number into the open arms in elevated plus-maze compared to the Aß group. The volatile oil also enhanced the swimming time and decreased the immobility time in Alzheimer rats. In addition, the antioxidant parameters such as catalase (CAT) activity decreased, and the total content of reduced glutathione (GSH) was increased in the hippocampal tissues of the rats (Cioanca et al. 2014).

Aqueous coriander extract's protective and therapeutic effects in aluminum chloride ($AlCl_3$) induced animal model of Alzheimer's disease and its effects on the cortical pyramidal cells were investigated in male albino rats. Treatment with coriander extract (0.5 mg/kg, p.o.) for one month after stopping $AlCl_3$ (300 mg/kg, p.o.) treatment restored deleterious effects on the pyramidal cells including the dilatation of blood capillaries and presence of many shrunk pyramidal cells, and detached and irregular fibers in thickness (Enas 2010).

Anti-amnestic and anti-stress properties of coriander extract on vanillylmandelic acid (VMA) and ascorbic acid urinary levels in scopolamine-induced amnesia model showed daily administration of 100, 200, and 300 mg/kg of the extract, one hour prior to exposure of animals to stress, decreased the stress-induced urinary levels of VMA, while increasing the excretion of ascorbic acid. Moreover, the plant extract dose-dependently reversed acquisition, retention, and recovery impairments induced by scopolamine (1 mg/kg, i.p.) accompanied by lipid peroxidation inhibition in both liver and brain tissues (Koppula and Choi 2012).

Oral administration of young and aged rats with leaf extract (5, 10, and 15 % w/w) for 45 days dose-dependently improved memory. In addition, the leaf extract also effectively reversed the memory impairments induced by scopolamine (0.4 mg/kg, i.p.) or diazepam (1 mg/kg, i.p.) administration in plus-maze and Hebb-Williams maze models for testing memory (Mani and Parle 2009, Mani et al. 2011). In learning and memory impairment due to seizures induced by pilocarpine (30 mg/kg, i.p.), administration of 200 mg/kg (i.p.) of the extract for 7 consecutive days increased the latencies of seizure onsets which was accompanied by a decreased level of latency to find the platform in Morris water maze (Elahdadi-Salmani, Khorshidi, and Ozbaki 2015) (Table 12.3).

12.5.2 LINALOOL

The beneficial effects of linalool on triple transgenic Alzheimer's disease model (3xTg-AD) in mice were studied. Both non-transgenic and 3xTg-AD mice received linalool (25 mg/kg, p.o.) every 48 h for 3 months and showed a significantly better retention performance and swam a longer distance in the platform location compared to 3xTg-AD vehicle-treated mice. In addition, linalool treatment successfully increased the time spent and the entries into the open arms compared to vehicle-treated mice. Linalool treatment also delayed deposits and reduced the abundance of β-amyloid peptide and significantly reduced the Aß (1-40), (1-42) protein and pair helical filaments (PHFs) levels in the hippocampal tissues. Additionally, an immunohistochemistry study showed that glial fibrillary acidic protein (GFAP) decreased in the CA1 area of the hippocampus, entorhinal cortex (EC), and amygdala in mice. Moreover, linalool significantly reduced the extracellular of β-amyloids,

TABLE 12.4

Memory Improving and Anti-Alzheimer's Disease Properties of Coriander and Linalool

Extract/Component	Type of Study	Doses	Results
Volatile oil	Induced Alzheimer's disease in rats	1 and 3%	Improved and reversed the deleterious effects of amyloid ß. Also increased GP_X levels and reduced SOD, lactate dehydrogenase (LDH), and MDA levels (Cioanca et al. 2013).
Volatile oil		1 and 3%	Enhanced the swimming time and decreased the immobility time in Alzheimer rats. In addition, the reduced glutathione (GSH) was increased in the hippocampal tissues (Cioanca et al. 2014).
Aqueous seed extract		0.5 mg/kg, p.o.	Restored deleterious effects on the pyramidal cells, including dilatation of blood capillaries and many shrunk pyramidal cells (Enas 2010).
Aqueous fruit extract		100, 200, and 300 mg/kg	Decreased VMA urinary levels in the stress-induced animal model while increasing ascorbic acid excretion (Koppula and Choi 2012).
Leaves mixed in the diet		5, 10, and 15% (w/w)	Improved memory scores in aged and young rats.
Ethanolic leaf extract		200 mg/kg, i.p.	Increased latencies of seizure onsets accompanied a decreased latency to find the platform.
Ethanolic seed extract		100 and 200 mg/kg, p.o.	Decreased VCM, TP, and OB behaviors induced by tacrine and improved locomotor activity and cognitive functions.
Linalool	Induced Alzheimer's disease in mice	25 mg/kg, p.o.	Reduced extracellular beta-amyloids, astroglia, and microglia and the pro-inflammatory marker levels, including mitogen-activated protein kinases, p38, nitric oxide (NO) synthase 2, cyclooxygenase-2, and interleukin 1 beta, in the AD treated mice (Sabogal-Guáqueta, Osorio, and Cardona-Gómez 2016).
Inhaled linalool		1%	Increased social interaction, reduced aggressive behavior, and impaired memory compared to control mice (Linck et al. 2010).

astroglia, and microglia, and the pro-inflammatory marker levels, including mitogen-activated protein kinases, p38 nitric oxide (NO) synthase 2, cyclooxygenase-2, and interleukin 1 beta, in the AD-treated mice (Sabogal-Guáqueta, Osorio, and Cardona-Gómez 2016).

The effects of linalool (1%) inhalation on acquisition memory in the step-down inhibitory avoidance test showed anxiolytic properties, increased social interaction, and reduced aggressive behavior as well as impaired memory in comparison with the control mice (Linck et al. 2010) (Table 12.4).

12.6   ANTI-PARKINSON EFFECTS

The effects of 100 and 200 mg/kg of the seed ethanol extract were evaluated on orofacial dyskinesia induced by tacrine. Administration of 2.5 mg/kg of tacrine (i.p) induced some orofacial dyskinesia symptoms, including tongue protrusions (TP), vacuous chewing movements (VCM), and orofacial bursts (OB) 1 h after administration. However, pre-treatment with 100 and 200 mg/kg (p.o.) of the seed extract for 15 days considerably diminished VCM, TP, and OB induced by tacrine and also improved locomotor activity and cognitive functions. In addition, administration of the extract significantly improved SOD, CAT, and GSH levels, while attenuating lipid peroxidation (Mohan, Yarlagadda, and Chintala 2015).

12.7 THE EFFECTS OF CORIANDER AND LINALOOL ON NEUROTOXICITY

12.7.1 CORIANDER

The beneficial effects of coriander seed hydroalcoholic extract on neurodegeneration due to lead exposure showed that daily treatment with the extract (250 and 500 mg/kg) for a 7-day period, which was started after 4 weeks of lead exposure, increased reactive oxygen species (ROS) production and lipid peroxidation accompanied by diminished total protein carbonyl content in the cerebellum, hippocampus, frontal cortex, and brain stem (Velaga et al. 2014).

Hydroalcoholic extract, WF, EAF, and NBF (0.1, 0.2, 0.4, and 0.8 mg/mL) had a protective activity on neurons (PC12 cells) in glucose/serum deprivation-induced cytotoxicity. The extract and the fractions showed no cytotoxicity effect in standard conditions, while EAF and NBF (1.6 mg/mL) decreased cell survival. On the other hand, glucose/serum deprivation-induced cytotoxicity was significantly reduced by WF at 0.4, 0.8, and 1.6 mg/mL (Ghorbani et al. 2011).

The leaves' methanol extract (200 mg/kg, p.o., per day) increased endogenous enzyme levels such as superoxide dismutase, glutathione, catalase, and total protein levels. At the same time, it reduced the infarcted volume of the brain and attenuated the lipid peroxidation and calcium concentration in the brains of the rats in ischemia conditions induced by bilateral common carotid arteries occlusion. Additionally, blood vessel congestions and neutrophil infiltration were observed, accompanied by an increasing level of intercellular spaces due to ischemia in rats. The extract improved all the damage. The infarction area in the caudal side of the hippocampus was also reduced after treatment by the extract (Vekaria et al. 2012) (Table 12.5).

12.7.2 LINALOOL

The possible neuroprotective effect of linalool was investigated in the glucose/serum deprivation-induced cytotoxicity model. Glucose/serum deprivation reduced cell viability when cultured for 8 h compared to the standard condition culture, but treatment with 16 μg/ml of linalool improved cell viability (Alinejad, Ghorbani, and Sadeghnia 2013).

The effect of linalool on neuronal damages induced by acrylamide (ACR) was also evaluated. ACR is potentially a neurotoxic water-soluble monomer and is extensively used in different industries. This monomer damages the central nervous system (CNS) and peripheral nervous system (PNS) in humans and animals. ACR-induced neurotoxicity has been suggested to be mainly due to oxidative stress. Administration of 12.5, 25, 50, and 100 mg/kg (i.p.) of linalool for 11 days significantly reduced severe gait abnormalities induced by ACR. Linalool also increased GSH content; however, it decreased lipid peroxidation in the cortical tissues of the rat brain (Mehri, Meshki, and Hosseinzadeh 2015).

Linalool significantly reduced neuronal injury in the cortical tissues due to oxygen-glucose deprivation/reoxygenation (OGD/R), while it could not inhibit N-methyl-D-aspartate (NMDA)-induced excitotoxicity. It also significantly scavenged peroxyl radicals and reduced intracellular oxidative stress during OGD/R. Monocyte-chemoattractant protein-1 (MCP-1), which is a chemokine that is released during OGD/R and induces microglial migration, was also inhibited by linalool (Park et al. 2016a).

Linalool (1 μM, 2.5 μM, and 5 μM) had no cytotoxic effects on SH-SY5Y cells and did not significantly induce NO. However, incubation of the cells with sodium nitroprusside (SNP) (2.5 mM) reduced cell viability. Linalool also increased neuronal cell viability (about 60% cell viability) and significantly prevented NO over-production. In addition, linalool significantly increased antioxidant levels and protected SH-SY5Y cells against SNP-induced cytotoxicity (Kim et al. 2015). It has been

TABLE 12.5

The Effects of Coriander and Linalool on Neurotoxicity

Extract/ Component	Type of Study	Doses	Results
Seed hydroalcoholic extract	Lead exposure	250 and 500 mg/kg	Reduced ROS production and lipid peroxidation in the cerebellum, hippocampus, frontal cortex, and brain stem (Velaga et al. 2014).
Leaf methanolic extract	Ischemic rat model	200 mg/kg, p.o.,	Increased total proteins, superoxide dismutase, catalase, and glutathione while attenuating cerebral infarcted volume prevented increasing lipid peroxidation and calcium levels in the ischemic rat brain (Vekaria et al. 2012).
WF, EAF, and NBF of coriander	*In vitro*	0.1, 0.2, 0.4, and 0.8 mg/ mL	Glucose/serum deprivation induced cytotoxicity were significantly reduced by WF at 0.4, 0.8, and 1.6 mg/ml (Ghorbani et al. 2011).
Linalool	*In vitro*	16 μg/mL	Significantly improved cell viability (Alinejad, Ghorbani, and Sadeghnia 2013).
Linalool	Acrylamide (ACR) induced neural damage in rat	12.5, 25, 50, and 100 mg/ kg i.p.	Significantly reduced severe gait abnormalities induced by ACR (Mehri, Meshki, and Hosseinzadeh 2015).
Linalool	*In vitro*	0.01, 0.1, 1, and 10 μm	Significantly reduced intracellular oxidative damage in the OGD/R model. It also scavenged peroxyl radicals and inhibited the monocyte-chemoattractant protein-1 (Park et al. 2016a).
Linalool	*In vitro*	1 μm, 2.5 μm, and 5 μm	Increased neuronal cell viability (60% cell viability) significantly prevented NO over-production and improved antioxidant status (Kim et al. 2015).
Linalool	*In vitro*	10 μM	Showed therapeutically effects in treating ischemia-induced cerebral neuronal injury by inhibiting oxidative stress and inflammatory responses (Park et al. 2016b).
Linalool	ACR-induced neural damage in rat	12.5 mg/kg, i.p.	Showed neuroprotective effects by reducing the progressive gait abnormalities promoted by acrylamide, oxidative stress, lipid peroxidation, and increasing glutathione (GSH) (Mehri, Meshki, and Hosseinzadeh 2015).
Beta-myrcene (MYR)	Induced ischemia in mice	200 mg/kg, i.p.	Decreased antioxidant factors including GSH, catalase, glutathione peroxidase (GPX), and SOD (Ciftci, Oztanir, and Cetin 2014).

reported that linalool (10 μM) had effects therapeutically in treating ischemia-induced cerebral neuronal injury by inhibiting oxidative stress and inflammatory responses in rat cortical cells (Park et al. 2016b). Treatment of rats with linalool (12.5 mg/kg, i.p.) showed neuroprotective effects by reducing the progressive gait abnormalities promoted by acrylamide, oxidative stress, lipid peroxidation, and increasing the levels of glutathione (GSH) (Mehri, Meshki, and Hosseinzadeh 2015) (Table 12.5).

12.7.3 BETA-MYRCENE

The effects of β-myrcene (a monoterpene derived from coriander) on oxidative and histological damages induced by global cerebral ischemia/reperfusion (I/R) in tissues of C57BL/J6 mice brain were investigated. Apoptosis in the tissues and other brain damage were revealed following I/R induction, while antioxidant factors, including GSH, catalase, glutathione peroxidase

(GPx), and SOD, decreased. Furthermore, treating animals with MYR (200 mg/kg, i.p.) for 10 days significantly decreased thiobarbituric acid reactive substances (TBARS) formation. Furthermore, MYR protected against oxidative stress factors and significantly increased GSH, GPx, and SOD. Additionally, MYR showed a neuroprotective effect and decreased incidence of histopathological damages and apoptosis in the brain tissue (Ciftci, Oztanir, and Cetin 2014) (Table 12.5).

12.8 CONCLUSION

Various parts of the plant, such as coriander seeds and leaves, possess antioxidant, anxiolytic, anti-depressive, learning and memory improving, neuroprotective, sedative-hypnotic, analgesic, and anticonvulsant activities. Furthermore, the plant's main components, including linalool, have different neuropharmacological effects, including anti-anxiety, sedative, anticonvulsant, and anti-Alzheimer activities. Finally, the neuroprotective effects of *C. sativum* and its main component may be their antioxidant and anti-inflammatory properties.

REFERENCES

Alinejad, Bagher, Ahmad Ghorbani, and Hamid Reza Sadeghnia. 2013. "Effects of combinations of curcumin, linalool, rutin, safranal, and thymoquinone on glucose/serum deprivation-induced cell death." *Avicenna Journal of Phytomedicine* 3(4):321–328.

Anaeigoudari, Akbar, Mahmoud Hosseini, Reza Karami, Farzaneh Vafaee, Toktam Mohammadpour, Ahmad Ghorbani, and Hamid Reza Sadeghnia. 2016. "The effects of different fractions of *Coriandrum sativum* on pentylenetetrazole-induced seizures and brain tissues oxidative damage in rats." *Avicenna Journal of Phytomedicine* 6(2):223–235.

Balick, Michael J., and Paul Alan Cox. 1996. *Plants, People, and Culture: The Science of Ethnobotany.* Scientific American Library.

Bhat, S., P. Kaushal, M. Kaur, and H. K. Sharma. 2014. "Coriander (*Coriandrum sativum* L.): Processing, nutritional and functional aspects." *African Journal of Plant Science* 8(1):25–33.

Brum, L. F. Silva, T. Emanuelli, D. O. Souza, and E. Elisabetsky. 2001. "Effects of linalool on glutamate release and uptake in mouse cortical synaptosomes." *Neurochemical Research* 26(3):191–194.

Ciftci, Osman, M. Namik Oztanir, and Aslı Cetin. 2014. "Neuroprotective effects of β-myrcene following global cerebral ischemia/reperfusion-mediated oxidative and neuronal damage in a C57BL/J6 mouse." *Neurochemical Research* 39(9):1717–1723.

Cioanca, Oana, Lucian Hritcu, Marius Mihasan, and Monica Hancianu. 2013. "Cognitive-enhancing and antioxidant activities of inhaled coriander volatile oil in amyloid β (1–42) rat model of Alzheimer's disease." *Physiology and Behavior* 120:193–202.

Cioanca, Oana, Lucian Hritcu, Marius Mihasan, Adriana Trifan, and Monica Hancianu. 2014. "Inhalation of coriander volatile oil increased anxiolytic-antidepressant-like behaviors and decreased oxidative status in beta-amyloid (1–42) rat model of Alzheimer's disease." *Physiology and Behavior* 131:68–74.

de Figueiredo, R. Oliveira, Márcia Ortiz Maio Marques, João Nakagawa, and Lin Chau Ming. 2004. "Composition of coriander essential oil from Brazil." *Acta Horticulturae*:135–138.

de Moura Linck, Viviane, Adriana Lourenço da Silva, Micheli Figueiró, Angelo Luis Piato, Ana Paula Herrmann, Franciele Dupont Birck, Elina Bastos Caramao, Domingos Savio Nunes, Paulo Roberto H. Moreno, and Elaine Elisabetsky. 2009. "Inhaled linalool-induced sedation in mice." *Phytomedicine* 16(4):303–307.

Dhanapakiam, P., J. Mini Joseph, V. K. Ramaswamy, M. Moorthi, and A. Senthil Kumar. 2007. "The cholesterol lowering property of coriander seeds (*Coriandrum sativum*): Mechanism of action." *Journal of Environmental Biology* 29(1):53.

Diederichsen, Axel, and Karl Hammer. 2003. "The infraspecific taxa of coriander (*Coriandrum sativum* L.)." *Genetic Resources and Crop Evolution* 50(1):33–63.

dos Santos, Éverton Renan Q., Cristiane Socorro F. Maia, Enéas A. Fontes Junior, Ademar S. Melo, Bruno G. Pinheiro, and José Guilherme S. Maia. 2018. "Linalool-rich essential oils from the Amazon display antidepressant-type effect in rodents." *Journal of Ethnopharmacology* 212:43–49.

Duke, James A. 2002. *Handbook of Medicinal Herbs.* CRC Press.

Elahdadi-Salmani, Mahmoud, Mahdi Khorshidi, and Jamile Ozbaki. 2015. "Reversal effect of *Coriandrum sativum* leaves extract on learning and memory deficits induced by epilepsy in male rat." *Zahedan Journal of Research in Medical Sciences* 17(3).

Elisabetsky, E., L. F. Silva Brum, and D. O. Souza. 1999. "Anticonvulsant properties of linalool in glutamate-related seizure models." *Phytomedicine* 6(2):107–113.

Emam, Ghoreyshi M., and Hamedani Ghazal Heydari. 2008. "Effect of extract and essential oil of *Coriandrum sativum* seed against pentylenetetrazole-induced seizure." *Pharmaceutical Sciences* 14(3):1–10.

Emamghoreishi, M., and G. Heidari-Hamedani. 2010. "Anticonvulsant effect of extract and essential oil of *Coriandrum sativum* seed in concious mice." *Iranian Journal of Pharmaceutical Research*:71.

Emamghoreishi, M., and G. Heidari-Hamedani. 2015. "Sedative-hypnotic activity of extracts and essential oil of coriander seeds." *Iranian Journal of Medical Sciences* 31(1):22–27.

Emamghoreishi, Masoumeh, Mohammad Khasaki, and Maryam Fath Aazam. 2005. "*Coriandrum sativum*: Evaluation of its anxiolytic effect in the elevated plus-maze." *Journal of Ethnopharmacology* 96(3):365–370.

Enas, A. Khalil. 2010. "Study of the possible protective and therapeutic influence of Coriander (*Coriandrum sativum* L.) against neurodegenerative disorders and Alzheimer's disease induced by aluminum chloride in cerebral cortex of male Albino rats." *Natural Science* 8(11):202–213.

Ertas, O. N., T. Guler, M. Ciftci, B. Dalkilic, and O. Yilmaz. 2005. "The effect of a dietary supplement coriander seeds on the fatty acid composition of breast muscle in Japanese quail." *Revue de médecine vétérinaire* 156(10):514.

Gastón, María Soledad, Mariana Paula Cid, Ana María Vázquez, María Florencia Decarlini, Gabriela I. Demmel, Laura I. Rossi, Mario Leandro Aimar, and Nancy Alicia Salvatierra. 2016. "Sedative effect of central administration of *Coriandrum sativum* essential oil and its major component linalool in neonatal chicks." *Pharmaceutical Biology*:1–8.

Ghorbani, Ahmad, Hassan Rakhshandeh, Elham Asadpour, and Hamid Reza Sadeghnia. 2011. "Effects of *Coriandrum sativum* extracts on glucose/serum deprivation-induced neuronal cell death." *Avicenna Journal of Phytomedicine* 2(1):4–9.

Gilani, Anwarul Hassan, and Attar-ur Rahman. 2005. "Trends in ethnopharmacology." *Journal of Ethnopharmacology* 100(1):43–49.

Hosseinzadeh, Hossein, and Mohammad Madanifard. 2000. "Anticonvulsant effects of *Coriandrum sativum* L. seed extracts in mice." *Archives of Iranian Medicine* 3:182–184.

Iwatani, Yoko, Jayashree Arcot, and Ashok K. Shrestha. 2003. "Determination of folate contents in some Australian vegetables." *Journal of Food Composition and Analysis* 16(1):37–48.

Kandlakunta, Bhaskarachary, Ananthan Rajendran, and Longvah Thingnganing. 2008. "Carotene content of some common (cereals, pulses, vegetables, spices and condiments) and unconventional sources of plant origin." *Food Chemistry* 106(1):85–89.

Kansal, Leena, Veena Sharma, Arti Sharma, Shweta Lodi, and S. H. Sharma. 2011. "Protective role of *Coriandrum sativum* (coriander) extracts against lead nitrate induced oxidative stress and tissue damage in the liver and kidney in male mice." *Journal of Applied Pharmaceutical Technology* 2(3):65–83.

Karami, Reza, Mahmoud Hosseini, Toktam Mohammadpour, Ahmad Ghorbani, Hamid Reza Sadeghnia, Hassan Rakhshandeh, Farzaneh Vafaee, and Mahdi Esmaeilizadeh. 2015. "Effects of hydroalcoholic extract of *Coriandrum sativum* on oxidative damage in pentylenetetrazole-induced seizures in rats." *Iranian Journal of Neurology* 14(2):59.

Kharade, S. M., D. S. Gumate, V. M. Patil, S. P. Kokane, and N. S. Naikwade. 2011. "Behavioral and biochemical studies of seeds of *Coriandrum sativum* in various stress models of depression." *International Journal of Current Research and Review* 3(3):4–11.

Kim, K. Y., H. S. Lee, S. S. Min, and G. H. Seol. 2015. "Neuroprotective effect of (-)-linalool against sodium nitroprusside-induced cytotoxicity." *Medicinal Chemistry* 5(4):178–182.

Kishore, K., and N. A. Siddiqui. 2003. "Anti-despair activity of green and dried plant extract of *Coriandrum sativum* in mice." *Indian Drugs* 40(7):419–421.

Koppula, Sushruta, and C. K. Choi. 2012. "Anti-stress and anti-amnesic effects of *Coriandrum sativum* Linn (Umbelliferae) extract-an experimental study in rats." *Tropical Journal of Pharmaceutical Research* 11(1):36–42.

Linck, V. M., A. L. Da Silva, M. Figueiró, E. B. Caramão, P. R. H. Moreno, and E. Elisabetsky. 2010. "Effects of inhaled linalool in anxiety, social interaction and aggressive behavior in mice." *Phytomedicine* 17(8):679–683.

Mahendra, Poonam, and Shradha Bisht. 2011. "Anti-anxiety activity of *Coriandrum sativum* assessed using different experimental anxiety models." *Indian Journal of Pharmacology* 43(5):574.

Mani, Vasudevan, and Milind Parle. 2009. "Memory-enhancing activity of *Coriandrum sativum* in rats." *Pharmacologyonline* 2:827–839.

Mani, Vasudevan, Milind Parle, Kalavathy Ramasamy, Abdul Majeed, and Abu Bakar. 2011. "Reversal of memory deficits by *Coriandrum sativum* leaves in mice." *Journal of the Science of Food and Agriculture* 91(1):186–192.

Mehri, Soghra, Mohammad Ali Meshki, and Hossein Hosseinzadeh. 2015. "Linalool as a neuroprotective agent against acrylamide-induced neurotoxicity in Wistar rats." *Drug and Chemical Toxicology* 38(2):162–166.

Mir, Heidar. 1992. "*Coriandrum sativum* in: Application of plants in prevention and treatment of illnesses." *Persian* 1:257–252.

Mohan, Mahalaxmi, Sanjyothi Yarlagadda, and Saritha Chintala. 2015. "Effect of ethanolic extract of *Coriandrum sativum* L. on tacrine induced orofacial dyskinesia." *Indian Journal of Experimental Biology* 53(5):292–296.

Msaada, Kamel, Karim Hosni, Mouna Ben Taarit, Thouraya Chahed, Mohamed Elyes Kchouk, and Brahim Marzouk. 2007. "Changes on essential oil composition of coriander (*Coriandrum sativum* L.) fruits during three stages of maturity." *Food Chemistry* 102(4):1131–1134.

Nejad Ebrahimi, Samad, Javad Hadian, and Hamid Ranjbar. 2010. "Essential oil compositions of different accessions of *Coriandrum sativum* L. from Iran." *Natural Product Research* 24(14):1287–1294.

Nurzyńska-Wierdak, Renata. 2013. "Essential oil composition of the coriander (*Coriandrum sativum* L.) herb depending on the development stage." *Acta Agrobotanica* 66(1):53–60.

Orav, Anne, Elmar Arak, and Ain Raal. 2011. "Essential oil composition of *Coriandrum sativum* L. fruits from different countries." *Journal of Essential Oil Bearing Plants* 14(1):118–123.

Pandey, A., P. Bigoniya, V. Raj, and K. K. Patel. 2011. "Pharmacological screening of *Coriandrum sativum* Linn. for hepatoprotective activity." *Journal of Pharmacy and Bioallied Sciences* 3(3):435.

Park, Hyeon, Geun Hee Seol, Sangwoo Ryu, and In-Young Choi. 2016a. "Neuroprotective effects of (-)-linalool against oxygen-glucose deprivation-induced neuronal injury." *Archives of Pharmacal Research*:1–10.

Park, Hyeon, Geun Hee Seol, Sangwoo Ryu, and In-Young Choi. 2016b. "Neuroprotective effects of (-)-linalool against oxygen-glucose deprivation-induced neuronal injury." *Archives of Pharmacal Research* 39(4):555–564.

Pathan, A. R., K. A. Kothawade, and Mohd Nadeem Logade. 2011. "Anxiolytic and analgesic effect of seeds of *Coriandrum sativum* Linn. " *International Journal of Research in Pharmacy and Chemistry* 1:1087–1099.

Pourzaki, Mojtaba, Mansour Homayoun, Saeed Sadeghi, Masoumeh Seghatoleslam, Mahmoud Hosseini, and Alireza Ebrahimzadeh Bideskan. 2017. "Preventive effect of *Coriandrum sativum* on neuronal damages in pentylentetrazole-induced seizure in rats." *Avicenna Journal of Phytomedicine* 7(2):116.

Rakhshandeh, Hassan, Hamid Reza Sadeghnia, and Ahmad Ghorbani. 2012. "Sleep-prolonging effect of *Coriandrum sativum* hydro-alcoholic extract in mice." *Natural Product Research* 26(22):2095–2098.

Sabogal-Guáqueta, Angélica Maria, Edison Osorio, and Gloria Patricia Cardona-Gómez. 2016. "Linalool reverses neuropathological and behavioral impairments in old triple transgenic Alzheimer's mice." *Neuropharmacology* 102:111–120.

Sahib, Najla Gooda, Farooq Anwar, Anwarul-Hassan Gilani, Azizah Abdul Hamid, Nazamid Saari, and Khalid M. Alkharfy. 2013. "Coriander (*Coriandrum sativum* L.): A potential source of high-value components for functional foods and nutraceuticals-A review." *Phytotherapy Research* 27(10):1439–1456.

Souto-Maior, Flávia Negromonte, Fabíola Lélis de Carvalho, Liana Clébia Soares Lima de Morais, Sueli Mendonça Netto, Damião Pergentino de Sousa, and Reinaldo Nóbrega de Almeida. 2011. "Anxiolytic-like effects of inhaled linalool oxide in experimental mouse anxiety models." *Pharmacology, Biochemistry, and Behavior* 100(2):259–263.

Sriti Eljazi, Jazia, Olfa Bachrouch, Nidhal Salem, Kamel Msaada, Jihad Aouini, Majdi Hammami, Emna Boushih, Manef Abderraba, Ferid Limam, and Jouda Mediouni Ben Jemaa. 2017. "Chemical composition and insecticidal activity of essential oil from coriander fruit against *Tribolium castaenum*, *Sitophilus oryzae*, and *Lasioderma serricorne*." *International Journal of Food Properties* 20(sup3):S2833–S2845.

Tashinen, J., and L. Nykänen. 1975. "Volatile constituents obtained by the extraction with alcohol-water mixture and by steam distillation of coriander fruit." *Acta Chemica Scandidavica* B29.

Vekaria, Rutvi H., Milan N. Patel, Payal N. Bhalodiya, Vishal Patel, Tushar R. Desai, and Pravin R. Tirgar. 2012. "Evaluation of neuroprotective effect of *Coriandrum sativum* linn. against ischemic-reperfusion insult in brain." *International Journal of Phytopharmacology* 3(2):186–193.

Velaga, Manoj Kumar, Prabhakara Rao Yallapragada, Dale Williams, Sharada Rajanna, and Rajanna Bettaiya. 2014. "Hydroalcoholic seed extract of *Coriandrum sativum* (Coriander) alleviates lead-induced oxidative stress in different regions of rat brain." *Biological Trace Element Research* 159(1–3):351–363.

Wangensteen, Helle, Anne Berit Samuelsen, and Karl Egil Malterud. 2004. "Antioxidant activity in extracts from coriander." *Food Chemistry* 88(2):293–297.

Xu, Pan, Kezhu Wang, Cong Lu, Liming Dong, Li Gao, Ming Yan, Silafu Aibai, Yanyan Yang, and Xinmin Liu. 2017. "Protective effects of linalool against amyloid beta-induced cognitive deficits and damages in mice." *Life Sciences* 174:21–27.

Yeung, Edward C., and Steve Bowra. 2011. "Embryo and endosperm development in coriander (*Coriandrum sativum*)." *Botany* 89(4):263–273.

Zargar-Nattaj, Seyed Sadegh, Pooya Tayyebi, Vahid Zangoori, Yasaman Moghadamnia, Hasan Roodgari, Seyed Gholamali Jorsaraei, and Ali Akbar Moghadamnia. 2011. "The effect of *Coriandrum sativum* seed extract on the learning of newborn mice by electric shock: interaction with caffeine and diazepam." *Psychology Research and Behavior Management* 4:13.

Zargari, Ali. 1991. *Medicinal Plants*. 5th ed. Tehran University Publications.

13 Pharmaceutical Applications of Coriander in Neurodegenerative Disorders

Swati Sahoo and Brijesh Sukumaran

CONTENTS

13.1 INTRODUCTION

Neurodegeneration is a complex process involving progressive damage to the central nervous system (CNS) neurons leading to loss of brain function. The selective loss of neuronal activity results in miscommunications between brain cells. The degeneration process is irreversible, resulting in cognitive and sensory impairments which affect an individual's movement, speech, memory, intelligence, etc. Further, a prolonged and significant neuronal loss also results in the development of neurodegenerative disorders (ND) such as Parkinson's disease (PD), Huntington's disease (HD), Alzheimer's disease (AD), spinocerebellar ataxia, epilepsy, cerebral ischemia, frontotemporal dementia (FTD), multiple sclerosis (MS), and amyotrophic lateral sclerosis (ALS) (Roos, 2010; Gitler et al., 2017; Lal and Lal, 2020; Naveen and Bhattacharjee 2021).

The global burden of neurological and mental disorders is massive and increasing, making them the second-largest cause of global deaths (Pakpoor and Goldacre, 2017). An estimated 6.8 million deaths are reported every year in the US from neurological disorders (Gooch et al., 2017). These disorders are age-dependent and typically found in people 65 years old and above. As the world population ages, these diseases become increasingly prevalent and present a significant threat to human health. In addition, neurodegenerative disorders are generally considered incurable, and the current treatments only provide symptomatic relief rather than stopping or reversing the progression of the disease.

Aging is a natural process wherein the brain starts developing neuropathological changes and could lose its structural and functional plasticity. Apart from aging, environmental factors including exposure to metals, pesticides, viral infections, smoking, pollutants, and microorganisms could lead to neurological disorders. Studies have shown that the most common pathophysiology of neurodegeneration is protein deposition with alteration in brain physicochemical traits. These pathologic changes are called protein miscoding or misfolding, which result in the accumulation and

DOI: 10.1201/9781003204626-14

aggregation of β-amyloid (Aβ) and neurofibrillary tangles (NFTs) in the case of AD (Huang et al., 2000), α-synuclein in PD (Stefanis L, 2012), huntingtin protein in HD (Schulte and Littleton, 2011), and TAR DNA binding protein-43 (TDP-43) in FTD and ALS (Bourbouli et al., 2017).

Coriander (*Coriandrum sativum* L.) is an aromatic plant that belongs to the family Umbelliferae (Apiaceae). Although indigenous to the Mediterranean region and the Middle East, it is now widely cultivated in central and eastern Europe and many Asian countries, including China, Iran, and India (Laribi et al., 2015). It has various common names such as Chinese parsley, phak chee (Thai), cilantro (Spanish), and dhaniya (Hindi) (Singletary, 2016). Coriander is a popular herb with versatile applications in perfumes, cosmetics, traditional medicines, and the flavoring of alcohol-based liquors. All parts of the plants are edible. The young leaves of coriander are used as a flavoring in the preparation of salads, soups, sauces, Mexican salsas, and chilies, whereas the aromatic seeds, either ground or whole, are used in the preparation of seafood dishes, meat-based curry dishes, puddings, bread, soups, stews, and several other ethnic foods. The roots are commonly used as an ingredient in Thai cuisines. The usage of coriander essential oil (CEO) in cosmetics and perfumes has been well documented and dates back to 2000 bc (Draelos, 2013). The traditional Indian cosmetic formula "Varnakarlepa" used coriander as an ingredient to normalize skin color (Yadav, 2016).

C. sativum contains various classes of bioactive compounds, including polyphenols, essential oils, lipids, flavonoids, isoprenoids, alkaloids, tannins, monoterpenoids, and glycosides (Prachayasittikul et al., 2018). Globally, CEO is among the top 20 essential oils and constitutes approximately 1% of the yield of the coriander fruit. Linalool is the major constituent of CEO, followed by α-pinene (0.2%–8%), γ-terpinene (1%–8%), geranyl acetate (0.1%–4.7%), and camphor (0.9%–4.9%) In this chapter, the discussion focuses on the CEO constituent linalool as it has been well studied due to its wide range of biological activities (Singletary, 2016).

Historical records indicate the usage of coriander as folk medicine since the Greek era (Hippocrates, 460–377 bc) (Prachayasittikul et al., 2018). In traditional medicine, coriander is used for the treatment of a wide variety of gastrointestinal diseases (diarrhea, jaundice, flatulence, dysentery, and vomiting) and loss of appetite (Platel and Srinivasan, 2004; Khan and Khatoon, 2008; Ugulu et al., 2009), and as a diuretic (Ugulu et al., 2009). Studies have also shown that coriander leaf extracts exhibit hepatoprotective properties by lowering the serum glutamate oxaloacetate transaminase and glutamate pyruvate transaminase levels. Furthermore, treatment with the coriander leaf extract has also shown a significant increase in the levels of hepatic enzymes such as superoxide dismutase (SOD), catalase, and glutathione peroxidase (GPX) (Sreelatha et al., 2009). In addition, the CEO from the seeds has traditionally been used as an anti-fungal and antibiotic agent, inhibiting a wide range of microorganisms, including Gram-negative and Gram-positive microorganisms (Chaudhry and Tariq, 2006; Silva et al., 2011a, Silva et al., 2011b). Other historical uses of coriander include its use as an aphrodisiac, hypoglycemic, hypotensive, myorelaxant, carminative, anti-inflammatory, and analgesic, a remedy for respiratory ailments, and improving memory (Momin et al., 2012).

Additionally, the plant contains a mixture of active phytoconstituents such as polyphenols, flavonoids, isoprenoids, alkaloids, and tannins which have neuroprotective properties. In Iranian traditional medicine, the plant has been indicated for anti-anxiety and anti-epileptic activities (Hūšang, 2011). Various parts of *C. sativum* have been traditionally associated with neuronal health, and these activities have also been demonstrated in *in vitro* and *in vivo* models (Enas, 2010; Enas, 2011; Mani et al., 2011; Ghorbani et al., 2012). The aqueous fraction of the aerial parts (leaves, stems, and twigs) of *C. sativum* has shown a significant decrease in glucose/serum deprivation-induced cell death in PC12 neuronal cells (Ghorbani et al., 2012). Treatment with *C. sativum* leaves (5, 10, and 15% w/w of diet) in mice has shown reversal of scopolamine and diazepam-induced memory deficits (Mani et al., 2011). Treatment with aqueous extract of coriander seeds has shown protective effects against aluminum chloride-induced neurodegeneration and AD in the pyramidal cells of the cerebral cortex in rats (Enas, 2010; Enas, 2011). Moreover, *C. sativum* volatile oil showed cognitive improvement in an Aβ42 rat AD model (Cioanca et al., 2014). The CEO from the seeds has also been reported to possess sedative and hypnotic activities (Emamghoreishi and Heidari-Hamedani, 2006).

This chapter gives a brief idea regarding neurodegenerative and affective disorders and highlights studies about potential neurological health benefits of *C. sativum* and its constituents.

13.2 METHODOLOGY

Multiple online interactive searches for original research and review articles were done across various databases including PubMed, Google Scholar, MEDLINE, and ResearchGate using keywords and terms such as neurodegenerative disorders, neurologic disorders, *Coriandrum sativum* in neurodegenerative disorders, anti-Alzheimer's activity of *C. sativum*, anti-Parkinson's activity of *C. sativum*, the anticonvulsant activity of *C. sativum*, the anti-epileptic activity of *C. sativum*, antiamyotrophic lateral sclerosis activity of *C. sativum*, anti-anxiety/anxiolytic activity of *C. sativum*, anti-depressant activity of *C. sativum*, etc. Fact sheets and updates from the websites of organizations such as the World Health Organization (WHO), Centers for Disease Control and Prevention (CDC), and Global Burden of Disease (GBD) were also searched. The obtained results were screened, full texts were obtained, and inclusion and exclusion criteria were applied to determine the suitability of articles to be used in this review. Studies that reported therapeutic activities of whole, parts, extracts, and phytoconstituents of *C. sativum* in treating neurodegenerative disorders were included.

13.3 NEUROPROTECTIVE POTENTIAL OF *C. SATIVUM*

13.3.1 ANTIDEPRESSION

Major depressive disorder (MDD) is one of the most prevalent affective disorders associated with broad cognitive deficits, as evidenced by difficulties in executive functioning, working memory, and processing speed (LeMoult and Gotlib, 2019). The non-cognitive symptoms include anxiety, pessimistic behavior, feelings of worthlessness, sleep deprivation or oversleeping, loss of appetite, exhaustion, and suicidal thoughts (Martins and Brijesh, 2020). Like anxiety, depression manifests as mental, physical, and/or emotional stress. Globally, with more than 300 million (4.4%) people suffering, depression is the leading cause of disability and poses an enormous cost at both individual and societal levels (WHO, 2017).

Pharmacological treatments remain the primary choice of intervention that works by modulating the neurotransmitters in the brain. Several anti-depressants classes are available, including tricyclic antidepressants, MAOIs, selective serotonin reuptake inhibitors, and serotonin and norepinephrine reuptake inhibitors (Lieberman, 2003). These drugs have moderate (nausea, diarrhea, weight loss, sweating, and dry mouth) to severe (coma, seizures, cardiac toxicity, sexual dysfunctions) side effects (Coleman, 2011; Isbister et al., 2004). Further, discontinuation of these drugs has been shown to cause withdrawal symptoms (Broekhoven et al., 2002). Therefore, complementary and alternative approaches such as psychotherapy, brain stimulation therapy, yoga, music therapy, and herbal medicines have been explored. Many herbal products with a traditional claim have been introduced into psychiatric practice. These herbal medicines are cost-effective, readily available, and have synergistic effects with synthetic drugs (Fajemiroye et al., 2016).

The antidepressant activity of CEO (1% and 3%) in the Aβ (1-42) rat model of AD has been reported. The antidepressant activity was assessed using the forced swimming test (FST), wherein animals are forced to swim in a confined space (inescapable stressful situation) which resembles a state of despair and mental depression (Tolardo et al., 2010). Animals exposed to CEO for 21 consecutive days showed significantly prolonged swimming time and decreased immobility time, especially at 3% concentration. The results suggest that CEO has an antidepressant-like effect in the AD model (Cioanca et al., 2014). Hence, coriander oil could be used as a potential medicine in comorbid disorders such as AD and depression. Pathan et al. (2015) investigated the antidepressant effects of

C. sativum ethanol extract (CSEE) seed in mice using FST. CSEE (200 mg/kg) elicited a significant decrease in immobility time compared to the standard treatment (imipramine, 10 mg/kg) group.

Linalool, one of the major phytoconstituents of volatile coriander oil, has been reported for its antidepressant-like activity. Mice treated with (-)-linalool (10, 50, 100, and 200 mg/kg, i.p.) were evaluated using the tail suspension test (TST). The results indicated that the animals presented a decrease in immobility times at 100 and 200 mg/kg doses compared to vehicle control (Coelho et al., 2013). Guzmán-Gutiérrez et al. (2012) have also reported antidepressant-like activity of linalool (100 mg/kg, i.p.) in the FST. Further, the mechanism of action of linalool was investigated using antagonists for serotonergic, dopaminergic, and noradrenergic pathways (Guzmán-Gutiérrez et al., 2015). WAY 100635, the $5HT_{1A}$ selective antagonist, could reverse the antidepressant effect of linalool, which indicates that the $5\text{-}HT_{1A}$ pathway is involved in the mechanism of action of linalool.

13.3.2 Anti-Anxiety Activity

Anxiety disorders are common and frequently occurring mental illnesses among children and adolescents worldwide (Polanczyk et al., 2015). Under chronic conditions (more than six months), it can become pathologic and interfere with a person's ability to perform routine tasks. In addition, anxiety disorders have been associated with an increased risk of suicide (Sareen, 2011; Ferrari et al., 2014). Conventional anxiolytic and antidepressant drugs are safe, effective, and broad-spectrum in action; however, their clinical uses are limited due to uncertain treatment outcomes and considerable side effects. Therefore, medicinal plants have become a great alternative and complementary therapy choice (Fajemiroye et al., 2016).

Numerous studies have reported on the anxiolytic properties of coriander and its phytoconstituents. Emamghoreishi et al. (2005) studied the anti-anxiety effects of *C. sativum* aqueous seed extract (CSAE) using the EPM model. Animals treated with 100 mg/kg extract showed an increase in the time spent on open arms and percentage of open arm entries on the model, compared with controls. Ravindran et al. (2014) studied the anxiolytic effect of CSAE with chronic dosing of ten days, wherein 25 and 50 mg/kg showed a significant effect on EPM, compared to controls. In a recent study, the effect of CSAE was studied using EPM and light-dark test (LDT) models following chronic restraint-induced stress. Further, the effect of the extract on the neurotransmitters such as GABA, glutamate, serotonin (5-HT), norepinephrine (NE), and dopamine (DA) was evaluated in the hippocampus, cerebral cortex, cerebellum, and brain stem region of the brain. The results indicated that CSAE reduced anxiety by reducing the GABA, 5-HT, NE, and DA levels in all brain regions. Also, treatment with CSAE brought down high levels of glutamate in the hippocampus region induced by chronic restraint stress (Sahoo and Brijesh, 2020). Mahendra and Bisht (2011) demonstrated the anxiolytic effect of coriander seeds ethanol extract (CSEE), wherein the animals treated with the extract (100 and 200 mg/kg) showed less anxiety-like behavior compared to the control animals in four anxiety models, viz., EPM, LDT, open field, and social interaction test. Similar results were observed by Pathan et al. (2011) with CSEE, wherein the highest dose tested (200 mg/kg, IP) demonstrated a significant anxiolytic-like effect in the EPM. Latha et al. (2015) also studied the anxiolytic effects of aqueous extract of coriander leaves (50, 100, and 200 mg/kg) in mice using the EPM model, wherein all the doses exhibited significant activity compared to vehicle control.

Linck et al. (2010) investigated the effect of linalool, when inhaled, on anti-anxiety, anti-aggressiveness, and social interaction in mice. Further, they investigated its effect on the acquisition phase of a step-down memory task in mice with short- and long-term memory (STM and LTM) of 90 min and 24 h, respectively. The results indicated that linalool (3% in Tween 80 v/v) elicited anxiolytic-like activity as observed on the LDT. Also, decreased aggressive behavior and increased social interaction were observed with the linalool-treated mice. Additionally, mice treated with linalool showed shorter latency time and better acquisition memory. The mechanism for linalool's anxiolytic action has been suggested to be either by inhibition of glutamate release (Elisabetsky et al., 1999) or monoamine (DA, 5-HT, and NE) modulation (Cheng et al., 2015). Cheng et al. (2015) also reported on the anxiolytic activity of S-(+)-enantiomer of linalool extracted from *Cinnamomum osphloeum* ct. leaves in orally treated mice. The treated mice showed reduced levels of monoamines (i.e., 5-HT, DA, and NE) in the frontal cortex and hippocampus region. Interestingly, the R-(-)-linalool has demonstrated a stress-relieving effect in a human model (de Sousa et al., 2010).

Besides coriander, linalool-rich essential oils from other plants, including *Piper guineense* (Tankam and Ito, 2013) and *Cananga odorata* (Zhang et al., 2016), have been shown to exhibit anxiolytic effects in mice models. These findings indicate that the anxiolytic property of these linalool-rich essential oils is mediated *via* the glutamatergic and nicotinic acid receptors (Elisabetsky et al., 1995; Re et al., 2000).

13.3.3 Anti-Alzheimer's Activity

AD is characterized by gradual onset and a progressive and irreversible loss of neurons in the brain leading to build up of plaques, tangles, and cognitive decline. The pathological hallmarks of the disease include the deposition of beta-amyloid (Aβ) plaques and aggregation of intracellular tau protein in the neurofibrillary tangles in the neocortex and cerebral vasculature (Huang et al., 2000; Hardy and Selkoe, 2002). Various animal models have shown that the accumulation of neurotoxic levels of Aβ results in neuronal damage through various mechanisms, including enhanced oxidative stress, increased inflammatory response, impaired energy metabolism, and disturbed cellular and calcium homeostasis (Mattson, 2004; Glass et al., 2010). Parallelly, genetic studies have supported the observations in animals and patients with AD by showing the involvement of mitogen-activated protein kinases (MAPK), wherein hyperactivation of the extracellular signal-regulated kinase (ERK) and c-Jun N-terminal kinase (JNK) was observed. Furthermore, genetic and pharmaceutical interventions in these pathways have been shown to ameliorate neurological phenotypes in animal models (Hong et al., 2012; Park et al., 2013).

Studies have shown neuroprotective effects of coriander, wherein treatment with volatile oil extracted from the seeds showed a reduction in Aβ deposits in Wistar rats treated with Aβ(1-42). The GC-MS and GS-FID analyses revealed monoterpenes as the major class of compounds, with linalool being the critical component. Other important compounds detected were γ-terpinene, α-pinene, pinocarvone, β-ocimene (E + Z), carvone, and geranyl acetate. In addition, animals treated with 1% and 3% of the volatile oil exhibited increased locomotor activity, swimming time in the FST, percentage time spent, and several open arm entries in the elevated plus-maze (EPM) test suggesting antidepressant- and anxiolytic-like effects. Furthermore, the coriander volatile oil has also shown an increase in glutathione level and a decrease in catalase activity in the hippocampus, as is expected for the antioxidant agents (Cioanca et al., 2014).

Mani et al. (2011) investigated fresh *C. sativum* leaves (CSL; 5, 10, and 15% w/w of diet) for their effects on cognitive functions in Swiss albino mice. The exteroceptive (i.e., EPM and passive avoidance apparatus) and the interoceptive (i.e., diazepam, scopolamine, and aging-induced amnesia) behavioral models were used for testing memory. CSL showed a dose-dependent improvement in memory scores in both the young and the aged mice. CSL also showed reversal of scopolamine- and diazepam-induced memory deficits. Interestingly, concomitant administration of CSL with the daily diet for 45 days significantly reduced cholinesterase activity in the brain and serum total cholesterol levels in young and aged mice. CSL can be a useful remedy for the management of AD.

The neuroprotective effects of an ethanol extract of *C. sativum* leaves (200 μg/mL) were investigated on Aβ cytotoxicity *in vitro* (human neuroblastoma cells, SH-SY5Y; and primary cultured mouse cortical neurons) and *in vivo* (Aβ 42-expressing *Drosophila*). Treatment with *C. sativum* extract (CSE) inhibited the death of both Aβ-treated SH-SY5Y cells and mouse cortical neurons. Similarly, CSE intake increased survival rate, shortened pupariation time, and rescued motor defects in Aβ 42-expressing flies compared to control AD flies. Further, the extract suppressed Aβ 42-induced cell death in the larval imaginal disc and brain without affecting Aβ 42 expression and accumulation. Interestingly, the extract suppressed H_2O_2-induced increase in reactive oxygen species (ROS) levels and decreased the glial cell number in AD model flies.

Additionally, CSE inhibited the epidermal growth factor receptor- and $A\beta$ -induced phosphorylation of ERK. The protective function of CSE against the $A\beta$ 42-induced eye defect phenotype in *Drosophila* was abolished by the constitutive form of ERK. Also, CSE intake restored the increased number of microglial cells to normal in the larval brains of the AD model to that of control. These results suggest that CSE has neuroprotective, antioxidant, anti-inflammatory, and ERK inhibitory properties (Liu et al., 2016). Collectively, these studies point towards the neuroprotective potential of *C. sativum* that can be explored for the treatment of patients with AD.

13.3.4 ANTI-EPILEPTIC AND ANTI-CONVULSANT ACTIVITY

Worldwide, about 50 million people are affected by epilepsy, making it one of the most common neurological disorders. Eighty percent of the burden of epilepsy is in developing countries, where 80–90% of people do not receive any treatment (Meyer et al., 2010; Cooper, 2019). Epilepsy is characterized by recurrent seizures, occurring in both males and females of all ages, from childhood to adolescence to old age (Beghi, 2020). Etiologies that have been generally associated with epilepsy include CNS infections, cerebrovascular disorders, antenatal and perinatal risk factors, and idiopathic etiology. The first line of antiepileptic drugs includes phenobarbital, carbamazepine, phenytoin, and valproic acid (WHO, 2005).

Karami et al. (2015) reported coriander's aerial parts' antioxidant and anticonvulsive effect (leaves, stems, and twigs). The convulsions/seizures in rats were induced by administration of pentylenetetrazole (PTZ), progressing from minimal clonic seizures (MCS) to clonic seizures followed by generalized tonic-clonic seizures (GTCS). These seizures accompanied oxidative damage of brain tissue, measured by increased malondialdehyde (MDA) and reduced total thiol (SH) group levels in the brain. The hydroalcoholic extract (100, 500, and 1000 mg/kg) treatment significantly increased the first MCS and the first GTCS latencies following PTZ-induced seizures. Also, the extract showed antioxidant properties by preventing PTZ-induced elevation of the MDA levels in the cortex and hippocampus.

Furthermore, the extract at 500 mg/kg significantly reversed the decrease in total thiol concentration in the cortical tissues. Anaeigoudari et al. (2016) studied the effect of different fractions (water, *n*-butanol, and ethyl acetate) of hydroalcoholic extract of aerial parts of *C. sativum* on PTZ-induced seizures. The water and ethyl acetate fractions (100 mg/kg) showed significant latency to GTCS compared to the *n*-butanol fraction, whereas only water effectively increased the MCS latency. In addition, the MDA levels in the hippocampus were significantly reduced by all three extracts, whereas the water and *n*-butanol fractions significantly elevated the thiol levels in the cortex and hippocampus regions, respectively.

Linalool, an important phytoconstituent of coriander, has been reported to possess anticonvulsant activity by various studies (Elisabetsky et al., 1999; Sampaio et al., 2012; Vatanparast et al., 2017). Elisabetsky et al. studied the effect of linalool in three convulsion models, i.e., the intracerebroventricular quinolinic acid (Quin)-induced convulsions, the N-methyl-D-aspartate (NMDA)-induced convulsions, and the PTZ-kindling model. Interestingly, linalool demonstrated protective effects in all three convulsion models. Mechanisms governing the anticonvulsant activity of linalool include modulation of glutamatergic transmission, activation of protein kinase C, and inhibition of adenylate cyclase activity (Elisabetsky et al., 1999; Sampaio et al., 2012; Vatanparast et al., 2017).

13.3.5 ANTI-PARKINSON ACTIVITY

PD is a common progressive age-related movement disorder of the CNS characterized by gradual loss of dopaminergic neurons in the substantia nigra pars compacta (SNc) region and their terminals in the corpus striatum (de Lau and Breteler 2006; Herrera et al., 2017). In 2020, an estimated 9.4 million people were affected globally by PD (Maserejian et al., 2020). Patients with PD often exhibit motor impairments, including muscle rigidity, resting tremor, flexed posture, bradykinesia, loss of postural reflexes, and freezing phenomenon. Among the non-motor symptoms, anxiety, depression, fatigue, sleep disturbances, and dementia are common occurrences. The etiology of PD is multifactorial and includes oxidative/nitrosative stress, inflammation, mitochondrial dysfunction, endoplasmic reticulum stress, microglial activation, excitotoxicity, and ubiquitin-proteasome system dysfunction (Moore et al., 2005; Im et al., 2006; Lashuel et al., 2013; Ur Rasheed et al., 2016; Gelders et al., 2018; Troncoso-Escudero et al., 2018). Chronically, these factors lead to the accumulation of abnormal or misfolded α-synuclein (α-syn) protein in the different brain regions and peripheral tissues, loss of neuromelanin containing dopamine, and appearance of Lewy bodies. Gradually, these abnormal proteins undergo fibrillation, oligomerization, and polymerization (Cheng et al., 2010, Carballo-Carbajal et al., 2010). The current therapeutics available for PD include dopamine agonists, cholinesterase, and monoamine oxidase inhibitors (MAOIs) (Ellis and Fell, 2017). However, these drugs have limited effectiveness, offer only symptomatic relief, and do not halt the disease progression. Also, long-term usage of these drugs leads to various unwanted side effects (Samii et al., 2004). Recently, various research groups have tried natural herbs such as *Bacopa monnieri, Cinnamon asiatica, Carthamus tinctorius, Crocus sativus, Chondrus crispus, Corema album,* and *Panax ginseng* for treatment of PD, the majority of which target α-syn (Javed et al., 2019).

C. sativum fruit extract (CSFE) has been investigated for its neuroprotective effects in both *in vitro* [lipopolysaccharide (LPS)-stimulated BV-2 microglia-mediated neuroinflammation] and *in vivo* [1-methyl-4 phenyl-1,2,3,6-tetrahydropyridine (MPTP)-induced PD in mice] models. The CSFE (25, 50, and 100 µg/mL) significantly inhibited LPS-stimulated increase in nitric oxide (NO), inducible NO synthase, cyclooxygenase-2, interleukin-6, tumor necrosis factor-alpha nuclear factor of kappa B (NF-κB) activation, and inhibitor of κB (IκB)-α phosphorylation in the BV-2 microglial cells in a dose-dependent manner. Furthermore, in the mouse PD model, treatment with CSFE (100, 200, and 300 mg/kg) significantly attenuated the MPTP-induced changes in locomotor, cognitive, and behavioral functions evaluated by rotarod, passive avoidance, and open field tests. Further, CSFE restored the levels of MPTP-induced oxidative enzymes in the brain, such as SOD, catalase, and lipid peroxides. In addition, the authors could identify quercetin and kaempferol in the extract using high-performance thin-layer chromatography. The study, therefore, indicated that coriander seed extract has therapeutic potential for the treatment of neuroinflammatory and oxidative stress-mediated neurodegeneration seen in PD (Koppula et al., 2021).

13.3.6 ANTI-AMYOTROPHIC LATERAL SCLEROSIS ACTIVITY

Amyotrophic lateral disease (ALS), or Lou Gehrig's disease, is a progressive and fatal neurological disease that causes dysfunctions of motor neurons of the cortex, brain stem, and spinal cord (Ravits and La Spada, 2009). Patients with ALS experience muscle weakness and cramps, progressive muscle atrophy, fasciculations, slurred speech, swallowing, chewing, hyperactive reflexes, cognitive impairment, and severe constipation. Eventually, the patient shows breathing difficulty leading to respiratory failure (Borasio and Miller, 2001; Phukan and Hardiman, 2009; Wijesekera and Leigh, 2009). Genetics (familial and sporadic) and aging are two major risk factors for ALS development. Familial ALS involves mutations in genes such as SOD-1, tar-DNA binding protein 43, Chromosome 9 open reading frame 72, and fused in sarcoma. Additionally, environmental factors such as toxins, metals, infections, traumatic head injury, and smoking increase the chances of developing ALS. Specifically, long-term exposure to methylmercury can produce symptoms such as the early onset of hind limb weakness similar to those observed in ALS (Johnson and Aitchison, 2009).

There are no specific studies of the effect of coriander on ALS treatment. However, coriander has been reported to remove mercury and other toxic metals from the CNS (Pandey et al., 2016). Also, coriander leaves have been reported to enhance the excretion of mercury following dental amalgam removal (Abascal and Yarnell, 2012; Sears, 2013). Further, treatment with coriander has been shown to increase the levels of SOD, catalase, and GPX in the liver and kidney of lead-exposed animals (Sharma et al., 2011).

Myricetin, the polyphenolic flavonoid in coriander, has been shown to clear protein aggregates in cell culture. Cos-7 cells transfected with plasmid constructs of WT and mutant SOD1 show spontaneous intracellular accumulation of mutant SOD1 and WT accumulation following the addition of proteasome inhibiting compound. Myricetin treatment (10 µM for 48 h) showed reduced intracellular aggregation of ubiquitin-positive SOD1 (Joshi et al., 2019).

13.4   CONCLUSION

C. sativum, an herbal plant belonging to the family Apiaceae, is a storehouse for bioactive compounds. Since ancient times, all parts of this plant have been useful and integrated into human life as food, fragrance, and medicine. There has been an increase in the demand and usage of essential oils from herbal plants. Coriander seeds contain a high amount of essential oil, and coriander essential oil is the second major annually produced oil following orange essential oil, indicating its importance as an economic spice (Lawrence, 1993). These essential oils, along with other phytoconstituents (i.e., phenolics, polyunsaturated fatty acids, coumarins, and flavonoids), have been reported for their antioxidant and neuroprotective activity. Linalool, the major monoterpenoid in the coriander seeds, has been well studied for its CNS properties (i.e., anxiolytic, anticonvulsant, antiepileptic, antidepressant, and anti-Alzheimer's). Apart from the seed essential oil, various fresh leaf extracts have been explored and studied for neurological activities. However, most of these studies are either *in vitro* or *in vivo*. Therefore, clinical studies are required for the authentication of these reported activities and their therapeutic dose. In summary, the *C. sativum* is a promising functional food that demonstrates therapeutic values against neurodegenerative and behavioral disorders.

REFERENCES

Abascal K, Yarnell Y. Cilantro-culinary herb or miracle medicinal plant? *Alternative and Complementary Therapies*. 2012;18(5):259–264.

Anaeigoudari A, Hosseini M, Karami R, Vafaee F, Mohammadpour T, Ghorbani A, Sadeghnia HR. The effects of different fractions of *Coriandrum sativum* on pentylenetetrazole-induced seizures and brain tissues oxidative damage in rats. *Avicenna Journal of Phytomedicine*. 2016;6(2):223–235.

Beghi E. The epidemiology of epilepsy. *Neuroepidemiology*. 2020;54(2):185–191.

Borasio GD, Miller RG. Clinical characteristics and management of ALS. *Seminars in Neurology*. 2001;21(2):155–166.

Bourbouli M, Rentzos M, Bougea A, Zouvelou V, Constantinides VC, Zaganas I, Evdokimidis I, Kapaki E, Paraskevas GP. Cerebrospinal fluid TAR DNA-binding protein 43 combined with tau proteins as a candidate biomarker for amyotrophic lateral sclerosis and frontotemporal dementia spectrum disorders. *Dementia and Geriatric Cognitive Disorders*. 2017;44(3–4):144–152.

Broekhoven F, Kan CC, Zitman FG. Dependence potential of antidepressants compared to benzodiazepines. *Progress in Neuro-Psychopharmacology and Biological Psychiatry*. 2002;26(5):939–943.

Carballo-Carbajal I, Laguna A, Romero-Giménez J, Cuadros T, Bové J, Martinez-Vicente M, Parent A, Gonzalez-Sepulveda M, Peñuelas N, Torra A, Rodríguez-Galván B. Brain tyrosinase overexpression implicates age-dependent neuromelanin production in Parkinson's disease pathogenesis. *Nature Communications*. 2010;10(1):1–9.

Chaudhry NM, Tariq P. Bactericidal activity of black pepper, bay leaf, aniseed and coriander against oral isolates. *Pakistan Journal of Pharmaceutical Sciences*. 2006;19(3):214–218.

Cheng BH, Sheen LY, Chang ST. Evaluation of anxiolytic potency of essential oil and S-(+)-linalool from *Cinnamomum osmophloeum* ct. linalool leaves in mice. *Journal of Traditional and Complementary Medicine*. 2015;5(1):27–34.

Cheng HC, Ulane CM, Burke RE. Clinical progression in Parkinson disease and the neurobiology of axons. *Annals of Neurology*. 2010;67(6):715–725.

Cioanca O, Hritcu L, Mihasan M, Trifan A, Hancianu M. Inhalation of coriander volatile oil increased anxiolytic-antidepressant-like behaviors and decreased oxidative status in beta-amyloid (1–42) rat model of Alzheimer's disease. *Physiology & Behavior.* 2014;131:68–74.

Coelho V, Mazzardo-Martins L, Martins DF, Santos AR, da Silva Brum LF, Picada JN, Pereira P. Neurobehavioral and genotoxic evaluation of (2)-linalool in mice. *Journal of Natural Medicine.* 2013;67(4):876–880.

Coleman E. Impulsive/compulsive sexual behavior: Assessment and treatment. In Grant, Jon E.; Potenza, Marc N. (eds), *The Oxford Handbook of Impulse Control Disorders.* New York: Oxford University Press; 2011, Chapter 28, p. 385.

Cooper C. Global, regional, and national burden of neurological disorders, 1990–2016: A systematic analysis for the Global Burden of Disease Study 2016. *The Lancet Neurolology.* 2019;18(4):357–375.

de Lau LML, Breteler MMB. Epidemiology of Parkinson's disease. *The Lancet Neurology.* 2006;5(6): 525–535.

de Sousa DP, Nóbrega FF, Santos CC, de Almeida RN. Anticonvulsant activity of the linalool enantiomers and racemate: Investigation of chiral influence. *Natural Product Communications.* 2010;5(12):1934578X1000501201.

Draelos ZD. To smell or not to smell? That is the question! *Journal of Cosmetic Dermatology.* 2013;12:1–2.

Elisabetsky E, Brum LS, Souza DO. Anticonvulsant properties of linalool in glutamate-related seizure models. *Phytomedicine.* 1999;6(2):107–113.

Elisabetsky E, Marschner J, Souza DO. Effects of linalool on glutamatergic system in the rat cerebral cortex. *Neurochemical Research.* 1995;20(4):461–465.

Ellis JM, Fell MJ. Current approaches to the treatment of Parkinson's disease. *Bioorganic and Medicinal Chemistry Letters.* 2017;27:4247–4255.

Emamghoreishi M, Heidari-Hamedani G. Sedative-hypnotic activity of extracts and essential oil of coriander seeds. *Iranian Journal of Medical Sciences.* 2006;31(1):22–27.

Emamghoreishi M, Khasaki M, Aazam MF. *Coriandrum sativum*: Evaluation of its anxiolytic effect in the elevated plus-maze. *Journal of Ethnopharmacology.* 2005;96:365–370.

Enas AK. Evaluation of the possible protective and therapeutic influence of coriander (*Coriandrum sativum* L.) seed aqueous extract on hippocampal pyramidal cells against Alzheimer's disease induced by aluminum chloride in adult male albino rats. *Researcher.* 2011;3(1):22–29.

Enas AK. Study the possible protective and therapeutic influence of coriander (*Coriandrum sativum* L.) against neurodegenerative disorders and Alzheimer's disease induced by aluminum chloride in cerebral cortex of male albino rats. *Nature and Science.* 2010;8(11):202–213.

Fajemiroye JO, da Silva DM, de Oliveira DR, Costa EA. Treatment of anxiety and depression: Medicinal plants in retrospect. *Fundamental and Clinical Pharmacology.* 2016;30(3):198–215.

Ferrari AJ, Norman RE, Freedman G, Baxter AJ, Pirkis JE, Harris MG, Page A, Carnahan E, Degenhardt L, Vos T, Whiteford HA. The burden attributable to mental and substance use disorders as risk factors for suicide: Findings from the Global Burden of Disease Study 2010. *PloS One.* 2014;9(4):e91936.

Gelders G, Baekelandt V, Van der Perren A. Linking neuroinflammation and neurodegeneration in Parkinson's disease. *Journal of Immunology Research.* 2018;2018:4784268.

Ghorbani A, Rakhshandeh H, Asadpour E, Sadeghnia HR. Effects of *Coriandrum sativum* extracts on glucose/serum deprivation-induced neuronal cell death. *Avicenna Journal of Phytomedicine.* 2012;2(1):4–9.

Gitler AD, Dhillon P, Shorter J. Neurodegenerative disease: Models, mechanisms, and a new hope. *Disease Model and Mechanisms.* 2017;10(5):499–502.

Glass CK, Saijo K, Winner B, Marchetto MC, Gage FH. Mechanisms underlying inflammation in neurodegeneration. *Cell.* 2010;140(6):918–934.

Gooch CL, Pracht E, Borenstein AR. The burden of neurological disease in the United States: A summary report and call to action. *Annals of Neurology.* 2017;81(4):479–484.

Guzmán-Gutiérrez SL, Bonilla-Jaime H, Gómez-Cansino R, Reyes-Chilpa R. Linalool and β-pinene exert their antidepressant-like activity through the monoaminergic pathway. *Life Sciences.* 2015;128:24–29.

Guzmán-Gutiérrez SL, Gómez-Cansino R, García-Zebadúa JC, Jiménez-Pérez NC, Reyes-Chilpa R. Antidepressant activity of *Litsea glaucescens* essential oil: Identification of β-pinene and linalool as active principles. *Journal of Ethnopharmacology.* 2012;143(2):673–679.

Hardy J, Selkoe DJ. The amyloid hypothesis of Alzheimer's disease: Progress and problems on the road to therapeutics. *Science.* 2002;297(5580):353–356.

Herrera A, Muñoz P, Steinbusch HWM, Segura Aguilar J. Are dopamine oxidation metabolites involved in the loss of dopaminergic neurons in the nigrostriatal system in Parkinson's disease? *ACS Chemical Neuroscience.* 2017;8(4):702–711.

Hong YK, Lee S, Park SH, Lee JH, Han SY, Kim ST, Kim YK, Jeon S, Koo BS, Cho KS. Inhibition of JNK/ dFOXO pathway and caspases rescues neurological impairments in Drosophila Alzheimer's disease model. *Biochemical and Biophysical Research Communications*. 2012;419(1):49–53.

Hūšang A. Coriander. In *Encyclopaedia Iranica*, 2011; VI/3, p. 273. Available online at: http://www.iranica-online.org/articles/coriander-coriandrum-sativum-l (accessed on 31st October 2021).

Hūšang A. Coriander. In *Encyclopaedia Iranica*, 2011; VI/3, p. 273. Available online at (31st Dec 2018).

Im HI, Nam E, Lee ES, Hwang YJ, Kim YS. Baicalein protects 6-OHDA-induced neuronal damage by suppressing oxidative stress. *The Korean Journal of Physiology and Pharmacology*. 2006;10(6): 309–315.

Isbister GK, Bowe SJ, Dawson A, Whyte IM. Relative toxicity of selective serotonin reuptake inhibitors (SSRIs) in overdose. *Journal of Toxicology: Clinical Toxicology*. 2004;42(3):277–285.

Javed H, Nagoor Meeran MF, Azimullah S, Adem A, Sadek B, Ojha SK. Plant extracts and phytochemicals targeting α-synuclein aggregation in Parkinson's disease models. *Frontiers in Pharmacology*. 2019;9:1555(1–27).

Johnson FO, Atchison WD. The role of environmental mercury, lead and pesticide exposure in development of amyotrophic lateral sclerosis. *Neurotoxicology*. 2009;30(5):761–765.

Joshi V, Mishra R, Upadhyay A, Amanullah A, Poluri KM, Singh S, Kumar A, Mishra A. Polyphenolic flavonoid (Myricetin) upregulated proteasomal degradation mechanisms: Eliminates neurodegenerative proteins aggregation. *Journal of Cellular Physiology*. 2019;234(11):20900–20914.

Karami R, Hosseini M, Mohammadpour T, Ghorbani A, Sadeghnia HR, Rakhshandeh H, Vafaee F, Esmaeilizadeh M. Effects of hydroalcoholic extract of *Coriandrum sativum* on oxidative damage in pentylenetetrazole-induced seizures in rats. *Iranian Journal of Neurology*. 2015;14(2):59–66.

Khan SW, Khatoon SU. Ethnobotanical studies on some useful herbs of Haramosh and Bugrote valleys in Gilgit, northern areas of Pakistan. *Pakistan Journal of Botany*. 2008;40(1):43–58.

Koppula S, Alluri R, Kopalli SR. *Coriandrum sativum* attenuates microglia mediated neuroinflammation and MPTP-induced behavioral and oxidative changes in Parkinson's disease mouse model. *EXCLI Journal*. 2021;20:835–850.

Lal UR, Lal S. Bioactive molecules from Indian medicinal plants as possible candidates for the management of neurodegenerative disorders. In: Sharma K, Mishra K, Senapati KK, Danciu C (Eds.), *Bioactive Compounds in Nutraceutical and Functional Food for Good Human Health* [Internet]. London: IntechOpen; 2020. Available online at: https://www.intechopen.com/chapters/72064; doi: 10.5772/intechopen.92043 (accessed on 31st October 2021).

Laribi B, Kouki K, M'Hamdi M, Bettaieb T. Coriander (*Coriandrum sativum* L.) and its bioactive constituents. *Fitoterapia*. 2015;103:9–26.

Lashuel HA, Overk CR, Oueslati A, Masliah E. The many faces of α-synuclein: From structure and toxicity to therapeutic target. *Nature Reviews Neuroscience*. 2013;14:38–48.

Latha K, Rammohan B, Sunanda BP, Maheswari MU, Mohan SK. Evaluation of anxiolytic activity of aqueous extract of *Coriandrum sativum* Linn. in mice: A preliminary experimental study. *Pharmacognosy Research*. 2015;7(Suppl 1):S47–S51.

Lawrence BM. A planning scheme to evaluate new aromatic plants for the flavor and fragrance industries. In: Janick J, Simon JE (Eds.), *New Crops*. New York: Wiley; 1993;620–627.

LeMoult J, Gotlib IH. Depression: A cognitive perspective. *Clinical Psychology Review*. 2019;69:51–66.

Lieberman JA. History of the use of antidepressants in primary care. *Journal of Clinical Psychiatry*. 2003;5(Suppl 7):6–10.

Linck VM, da Silva AL, Figueiro M, Caramao EB, Moreno PR, Elisabetsky E. Effects of inhaled Linalool in anxiety, social interaction and aggressive behavior in mice. *Phytomedicine*. 2010;17(8–9):679–683.

Liu QF, Jeong H, Lee JH, Hong YK, Oh Y, Kim YM, Suh YS, Bang S, Yun HS, Lee K, Cho SM. *Coriandrum sativum* suppresses Aβ42-induced ROS increases, glial cell proliferation, and ERK activation. *The American Journal of Chinese Medicine*. 2016;44(07):1325–1347.

Mahendra P, Bisht S. Anti-anxiety activity of *Coriandrum sativum* assessed using different experimental anxiety models. *Indian Journal of Pharmacology*. 2011;43(5):574–577.

Mani V, Parle M, Ramasamy K, Abdul Majeed AB. Reversal of memory deficits by *Coriandrum sativum* leaves in mice. *Journal of the Science of Food and Agriculture*. 2011;91(1):186–192.

Martins J, Brijesh S. Anti-depressant activity of *Erythrina variegata* bark extract and regulation of monoamine oxidase activities in mice. *Journal of Ethnopharmacology*. 2020;248:112280.

Maserejian N, Vinikoor-Imler L, Dilley A. Estimation of the 2020 global population of Parkinson's disease (PD). *Movement Disorder*. 2020;35(suppl 1).

Mattson MP. Pathways towards and away from Alzheimer's disease. *Nature*. 2004;430(7000):631–639.

Meyer AC, Dua T, Ma J, Saxena S, Birbeck G. Global disparities in the epilepsy treatment gap: A systematic review. *Bulletin of the World Health Organization.* 2010;88(4):260–266.

Momin AH, Acharya SS, Gajjar AV. *Coriandrum sativum*-review of advances in phytopharmacology. *International Journal of Pharmaceutical Sciences and Research.* 2012;3(5):1233–1239.

Moore DJ, West AB, Dawson VL, Dawson TM. Molecular pathophysiology of Parkinson's disease. *Annual Review of Neuroscience.* 2005;28:57–87.

Naveen KL, Bhattacharjee A. Medicinal herbs as neuroprotective agents. *World Journal of Pharmacy and Pharmaceutical Sciences.* 2021;10(4):675–689.

Pakpoor J, Goldacre M. The increasing burden of mortality from neurological diseases. *Nature Reviews Neurology.* 2017;13(9):518–519.

Pandey S, Vinamra S, Sharma V, Chaudhary AK. Chelation therapy and chelating agents of ayurveda. *International Journal of Green Pharmacy.* 2016;10(03):143–150.

Park SH, Lee S, Hong YK, Hwang S, Lee JH, Bang SM, Kim YK, Koo BS, Lee IS, Cho KS. Suppressive effects of SuHeXiang Wan on amyloid-β42-induced extracellular signal-regulated kinase hyperactivation and glial cell proliferation in a transgenic Drosophila model of Alzheimer's disease. *Biological and Pharmaceutical Bulletin.* 2013;36(3):390–398.

Pathan A, Alshahrani A, Al-Marshad F. Neurological assessment of seeds of *Coriandrum sativum* by using antidepressant and anxiolytic like activity on albino mice. *Inventi Impact: Ethnopharmacology.* 2015;3:102–105.

Pathan AR, Kothawade KA, Logade MN. Anxiolytic and analgesic effect of seeds of *Coriandrum sativum* Linn. *International Journal of Research in Pharmacy and Chemistry.* 2011;1(4):1087–1099.

Phukan J, Hardiman O. The management of amyotrophic lateral sclerosis. *Journal of Neurology.* 2009;256(2):176–186.

Platel K, Srinivasan K. Digestive stimulant action of spices: A myth or reality? *Indian Journal of Medical Research.* 2004;119(5):167–179.

Polanczyk GV, Salum GA, Sugaya LS, Caye A, Rohde LA. Annual research review: A meta-analysis of the worldwide prevalence of mental disorders in children and adolescents. *Journal of Child Psychology and Psychiatry.* 2015;56(3):345–365.

Prachayasittikul V, Prachayasittikul S, Ruchirawat S, Prachayasittikul V. Coriander (*Coriandrum sativum*): A promising functional food toward the well-being. *Food Research International.* 2018;105:305–323.

Ravindran A, Rai M, Raveendran N, Naik H. Chronic anxiolytic-like activity of aqueous extract of *Coriandrum sativum* seeds using elevated plus maze test in Swiss albino mice. *International Journal of Pharmacy and Pharmaceutical Sciences.* 2014;6(2):93–95.

Ravits JM, La Spada AR. ALS motor phenotype heterogeneity, focality, and spread: Deconstructing motor neuron degeneration. *Neurology.* 2009;73:805–811.

Re L, Barocci S, Sonnino S, Mencarelli A, Vivani C, Paolucci G, Scarpantonio A, Rinaldi L, Mosca E. Linalool modifies the nicotinic receptor-ion channel kinetics at the mouse neuromuscular junction. *Pharmacological Research.* 2000;42(2):177–181.

Roos RA. Huntington's disease: A clinical review. *Orphanet Journal of Rare Diseases.* 2010;5(1):1–8.

Sahoo S, Brijesh S. Anxiolytic activity of *Coriandrum sativum* seeds aqueous extract on chronic restraint stressed mice and effect on brain neurotransmitters. *Journal of Functional Foods.* 2020;68:103884.

Samii A, Nutt JG, Ransom BR. Parkinson's disease. *Lancet.* 2004;363:1783–93.

Sampaio LD, Maia JG, de Parijós AM, de Souza RZ, Barata LE. Linalool from rosewood (Aniba rosaeodora Ducke) oil inhibits adenylate cyclase in the retina, contributing to understanding its biological activity. *Phytotherapy Research.* 2012;26(1):73–77.

Sareen J. Anxiety disorders and risk for suicide: Why such controversy?. *Depression and Anxiety.* 2011;28(11):941–945.

Schulte J, Littleton JT. The biological function of the Huntingtin protein and its relevance to Huntington's Disease pathology. *Current Trends in Neurology.* 2011;5:65–78.

Sears ME. Chelation: Harnessing and enhancing heavy metal detoxification-A review. *The Scientific World Journal.* 2013;2013:13.

Sharma V, Kansal L, Sharma A, Lodi S, Sharma SH. Protective role of *Coriandrum sativum* (coriander) extracts against lead nitrate induced oxidative stress and tissue damage in the liver and kidney in male mice. *International Journal of Applied Biology Pharmaceutical Technology.* 2011;2:65–83.

Silva F, Ferreira S, Duarte A, Mendonca DI, Domingues FC. Antifungal activity of *Coriandrum sativum* essential oil, its mode of action against Candida species and potential synergism with amphotericin B. *Phytomedicine.* 2011a;19(1):42–47.

Silva F, Ferreira S, Queiroz JA, Domingues FC. Coriander (*Coriandrum sativum* L.) essential oil: Its antibacterial activity and mode of action evaluated by flow cytometry. *Journal of Medical Microbiology.* 2011b;60(10):1479–1486.

Singletary K. Coriander: Overview of potential health benefits. *Nutrition Today.* 2016;51(3):151–161.

Sreelatha S, Padma PR, Umadevi M. Protective effects of *Coriandrum sativum* extracts on carbon tetrachloride-induced hepatotoxicity in rats. *Food and Chemical Toxicology.* 2009;47(4):702–708.

Stefanis L. α-Synuclein in Parkinson's disease. *Cold Spring Harbor Perspectives in Medicine.* 2012;2(2): a009399.

Tankam JM, Ito M. Inhalation of the essential oil of Piper guineense from Cameroon shows sedative and anxiolytic-like effects in mice. *Biological and Pharmaceutical Bulletin.* 2013;36(10):1608–1614.

Tolardo R, Zetterman L, Bitencourtt DR, Mora TC, de Oliveira FL, Biavatti MW, Amoah SK, Bürger C, de Souza MM. Evaluation of behavioral and pharmacological effects of *Hedyosmum brasiliense* and isolated sesquiterpene lactones in rodents. *Journal of Ethnopharmacology.* 2010;128(1):63–70.

Troncoso-Escudero P, Parra A, Nassif M, Vidal RL. Outside in: Unraveling the role of neuroinflammation in the progression of Parkinson's disease. *Frontiers in Neurology.* 2018;9:860.

Ugulu I, Baslar S, Yorek N, Dogan Y. The investigation and quantitative ethnobotanical evaluation of medicinal plants used around Izmir province, Turkey. *Journal of Medicinal Plants Research.* 2009;3(5):345–367.

Ur Rasheed MS, Tripathi MK, Mishra AK, Shukla S, Singh MP. Resveratrol protects from toxin-induced Parkinsonism: A plethora of proofs hitherto petty translational value. *Molecular Neurobiology.* 2016;53:2751–2760.

Vatanparast J, Bazleh S, Janahmadi M. The effects of linalool on the excitability of central neurons of snail *Caucasotachea atrolabiata. Comparative Biochemistry and Physiology Part C: Toxicology and Pharmacology.* 2017;192:33–39.

Wijesekera LC, Leigh PN. Amyotrophic lateral sclerosis. *Orphanet Journal of Rare Disease.* 2009;4(1):3.

World Health Organization. *Atlas: Epilepsy Care in the World 2005.* World Health Organization, International Bureau for Epilepsy and the International League against Epilepsy, Geneva; 2005.

World Health Organization. *Depression and Other Common Mental Disorders: Global Health Estimates.* World Health Organization, Geneva; 2017.

Yadav KD. Cosmeceutical assets of ancient and contemporary ayurvedic astuteness. *International Journal of Green Pharmacy.* 2016;9(4): S1–S6.

Zhang N, Zhang L, Feng L, Yao L. The anxiolytic effect of essential oil of *Cananga odorata* exposure on mice and determination of its major active constituents. *Phytomedicine.* 2016;23(14):1727–1734.

14 Pharmaceutical Applications of Coriander

Nazish Jahan

CONTENTS

14.1 INTRODUCTION

Coriandrum sativum L. is an aromatic herb; its leaves, stems, and seeds are commonly used as spices for taste and fragrance in cooking worldwide. Fresh coriander leaves are rich in vitamin C, vitamin A, α-tocopherol, vitamin K, niacin, folate, and protein. Coriander also contains micronutrients such as potassium, calcium, magnesium, manganese, choline phosphorous, and zinc. The seeds and the aerial parts of the plant are a potential source of lipids and essential oil. The amount of bioactive phytotherapeutics and essential oil varies in different varieties and chemotypes growing under different soil conditions due to different geometrical conditions. However, coriander is very low in saturated lipids and contains linoleic acid. Dietary supplementation of coriander seed greatly affects the lipid composition of the carcass by decreasing saturated fatty acid (SFA) contents (palmitic and stearic acids) and by increasing monounsaturated and polyunsaturated fatty acid. The seeds contain 1–2% essential oil, and the monoterpenoid, linalool, is the main component and ranges from 35 to 85%. Other important components with range are γ-terpinene (0.1–14%), α-pinene (0.1–10.9%), *p*-cymene (0.1–8.1%), camphor (2.0–5.5%), geranyl acetate (0.2–5.4%), and myrcene (0.1–3%) (Ebrahimi et al., 2010; Mandal and Manda 2015). Coriander seed oil is a rich source of two bioactive phytosterols, stigmasterol and sitosterol.

Phenolic compounds are responsible for many physiological effects of coriander. Bioactive phenolics and flavonoids reported in coriander are caffeic acids, chlorogenic acid, *p*-coumaric acid, ferulic acid, gallic acid, synergic acid, vanillic acid, quercetin, luteolin, and kaempferol (Debouch et al., 2020; Mahleyuddin et al., 2022).

Coriander seed is famous for its extensive healing applications in gastrointestinal disorders. Various studies have proven its efficacy against intestinal parasites, and it also possesses larvicidal, bactericidal, and fungicidal potential. Coriander is also well known for its antioxidant, anti-diabetic, anti-mutagenic, anti-anxiety, and antimicrobial activity, along with analgesic and hormone balancing effect. Its nutritional and extraordinary health benefits promote its use as a nutraceutical. In this chapter the pharmaceutical potential of coriander is explained.

DOI: 10.1201/9781003204626-15

14.2 PHARMACEUTICAL POTENTIAL OF CORIANDER

14.2.1 CARDIOPROTECTIVE POTENTIAL

A high level of serum lipids, also known as hyperlipidemia, leads to an increased risk of lipid oxidation and production of oxidized low-density lipoprotein (LDL) and other free radical products stored in the subendothelial layers, vasculature, and bone. The consumption of thermogenic high-fat and accumulation of oxidized products in the endothelium is crucial for initiating chronic diseases atherosclerosis. The researchers studied the antilipidemic effect of coriander (*Coriandrum sativum* L.). In an animal trial, coriander was administered at a 1 g/kg dose to triton-induced hyperlipidemic rats. It was found that coriander decreases the uptake and enhances the breakdown of lipids. Results were compared with commercially available herbal drugs for hypolipidemia (Mahleyuddin et al., 2022).

Coriander seeds contain beneficial fatty acids like linoleic acid, oleic acid, palmitic acid, stearic acid, and ascorbic acid. A combination of these acids may be beneficial in decreasing high levels of cholesterol in the blood. They are also effective against high levels of bad cholesterol (LDL) and reduce their accumulation along the inner walls of the arteries and veins. High LDL level is a leading cause of cardiovascular disorders like atherosclerosis and acute myocardial cardiac arrest. More importantly, coriander could also increase the levels of "good" cholesterol (HDL), which promotes defense against some cardiac disorders.

Coriander seeds are very effective in the treatment of cardiac dysrhythmia. The cardiac anti-arrhythmic (anti-tachycardial and anti-bradycardial) potential of *C. sativum* seeds was reported through both curative and preventive modes of treatment in $BaCl_2$-induced tachycardial and KCL-induced bradycardia in rats. Coriander seeds were very efficacious in dysrhythmia and maintained the heartbeat rate. Results indicated that the oral administration of *C. sativum* seeds (300 mg/kg) had a profound effect on the level of cardiac biomarkers in both curative and preventive modes of treatment. This treatment significantly reduced the $BaCl_2$- and KCl-induced secretion of cardiac enzymes (CK-MB, AST, ALT, and LDH). ECG patterns of KCl and $BaCl_2$ treated rats showed abnormalities; however, the normal ECG pattern of curative and preventive groups confirmed the anti-arrhythmic potential of *C. sativum* seeds. The anti-arrhythmic potential of coriander seeds may be attributed to important bioactive phenolic compounds (Rehman et al., 2016). Certain studies also reported that alkaloids in coriander seeds are responsible for their curative and preventive potential in dysrhythmia (Mahleyuddin et al., 2022). Therefore, it is concluded that oral administration of *C. sativum* seeds could attenuate both types of cardiac arrhythmias.

14.2.2 ANTIHYPERTENSIVE POTENTIAL

Coriander is very helpful in lowering blood pressure. Many medicines are used to lower blood pressure, such as calcium channel blockers, angiotensin-converting enzyme (ACE) inhibitors, beta-blockers, and diuretics. In addition, bioactive phytochemicals enhance the interaction of calcium ions and acetylcholine, a neurotransmitter in the peripheral and central nervous system, relax blood vessel tension, and inhibit enzymes' activity, thereby reducing the chances of several cardiovascular conditions, including heart attacks and strokes.

Hussain et al. (2018) identified and characterized the bioactive compounds of *Coriandrum sativum* responsible for treating hypertension and explored their mechanism of action as ACE inhibitors. Four bioactive rich fractions, including alkaloids, flavonoids, steroids, and tannins, were extracted from the coriander seed and evaluated for their ACE inhibition potential. The flavonoid-rich fraction showed high ACE inhibition potential with an IC_{50} value of 28.9 μg/mL. The flavonoids were characterized through LC-ESI-MS/MS. Seventeen flavonoids were identified in this fraction of *Coriandrum sativum*, including pinocembrin, apigenin, pseudobaptigenin, galangin-5-methyl ether, quercetin, baicalein trimethyl ether, kaempferol dimethyl ether, pinobanksin-5-methylether-3-*O*-acetate, pinobanksin-3-*O*-pentenoate, pinobanksin-3-*O*-phenylpropionate, pinobanksin-3-*O*-pentanoate, apigenin-7-*O*-glucuronoide, quercetin-3-*O*-glucoside, apigenin-3-*O*-rutinoside, rutin, isorhamnetin-3-*O*-rutinoside, and quercetin dimethyl ether-3-*O*-rutinoside, daidzein, luteolin, pectolinarigenin, apigenin-C-glucoside, kaempferol-3-7-dimethyl ether-3-*O*-glucoside, and apigenin-7-*O*-(6-methyl-beta-D-glucoside). The results of this study revealed that *Coriandrum sativum* is a valuable functional food that possesses several therapeutic flavonoids with ACE inhibition potential that can manage blood pressure very efficiently (Hussain et al., 2018).

Jabeen et al. (2009) studied the diuretic activity of coriander in Wistar rats; the animals were divided into negative and positive control and treatment groups, which were treated with different doses of coriander seeds (50 mL/kg). This study concluded that the diuretic effect of coriander was confirmed due to a significant increase in urine output (diuresis) in rats, similar to furosemide, a standard diuretic. Therefore, a diuretic is considered one of the best choices for treating and managing uncomplicated hypertension.

14.2.3 Anti-Hyperglycemic Activity

The coriander seed extract is used in complementary and traditional medicine to treat diabetic patients. The coriander leaves also contain high dietary fibers and proteins, which help reduce high blood sugar levels. A study in rats confirmed that incorporating coriander seed extract into the diet controlled the blood glucose and significantly decreased insulin levels in diabetic rats (Chithra and Leelamma, 1999; Eidi et al., 2009, Zamany et al., 2021). In addition, Gray and Flatt (1999) studied coriander's insulin-releasing and insulin-like activity and reported that coriander consumption stimulated the secretion of insulin colon B-cell line.

14.2.4 Anti-Anthelmintic Activity

Many *in vitro* and *in vivo* studies proved the anthelmintic potential of coriander seeds against the parasite *Haemonchus contortus* (Debella et al., 2007).

14.2.5 Anti-Bacterial Activity

Coriander is a potent inhibitor of *E. coli* and other bacteria and fungi. Fresh coriander leaves containing a high concentration of aliphatic (2*E*)-alkenals and alkanals showed bactericidal activity against Gram-negative strain *Salmonella choleraesuis* spp. Coriander seeds possess a complex combination of different bioactive compounds in different amounts. Seeds extracts showed potential against *S. aureus*, *S. Typhi*, *Salmonella choleraesuis*, and *P. aeruginosa* (Ali and Malik 2021). In addition, coriander essential oil has been reported to inhibit a broad spectrum of microorganisms like *E. coli*, *B. cereus*, *A. baumannii* and *P. aeruginosa*, *K. pneumoniae*, and *Listeria monocytogenes* (Silva and Domingues 2017; Duarte et al., 2012). These studies confirmed that the potential of coriander essential oil was greater than its isolated main constituent, linalool.

14.2.6 Anti-Fungal Activity

Coriander showed excellent potential against fungal strains such as *Pyricularia oryzae*, *Bipolaris oryzae*, *Alternaria alternata*, *Tricoconis padwickii*, *Drechslera tetramera*, *Drechslera halodes*, *Curvularia lunata*, *Fusarium moniliforme*, *Fusarium oxysprorum*, *Candida tropicals*, *Aspergillus flavus*, *Mucor* sp., and *Emericella nidulans* (Mandal and Mandal 2015).

14.2.7 Antioxidant Activity

Coriander contains a galaxy of antioxidants, which avert the cellular damage instigated by free radicals. In addition, its antioxidants have helped increase the ability of the body to fight against inflammation (Sriti et al., 2019: Sittikul et al., 2017).

14.3 CONCLUSION

Coriander is a rich source of bioactive compounds with versatile pharmacological benefits. The information reported in this chapter is helpful to generate interest in this medicinal and nutritional herb and, hence, may help in discovering innovative drugs and nutraceuticals in the future.

REFERENCES

Amin, S., Q. Ali, and A. Malik. 2021. Antimicrobial activity of *Coriander sativum*. *Journal of Pharmaceutical Research International* 32(47): 74–81.

Bettaieb, S. J. I., O. Bachrouch, and T. Talou. 2019. Chemical composition and antioxidant activity of the coriander cake obtained by extrusion. *Arabian Journal of Chemistry* 12(7): 1765–1773.

Chithra, V., and S. Leelamma. 1999. *Coriandrum sativum*-mechanism of hypoglycemic action. *Food Chemistry* 67: 229–231.

Derouich, M., E. D. T. Bouhlali, M. Bammou, A. Hmidani, K. Sellam, and C. Alem. 2020. Bioactive compounds and antioxidant, antiperoxidative, and antihemolytic properties investigation of three *Apiaceae* species grown in the southeast of Morocco. *Scientifica* 2020: 3971041. https://doi.org/10.1155/2020/3971041

Duarte, A., S. Ferreira, F. Silva, and F. C. Domingues. 2012. Synergistic activity of coriander oil and conventional antibiotics against *Acinetobacter baumannii*. *Phytomedicine* 19: 236–238.

Ebrahimi, S. N., J. Hadian, and H. Ranjbar. 2010. Essential oil compositions of different accessions of *Coriandrum sativum* L. from Iran. *Natural Product Research* 24(14): 1287–1294.

Eidi, M., A. Eidi, A. Saeidi, S. Molanaei, A. Sadeghipour, M. Bahar, and K. Baha. 2009. Effect of coriander seed (*Coriandrum sativum* L.) ethanol extract on insulin release from pancreatic beta cells in streptozotocin-induced diabetic rats. *Phytotherapy Research* 23(3): 404–406. https://doi.org/10.1002/ptr.2642

Gray, A. M., and P. R. Flatt. 1999. Insulin-releasing and insulin-like activity of the traditional anti-diabetic plant *Coriandrum sativum* (coriander). *British Journal of Nutrition* 81(3): 203–209. https://doi.org/10.1017/s0007114599000392

Hussain, F., N. Jahan, Khalil-ur-Rahman, B. Sultana, and S. Jamil. 2018. Identification of hypotensive biofunctional compounds of Coriandrum sativum and evaluation of their angiotensin-converting enzyme (ACE) inhibition potential. *Oxidative Medicine and Cellular Longevity* 2018: 4643736. https://doi.org/10.1155/2018/4643736

Mahleyuddin, N. N., S. Moshawih, L. C. Ming, H. H. Zulkifly, N. Kifli, M. J. Loy, M. M. R. Sarker, Y. Mohammed Al-Worafi, B. H. Goh, S. Thuraisingam, and H. P. Goh. 2022. *Coriandrum sativum* L.: A review on ethnopharmacology, phytochemistry, and cardiovascular benefits. *Molecules* 27: 209. https://doi.org/10.3390/molecules27010209

Mandal, S., and M. Mandal. 2015. Coriander (*Coriandrum sativum* L.) essential oil: Chemistry and biological activity. *Asian Pacific Journal of Tropical Biomedicine* 5(6): 421–428.

Rehman, N., N. Jahan, Khalil-ul-Rahman, K. Mahmood Khan, and F. Zafar. 2016. Anti-arrhythmic potential of seeds in salt induced arrhythmic rats. *Pakistan Veterinary Journal* 36(4): 465–471.

Silva, F., and F. C. Domingues. 2017. Antimicrobial activity of coriander oil and its effectiveness as food preservative. *Critical Reviews in Food Science and Nutrition* 57(1): 35–47. https://doi.org/10.1080/10408398.2013.847818

Sittikul, V. P., Supaluk Prachayasittikul, Somsak Ruchirawat, and Virapong Prachayasittikul. 2017. Coriander (*Coriandrum sativum*): A promising functional food toward the well-being. *Food Research International* 105: 305–323.

Zamany, S., M. Mahdavi, S. Pirouzpanah, and A. Barzegar. 2021. The effects of coriander seed supplementation on serum glycemic indices, lipid profile and parameters of oxidative stress in patients with type 2 diabetes mellitus: A randomized double-blind placebo-controlled clinical trial. https://doi.org/10.21203/rs.3.rs-262149/v1.

15 Coriander and Physiological Systems of the Body

Haseeb Anwar, Ghulam Hussain, Humaira Muzaffar,
Bazla Naseer Hashmi, Asma Mukhtar, and Liza Malik

CONTENTS

15.1 INTRODUCTION

15.1.1 BACKGROUND

Coriander (*C. sativum* L.) is well-recognized as cilantrillo, cilantro, Chinese parsley, Arabian parsley, Dhania, Yuen sai, and Mexican parsley. It belongs to the family of Apiaceae (previously known as Umbelliferae). It is a perennial herb popular in African, Mediterranean, Middle Eastern, Southeast Asia, Latin American, and Indian cuisine. Though coriander and cilantro appear to be interchangeable, they have distinct meanings. Cilantro refers to the green leafy part of the plant when freshly gathered; the herb is called coriander when the dried fruits are consumed (Önder 2018).

Coriander possesses many aromatic components and bioactive components possessing antifungal, antibacterial, and antioxidant properties, and therefore *C. sativum* is beneficial in food preparation, i.e., as a flavoring agent, for preservation, and for the prevention of foodborne illnesses and decay (Mandal and Mandal 2015). In Sanskrit literature, coriander is known as "kusthumbari" or "dhanayaka"; in India, it is known as Dhania; and in Bengal, it is known as Dhane. It is native to the Mediterranean and spread to China, India, and the whole world (Coşkuner and Karababa 2007).

Coriandrum sativum has been known for its health benefits and usages since 1550 BC in Greek Eber Papyrus, and it is also referred to in the literature of Sanskrit as far back as *ca.* 5000 BC. In order to give the wine a pungent taste and odor, the Greeks widely used coriander as a flavoring agent. It is widely cultivated in the Indo-Pak subcontinent and Asia, Central Europe, South and Western Australia, and Morocco. Coriander also contains 0.03 to 2.6% of essential oils, having numerous health and medicinal purposes. Coriander is referred to as a functional food that has promising health benefits for the well-being of humanity. Coriander is a vast reservoir of essential bioactive compounds that have created new avenues in drug discovery. *Coriandrum sativum* contains flavonoids and phenolic constituents, which are the reason for the antioxidant potential of coriander (Foudah et al. 2021). The phenolic compounds present in *Coriandrum sativum* are caffeic acid, cinnamic acid, and coumarins, which provide the cytoprotective potential of coriander (Mishra, Sasmal, and Kumar 2017).

15.1.2 OCCURRENCE

The plant was cultivated first in the Mediterranean region and now is mainly cultivated in places such as Russia, Ukraine, Morocco, Romania, Mexico, Argentina, and India (Priyadarshi et al. 2016).

15.1.3 OPTIMUM TEMPERATURE

The optimal temperature for the germination of coriander is 20–25°C. Although most preferential regions are arid, it can survive in different soil types. Dampening of soil, loamy, light and heavy dark soil, light, and well-drained soils can affect its cultivation (Bhat et al. 2014).

15.2 CONSTITUENTS AND MEDICINAL USES

The annual herb coriander (*Coriandrum sativum* L.) is mainly used as a flavoring agent. Its plant roots, seeds, and leaves are edible; however, their tastes and applications are somewhat different. The flavor of the herb is mild and refreshing. Its roots and leaves can enhance the sweetness of ripened fruits. It can be utilized as a flavoring component in different culinary preparations because of the delicate nature of its leaves. Coriander is used in cooking as a whole plant, mostly fresh leaves, and ripe fruits. Coriander leaves have a distinct flavor from coriander seeds, having citrus undertones. Its plant is a rich source of nutritional and micronutrients components. Coriander has less saturated fat content, but linoleic acid content is high which is an excellent supplier of vitamin K and tocopherol. Its leaves are rich in vitamins, while its seeds have high amounts of essential oils and polyphenols. Its flavor comes from its essential oil, containing many furanocoumarins (dihydrocoriandrine and coriandrine) and linoleic acid. *C. sativum* is well recognized for its anti-diabetic, antioxidant, anti-anxiety, anti-mutagenic, and analgesic activity, as well as antimicrobial and hormone modulating activities, which encourages its consumption in food because of various health benefits as well as its protective action in preserving food for long durations (Bhat et al. 2014).

Chromatography-mass spectroscopy was used to examine essential oils present in the fruits and leaves of coriander. See Table 15.1 for major chemicals present in essential oils of coriander leaves

TABLE 15.1
Constituents of Essential Oils

Constituent	%	Constituent	%
2-decenoic acid	30.8	*E*-11-tetradecenoic acid	13.4
Capric acid	12.7	Undecyl alcohol	6.4
Tridecanoic acid	5.5	Undecanoic acid	7.1

Source: Bhuiyan, Begum, and Sultana 2009.

TABLE 15.2
Constituents of Essential Oils of Seeds

Constituent	%	Constituent	%
Linalool	37.7	Geranyl acetate	17.6
Terpinene	14.4		

Source: Bhuiyan, Begum, and Sultana 2009.

and Table 15.2 for seeds. Both oils had different qualitative and quantitative constituents (Bhuiyan, Begum, and Sultana 2009).

Essential oil (linalool-rich) and lipids (petroselinic acid-rich) can be extracted from the seeds and aerial portions of the plant. A broad range of pharmacological functions has been attributed to various plant portions because of various biologically active components, including antioxidant, antimicrobial, anti-epileptic, anxiolytic, anti-diabetic, anti-mutagenic, anti-depressant, anti-dyslipidemic, diuretic, anti-hypertensive neuroprotective and anti-inflammatory properties. *C. sativum*, it turned out, has lead-detoxifying properties (Sahib et al. 2013). In addition, many therapeutic qualities have been attributed to *Coriandrum sativum* Linn. It has been used as a pain reliever, carminative, digestive aid, anti-rheumatic, and antispasmodic (Mahendra and Bisht 2011).

15.3 PHYSIOLOGICAL SYSTEMS OF THE BODY

Tissues and organs work together to form organ systems. The body's physiology depends upon the normal functioning of all the organ systems efficiently. Therefore, organs work together in interdependent systems that serve a larger role for the whole body. Our body comprises ten major physiological systems that interact with each other in a well-defined manner, which means that each organ system has its specialty, but still, all of the organ systems interact with each other in a variety of ways. The interdependency of one physiological system with another also significantly impacts the diagnosis and treatment of many ailments. Every physiological system of the body plays a specific physiological role in order to maintain the normal homeostatic conditions of the body; for example, the respiratory system is comprised of nasal passages, trachea/wind pipe, and lungs/alveoli, and it deals with the procedure of air entry from the nostrils and its passage through the trachea to reach the alveoli where gaseous exchange takes place (Sherwood 2015). See Figure 15.1 for a visual representation of the overall physiological systems of the body.

The cardiovascular system comprises the heart and major blood vessels, which transport the blood between the heart and different tissues of the body. Both respiratory and cardiovascular systems work in close coordination with the body to carry out the phenomenon of pulmonary circulation. Another important physiological system of the body is the endocrine system which consists of major endocrine glands responsible for secreting chemical messengers in the body known as hormones. These hormones serve a vital role in the coordination of growth, metabolism, blood pressure regulation, maintenance of electrolytes concentration, the phenomenon of reproduction, etc. The reproductive system is also one of the major physiological systems of the human body; it is responsible for the sexual traits of both the sexes, i.e., male and female; the male reproductive system is comprised of testes (storage and production of male gametes/sperms), penis (copulatory organ). and other accessory organs like vas deferens, epididymis, prostate gland, etc. At the same time, the female reproductive system is comprised of ovaries (site of production of female gametes/egg), fallopian tubes, uterus (implanting region for the embryo), and mammary tissues (site of milk production). Finally, the nervous system is the controlling region of the body; all the body's physiological systems work under the command of the nervous system. The nervous system comprises

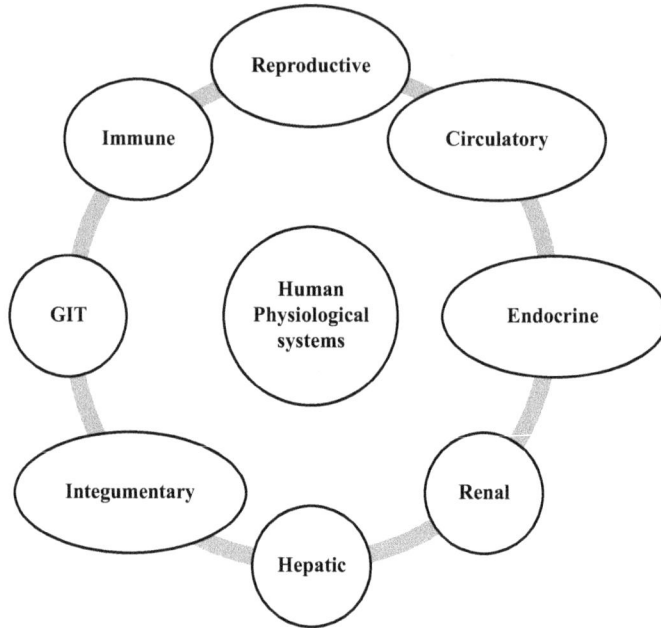

FIGURE 15.1 Physiological systems operating inside the body.

nerves, the brain, and the spinal cord, which messages between body tissues and the brain or spinal cord for the actions required in the body. Five basic senses make up the sensory system inside the nervous system, accountable for getting messages from outside the body and carrying them to the brain for integration. The integumentary system protects the body, composed of skin, nails, hairs, sweat, and sebaceous glands. This system covers the body in a defensive barrier with many functions, including protection against ultraviolet rays and temperature regulation. The immune system is associated with the body's defense mechanism, protecting the body against foreign invaders or any disease-causing pathogen. At the same time, the lymphatic system is associated with returning lymphatic fluid to the blood (Sherwood 2015).

The excretory or urinary system comprises excretory organs, i.e., kidneys (functional unit; nephrons), ureters, urinary bladder, and urethra. The excretory system is responsible for blood filtration, waste removal, and the maintenance of plasma electrolytes. Kidneys are called blood purifiers and filter the bloodstream, remove undesirable substances, toxins, from the body, and return those substances (electrolytes) that are still required to the blood. The body's musculoskeletal system comprises bones, cartilage, tendons, and ligaments that provide movement and support to the whole body, providing the framework for the whole body. The digestive system serves to digest food by breaking down food matter into digestible components to allow them to be absorbed into the blood to nourish the body's tissues (Boron and Boulpaep 2012).

15.4 CORIANDER AND PHYSIOLOGICAL SYSTEMS OF THE BODY

Plant polyphenolic compounds have sparked much attention due to their efficacy in fighting cancer as chemopreventive agents and chronic disease defenders. Overall positive effects are primarily attributable to their antioxidant and chelating functions. Flavonoids are polyphenolic chemicals found in various plant foods, including vegetables, fruits, beverages, and cereals (Anila and Vijayalakshmi 2002). Antioxidants react differently in different situations. They operate as antioxidants at physiological levels that are characteristic of a healthy food intake, but at pharmaceutical dosages, some of them work as pro-oxidants, promoting the formation of free radicals (Hashim et al. 2005).

15.4.1 CIRCULATORY SYSTEM

In male Wistar rats, the impact of coriander on heart pathology was assessed using a cardiotoxicity model induced by isoproterenol. According to the findings, a methanolic extract of coriander can avert myocardial infarction by reducing myofibrillar deterioration. Furthermore, coriander extract possesses high phenolic concentrations, also accountable for reducing oxidative stress by efficiently scavenging the isoproterenol-produced reactive oxygen species, according to the findings (Patel et al. 2012).

The therapeutic and preventive potential of *Coriandrum sativum* extract in Wistar rats with isoproterenol-induced heart failure (HF) was also assessed. Using *C. sativum* extract as a therapeutic and preventive therapy enhanced left ventricular function and hemodynamic parameters while increasing baroreflex sensitivity. It also reduced lipid peroxidation, improved lipid profile, and decreased endothelin receptor expression. Treatment with simvastatin had a comparable cardioprotective effect. Findings showed that *C. sativum* extract protects against heart failure, perhaps due to its capacity to enhance left ventricular function and baroreflex sensitivity, reduce lipid peroxidation, and modify endothelin receptor expression (Dhyani et al. 2020).

Supplementation of coriander in type-II diabetic individuals lowered LDL cholesterol, total cholesterol, triglycerides, and HDL cholesterol while not affecting plasma urea and creatinine. These data support its use in diabetes therapy and control. Frequent use of CSP (which is generally innocuous) in type-II diabetes patients might lower plasma glucose and prevent or minimize CV problems caused by hyperlipidemia (Parsaeyan 2012). Coriander fruit hydro-methanolic extract has been discovered to have cardioprotective properties. This impact should also be attributed to the fruits' high polyphenol content (Patel et al. 2012).

15.4.2 NERVOUS SYSTEM

From the medical point of view, coriander has the following properties: antibilious, anti-inflammatory, galactagogue, antioxidant, aphrodisiac, deodorant, aromatic, digestive, antispasmodic, and refrigerant aromatic, stimulant, stomachic, appetizer, lypolytic, diuretic, fungicidal agent. As a result, it reduces intestinal cramping, stomach problems, dyspepsia, cold, gastrointestinal problems, fever, anxiety, nervous tension, and cholesterol; helps the removal of fluid wastes and toxins, the flow of gastric juice, and migraine; and stimulates the ease of fatigue and mind (Balasubramanian et al. 2016).

In Iran, coriander is used as a traditional medicine with anti-depressant, sedative, anticonvulsant, and anxiolytic properties. Coriander seeds have anti-seizure potential too. Ethanol and aqueous coriander seed extract may delay the beginning of clonic convulsion that showed anticonvulsant activity. It was observed that there was a reduction in the period of tonic seizures as checked by maximum electric shock. A 5 mg/kg dose showed the defensive effect of 20 mg/kg of phenobarbitural. Another finding suggested that both alcoholic and aqueous extracts and oil of coriander seed supply dose-dependent defensive mechanisms against PTZ-induced death and tonic convulsions. Aqueous extract of coriander seed works actively compared to oil of coriander. It is also involved in the sedative activity and induction of sleeping time. Although coriander seed oil at a dose of 600 mg/kg showed a soothing effect, it is suggested that sedative components may be present in higher amounts (Sahib et al. 2013).

Coriander is a rich source of vitamin A, B2, and C. Dried seed and fresh coriander leaves are mostly used as pleasantly aromatic spices, as traditional medicine, and the fresh juice of coriander leaves and powder fruit tea are suggested for the improvement of anxiety and insomnia. Apart from these uses, it is also used to reduce the loss of appetite, pain, galactagogue, and flatulence, and used in digestive problems, urinary system, and respiratory system. According to the medical point of view, coriander has anti-convulsant, hypnotic, and sedative effects on the central nervous system. Oil of coriander seed and extract may also cause depression in the central nervous system.

Coriander has antioxidant properties; it showed a neuroprotective effect and prevented apoptotic cell production and dark neurons in the hippocampal part. It also affects memory due to its anxiolytic property (Pachiappan and Institutions, 2020).

It is reported that coriander has a protective effect on pentylentetrazole and induces seizures and oxidative damage in brain tissues. For seizures in the hippocampal tissues of the rats, it was shown that extract of coriander plant prevented seizure by preventing lipid peroxidation (Hosseini et al. 2015). Due to seizure, oxidative damage is an essential factor in memory loss and neural damage, which may be avoided using antioxidant reagents (Seghatoleslam et al. 2013). Considering the well-known antioxidant effects of coriander, it seems that the hydro-alcoholic extract of the plant can prevent neuronal damage and apoptotic cell production in the hippocampal regions. Thus, there are neuroprotective, anti-apoptotic, and anti-convulsive effects of hydro-alcoholic extract of *C. sativum* in seizures (Pourzaki et al. 2016).

Alzheimer's disease (AD) is a primary degenerative disease characterized by memory loss, unusual behavior, and declining cognitive function. The main pathological culprits are the extracellular deposition of beta-amyloid-Aβ and intracellular accumulation of tau protein and some risk factors. The inhalation of volatile coriander oil increased anxiolytic-antidepressant-like behaviors and decreased oxidative status in the beta-amyloid rat model of Alzheimer's disease. *Coriandrum sativum* leaves appear to be a promising candidate for improving memory, and it would be worthwhile to explore the potential of this plant in the management of Alzheimer's patients (Manuha, 2018).

Many studies indicated that coriander seeds affect neuromuscular management and locomotion and have anxiolytic and antidepressant effects. It was seen that ethanol extract of coriander is more effective than aqueous and methanol extract; therefore, the sedative-hypnotic efficacy of ethanol extract of *Coriander sativum* was assessed. Furthermore, induction of pentobarbital sodium was used to evaluate the possible sedative-hypnotic effect of coriander seeds. Barbiturates and benzodiazepines enhance the period of chloride channel opening. GABA receptors exert a direct stimulatory effect by acting as a GABA agonist and directly causing inhibitory action (Alshahrani 2016).

Extensive study of phytochemicals led to coriander separation and detection of seven new compounds such as phenolic glycoside and isocoumarin. Analysis and chemical structures of these compounds are explained by 1D and 2D nuclear magnetic resonance. Separated compounds were described not based on their neurotropic activity but by induction of growth factors in C6 glioma cells. These compounds were stimulated to release the nerve growth factors. Furthermore, the aglycones showed more effective nerve growth factor secretion activity and anti-neuroinflammatory effect than glycosides (Cha et al. 2020).

Administration of 200 mg/kg body weight *Coriandrum sativum* seed extract exhibits better effects on memory destruction in SAMP8 mice. Coriander seed extract's mechanisms involved in repairing memory deficits were connected with increased NF-L mRNA and decreased nNOS mRNA in the brain's frontal lobe (Mima et al. 2020).

Different parts of the plant, like leaves and seeds, possess antioxidant, anxiolytic, anti-depressive, learning and memory improving, neuroprotective, sedative-hypnotic, analgesic, and anticonvulsant activities. Furthermore, linalool has various neuropharmacological effects, including sedative, anti-anxiety, anticonvulsant, and anti-Alzheimer's activities. The neuroprotective effects of coriander and its major component may be due to their anti-inflammatory and antioxidant properties (Hosseini, Boskabady, and Khazdair 2021). Coriander shows a protective effect due to its antioxidant property. Linalool is the main component of the coriander seeds and accounts for central nervous system modulation, i.e., migraine-relieving, anxiolytic, anticonvulsant, and antimicrobial activity, therapeutic potential against metabolic syndrome, and gut modulatory effect (Prachayasittikul et al. 2018). The aqueous extract of leaves of CS Linn. possesses dose-dependent anxiolytic activity. Aqueous extract remarkably reduced spontaneous activity and neuromuscular coordination compared to the normal group. These results showed that the aqueous extract of coriander seed has an anxiolytic effect and may have potential sedative and muscle relaxant effects (Latha et al. 2015).

15.4.3 GROWTH AND MUSCULOSKELETAL SYSTEM

Feed supplemented with coriander and cinnamon powder improved growth performance, feed conversion ratio, protein retention, and protein productive value. Furthermore, the edible parts of fish, when fed with both types of medicinal plants, were compared to the negative control feed, which showed a significant reduction in fat percentage (Al-Shakarchi and Mohammad 2021). In addition, the body weight, weight gain, and feed consumption were significantly increased in broiler ration with the addition of coriander. Hence, coriander powder could be used to substitute for antibiotics and a natural feed additive that promotes growth in poultry diets (Ahmad, Kumar, and Singh 2016).

Essential oils of coriander also exerted a strong antioxidant effect on chicken tissues. It reduced malondialdehyde significantly (Rahiminiat et al. 2017). Coriander fruit oil has abundant fatty acids. In anti-aging products the use of the fatty acids (long-chain) is potentially effective, serving as repair barrier with characteristics of prevention of moisture loss and epidermis. That is why fatty acids (long-chain) are thought of as potential anti-aging agents. Its fruit oil is used as topical therapy for several skin conditions (Abascal and Yarnell 2012).

15.4.4 RENAL SYSTEM

Coriander also acts as a nephron-protective agent to prevent heavy metal deposition in kidneys by the phenomenon of chelation (Téllez-López et al. 2017). Herbal medicines or herbal tinctures contain coriander as an active ingredient that protects against renal damage (El-Masry et al. 2016). The aqueous and ethanol extract of coriander exhibited a protective effect countering lead-induced oxidative damage in mice as the treatment with coriander extracts led to an increase in antioxidants and a further decrease in lipid peroxidation. Furthermore, in the kidneys of mice which were damaged due to ingestion of lead, when treated with ethanol and aqueous extract of *Coriandrum sativum*, the kidney tubules were reversed to near normal (Dutta, Harikumar, and Grewal 2013). Therefore coriander extract can have protective effects on kidneys against lead toxicity.

Marked enhancement in renal function tests (RFTs) was seen as a result of the reduction in creatinine-clearance and an increase in serum sodium and potassium ion concentration with the decrease in urine concentrations in rats exposed to lead induction. The previous studies thus suggested that coriander has beneficial effects in renal protection. Lead accumulation disrupts tubules, decreasing reabsorption and increasing urinary protein excretion. Administration of coriander extract in rats showed a significant improvement in all vital functions of the kidney and regulation of electrolyte balance (potassium and sodium) in serum and urine (El-Masry et al. 2016). Lead accumulation disrupts tubules, decreasing reabsorption and increasing urinary protein excretion. Administration of coriander extract in rats showed a significant improvement in all vital functions of the kidney and regulation of electrolyte balance (potassium and sodium) in serum and urine. These potential effects of coriander are due to the diuretic and natriuretic effects of coriander which lead to a higher glomerular filtration rate (GFR) (Chahal et al. 2018).

Herbal plants like coriander enhance a variety of parameters in a dose-dependent manner, including the normalization of serum uric acid and creatinine, restoration of Delta-aminolevulinic acid dehydratase enzyme activity, rise in enzymatic and non-enzymatic antioxidant concentration levels, and reduction of pro-oxidant concentrations in the kidney and plasma. Following coriander treatment, antioxidants such as catalase, superoxide dismutase, and glutathione rise, while oxidative damage in the kidney and liver of animal studies declines. As a result, suppression of increased LPO levels in organs could be a plausible mechanism for coriander's ameliorative actions. The higher quantities of carotenoids, tocopherols, and phospholipids in *Coriandrum sativum* contribute to its antioxidative properties (Al-Marzoqi, Hameed, and Idan 2015). These substances use different modes of action. Secondary antioxidants work by decreasing singlet oxygen, while primary antioxidants work by reducing oxidative stress (Al-Marzoqi, Hameed, and Idan 2015).

Lead poisoning also raises AST, ALT, ALP, and ACP activity, cholesterol, and protein levels. Because of their inherent harmful effect, several antibacterial and anticancer medicines and synthetic compounds cause nephrotoxicity. Chronic kidney impairment is connected to oxidative damage, greater levels of cytokine production, and a variety of lipid and homocysteine metabolites (Kannan and Quine 2013). Patients should be treated with anti-inflammatory or antioxidant medicines to prevent such disorders, but organic herb therapy would be preferable because the drugs are manufactured and have side effects. Because chronic kidney illness is associated with increased mortality and morbidity, the link between coriander and renal function may provide a novel avenue for treating the disease.

15.4.5 Integumentary System

The largest organ covering our whole body is the skin, the outer layer called the integumentary system. It is composed of three different layers, the epidermis, dermis, and hypodermis. The chief function of the skin is to protect the body against pathogens, regulate excessive water loss and temperature, and vitamin D synthesis. The imbalance in the perspiration of skin cells with the environment causes skin diseases. In addition, the sweat which is excreted from the skin contains urea which is responsible for body temperature (Guna et al.). Skin diseases are categorized into two classes:

Acute skin diseases: These diseases are easily curable and do not create patches or marks after treatment.
Chronic skin diseases: These disease conditions take long-term treatment to recover from; also, they cause scars or marks on the skin (Baghel, Gidwani, and Kaur 2017).

Medicinal plants are still used to treat and prevent disease. See Table 15.3 for some uses of coriander leaves in skin conditions. Some studies showed that people in Anatolia had used plants to treat many diseases. Plant-based medicines have been used to treat disease since ancient times. Topically, coriander seed oil has good antibacterial and antimicrobial activity for shallow skin infections and oozing dermatitis. Cilantro leaf can reduce urticaria, rashes, itching, and burns because it is used as a cooling agent (Wood 2008). Cilantro leaf can be adequately chopped and applied to the affected area after disinfecting for burns, rashes, and itching.

Photodamage to the skin is caused by ultraviolet (UV) radiation which leads to the diminution of the extracellular matrix present in the dermal region and causes chronic alterations in skin structure. The synthesis of matrix metalloproteinase-1 (MMP-1) and collagen leads to skin wrinkle development. *Coriandrum sativum* is commonly used as herbal medicine to treat dermatitis. A previous study showed UV radiation-induced skin damage in mice with hairless skin and dermal fibroblasts of humans with normal skin. The cells treated with *Coriandrum sativum* showed an elevated level of procollagen type-I and decreased MMP-1. The results indicated that *Coriandrum sativum* inhibits the transcription factor activator protein-1 activity. Furthermore, it was reported that the mice treated with *Coriandrum sativum* exhibited thinner layers of the epidermis and denser dermal collagen fibers than untreated mice. Therefore, that study revealed the beneficial effects of *Coriandrum sativum* to prevent skin dermatitis caused by exposure to UV radiation (Hwang et al. 2014).

TABLE 15.3
Coriander Used in Skin Conditions

Herb	Scientific Name	Family	Active Constituent	Uses	Parts Used
Coriander	*Coriandrum Sativum* Linn.	Apiaceae	*p*-Cymene, linalool, geraniol, hydroxy-coumarins	Antibacterial Antifungal	Leaves

Source: Baghel, Gidwani, and Kaur 2017.

The protective effect of *Coriandrum sativum* also proved to be protective against contact dermatitis, an inflammatory condition of the skin caused by the contact of external factors with the characteristic features of redness, itching, scaling, and erythema. *Coriandrum sativum*, commonly known as coriander or Dhanya, is a member of the family Apiaceae, and it is cultivated all around the globe due to its nutritional and culinary value. Linoleic acid is the major component in *Coriandrum sativum*, with multiple pharmacological aspects. So, the protective effects of CS against 2,4-dinitrochlorobenzene-induced CD-like skin lesions were assessed. When coriander dosages of 0.5–1% were administered to dorsal skin, the establishment of contact dermatitis-like skin lesions was suppressed. In addition, immunoglobulin E, tumor necrosis factor, interleukin-1, interferon, interleukin-1, 13, and interleukin-1,-4, all Th2-mediated pro-inflammatory cytokines, were also significantly reduced. Furthermore, coriander increased the amounts of glutathione and heme-oxygenase-1 proteins. Thus, by modifying immune mediators, coriander can suppress the growth of skin irritation in mice, including skin lesions, and can be utilized as an alternative treatment for contact illnesses (Park et al. 2014).

Sensitive skin is a common problem affecting many people all over the world. The occurrence of painful sensations such as scorching and blistering in response to specific stimuli that should not normally evoke such sensations is a symptom of this disease. The primary ideas linking the emergence of sensitive skin to the rupture of the epidermal membrane and greater penetration of chemicals such as allergens are the rupture of the epidermal membrane and increased penetration of substances such as irritants. The necrosis factor-B cascade, which is important in regulating inflammatory reactions, is triggered in keratinocytes. Regulating this activity can help to manage inflammation. Neurosensory impairment is an important mechanism to investigate because of the wide range of auditory processing disorders. Ankyrin-1, an ion channel best recognized as a sensor for discomfort, cold, and irritation or a sensor of barrier function injury, is produced by some dermal nerve endings and is engaged in a wide of physiological or biological processes involving nociception, itching, and neurogenic swelling. Treating neurosensory malfunction necessitates an understanding of how sensory function is modulated. Coriander seed oil is pure cilantro seed oil with unique fatty acid content. The efficacy of cilantro seed oil in inhibiting necrosis factor-B and transient receptor potential ankyrin-1 (TRPA1) activation was investigated to determine if it could soothe irritated skin. Coriander seed oil modulated necrosis factor-B activation caused by tumor necrosis factor-b in an *in vitro* analysis of inflammatory response in keratinocytes. It also modulated the activation of TRPA1 caused by allyl iso-thiocyanate in an *in vitro* study of keratinocytes-neurons co-culture. These data confirmed the soothing effect of cilantro seed oil (Kern et al. 2020).

Because several chemicals found in seed oils have been shown to have nutritional benefits, they can create novel functional vegetable oils. Given the amount ingested in most developed nations, vegetable oils with more beneficial active components could significantly impact human health. This is already happening in Japan, where oils with higher vitamin E and phytosterols are now accessible. Fortifying conventional vegetable oils with increased levels of certain functional additives is one technique to generate functional oils. This notion is analogous to the fortification of white flour, effectively used for decades. This method enables the precise inclusion of specific beneficial components while keeping the original sensory aspects of the meal that customers are already familiar with and love (Athar and Nasir 2005). For example, coriander oil promotes circulation, reduces cellulite, relieves face neuralgia, and aids in the battle against fungal diseases. It is also good for rheumatism, broken capillaries, psoriasis, eczema, musculoskeletal soreness, rheumatism, cramps, rigidity, and sweaty feet (Shaheen et al. 2003).

15.4.6 Hepatic System

The effect of pretreatment with 10% coriander powder was assessed against oxidative stress induced by hexachlorohexane in rat's liver. Coriander powder prefeeding reduced the HCH-induced increase in conjugated dienes, hydroperoxide, and malondialdehyde (MDA) in the liver of rats. Hepatic superoxide dismutase, glutathione S-transferase, catalase, glutathione reductase, and glucose-6-phosphate dehydrogenase activities were reduced by intraperitoneal injection of HCH hepatic glutathione peroxidase and renal glutamyltranspeptid. Hepatic antioxidant system alterations occur due to CSP prefeeding, otherwise attenuated by HCH injection. The findings show that coriander seeds have an antioxidative impact against HCH-induced free radical production in rat liver (Anilakumar, Nagaraj, and Santhanam 2001).

The hepatotoxin thioacetamide (TAA) produces centrilobular necrosis and nephrotoxic damage. In addition, TAA exposure can produce bile duct growth and liver cirrhosis that is histopathologically identical to that caused by viral hepatitis. Coriander in diet boosts antioxidant levels, acting as a natural antioxidant and preventing harmful oxidation reactions. Coriander leaves and seeds were tested for phenolic content and antioxidant activity. The use of coriander leaves and seeds can help reduce the adverse effects of TAA-induced hepatotoxicity (Moustafa et al. 2014).

The goal of one study was to see if *Coriander sativa* has any antioxidant effect against CCl4-induced oxidative damage in Wistar healthy mice. The injection of CCl_4 resulted in a considerable increase in serum marker enzymes and thiobarbituric acid reactive substances (TBARS), as well as a decrease in antioxidant enzymes, indicating cell damage. The activity of enzymes such as ALP, ACP, protein, and bilirubin was measured in serum. At a dose of 200 mg/kg, the activity of leaf extract was comparable to that of the standard medication, silymarin. According to these findings, coriander extract protects the liver from oxidative damage induced by CCl_4 and so aids in the assessment of traditional claims about this herb. The beneficial effect of the coriander extract could be characterized by the presence of phenolic components with membrane-stabilizing properties. These findings indicate that the chemical found in the crude extracts effectively works on the liver to keep it generally operating while reducing cell membrane disruptions (Sreelatha, Padma, and Umadevi 2009). Fast-food consumption, along with exposure to environmental toxins, has increased the occurrence of several life-threatening illnesses, including renal and hepatic toxicity, and the ability to stimulate inflammatory pathways that lead to various carcinogenic events. The use of raw foods as a therapeutic cure for many illnesses has been the subject of converging scientific research. *Coriandrum sativum* is its scientific name. Coriander seeds contain a volatile oil that is both necessary and volatile. The principal constituents of coriander essential oil are alcoholic monoterpenes, with linalool being the most prevalent.

Furthermore, freshly coriander leaves and stems contain polyphenols such as phenolic acids, flavonoids (particularly quercetin), and essential oil to a degree. Coriander polyphenols and linalool can boost the body's enzymatic and non-enzymatic defenses. In addition, they help to regulate liver enzymes like aspartate aminotransferase and alanine aminotransferase, which reduces the harmful effects of free radicals in CCl_4-induced hepatotoxic rats. In summary, the protective effects of coriander against CCl_4 are due to a variety of mechanisms, the most prominent of which is the antioxidant activity provided by the polyphenolic ingredient of coriander. As seen by a sudden increase in their biomarkers, carbon tetrachloride poisoning can harm both hepatic and renal organs. Treatment with coriander significantly reduced the adverse effects of carbon tetrachloride by conserving antioxidants and free radical scavengers, whereas toxicity reduced intrinsic antioxidant enzymes. It has been proven that eating a diet rich in cilantro seeds and leaves can protect humans from xenobiotic-induced hepato-renal damage. Moreover, cilantro leaves have a higher hepatoprotective potential than cilantro seeds (Iqbal et al. 2018).

Coriander is one of the most extensively used spices, with nutritional and therapeutic benefits. It is widely spread and cultivated primarily for its seeds, including essential oil and the monoterpenoid linalool. Many household medications contain coriander, which is used to treat bed colds, seasonal fevers, nausea, vomiting, stomach diseases, indigestion, parasites, arthritis, and joint pain. Coriander's excellent phytonutrients are responsible for many of its healing qualities, and it is sometimes referred to as a bioactive chemical storehouse. The study of antioxidants found in spices is gaining traction in human health because the body quickly absorbs them. *Coriandrum sativum*, for example, is well known for its antioxidant capabilities, and some of its active ingredients have been discovered. Caffeic and chlorogenic acid are active phenolic acid molecules found in cilantro. When these substances are obtained from the diet, they suppress the free radicals created in the digestive system. The volatile oil of coriander is high in beneficial phytochemicals, and the seeds have a health-supporting reputation. It has been used as an antispasmodic, carminative, stimulant, cytotoxic, lipolytic, fungicidal, and stomachic chemical. Coriander has hypoglycemic, antimicrobial, hypolipidemic, insecticidal, antimutagenic, and aflatoxin-controlling properties, among other things. In addition, coriander has a long list of traditional health advantages. Coriander's medicinal benefits can be linked to its high phytonutrient concentration (Rajeshwari and Andallu 2011).

15.4.7 GASTROINTESTINAL SYSTEM

15.4.7.1 Glycemic Index and Lipid Profile

Coriander is proved to be anti-diabetic as it can lower the glycemic index. Experimental trials on animals have shown a significant reduction of glucose levels in plasma (Sá-Caputo et al. 2016). Another study elucidated the effectiveness of methanolic extract of seeds of *Coriandrum sativum* in attenuating diabetes-related secondary complications like oxidative damage and renal insufficiency. Such studies have shown that coriander has both anti-diabetic and antioxidant properties. The results were gained by assessing assays relevant to glucose parameters and antioxidant/oxidant assays. Results showed a significant increase in antioxidant enzymes like catalase (CAT), superoxide dismutase (SOD), glutathione peroxidase (GPx), and glutathione-s-transferase (GST), along with an increase in glutathione (GSH) level when compared to the control positive or diabetic group. The study results proved that methanolic extract of the seeds of *Coriandrum sativum* L. has a prospective ability to attenuate hyperglycemia and diabetes-related secondary complications (Mishra et al.).

Another study evaluated the hypoglycemic effects of aqueous extract of coriander fruits in terms of improvement in glucose secretion of insulin by increasing the metabolism of muscles. These results suggested that *Coriandrum sativum* may act as a novel oral mediator for diabetes (Al-Mofleh et al. 2006). Patients with type-II diabetes mellitus, when supplemented with coriander seed as a complementary treatment, significantly improved their glycemic index, oxidative stress, and serum lipid profile. In addition, the results showed reduced serum lipid profile markers like triglycerides, low-density lipoproteins, improved insulin secretion, improved glucose uptake, and increased level of antioxidant enzymes (Zamany et al. 2021).

Further experimentation was done on the Awassi sheep diet in which powdered seeds of *Coriandrum sativum* were incorporated to investigate the effects of coriander seed powder on inviter digestible co-efficient rumen fermentation body weight and serum lipid profile. The study showed significant results regarding low-density lipoproteins and triglyceride levels. In addition, the Awassi sheep diet incorporating coriander seed powder showed a significant reduction in serum lipid profile parameters compared to those ewes fed with the Awassi sheep diet without coriander. These results clarified that the consumption of coriander is safe and healthy, having no detrimental side effects (Mohammed et al. 2018). Furthermore, recent studies have shown that polyphenolic fractions derived from *Coriandrum sativum* served as a complementary treatment against type-I and type-II diabetes mellitus as they possess hypoglycemic and anti-inflammatory properties (Mechchate et al. 2021).

15.4.7.2 Gastric Ulcer and Heavy Metal Absorption

Coriander helps prevent heavy metals and other toxic chemicals from being absorbed by the body. Furthermore, coriander helps to prevent *Helicobacter pylori* from causing gastric ulcers. The potential of coriander to prevent gastric ulcers may be due to the antioxidant potential of various constituents present in coriander which helps in scavenging the reactive oxygen species attacking the mucosal lining of the gastrointestinal tract and also by forming a protective shield around gastric mucosa by forming hydrophobic linkages, and thus in this way coriander helps to protect the enteric cells (Al-Mofleh et al. 2006).

Al-Mofleh et al. (2006) stated that gastric lining was injured in the experimental animal (rat) model using indomethacin, NaOH, NaCl, ethanol, and pylorus ligation, which ultimately caused the accumulation of gastric secretions. The treatment with different doses of coriander, i.e., 250 mg per kg body weight and 500 mg per kg body weight, was beneficial in compensating for and protecting from the damage caused by injurious agents. Coriander prevents ulceration caused by various necrosis-provoking agents and helps prevent the histopathological lesions caused *via* ethanol injury. Furthermore, coriander prevents the accumulation of gastric secretions caused by pylorus ligation. Furthermore, the study results suggested coriander served as a protective shield against ethanol damage of gastric linings due to the presence of several essential constituents having antioxidant potentials like phenolic compounds, coumarins, and terpenes flavonoids, catechins, and linalool (Al-Mofleh et al. 2006).

15.4.8 HEMATOLOGICAL SYSTEM

Some of the previous studies evaluated the beneficial health impacts of coriander by incorporating the seed powder of coriander into the diets of broiler in order to evaluate the effects of coriander on various biochemical, hematological, and growth parameters. One relevant study was done by Taha et al. (2019) and suggested that 0.4% of coriander seed powder significantly improved the parameters regarding growth performance along with several other body characteristics. The number of harmful bacteria residing in the ileum region was also reduced with coriander seed powder inculpation in broiler diets. The hematological parameters were also evaluated, which showed an overall increase in production of broilers chicks which further improved the economic efficiency (Taha et al. 2019).

Other studies revealed that treatment with coriander has an excellent power to improve the body weight gain along with improved feed intake and renal and hepatic efficiency when rats were infected with lead and mercury. Furthermore, another study revealed that when rats were treated with coriander, all the tested parameters improved significantly. Furthermore, significant enhancements were found during histopathological evaluation.

15.4.9 IMMUNE SYSTEM

Coriander is equipped with several health benefits like immune-stimulatory, antiviral, antioxidant, antibacterial, and growth-stimulating activities. In addition, coriander also plays a vital role in the chelation of heavy metals (Abdou Said, M Reda, and M Abd El-Hady 2021). One of the previous studies evaluated the impact of various medicinal herbaceous plants, which were selected from different regions of Pakistan in order to figure out their cytoprotective and immune-stimulatory properties in epithelial cells of gastric lining against oxidative stress and during the increased expression of inflammatory cytokine-like interleukin-8 (IL-8) *via* infection caused by *Helicobacter pylori*. Thus, after treating with coriander, fruitful results were gained from suppressing the free radicals/ ROS production (Al-Snafi 2016).

15.5 CONCLUSION

The regenerative and healing potential of coriander is due to bioactive phytochemicals. These bioactive metabolites, when ingested, are taken actively by all systems. Many findings have supported that coriander extract and oils positively impact all systems. By considering these potentials of *C. sativum*, its metabolites, essential oils, and powder extract, this plant is still worthy of future research, utilization, and investigations.

REFERENCES

Abascal, Kathy, and Eric Yarnell. 2012. "Cilantro-culinary herb or miracle medicinal plant?" *Alternative and Complementary Therapies* 18(5):259–264. https://doi.org/10.1089/act.2012.18507

Abdou Said, Ahmed, Rasha M. Reda, and Heba M. Abd El-Hady. 2021. "Overview of herbal biomedicines with special reference to coriander (*Coriandrum sativum*) as new alternative trend for the development of aquaculture." *Egyptian Journal of Aquatic Biology and Fisheries* 25(2):539–550. https://doi.org/10.21608/EJABF.2021.165900

Ahmad, Maroof, Amit Kumar, and Pragati Singh. 2016. "Effects of coriander (*Coriandrum sativum* L.) seed powder on growth performance of broiler chickens." *Journal of Krishi Vigyan* 5(1):57–59. https://doi.org/10.5958/2349-4433.2016.00034.9

Al-Marzoqi, Ali Hussein, Imad Hadi Hameed, and Salah Ali Idan. 2015. "Analysis of bioactive chemical components of two medicinal plants (*Coriandrum sativum* and *Melia azedarach*) leaves using gas chromatography-mass spectrometry (GC-MS)." *African Journal of Biotechnology* 14(40):2812–2830. https://doi.org/10.5897/AJB2015.14956

Al-Mofleh, I. A., A. A. Alhaider, J. S. Mossa, M. O. Al-Sohaibani, S. Rafatullah, and S. Qureshi. 2006. "Protection of gastric mucosal damage by *Coriandrum sativum* L. pretreatment in Wistar albino rats." *Environmental Toxicology and Pharmacology* 22(1):64–69. https://doi.org/10.1016/j.etap.2005.12.002

Al-Shakarchi, Hani Hashim, and Mahmoud A. Mohammad. 2021. "Effects of different levels of cinnamon and coriander powders on growth performance, feed utilization, survival rate and body chemical composition of common carp *Cyprinus carpio* l." *Plant Archives* 21(1):46–54. https://doi.org/10.51470/PLANTARCHIVES.2021.v21.S1.010

Al-Snafi, Ali Esmail. 2016. "Beneficial medicinal plants in digestive system disorders (part 2): plant based review." *IOSR Journal of Pharmacy* 6(7):85–92.

Alshahrani, Abdulrahman M. 2016. "In vivo neurological assessment of sedative hypnotic effect of *Coriandrum sativum* L. seeds in mice." *International Journal of Phytomedicine* 8:113–116.

Anila, L., and N. R. Vijayalakshmi. 2002. "Flavonoids from *Emblica officinalis* and *Mangifera indica*-effectiveness for dyslipidemia." *Journal of Ethnopharmacology* 79(1):81–87. https://doi.org/10.1016/S0378-8741(01)00361-0

Anilakumar, K. R., N. S. Nagaraj, and K. Santhanam. 2001. "Effect of coriander seeds on hexachlorocyclohexane induced lipid peroxidation in rat liver." *Nutrition Research* 21(11):1455–1462. https://doi.org/10.1016/S0271-5317(01)00338-4

Athar, Mohammad, and Syed Mahmood Nasir. 2005. "Taxonomic perspective of plant species yielding vegetable oils used in cosmetics and skin care products." *African Journal of Biotechnology* 4(1):36–44.

Baghel, Surekha, Bina Gidwani, and Chanchal Deep Kaur. 2017. "Novel drug delivery systems of herbal constituents used in acne." *Seed* 10:2. https://doi.org/10.5958/2231-5659.2017.00009.1

Balasubramanian, S., Roselin, P., Singh, K. K., Zachariah, J., & Saxena, S. N. (2016). Postharvest processing and benefits of black pepper, coriander, cinnamon, fenugreek, and turmeric spices. *Critical Reviews in Food Science and Nutrition*, 56(10), 1585–1607.

Bhat, S., P. Kaushal, M. Kaur, and H. K. Sharma. 2014. "Coriander (*Coriandrum sativum* L.): Processing, nutritional and functional aspects." *African Journal of Plant Science* 8(1):25–33. https://doi.org/10.5897/AJPS2013.1118

Bhuiyan, Md Nazrul Islam, Jaripa Begum, and Mahbuba Sultana. 2009. "Chemical composition of leaf and seed essential oil of *Coriandrum sativum* L. from Bangladesh." *Bangladesh Journal of Pharmacology* 4(2):150–153. https://doi.org/10.3329/bjp.v4i2.2800

Boron, Walter F., and Emile L. Boulpaep. 2012. *Medical physiology, 2e updated edition e-book: With student consult online access.* Elsevier Health Sciences.

Cha, J. M., Yoon, D., Kim, S. Y., Kim, C. S., & Lee, K. R. (2020). Neurotrophic and anti-neuroinflammatory constituents from the aerial parts of Coriandrum sativum. *Bioorganic Chemistry*, 105, 104443.

Chahal, K. K., Ravinder Singh, Amit Kumar, and Urvashi Bhardwaj. 2018. "Chemical composition and biological activity of *Coriandrum sativum* L.: A review." *Indian Journal of Natural Products and Resources (IJNPR)[Formerly Natural Product Radiance (NPR)]* 8(3):193–203.

Coşkuner, Yalçın, and Erşan Karababa. 2007. "Physical properties of coriander seeds (*Coriandrum sativum* L.)." *Journal of Food Engineering* 80(2):408–416. https://doi.org/10.1016/j.jfoodeng.2006.02.042

Dhyani, Neha, Adıla Parveen, Aisha Siddiqi, M. Ejaz Hussain, and Mohammad Fahim. 2020. "Cardioprotective efficacy of *Coriandrum sativum* (L.) seed extract in heart failure rats through modulation of endothelin receptors and antioxidant potential." *Journal of Dietary Supplements* 17(1):13–26. https://doi.org/10.1080/19390211.2018.1481483

Dutta, Deepika, S. L. Harikumar, and Amarjot Kaur Grewal. 2013. "Role of herbal drugs in nephrotoxicity and neurological disorder: Connecting a bridge between kidney and brain." *Research Journal of Pharmacy and Technology* 6(10):1165–1168.

El-Masry, Shimaa, H. A. Ali, Nora M. El-Sheikh, and Safaa Mostafa Awad. 2016. "Dose-dependent effect of coriander (*Coriandrum sativum* L.) and fennel (*Foeniculum vulgare* M.) on lead nephrotoxicity in rats." International Journal of Research Studies in Biosciences 4:36–45. https://doi.org/10.20431/2349-0365 .0406006

Foudah, Ahmed I., Mohammad H. Alqarni, Aftab Alam, Mohammad Ayman Salkini, Elmutasim O. Ibnouf Ahmed, and Hasan S. Yusufoglu. 2021. "Evaluation of the composition and in vitro antimicrobial, antioxidant, and anti-inflammatory activities of Cilantro (*Coriandrum sativum* L. leaves) cultivated in Saudi Arabia (Al-Kharj)." *Saudi Journal of Biological Sciences*. https://doi.org/10.1016/j.sjbs.2021.03.011

Guna, Ada, Ashley Giordano, Petra Sovcovova, and David Shaw. "Understanding herbs coriander/cilantro-*Coriandrum sativum*."

Hashim, M. S., S. Lincy, V. Remya, M. Teena, and L. Anila. 2005. "Effect of polyphenolic compounds from *Coriandrum sativum* on H_2O_2-induced oxidative stress in human lymphocytes." *Food Chemistry* 92(4):653–660. https://doi.org/10.1016/j.foodchem.2004.08.027

Hosseini, Mahmoud, Mohammad Hossein Boskabady, and Mohammad Reza Khazdair. 2021. "Neuroprotective effects of *Coriandrum sativum* and its constituent, linalool: A review." *Avicenna Journal of Phytomedicine* 11(5):436–450.

Hwang, Eunson, Do-Gyeong Lee, Sin Hee Park, Myung Sook Oh, and Sun Yeou Kim. 2014. "Coriander leaf extract exerts antioxidant activity and protects against UVB-induced photoaging of skin by regulation of procollagen type I and MMP-1 expression." *Journal of Medicinal Food* 17(9):985–995. https://doi.org /10.1089/jmf.2013.2999

Iqbal, Muhammad Jawad, Masood Sadiq Butt, Aamir Shehzad, and Muhammad Asghar. 2018. "Evaluating therapeutic potential of coriander seeds and leaves (*Coriandrum sativum* L.) to mitigate carbon tetrachloride-induced hepatotoxicity in rabbits." *Asian Pacific Journal of Tropical Medicine* 11(3):209. https://doi.org/10.4103/1995-7645.228435

Kannan, M. Mari, and S. Darlin Quine. 2013. "Ellagic acid inhibits cardiac arrhythmias, hypertrophy and hyperlipidaemia during myocardial infarction in rats." *Metabolism* 62(1):52–61. https://doi.org/10.1016 /j.metabol.2012.06.003

Karami, R., Hosseini, M., Mohammadpour, T., Ghorbani, A., Sadeghnia, H. R., Rakhshandeh, H., … Esmaeilizadeh, M. (2015). Effects of hydroalcoholic extract of Coriandrum sativum on oxidative damage in pentylenetetrazole-induced seizures in rats. *Iranian Journal of Neurology*, 14(2), 59.

Kern, Catherine, Christian Gombert, Alicia Roso, and Christine Garcia. 2020. "Soothing effect of virgin coriander seed oil on sensitive skin." *OCL* 27:49. https://doi.org/10.1051/ocl/2020043

Latha, K., B. Rammohan, B. P. V. Sunanda, M. S. Uma Maheswari, and Surapaneni Krishna Mohan. 2015. "Evaluation of anxiolytic activity of aqueous extract of *Coriandrum sativum* Linn. in mice: A preliminary experimental study." *Pharmacognosy Research* 7(Suppl 1):S47. https://doi.org/10.4103%2F0974 -8490.157996

Mahendra, Poonam, and Shradha Bisht. 2011. "*Coriandrum sativum*: A daily use spice with great medicinal effect." *Pharmacognosy Journal* 3(21):84–88. https://doi.org/10.5530/pj.2011.21.16

Mandal, Shyamapada, and Manisha Mandal. 2015. "Coriander (*Coriandrum sativum* L.) essential oil: Chemistry and biological activity." *Asian Pacific Journal of Tropical Biomedicine* 5(6):421–428. https:// doi.org/10.1016/j.apjtb.2015.04.001

Mechchate, Hamza, Imane Es-Safi, Amal Amaghnouje, Smahane Boukhira, Amal A. Alotaibi, Mohammed Al-Zharani, Fahd A. Nasr, Omar M. Noman, Raffaele Conte, and El Hamsas El Youbi Amal. 2021. "Antioxidant, anti-inflammatory and antidiabetic proprieties of LC-MS/MS identified polyphenols from coriander seeds." *Molecules* 26(2):487. https://doi.org/10.3390/molecules26020487

Mima, Y., Izumo, N., Chen, J. R., Yang, S. C., Furukawa, M., & Watanabe, Y. (2020). Effects of Coriandrum sativum seed extract on aging-induced memory impairment in Samp8 mice. *Nutrients*, 12(2), 455.

Mishra, Bipin B., Shree R. Padmadeo, Kumud R. Thakur, Deepak K. Jha, Kumar Pranay VyomeshVibhaw, and Pankaj Kumar. 2021. "Hypoglycemic and antioxidative potential of *Coriandrum sativum* seed extract in alloxan induced diabetic rats." *Bioscience Biotechnology Research Communications* 14(1). https://doi.org/10.21786/bbrc/14.1/39

Mishra, Chanchal Kumar, Dinakar Sasmal, and Dhiraj Kumar. 2017. "In vitro antioxidant activity of chloroform and ethanolic fruit and root extracts of *Carissa carandas* Linn." Journal of Bio Innovation 6(5):741–748.

Mohammed, S. F., A. A. Saeed, O. S. Al-Jubori, and A. A. Saeed. 2018. "Effect of daily supplement of coriander seeds powder on weight gain, rumen fermentation, digestion and some blood characteristics of Awassi ewes." *Journal of Research in Ecology* 6(2):1762–1770.

Moustafa, Abdel Halim A., Ehab Mostafa M. Ali, Said S. Moselhey, Ehab Tousson, and Karim S. El-Said. 2014. "Effect of coriander on thioacetamide-induced hepatotoxicity in rats." *Toxicology and Industrial Health* 30(7):621–629. https://doi.org/10.1177%2F0748233712462470

Önder, Alev. 2018. "Coriander and its phytoconstituents for the beneficial effects." *Potential of Essential Oils*:165–185.

Park, Gunhyuk, Hyo Geun Kim, Soonmin Lim, Wonil Lee, Yeomoon Sim, and Myung Sook Oh. 2014. "Coriander alleviates 2, 4-dinitrochlorobenzene-induced contact dermatitis-like skin lesions in mice." *Journal of Medicinal Food* 17(8):862–868. https://doi.org/10.1089/jmf.2013.2910

Parsaeyan, Nayereh. 2012. "The effect of coriander seed powder consumption on atherosclerotic and cardio-protective indices of type 2 diabetic patients." *Iranian Journal of Diabetes and Obesity 4*, 2.

Patel, Dipak K., Swati N. Desai, Hardik P. Gandhi, Ranjitsinh V. Devkar, and A. V. Ramachandran. 2012. "Cardio protective effect of *Coriandrum sativum* L. on isoproterenol induced myocardial necrosis in rats." *Food and Chemical Toxicology* 50(9):3120–3125. https://doi.org/10.1016/j.fct.2012.06.033

Pourzaki, M., Homayoun, M., Sadeghi, S., Seghatoleslam, M., Hosseini, M., & Bideskan, A. E. (2017). Preventive effect of Coriandrum sativum on neuronal damages in pentylentetrazole-induced seizure in rats. *Avicenna Journal of Phytomedicine*, 7(2), 116. (Pachiappan and Institutions, 2020)

Prachayasittikul, Veda, Supaluk Prachayasittikul, Somsak Ruchirawat, and Virapong Prachayasittikul. 2018. "Coriander (*Coriandrum sativum*): A promising functional food toward the well-being." *Food Research International* 105:305–323. https://doi.org/10.1016/j.foodres.2017.11.019

Priyadarshi, Siddharth, Hafeeza Khanum, Ramasamy Ravi, Babasaheb Baskarrao Borse, and Madeneni Madhava Naidu. 2016. "Flavour characterisation and free radical scavenging activity of coriander (*Coriandrum sativum* L.) foliage." *Journal of Food Science and Technology* 53(3):1670–1678. https://doi.org/10.1007/s13197-015-2071-1

Rahiminiat, F., Ghazanfari, S., Mohammadi, Z., & Sharifi, S. D. (2017). Feeding artemisia sieberi, coriander and clove essential oils alters muscle lipid oxidation in broiler chicken. *Bulgarian Journal of Agricultural Science*, 23(4), 625–631.

Rajeshwari, Ullagaddi, and Bondada Andallu. 2011. "Medicinal benefits of coriander (*Coriandrum sativum* L.)." *Spatula DD* 1(1):51–58. https://doi.org/10.5455/spatula.20110106123153

Sá-Caputo, D. C., C. F. Dionello, D. S. Morel, and L. L. Paineiras-Domingos. 2016. "Possible benefits of the *Coriandrum sativum* in the management of diabetes in animal model: A systematic review." *ARCHIVOS DE MEDICINA* 2(1):10.

Sahib, Najla Gooda, Farooq Anwar, Anwarul-Hassan Gilani, Azizah Abdul Hamid, Nazamid Saari, and Khalid M. Alkharfy. 2013. "Coriander (*Coriandrum sativum* L.): A potential source of high-value components for functional foods and nutraceuticals-A review." *Phytotherapy Research* 27(10):1439–1456. https://doi.org/10.1002/ptr.4897

Seghatoleslam, M., Alipour, F., Shafieian, R., Hassanzadeh, Z., Edalatmanesh, M. A., Sadeghnia, H. R., & Hosseini, M. (2016). The effects of Nigella sativa on neural damage after pentylenetetrazole induced seizures in rats. *Journal of Traditional and Complementary Medicine*, 6(3), 262–268.

Shaheen, Farzana, Attaur Rahman, K. Vasisht, and M. Iqbal Choudhary. 2003. "The status of medicinal and aromatic plants in Pakistan." In *Medicinal plants and their utilization*, UNIDO:77–87.

Sherwood, Lauralee. 2015. *Human physiology: from cells to systems.* Cengage Learning.

Sreelatha, S., P. R. Padma, and M. Umadevi. 2009. "Protective effects of *Coriandrum sativum* extracts on carbon tetrachloride-induced hepatotoxicity in rats." *Food and Chemical Toxicology* 47(4):702–708. https://doi.org/10.1016/j.fct.2008.12.022

Taha, Ayman E., Saber S. Hassan, Ramadan S. Shewita, Ahmed A. El-seidy, Mohamed E. Abd El-Hack, El-sayed O. S. Hussein, Islam M. Saadeldin, Ayman A. Swelum, and Mohamed A. El-Edel. 2019. "Effects of supplementing broiler diets with coriander seed powder on growth performance, blood haematology, ileum microflora and economic efficiency." *Journal of Animal Physiology and Animal Nutrition* 103(5):1474–1483. https://doi.org/10.1111/jpn.13165

Téllez-López, Miguel Ángel, Gabriela Mora-Tovar, Iromi Marlen Ceniceros-Méndez, Concepción García-Lujan, Cristo Omar Puente-Valenzuela, María del Carmen Vega-Menchaca, Luis Benjamín Serrano-Gallardo, Rubén García Garza, and Javier Morán-Martínez. 2017. "Evaluation of the chelating effect of methanolic extract of *Coriandrum sativum* and its fractions on Wistar rats poisoned with lead acetate." *African Journal of Traditional, Complementary and Alternative Medicines* 14(2):92–102.

Wood, Matthew. 2008. *The earthwise herbal: A complete guide to old world medicinal plants.* Vol. 1. North Atlantic Books.

Zamany, Sanaz, Aida Malek Mahdavi, Saeed Pirouzpanah, and Ali Barzegar. 2021. "The effects of coriander seed supplementation on serum glycemic indices, lipid profile and parameters of oxidative stress in patients with type 2 diabetes mellitus: A randomized double-blind placebo-controlled clinical trial." https://doi.org/10.21203/rs.3.rs-262149/v1

16 Non-Food Applications of Coriander Seeds and Leaves

Abdul Hameed Kori, Sarfaraz Ahmed Mahesar,
Syed Tufail Hussain Sherazi, Zahid Hussain Laghari,
and Aijaz Ahmed Otho

CONTENTS

ABBREVIATIONS

Coriandrum sativum L. (*C. sativum*), diphenylpicryl hydrazine (DPPH·), *Escherichia coli* (*E. coli*), high-density lipoprotein (HDL), interferon gamma (IFN-ɣ), low-density lipoprotein (LDL), mono-unsaturated fatty acid (MUFA), polyunsaturated fatty acid (PUFA), saturated fatty acids (SFA), nanoparticles (NPs), *Salmonella typhi* (*S. typhi*), silver nanoparticles (Ag-NPs), *Staphylococcus aureus* (*S. aureus*).

DOI: 10.1201/9781003204626-17

16.1 INTRODUCTION

Coriander (*Coriandrum sativum* L.) is a spice and an annual herb that belongs to the carrot family (Umbelliferae). Its leaves and seeds are used to flavor foods. Fresh coriander leaves are also known as cilantro, and they look a lot like Italian (*Eryngium foetidum*) flat-leaf parsley (Lohwasser et al., 2008). The seeds, leaves, stem, and root all have a good aromatic odor and are utilized in various cuisines.

16.1.1 ORIGIN

Coriander's origin is unknown, though most authors believe it is native to the Mediterranean and the Middle East and Southwestern Asia, North Africa, and Southern Europe. There are two species of coriander in the tropic region. Only *Coriandrum sativum* L. (*C. sativum*) is commonly cultivated, while the wild species is *Coriander tordylium* (Ifenzl) Bomm. In 2020, India was the world's biggest producer of coriander, with 755 tons accounting for 39% of total global exports (Silva et al., 2020). Coriander is also grown commercially in different countries, including Mexico, Morocco, Italy, Argentina, Pakistan, Romania, Myanmar, Bulgaria, the Netherlands, France, Turkey, Spain, Canada, and a little bit in the USA and UK. Coriander is among the world's oldest spices grown since prehistoric times (Telci et al., 2006). The young plant as a whole is used in a variety of food and non-food applications. Coriander plants, seeds, leaves, and flowers are shown in Figure 16.1.

FIGURE 16.1   Coriander plant, seeds, and leaves.

16.2 PHYSICOCHEMICAL COMPOSITION

16.2.1 Leaves and Seeds

Oil, fibers, carbohydrates, proteins, minerals, and vitamins are all present in varying amounts in coriander. Although they are used in small amounts in food, none of them contribute considerably to nutrient needs despite the importance of the essential oils of coriander seeds.

Table 16.1 provides the composition of coriander green leaves and seeds. It has a range of moisture content of 6.3–87.9%, carbohydrate content of 3.6–56.5%, lipids content of 0.22–30.0%, protein content of 2.13–21.3%, crude fiber of 2.8–41.9%, ash 1.2–6.0%, and energy 279–298 kcal of seeds and leaves. Coriander seeds also have the highest selenium level (23.5 ppm) of any spice (Ozcan et al., 2008). In addition, other minerals such as magnesium, sulfur, aluminum, potassium, silicon, tin, phosphorus, copper, zinc, calcium, iron, and manganese were reported (Al-Bataina et al., 2003).

Coriander nutrition comes primarily from its green leaves and dried fruits, which include considerable amounts of carotene, riboflavin, sodium, thiamine, tryptophan, folate, niacin, and vitamin A, C, D, E, and B6 and B12 (Nimish et al., 2011; Bhat et al., 2014; USDA, 2016).

16.2.2 Essential Oil

Coriander seeds have a fragrant odor and flavor from their volatile oil, transparent and clear to light yellow liquid. The oil has a spicy, warm, sweet, and fruity aroma. When compared to immature fruits, it has been reported that essential oil in ripe fruits is minimal (usually less than 1%), and a maximum of 2% was reported by Telci et al. (2006). Monoterpene hydrocarbons make up most of the essential oil, with oxygenated monoterpenes accounting for around 20% of the volatile oil (Bandoni et al., 1998). Linalool, commonly known as coriandrol, is the primary and most important oxygenated molecule. Its content in essential oils ranges between 19.8 and 82%. Major components in the essential oil of leaves and seeds are (E)-2-decenal 32.2% and linalool 55.4%, respectively (Sharma & Sharma, 2012). Different components of essential oils in the various percentage of leaves and seeds are given in Table 16.2.

The chemical profile of coriander oil has been demonstrated to be influenced by location, fertilization, weeds, cultivar, and seeding date (Gil et al., 2002; Zheljazkov et al., 2008). The chemical structure and relative amounts of the constituents in coriander seed essential oil are affected by the

TABLE 16.1
Composition of Coriander Seeds and Leaves

Parameters	Seeds	Leaves	Reference
Moisture (%)	6.3–8.0	45–87	(Mandal & Mandal, 2015; Sharma & Sharma, 2012; Kassahun, 2020; Sliva, et al., 2020).
Protein (%)	11.5–21.3	2.13–12.3	(Mandal & Mandal, 2015; Sharma & Sharma, 2012; Kassahun, 2020; Sliva, et al., 2020).
Carbohydrate (%)	54.9–56.5	3.6–52.1	(Mandal & Mandal, 2015; Sharma & Sharma, 2012; Kassahun, 2020; Sliva, et al., 2020).
Lipids (%)	4.78–30.0	0.22–17.7	(Mandal & Mandal, 2015; Sharma & Sharma, 2012; Kassahun, 2020; Sliva, et al., 2020).
Fiber (%)	28.4–29.1	2.8–41.9	(Mandal & Mandal, 2015; Sharma & Sharma, 2012; Kassahun, 2020; Sliva, et al., 2020).
Ash (%)	4.9–6.0	1.2–1.7	(Mandal & Mandal, 2015; Sharma & Sharma, 2012; Kassahun, 2020; Sliva, et al., 2020).
Energy (kcal)	298.0	279.0	(Mandal & Mandal, 2015; Sharma & Sharma, 2012; Kassahun, 2020; Sliva, et al., 2020).

TABLE 16.2
Essential Oils Composition (%) of Coriander Leaves and Seeds

Components	Leaves*	Seeds*
α-pinene	1.9	7.14
Decanal	1.73	4.69
(E)-2-decenal	32.23	-
2-Decen-l-ol	5.45	-
1-Decanol	2.18	-
Linalool	13.9	55.4
Undecanal	2.43	-
(E)-2-undecenal	4.31	-
Dodecanal	4.07	-
1-Eicosanol	1.35	-
(E)-2-dodecenal	7.51	-
Tridecanal	1.09	-
(E)-2-tridecenal	3.0	-
Tetradecanal	1.01	-
(E)-2-tetradecenal	6.56	-
Carvone	1.40	-
Pentadecenal	2.47	-
(E)-2-hexadecenal	2.94	-
Camphene	-	1.78
D-limonene	-	1.36
Limonene	-	3.10
γ-Terpinene	-	7.47
Geraniol	-	2.23
Camphor	-	5.59
Geranyl acetate	-	4.24

* Silva et al. (2020)

stage of maturity (Msaada et al., 2017). Fresh herbage has a very distinct odor and flavor from the mature seed. Otherwise, the leaf oil composition differs significantly from the seed oil.

16.2.3 Fatty Acids

Apart from the essential oil, the seeds contain fatty oils that range from 4.9 to 30.1%, as shown in Figure 16.2. The fatty oil is dark brownish-green and has a scent quite close to the essential oil. Petroselinic acid (54.6–80.9%), linoleic acid (14.6–17.3%), oleic acid (2.94–16.6%), palmitic acid (3.74–7.51%), and stearic acid (0.8–5.45%) are the main components of the fruit's fatty acids. In addition, the unsaturation and the high amount of petroselinic acid, which are uncommon among octadecenoic acids, allow for the production of chemical derivatives that are not possible with other oils (Kassahun, 2020).

16.2.4 Uses in the Food Industry

Nutrition-based therapy against numerous lifestyle-related illnesses has grown in popularity due to the link between diet and health. The demand for natural ingredients increases every day as people become more conscious of diet-related health issues (Sharma & Sharma, 2012). Coriander is a fresh green herb and spice frequently used as a flavoring ingredient. These two products have

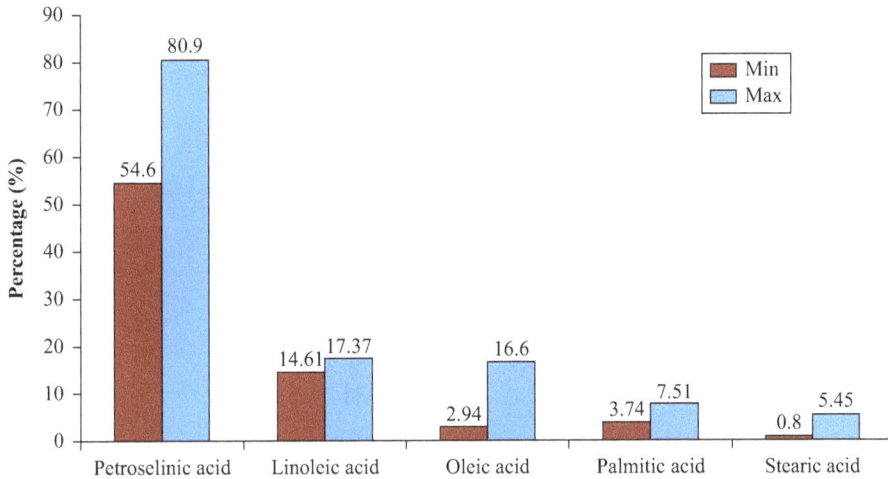

FIGURE 16.2 Fatty acid compositions of coriander seed oil.

very different scents and flavors. In Asia, Central and South America, and the Middle East, the plant flavors food. When the plant is young, the entire plant is used to make chutneys, and the leaves are used to spice curries, sauces, and soups.

Fresh leaves can be used as a garnish for curries and other foods, and certain ethnic groups eat them as green vegetables. The coriander seeds are a key element in some curry powders and are utilized in pickling, pastries, seasonings, meats, sausages, cakes, buns, tobacco items, and other confectionaries. Coriander essential oils are significant chemicals found in the seeds and leaves of the coriander plant employed in various industrial applications (Shahwar et al., 2012). Canned soups, baked dishes, candy, meat products, chocolate, cocoa, and alcoholic beverages like gin are all prepared with the distilled essential oil from the fruits (Caballero, 2003; Singh et al., 2006; Meena et al., 2014; Dussort et al., 2014).

16.3 NON-FOOD APPLICATIONS

Many products can be obtained from coriander besides food. The non-food applications of coriander include use in medicine, cosmetics/perfume, soaps, green synthesis of metal nanoparticles, biofuels, etc. (Figure 16.3).

16.3.1 TRADITIONAL MEDICINE

Coriander seeds, leaves, and oil are used in traditional and folk medicine for various purposes. Coriander is an excellent plant for aiding digestion and treating gastrointestinal problems like vomiting, flatulence, dyspepsia, loss of appetite, and griping pain (Jabeen et al., 2009). When fresh buttermilk is added with one or two spoons of coriander juice, it is thought to help indigestion, ulcerative, colitis, nausea, dysentery, and hepatitis. Coriander is also effective for the treatment of typhoid fever.

Dried coriander is used to cure diarrhea and chronic dysentery, as well as to minimize acidity. In addition, dried coriander, green chilies, grated coconut, ginger, and black grapes without seeds can be used to make a chutney for indigestion-related abdominal pain. Coriander is often used as a diuretic in some parts of the world, particularly Morocco (Aissaoui et al., 2008). Coriander water helps decrease blood cholesterol by activating the kidney's process when consumed regularly.

Coriander has anti-inflammatory and analgesic qualities, as well as antibacterial characteristics. The seeds' extract, mixed with castor oil, treats rheumatism and joint pain (Ghani, 2003). A

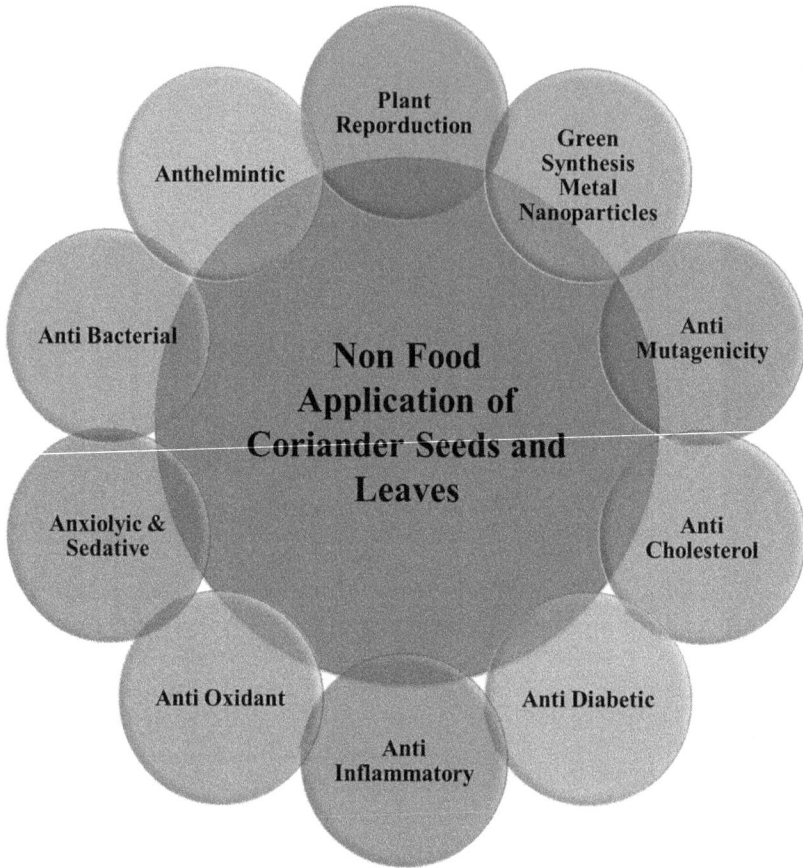

FIGURE 16.3 Non-food application of coriander seeds and leaves.

decoction prepared from freshly dried coriander is an effective eyewash for treating conjunctivitis, in addition to the effects on diarrhea and chronic dysentery. It soothes pain and inflammation while reducing burning. A teaspoon of coriander juice mixed with a pinch of turmeric powder is said to be a good cure for pimples, blackheads, and other skin problems.

Anxiety and insomnia are relieved using the seeds, particularly in Iran (Emam et al., 2005). Excessive menstrual flow can also be controlled with the coriander plant. Coriander leaves are an aphrodisiac that can also cure stomatitis and foul breath and inhibit pyorrhea. They also aid in evacuating catarrh and phlegm from the bronchial tubes, alleviating spasmodic symptoms. Furthermore, the juice of the leaf can be applied to the forehead for migraine and other types of headache treatment. Coriander is also thought to offer antidiabetic benefits in various parts of Europe.

Coriander is high in dietary fiber and contains significant manganese, magnesium, and iron. Coriander leaves are exceptionally high in vitamin A (5200 IU/100 g) and vitamin C (250 mg/100 g), making them one of the most nutritious foods. Coriander juice is therefore highly effective in the treatment of iron, vitamin A, B1, B2, and C deficiency. Coriander is a popular ingredient in perfumes and cosmetics, soaps, detergents, and other personal care products (Smith et al., 1997). Its soft, pleasant, slightly spicy aroma fits well with oriental aromas. The leaf juice can be utilized as a cosmetic, and the essential oil can be employed in aromatherapy. However, it is not as widely used as other seed oils. Coriander is interesting since it contains much petroselinic acid, an isomer of oleic acid utilized as a plastics lubricant in nylon production and cosmetics.

TABLE 16.3

Non-Food Applications of Different Parts of the Coriander Plant

Application	Parts of Plant	References
Vitamin deficiencies	Leaves, stem, and seeds	(EFSA Panel, 2013)
Surfactants and emulsifiers	Seeds	(Smith et al., 1997)
Plastics lubricant and nylon	Seeds	(Vollmann & Laimer, 2013; Isbell et al., 2016)
Biological purposes such as pollination and reproduction	Fruit/seeds	(Bhalchandra et al., 2014; Bou-Shaara, 2015; Moise, 2015)
Insecticidal	Leaves and seeds	(Eljazi et al., 2018)
Antibiotic in feed	Leaves, stems, and seeds	(Hosseinzadeh et al., 2014)
Heavy metals removal	Leaves powder, coriander extracts as nanoparticles	(Kadiri et al., 2017; Winarti et al., 2018).
Perfumes and cosmetics	Leaves and seeds	(Meena et al., 2014; Dussort et al., 2014; Caballero., 2003)
Antioxidants	Leaves and seeds extracts. EO of seeds and leaves	(Shyamala et al., 2005; Yepez et al., 2002; Mousavizadeh & Sedaghathoor, 2011)
Antibacterial	EO of coriander and extract of leaves and seeds	(Saeed & Tariq, 2007; Wong & Kitts, 2006; Kubo et al., 2004; Delaquis et al., 2002)
Antifungal	EO of leaves and seeds	(Delaquis et al., 2002)
Anthelmintic activity	Leaves and seeds extract	(Eguale et al., 2007)
Antidiabetic	Leaves and seeds extract	(Chitra & Leelamma, 1999)
Anticholesterol	Leaves and seeds extract	(Lal et al., 2004)
Anticancer	Leaves and seeds extract	(Chitra & Leelamma, 2000)
Anxiolytic and sedative effects	Seeds extract	(Emam et al., 2005)
Antifertility activity	Seeds extract	(Al-Said et al., 1987)
Diuretic effect	Leaves and seeds extract	(Aissaoui et al., 2008)
Prevention of heavy metals deposition	Seeds extract	(Aga et al., 2001)
Bowel disorders	Leaves and seeds extract	(Jagtap et al., 2004)
Biodiesel	Seed oil	(Moser & Vaughn, 2010; Babu et al., 2020)
Green synthesis of NPs	Leaves and seeds extract	(Kakarla & Rama, 2015; Senthilkumar et al., 2018; Asma Ashraf et al., 2019; Hu et al., 2020; Abbas et al., 2020; Ruiz-Torres et al., 2021)

During oxidation cleavage, petroselinic breaks up into two chemicals. One is lauric acid, which is utilized to make surfactants and food products, while the other is adipic acid which is employed to make polymers and nylon (Lopez et al., 2008. Isbell et al., 2006). Adipic acid is currently produced from mineral oil using a method that emits gases such as N_2O, which harm the ozone layer and contribute to global warming. Instead, adipic acid is a more environmentally friendly substance generated from coriander seeds (Vollmann & Laimer, 2013).

Coriander oil has been applied as a finish for wood goods, including furniture and construction materials for indoor applications. Coriander oil preserves the construction from pests and enhances the air quality and environment (Uitterhaegen et al., 2016). Various non-food applications of coriander in different fields have been summarized in Table 16.3.

16.3.2 BIOLOGICAL USES OF CORIANDER

The reproductive biology of plants is another advantage of coriander. Coriander generates a large amount of nectar, which attracts various insects for pollination, indicating an excellent melliferous plant (Jerkovic et al., 2013; Bhalchandra et al., 2014; Bou-Shaara, 2015). One hectare of coriander is utilized to obtain 500 kg of honey by honeybees in Russia (Moise, 2015).

After harvest, the stubble can be utilized as fuelwood, and after distillation, the seed waste can be utilized in calf fodder. In addition, insecticidal efficacy of coriander essential oil against stored product beetle pests was also reported (Pascual-Villalobos, 2002; Eljazi et al., 2018).

It was recently demonstrated that coriander leaf extract could minimize heavy metal contamination of Cu, Hg, and Pb in rod shellfish (Kadiri et al., 2017; Winarti et al., 2018). Coriander powder could be substituted for synthetic antibiotics used in the diet and coriander extract in water (Naeemasa et al., 2015, Ahmad et al., 2016). Coriander powder could be used as natural feed additives and growth promoters in chicken diets (Hosseinzadeh et al., 2014).

16.3.3 USES AS BIODIESEL

Few studies prepared and assessed as substitute biodiesel fuel for methyl esters of coriander seed oil (CSME) (Moser & Vaughn, 2010; Selva et al., 2020). Coriander seed oil contains a unique fatty acid as the main component, petroselinic (6Z-octadecenoic; 68.5 wt percent), which had never been reported before in biodiesel fuels. CSME surpassed soybean oil methyl ester in oxidative stability, low-temperature characteristics, and iodine value. As a result of its particular fatty acid constitution, coriander oil methyl ester offers outstanding fuel qualities (Moser & Vaughn, 2010). Another study by Babu et al. (2020) found that biodiesel prepared from coriander had the least exhaust gas temperature to emit toxic gases than diesel.

16.3.4 USES IN THE GREEN SYNTHESIS OF NANOPARTICLES

Nowadays, green chemistry has attracted the world's attention to preparing nanoparticles (NPs) in many fields due to its low cost, lower toxicity, and being environmentally friendly. The different NPs such as silver, gold, iron, zinc, titanium, and copper have been prepared and reported with various plants extracts such as *Aloe vera*, blue and blackberry, pomegranate, turmeric, alcea and thyme, rosella, reishi mushroom, and black cumin (Jadoun et al., 2021). The coriander extract of different parts (roots, stem, leaves, and seeds) has also been utilized to prepare environmentally friendly NPs.

Nanomaterial has a smaller particle size from 1 nm to 100 nm due to its unique size and shape. As a result, it has a large surface area to volume ratio, resulting in excellent materials and higher efficiency. As a result, nanomaterial is being applied in various fields such as medicine, the environment, and engineering. Silver nanoparticles (Ag-NPs) were successfully prepared from coriander leaf extract and utilized in many applications such as catalysis, chemical analysis, and photographic reactions on a commercial scale (Kakarla & Rama, 2015).

Senthilkumar et al. synthesized Ag-NPs using a green approach by applying *Coriandrum* leaves, roots, seeds, and stem extracts. They observed good antibacterial activity (Senthilkumar et al., 2018).

Recently, Asma Ashraf et al. (2019) constructed biologically friendly Ag-NPs with coriander leaf extract and tested them against gram-negative and positive bacteria *viz* gentamicin antibiotic to check its impact. It was reported that particles with smaller sizes having a higher surface area ratio were more effective in inhibiting the growth and proliferation of bacteria than synthetic antibiotics. Similarly, some other metallic NPs such as iron, copper, zinc, and titanium dioxide were also reported using green synthetical approaches with coriander leaf extract and produced excellent results as an antibacterial against pathogens (Hu et al., 2020; Abbas et al., 2020; Ruiz-Torres et al., 2021).

To enhance the antimicrobial activity of Ag-NPs, coriander essential oil was added, resulting in tiny NPs with mean diameters between 2.5 and 5 nm, which had good antibacterial characteristics (Saygi et al., 2021).

16.4 MODERN RESEARCH IN THE MEDICINAL TRAITS OF CORIANDER

16.4.1 Antioxidant Activity

Herbs and spices have been utilized for health-promoting and disease-prevention purposes since prehistoric times due to bioactive compounds. Polyphones, also known as phenolic chemicals, are key bioactive components with substantial antioxidant potential and favorable physiological benefits. Phytochemicals are called natural antioxidants because of their bioactivity and broad distribution in vegetables. Phenolic acids and flavonoids are bioactive components with significant antioxidant potential (Melo et al., 2005).

Borneol, geraniol, elemol, linalool, carvone, limonene, and camphor are the beneficial phytonutrients found in the seeds and volatile oil. Rhamnetin, quercitin, epigenin, and kaempferol are some of the flavonoids found in coriander. In addition, chlorogenic and caffeic acid are two active phenolic acid molecules found in coriander.

According to the World Health Organization (2002), traditional herbal remedies are used by many of the world's population. Herbs and spices have high antioxidant content and have been used for medicinal purposes for ages. Various studies have been undertaken in recent decades to investigate herbs and spices' antioxidant and health-promoting properties.

Shyamala et al. (2005) reported that the extract of *C. sativum* leaves in ethanol is a powerful and stable antioxidant at elevated temperatures that can replace the synthetic antioxidants. Diphenyl picryl hydrazine (DPPH·) solution in methanol showed the main antioxidant activities of extracts of coriander seeds acquired with supercritical CO_2 equivalent to commercial antioxidants (Yepez et al., 2002). In three independent bioassays, the antioxidant potential of phenolic compounds in coriander leaves was higher than that of seeds (Wong & Kitts, 2006).

Coriander oil has been utilized as an antioxidant to minimize fruit and green vegetable browning by lowering the peroxidase activity (Mousavizadeh & Sedaghathoor, 2011) and to minimize lipid peroxidation in meat products (Marangoni & Moura, 2011). Other research has used the benefits of *C. sativum* dual-action and applied it as a food preservative to prevent oxidation and microbiological decomposition. Coriander oil has also been evaluated in this situation, either used according to the food product or mixed into packaging films.

In addition to the food product, coriander EO/extracts were found to reduce the number of gram-negative bacteria (Enterobacteriaceae, *C. jejuni*, and *A. hydrophila*), Gram-positive bacteria (*L. monocytogenes*, *S. aureus*), and fungi (*Byssochlamys fulva*), preventing lipid oxidation when added directly to food. Many bio-based films are produced using coriander extracts, including chicken foot protein, bone meal, whey protein isolate, hake protein, and pig meat, as well as more standard polyamide/polyethylene films (Silva et al., 2020).

16.4.2 Antibacterial Properties

Essential oils or extracts of *C. sativum* from seeds, leaves, or the entire plant have exhibited antibacterial efficacy against molds, yeasts, gram-negative and gram-positive bacteria, parasite *Leishmania donovani*, and even bacteriophages (Chao et al., 2000). The activity of bacteria such as *Escherichia coli* (*E. coli*), *Salmonella typhi* (*S. typhi*), and *Staphylococcus aureus* (*S. aureus*) are all prevented with coriander due to its potent antioxidant activity (Al-Jedah et al., 2000). In another study, Delaquis et al. (2002) found that coriander oil suppressed gram-positive bacteria "*Listeria monocytogenes* and *S. aureus*" but had minimal effect on gram-negative bacteria (*Pseudomonas fragi*, *E. coli*, and *S. typhi*). Dodecenal, an antibacterial compound in coriander, is twice as effective in destroying *Salmonella* as the regularly used antibiotic drug gentamicin (Kubo et al., 2004).

Eight more antibiotic compounds were discovered from fresh coriander in addition to dodecenal, leading food scientists to believe that dodecenal could be used as a flavorless food additive to inhibit foodborne illness. To measure the cell damage, methanol and water extracts of coriander leaves and stems were evaluated for antibacterial efficacy against *E. coli* and *Bacillus subtilis*.

The methanol stem extracts have done more considerable bacterial cell damage, resulting in more bacterial growth suppression, which correlated to iron sequestering capacity (Wong & Kitts, 2006). However, several investigations have discovered that coriander decoction has no antibacterial activity against urinary pathogens, gram-negative bacteria, and gram-positive bacteria, as well (Chaudhry & Tariq, 2006; Saeed & Tariq, 2007).

16.4.3 ANTHELMINTIC ACTIVITY

At concentrations less than 0.5 mg/mL *C. sativum* prepared in aqueous or hydroalcoholic extracts completely stopped the hatching of nematode eggs and no significant difference was observed among extracts. On the other hand, *in vitro* activity of the hydroalcoholic extract outperformed the aqueous extract against adult parasites (Eguale et al., 2007).

16.4.4 ANTIDIABETIC ACTIVITY

In rats given high-cholesterol feed, *C. sativum* had a substantial hypoglycemic effect. Glycogen synthase activity was increased, indicating an increase in hepatic glycogen content, but gluconeogenic and glycogen phosphorylase enzyme activity suggested a decrease in gluconeogenesis and glycogenolysis rates. The activities of glycolytic and glucose-6-phosphate dehydrogenase enzymes were also improved, indicating that glucose is utilized through the pentose phosphate route and glycolysis, respectively (Silva et al., 2020).

16.4.5 ANTICHOLESTEROL ACTIVITY

Research reported that *C. sativum* (1.0 g/kg body weight) lowered triglyceride and cholesterol levels in triton-induced hyperlipidemic rats. Coriander reduces lipid intake while increasing lipid breakdown, a potential preventative and therapeutic treatment for hypolipidemia (Lal et al., 2004). In another study, it has been found that coriander reduced the quantity of lipid peroxides in cell membranes, as well as decreasing total and LDL levels, while raising HDL levels (Chitra & Leelamma, 1999).

16.4.6 ANTIMUTAGENICITY

Chitra and Leelamma (2000) reported that coriander protected rats from the adverse effects of 1, 2-dimethyl hydrazine on lipid metabolism in an experimental colon cancer model. Using the Ames reversion mutagenesis experiment (his- to his+) with the *S. typhi* TA98 strain as the indicator organism, the antimutagenicity of coriander juice against the mutagenic activity of 4-nitro-o-phenylenediamine, m-phenylenediamine, and 2-aminofluorene was examined. The mutagenicity of metabolized amines was considerably reduced by aqueous crude coriander juice.

16.4.7 IMMUNOMODULATORY ACTIVITY

At 50 to 200 g/mL concentration, aqueous crude extracts of *C. sativum* enhanced the development of human peripheral blood mono nucleus cells (PBMC) and the production of interferon-gamma (IFN-γ). Furthermore, research on numerous bioactive components of coriander in the extract revealed that the flavonoid quercetin increased human PBMC growth and IFN-y release. The flavonoid rutin and the coumarins bergapten and xanthotoxin were found to regulate IFN-y secretion but not human PBMC proliferation. In contrast, the coumarin isopimpinellin increased PBMC proliferation but did not change IFN-y secretion (Cherng et al., 2008).

16.4.8 ANXIOLYTIC AND SEDATIVE EFFECTS

C. sativum seed aqueous extract has anxiolytic properties and could function as a tranquilizer and muscle relaxant. Using the elevated plus maze technique, the aqueous extract (100 mg/kg) had an anxiolytic effect in male albino mice. Moreover, compared to the control group, the aqueous extract (500, 100, and 50 mg/kg) drastically decreased spontaneous activity and neuromuscular coordination (Emam et al., 2005). Coriander seed extract and essential oil have sedative-hypnotic (unknown) properties. The key active constituent(s) in the aqueous extract is responsible for the hypnotic action (Emam & Heidari, 2006).

16.4.9 DIURETIC EFFECT

In anesthetized Wister rats, the aqueous extract of coriander at doses of 40 and 100 mg/kg delivered by intravenous infusion (120 min) increased diuresis, urine excretion of sodium-potassium chloride, and glomerular filtration rate. Coriander diuretic is similar to furosemide in terms of mechanism (Aissaoui et al., 2008). On the other hand, furosemide was more effective as a diuretic and saluretic.

16.4.10 ANTIFERTILITY ACTIVITY

Al-said et al. (1987) examined female fertility in rats using an aqueous extract of fresh coriander seeds. The extract (250 and 500 mg/kg orally) had a dose-dependent strong anti-implantation impact; however, it failed to generate complete infertility.

16.4.11 PREVENTION OF HEAVY METAL DEPOSITION

In mice, it has been reported that *C. sativum* has a protective effect against localized lead accumulation when fed lead (1000 ppm) as lead acetate trihydrate in drinking water for 32 days. The *C. sativum* administration in mice significantly reduced lead deposition in the femur and severe lead-induced damage in the kidney (Aga et al., 2001).

16.4.12 EFFECTS ON BOWEL DISORDERS

Two separate experimental animal models of inflammatory bowel disease "indomethacin-induced enterocolitis and acetic acid-induced colitis in mice in rats" were used to test a polyherbal ayurvedic formulation including ripe coriander fruits as one of its key constituents. The findings revealed that the formulation was a successful treatment for inflammatory bowel disease (Jagtap et al., 2004). In a pilot investigation of individuals with irritable bowel syndrome, carmint herbal medicine containing *Menta spicata*, *Melissa officinalis*, and whole extract of *C. sativum* was found to exhibit sedative, carminative, and antispasmodic properties (Vejdani et al., 2006).

16.4.13 EFFECTS ON THE LIPID COMPOSITION OF MEAT

In comparison to the control group, dietary supplementation with coriander seed significantly changed the lipid composition of meat by decreasing saturated fatty acid (plasmic and stearic acids) concentration and increasing monounsaturated and polyunsaturated fatty acid (MUFA and PUFA) proportions. The greatest dose of coriander seed (4% added to the ration) has consistently resulted in the most substantial change in fatty acid profile. As a result, coriander seed supplementation would increase the quality of quintal lipids by decreasing the saturated fatty acids (SFA) proportion and increasing the PUFA content, notably of n3-PUFA (Ertas et al., 2005).

16.5 CONCLUSION

Herbs and spices have been utilized in foods since ancient times for flavor, increasing shelf life, and restoring health. Coriander is an aromatic plant applied in cuisines, pharmaceuticals, cosmetics, agriculture, biology, and the environment. Because of the phytochemicals qualities of *C. sativum* EO and extracts, they could be employed as natural antioxidants. Their health benefits include antibacterial and anticancer properties. In addition, the leaves and fruits have a strong fragrance and hold nutrients like lipids, proteins, vitamins, minerals, and other nutrients that are useful in functional food and improve a healthy lifestyle.

REFERENCES

Abbas, S., Nasreen, S., Haroon, A., & Ashraf, M. A. (2020). Synhesis of silver and copper nanoparticles from plants and application as adsorbents for naphthalene decontamination. *Saudi Journal of Biological Sciences*, 27(4), 1016–1023. https://doi.org/10.1016/j.sjbs.2020.02.011

Abou-Shaara, H. F. (2015). Potential honey bee plants of Egypt. *Cercetări Agronomice în Moldova*, 3(2), 99–108. https://doi.org/10.1515/cerce-2015-0034

Aga, M., Iwaki, K., Ueda, Y., Ushio, S., Masaki, N., Fukuda, S., & Kurimoto, M. (2001). Preventive effect of *Coriandrum sativum* (Chinese parsley) on localized lead deposition in ICR mice. *Journal of Ethnopharmacology*, 77(2–3), 203–208. https://doi.org/10.1016/S0378-8741(01)00299-9

Agostoni, C., Canani, R. B., Fairweather-Tait, S. J., Heinonen, M. I., Korhonen, H. J., Vieille, S. L., Marchelli, R., Martin, A., Naska, A., Neuhäuser-Berthold, M., Nowicka, G., Sanz, Y., Siani, A., Sjödin, A. M., Stern, M., Tetens, I., Tomé, D., Turck, D., Engel, H., Moseley, B., Pöting, A., Poulsen, M., & Salminen, S. (2013). Scientific opinion on the safety of "coriander seed oil" as a novel food ingredient. *EFSA Journal*, 11(10), 3422. https://doi.org/10.2903/j.efsa.2013.3422

Ahmad, M., Kumar, A., & Singh, P. (2016). Effects of coriander (*Coriandrum sativum* L.) seed powder on growth performance of broiler chickens. *Journal of Krishi Vigyan*, 5(1), 57–59. https://doi.org/10.5958/2349-4433.2016.00034.9

Aissaoui, A., El-Hilaly, J., Israili, Z. H., & Lyoussi, B. (2008). Acute diuretic effect of continuous intravenous infusion of an aqueous extract of *Coriandrum sativum* L. in anesthetized rats. *Journal of Ethnopharmacology*, 115(1), 89–95. https://doi.org/10.1016/j.jep.2007.09.007

Al-Bataina, B. A., Maslat, A. O., & Al-Kofahi, M. M. (2003). Element analysis and biological studies on ten oriental spices using XRF and Ames test. *Journal of Trace Elements in Medicine and Biology*, 17(2), 85–90. https://doi.org/10.1016/S0946-672X(03)80003-2

Al-Jedah, J. H., Ali, M. Z., & Robinson, R. K. (2000). The inhibitory action of spices against pathogens that might be capable of growth in a fish sauce (mehiawah) from the Middle East. *International Journal of Food Microbiology*, 57(1–2), 129–133. https://doi.org/10.1016/S0168-1605(00)00231-2

Al-Said, M. S., Al-Khamis, K. I., Islam, M. W., Parmar, N. S., Tariq, M., & Ageel, A. M. (1987). Post-coital antifertility activity of the seeds of *Coriandrum sativum* in rats. *Journal of Ethnopharmacology*, 21(2), 165–173. https://doi.org/10.1016/0378-8741(87)90126-7

Ashraf, A., Zafar, S., Zahid, K., Shah, M. S., Al-Ghanim, K. A., Al-Misned, F., & Mahboob, S. (2019). Synthesis, characterization, and antibacterial potential of silver nanoparticles synthesized from *Coriandrum sativum* L. *Journal of Infection and Public Health*, 12(2), 275–281. https://doi.org/10.1016/j.jiph.2018.11.002

Babu, B. S., Kumar, D. B., & Sathiyaraj, S. (2020). Experimental investigation and performance of diesel engine using biodiesel coriander seed oil. *Materials Today Proceedings*, 33(1), 1044–1048. https://doi.org/10.1016/j.matpr.2020.07.055

Bandoni, A. L., Mizrahi, I., & Juárez, M. A. (1998). Composition and quality of the essential oil of coriander (*Coriandrum sativum* L.) from Argentina. *Journal of Essential Oil Research*, 10(5), 581–584. https://doi.org/10.1080/10412905.1998.9700977

Bhalchandra, W., Baviskar, R. K., & Nikam, T. B. (2014). Diversity of nectariferous and polleniferous bee flora at Anjaneri and Dugarwadi hills of Western Ghats of Nasik district (MS) India. *Journal of Entomology and Zoology Studies*, 2(4), 244–249. https://www.entomoljournal.com/vol2Issue4/pdf/87.1.pdf

Bhat, S., Kaushal, P., Kaur, M., & Sharma, H. K. (2014). Coriander (*Coriandrum sativum* L.): Processing, nutritional and functional aspects. *African Journal of Plant Science*, 8(1), 25–33. https://doi.org/10.5897/AJPS2013.1118

Caballero, B., Trugo, L., & Finglas, P. (2003). *Encyclopedia of Food Sciences and Nutrition: Volumes 1–10* (2nd ed.).Oxford and Waltham, MA: Academic Press of Elsevier.

Chao, S. C., Young, D. G., & Oberg, C. J. (2000). Screening for inhibitory activity of essential oils on selected bacteria, fungi and viruses. *Journal of Essential Oil Research*, 12(5), 639–649. https://doi.org/10.1080/10412905.2000.9712177

Chaudhry, N. M., & Tariq, P. (2006). Bactericidal activity of black pepper, bay leaf, aniseed and coriander against oral isolates. *Pakistan Journal of Pharmaceutical Sciences*, 19(3), 214–218.

Cherng, J. M., Chiang, W., & Chiang, L. C. (2008). Immunomodulatory activities of common vegetables and spices of umbelliferae and its related coumarins and flavonoids. *Food Chemistry*, 106(3), 944–950. https://doi.org/10.1016/j.foodchem.2007.07.005

Chithra, V., & Leelamma, S. (1999). Coriandrum sativum mechanism of hypoglycemic action. *Food Chemistry*, 67(3), 229–231. https://doi.org/10.1016/S0308-8146(99)00113-2

Chithra, V., & Leelamma, S. (2000). *Coriandrum sativum* effect on lipid metabolism in 1, 2-dimethyl hydrazine induced colon cancer. *Journal of Ethnopharmacology*, 71(3), 457–463. https://doi.org/10.1016/s0378-8741(00)00182-3

Delaquis, P. J., Stanich, K., Girard, B., & Mazza, G. (2002). Antimicrobial activity of individual and mixed fractions of dill, cilantro, coriander and eucalyptus essential oils. *International Journal of Food Microbiology*, 74(1–2), 101–109. https://doi.org/10.1016/s0168-1605(01)00734-6

Dussort, P., Deprêtre, N., Bou-Maroun, E., Brunerie, P., Guichard, E., Le Fur, Y., & Le Quéré, J. L. (2014). Identification of key gin aroma compounds: An integrative approach based on an original selection procedure. In *Flavour Science, Proceedings from XIII Weurman Flavour Research Symposium* (pp. 367–370). Elsevier.

Eguale, T., Tilahun, G., Debella, A., Feleke, A., & Makonnen, E. (2007). In vitro and in vivo anthelmintic activity of crude extracts of *Coriandrum sativum* against haemonchus contortus. *Journal of Ethnopharmacology*, 110(3), 428–433. https://doi.org/10.1016/j.jep.2006.10.003

Eljazi, J. S., Bachrouch, O., Salem, N., Msaada, K., Aouini, J., Hammami, M., Boushih, E., Abderraba, M., Limam, F., & Jemaa, J. M. B. (2018). Chemical composition and insecticidal activity of essential oil from coriander fruit against *tribolium castaenum, sitophilus oryzae*, and *lasioderma serricorne*. *International Journal of Food Properties*, 21(1), 2833–2845. https://doi.org/10.1080/10942912.2017.1381112

Emam, G. M., & Heidari, H. G. (2006). Sedative & hypnotic activity of extracts and essential oil of coriander seeds. *Iran Journal of Medicine Science*, 31(1), 22–27.

Emam, G. M., Khasaki, M., & Aazam, M. F. (2005). *Coriandrum sativum*: Evaluation of its anxiolytic effect in the elevated plus maze. *Journal of Ethnopharmacology*, 96, 365–70. https://doi.org/10.1016/j.jep.2004.06.022

Ertas, O. N., Guler, T., Ciftci, M., Dalkilic, B., & Yilmaz, O. (2005). The effect of a dietary supplement coriander seeds on the fatty acid composition of breast muscle in Japanese quail. *Review Medicine Veterinary*, 156(10), 514. https://doi.org/10.2754/avb201079010033

Ghani, A. (2003). *Medicinal Plants of Bangladesh: Chemical Constituents & Uses* (2nd ed.). Dhaka: Asiatic Society of Bangladesh, pp. 1–16. https://www.scirp.org/(S(351jmbntvnsjt1aadkposzje))/reference/ReferencesPapers.aspx?ReferenceID=1002529

Gil, A., De La Fuente, E. B., Lenardis, A. E., López Pereira, M., Suárez, S. A., Bandoni, A., & Ghersa, C. M. (2002). Coriander essential oil composition from two genotypes grown in different environmental conditions. *Journal of Agricultural and Food Chemistry*, 50(10), 2870–2877. https://doi.org/10.1021/jf011128i

Hosseinzadeh, H., Alaw Qotbi, A. A., Seidavi, A., Norris, D., & Brown, D. (2014). Effects of different levels of coriander (*Coriandrum sativum*) seed powder and extract on serum biochemical parameters, microbiota, and immunity in broiler chicks. *The Scientific World Journal*, 2014, 628979. https://doi.org/10.1155/2014/628979

Hu, J., Wu, X., Wu, F., Chen, W., White, J. C., Yang, Y., & Wang, X. (2020). Potential application of titanium dioxide nanoparticles to improve the nutritional quality of coriander (*Coriandrum sativum* L.). *Journal of Hazardous Materials*, 389, 121837. https://doi.org/10.1016/j.jhazmat.2019.121837

Isbell, T. A., Green, L. A., DeKeyser, S. S., Manthey, L. K., Kenar, J. A., & Cermak, S. C. (2006). Improvement in the gas chromatographic resolution of petroselinate from oleate. *Journal of the American Oil Chemists' Society*, 83(5), 429–434. https://doi.org/10.1007/s11746-006-1222-y

Jabeen, Q., Bashir, S., Lyoussi, B., & Gilani, A. H. (2009). Coriander fruit exhibits gut modulatory, blood pressure lowering and diuretic activities. *Journal of Ethnopharmacology*, 122(1), 123–130. https://doi.org/10.1016/j.jep.2008.12.016

Jadoun, S., Arif, R., Jangid, N. K., & Meena, R. K. (2021). Green synthesis of nanoparticles using plant extracts: A review. *Environmental Chemistry Letters*, 19(1), 355–374. https://doi.org/10.1007/s10311 -020-01074-x

Jagtap, A. G., Shirke, S. S., & Phadke, A. S. (2004). Effect of polyherbal formulation on experimental models of inflammatory bowel diseases. *Journal of Ethnopharmacology*, 90(2–3), 195–204. https://doi.org/10 .1016/j.jep.2003.09.042

Jerković, I., Obradović, M., Kuś, P. M., & Šarolić, M. (2013). Bioorganic diversity of rare *Coriandrum sativum* L. honey: Unusual chromatographic profiles containing derivatives of linalool/oxygenated methoxyben-zene. *Chemistry & Biodiversity*, 10(8), 1549–1558. https://doi.org/10.1002/cbdv.201300074

Kadiri, L., Lebkiri, A., Rifi, E. H., Essaadaoui, Y., Ouass, A., Lebkiri, I., & Hamad, H. (2017). Characterization of coriander seeds "*Coriandrum sativum*". *International Journal of Scientific and Engineering Research*, 8(7), 2303–2308.

Kakarla, L., & Rama, C. (2015). Synthesis of silver nanoparticles from different plant leaf extracts and its critical analysis using UV-spectroscopy. *International Journal of Nanotechnology and Application*, 5(3), 21–26. https://doi.org/10.1155/2019/8642303

Kassahun, B. M. (2020). Unleashing the exploitation of coriander (*Coriander sativum* L.) for biological, industrial and pharmaceutical applications. *Academic Research Journal of Agricultural Science and Research*, 8(6), 552–564. https://doi.org/10.14662/ARJASR2020.555

Kubo, I., Fujita, K. I., Kubo, A., Nihei, K. I., & Ogura, T. (2004). Antibacterial activity of coriander vola-tile compounds against *Salmonella choleraesuis*. *Journal of Agricultural and Food Chemistry*, 52(11), 3329–3332. https://doi.org/10.1021/jf0354186

Lal, A., Kumar, T., Murthy, P. B., & Pillai, K. S. (2004). Hypolipidemic effect of *Coriandrum sativum* L. in triton-induced hyperlipidemic rats. *Indian Journal of Experimental Biology*, 42(9), 909–912.

Lohwasser, U., Borner, A., & Kruger, H. (2008). Intraspecific taxonomy of *Coriandrum sativum* L.: Comparison of morphological and phytochemical data. *Acta Horticulturae*, 799, 111–113. https://doi .org/10.17660/ActaHortic.2008.799.13

López, P. A., Widrlechner, M. P., Simon, P. W., Rai, S., Boylston, T. D., Isbell, T. A., & Wilson, L. A. (2008). Assessing phenotypic, biochemical, and molecular diversity in coriander (*Coriandrum sativum* L.) germplasm. *Genetic Resources and Crop Evolution*, 55(2), 247–275. https://doi.org/10.1007/s10722-007 -9232-7

Mandal, S., & Mandal, M. (2015). Coriander (*Coriandrum sativum* L.) essential oil: Chemistry and biological activity. *Asian Pacific Journal of Tropical Biomedicine*, 5(6), 421–428. https://doi.org/10.1016/j.apjtb .2015.04.001

Marangoni, C., & Moura, N. F. D. (2011). Antioxidant activity of essential oil from *Coriandrum sati-vum* L. in Italian salami. *Food Science and Technology*, 31, 124–128. https://doi.org/10.1590/S0101 -20612011000100017

Meena, S. K., Jat, N. L., Sharma, B., & Meena, V. S. (2014). Effect of plant growth regulators and sulphur on productivity of coriander (L.) in Rajasthan. *International Journal of Environmental Science*, 6, 69–73.

Melo, E. A, Mancini, F. J., & Guerra, N. B. (2005). Characterization of antioxidant compounds in aqueous coriander extract (*Coriandrum sativum* L.). *LWT-Food Science and Technology*, 38(1), 15–19. https:// doi.org/10.1016/j.lwt.2004.03.011

Moise, G. (2015). Research on quality analysis of an assortment of five types of honey in Romania. *Scientific Papers Series Management, Economic Engineering in Agriculture and Rural Development*, 15(3), 195–200.

Moser, B. R., & Vaughn, S. F. (2010). Coriander seed oil methyl esters as biodiesel fuel: Unique fatty acid composition and excellent oxidative stability. *Biomass and Bioenergy*, 34(4), 550–558. https://doi.org/10 .1016/j.biombioe.2009.12.022

Mousavizadeh, S. J., & Sedaghathoor, S. (2011). Apple and quince peroxidase activity in response to essential oils application. *African Journal of Biotechnology*, 10(57), 12319–12325. https://www.ajol.info/index .php/ajb/article/view/96410

Msaada, K., Jemia, M. B., Salem, N., Bachrouch, O., Sriti, J., Tammar, S., & Marzouk, B. (2017). Antioxidant activity of methanolic extracts from three coriander (*Coriandrum sativum* L.) fruit varieties. *Arabian Journal of Chemistry*, 10, S3176–S3183. https://doi.org/10.1016/j.arabjc.2013.12.011

Naeemasa, M., Qotbi, A. A., Seidavi, A., Norris, D., Brown, D., & Ginindza, M. (2015). Effects of coriander (*Coriandrum sativum* L.) seed powder and extract on performance of broiler chickens. *South African Journal of Animal Science*, 45(4), 371–378. https://doi.org/10.4314/sajas.v45i4.3

Nimish, P. L., Sanjay, K. B., Nayna, B. M., & Jaimik, R. D. (2011). Phyto pharmacological properties of cori-ander sativum as a potential medicinal tree: An overview. *Journal of Applied Pharmaceutical*, 01(04), 20–25. http://japsonline.com/admin/php/uploads/55_pdf.pdf

Özcan, M. M., Ünver, A., Uçar, T., & Arslan, D. (2008). Mineral content of some herbs and herbal teas by infusion and decoction. *Food Chemistry*, 106(3), 1120–1127. https://doi.org/10.1016/j.foodchem.2007.07.042

Pascual-Villalobos, M. J. (2002). Volatile activity of plant essential oils against stored product beetle pests. In Credland, P. F., Armitage, D. M., Bell, C. H., Cogan, P., & Highley, E. (Eds.), *Advances in Stored Product Protection. Proceedings of the 8th International Working Conference on Stored Product Protection* (pp. 648–650). York, UK, 22–26 July 2002. https://doi.org/10.1155/2018/6906105

Ruiz-Torres, N., Flores-Naveda, A., Barriga-Castro, E. D., Camposeco-Montejo, N., Ramírez-Barrón, S., Borrego-Escalante, F., & García-López, J. I. (2021). Zinc oxide nanoparticles and zinc sulfate impact physiological parameters and boosts lipid peroxidation in soil grown coriander plants (*Coriandrum sativum*). *Molecules*, 26(7), 1998. https://doi.org/10.3390/molecules26071998

Saeed, S., & Tariq, P. (2007). Antibacterial activities of *Emblica officinalis* and *Coriandrum sativum* against gram negative urinary pathogens. *Pakistan Journal of Pharmaceutical Sciences*, 20(1), 32–35.

Saygi, K. O., Kacmaz, B., & Gul, S. (2021). Antimicrobial activities of coriander seed essential oil and silver nanoparticles. *Research Square*. https://doi.org/10.21203/rs.3.rs-526332/v1

Senthilkumar, N., Aravindhan, V., Ruckmani, K., & Potheher, I. V. (2018). *Coriandrum sativum* mediated synthesis of silver nanoparticles and evaluation of their biological characteristics. *Materials Research Express*, 5(5), 055032. https://doi.org/10.1088/2053-1591/aac312

Shahwar, M. K., El-Ghorab, A. H., Anjum, F. M., Butt, M. S., Hussain, S., & Nadeem, M. (2012). Characterization of coriander (*Coriandrum sativum* L.) seeds and leaves: Volatile and non volatile extracts. *International Journal of Food Properties*, 15(4), 736–747. https://doi.org/10.1080/10942912.2010.500068

Sharma, M. M., & Sharma, R. K. (2012). Coriander. In Peter, K. V. (Ed.), *Handbook of Herbs and Spices* (pp. 216–249). Cambridge, UK: Woodhead Publishing.

Shyamala, B. N., Gupta, S., Lakshmi, A. J., & Prakash, J. (2005). Leafy vegetable extracts-antioxidant activity and effect on storage stability of heated oils. *Innovative Food Science & Emerging Technologies*, 6(2), 239–245. https://doi.org/10.1016/j.ifset.2004.12.002

Silva, F., Domeño, C., & Domingues, F. C. (2020). *Coriandrum sativum* L.: Characterization, biological activities, and applications. In Preedy, V. R., & Watson, R. R. (Eds.), *Nuts and Seeds in Health and Disease Prevention* (2nd ed.) (pp. 497–519). London: Academic Press.

Singh, G., Maurya, S., De Lampasona, M. P., & Catalan, C. A. (2006). Studies on essential oils, Part 41. Chemical composition, antifungal, antioxidant and sprout suppressant activities of coriander (*Coriandrum sativum*) essential oil and its oleoresin. *Flavour and Fragrance Journal*, 21(3), 472–479. https://doi.org/10.1002/ffj.1608

Smith, N. O., Maclean, I., Miller, F. A., & Carruthers, S. R. (1997). *Crops for Industry and Energy in Europe*. University of Reading, European Commission, Directorate General XII E-2, Agro-Industrial Research Unit, 11–19. https://www.feedipedia.org/node/13677

Telci, I., Bayram, E. M. İ. N. E., & Avci, B. (2006). Changes in yields, essential oil and linalool contents of *Coriandrum sativum* varieties (var. vulgare Alef. and var. microcarpum DC.) harvested at different development stages. *European Journal of Horticultural Science*, 71(6), 267–271. https://hdl.handle.net/20.500.12881/7183

Uitterhaegen, E., Sampaio, K. A., Delbeke, E. I., De Greyt, W., Cerny, M., Evon, P., & Stevens, C. V. (2016). Characterization of French coriander oil as source of petroselinic acid. *Molecules*, 21(9), 1202. https://doi.org/10.3390/molecules21091202

USDA. (2016). *National Nutrient Database for Standard Reference Release 28 Full Report (All Nutrients)*. Nutrient data for, spices, coriander seed.

Vejdani, R., Shalmani, H. R. M., Mir-Fattahi, M., Sajed-Nia, F., Abdollahi, M., Zali, M. R., & Amin, G. (2006). The efficacy of an herbal medicine, carmint, on the relief of abdominal pain and bloating in patients with irritable bowel syndrome: A pilot study. *Digestive Diseases and Sciences*, 51(8), 1501–1507. https://doi.org/10.1007/s10620-006-9079-3

Vollmann, J., & Laimer, M. (2013). Novel and traditional oil crops and their biorefinery potential. In Yang, S.-T., El-Enshasy, H. A., & Thongchul, N. (Eds.), *Bioprocessing Technologies in Biorefinery for Sustainable Production of Fuels, Chemicals, and Polymers* (pp. 47–60). https://doi.org/10.1002/9781118642047.ch3

Winarti, S., Pertiwi, C. N., Hanani, A. Z., Mujamil, S. I., Putra, K. A., & Herlambang, K. C. (2018). Beneficial of coriander leaves (*Coriandrum sativum* L.) to reduce heavy metals contamination in rod shellfish. *Journal of Physics Conference Series*, 953(1), 012237. https://doi.org/10.1088/1742-6596/953/1/012237

Wong, P. Y., & Kitts, D. D. (2006). Studies on the dual antioxidant and antibacterial properties of parsley (*Petroselinum crispum*) and cilantro (*Coriandrum sativum*) extracts. *Food Chemistry*, 97(3), 505–515. https://doi.org/10.1016/j.foodchem.2005.05.031

World Health Organization, Tufts University. School of Nutrition Science, Policy, & Tufts University. (2002). *Keep fit for life: Meeting the nutritional needs of older persons.* World Health Organization.

Yepez, B., Espinosa, M., López, S., & Bolanos, G. (2002). Producing antioxidant fractions from herbaceous matrices by supercritical fluid extraction. *Fluid Phase Equilibria,* 194, 879–884. https://doi.org/10.1016/S0378-3812(01)00707-5

Zheljazkov, V. D., Pickett, K. M., Caldwell, C. D., Pincock, J. A., Roberts, J. C., & Mapplebeck, L. (2008). Cultivar and sowing date effects on seed yield and oil composition of coriander in Atlantic Canada. *Industrial Crops and Products,* 28(1), 88–94. https://doi.org/10.1016/j.indcrop.2008.01.011

17 Heavy Metals Detoxification Using Coriander

Ahmad Cheikhyoussef, Natascha Cheikhyoussef,
Aiman K. H. Bashir, and Ahmed A. Hussein

CONTENTS

17.1 INTRODUCTION

Anthropogenic activities have been rapidly increased and have resulted in environmental pollution with heavy metals (HMs), which, if accumulated in higher concentrations than what is recommended, or naturally found on earth, will accumulate in plants and animals as a result of such activities (Yan et al., 2021). Heavy metal pollution (HMP) is a deleterious stress factor threatening human health, disturbing plant growth, and negatively impacting the overall status of the environment. More specifically, HMP affects food security and can severely lead to food quality decline and threaten public health (Sardar et al., 2022). HMs are a part of hazardous waste substances that originate from chemical mining, metal processing, and other related industries (Nabi et al., 2021). They are linked to non-communicable chronic diseases (NCDs) without contributing to human homeostasis (Sears, 2013). The US Environmental Protection Agency (EPA) indicated that the permissible limits for Pb(II), Zn(II), and Cu(II) ions in drinking water should not exceed 0.015, 1.3, and 5.00 mg/L, respectively (Rao & Kashifuddin, 2012). Among the technologies and processes that have been used efficiently in detoxifying HMs such as arsenic, cadmium, copper, lead, and mercury are adsorption, chelation, coagulation, ion exchange, and photocatalysis (Kenduzler and Turker 2005; Sears, 2013; Susan et al., 2019; Ajiboye et al., 2021) with several reported advantages and functions such as efficiency, low cost, precision and high selectivity, rapidity, versatility, and regeneration for multiple uses (Sears, 2013; de Freitas et al., 2019; Ajiboye et al., 2021).

Coriander is a tropical herb grown under a wide range of conditions in different regions, with India being the world's largest producer (CSIR 2001). This plant is also cultivated commercially in Morocco, Poland, Romania, Mexico, the USA, and Russia (Rao & Kashifuddin, 2012). Coriander belongs to the Apiaceae family, containing around 300 genera and over 3000 species (Asgarpanah et al., 2012). There are several common names for coriander; these include cilantro, dhaniya, Arab parsley, Kasbour, and Chinese parsley. *Coriandrum sativum* L. belongs to the Umbelliferae family. It is one member of the natural adsorbent plants that have been successfully reported to remove aluminum (Abidhusen, Sawapnil, and Amit 2012), mercury (Karunasagar et al., 2005; Abascal & Yarnell, 2012), lead, copper, and zinc (Rao & Kashifuddin, 2012), aluminum and lead (Aga et al., 2001; Deldar et al., 2008; Téllez-López et al., 2017), cadmium (Jia et a., 2009; Ren et al., 2009; Nicula et al. 2016; Martin et al., 2020), copper (Kadiri et al., 2018), iron (Mirzaei & Khatami, 2013; Talaei et al., 2018), and other HMs (Nishio et al., 2019). The chapter focuses on the latest developments and use of coriander plant parts to detoxify HMs.

17.2 HEAVY METALS PHYTOREMEDIATION BY CORIANDER

Coriandrum sativum is commonly called coriander, cilantro, or Chinese parsley. It has a role in the apothecary of most Western practitioners (Abascal & Yarnell, 2012). The name changes depending on the use status of the plant materials whereby if it is freshly harvested and green in color, and the leaves are used for different purposes, it is called cilantro; meanwhile, when the dried fruits are used, it is called coriander (Abascal & Yarnell, 2012). The administration of chemical chelators is the most common method for heavy metals removal from the body, and eliminating the source of exposure alongside keeping the patient away from the exposure source of HMs are the first steps in the treatment of heavy metals poisoning (Mehrandish, Rahimian, & Shahriary, 2019). Chelation therapy is well-known and has widely been used to detoxify HMs (Flora & Pachauri, 2010; Aaseth, Crisponi & Andersen, 2016), such as arsenic (Flora et al., 2007; Susan et al., 2019) and mercury (Clarke et al., 2012). It detoxifies HMs *via* the formation of complexes, particularly with glutathione and other small molecules, and then facilitates their excretion outside the body (Flora, 2009), e.g., it prevents arsenic from interacting with biological targets (proteins and DNA), and then, it facilitates the elimination of arsenic from the body (Flora et al., 2007). *C. sativum* has strong chelating properties; however, there is insufficient scientific evidence (Gurib-Fakim, 2006). Phytoremediation is another effective method and a sustainable green solution for cleaning the environment of HMs and their toxicological effects due to its economic feasibility, eco-friendliness, and applicability (Ashraf et al., 2019; Yan et al., 2020; Babu et al., 2021).

17.2.1 CORIANDER AND LEAD (PB) DETOXIFICATION

Lead (Pb) is a well-known multi-organ toxicant that can damage the liver and kidney (Manoj-Kumar et al., 2013). The preventive effect of *C. sativum* against lead (Pb) deposition in male ICR mice administered with lead (1000 ppm) as lead acetate trihydrate in drinking water over 32 days was studied (Aga et al., 2001). They further highlighted that a decrease in the lead absorption into bone was obtained, and inhibited delta-aminolevulinic acid dehydratase (ALAD) activity *in vitro* was observed (Aga et al., 2001). They suggested that coriander extracts contain certain substances that can chelate lead and reduce tissue lead accumulation (Aga et al., 2001).

A trial was undertaken in 3- to 7-year-old children exposed to lead; the coriander extract was as effective as the placebo group in increasing renal excretion (Deldar et al., 2008). Furthermore, Pb(II) ions from an aqueous solution have been removed by coriander seed powder in a pH-dependent model (Rao & Kashifuddin, 2012). Finally, a study by Manoj-Kumar et al. (2013) investigated the therapeutic role of *C. sativum* seed extract against lead-induced oxidative stress in rat liver and kidney and reported a reduction in the lead-induced oxidative stress in organs (kidney and liver) due to its antioxidant and metal chelating activity.

The aqueous extract of coriander seeds was effective in normalizing the adverse effects of lead-induced nephrotoxicity on renal functions (El-Masry et al., 2016). Several natural products involved in nephroprotection (Sri Laasya et al., 2020), such as phenolics and flavonoids, can reduce renal damage, lipid peroxidation, urea, and creatinine reduction (Javaid et al., 2012; Sharma et al., 2011; Gholamine et al., 2021), decrease oxidative stress (Ahmed, Zaki, & Nabil 2015; Kükner et al., 2021), and decrease SCr, urea, uric acid, and liver enzymes levels (Adeneye & Benebo, 2008; Mashayekhi, 2012; Heidari-Soreshjani et al., 2017; Altınok-Yipel et al., 2020) and their potential antioxidant activities (Wardle, 1999; Abou El-Soud et al., 2014; Ratliff et al., 2016; Esmail and Ali, 2019; Neha et al., 2019). The presence of catechin, epicatechin, epigallocatechin, epigallocatechin-3-gallate, resveratrol, quercetin, luteolin, and kaempferol (Demir & Korukluoglu, 2020), gallic acid, and rutin (Laribi et al., 2015; Singletary, 2016), and caffeic, ferulic, gallic, and chlorogenic acids (Oganesyan, Nersesyan, & Parkhomenko, 2007) in coriander plant parts could explain the nephroprotection for coriander extracts.

The chelating effect of coriander methanolic extract (LD_{50} > 1000 mg/dL) and its fractions was reported on Wistar rats intoxicated with lead (Pb) by Téllez-López et al. (2017), and it was confirmed that coriander extracts could protect the liver and lower lead concentration in rats intoxicated with lead. The study reported that histological evaluation of tissues revealed less damage in groups administered with methanol extract and its fractions (significant difference in hemoglobin and hematocrit levels) compared to the positive control, which presented structural alterations due to lead toxicity. Previous studies have also reported and confirmed chelation abilities and poisoning reduction for *C. sativum* extracts in the testis of lead-exposed mice by reducing the negative effect of lead on liver enzymes, testosterone levels, sperm density, and concentration of lead (Sharma et al., 2010) by reducing lead-induced oxidative stress, and male Wistar strain rats challenged with lead-induced oxidative stress (Velaga et al., 2014). The lead toxicity enzymatic indicator, delta-amino levulinic acid dehydratase (δ-ALAD), decreased together with significantly increased reactive oxygen species (ROS), lipid peroxidation products (LPP), and total protein carbonyl content (TPCC) levels in exposed rat brain regions indicating lead-induced oxidative stress could be the mechanism for the hydroalcoholic seed extract of *C. sativum* in the tissue-specific amelioration of oxidative stress as a result of lead toxicity (Velaga et al., 2014).

Kansal et al. (2011) reported that two extracts of coriander seeds (aqueous and ethanolic) could protect against oxidative damages in the liver and kidney due to lead toxicity in mice. They used oral administration to feed the mice and reported ameliorative effects for the lead-induced oxidative damage through increasing the antioxidant enzymatic activities such as superoxide dismutase (SOD) and catalase (CAT), and glutathione level, and decreasing the lipid peroxidation, alongside modulating the hepatic and renal biochemical parameters. Furthermore, the histological investigation confirmed that coriander extracts could protect the liver and renal tissues from oxidative damage, with flavonoids and ascorbic acid being assumed to be the main bioactive compounds responsible for the vital role (Kansal et al., 2011).

In addition, coriander extracts reduced the negative effect of lead on liver enzymes, testosterone levels, sperm density, and concentration of lead in the mice's testis (Sharma et al., 2010). Furthermore, several reports confirmed the protective roles of flavonoids and their enhancing effect on the antioxidant functions of the liver (Meunier, de Visser, & Shaik, 2004; Wu et al., 2006; Noh et al., 2011; Chen et al., 2015; Aćimović, & Milić, 2017; Wan & Jiang, 2018; Huang et al., 2021; Komolafe et al., 2021) via different mechanisms:

1. Increasing enzymatic activities of superoxide dismutase, glutathione s-transferase, and glutathione peroxidase.
2. Improving insulin sensitivity and inhibiting hepatic stellate cell activation *via* regulating the following enzymatic activities: heme oxygenase-1, cytochrome P450, and telomerase.
3. Reducing inflammatory reaction by restraining the expression of tumor necrosis factor-α, interferon-γ, and interleukin-6.
4. Mediating apoptosis and autophagy via controlling the pathways of genes-p 53-genetics, nuclear factor κB, and phosphatidylinositol 3-kinase/protein kinase B signaling.

17.2.2   CORIANDER AND MERCURY (HG) DETOXIFICATION

There are three forms of mercury; these include inorganic salts, metallic elements, and organic compounds with different toxicity and bioavailability for each form (Abdel-Salam et al., 2018). At the beginning of the 21st century, there were more than 2500 tons of mercury produced on an annual basis from different global anthropogenic sources, which contributes to the high levels of mercury worldwide with organic mercury compounds (methyl mercury: MeHg) being 50–100 times more toxic than the inorganic mercury species (de Wuilloud et al., 2002; Krishna & Karunasagar, 2015).

Generally, the brain is the main target organ for mercury toxicity; however, it can harm any organ and cause malfunctioning of nerves, kidneys, and muscles, whereby it disrupts membrane potential and interrupts the intracellular calcium homeostasis (Patrick, 2002; Fernandes Azevedo et al., 2012; Abdel-Salam et al., 2018).

Omura et al. (1996) developed a bidigital *O*-ring test (BDORT) to measure HMs after they observed that mercury levels were higher in a patient after dental amalgam removal; then they recommended the use of a 100 mg coriander leaf tablet, four times per day, together with some enhancement method using their BDORT to remove Hg from the dental patient. They further added that a combination of coriander leaf tablets and antibiotics could eliminate *Chlamydia trachomatis*, herpes viral infections, and other infections (Omura & Beckman, 1995).

Karunasagar et al. (2005) prepared coriander-based sorbent with the ability to remove inorganic (Hg^{2+}) and methyl mercury (CH_3Hg^+) from aqueous solutions due to the carboxylic acid group's property of binding mercury in contaminated waters. Mercury-induced neurotoxicity in male rats was treated with hot water extracts of dandelion (DA), coriander (CO), date palm seeds (DS), probiotic supernatant (PS), and their combined mixture (Abdel-Salam et al., 2018). The authors reported that due to the antioxidant treatment with the three herbal extracts, probiotics, and their mixture, they were completely able to invert the cytotoxicity induced by $HgCl_2$ with variable degrees and suggested that a probiotic-herbal mixture neutralizing mercury-induced neurotoxicity represents a promising approach that deserves further confirmation.

17.2.3 CORIANDER AND ARSENIC (AS) DETOXIFICATION

Susan et al. (2019) reported that phytochemicals might aid in the elimination of arsenic (As) from the biological system and more effectively than conventional therapeutic agents. Since coriander leaves have been reported to contain good levels of caffeic, ferulic, gallic, and chlorogenic acids (Sreelatha, Padma, & Umadevi, 2009), these phenolic acids might mediate the detoxification mechanism; however, further investigations are needed to elucidate its mode of action. It is well known that oxidative stress and pro-inflammatory actions are among the most common mechanisms for HM toxicities; however, on the other side of the process, phytonutrients such as phenolic acids counteract these cellular invectives through the anti-oxidation action, upregulation of anti-inflammatory pathways, and chelation (Chung, 2017). The phytoremediation ability of coriander was confirmed in soil and water contaminated with lead and arsenic (Gaur et al., 2017), and the whole coriander plant seems to be an excellent natural chelating agent for toxic HM detoxification (Nishio et al., 2019).

17.2.4 CORIANDER AND NICKEL (NI) DETOXIFICATION

The phytoremediation of nickel (Ni) by coriander was reported by Tagharobiyan and Poozesh (2015) in hydroponic design. They reported that treatment with nickel nitrate (0, 100, 200, and 500 µM) resulted in a significant increase in flavonoids, hydrogen peroxide, MDA, and other aldehydes in the coriander plant to Ni has changed the catalase enzymes and led to a significant decrease in their contents.

17.2.5 CORIANDER AND CADMIUM (CD) DETOXIFICATION

Cadmium (Cd) is a non-essential metal that has polluted the environment heavily due to human activities. Cd can be absorbed into the human body *via* the gastrointestinal tract, respiratory tract, and skin, and it can cause chronic damage within the kidneys' nephrons (Yan & Allen, 2021). Fish feed formula containing coriander was prepared and administered to rainbow trout, *Oncorhynchus mykiss* (Jia et al., 2009; Ren et al., 2009). The developed formula did not negatively affect the growth or health of farmed rainbow trout; it reduced their cadmium uptake by 12–48%. The fish were divided into groups, with one group fed with a formula equal to 20% fresh coriander leaf in freeze-dried form and cadmium. They reported that the coriander-supplemented formula increased the amount of cadmium bound to metallothionein (MT), significantly reducing the cadmium toxicity in the fish. They further emphasized that the synergetic actions of several compounds in coriander leaf may contribute to binding MT, preventing the accumulation of harmful HMs in fish (Jia et al., 2009; Ren et al., 2009).

Aqueous extract of coriander leaf treatment led to the normalization of the enzymatic and non-enzymatic parameters in cadmium-induced toxicity in albino rats (John, Shobana, & Keerthana, 2014). Moreover, when coriander was introduced as a supplement, together with garlic and chlorella algae in the diets (2%) of Prussian carp (*Carassius gibelio*), protection against kidney damage from cadmium exposure at 10 mg/L was observed (Nicula et al., 2016).

A recent study recommended the addition of 10–15 g/kg of coriander leaf powder to the beluga diet for the optimal growth and health of the fish regardless of whether the water is contaminated with HMs, whereby it enhances the metabolism of protein and lipid, and reduces the adverse hematological effects of HMs (Bahrekazemi, Eslami, and Nikbakhsh, 2020). Martin et al. (2020) investigated the adsorption of Cd^{2+} ions by fresh and dry cilantro leaves and stems, whereby the plant materials were sun-dried at room temperature until their weighed dry masses remained constant, and after that, they were ground to fine powders that were used to remove Cd^{2+} ions from contaminated water. They further reported that fresh leaves and fresh stems powder could remove more dry parts powder as per the obtained Langmuir isotherm with Qmax values of 250.00, 51.28 mg/g compared to 4.68, 15.06 mg/g, respectively.

17.2.6 CORIANDER AND IRON (FE) DETOXIFICATION

Toxic levels of iron can cause several types of physiological damage to the kidney and heart via the increasing activity of the oxidative stress pathway that may lead to the dysfunction of these two important organs (Aziza, Azab, & El-Shall, 2014; Sarkar, Hazra, & Mandal, 2012; Sengsuk et al., 2014). The hydro-alcoholic extract of *C. sativum* whole plant powder has iron-chelating and liver-protective properties comparative to deferoxamine against experimental iron in an overload condition in Wistar rats (Talaei et al., 2018). They further indicated that *C. sativum* extract treatment significantly decreased biochemical parameters, such as alanine aminotransferase (ALT), aspartate aminotransferase (AST), alkaline phosphatase (ALP), lactate dehydrogenase (LDH), and creatine phosphokinase (CPK), alongside improving the tissue damage and decreasing the iron accumulation in the liver compared to the iron overload group. Furthermore, the histological injuries inside the liver were alleviated significantly after treatment with coriander extract, and a reduction in liver iron content was also observed (Talaei et al., 2018). Mirzaei and Khatami (2013) reported that coriander is considered a potent iron chelator in *in vitro* conditions; they reported a positive correlation between total phenolic and flavonoid contents of the plant and the iron-chelating effects.

Several reports emphasized the biological activity of flavonoids and their function of interfering in the iron absorption and acting as iron chelator agents (Kumar & Pandey, 2013; Ullah et al., 2020; Hawula et al., 2021; Wang et al., 2021) with a direct relationship between iron-chelating properties of a plant and its flavonoids content (Ebrahimzadeh, Pourmorad, & Bekhradnia, 2008; Najafzadeh et al., 2010) and phenol and flavonoid contents (Ebrahimzadeh, Nabavi, & Nabavi, 2009).

17.2.7 CORIANDER AND OTHER HMS DETOXIFICATION

The leaves of *C. sativum* gained attention when a coriander leaf soup was prepared and observed to enhance urinary excretion of mercury, lead, and aluminum, whereby it may reduce the uptake of HMs if taken during exposure or, if taken in advance, it may protect the liver from some other toxins (Abascal & Yarnell, 2012). Furthermore, Rao and Kashifuddin (2012) indicated that coriander seed powder has excellent potential to remove Cu(II) and Zn(II) ions from aqueous solution with the adsorption process being pH-dependent whereby pH 4–6 yielded the maximum adsorption of these ions.

Winarti et al. (2018) reported the beneficial effects of coriander leaves in reducing HM contamination (Pb, Hg, and Cu) in rod shellfish and reported that the longer soaking time resulted in decreasing the Pb level from 4.4 ppb, Hg from 4.11, and Cu from 433.7 ppb to 1.7, 1.12 and 117 ppb respectively with no significant effect on decreasing the protein content. Kadiri et al. (2018) successfully removed Cu(II) from an aqueous solution with coriander seeds using adsorption to treat wastewaters containing copper. Oxidative stress and pro-inflammatory actions are the most popular mechanisms for HM toxicity, whereby coriander active constituents neutralize this toxic effect *via* anti-oxidation, anti-inflammatory pathway upregulation, and chelation (Chung, 2017). Among these bioactive constituents of coriander is epigallocatechin (Demir & Korukluoglu, 2020) which is believed to play a pivotal role in HM detoxification *via* scavenging reactive oxygen species, chelating HMs, nuclear factor erythroid 2-related factor 2 (Nrf2) activation, anti-inflammatory effects, and mitochondria protection (Zwolak, 2021).

Nishio et al. (2019) studied the impact of heated leaf extract of *C. sativum* on the concentrations of zinc, iron, copper, arsenic, and cadmium in the liver and kidney of male ddY mice. There was no effect on the heavy metals tested in the liver; however, the concentrations of iron, arsenic, and cadmium in the kidney were decreased due to its contribution to resistance to oxidative stress caused by hydrogen peroxide. As a result, a maximum tolerable intake for some heavy metals such as As, Cd, Pb, and Cr of 2, 1, 6.03, and 3.32 µg/kg body weight/day, respectively, has been recommended by a joint committee of the World Health Organization (WHO) and the Food and Agriculture Organization (FAO) on food additives (JCEFA) (Dubey et al., 2018).

17.3   FUTURE PERSPECTIVES AND CONCLUSIONS

Coriander has chelating properties and the ability to suppress lead deposition through lead chelating by specific compounds contained in coriander's different plant parts, which need further purification, isolation, and identification via bio-guided fractionation and mass spectrometry techniques (Badria et al., 2018). The protective effects of coriander against HM toxicity are significant for food toxicology to develop natural remedies for HM toxicity. Even though coriander leaves are amazingly considered a popular mercury detoxifier among the public and some practitioners, follow-up studies are needed using accepted techniques to evaluate the effects and doses on excretion of HMs (Millet, 2006; Abascal & Yarnell, 2012). Toxic HM excretion from the body is essential to improve health and prevent kidney and liver failure (hepatorenal syndrome and acute liver failure). Considering the increasing HM contamination in food chains worldwide (Narendrula-Kotha et al., 2020; Gong et al.,2020; Angulo-Bejarano, Puente-Rivera, & Cruz-Ortega, 2021), it is very encouraging and beneficial to incorporate coriander leaf in diets in forms of pesto, salsas, salads, and soup (Chen et al., 2011; Abascal & Yarnell, 2012). *C. sativum* extracts possess several beneficial effects thanks to their phenolic compounds that positively affect liver and kidney functions. Coriander powder/extracts can be used as prophylactic or therapeutic agents to treat HM toxicity in fish, whereby it can be added to the feeding materials with several overall health benefits. Aluminum, arsenic, cadmium, copper, iron, lead, mercury, nickel, zinc, and other HMs have been successfully removed/detoxified by coriander plant parts powder/extracts. Chelation therapy and phytoremediation technology are the most common methods for HM removal from the body and environment to eliminate/reduce their toxicities.

REFERENCES

Aaseth, J., Crisponi, G., & Andersen, O. (2016). *Chelation Therapy in the Treatment of Metal Intoxication* (1st ed.). Oxford: Academic Press, Elsevier, 35–57.

Abascal, K., & Yarnell, E. (2012). Cilantro-culinary herb or miracle medicinal plant? *Alternative and Complementary Therapies*, *18*(5), 259–264. https://doi.org/10.1089/act.2012.18507

Abdel-Salam, A. M., Al Hemaid, W. A., Afifi, A. A., Othman, A. I., Farrag, A., & Zeitoun, M. M. (2018). Consolidating probiotic with dandelion, coriander and date palm seeds extracts against mercury neurotoxicity and for maintaining normal testosterone levels in male rats. *Toxicology Reports*, *5*, 1069–1077. https://doi.org/10.1016/j.toxrep.2018.10.013

Abidhusen, H. M., Sawapnil, S. A., & Amit, V. G. (2012). *Coriandrum sativum*: Review of advances in phytopharmacology. *International Journal of Pharmaceutical Sciences and Research*, *3*(5), 1233–1239. https://doi.10.13040/IJPSR.0975-8232

Abou El-Soud, N. H., El-Lithy, N. A., El-Saeed, G., Wahby, M. S., Ykhalil, M., Morsy, F., & Shaffie, N. (2014). Renoprotective effects of caraway (*Carum carvi* L.) essential oil in streptozotocin induced diabetic rats. *Journal of Applied Pharmaceutical Science*, *4*(2), 27–33. DOI. 10.7324/JAPS.2014.40205

Aćimović, M. G., & Milić, N. B. (2017). Perspectives of the Apiaceae Hepatoprotective Effects - A Review. *Natural product communications*, *12*(2), 309–317.

Adeneye, A. A., & Benebo, A. S. (2008). Protective effect of the aqueous leaf and seed extract of *Phyllanthus amarus* on gentamicin and acetaminophen-induced nephrotoxic rats. *Journal of Ethnopharmacology* *118*(2), 318–323. https://doi.org/10.1016/j.jep.2008.04.025

Aga, M., Iwaki, K., Ueda, Y., Ushio, S., Masaki, N., Fukuda, S., Kimoto, T., Ikeda, M., & Kurimoto, M. (2001). Preventive effect of *Coriandrum sativum* (Chinese parsley) on localized lead deposition in ICR mice. *Journal of Ethnopharmacology*, *77*(2–3), 203–208. https://doi.org/10.1016/s0378-8741(01)00299-9

Ahmed, W., Zaki, A., & Nabil, T. (2015). Prevention of methotrexate-induced nephrotoxicity by concomitant administration of garlic aqueous extract in rat. *Turkish Journal of Medical Sciences*, *45*(3), 507–516. https://doi.org/10.3906/sag-1408-121

Ajiboye, T. O., Oyewo, O. A., & Onwudiwe, D. C. (2021). Simultaneous removal of organics and heavy metals from industrial wastewater: A review. *Chemosphere*, *262*, 128379. https://doi.org/10.1016/j.chemosphere.2020.128379

Altınok-Yipel, F., Tekeli, I. O., Özsoy, S. Y., Güvenç, M., Kaya, A., & Yipel, M. (2020). Hepatoprotective activity of linalool in rats against liver injury induced by carbon tetrachloride. *International Journal for Vitamin and Nutrition Research*, *90*(3–4), 302–308. https://doi.org/10.1024/0300-9831/a000581

Angulo-Bejarano, P. I., Puente-Rivera, J., & Cruz-Ortega, R. (2021). Metal and metalloid toxicity in plants: An overview on molecular aspects. *Plants (Basel, Switzerland)*, *10*(4), 635. https://doi.org/10.3390/plants10040635

Asgarpanah, J., Dadashzadeh, M. G., Ahmadi, M., Ranjbar, R., & Safialdin-Ardebily, M. (2012). Chemistry, pharmacology and medicinal properties of *Heracleum persicum* Desf. Ex Fischer: A review. *Journal of Medicinal Plants Research*, *6*(10), 1813–1820. https://doi.org/10.5897/JMPR11.1716

Ashraf, S., Ali, Q., Zahir, Z. A., Ashraf, S., & Asghar, H. N. (2019). Phytoremediation: Environmentally sustainable way for reclamation of heavy metal polluted soils. *Ecotoxicology and Environmental Safety*, *174*, 714–727. https://doi.org/10.1016/j.ecoenv.2019.02.068

Aziza, S. A., Azab, M., & El-Shall, S. K. (2014). Ameliorating role of rutin on oxidative stress induced by iron overload in hepatic tissue of rats. *Pakistan Journal of Biological Sciences: PJBS*, *17*(8), 964–977. https://doi.org/10.3923/pjbs.2014.964.977

Babu, S. M. O. F., Hossain, M. B., Rahman, M. S., Rahman, M., Ahmed, A. S. S., Hasan, M. M., Rakib, A., et al. (2021). Phytoremediation of toxic metals: A sustainable green solution for clean environment. *Applied Sciences*, *11*(21), 10348. http://dx.doi.org/10.3390/app112110348

Badria, F. A., Suliman, S. N., Elsbaey, M., & El-Naggar, M. H. (2018). Bio-guided fractionation and iron chelating activity of agricultural residues. *Drug Discoveries & Therapeutics*, *12*(5), 299–303. https://doi.org/10.5582/ddt.2018.01046

Bahrekazemi, M., Eslami, M., & Nikbakhsh, J. (2020). The effect of dietary coriander supplementation on growth performance, biochemical responses, carcass proximate composition, and heavy metal accumulation in beluga, Huso huso. *Journal of Applied Aquaculture*, *32*. https://doi.org/10.1080/10454438.2020.1782798

Chen, C., Qian, Y., Chen, Q., & Li, C. (2011). Assessment of daily intake of toxic elements due to consumption of vegetables, fruits, meat, and seafood by inhabitants of Xiamen, China. *Journal of Food Science*, *76*(8), T181–T188. https://doi.org/10.1111/j.1750-3841.2011.02341.x

Chen, S., Zhao, X., Wan, J., Ran, L., Qin, Y., Wang, X., Gao, Y., Shu, F., Zhang, Y., Liu, P., Zhang, Q., Zhu, J., & Mi, M. (2015). Dihydromyricetin improves glucose and lipid metabolism and exerts anti-inflammatory effects in nonalcoholic fatty liver disease: A randomized controlled trial. *Pharmacological Research*, *99*, 74–81. https://doi.org/10.1016/j.phrs.2015.05.009

Chung, R. T. (2017). Detoxification effects of phytonutrients against environmental toxicants and sharing of clinical experience on practical applications. *Environmental Science and Pollution Research International, 24*(10), 8946–8956. https://doi.org/10.1007/s11356-015-5263-3

Clarke, D., Buchanan, R., Gupta, N., & Haley, B. (2012). Amelioration of acute mercury toxicity by a novel, non-toxic lipid soluble chelator N,N'bis-(2-mercaptoethyl)isophthalamide: Effect on animal survival, health, mercury excretion and organ accumulation. *Toxicological and Environmental Chemistry, 94*(3), 616–640. https://doi.org/10.1080/02772248.2012.657199

CSIR. (2001). *The Wealth of India - a Dictionary of Indian Raw Materials and Industrial Products.* New Delhi: National Institute of Science Communication, CSIR. Vol. 2, pp. 203–206.

de Freitas, G. R., da Silva, M., & Vieira, M. (2019). Biosorption technology for removal of toxic metals: A review of commercial biosorbents and patents. *Environmental Science and Pollution Research International, 26*(19), 19097–19118. https://doi.org/10.1007/s11356-019-05330-8

de Wuilloud, J. C. A., Wuilloud, R. G., Olsinaa, R. A., & Martinez, L. D. (2002). Separation and preconcentration of inorganic and organomercury species in water samples using a selective reagent and an anion exchange resin and determination by flow injection-cold vapor atomic absorption spectrometry. *Journal of Analytical Atomic Spectrometry, 17*(2), 389–394. https://doi.org/10.1039/B110853K

Deldar, K., Nazemi, E., Mood, M. B., Emami, S., MohammadPour, A., Tafaghodi, M., & Afshari, R. (2008). Effect of *Coriandrum sativum* L. extract on lead excretion in 3-7 year old children. *Journal of Birjand University of Medical Sciences, 15*(3), 11–19. http://journal.bums.ac.ir/article-1-337-en.html

Demir, S., & Korukluoglu, M. (2020). A comparative study about antioxidant activity and phenolic composition of cumin (*Cuminum cyminum* L.) and coriander (*Coriandrum sativum* L.). *Indian Journal of Traditional Knowledge, 19*(2), 383–393.

Dubey, S., Shri, M., Gupta, A., Rani, V., & Chakrabarty, D. (2018). Toxicity and detoxification of heavy metals during plant growth and metabolism. *Environmental Chemistry Letters, 16*, 1169–1192. https://doi.org/10.1007/s10311-018-0741-8

Ebrahimzadeh, M. A., Nabavi, S. M., & Nabavi, S. F. (2009). Correlation between the in vitro iron chelating activity and poly phenol and flavonoid contents of some medicinal plants. *Pakistan Journal of Biological Sciences: PJBS, 12*(12), 934–938. https://doi.org/10.3923/pjbs.2009.934.938

Ebrahimzadeh, M. A., Pourmorad, F., & Bekhradnia, A. R. (2008). Iron chelating activity, phenol and flavonoid content of some medicinal plants from Iran. *African Journal of Biotechnology, 7*(18), 3188–3192.

El-Masry, S., Ali, H. A., El-Sheikh, N. M., & Awad, S. M. (2016). Dose-dependent effect of coriander (*Coriandrum sativum* L.) and fennel (*Foeniculum vulgare* M.) on lead nephrotoxicity in rats. *Journal of Research Studies in Biosciences, 4*, 36–45.

Esmail, A.-S. A., & Ali, T. T., (2019). A review of medicinal plants with nephroprotective effects. *GSC Biological and Pharmaceutical, 8*(1), 114–122. https://doi.org/10.30574/gscbps.2019.8.1.0108

Flora, S. J., Bhadauria, S., Kannan, G. M., & Singh, N. (2007). Arsenic induced oxidative stress and the role of antioxidant supplementation during chelation: A review. *Journal of Environmental Biology, 28*(2 Suppl), 333–347.

Flora, S. J., & Pachauri, V. (2010). Chelation in metal intoxication. *International Journal of Environmental Research and Public Health, 7*(7), 2745–2788. https://doi.org/10.3390/ijerph7072745

Flora, S. J. S. (2009). Metal poisoning: Threat and management. *Al Ameen Journal of Medical Science, 2*(2), 4–26.

Fernandes Azevedo, B., Barros Furieri, L., Peçanha, F. M., Wiggers, G. A., Frizera Vassallo, P., Ronacher Simões, M., Fiorim, J., Rossi de Batista, P., Fioresi, M., Rossoni, L., Stefanon, I., Alonso, M. J., Salaices, M., & Valentim Vassallo, D. (2012). Toxic effects of mercury on the cardiovascular and central nervous systems. *Journal of Biomedicine & Biotechnology, 2012*, 949048. https://doi.org/10.1155/2012/949048

Gaur, N., Kukreja, A., Yadav, M., & Tiwari, A. (2017). Assessment of phytoremediation ability of *Coriander sativum* for soil and water co-contaminated with lead and arsenic: A small-scale study. *3 Biotech, 7*(3), 196. https://doi.org/10.1007/s13205-017-0794-6

Gholamine, B., Houshmand, G., Hosseinzadeh, A., Kalantar, M., Mehrzadi, S., & Goudarzi, M. (2021). Gallic acid ameliorates sodium arsenite-induced renal and hepatic toxicity in rats. *Drug and Chemical Toxicology, 44*(4), 341–352. https://doi.org/10.1080/01480545.2019.1591434

Gong, Z., Xiong, L., Shi, H., Yang, S., Herrera-Estrella, L. R., Xu, G., Chao, D. Y., Li, J., Wang, P. Y., Qin, F., Li, J., Ding, Y., Shi, Y., Wang, Y., Yang, Y., Guo, Y., & Zhu, J. K. (2020). Plant abiotic stress response and nutrient use efficiency. *Science China. Life Sciences, 63*(5), 635–674. https://doi.org/10.1007/s11427-020-1683-x

Gurib-Fakim, A. (2006). Medicinal plants: Traditions of yesterday and drugs of tomorrow. *Molecular Aspects of Medicine, 27*(1), 1–93. https://doi.org/10.1016/j.mam.2005.07.008

Hawula, Z. J., Secondes, E. S., Wallace, D. F., Rishi, G., & Subramaniam, V. N. (2021). The effect of the flavonol rutin on serum and liver iron content in a genetic mouse model of iron overload. *Bioscience Reports*, *41*(7), BSR20210720. https://doi.org/10.1042/BSR20210720

Heidari-Soreshjani, S., Asadi-Samani, M., Yang, Q., & Saeedi-Boroujeni, A. (2017). Phytotherapy of nephrotoxicity-induced by cancer drugs: An updated review. *Journal of Nephropathology*, *6*(3), 254–263. https://doi.org/10.15171/jnp.2017.41

Huang, L., Zeng, X., Li, B., Wang, C., Zhou, M., Lang, H., Yi, L., & Mi, M. (2021). Dihydromyricetin attenuates palmitic acid-induced oxidative stress by promoting autophagy via SIRT3-ATG4B signaling in hepatocytes. *Nutrition & Metabolism*, *18*(1), 83. https://doi.org/10.1186/s12986-021-00612-w

Javaid, R., Aslam, M., Nizami, Q., & Javaid, R. (2012). Role of antioxidant herbal drugs in renal disorders: An overview. *Free Radicals Antioxidants* 2(1), 2–5. https://doi.org/10.5530/ax.2012.2.2

Jia, H. J., Ren, H. F., Endo, H., & Hayashi, T. (2009). Effect of Chinese parsley *Coriandrum sativum* on cadmium binding to proteins from the liver and kidney of rainbow trout *Oncorhynchus mykiss*. *Marine and Freshwater Behaviour and Physiology*, *42*(3), 187–199. https://doi.org/10.1080/10236240903036371

John, N. A. A., Shobana, G., & Keerthana, K. (2014). Protective effect of *Coriander sativum* L. on cadmium induced toxicity in albino rats. *World Journal of Pharmaceutical Sciences*, *3*, 525–534.

Kadiri, L., Lebkiri, A., Rifi, E. H., Ouass, A., Essaadaoui, Y., Lebkiri, I., & Hamad, H. (2018). Kinetic studies of adsorption of Cu (II) from aqueous solution by coriander seeds (*Coriandrum sativum*). *E3S Web of Conferences 37*, 02005. https://doi.org/10.1051/e3sconf/20183702005

Kansal, L., Sharma, V., Sharma, A., Lodi, S., & Sharma, S. H. (2011). Protective role of *Coriandrum sativum* (coriander) extracts against lead nitrate induced oxidative stress and tissue damage in the liver and kidney in male mice. *International Journal of Applied Biology and Pharmaceutical Technology*, *2*, 65–83.

Karunasagar, D., Krishna, M. V., Rao, S. V., & Arunachalam, J. (2005). Removal and preconcentration of inorganic and methyl mercury from aqueous media using a sorbent prepared from the plant *Coriandrum sativum*. *Journal of Hazardous Materials*, *118*(1–3), 133–139. https://doi.org/10.1016/j.jhazmat.2004.10.021

Kendüzler, E., & Türker, A. R. (2005). Optimization of a new resin, Amberlyst 36, as a solid-phase extractor and determination of copper(II) in drinking water and tea samples by flame atomic absorption spectrometry. *Journal of Separation Science*, *28*(17), 2344–2349. https://doi.org/10.1002/jssc.200500212

Komolafe, O., Buzzetti, E., Linden, A., Best, L. M., Madden, A. M., Roberts, D., Chase, T. J., Fritche, D., Freeman, S. C., Cooper, N. J., Sutton, A. J., Milne, E. J., Wright, K., Pavlov, C. S., Davidson, B. R., Tsochatzis, E., & Gurusamy, K. S. (2021). Nutritional supplementation for nonalcohol-related fatty liver disease: A network meta-analysis. *The Cochrane Database of Systematic Reviews*, *7*(7), CD013157. https://doi.org/10.1002/14651858.CD013157.pub2

Krishna, M. V. B., & Karunasagar, D. (2015). Robust ultrasound assisted extraction approach using dilute TMAH solutions for the speciation of mercury in fish and plant materials by cold vapour atomic absorption spectrometry (CVAAS). *Analytical Methods*, *7*(5), 1997–2005. https://doi.org/10.1039/C4AY02114B

Kükner, A., Soyler, G., Toros, P., Dede, G., Meriçli, F., Işık, S., Edebal, O., & Özoğul, C. (2021). Protective effect of *Coriandrum sativum* extract against inflammation and apoptosis in liver ischaemia/reperfusion injury. *Folia Morphologica*, *80*(2), 363–371. https://doi.org/10.5603/FM.a2020.0060

Kumar, S., & Pandey, A. K. (2013). Chemistry and biological activities of flavonoids: An overview. *The Scientific World Journal*, *2013*, 162750. https://doi.org/10.1155/2013/162750

Laribi, B., Kouki, K., M'Hamdi, M., & Bettaieb, T. (2015). Coriander (*Coriandrum sativum* L.) and its bioactive constituents. *Fitoterapia*, *103*, 9–26. https://doi.org/10.1016/j.fitote.2015.03.012

Manoj Kumar, V., Dale, W., Prabhakara Rao, Y., Sharada, R., & Bettaiya, R. (2013). Protective role of *Coriandrum sativum* seed extract against lead-induced oxidative stress in rat liver and kidney. *Current Trends in Biotechnology and Pharmacy*, *7*(2), 650–64.

Martin, N., Harun, M., Nga'nga', M., & Gerald, M. (2020). Using cilantro (*Coriandrum sativum*) to remove cadmium from contaminated water. *World Environment*, *10*(1), 1–9. https://doi.org/10.5923/j.env.20201001.01

Mashayekhi, M. (2012). Renoprotective effect of silymarin on gentamicin-induced nephropathy. *African Journal of Pharmacy and Pharmacology*, *6*(29), 2241–2246. https://doi.org/10.5897/AJPP12.628

Mehrandish, R., Rahimian, A., & Shahriary, A. (2019). Heavy metals detoxification: A review of herbal compounds for chelation therapy in heavy metals toxicity. *Journal of Herbmed Pharmacology*, *8*(2), 69–77. https://doi.org/10.15171/jhp.2019.12

Meunier, B., de Visser, S. P., & Shaik, S. (2004). Mechanism of oxidation reactions catalyzed by cytochrome p450 enzymes. *Chemical Reviews*, *104*(9), 3947–3980. https://doi.org/10.1021/cr020443g

Millet, J. (2006). Cilantro, chlorella, and heavy metals. *Med Herbalism*, *14*, 17–19.

Mirzaei, A., & Khatami, R. (2013). Antioxidant and iron chelating activity of *Coriandrum sativum* and *Petroselinum crispum*. *Pharmacology and Life Sciences*, *2*, 27–31.

Nabi, A., Naeem, M., Aftab, T., Khan, M., & Ahmad, P. (2021). A comprehensive review of adaptations in plants under arsenic toxicity: Physiological, metabolic and molecular interventions. *Environmental Pollution*, *290*, 118029. https://doi.org/10.1016/j.envpol.2021.118029

Najafzadeh, H., Jalali, M. R., Morovvati, H., & Taravati, F. (2010). Comparison of the prophylactic effect of silymarin and deferoxamine on iron overload-induced hepatotoxicity in rat. *Journal of Medical Toxicology*, *6*(1), 22–26. https://doi.org/10.1007/s13181-010-0030-9

Narendrula-Kotha, R., Theriault, G., Mehes-Smith, M., Kalubi, K., & Nkongolo, K. (2020). Metal Toxicity and resistance in plants and microorganisms in terrestrial ecosystems. *Reviews of Environmental Contamination and Toxicology*, *249*, 1–27. https://doi.org/10.1007/398_2018_22

Neha, K., Haider, M. R., Pathak, A., & Yar, M. S. (2019). Medicinal prospects of antioxidants: A review. *European Journal of Medicinal Chemistry*, *178*, 687–704. https://doi.org/10.1016/j.ejmech.2019.06.010

Nicula, M. G., N. P. Dumitrescu, L. Stef, C. Tulcan, M. Dragomirescu, I. Bencsik, S. Patruica, D. Dronca, L. Petculescu-Ciochina, E. Simiz, et al. (2016). Garlic, cilantro and chlorella's effect on kidney histoarchitecture changes in Cd-intoxicated Prussian carp (*Carassius gibelio*). *Animal Science and Biotechnologies*, *49*, 168–177.

Nishio, R., Tamano, H., Morioka, H., Takeuchi, A., & Takeda, A. (2019). Intake of Heated leaf extract of *Coriandrum sativum* contributes to resistance to oxidative stress via decreases in heavy metal concentrations in the kidney. *Plant Foods for Human Nutrition (Dordrecht, Netherlands)*, *74*(2), 204–209. https://doi.org/10.1007/s11130-019-00720-2

Noh, B. K., Lee, J. K., Jun, H. J., Lee, J. H., Jia, Y., Hoang, M. H., Kim, J. W., Park, K. H., & Lee, S. J. (2011). Restoration of autophagy by puerarin in ethanol-treated hepatocytes via the activation of AMP-activated protein kinase. *Biochemical and Biophysical Research Communications*, *414*(2), 361–366. https://doi.org/10.1016/j.bbrc.2011.09.077

Oganesyan, E. T., Nersesyan, Z. M., & Parkhomenko, A. Y. (2007). Chemical composition of the aboveground part of *Coriandrum sativum*. *Pharmaceutical Chemistry Journal*, *41*(3), 149–153. https://doi.org/10.1007/s11094-007-0033-2.

Omura, Y., & Beckman, S. L. (1995). Role of mercury (Hg) in resistant infections & effective treatment of *Chlamydia trachomatis* and Herpes family viral infections (and potential treatment for cancer) by removing localized Hg deposits with Chinese parsley and delivering effective antibiotics using various drug uptake enhancement methods. *Acupuncture & Electro-therapeutics Research*, *20*(3–4), 195–229. https://doi.org/10.3727/036012995816357014

Omura, Y., Shimotsuura, Y., Fukuoka, A., Fukuoka, H., & Nomoto, T. (1996). Significant mercury deposits in internal organs following the removal of dental amalgam, & development of pre-cancer on the gingiva and the sides of the tongue and their represented organs as a result of inadvertent exposure to strong curing light (used to solidify synthetic dental filling material) & effective treatment: A clinical case report, along with organ representation areas for each tooth. *Acupuncture & Electro-Therapeutics Research*, *21*(2), 133–160. https://doi.org/10.3727/036012996816356915

Patrick, L. (2002). Mercury toxicity and antioxidants: Part 1: Role of glutathione and alpha-lipoic acid in the treatment of mercury toxicity. *Alternative Medicine Review*, *7*(6), 456–471.

Rao, R. A. K., & Kashifuddin, M. (2012). Adsorption properties of coriander seed powder (*Coriandrum sativum*): Extraction and pre-concentration of Pb(II), Cu(II) and Zn(II) ions from aqueous solution. *Adsorption Science & Technology*, *30*(2), 127–146. https://doi.org/10.1260/0263-6174.30.2.127

Ratliff, B. B., Abdulmahdi, W., Pawar, R., & Wolin, M. S. (2016). Oxidant mechanisms in renal injury and disease. *Antioxidants & Redox Signaling*, *25*(3), 119–146. https://doi.org/10.1089/ars.2016.6665

Ren, H. F., Jia, H. J., Endo, H., & Hayashi, T. (2009). Cadmium detoxification effect of Chinese parsley *Coriandrum sativum* in liver and kidney of rainbow trout *Oncorhynchus mykiss*. *Fisheries Science*, *75*(3), 731–741. https://doi.org/10.1007/s12562-009-0066-4

Sardar, R., Ahmed, S., Shah, A. A., & Yasin, N. A. (2022). Selenium nanoparticles reduced cadmium uptake, regulated nutritional homeostasis and antioxidative system in *Coriandrum sativum* grown in cadmium toxic conditions. *Chemosphere*, *287*(Pt 3), 132332. Advance online publication. https://doi.org/10.1016/j.chemosphere.2021.132332

Sarkar, R., Hazra, B., & Mandal, N. (2012). Hepatoprotective potential of *Caesalpinia crista* against iron-overload-induced liver toxicity in mice. *Evidence-Based Complementary and Alternative Medicine: eCAM*, *2012*, 896341. https://doi.org/10.1155/2012/896341

Sears, M. E. (2013). Chelation: Harnessing and enhancing heavy metal detoxification-a review. *The Scientific World Journal*, *2013*, 219840. https://doi.org/10.1155/2013/219840

Sengsuk, C., Tangvarasittichai, O., Chantanaskulwong, P., Pimanprom, A., Wantaneeyawong, S., Choowet, A., & Tangvarasittichai, S. (2014). Association of iron overload with oxidative stress, hepatic damage and dyslipidemia in transfusion-dependent β-Thalassemia/HbE patients. *Indian Journal of Clinical Biochemistry: IJCB*, *29*(3), 298–305. https://doi.org/10.1007/s12291-013-0376-2

Sharma, R. K., Rajani, G., Sharma, V., Komala, N. (2011). Effect of ethanolic and aqueous extracts of *Bauhinia variegata* Linn. on gentamicin-induced nephrotoxicity in rats. *Indian Journal of Pharmaceutical Education and Research*, *45*(2), 192–198.

Sharma, V., Kansal, L., & Sharma, A. (2010). Prophylactic efficacy of *Coriandrum sativum* (Coriander) on testis of lead-exposed mice. *Biological Trace Element Research*, *136*(3), 337–354. https://doi.org/10.1007/s12011-009-8553-0

Singletary, K. (2016). Coriander: Overview of potential health benefits. *Nutrition Today*, *51*(3), 151–161. https://doi.org/10.1097/NT.0000000000000159

Sri Laasya, T. P., Thakur, S., Poduri, R., & Joshi, G. (2020). Current insights toward kidney injury: Decrypting the dual role and mechanism involved of herbal drugs in inducing kidney injury and its treatment. *Current Research in Biotechnology*, *2*, 161–175. https://doi.org/10.1016/j.crbiot.2020.11.002

Sreelatha, S., Padma, P. R., & Umadevi, M. (2009). Protective effects of *Coriandrum sativum* extracts on carbon tetrachloride-induced hepatotoxicity in rats. *Food and Chemical Toxicology*, *47*(4), 702–708. https://doi.org/10.1016/j.fct.2008.12.022

Susan, A., Rajendran, K., Sathyasivam, K., & Krishnan, U. M. (2019). An overview of plant-based interventions to ameliorate arsenic toxicity. *Biomedicine & Pharmacotherapy*, *109*, 838–852. https://doi.org/10.1016/j.biopha.2018.10.099

Tagharobiyan, M., & Poozesh, V. (2015). Hydroponic phytoremediation of nickel by coriander (*Coriandrum sativum*). *Journal of Chemical Health Risks*, *5*(4), 273–284. https://doi.org/10.22034/JCHR.2015.544117

Talaei, R., Kheirollah, A., Rezaei, H. B., Mansouri, E., & Mohammadzadeh, G. (2018). The protective effects of hydro-alcoholic extract of *Coriandrum sativum* in rats with experimental iron-overload condition. *Jundishapur Journal of Natural Pharmaceutical Products*, *13*(2):e65028. https://doi.org/10.5812/jjnpp.65028

Téllez-López, M. Á., Mora-Tovar, G., Ceniceros-Méndez, I. M., García-Lujan, C., Puente-Valenzuela, C. O., Vega-Menchaca, M., Serrano-Gallardo, L. B., Garza, R. G., & Morán-Martínez, J. (2017). Evaluation of the chelating effect of methanolic extract of Coriandrum *sativum* and its fractions on Wistar rats poisoned with lead acetate. *African* Journal *of* Traditional, Complementary, *and* Alternative Medicines*: AJTCAM*, *14*(2), 92–102. https://doi.org/10.21010/ajtcam.v14i2.11

Ullah, A., Munir, S., Badshah, S. L., Khan, N., Ghani, L., Poulson, B. G., Emwas, A. H., & Jaremko, M. (2020). Important flavonoids and their role as a therapeutic agent. *Molecules (Basel, Switzerland)*, *25*(22), 5243. https://doi.org/10.3390/molecules25225243

Velaga, M. K., Yallapragada, P. R., Williams, D., Rajanna, S., & Bettaiya, R. (2014). Hydroalcoholic seed extract of *Coriandrum sativum* (Coriander) alleviates lead-induced oxidative stress in different regions of rat brain. *Biological Trace Element Research*, *159*(1–3), 351–363. https://doi.org/10.1007/s12011-014-9989-4

Wan, L., & Jiang, J. G. (2018). Protective effects of plant-derived flavonoids on hepatic injury. *Journal of Functional Foods*, *44*, 283–291. https://doi.org/10.1016/j.jff.2018.03.015

Wang, X., Li, Y., Han, L., Li, J., Liu, C., & Sun, C. (2021). Role of flavonoids in the treatment of iron overload. *Frontiers in Cell and Developmental Biology*, *9*, 685364. https://doi.org/10.3389/fcell.2021.685364

Wardle, E. N. (1999). Antioxidants in the prevention of renal disease. *Renal Failure*, *21*(6), 581–591. https://doi.org/10.3109/08860229909094152

Winarti, S., Pertiwi, C. N., Hanani, A. Z., Mujamil, S. I., Putra, K. A., & Herlambang, K. C. (2018). Beneficial of coriander leaves (*Coriandrum sativum* L.) to reduce heavy metals contamination in rod shellfish. *Journal of Physics: Conference Series*, *953*, 012237.

Wu, Y., Wang, F., Zheng, Q., Lu, L., Yao, H., Zhou, C., Wu, X., & Zhao, Y. (2006). Hepatoprotective effect of total flavonoids from *Laggera alata* against carbon tetrachloride-induced injury in primary cultured neonatal rat hepatocytes and in rats with hepatic damage. *Journal of Biomedical Science*, *13*(4), 569–578. https://doi.org/10.1007/s11373-006-9081-y

Yan, A., Wang, Y., Tan, S. N., Mohd Yusof, M. L., Ghosh, S., & Chen, Z. (2020). Phytoremediation: A promising approach for revegetation of heavy metal-polluted land. *Frontiers in Plant Science*, *11*, 359. https://doi.org/10.3389/fpls.2020.00359

Yan, L. J., & Allen, D. C. (2021). Cadmium-induced kidney injury: Oxidative damage as a unifying mechanism. *Biomolecules*, *11*(11), 1575. https://doi.org/10.3390/biom11111575

Yan, S., Wu, F., Zhou, S., Yang, J., Tang, X., & Ye, W. (2021). Zinc oxide nanoparticles alleviate the arsenic toxicity and decrease the accumulation of arsenic in rice (*Oryza sativa L.*). *BMC Plant Biology, 21*(1), 150. https://doi.org/10.1186/s12870-021-02929-3

Zwolak, I. (2021). Epigallocatechin gallate for management of heavy metal-induced oxidative stress: Mechanisms of action, efficacy, and concerns. *International Journal of Molecular Sciences, 22*(8), 4027. https://doi.org/10.3390/ijms22084027

18 Coriander in Poultry Feed

Kok Ming Goh, Li Ann Lew, and Kar Lin Nyam

CONTENTS

18.1 INTRODUCTION

18.1.1 DESCRIPTION OF CORIANDER

Coriander (*Coriandrum sativum* L.), also known as cilantro or Chinese parsley, is regarded as one of the famous culinary and medicinal spices which belong to the family Apiaceae or Umbelliferae. It is considered an annual herbaceous plant that originated in Mediterranean regions and southern Europe but is now widely grown in many other parts of the world such as Asia, Africa, America, Europe, and the United States. As a photophilic plant, coriander requires intense solar energy (around 900 to 1000 h of light) from germination to flowering. Therefore, it thrives in a high humidity environment and well-drained soil with a pH of 6.2 to 6.8 (Acimovic, Oljaca, Jacimovic, Drazic, & Tasic, 2011).

The whole coriander plant, including leaves, stems, seeds (fruits), and roots, is edible, but the dried seeds and fresh leaves are the most commonly utilized portions, primarily in preparing many different dishes. Coriander has a long history of use. In Europe, coriander is usually used in making pastries, bread, cheese, and processed meat products (sausage and bologna). In Asia, coriander is a staple spice used in Chinese, Indian, and Indonesian cuisines. As a spice, the lemony and floral flavor from coriander leaves contributes an amazing taste and flavor to the dishes. Besides, coriander is also applied as a folk medicine due to its medical capabilities in treating different disorders (Kumar, Marković, Emerald, & Dey, 2016). Coriander has antioxidant activity, anti-inflammatory activity,

DOI: 10.1201/9781003204626-19

antimicrobial activity, antifungal activity, cholesterol-lowering activity, and antidiabetic activity (Bhat, Kaushal, Kaur, & Sharma, 2014; Nadeem et al., 2013). Thus, coriander is also involved in prescription drugs as an herbal alternative.

18.1.2 PHYSICOCHEMICAL PROPERTIES OF CORIANDER

Many people switch from modern medicines to herbal remedies due to lower expenses and lesser side effects. The growing use of herbal medicines has attracted much attention from researchers to investigate their biological activities and health-enhancing potential. There are varying amounts of physicochemical properties in different parts of coriander. The study from Hameed et al. (2017) reported that fresh coriander leaves comprise 85.1% of moisture, 4.29% of carbohydrate, 14.9% of total solids, 3.06% of crude protein, 2.01% of crude fiber, and 4.50% of ash (Hameed, Shafi, & Jan 2017). Fresh coriander leaves also contain trace amount of minerals including calcium (0.66%), potassium (1.62%), iron (0.03%), magnesium (0.52%), and sodium (0.20%). Similar to other green leafy vegetables, coriander leaves are rich in vitamin C (135 mg per 100 g), vitamin B2 (60 mg per 100 g), and vitamin A (12 mg per 100 g). However, it is difficult for them to contribute to nutritional requirements due to the small quantities of usage in food. On the other hand, every 100 g of coriander seed contains nearly 11.3% of moisture, 11.5% of crude protein, 28.4% of crude fiber, 19.1% of lipid, 10.5% of starch, 4.98% of minerals constituents, and 0.84% of essential oil (Nadeem et al., 2013). The high amount of essential oils in coriander (0.23 mL per 100 g) also makes coriander beneficial in different health aspects such as purifying the blood and eliminating gas build up in the body. The essential oil extracted from fully ripe coriander dried seeds is a pale yellow liquid with aromatic flavor contributed by the oil gland from the mericarp. The fatty acid composition that is present in the essential oil mainly includes petroselinic acid (68.8%), linoleic acid (16.6%), oleic acid (7.5%), and palmitic acid (3.8%) (Mandal & Mandal, 2015).

18.2 POULTRY FEED AND CHALLENGES TO THE POULTRY INDUSTRY

Poultry production has been a feature of human agriculture for a long time. With the increasing population size globally and rising individual consumption, the global demand for poultry products has also increased significantly. The poultry industry must ensure a high production volume to support this substantial rise in market demand for poultry meat. Therefore, greater feed volume in both raw ingredients and concentrate compound is required to increase poultry production.

18.2.1 COMPONENTS OF POULTRY FEED

Generally, poultry diets are largely made up of cereal grains, soybean meal, animal by-products, fats, and vitamin and mineral premixes. All the feed ingredients are chosen based on their nutritional content. In addition, a suitable feed should be free from hazardous chemicals and palatable to increase poultry feed intake. The diets, together with water, tend to supply energy for maintaining the body's general metabolism and essential nutrients required for the poultry's growth, reproduction, and health.

Poultry feedstuffs also contain other feed additives such as natural coloring (xanthophylls), antibiotics, and antimicrobial agents. Over the years, antibiotics at subtherapeutic doses have helped the poultry industry grow by eliminating the detrimental effects of various avian diseases. They protect against infectious diseases whose causative agents are bacteria (*Salmonella* and *Escherichia coli*) and as metaphylaxis to minimize an expected disease outbreak. Besides, the administration of antibiotics through poultry feed helps promote poultry's growth rate and optimize feed utilization. Some antibiotics such as tetracycline, ß-lactam penicillin, tylosin, and bacitracin are often used (Alhotan, 2021).

18.2.2 Effects of Antibiotics in Poultry Feed

18.2.2.1 Antibiotic Resistance

Even though antibiotics seemed to benefit poultry farming, many countries have prohibited antibiotics in poultry feed, mainly European countries. The European Union banned feed-grade antibiotic growth promoters in 2006 due to the possibility of drug resistance in pathogenic bacteria after ingestion of antibiotics for a period of time and transference of antibiotic resistance genes from poultry to human microbiota (Castanon, 2007). Antibiotic resistance is the ability of some bacteria to protect themselves against the effects of antibiotics. These resistant variants can become dominant and will proliferate in the poultry. The resistant bacteria are transmitted to other poultry through interaction, thus forming the colonization of antibiotic-resistant bacteria. Skockova et al. (2015) discovered that *Escherichia coli* strains isolated from poultry were resistant to a number of antimicrobial agents, including β-lactams, tetracycline, and quinolones (Skočková, Koláčková, Bogdanovičová, & Karpíšková, 2015). Antibiotic resistance is more likely to develop among commensal bacteria in an exposed population when the antibiotic is used more frequently.

18.2.2.2 Antibiotic Residues in Poultry Products

When poultry is administrated with antibiotics for an extended period, the residue can accumulate in the poultry's tissue and carry over to poultry products such as eggs. Long-term exposure to antibiotic residues through food consumption may affect the human intestinal microbiota, but the tolerance to antibiotic residues is still unclear. For example, a study that evaluated the presence of antibiotics residues in poultry meat revealed that 37 out of 40 samples of retail poultry meat from different markets in Estonia, Latvia, Lithuania, Poland, and France contained enrofloxacin levels ranging from 3.3 to 1126 ng/kg (Pugajeva et al., 2018). Similarly, another survey from Turkey indicated that quinolones were identified in 58 out of 127 chicken samples (45.7%), with a mean level of 30.8 μg/kg (Er et al., 2013). All the obtained findings stress that the poultry industry needs to take necessary actions to limit the widespread use of antibiotics.

18.2.2.3 Environment Contamination

Extensive use of antibiotics in poultry feed has raised environmental concerns because they can enter the environment through various routes, resulting in the contamination of groundwater, surface water, drinking water, and soil. Most of the antibiotics are poorly absorbed by the poultry gut. Subsequently, the antibiotics and their metabolites are eliminated in urine or feces and disposed into the environment. The research found that approximately 75% of ingested antibiotics are excreted as parent compounds or active form through animal wastes. Once they are discharged into the environment, this group of drugs can penetrate the soil and enter groundwater.

Meanwhile, crops also may be contaminated with antibiotic residue when people use contaminated manure as fertilizer. Accumulation of antibiotic residues in crops favors antibiotic resistance among consumers, affecting the germination process and consequently decreasing the yield in farmland (Polianciuc, Gurzău, Kiss, Ştefan, & Loghin, 2020). In addition, their entry rate into the environment is higher than the elimination rate, and it is difficult to remove antibiotics 100% in wastewater treatment plants.

Antibiotic degradation in the environment is very much dependent on their functional groups and interaction with the environmental conditions such as soil pH, water content, temperature, and organic carbon content. The mechanisms of antibiotic degradation can be biotic or non-biotic. Biodegradation involves the use of microorganisms such as fungi and bacteria. The microbial population breaks down macromolecules into smaller molecules that are non-hazardous. The non-biotic process involves hydrolysis, photolysis, reduction, and oxidation reactions to dissipate the active compounds (Kumar et al., 2019). For instance, tetracyclines are light-sensitive and efficiently degrade tetracyclines by photolysis technology (Ahmad, Zhu, & Sun, 2021; Huang et al., 2019).

18.2.2.4   Human Health Concerns

Antibiotic resistance and environmental contamination can lead to severe consequences for public health. Transfer of the bacteria from animal to human is possible through many practices. The primary exposure of humans to resistant bacteria occurs across the food chain, especially in farms and slaughterhouses. Humans may be involved in the cleaning of the feces, which contain the bacteria of the animals in farms. During the cleaning process, humans come into contact with the bacteria on their bodies and hands. If it is not properly cleaned, the person could ingest the bacteria.

Along with the previous sources of contamination, humans can get infected by eating meat from animals with resistant bacteria. Even though adequate cooking reduces the survival of the bacteria, some may still survive and infect a human. Antibiotic-resistant bacteria still represent a concern to human health, especially those sensitive to antibiotics or allergic reactions. However, most antibiotics are relatively safe and there is no conclusive evidence to show that antibiotic residue in poultry meat negatively influences human health. Many previous papers had reported cases of allergic reactions to antibiotics. Antibiotics that include β-lactam, mainly penicillin and cephalosporins, can cause mild allergic reactions such as itchy skin rash and hives. Sometimes, they can also trigger a life-threatening allergic reaction and lead to anaphylaxic conditions (Kyuchukova, 2020).

Baynes et al. (2016) also confirmed the risk of developing an allergic reaction after consuming meat products with penicillin residues, manifesting as allergic dermatitis (Baynes et al., 2016). In addition, a few antibiotic-resistant bacteria known as indicator organisms cause no disease in poultry but cause foodborne disease to humans (Phillips, 2003). Apart from spreading infectious foodborne diseases, previous studies indicated that long-term exposure to tetracyclines may cause mutagenic and carcinogenic effects and interrupt teeth development in young children (Jayalakshmi, Paramasivam, Sasikala, Tamilam, & Sumithra, 2017).

18.2.3   Ban on Antibiotics in Poultry Feed

The use of antibiotics is strictly regulated by the Food and Drug Administration (FDA) and the United States Department of Agriculture (USDA). Due to the emergence of bacterial resistance to antibiotics, which is considered a worldwide public health problem, many countries have prohibited in-feed antibiotics in the poultry industry. Since 2006, the European Union has prohibited the use of feed-grade antibiotic growth promoters, and the FDA also has withdrawn the use of certain antibiotics for enhanced food production since 2000 (Castanon, 2007). When antibiotics are phased out from poultry production, other problems typically occur. One of the biggest challenges is preventing infectious disease and ensuring sustainable poultry production without in-feed antibiotics.

18.2.4   Future Trend of Antibiotic Use in Poultry Feed

Nowadays, consumers are more health-conscious and concerned about their diets. This is indicated by a consumer survey conducted in Thailand. The results revealed that 62% of participants (313 participants) would look closely at product labels, shift toward products "grown without antibiotics", and inquire about the usage of phytogenic feed additives. Moreover, the survey found that 9 out of 10 consumers preferred poultry fed with natural ingredients instead of antibiotics (Delacon, 2018). With the increasing global demand for poultry meat and products, there is an urgent need for poultry feed manufacturers to replace synthetic agents with similar or better beneficial effects. The trend toward reducing antibiotic growth promoters has made phytogenic feed additives generalized and well accepted. Phytogenic feed additives consist of substances of plant origin, such as plant extracts and their bioactive compounds.

18.3 BENEFITS OF USING CORIANDER IN POULTRY FEED

Researchers also found that coriander may emerge as an alternative feed additive and potential antibiotic replacement in poultry diets as a multi-purpose plant. Researchers discovered that these phytogenic feed additives are effective in maintaining poultry health. Moreover, the high content of fats and protein in the coriander seed makes distillation residues suitable for animal feed. Coriander powder or its extract is also found to contribute multiple health benefits.

18.3.1 IMPROVE GASTROINTESTINAL HEALTH

Coriander has been discovered to be associated with intestinal health enhancement. Many investigations showed that coriander is useful in modulating gut microbiota. Gut microbiota plays a significant functional role in maintaining intestinal integrity and influencing poultry nutrition and development. In addition, a healthy gut microflora may serve as an ideal barrier to protect against pathogenic microorganisms by reducing the opportunity for colonization of pathogenic microorganisms.

The study from Hosseinzadeh et al. (2014) observed that broiler chicks fed with a diet containing 1.5% coriander powder had a significantly low population of *E.coli* in their ileum. On the other hand, the population of beneficial bacteria such as *Lactobacillus* bacteria was numerically higher at 21 and 42 days of age (Hosseinzadeh, Alaw Qotbi, Seidavi, Norris, & Brown, 2014). The results are in agreement with an analysis which revealed that coriander essential oil showed effective antibacterial activity against ten bacterial strains (*Staphylococcus aureus* SA08, *Staphylococcus aureus* ATCC 25923, MRSA 10/08, MRSA 12/08, *Pseudomonas aeruginosa* ATCC 27853, *Klebsiella pneumoniae* ATCC 13883, *Escherichia coli* ATCC 25922, *Salmonella typhimurium* ATCC 13311, *Acinetobacter baumannii* 2/10, *Acinetobacter baumannii* 3/10). The excellent antimicrobial activity of coriander is contributed by the presence of phytochemicals that have a defensive function against pathogenic microorganisms such as linalool (Silva, Ferreira, Queiroz, & Domingues, 2011). Linalool is an acyclic monoterpene tertiary alcohol and is one of the most studied odorant molecules. It helps to inhibit the growth of certain bacteria by disrupting the bacteria membrane and suppressing the regulatory gene products. According to research, linalool is more effective against Gram-positive bacteria compared to Gram-negative bacteria (Aprotosoaie, Hăncianu, Costache, & Miron, 2014). In addition, linalool has been explored to inhibit fungal activities such as dermatophytes, which cause skin infections (Costa et al., 2013).

Furthermore, dietary supplementation of coriander was reported to improve gut morphology. Research from Ghazanfari et al. (2015) showed that 100–300 mg/kg of coriander oil increased villus height and crypt depth, and reduced epithelial thickness in the small intestine of the broiler chicken (Ghazanfari, Mohammadi, & Adib Moradi, 2015). Another study also confirmed that broiler chickens fed with coriander additives had improved villus length, crypt depth, and jejunum morphology compared to the control group (Hady, Zaki, Abd El-Ghany, & Korany Reda, 2016). The gut mucosa is the first tissue to contact with food particles, and the intestine can adapt to react morphologically in response to dietary changes. Long villi and deep intestinal crypt can help digestion and optimize nutrient absorption. In addition, long villi and crypts provide a large absorptive surface area for nutrients to be transferred into the bloodstream (Mohiti-Asli & Ghanaatparast-Rashti, 2017). Therefore, intestine structure is highly associated with nutrient uptake and may affect the growth performance of poultry.

Hosseinzadeh et al. (2014) also noticed that dietary inclusion of coriander powder has better results in gut health than including the coriander extract in drinking water (Hosseinzadeh et al., 2014). The factor could be due to environmental factors that may influence plant extract's chemical composition (Naeemasa et al., 2015).

18.3.2 IMPROVE GROWTH PERFORMANCE

Coriander exhibited a positive impact on growth parameters. This statement was supported by several studies that indicated that final body weights and total weight gains of broiler chicks in coriander groups were higher than control groups by the end of the experiment (Ghazanfari et al., 2015; Hady et al., 2016; Hosseinzadeh et al., 2014; Taha et al., 2019).

Improvement in growth metrics can be related to the increase in feed intake. According to a study by Naeemasa et al. (2015), total feed intake (TFI) of broilers fed with 1000 mg/kg coriander was significantly higher when compared to broilers fed with a regular diet (108 g/bird/day versus 102 g/bird/day) (Naeemasa et al., 2015). A similar finding was reported by Taha et al. (2019), which reported that broilers that received a diet with 0.2% and 0.4% coriander seed powder had significantly higher TFI than broilers that received a regular diet with 0.1% coriander seed powder (Taha et al., 2019). These positive effects were also applied on broilers grown under high ambient temperature. A study from Iraq observed that growth performance parameters and feed consumption of broilers during summer months were significantly higher in 2% coriander seed diets compared with other treatment groups (Hamodi, Al-Mashhad, Al-Jaff, & Al-Mashhad, 2010). Heat stress is one of the common problems in poultry farms. It occurs when birds are exposed to high temperatures and humidity. Birds cannot balance body heat production and the elimination of heat (Wasti, Sah, & Mishra, 2020). When birds are under heat stress, the research found a 28.6% reduction in daily feed intake (Lara & Rostagno, 2013).

Despite that, a study from EL-Shoukary et al. (2014) explained that heat stressed broilers supplemented with 2% coriander seeds manifested significantly improved feeding behavior (EL-Shoukary, Darwish, & Abdel-Rahman, 2014). Researchers observed that phytogenic feed additives induced an appetizing effect in the poultry diet. The linalool compounds and other monoterpenoids can enhance flavor and increase diet consumption. As mentioned earlier, linalool not only possesses beneficial effects on gut morphology, resulting in higher nutrient absorption. Linalool compounds may also stimulate the digestive system activities by improving liver function and increasing digestive enzymes, which helps accelerate carbohydrate and protein metabolism (Almremdhy, 2020). An animal study revealed that lipase and amylase activities were 26% and 36% higher than control when Wistar rats were administrated with 1.0 mg/mL of coriander powder (Ramakrishna Rao, Platel, & Srinivasan, 2003). Thus, the results proved that coriander powder or extract could replace synthetic antibiotics as growth promoters.

18.3.3 INCREASE CARCASS YIELD

Some studies reported the influence of coriander as a feed additive on carcass yield. Farag (2013) revealed that adding 0.6% coriander seed in diets obtained a higher carcass yield compared to other experiment groups (0.2% and 0.4% coriander seed). The result agrees with another study that fed broilers with coriander seed diets for 42 days. Broilers fed with a 0.3% coriander seed diet exhibited larger carcass yields than the control group (Saeid & Al-Nasry, 2010). The results indicated that supplement coriander in poultry diets effectively promotes poultry growth. The increasing effect in carcass yield could be attributed to the high feed conversion ratio and daily feed intake in birds. Improved carcass yield is also induced by the stimulating effect of phytoconstituents on fat and protein metabolism (Naeemasa et al., 2015).

18.3.4 STIMULATE IMMUNE RESPONSE

Coriander has been reported to have noticeable immunomodulatory properties. It is crucial to strengthen the immunity system of poultry to provide them with a better defense mechanism for reducing the occurrence of illness.

A study examining the effects of coriander extract in drinking water at concentrations of 750 ppm, 1000 ppm, and 1250 ppm on the immune response observed a significant enhancement in antibody titer against Newcastle disease at 42 days of age. However, there were no significant differences between the treatment diet with coriander powder and the control diet without coriander (Hosseinzadeh et al., 2014). The non-significant effect may be due to biosecurity measures throughout the experiment, but further research was required to confirm this observation.

Linalool has been found to stimulate interferon-gamma (INF-γ) secretion. INF-γ is a proinflammatory cytokine mainly secreted by natural killer cells and T cells. Its major role is to promote cellular immunity activation, such as activating macrophages and stimulating the antitumor response (Castro, Cardoso, Gonçalves, Serre, & Oliveira, 2018). Furthermore, the researchers discovered that linalool has excellent suppression effects against cancer cells. Based on the results obtained, linalool suppressed the growth of A549 (lung cancer cells), SW620 (colon cancer cells), and Hep G2 (liver cancer cells) effectively with IC_{50} values of 438, 222, and 290 µM, respectively (Chang & Shen, 2014). Researchers have recently conducted research to determine the possibility of adding natural anti-cancer agents such as coriander to poultry feed to produce anti-cancer poultry meat and eggs. The idea is that if consumers consumed meat or eggs, they would benefit from the anti-cancer properties.

In addition, a healthy gut environment can contribute to a strong immune system. Balanced gut microbiota has profound effects on innate and adaptive immune systems and achieves immune homeostasis (Wu & Wu, 2014). For example, bacteria such as *Roseburia intestinalis* and *Faecalibacterium prausnitzii* are responsible for fermenting carbohydrates and producing short-chain fatty acids that modulate host immune cells and regulate T cell development (Yoo, Groer, Dutra, Sarkar, & McSkimming, 2020).

18.3.5 IMPACTS ON HEMATOLOGICAL PARAMETERS

Coriander has pronounced effects on specific hematological parameters. According to a study by Taha et al. (2019), broilers administrated with coriander seed powder at concentrations of 0.1%, 0.2%, and 0.4% had significantly higher hemoglobin concentration (Hb), hematocrit value (PCV), and platelet counts (PLT) than those broilers in the non-supplemented group. Besides, red blood cell (RBC) counts significantly increased when broilers were fed with diets containing 0.2% and 0.4% coriander seed powder (Taha et al., 2019). The results obtained agree with other studies that stated the PCV, Hb, and RBC count of chicken supplemented with coriander diets were improved significantly compared to the control group (Farag, 2013; Saeid & Al-Nasry, 2010).

The improvement in these hematological parameters was related to the improvement of metabolism and enhancement of nutritional absorption as the gut morphology is improved (Taha et al., 2019). Moreover, coriander seed is abundant in iron, zinc, and copper, the essential minerals for RBC production. Based on the USDA report, coriander leaf and seed contain 16.3 mg and 42.4 mg of iron, respectively (USDA, 2015). Therefore, in some parts of the world, coriander is consumed to reduce the risk of anemia.

For white blood cells (WBCs), a study from Khubeiz and Shirif (2020) showed that the inclusion of 2.5% coriander seed powder in diets increased the basophil and eosinophil cell count (Khubeiz & Shirif, 2020). Likewise, broilers fed with 1.5% coriander seed powder indicated higher lymphocytes than those without coriander seed powder. Nevertheless, there was no significant effect in monocyte cells. Dietary coriander seeds showed a positive correlation on basophils, eosinophils, and lymphocytes due to sterol and tocopherol compounds that coriander seeds contain. The sterol and tocopherol compounds have antioxidant ability, which suppresses the free radicals (Khamisabadi & Ahmadpanah, 2020).

Apart from that, broilers fed 1.5% coriander seeds showed a lower heterophils/lymphocytes (H/L) ratio than those on a control diet. These findings correspond with research from Farag (2013), which revealed coriander seed treatment groups (0.2%, 0.4%, and 0.6%) had a lower H/L ratio than the control group (Farag, 2013). Measuring the ratio of H/L is a standard tool for assessing long-term stress in birds. The greater H/L ratio indicated the birds had a higher level of stress.

18.3.6 IMPACTS ON SERUM BIOCHEMISTRY PARAMETER

Also, dietary coriander has been shown to affect serum biochemical parameters. For example, several studies demonstrated that the inclusion of coriander seed or coriander oil effectively reduced the level of blood glucose compared to the control group (Al-Jaff, 2011; Al-Mashhadani, Al-Jaff, Hamodi, & Al-Mashhadani, 2011; Hosseinzadeh et al., 2014). For example, according to Al-Jaff's findings, blood glucose levels in broilers receiving diets plus 2% and 3% coriander seed were reduced by 7.88% and 4.72%, respectively (Al-Jaff, 2011).

The hypoglycemic effect of coriander is related to its phytoconstituents, particularly phenolics. Phenolic compounds aid the regeneration of β-cells and restore insulin response, thereby suppressing the glucose level. Another animal study that examined the hypoglycemic action of coriander leaf extract in alloxan-induced diabetic rats had proven that 200 mg/kg of coriander leaf extract reduced the glucose level significantly compared to a diabetic control group (128 mg/dl versus 260 mg/dl). Furthermore, coriander leaf extract had a hypoglycemic effect comparable to the 5 mg/kg glibenclamide treated group (Sreelatha & Inbavalli, 2012). These results were in line with the findings from Aligita et al. (2018). In addition, coriander leaf extract is effective in inhibiting the α-glucosidase enzymes. Based on an experiment that analyzed the inhibition of α-glucosidase enzyme activity, the IC_{50} value for coriander leaf extract was 32,376 ppm, while the standard drug (acarbose) value was 82,272 ppm (Aligita, Susilawati, Septiani, & Atsil, 2018). The results indicated that a lower coriander leaf extract concentration is needed to inhibit 50% of the enzyme activity.

The inclusion of coriander in poultry diets showed a hypolipidemic effect as well. Research from Hosseinzadeh et al. (2014) found that birds that received diets containing 2% coriander powder had a lower total cholesterol (TC) level and low-density lipoproteins (LDL) cholesterol than birds that received normal diets (Hosseinzadeh et al., 2014). These findings are consistent with several previous studies. According to Al-Jaff (2011), the addition of 2% coriander seed reduced cholesterol levels by about 8.27% (Al-Jaff, 2011). The cholesterol-reducing effect is also shown in heat-stressed broilers. A study from Al-Mashhadani et al. (2011) mentioned that administration of 0.5% and 1.0% coriander oil resulted in lower cholesterol levels. The cholesterol levels for the control group, 0.5% coriander group, and 1.0% coriander group after 42 days were 187 mg/100 mL, 176 mg/100 mL, and 175 mg/100 mL, respectively (Al-Mashhadani et al., 2011). Farag (2013) reported a similar finding in which the levels of serum triglycerides and cholesterol in broiler chickens were shown to decrease after being fed with 0.2%, 0.4%, and 0.6% coriander seed diets (Farag, 2013).

The use of coriander in animal feed is observed to influence lipid metabolism. The phytoconstituents from the coriander can decrease the activity of HMG-CoA reductase in the liver. This enzyme is recognized as a rate-limiting enzyme, and it is responsible for liver cholesterol synthesis (Kostner et al., 1989). In poultry, suppressing HMG-CoA reductase by 5% resulted in a 2% reduction in blood cholesterol (Case, He, Mo, & Elson, 1995). In rat research, supplementation with 10% coriander seed also caused a significant increase in plasma lecithin cholesterol acyl transferase (LCAT) activity (Dhanapakiam, Joseph, Ramaswamy, Moorthi, & Kumar, 2007). Plasma LCAT is an enzyme that removes excess cholesterol from blood and tissues through cholesterol esterification. Plasma LCAT activity is negatively correlated with total cholesterol, triglycerides, and LDL cholesterol (Dullaart, Perton, Sluiter, de Vries, & van Tol, 2008).

Furthermore, there is an association between free radicals and increased blood glucose and blood cholesterol. High ambient temperature may also elevate oxidative stress in birds. Hence, coriander can be a useful feed additive in poultry diets to reduce thermal stress as it is an antioxidant-rich herb (Abd El-Hack et al., 2020). According to Sreelatha and Inbavalli (2012), coriander seeds ameliorated oxidative stress in hyperlipidemia and diabetic rats (Sreelatha & Inbavalli, 2012). This statement can be explained by the increased level of superoxide dismutase (SOD), and malondialdehyde (MDA) concentration decreased after the diabetic rats received coriander extract treatment.

18.4 CONCLUSION

Based on our findings, coriander used as a feed additive in poultry diets significantly improves gut health, growth performance, and immunity in poultry cultivation. In addition, coriander has an equal or better impact than synthetic antibiotics as a phytogenic feed component. Meanwhile, this natural agent's use also protects the environment and consumer health. Overall, coriander offers sustainable therapeutic and performance aids to poultry health and production.

REFERENCES

Abd El-Hack, M. E., Abdelnour, S. A., Taha, A. E., Khafaga, A. F., Arif, M., Ayasan, T., … Aleya, L. (2020). Herbs as thermoregulatory agents in poultry: An overview. *Science of the Total Environment, 703*, 134399.

Acimovic, M., Oljaca, S., Jacimovic, G., Drazic, S., & Tasic, S. (2011). Benefits of environmental conditions for growing coriander in Banat Region, Serbia. *Natural Product Communications, 6*(10), 1934578X1100601014.

Ahmad, F., Zhu, D., & Sun, J. (2021). Environmental fate of tetracycline antibiotics: Degradation pathway mechanisms, challenges, and perspectives. *Environmental Sciences Europe, 33*(1). doi: 10.1186/s12302-021-00505-y

Al-Jaff, F. K. (2011). Effect of coriander seeds as diet ingredient on blood parameters of broiler chicks raised under high ambient temperature. *International Journal of Poultry Science, 10*(2), 82–86.

Al-Mashhadani, E. H., Al-Jaff, F. K., Hamodi, S. J., & Al-Mashhadani, H. E. (2011). Effect of different levels of coriander oil on broiler performance and some physiological traits under summer condition. *Pakistan Journal of Nutrition, 10*(1), 10–14.

Alhotan, R. (2021). Commercial poultry feed formulation: Current status, challenges, and future expectations. *World's Poultry Science Journal, 77*, 1–21.

Aligita, W., Susilawati, E., Septiani, H., & Atsil, R. (2018). Antidiabetic activity of coriander (Coriandrum *sativum* L.) leaves' ethanolic extract. *International Journal of Pharmaceutical and Phytopharmacological Research, 8*, 59–63.

Almremdhy, H. A. A.-e. (2020). The effect of coriander oil in drinking water of broiler chickens in growth performance and immune response. *Plant Archives 20*(Supplement 1), 1999–2002.

Aprotosoaie, A. C., Hăncianu, M., Costache, I.-I., & Miron, A. (2014). Linalool: A review on a key odorant molecule with valuable biological properties. *Flavour and Fragrance Journal, 29*(4), 193–219. doi: 10.1002/ffj.3197

Baynes, R. E., Dedonder, K., Kissell, L., Mzyk, D., Marmulak, T., Smith, G., … Riviere, J. E. (2016). Health concerns and management of select veterinary drug residues. *Food and Chemical Toxicology, 88*, 112–122. doi: 10.1016/j.fct.2015.12.020

Bhat, S., Kaushal, P., Kaur, M., & Sharma, H. (2014). Coriander (*Coriandrum sativum* L.): Processing, nutritional and functional aspects. *African Journal of Plant Science, 8*(1), 25–33.

Case, G. L., He, L., Mo, H., & Elson, C. E. (1995). Induction of geranyl pyrophosphate pyrophosphatase activity by cholesterol-suppressive isoprenoids. *Lipids, 30*(4), 357–359.

Castanon, J. I. R. (2007). History of the use of antibiotic as growth promoters in European poultry feeds. *Poultry Science, 86*(11), 2466–2471. doi: 10.3382/ps.2007-00249

Castro, F., Cardoso, A. P., Gonçalves, R. M., Serre, K., & Oliveira, M. J. (2018). Interferon-gamma at the crossroads of tumor immune surveillance or evasion. *Frontiers in Immunology, 9*. doi: 10.3389/fimmu.2018.00847

Chang, M.-Y., & Shen, Y.-L. (2014). Linalool exhibits cytotoxic effects by activating antitumor immunity. *Molecules, 19*(5), 6694–6706. doi: 10.3390/molecules19056694

Costa, D. C. M., Vermelho, A. B., Almeida, C. A., Dias, E. P. d. S., Cedrola, S. M. L., Arrigoni-Blank, M. d. F., … Alviano, D. S. (2013). Inhibitory effect of linalool-rich essential oil from *Lippia alba* on the peptidase and keratinase activities of dermatophytes. *Journal of Enzyme Inhibition and Medicinal Chemistry, 29*(1), 12–17. doi: 10.3109/14756366.2012.743537

Delacon. (2018). Survey suggests Thai Millennials would choose meat raised with phytogenic feed additives. Retrieved 7 August 2021. https://benisonmedia.com/survey-suggests-thai-millennials-would-choose-meat-raised-with-phytogenic-feed-additives/

Dhanapakiam, P., Joseph, J. M., Ramaswamy, V., Moorthi, M., & Kumar, A. S. (2007). The cholesterol lowering property of coriander seeds (*Coriandrum sativum*): Mechanism of action. *Journal of Environmental Biology, 29*(1), 53.

Dullaart, R. P. F., Perton, F., Sluiter, W. J., de Vries, R., & van Tol, A. (2008). Plasma lecithin: Cholesterol acyltransferase activity is elevated in metabolic syndrome and is an independent marker of increased carotid artery intima media thickness. *The Journal of Clinical Endocrinology & Metabolism, 93*(12), 4860–4866. doi: 10.1210/jc.2008-1213

El-Shoukary, R. D., Darwish, M. H., & Abdel-Rahman, M. A. (2014). Behavioral, performance, carcass traits and hormonal changes of heat stressed broilers feeding black and coriander seeds. *Journal of Advanced Veterinary Research*, *4*(3), 97–101.

Er, B., Onurdağ, F. K., Demirhan, B., Özgacar, S. Ö., Öktem, A. B., & Abbasoğlu, U. (2013). Screening of quinolone antibiotic residues in chicken meat and beef sold in the markets of Ankara, Turkey. *Poultry Science*, *92*(8), 2212–2215. doi: 10.3382/ps.2013-03072

Farag, S. A. (2013). The efficiency of coriander seeds as dietary additives in broiler chicken's diets. Egyptian Journal of Nutrition and Feeds, *16*, 491–501.

Ghazanfari, S., Mohammadi, Z., & Adib Moradi, M. (2015). Effects of coriander essential oil on the performance, blood characteristics, intestinal microbiota and histological of broilers. *Revista Brasileira de Ciência Avícola*, *17*(4), 419–426. doi: 10.1590/1516-635x1704419-426

Hady, M., Zaki, M., Abd El-Ghany, W., & Korany Reda, M. (2016). Assessment of the broilers performance, gut healthiness and carcass characteristics in response to dietary inclusion of dried coriander, turmeric and thyme. *International Journal of Environmental & Agriculture Research*, *2*, 153–159.

Hameed, O. B., Shafi, F., & Jan, N. (2017). Studies on nutritional composition of coriander leaves by using sun and cabinet drying methods. *International. Journal of Chemical Studies*, *5*(6), 12–14.

Hamodi, S. J., Al-Mashhad, E. H., Al-Jaff, F. K., & Al-Mashhad, H. E. (2010). Effect of coriander seed (*Coriandrum sativum* L.) as diet ingredient on broilers performance under high ambient temperature. *International Journal of Poultry Science*, *9*(10), 968–971. doi: 10.3923/ijps.2010.968.971

Hosseinzadeh, H., Alaw Qotbi, A. A., Seidavi, A., Norris, D., & Brown, D. (2014). Effects of different levels of coriander (*Coriandrum sativum*) seed powder and extract on serum biochemical parameters, microbiota, and immunity in broiler chicks. *The Scientific World Journal*, *2014*, 1–11. doi: 10.1155/2014/628979

Huang, S.-T., Lee, S.-Y., Wang, S.-H., Wu, C.-Y., Yuann, J.-M. P., He, S., … Liang, J.-Y. (2019). The influence of the degradation of tetracycline by free radicals from riboflavin-5′-phosphate photolysis on microbial viability. *Microorganisms*, *7*(11), 500. doi: 10.3390/microorganisms7110500

Jayalakshmi, K., Paramasivam, M., Sasikala, M., Tamilam, T., & Sumithra, A. (2017). Review on antibiotic residues in animal products and its impact on environments and human health. *Journal of Entomology and Zoology Studies*, *5*(3), 1446–1451.

Khamisabadi, H., & Ahmadpanah, J. (2020). The effect of diets supplemented with *Coriandrum sativum* seeds on carcass performance, immune system, blood metabolites, rumen parameters and meat quality of lambs. *Acta Scientiarum. Animal Sciences*, *43*, e52048. doi: 10.4025/actascianimsci.v43i1.52048

Khubeiz, M. M., & Shirif, A. M. (2020). Effect of coriander (*Coriandrum sativum* L.) seed powder as feed additives on performance and some blood parameters of broiler chickens. *Open Veterinary Journal*, *10*(2). doi: 10.4314/ovj.v10i2.9

Kostner, G. M., Gavish, D., Leopold, B., Bolzano, K., Weintraub, M. S., & Breslow, J. L. (1989). HMG CoA reductase inhibitors lower LDL cholesterol without reducing Lp (a) levels. *Circulation*, *80*(5), 1313–1319.

Kumar, M., Jaiswal, S., Sodhi, K. K., Shree, P., Singh, D. K., Agrawal, P. K., & Shukla, P. (2019). Antibiotics bioremediation: Perspectives on its ecotoxicity and resistance. *Environment International*, *124*, 448–461. doi: 10.1016/j.envint.2018.12.065

Kumar, V., Marković, T., Emerald, M., & Dey, A. (2016). Herbs: Composition and dietary importance. 332–337. doi: 10.1016/b978-0-12-384947-2.00376-7

Kyuchukova, R. (2020). Antibiotic residues and human health hazard-review. *Bulgarian Journal of Agricultural Science*, *26*(3), 664–668.

Lara, L., & Rostagno, M. (2013). Impact of heat stress on poultry production. *Animals*, *3*(2), 356–369. doi: 10.3390/ani3020356

Mandal, S., & Mandal, M. (2015). Coriander (*Coriandrum sativum* L.) essential oil: Chemistry and biological activity. *Asian Pacific Journal of Tropical Biomedicine*, *5*(6), 421–428. doi: 10.1016/j.apjtb.2015.04.001

Mohiti-Asli, M., & Ghanaatparast-Rashti, M. (2017). Comparing the effects of a combined phytogenic feed additive with an individual essential oil of oregano on intestinal morphology and microflora in broilers. *Journal of Applied Animal Research*, *46*(1), 184–189. doi: 10.1080/09712119.2017.1284074

Nadeem, M., Anjum, F. M., Khan, M. I., Tehseen, S., El-Ghorab, A., & Sultan, J. I. (2013). Nutritional and medicinal aspects of coriander (*Coriandrum sativum* L.): A review. *British Food Journal*. 115(5), 743–755.

Naeemasa, M., Qotbi, A. A. A., Seidavi, A., Norris, D., Brown, D., & Ginindza, M. (2015). Effects of coriander (*Coriandrum sativum* L.) seed powder and extract on performance of broiler chickens. *South African Journal of Animal Science*, *45*(4), 371. doi: 10.4314/sajas.v45i4.3

Phillips, I. (2003). Does the use of antibiotics in food animals pose a risk to human health? A critical review of published data. *Journal of Antimicrobial Chemotherapy*, *53*(1), 28–52. doi: 10.1093/jac/dkg483

Polianciuc, S. I., Gurzău, A. E., Kiss, B., Ştefan, M. G., & Loghin, F. (2020). Antibiotics in the environment: Causes and consequences. *Medicine and Pharmacy Reports, 93*(3), 231.

Pugajeva, I., Avsejenko, J., Judjallo, E., Bērziņš, A., Bartkiene, E., & Bartkevics, V. (2018). High occurrence rates of enrofloxacin and ciprofloxacin residues in retail poultry meat revealed by an ultra-sensitive mass-spectrometric method, and antimicrobial resistance to fluoroquinolones in *Campylobacter* spp. *Food Additives & Contaminants: Part A, 35*(6), 1107–1115. doi: 10.1080/19440049.2018.1432900

Ramakrishna Rao, R., Platel, K., & Srinivasan, K. (2003). In vitro influence of spices and spice-active principles on digestive enzymes of rat pancreas and small intestine. *Food/Nahrung, 47*(6), 408–412.

Saeid, J., & Al-Nasry, A. (2010). Effect of dietary coriander seeds supplementation on growth performance carcass traits and some blood parameters of broiler chickens. *International Journal of Poultry Science, 9*(9), 867–870.

Silva, F., Ferreira, S., Queiroz, J. A., & Domingues, F. C. (2011). Coriander (*Coriandrum sativum* L.) essential oil: Its antibacterial activity and mode of action evaluated by flow cytometry. *Journal of Medical Microbiology, 60*(10), 1479–1486. doi: 10.1099/jmm.0.034157-0

Skočková, A., Koláčková, I., Bogdanovičová, K., & Karpíšková, R. (2015). Characteristic and antimicrobial resistance in *Escherichia coli* from retail meats purchased in the Czech Republic. *Food Control, 47*, 401–406. doi: 10.1016/j.foodcont.2014.07.034

Sreelatha, S., & Inbavalli, R. (2012). Antioxidant, antihyperglycemic, and antihyperlipidemic effects of *Coriandrum sativum* leaf and stem in alloxan-induced diabetic rats. *Journal of Food Science, 77*(7), T119–T123. doi: 10.1111/j.1750-3841.2012.02755.x

Taha, A. E., Hassan, S. S., Shewita, R. S., El-seidy, A. A., Abd El-Hack, M. E., Hussein, E.-s. O. S., … El-Edel, M. A. (2019). Effects of supplementing broiler diets with coriander seed powder on growth performance, blood haematology, ileum microflora and economic efficiency. *Journal of Animal Physiology and Animal Nutrition, 103*(5), 1474–1483. doi: 10.1111/jpn.13165

USDA. (2015). National nutrient database for standard reference, release 28. Nutrients: Vitamin C, total ascorbic acid.

Wasti, S., Sah, N., & Mishra, B. (2020). Impact of heat stress on poultry health and performances, and potential mitigation strategies. *Animals, 10*(8), 1266. doi: 10.3390/ani10081266

Wu, H.-J., & Wu, E. (2014). The role of gut microbiota in immune homeostasis and autoimmunity. *Gut Microbes, 3*(1), 4–14. doi: 10.4161/gmic.19320

Yoo, J., Groer, M., Dutra, S., Sarkar, A., & McSkimming, D. (2020). Gut microbiota and immune system interactions. *Microorganisms, 8*(10), 1587. doi: 10.3390/microorganisms8101587

19 Coriander in Fish Feed

Nakul Ravishankar, Anant Kumar, and Gayathri Mahalingam

CONTENTS

19.1 A BRIEF INTRODUCTION TO FISH FEED

Aquaculture, also referred to as aquatic agriculture, is the culture of aquatic animals and plant species in fresh, brackish, marine, and hypersaline waters, aiming to increase seafood production. Aquaculture is not traditional fishing as it is more focused on manipulating parameters to boost production. About 50% of aquaculture consists of fisheries (Lucas, 2015).

Aquaculture is an integral part of the agricultural economy. Global production of aquaculture has witnessed a great boom in the last three decades (Kong et al., 2020). According to the Food and Agriculture Organization of the United Nations, global aquaculture production witnessed an increase of 521% between 1990 and 2018. During the same period, global capture-fisheries production increased by 14%, and consumption has increased up to 122% (*The State of World Fisheries and Aquaculture 2020*). It is evident from the data that capture-fisheries production exhibited an increasing trend, but it could not match the increase in the consumption growth rate. Thus, the sustainability and effectiveness of fisheries development are the key parameters to balance these trends.

Fisheries is not a worldwide industry; it is mostly limited to a few countries like China, Indonesia, Peru, India, Russia, the USA, and Vietnam (*The State of World Fisheries and Aquaculture 2020*). However, these eight countries account for almost 50% of the total capture production.

All these countries are promoting this industry to meet consumption demands and due to the number of jobs it generates. In a country like India, the fisheries sector constitutes 7.28% of the total profit of the agriculture economy.

The industry has focused more on finding and developing ways to increase production. However, the type and quality of the substance used in fish feed are prerequisites for ensuring sustainable

DOI: 10.1201/9781003204626-20

production. Therefore, although industries always try to achieve exponential production, our discussions in this chapter are restricted to sustainability. In that sense, high-quality fish feed is essential for developing fish and reducing the harmful environmental impacts induced by intensive aquaculture.

Indeed the aquaculture industry is dominated by fisheries. Moreover, the production quality mainly depends on nutrient supply to the system. The aquafeed industry is growing rapidly to maintain the aquaculture industry's production trends, which is backed by the quality and type of the aquafeed. Table 19.1 presents the world employment for fishers and fish farmers.

19.2 CONTENTS PRESENT IN FISH FEED

The fish feed has developed tremendously in recent years due to developing a novel well-balanced diet that ensures optimal fish health and growth. Most fish need more than 40 essential nutrients: proteins, fats, carbohydrates, vitamins, and minerals (Oliva-Teles, 2012).

The main available commercial fish feed has the following ingredients: fishmeal, animal trimmings, fish oil, and plants. Fish oil and fishmeal are essential ingredients to most fisheries due to their high proteins, fats, and carbohydrates. Table 19.2 provides information on various components of commercial fish feed with their ranges (Craig & Helfrich, 2017).

19.2.1 PROTEIN

Protein is one of the essential components of fish feed. It is essential for the development of fish muscles and also plays a vital role in supporting fish growth (Cacho et al., 1990). Proteins can also be utilized as a source of energy. Proteins are the most expensive component of the fish feed. Thus, it becomes crucial to determine each species' protein requirements as one wants to achieve optimum growth with the least investment. The main essential amino acids (EAA) are histidine, phenylalanine, threonine, isoleucine, leucine, lysine, methionine, tryptophan, and valine (Hou & Wu, 2018). The body cannot make them, and these EAA have to be supplemented externally through food (Wu, 2013). Among all EAAs, lysine and methionine are often the first limiting amino acids (Schwab & Whitehouse, 2022). Thus, estimating and fulfilling sufficient protein and specific amino acid requirements is essential to promote optimal growth and health. Fish feeds prepared with bacterial and yeast proteins contain low sources of lysine and methionine (Agboola et al., 2020, Craig & Helfrich, 2017). Hence, the feeds must be supplemented externally with these amino acids to ensure good growth (Craig & Helfrich, 2017).

Protein content in the fish feed varies from one fish type to another. The average range of protein used for catfish is 28–32%, 38–42% for hybrid striped bass, and 40–45% for marine finfish and trout. It is observed that herbivorous and omnivorous fish require less protein content than carnivorous fish (Craig & Helfrich, 2017).

TABLE 19.1
World Employment for Fishers and Fish Farmers, by Region (Thousands)

Fisheries	1995	2000	2005	2010	2015	2018
Africa	2743	3247	3736	4228	4712	5021
Americas	1793	1982	2013	2562	2816	2455
Asia	24,205	28,079	29,890	31,517	30,436	30,768
Europe	378	679	558	530	338	272
Oceania	460	451	458	467	469	460
Total	29,579	34,439	36,655	39,305	38,771	38,976

Source: The State of World Fisheries and Aquaculture 2020.

TABLE 19.2

Various Components to Prepare Fish Food Commercially

Component	Range (%)
Protein	18–50
Lipids	10–25
Carbohydrates	15–20
Ash	< 8.5
Water	< 10
Phosphorous	< 1.5
Vitamins and minerals	< 1

19.2.2 CARBOHYDRATES

Carbohydrates are not required in the fish diet, but they reduce the overall cost and increase the feed volume. Indeed carbohydrates are the main energy source in mammals and can be utilized as an energy source in fish too. Fish store carbohydrates as glycogen in their body but cannot utilize them efficiently. The energy extraction process is very inefficient as fish can utilize only about 20% of the dietary carbohydrates (Craig & Helfrich, 2017).

19.2.3 LIPIDS

Lipids are highly energy-dense molecules, better than protein and carbohydrates (Poian & Luz, n.d.). They usually make up about 7–15% of fish diets. They supply essential fatty acids and serve as transporters for fat-soluble vitamins. Since proteins are expensive, feed manufacturers tend to add more lipids than required. Nevertheless, this comes with a compromise as excessive fat in the fish diet can negatively affect its health and reduce its survival rate (Craig & Helfrich, 2017).

Long-chain fatty acids are not a part of freshwater fish diets, but they require linoleic acid in quantities ranging from 0.5 to 1.5% for optimal growth. Therefore, these freshwater fish could synthesize long-chain *omega*-3 fatty acids, like eicosapentaenoic acid (EPA) and docosahexaenoic acid (DHA), required for their metabolism by elongating desaturating linoleic acid *via* enzymatic systems. On the other hand, marine fish do not possess this enzyme system and thus directly depend on *omega*-3 fatty acids. Therefore, their feeds consist of 0.5–2% *omega*-3 fatty acids for good health and better growth (Craig & Helfrich, 2017).

19.2.4 VITAMINS

Vitamins are micronutrients that are necessary for optimal fish growth. A vitamin-rich and diverse diet must be provided to fish as fish metabolism does not synthesize them. Vitamins are classified based on the medium in which they are soluble. They can be fat- or water-soluble. (Lillemoen & Bjørke-Monsen, 2020).

Water-soluble vitamins (Said, 2011, Combs & McClung, 2017): vitamin B family, choline, inositol, and vitamin C (ascorbic acid). Vitamin C has high importance because it enhances fish's immune system and is an antioxidant.

Fat-soluble vitamins (Dobreva et al., 2011, Combs & McClung, 2017): vitamin A, vitamin D, vitamin E, and vitamin K. Among these vitamins, vitamin E is an excellent antioxidant.

Vitamin E and vitamin C are predominantly used in most fish feeds by manufacturers due to their antioxidant properties that help inhibit dietary oxidation of lipids and thereby prolong the shelf life of feeds.

Research showed that yellow perch fed with both vitamin C and vitamin E exhibited better gain in weight, higher intake of fish feed, and higher feed efficiency compared to groups fed vitamin C-deficient diets (Lee & Dabrowski, 2003). The same study also showed that supplemental vitamin C increased the concentration of α-tocopherol in the liver in the fish group which was fed with vitamin E-deficient fish feed. This showed that vitamin E was either spared or regenerated with the help of vitamin C.

Vitamin deficiency can cause ill effects to the fish, thereby decreasing the life span and decreasing the production of well-suited offspring (Zhang et al., 2017). In addition, vitamin A deficiency is known to cause oxidative damage and apoptosis of intestinal cells (Jiang et al., 2019). The most common deficiencies observed in fish due to lack of appropriate vitamins are scoliosis from vitamin C deficiency and dark coloration of fish skin from vitamin B (folic acid) deficiency (Craig & Helfrich, 2017).

19.2.5 Minerals

Minerals are essential micronutrients that help regulate the fish's normal body functions. Minerals can be divided into macrominerals and microminerals based on the amount required by the fish and the amount already present in it. Common examples of minerals include sodium, magnesium, calcium, chloride, sulfur, etc. (Craig & Helfrich, 2017).

Although fish can absorb many minerals directly from the water through their gills and skin to some extent, these nutrients must also be provided in the diet for normal body functions. Some minerals (macrominerals) are required for osmotic balance regulation, bone formation, and bone integrity. Phosphorus is one of the essential macrominerals that fish require as it plays a key role in the metabolism of fish and is not available in the water directly (Sugiura et al., 2004). Some minerals are required in trace amounts (microminerals or trace minerals) as components in enzymes and hormone systems as cofactors (Watanabe et al., 1997).

19.2.6 Computer Formulation of Fish Feed

Formulation of fish feed is a technique that couples several ingredients to deliver the essential nutrition that fish require to grow (Porchelvi et al., 2018). Using technology, one can enhance the quality of fish feed by knowing the following information (Lovell, 1989):

a. How much nutrient is present in the feed: balancing all the required nutrients
b. The cost of each ingredient to be put in the feed
c. The availability of each ingredient present in the feed from various sources
d. The fish species of interest, and what is its nutrient requirement
e. The maximum and minimum levels of intake for a particular ingredient

There are adequate resources to understand factors (a) and (b). Precise knowledge of the type of organism and its nutritional requirements is an absolute must before engaging in computer-based feed formulation (Ghosh et al., 2011). While formulating feeds, one must always keep in mind the different factors like the physical nature of the feed, the toxicological properties, balanced diet, etc.

There are certain limitations to this technique. For example, one cannot program certain features due to different milling characteristics. One good example of this case is the expulsion of sorghum. During the processing of floating catfish feeds, corn extrudes easier than sorghum. Another limitation could be the availability of nutrients, the logistics and handling factors, and the cost of different ingredients.

19.3 PLANT-BASED FISH FEED

Fisheries are moving towards plant-based fish feed in a slowly growing manner. One of the major reasons for this change is that fishmeal is very expensive. As a result, most fisheries close up within one to three years due to high losses mainly incurred from purchasing fishmeal-based feed (Kaushik & Hemre, 2008). Hence, many fisheries began a comprehensive search to find an alternative protein source to replace the existing fishmeal to sustain their place in the aquaculture industry and get ahead of their competitors (Francis et al. 2001, Tacon et al., 1983).

Plant-based fish feed is being employed as an alternative to fishmeal (Bhosale et al., 2010). The most commonly used plant products in fish feed include:

a) Soybean
b) Wheat
c) Canola oil
d) Corn
e) Ground rice

Upon experimenting with various plant-based fish feeds, most results indicated soybean as one of the best alternatives to fishmeal. Soybean is highly nutritious to the fish, and hence most fisheries have begun using it as the alternative source of quality protein in a variety of fish feeds (Lovell 1989, Gatlin et al., 2007). In addition, it has approximately 48% proteins and shows an excellent amino acid profile (El-Sayed, 1999). Another advantage of using soybean for fish feed is its cost. Soybean is highly cost-effective compared to quality fishmeal. It also has a steady supply and is readily available compared to other plant protein sources (Storebakken et al., 2000).

Another trend that was observed is the usage of vegetable oils which have replaced conventional fish oils. Research showed that switching to vegetable oils like soybean oil, rapeseed oil, etc., could be better sources of fats for *Macrobrachium rosenbergii* than traditional fish oil. This was due to the NF-κB-NO signal pathway, which induced an oxidative status that was modified compared to normal fish (Sun et al., 2020).

We can also improve the efficiency of plant-based fish feeds by nutritional supplementation. For example, it was reported that supplementing tryptophan in plant-based fish feed showed beneficial effects. The fish grown with tryptophan supplementation showed a better immune response and coped better with stress-causing agents commonly observed in most fisheries (Cerqueira et al., 2020).

19.4 CORIANDER AND ITS PROPERTIES

Coriander is an aromatic annual herb and belongs to the Apiaceae family. One of the striking features of coriander is that all parts of this plant are edible. The parts of the plant most commonly used for consumption are the fresh leaves and dried seeds (Rodriguez, 2021). It is an excellent source of carbohydrates, proteins, and vitamins like vitamin A, provitamin A, and vitamin K. It also contains folate, potassium, manganese, choline, β-carotene, lutein, and zeaxanthin in trace amounts. Coriander is a tender herb that is relatively easy to grow. In addition to the health benefits, it also enhances the flavor of the food. The specific fragrance of coriander is due to a terpenoid called linalool (Burdock & Carabin, 2009). Coriander seeds generally have lower vitamin content but significant calcium, iron, magnesium, and dietary fibers (Ware, 2019). Table 19.3 presents the nutrient content in 100 g of coriander seeds.

19.5 CORIANDER IN FISH FEED

The effect of different botanical additives (mint, coriander, and amaranth leaves) on the growth and coloration of ornamental goldfish was reported (Jegan et al., 2008). Attractive colors began to appear in the sixth or seventh week of a goldfish fry (Hervey & Hems, 1968). The amount of carotenoids present in the diet is directly proportional to the redness of the goldfish skin (Iwahashi & Wakui, 1976). Adult goldfish fed at a 3% level exhibited a higher mean weight gain (1.939 g)

TABLE 19.3
Nutrient Content in 100 g of Coriander Seeds

Amounts Per Selected Serving		Protein & Amino Acids	
Calories	298 (1248 kJ)	Protein	12.4 g
Carbohydrate	108 (452 kJ)	**Carbohydrates**	
Fat	149 (624 kJ)	Total carbohydrate	55.0 g
Protein	41.6 (174 kJ)	Dietary fiber	41.9 g
		Minerals	
Vitamins		Calcium	709 mg
Vitamin C	21.0 mg	Iron	16.3 mg
Thiamin	0.2 mg	Magnesium	330 mg
Riboflavin	0.3 mg	Phosphorus	409 mg
Niacin	2.1 mg	Potassium	1267 mg
		Sodium	35.0 mg
Fats & Fatty Acids		Zinc	4.7 mg
Total fat	17.8 g	Copper	1.0 mg
Saturated fat	1.0 g	Manganese	1.9 mg
Monounsaturated fat	13.6 g	Selenium	26.2 mcg
Polyunsaturated fat	1.8 g	**Sterols**	
Total *omega*-6 fatty acids	1750 mg	Cholesterol	0.0 mg
		Phytosterols	46.0 mg

Source: FoodData Central, n.d.

and specific growth rate (0.668 g) compared to other coriander feeds. The highest percentage of coloration among juvenile goldfish came from the group fed with coriander, followed by mint and amaranth. Coriander leaves contained 36.72 mg of total carotenoids for 100 g of coriander (with respect to dry weight, sundried), while mint leaves contained 29.06 mg/100 g and amaranth leaves contained 29.05 mg/100 g (Kowsalya et al., 2001). The nature and amount of carotenoids present in the feed or ingredient determine the rate of color development (Boonyaratpalin et al., 2001). Fish feed that contained coriander leaves, as an ingredient, had a higher amount of carotenes present than feeds with mint and amaranth leaves and hence displayed a higher percentage of color development. It was concluded that coriander had a constructive effect on the growth and coloration of ornamental goldfish.

Researchers have investigated the effects of additive coriander powder on the growth, feed utilization, survival rate, and body chemical composition of common carp *Cyprinus carpio*, which demonstrated an increase in protein efficiency ratio and protein productive values. The selected fish were fed with coriander and cinnamon powders at levels ranging from 1.25% to 1.75%, increasing by 0.25%, and were compared to a control that consisted of fish that were not fed this diet mixture. The result obtained showed significant differences in weight gain, live body weight, protein retention, and feed conversion ratio among both groups. In addition, it was also observed that there was an increase in the percentage of dry matter and raw protein and a decrease in fat percentage (Al-Shakarchi & A. Mohammad, 2021).

Coriander seed extract (CSE) has also been studied for its anxiolytic effect in adult zebrafish. Adult zebrafish were injected with CSE (25, 50, or 100 mg/kg) and were placed in a colorful alarm substance/saline was added to the tank to induce an anxiety-like response in the fish, and this was recorded. After 24 h, the fish underwent a light/dark test. It was observed that CSE prevented the anxiety-causing effect of the alarm substance. Clonazepam was also used in this study, and it was observed that similar results were obtained between clonazepam and CSE (Zenki et al., 2020).

19.6 IMPACT OF CORIANDER ON THE IMMUNE SYSTEM OF THE FISH

Researchers in Iran have discovered that coriander seed extracts in fish feed can lead to their high specific growth rate and final weight. According to them, coriander seed extracts have immunostimulatory effects, improving growth and increasing the overall survival rate against pathogens. In addition, the seeds contain 40–50% linalool, which is well known for its antioxidant, antimicrobial, and anti-inflammatory properties. It can also enhance lysozyme activity and complement system activation.

Yersinia ruckeri is a Gram-negative rod-shaped bacteria that causes enteric redmouth disease in some fish species. Significant economic losses have been observed in the aquaculture industry due to *Yersinia ruckeri* (Barnes & Horne, 2011). Rainbow trout are most susceptible to *Y. ruckeri* (Ross et al., 1966). For eight weeks, a study was done to determine the efficacy of coriander seed extract on physiological responses, immunity, and disease resistance of rainbow trout, *Oncorhynchus mykiss*. It was reported that fish feed with coriander seed extract (especially 2%) improved the survival rate against *Y. ruckeri*.

Further studies confirmed that the dietary incorporation of coriander extract could improve growth factors, immunological indices, and resistance of rainbow trout against *Y. ruckeri* infection. Although the research was conducted on rainbow trout against *Y. ruckeri* infection, it gives solid evidence of immune-enhancing properties of coriander seed extract in freshwater fishes (Naderi Farsani et al., 2019).

Another study conducted on the effect of *Oreochromis niloticus* fed with *Coriandrum sativum* seed powder and extract against exposure to immunotoxic metal (lead) also provided strong evidence for using coriander seed extract (CSE) to enhance fish immunity (Ahmed et al., 2019). Furthermore, it was also observed that coriander reduced the accumulation and toxicity of cadmium in rainbow trout. This was done by concerted interaction between various molecules present in coriander, like citric acid, chelating material (phytic acid), diacetate amino acid, etc., which led to the speeding up of cadmium excretion and decreasing its uptake (Jia et al., 2009, Jia et al., 2006).

It is well known that heavy metals are one of the most common aquatic pollutants among all the environmental pollutants. They pose a significant threat to the survival of all aquatic life. Among the heavy metals, lead is very pernicious as it is toxic even at low concentrations. Lead exposure is lethal for both fish and humans, and many studies have proved this. Coriander has been used as a medicinal herb for centuries. Its antioxidant, anti-microbial, and anti-toxicant properties have been demonstrated in many animals (Önder, 2018). However, this study focused on CSE, and finally concluded that CSE (30 mg/kg diet) enhances the immune responses of fish to neutralize the immunotoxic effects of lead exposure (Ahmed et al., 2019). The one unique piece of information obtained from this study is that CSE makes the fish more tolerant to pathogens and increases the tolerance of fish to abiotic toxins like heavy metals.

19.7 IMPACT OF CORIANDER ON THE GROWTH PERFORMANCE AND SURVIVAL RATE OF FISH

Researchers at the University of Saskatchewan have shown that coriander in the fish diet can increase the concentration of long-chain *omega*-3 fatty acids like eicosapentaenoic acid (EPA) and docosahexaenoic acid (DHA), which would directly benefit the consumer's health in many ways. In addition, as stated earlier, coriander seed extracts improve fish immunity and increase the overall survival rate. Coriander seed oil contains petroselinic acid (PSA). PSA is a monounsaturated fatty acid usually present in the seeds of the Apiaceae family. A study demonstrated that petroselinic acid positively impacts fish growth (Teoh & Ng, 2013).

Most fatty acids contain an even number of carbon atoms. Saturated fatty acids contain carbon atoms bonded via a single bond. Unsaturated fatty acids contain double and/or triple bonds between the carbon atoms in the fatty acid chain (Ruxton et al., 2004). *Omega*-3 fatty acids or polyunsaturated fatty acids (PUFAs) contain the first double bond in the third position when counting is initiated from the end containing the methyl-group (de Roos et al., 2009). This type of counting, which starts from the methyl end, is denoted by "n-3". The nomenclature of PUFAs is based on the position of the double bonds and the total chain length (Cuyamendous et al., 2016, Wallis et al., 2002). For example, EPA is also written as 20:5(n-3), describing a fatty acid with 20 carbon atoms and 5 double bonds.

A study observed that the addition of PSA from coriander oil increases the anti-inflammatory precursor (22:6n-3) and decreases the pro-inflammatory precursor (20:4n-6) in radiolabeled rainbow trout hepatocytes (Randall et al., 2013). PSA makes up more than 50% of the total fatty acid present in seeds. Researchers have found that PSA increased the fatty acid composition of fish tissues. Fish fed with the 20% PSA diet showed the highest content of DPA (long-chain *omega*-3 fatty acids) (Nguyen et al., 2015).

19.8   CONSUMER PERSPECTIVE ON PLANT-BASED FISH FEED

The diet to be provided to fish is not only for improving the overall health, quality of meat, and weight of the fish, but it should also enhance the overall nutrient value of the fish. For example, one of the main reasons people consume fish is to obtain *omega*-3 fatty acids since our bodies do not produce *omega*-3 fatty acids (Hjalmarsdottir, 2018). Some of the potential benefits of *omega*-3 fatty acids are (Washington State Department of Health, n.d.):

1. It reduces the risk of heart attacks, strokes, and low blood pressure in humans (Illingworth & Ullmann, 2020). It also decreases the risk of events related to total coronary heart disease (Kris-Etherton et al., 2019)
2. It prevents inflammation (Isobe & Arita, 2015)
3. It reduces the risk of macular degeneration (the leading cause of permanent eye damage and blindness)
4. *Omega*-3 fatty acids have also been shown to improve overall mental health (Lange, 2020, Shahidi & Miraliakbari, 2005)
5. An *Omega*-3 fatty acids-rich diet reduces liver fat and inflammation in people suffering from non-alcoholic fatty liver disease (NAFLD) (Lu et al., 2016, Li & Chen, 2012)
6. In cases of acute neurological activity, *omega*-3 fatty acids act as neuroprotective agents (Dyall & Michael-Titus, 2008)
7. They are also involved in the GPR40 receptor and other specific fatty acid receptors, which display comprehensive expression in the central nervous system (Ma et al., 2007)
8. They can enhance anti-inflammatory properties and help patients with diseases like lupus, rheumatoid arthritis, etc. (Calder & Yaqoob, 2009, Calder, 2008)

As discussed earlier, coriander seed extracts enhance *omega*-3 fatty acids content in fish. Therefore, people consume fish as they are rich in calcium and phosphorus and a great source of minerals, such as iron, zinc, iodine, magnesium, and potassium.

Coriander essential oil can control and inhibit the growth of *Campylobacter coli* and *Campylobacter jejuni*, bacterial species that cause campylobacteriosis (Duarte et al., 2016). In addition, coriander essential oil is cheap to obtain, possesses a higher availability, and has wide-range temperature effectiveness and thus is a very good natural antimicrobial agent compared to commonly used chemical preservatives (Rattanachaikunsopon & Phumkhachorn, 2010).

Coriander seed extracts are also a good source of phosphorus, calcium, potassium, and magnesium; 100 g of coriander seed contains 1267 mg potassium, 709 mg calcium, 409 mg phosphorus, and 330 mg magnesium. Phosphorus and calcium work together to help build the bones. Phosphorus plays a structural role in nucleic acid and cell membrane (Heaney & Graeff-Armas, 2018). It is involved in the body's energy metabolism. Similarly, calcium is also essential for our muscle functioning (Kraft, 2014; Washington State Department of Health, n.d.).

19.9 FUTURE DIRECTIONS

One of the main drawbacks observed with plant-based fish feed is environmental concerns. Due to avid farming, the usage of the land, water, soil, and other associated resources is causing detrimental effects to the environment. Growing coriander is easier than other plants incorporated in fish feeds. It will also be easier if the fisheries use some of their lands to cultivate coriander. Coriander could also be extensively studied for its antioxidative property and thereby help in increasing the shelf-life of fish feed.

Another direction is to study the effect of coriander in promoting brain activity in fish. Better brain functioning can decrease fish mortality rate, increase offspring survival chances, etc.

A new direction of study can be undertaken that explores the effect of coriander in increasing the levels of essential vitamins like vitamin B12. Vitamin B12 plays a major role in the functioning of the central nervous system and the maturation of red blood cells in the bone marrow, being a cofactor in DNA synthesis, and myelin synthesis or myelinogenesis. Fish is a major source of vitamin B12, and the consumption of fish meat can help people with vitamin B12 deficiency make a significant recovery.

A suggested direction of study could be performed on the effect of coriander on the fish's digestive system. The chances of fish becoming a good quality food source increase based on the fish's ability to digest and absorb the nutrients in the feed. Studies can also be done on specific sections of the digestive system concerning pathogenic activity and how coriander seed extract can decrease the rate of infection.

19.10 CONCLUSION

The aquaculture industry is developing rapidly as the quantity, quality, availability, and the different varieties of the aquafeed are driving the industry. The industry is also generating massive employment, thereby boosting the associated industries' growth. Excellent aquafeed quality is the prerequisite for maintaining sustainable growth in the fisheries. Indeed, fishmeal is quite expensive, and hence to maintain a sustainable business model, the aquaculture industry is moving towards plant-based fish feed.

A plethora of research data is already available, favoring plant-based fish feed. Creating a balanced mixture of nutrients could increase the quantity and enhance the quality of fish feed. In addition, fish feed and specific nutrient supplementation could also boost the quality of the feed and improve the average fish size and lifespan.

Although more research is required, our current knowledge in this area is valid enough to put forward the idea that coriander seed extract (CSE) in fish feed is highly favorable to the aquaculture industries and feed manufacturers. Coriander is an annual herb that could be grown worldwide. CSE-rich diets could increase fish weight, boost their immunity against pathogens and heavy metal toxicity, and increase the overall survival rate. It could also increase the concentration of *omega*-3 fatty acids in the tissues, making them more desirable for consumers. Researchers have already proven this concept in many fresh and marine water fish. Thus, we conclude that coriander and its seed extract could be an excellent choice for fish consumption, thereby improving the quality of meat obtained from fish required for human consumption, although more research could give better insights.

REFERENCES

Agboola, J. O., Øverland, M., Skrede, A., & Hansen, J. Ø. (2020). Yeast as major protein-rich ingredient in aquafeeds: A review of the implications for aquaculture production. *Reviews in Aquaculture, 13*(2), 949–970. https://doi.org/10.1111/raq.12507

Ahmed, S. A. A., Reda, R. M., & ElHady, M. (2019). Immunomodulation by *Coriandrum sativum* seeds (coriander) and its ameliorative effect on lead-induced immunotoxicity in Nile tilapia (*Oreochromis niloticus* L.). *Aquaculture Research, 51*(3), 1077–1088. https://doi.org/10.1111/are.14454

Al-Shakarchi, H. H., & Mohammad, M. (2021). Effects of different levels of cinnamon and coriander powders on growth performance, Feed Utilization, survival rate and body chemical composition of common carp cyprinus carpio L. *Plant Archives, 21*(Supplement 1), 46–54. https://doi.org/10.51470/plantarchives.2021.v21.s1.010

Barnes, A. C., & Horne, M. T. (2011). Enteric redmouth disease (ERM) (Yersinia ruckeri). In P. T. K. Woo & D. W. Bruno (Eds.), *Fish diseases and disorders. Volume 3: Viral, bacterial and fungal infections* (pp. 445–477). CABI. https://doi.org/10.1079/9781845935542.0484

Bhosale, S., Bhilave, M., & Nadaf, S. (2010). Formulation of fish feed using ingredients from plant sources. *Research Journal of Agricultural Sciences, 1*, 284–287.

Boonyaratpalin, M., Thongrod, S., Supamattaya, K., Britton, G., & Schlipalius, L. E. (2001). Effects of β-carotene source *Dunaliella salina*, and astaxanthin on pigmentation, growth, survival and health of *Penaeus monodon*. *Aquaculture Research, 32*(1), 182–190. https://doi.org/10.1046/j.1355-557x.2001.00039.x

Burdock, G. A., & Carabin, I. G. (2009). Safety assessment of coriander (*Coriandrum sativum* L.) essential oil as a food ingredient. *Food and Chemical Toxicology, 47*(1). https://doi.org/10.1016/j.fct.2008.11.006

Cacho, O. J., Hatch, U., & Kinnucan, H. (1990). Bioeconomic analysis of fish growth: Effects of dietary protein and ration size. *Aquaculture, 88*(3–4), 223–238. https://doi.org/10.1016/0044-8486(90)90150-1

Calder, P. C. (2008). Session 3: Joint Nutrition Society and Irish Nutrition and Dietetic Institute Symposium on 'Nutrition and autoimmune disease' PUFA, inflammatory processes and rheumatoid arthritis. *Proceedings of the Nutrition Society, 67*(4), 409–418. https://doi.org/10.1017/s0029665108008690

Calder, P. C., & Yaqoob, P. (2009). Omega-3 polyunsaturated fatty acids and human health outcomes. *BioFactors, 35*(3), 266–272. https://doi.org/10.1002/biof.42

Cerqueira, M., Schrama, D., Silva, T. S., Colen, R., Engrola, S. A. D., Conceição, L. E. C., Rodrigues, P. M. L., & Farinha, A. P. (2020). How tryptophan levels in plant-based aquafeeds affect fish physiology, metabolism and proteome. *Journal of Proteomics, 221*, 103782. https://doi.org/10.1016/j.jprot.2020.103782

Combs Jr., G. F., & McClung, J. P. (2017). General properties of vitamins. In *The vitamins* (pp. 33–58). Elsevier. https://doi.org/10.1016/b978-0-12-802965-7.00003-4

Craig, S., & Helfrich, L. (2017). Understanding fish nutrition, feeds, and feeding. *Virginia Cooperative Extension, 420*(256).

Cuyamendous, C., de la Torre, A., Lee, Y. Y., Leung, K. S., Guy, A., Bultel-Poncé, V., Galano, J.-M., Lee, J. C.-Y., Oger, C., & Durand, T. (2016). The novelty of phytofurans, isofurans, dihomo-isofurans and neurofurans: Discovery, synthesis and potential application. *Biochimie, 130*, 49–62. https://doi.org/10.1016/j.biochi.2016.08.002

de Roos, B., Mavrommatis, Y., & Brouwer, I. A. (2009). Long-chain n-3 polyunsaturated fatty acids: New insights into mechanisms relating to inflammation and coronary heart disease. *British Journal of Pharmacology, 158*(2), 413–428. https://doi.org/10.1111/j.1476-5381.2009.00189.x

Dobreva, D., Galunska, B., & Stancheva, M. (2011). Liquid chromatography method for the simultaneous quantification of fat soluble vitamins in fish tissue. *Scripta Scientifica Medica, 43*(1), 35. https://doi.org/10.14748/ssm.v43i1.405

Duarte, A., Luís, Â., Oleastro, M., & Domingues, F. C. (2016). Antioxidant properties of coriander essential oil and linalool and their potential to control *Campylobacter* spp. *Food Control, 61*, 115–122. https://doi.org/10.1016/j.foodcont.2015.09.033

Dyall, S. C., & Michael-Titus, A. T. (2008). Neurological benefits of omega-3 fatty acids. *NeuroMolecular Medicine, 10*(4), 219–235. https://doi.org/10.1007/s12017-008-8036-z

El-Sayed, A.-F. M. (1999). Alternative dietary protein sources for farmed tilapia, Oreochromis spp. *Aquaculture, 179*(1–4), 149–168. https://doi.org/10.1016/s0044-8486(99)00159-3

FoodData central. (n.d.). Retrieved November 3, 2021, from https://fdc.nal.usda.gov/index.html

Francis, G., Makkar, H. P. S., & Becker, K. (2001). Antinutritional factors present in plantderived alternate fish feed ingredients and their effects in fish. *Aquaculture*, 199(3–4), 197–227. https://doi.org/10.1016/s0044-8486(01)00526-9

Gatlin, D. M., Barrows, F. T., Brown, P., Dabrowski, K., Gaylord, T. G., Hardy, R. W., Herman, E., Hu, G., Krogdahl, Å., Nelson, R., Overturf, K., Rust, M., Sealey, W., Skonberg, D., Souza, E. J., Stone, D., Wilson, R., & Wurtele, E. (2007). Expanding the utilization of sustainable plant products in Aquafeeds: A review. *Aquaculture Research*, 38(6), 551–579. https://doi.org/10.1111/j.1365-2109.2007.01704.x

Ghosh, D., Sathianandan, T. V., & Vijayagopal, P. (2011). Feed formulation using linear programming for fry of catfish, milkfish, tilapia, Asian sea bass, and grouper in India. *Journal of Applied Aquaculture*, 23(1), 85–101. https://doi.org/10.1080/10454438.2011.549781

Heaney, R. P., & Graeff-Armas, L. A. (2018). Vitamin D role in the calcium and phosphorus economies. In *Vitamin D* (pp. 663–678). Elsevier. https://doi.org/10.1016/b978-0-12-809965-0.00038-0

Hervey, G. F., & Hems, J. (1968). *The goldfish* (pp. 196–229). Latimer Trend Co. Ltd.

Hjalmarsdottir, F. (2018, October 15). *17 science-based benefits of omega-3 fatty acids*. Healthline Media. https://www.healthline.com/nutrition/17-health-benefits-of-omega-3#TOC_TITLE_HDR_3

Hou, Y., & Wu, G. (2018). Nutritionally essential amino acids. *Advances in Nutrition*, 9(6), 849–851. https://doi.org/10.1093/advances/nmy054

Illingworth, D. R., & Ullmann, D. (2020). Effects of omega-3 fatty acids on risk factors for cardiovascular disease. In *Omega-3 fatty acids in health and disease* (pp. 39–70). CRC Press. https://doi.org/10.1201/9781003066453-2

Isobe, Y., & Arita, M. (2015). Omega-3 fatty acid metabolism and regulation of inflammation. In *Bioactive lipid mediators* (pp. 155–162). Springer Japan. https://doi.org/10.1007/978-4-431-55669-5_11

Iwahashi, M., & Wakui, H. (1976). *Nippon Suisan Gakkaishi*, 42(12), 1339–1344. https://doi.org/10.2331/suisan.42.1339

Jegan, K., Felix, N., & R, J. (2008). Influence of botanical additives on the growth and colouration of adult goldfish. *Veterinary and Animal Science*, 4.

Jia, H., Ren, H., Endo, H., & Hayashi, T. (2009). Effect of Chinese parsley *Coriandrum sativum* on cadmium binding to proteins from the liver and kidney of rainbow trout *Oncorhynchus mykiss*. *Marine and Freshwater Behaviour and Physiology*, 42(3), 187–199. https://doi.org/10.1080/10236240903036371

Jia, H., Ren, H., Maita, M., Satoh, S., Endo, H., & Hayashi, T. (2006). Development of functional fish feed with natural ingredients to control heavy metals. *Toxicology Mechanisms and Methods*, 16(8), 411–417. https://doi.org/10.1080/15376520600666995

Jiang, W.-D., Zhou, X.-Q., Zhang, L., Liu, Y., Wu, P., Jiang, J., Kuang, S.-Y., Tang, L., Tang, W.-N., Zhang, Y.-A., Shi, H.-Q., & Feng, L. (2019). Vitamin A deficiency impairs intestinal physical barrier function of fish. *Fish & Shellfish Immunology*, 87, 546–558. https://doi.org/10.1016/j.fsi.2019.01.056

Kaushik, S. J., & Hemre, G.-I. (2008). Plant proteins as alternative sources for fish feed and farmed fish quality. *Improving Farmed Fish Quality and Safety*, 300–327. https://doi.org/10.1533/9781845694920.2.300

Kong, W., Huang, S., Yang, Z., Shi, F., Feng, Y., & Khatoon, Z. (2020). Fish feed quality is a key factor in impacting aquaculture water environment: Evidence from incubator experiments. *Scientific Reports*, 10(1), 187. https://doi.org/10.1038/s41598-019-57063-w

Kowsalya, S., Chandrasekhar, U., & Balasasirekha, R. (2001). Beta carotene retention in selected green leafy vegetables subjected to dehydration. *The Indian Journal of Nutrition and Dietetics*, 38, 374–383.

Kraft, M. D. (2014). Phosphorus and calcium. *Nutrition in Clinical Practice*, 30(1), 21–33. https://doi.org/10.1177/0884533614565251

Kris-Etherton, P. M., Richter, C. K., Bowen, K. J., Skulas-Ray, A. C., Jackson, K. H., Petersen, K. S., & Harris, W. S. (2019). Recent clinical trials shed new light on the cardiovascular benefits of omega-3 fatty acids. *Methodist DeBakey Cardiovascular Journal*, 15(3), 171–178. https://doi.org/10.14797/mdcj-15-3-171

Lange, K. W. (2020). Omega-3 fatty acids and mental health. *Global Health Journal*, 4(1), 18–30. https://doi.org/10.1016/j.glohj.2020.01.004

Lee, K.-J., & Dabrowski, K. (2003). Interaction between vitamins C and E affects their tissue concentrations, growth, lipid oxidation, and deficiency symptoms in yellow perch (*Perca flavescens*). *British Journal of Nutrition*, 89(5), 589–596. https://doi.org/10.1079/bjn2003819

Li, Y., & Chen, D. (2012). The optimal dose of omega-3 supplementation for non-alcoholic fatty liver disease. *Journal of Hepatology*, 57(2), 468–469. https://doi.org/10.1016/j.jhep.2012.01.028

Lillemoen, P. K. S., & Bjørke-Monsen, A.-L. (2020). Ernæringsstatus av vitamin og sporelement. *Tidsskrift for Den Norske Legeforening*. https://doi.org/10.4045/tidsskr.19.0587

Lovell, R. T. (1989). *Nutrition and Feeding of Fish*. Van Nostrand Reinhold, New York. https://doi.org/10.1007/978-1-4757-1174-5

Lu, W., Li, S., Li, J., Wang, J., Zhang, R., Zhou, Y., Yin, Q., Zheng, Y., Wang, F., Xia, Y., Chen, K., Liu, T., Lu, J., Zhou, Y., & Guo, C. (2016). Effects of omega-3 fatty acid in nonalcoholic fatty liver disease: A meta-analysis. *Gastroenterology Research and Practice*, *2016*(1459790), 1–11. https://doi.org/10.1155 /2016/1459790

Lucas, J. (2015). Aquaculture. *Current Biology*, *25*(22), R1064–R1065. https://doi.org/10.1016/j.cub.2015.08.013

Ma, Q.-L., Teter, B., Ubeda, O. J., Morihara, T., Dhoot, D., Nyby, M. D., Tuck, M. L., Frautschy, S. A., & Cole, G. M. (2007). Omega-3 fatty acid docosahexaenoic acid increases sorla/lrl1, a sorting protein with reduced expression in sporadic Alzheimer's disease (AD): Relevance to AD prevention. *Journal of Neuroscience*, *27*(52), 14299–14307. https://doi.org/10.1523/jneurosci.3593-07.2007

Naderi Farsani, M., Hoseinifar, S. H., Rashidian, G., Ghafari Farsani, H., Ashouri, G., & Van Doan, H. (2019). Dietary effects of *Coriandrum sativum* extract on growth performance, physiological and innate immune responses and resistance of rainbow trout (*Oncorhynchus mykiss*) against *Yersinia ruckeri*. *Fish & Shellfish Immunology*, *91*, 233–240. https://doi.org/10.1016/j.fsi.2019.05.031

Nguyen, Q.-H., Talou, T., Cerny, M., Evon, P., & Merah, O. (2015). Oil and fatty acid accumulation during coriander (*Coriandrum sativum* L.) fruit ripening under organic cultivation. *The Crop Journal*, *3*(4), 366–369. https://doi.org/10.1016/j.cj.2015.05.002

Oliva-Teles, A. (2012). Nutrition and health of aquaculture fish. *Journal of Fish Diseases*, *35*(2), 83–108. https://doi.org/10.1111/j.1365-2761.2011.01333.x

Önder, A. (2018). Coriander and its phytoconstituents for the beneficial effects. In *Potential of essential oils*. InTech. https://doi.org/10.5772/intechopen.78656

Poian, A. T. D. P., & Luz, M. R. M. P. (n.d.). *Human metabolism, energy, nutrients*. Learn Science at Scitable. Retrieved November 4, 2021, from https://www.nature.com/scitable/topicpage/nutrient-utilization-in -humans-metabolism-pathways-14234029/

Porchelvi, S., Irine, J., & Regupathi, R. (2018). Linear programming method for solving optimized nutrients feed formulation in GIFT Tilapia. *OSR Journal of Humanities and Social Science (IOSR-JHSS)*, *23*(10), 28–33.

Randall, K. M., Drew, M. D., Øverland, M., Østbye, T.-K., Bjerke, M., Vogt, G., & Ruyter, B. (2013). Effects of dietary supplementation of coriander oil, in canola oil diets, on the metabolism of [1–14C] 18:3n-3 and [1-14C] 18:2n-6 in rainbow trout hepatocytes. *Comparative Biochemistry and Physiology Part B: Biochemistry and Molecular Biology*, *166*(1), 65–72. https://doi.org/10.1016/j.cbpb.2013.07.004

Rattanachaikunsopon, P., & Phumkhachorn, P. (2010). Potential of coriander (*Coriandrum sativum*) oil as a natural antimicrobial compound in controlling *Campylobacter jejuniin* raw meat. *Bioscience, Biotechnology, and Biochemistry*, *74*(1), 31–35. https://doi.org/10.1271/bbb.90409

Rodriguez, E. (2021, May 5). *Coriander herb and spice*. Encyclopedia Britannica. https://www.britannica .com/plant/coriander

Ross, A. J., Rucker, R. R., & Ewing, W. H. (1966). Description of a bacterium associated with redmouth disease of rainbow trout (*Salmo gairdneri*). *Canadian Journal of Microbiology*, *12*(4), 763–770. https://doi .org/10.1139/m66-103

Ruxton, C. H. S., Reed, S. C., Simpson, M. J. A., & Millington, K. J. (2004). The health benefits of omega-3 polyunsaturated fatty acids: A review of the evidence. *Journal of Human Nutrition and Dietetics*, *17*(5), 449–459. https://doi.org/10.1111/j.1365-277x.2004.00552.x

Said, H. M. (2011). Intestinal absorption of water-soluble vitamins in health and disease. *Biochemical Journal*, *437*(3), 357–372. https://doi.org/10.1042/bj20110326

Schwab, C. G., & Whitehouse, N. L. (2022). Feed supplements: Ruminally protected amino acids. In *Encyclopedia of dairy sciences* (pp. 540–547). Elsevier. https://doi.org/10.1016/b978-0-08-100596-5 .23055-2

Shahbandeh, M. (n.d.). *World fish production 2019*. Statista. Retrieved November 3, 2021, from http://statista .com/statistics/264577/total-world-fish-production-since-2002/

Shahidi, F., & Miraliakbari, H. (2005). Omega-3 fatty acids in health and disease: Part 2-health effects of omega-3 fatty acids in autoimmune diseases, mental health, and gene expression. *Journal of Medicinal Food*, *8*(2), 133–148. https://doi.org/10.1089/jmf.2005.8.133

Storebakken, T., Refstie, S., & Ruyter, B. (2000). Soy products as fat and protein sources in fish feeds for intensive aquaculture. In J. K. Drackley (Ed.), *Soy in animal nutrition* (pp. 127–170). Champaign: Federation of Animal Science Societies.

Sugiura, S. H., Hardy, R. W., & Roberts, R. J. (2004). The pathology of phosphorus deficiency in fish - A review. *Journal of Fish Diseases*, *27*(5), 255–265. https://doi.org/10.1111/j.1365-2761.2004.00527.x

Sun, C., Liu, B., Zhou, Q., Xiong, Z., Shan, F., & Zhang, H. (2020). Response of *Macrobrachium rosenbergii* to vegetable oils replacing dietary fish oil: Insights from antioxidant defense. *Frontiers in Physiology*, *11*. https://doi.org/10.3389/fphys.2020.00218

Tacon, A. G. J., Stafford, E. A., & Edwards, C. A. (1983). A preliminary investigation of the nutritive value of three terrestrial Lumbricid worms for Rainbow Trout. *Aquaculture*, 35, 187–199. https://doi.org/10.1016/0044-8486(83)90090-x

Teoh, C.-Y., & Ng, W.-K. (2013). Evaluation of the impact of dietary petroselinic acid on the growth performance, fatty acid composition, and efficacy of long chain-polyunsaturated fatty acid biosynthesis of farmed *Nile tilapia*. *Journal of Agricultural and Food Chemistry*, *61*(25), 6056–6068. https://doi.org/10.1021/jf400904j

The state of world fisheries and aquaculture 2020. (n.d.). FAO. Retrieved November 3, 2021, from https://www.fao.org/state-of-fisheries-aquaculture

Wallis, J. G., Watts, J. L., & Browse, J. (2002). Polyunsaturated fatty acid synthesis: What will they think of next? *Trends in Biochemical Sciences*, *27*(9), 467–473. https://doi.org/10.1016/s0968-0004(02)02168-0

Ware, M. (2019, November 15). Why is cilantro (coriander) good for you? *Medical News Today*. https://www.medicalnewstoday.com/articles/277627

Washington State department of health. (n.d.). Retrieved November 3, 2021, from https://www.doh.wa.gov/communityandenvironment/food/fish/healthbenefits

Watanabe, T., Kiron, V., & Satoh, S. (1997). Trace minerals in fish nutrition. *Aquaculture*, *151*(1–4), 185–207. https://doi.org/10.1016/s0044-8486(96)01503-7

Wu, G. (2013). *Amino acids: Biochemistry and nutrition*. CRC Press.

Zenki, K. C., Souza de, L. S., Góis, A. M., Lima, B. dos S., Araújo, A. A. de S., Vieira, J. S., Camargo, E. A., Kalinine, E., Oliveira de, D. L., & Walker, C. I. B. (2020). *Coriandrum sativum* extract prevents alarm substance-induced fear- and anxiety-like responses in adult zebrafish. *Zebrafish*, *17*(2), 120–130. https://doi.org/10.1089/zeb.2019.1805

Zhang, L., Feng, L., Jiang, W.-D., Liu, Y., Wu, P., Kuang, S.-Y., Tang, L., Tang, W.-N., Zhang, Y.-A., & Zhou, X.-Q. (2017). Vitamin A deficiency suppresses fish immune function with differences in different intestinal segments: The role of transcriptional factor NF-κBandp38 mitogen-activated protein kinase signalling pathways. *British Journal of Nutrition*, *117*(1), 67–82. https://doi.org/10.1017/s0007114516003342

20 Coriander Straw and Press Cake from Seeds

Compositions and Possible Uses in the Field of Bio-Sourced Materials

Philippe Evon, Laurent Labonne, Valérie Simon, Thierry Talou, Stéphane Ballas, Thierry Véronèse, and Othmane Merah

CONTENTS

20.1 INTRODUCTION

20.1.1 GENERALITIES

Frequently used as a condiment or spice, coriander (*Coriandrum sativum* L.) is an annual herb. Many pathologies can be treated with the fruit in traditional medicine, *e.g.*, indigestion, worms, rheumatism, and joint pain. The claimed biological activities of coriander are particularly numerous, *e.g.*, alterative, antibiliary, antispasmodic, aphrodisiac, appetite stimulant, aromatic, carminative,

DOI: 10.1201/9781003204626-21

diaphoretic, diuretic, refrigerant, stimulant, stomachic, and tonic activities. Nowadays, the global production of coriander fruits is estimated to be around 600,000 tons per year (Sharma et al., 2014; Kamrozzaman et al., 2016).

Coriander seeds are exciting and have seen a clear revival in recent years, containing vegetable oil and a fraction of essential oil. Various factors influence their chemical composition. These include, for example, the geographical area of cultivation and the maturity stages of the seeds (Msaada et al., 2009). Coriander vegetable oil has a particular fatty acid profile. Its main fatty acid (up to 75%) is petroselinic acid. This is a rare isomer of oleic acid, present at high levels in a narrow range of plant oils, primarily those extracted from seeds of the Apiaceae family (Gunstone et al., 2007). The potential of petroselinic acid is strong and has recently attracted the interest of many industries (*e.g.*, food, cosmetics, and pharmaceuticals). Moreover, it could also be used in oleochemistry to synthesize many platform molecules in the coming years. The vegetable oil derived from coriander seeds was also certified in 2013 by the European Food Safety Authority (EFSA) as a Novel Food Ingredient (NFI), allowing its use as a food supplement for healthy adults (EFSA, 2013).

The benefits of coriander vegetable oil in general, particularly the petroselinic acid it contains, are therefore giving rise to a new interest in its extraction. On an industrial scale, the extraction of vegetable oils is generally performed in two successive steps, *i.e.*, a first step of pressing by mechanical action followed by organic solvent extraction of the residual oil contained in the oily cake. Nevertheless, extrusion technology as a single tool for the extraction of vegetable oils by mechanical pressing of oleaginous seeds is gaining more and more importance (Isobe et al., 1992; Crowe et al., 2001; Singh et al., 2002; Zheng et al., 2003; Uitterhaegen & Evon, 2017). In addition to single-screw extrusion, twin-screw extrusion could also be used. It has many advantages compared to the single-screw technology, including increased conveying force, better mixing, better seed crushing, and higher energy efficiency (Isobe et al., 1992; Bouvier & Guyomard, 1997; Uitterhaegen & Evon, 2017). Twin-screw extrusion has been used recently to extract vegetable oils from a wide range of oleaginous seeds (Isobe et al., 1992; Bouvier & Guyomard, 1997; Amalia Kartika et al., 2006; Evon et al., 2007, 2009, 2010a, 2013; Uitterhaegen & Evon, 2017). For this reason, the twin-screw technology could be implemented for the extraction of the vegetable oil contained in coriander seeds.

At the same time, the materials industry is focusing on more sustainable products and processes in the coming years in order to face several challenges:

- The depletion of fossil resources and changes in the world's forest resources have led to the search for alternative adhesives, polymer matrices with thermoplastic behavior, and lignocellulosic fibers.
- A constantly growing population.
- Ever-changing environmental concerns and regulations.

In order to effectively replace conventional materials not derived from renewable resources, the bio-sourced materials developed must still be competitive in terms of mechanical properties, durability, efficiency, cost, etc.

Crop residues and other co-products from the primary processing of agricultural resources are attractive alternative feedstocks because of their low cost, wide availability, and lack of competition with the food industry for land use. For these reasons, straw and press cake from seeds are two coriander co-products of interest to the materials sector.

20.1.2 POTENTIAL OF PRESS CAKE FROM CORIANDER SEEDS IN THE FIELD OF BIO-SOURCED MATERIALS

The press cake obtained by mechanical treatment in a twin-screw extruder will contain proteins and lignocellulosic fibers. As such, it will constitute a composite bio-sourced mixture of interest that could be used for the manufacture by thermopressing of self-bonded particleboards of high environmental value since they are both bio-sourced and biodegradable. Indeed, although they can also be used in the animal feed industry or as a source of energy, cakes from the mechanical pressing of other oleaginous seeds have already been used to manufacture such panels: sunflower (Evon et al., 2012, 2014a, 2015a), jatropha (Evon et al., 2014b), castor (Kurniati et al., 2015), etc. The proteins then act as natural binders, and cohesive panels are obtained without adding an exogenous binder of synthetic origin such as formaldehyde-based thermosetting resins. This greatly simplifies the manufacturing process (Tajuddin et al., 2016). On the other hand, lignocellulosic fibers can act as mechanical reinforcement inside the panels.

Often representing more than 30% of the production cost of wood-based panels available on the market (Van Dam et al., 2004a), formaldehyde-based resins are also harmful to human health (Okuda & Sato, 2004). The indoor use of panels glued with formaldehyde resins reduces the indoor air quality due to toxic formaldehyde emissions, and new regulations have thus been recently established on this subject (Salthammer et al., 2010).

20.1.3 POTENTIAL OF CORIANDER STRAW IN THE FIELD OF BIO-SOURCED MATERIALS

Generally representing between 60% and 85% of the weight of the aerial part of the plant, the straw constitutes for its part the lignocellulosic co-product of the culture of the coriander. Therefore, it could be used to manufacture fiberboards by thermopressing, contributing to their mechanical reinforcement. In the literature, self-bonded panels with promising performances have even been obtained without the addition of exogenous binder, and this from different co-products rich in lignocellulosic fibers such as wheat straw (Halvarsson et al., 2009), *Miscanthus sinensis* (Velásquez et al., 2003), sugarcane bagasse (Nonaka et al., 2013), or kenaf core (Okuda et al., 2006a, 2006b; Xu et al., 2006).

Under the high pressure and temperature conditions used during hot pressing, the glass transition of lignins can be exceeded (Bouajila et al., 2005). Due to their plasticization, lignins then act as a natural binder, having a matrix effect by embedding the cellulose fibers (Van Dam et al., 2004b). Due to the applied thermal stress, lignins and hemicelluloses also undergo some degree of degradation. The resulting monomers can crosslink in subsequent condensation reactions, contributing to self-bonding (Okuda et al., 2006a, 2006b).

Prior fiber refining can improve the mechanical and water-resistance properties of fiberboards. Besides a simple grinding step that increases the fiber surface and the accessibility to the inner cell wall components (Okuda & Sato, 2004), refining is more often conducted by steam treatment (or steam explosion) in a thermomechanical refiner or a digester. The dimensional stability of the boards is then significantly improved (Anglès et al., 2001; Velásquez et al., 2003; Xu et al., 2006). During the steam explosion, lignins are released from the inner cell wall to the fiber surface: their accessibility is thus significantly increased during hot pressing, and their mobilization as a natural binder is thus largely exacerbated. Hydrolysis reactions are also likely to occur in the presence of steam, especially in hemicelluloses (Tajuddin et al., 2016). Their content then decreases. Since hemicelluloses are highly hygroscopic by nature, the water-resistance of fiberboards can be significantly improved (Velásquez et al., 2003; Anglès et al., 2001).

Last but not least, the morphology of the cellulose fibers is considerably enhanced after the steam treatment, which allows them to be separated from each other. This increases their specific surface area and aspect ratio, defined as the ratio between their length and width (Xu et al., 2006). This improves their ability to be mechanically reinforced. Therefore, in addition to refiners and other digesters, the twin-screw extrusion technology may also be suitable for the refining (*i.e.*, the defibration) of lignocellulosic fibers in the presence of water before board forming, as twin-screw extruders are effectively thermo-mechano-chemical reactors (Evon et al., 2018).

Whether ground, micronized as flour, or (extrusion-)refined, lignocellulose fibers can also be dispersed in plastic matrices for use as reinforcing fillers. In particular, the automotive and construction industries are important users of natural fiber composites (Faruk et al., 2014; Pickering et al., 2016). The additional advantages of natural fibers for this type of use are low density and reinforcement efficiency comparable to traditional fillers such as E-glass fibers (John & Thomas, 2008). In order to optimize the mechanical performance of the composite material, it is still necessary to ensure excellent dispersion of the fibers in the polymer matrix. In particular, the versatility of twin-screw extruders gives them excellent mixing capability (Evon et al., 2018). For this reason, these industrial tools are frequently implemented for compounding, whether plant fibers are dispersed in petroleum-based polymer matrices such as polypropylene (PP) (Bledzki et al., 2005; Yang et al., 2006) or bio-based ones such as poly(lactic acid) (PLA) (Gamon et al., 2013). In addition to obtaining fiber panels already mentioned above, coriander straw could also be used for the mechanical reinforcement of plastic composites.

Finally, due to its high lignocellulose content, coriander straw could also be used for thermal insulation of buildings (*i.e.*, walls, ceilings, attic spaces of houses, etc.). Indeed, lignocellulosic fibers naturally have good thermal insulation abilities (Saiah et al., 2010). Over the last 20 years, a multitude of agricultural co-products have been used for such applications, in the form of loose insulation or insulating panels, *e.g.*, kenaf (Ardente et al., 2008), flax and hemp (Kymäläinen & Sjöberg, 2008; Korjenic et al., 2011; Benfratello et al., 2013), cotton stalk (Zhou et al., 2010), jute (Korjenic et al., 2011), etc. Pith from sunflower stem (Sabathier et al., 2017; Verdier et al., 2021) and a cake from the whole sunflower plant (Evon et al., 2014a, 2015b) could also provide insulation boards.

Their density strongly influences the thermal conductivity of insulation boards (Zhou et al., 2010; Benfratello et al., 2013; Evon et al., 2014a, 2015b; Verdier et al., 2021). Therefore, the least dense materials have the lowest thermal conductivity values and thus the best insulating character. For example, a sunflower pith insulation board has a thermal conductivity of only 35 mW/m.K at 25°C for an extremely low density, equal to only 50 kg/m^3 (Verdier et al., 2021). This is a thermal conductivity perfectly comparable to those of conventional materials also used for the thermal insulation of buildings, *e.g.*, expanded polystyrene (37 mW/m.K), rock wool (36 mW/m.K), and glass wool (35 mW/m.K).

20.1.4 STRUCTURE OF THE CHAPTER

This chapter focuses on the possible uses already investigated at the laboratory scale of coriander straw and press cake from seeds in the field of bio-sourced materials. It is organized in three successive parts:

- First, the vegetable oil extraction process by mechanical pressing coriander seeds in a twin-screw extruder will be briefly described. Then, the influence of the extraction operating conditions on the vegetable oil yield will be discussed. The characteristics of the two liquid fractions resulting from this process, namely the fragrant vegetable oil and aromatic water, will also be described as well as their possible future uses. Finally, the solid raffinate resulting from the seeds' fractionation constitutes the press cake. It is precisely the latter that will be the object of valorization in the field of bio-sourced materials. Such valorization will be discussed exhaustively in the following part.
- In the second part devoted exclusively to the press cake, its chemical composition will first be mentioned. Concerning this, possible uses will be suggested. The thermoplastic and adhesive character of the proteins of the press cake, also briefly described in this part, will justify its use in the field of bio-sourced materials. Its transformation into particleboards by thermopressing will be discussed at this level. The properties of use of particleboards thus obtained will also be exposed.
- Finally, in the third and last part, the obtaining of bio-sourced materials based on coriander straw will be described. Four different forms of materials will be presented as well as their characteristics of use and the possible methods of valorization for these materials:

- Hot-pressed 100% coriander fiberboards combining straw as a mechanical reinforcement and press cake as a natural binder.
- Insulating materials in the form of loose straw or low-density blocks in which the straw particles are glued together with a starch-based binder.
- Injectable or extrudable thermoplastic composites in which the straw will act again as a mechanical reinforcement of the plastic matrices considered.

20.2 GENERATION OF CORIANDER PRESS CAKE FROM SEEDS WHEN EXTRACTING VEGETABLE OIL THROUGH MECHANICAL PRESSING

The coriander press cake is generated from seeds when continuously extracting vegetable oil through mechanical pressing using a twin-screw extruder (Uitterhaegen et al., 2015). This part of the chapter describes the conditions of oil extraction, the operating parameters that can influence the yield obtained, and the chemical characteristics of the three fractions generated, especially the cake.

20.2.1 OIL EXTRACTION EFFICIENCY AS A FUNCTION OF THE OPERATING CONDITIONS

In Uitterhaegen et al. (2015), the screw profile was first optimized, and the one that showed the best efficiency is shown in Figure 20.1. Then, it was also found during extraction that the filling ratio of the machine on the one hand and the temperature at the pressing zone on the other hand strongly influenced the efficiency of the oil extraction. In particular, with the optimized screw profile, an improvement in the efficiency of the liquid/solid pressing in the counter-threads (or reverse-pitch elements) (CF2C) positioned at the end of the screw profile is observed by increasing the filling ratio of the twin-screw extruder (from 31 g/h.rpm to 47 g/h.rpm). This increase is made possible by increasing the incoming flow rate of seeds (from 3.1 kg/h to 4.7 kg/h) for a constant screw rotation speed (100 rpm). This results in both a greater decrease in the residual lipid content in the cake (from 16.5% to 16.1% of its dry mass), while at the same time, a slightly higher R_C oil yield (from 47% to 49%) (Table 20.1). The R_L oil yield, which corresponds to the oil separated from the foot after centrifugation of the filtrate, increases significantly (from 40% to 46%). This is explained by the fact that the foot content in the filtrate is much lower in the case of the highest filling rate (8% instead of 16%).

A reduction of the pressing temperature in module 7 (from 120°C to 65°C) was considered in a second step using the previously optimized filling ratio (47 g/h.rpm). The compression effect in the pressing zone is then increased, which is explained by the gradual increase in the viscosity of the processed mixture. As a result, the residual oil content of the cake is reduced from 16.5% to 15.0% of its dry mass, and an increase in R_C yield from 47% to 54% is observed (Table 20.1). However, when the pressing temperature is less than or equal to 80°C, it is essential to note that the filtrate's

1		2						3		4		5		6		7		
T2F 50	T2F 50	C2F 33	C2F 25	DM 10×10 (45°)	C2F 33	C2F 25	BB 10×10 (90°)	C2F 33	C2F 33	C2F 25	C2F 25	C2F 16	C2F 16	CF2C -33	C2F 25	C2F 33

FIGURE 20.1 Optimized screw profile for extracting vegetable oil from coriander seeds in a Clextral BC 21 twin-screw extruder (reproduced from Uitterhaegen et al. (2015) with the publisher's permission). *T2F, trapezoidal double-thread screw; C2F, conveying double-thread screw; DM, monolobe paddle-screw; BB, bilobe paddle-screw; CF2C, reverse screw. The numbers following the type of screw indicate the pitch (in mm) of T2F, C2F, and CF2C screws and the number of the DM and BB paddles, each 5 mm in length.*

TABLE 20.1

Mechanical Pressing of Coriander Seeds in a Clextral BC 21 Twin-Screw Extruder: Operating Conditions and Vegetable Oil Extraction Yields

Trial Number	1	2	3	4	5	6
Operating conditions						
S_S (rpm)	100	100	100	100	100	100
Q_S (kg/h)	3.15	3.94	4.71	4.71	4.71	4.71
θ_{c7} (°C)	120	120	120	100	80	65
	(120.1 ± 0.7)	(118.1 ± 1.1)	(119.5 ± 0.7)	(101.0 ± 1.1)	(79.7 ± 0.9)	(64.8 ± 0.5)
Filtrate						
Q_F (kg/h)	0.38	0.52	0.59	0.59	0.59	0.59
T_L (%)	83.6	89.3	91.5	91.7	82.1	84.1
T_F (%)	16.4	10.7	8.5	8.3	17.9	15.9
Cake						
Q_C (kg/h)	2.61	3.31	3.92	4.00	3.97	4.00
H_C (%)	2.32 ± 0.14^f	3.73 ± 0.08^e	4.05 ± 0.06^d	4.88 ± 0.22^c	8.62 ± 0.27^b	8.99 ± 0.19^a
L_C (% of dry matter)	16.48 ± 0.21^b	16.82 ± 0.29^a	16.06 ± 0.08^c	16.46 ± 0.11^b	14.99 ± 0.05^d	15.23 ± 0.14^d
Vegetable oil extraction yield (%)						
R_L	40.5	46.9	45.8	46.3	41.1	41.8
R_C	46.7	45.6	48.7	46.7	53.7	52.8

Source: reproduced from Uitterhaegen et al. (2015) with the publisher's permission.

S_S is the screw rotation speed. Q_S is the inlet flow rate of coriander seeds. θ_{c7} is the temperature of the pressing zone (i.e., module 7) (set value first mentioned, plus temperature measured during sampling in parentheses). Q_F is the outlet flow rate of the filtrate. T_L is the oil proportion in the filtrate. T_F is the foot proportion in the filtrate. Q_C is the outlet flow rate of the cake. H_C is the cake moisture content. L_C is the residual oil content in the cake. R_L is the oil extraction yield in proportion to the seed oil content. R_C is the oil extraction yield in proportion to the residual oil in the cake. Means in the same line and with different superscript letters (a–f) are significantly different at $P < 0.05$.

foot content increases (from 8% to 18%). This is because the proportion of solid particles accumulating on top of the sieves is exceptionally high under such temperature conditions. As a result, a slight decrease in the R_L yield (from 46% to 41–42%) is observed. Therefore, it is reasonable to assume that a median temperature (100°C) is the optimum pressing temperature. In such a situation, the foot content of the filtrate is effectively reduced (only 8%), and the highest R_L yield in the study (46%) is then obtained (Uitterhaegen et al., 2015). In the future, in addition to adjusting the operating conditions of the twin-screw extruder, the influence of the dryness of the coriander seeds at the inlet of the machine on the extraction yield of vegetable oil by mechanical pressing can be studied. More dry seeds are expected for the increase in this yield (Evon et al., 2013).

20.2.2 LIQUID FRACTIONS OBTAINED DURING THE MECHANICAL PRESSING OF CORIANDER SEEDS, AND THEIR POSSIBLE USES

In addition to the solid raffinate (*i.e.*, press cake), two distinct liquid extracts are generated during the thermo-mechanical fractionation of coriander seeds in the twin-screw extruder. These are not only the fragrant vegetable oil fraction mentioned above but also a condensate enriched with essential oil (aromatic water) (Figure 20.2).

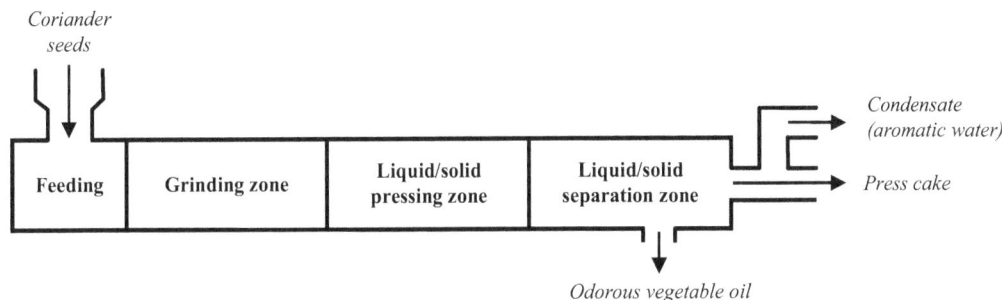

FIGURE 20.2 General diagram of the fractionation of coriander seeds in a twin-screw extruder through mechanical pressing (Evon, 2019).

From the point of view of its fatty acid composition, coriander vegetable oil is particularly rich in petroselinic acid (75%), a rare isomer of oleic acid. It thus has an original composition. Also rich in antioxidants, its virtues for human health are recognized (Uitterhaegen et al., 2015). During thermo-mechanical pressing in a twin-screw extruder, the vegetable and essential oils contained in the seed are co-extracted, which allows the vegetable oil generated to be pleasantly scented. Furthermore, the extraction conditions' choice allows the odorant level of the vegetable oil produced to be precisely adjusted. For standard extraction conditions, its concentration in volatile organic compounds (VOC) has been estimated at 35.3 mg/g. Among the fragrant constituents identified, linalool is by far the most prevalent, with a concentration of 25.3 mg/g (72%) (Figure 20.3). The other molecules identified are α-pinene, limonene, γ-terpinene, etc.

Concerning its characteristic profile in fatty acids on the one hand and the presence of odorous molecules, on the other hand, the vegetable oil of coriander generated during the mechanical pressing of the seeds in a twin-screw extruder could find applications in many sectors. It is already used today in nutraceuticals (food supplements) but could soon also be used in the formulation of cosmetic creams, in nutrition, or in pharmaceuticals (Uitterhaegen et al., 2015; Uitterhaegen, 2018). Furthermore, it could also be used as an essential source of petroselinic acid, which, once chemically modified, will be used to manufacture new reaction intermediates.

In addition to the pressed oil, aromatic water is also generated during the twin-screw fractionation (Uitterhaegen, 2018). This can be achieved by collecting the vapors escaping from the solid outlet of the twin-screw extruder. A coolant is then used to condense them. The analysis of its composition in odorous molecules was carried out in the same way as for the vegetable oil by a headspace sampling, followed by a chromatographic analysis (GC-FID/MS). It showed that linalool was also the majority constituent (Figure 20.3). The other molecules present are α-pinene, γ-terpinene, camphor, geranyl acetate, etc. These are precisely the major compounds present in the essential oil of coriander seeds when it is obtained in the classical hydrodistillation way. This highlights that the condensate contains part of this essential oil. Therefore, aromatic water can be used in the field of perfumery.

On the other hand, the transformation of the press cakes into bio-sourced materials can be envisaged. The following section describes the method to adopt and the characteristics of the materials obtained.

20.3 BIO-SOURCED MATERIALS FROM THE CORIANDER PRESS CAKE

20.3.1 CHEMICAL COMPOSITION

During the agro-industrial processes of biomass fractionation, the objective is to reduce as much as possible the production of waste and to try to valorize all the fractions generated. Therefore, it

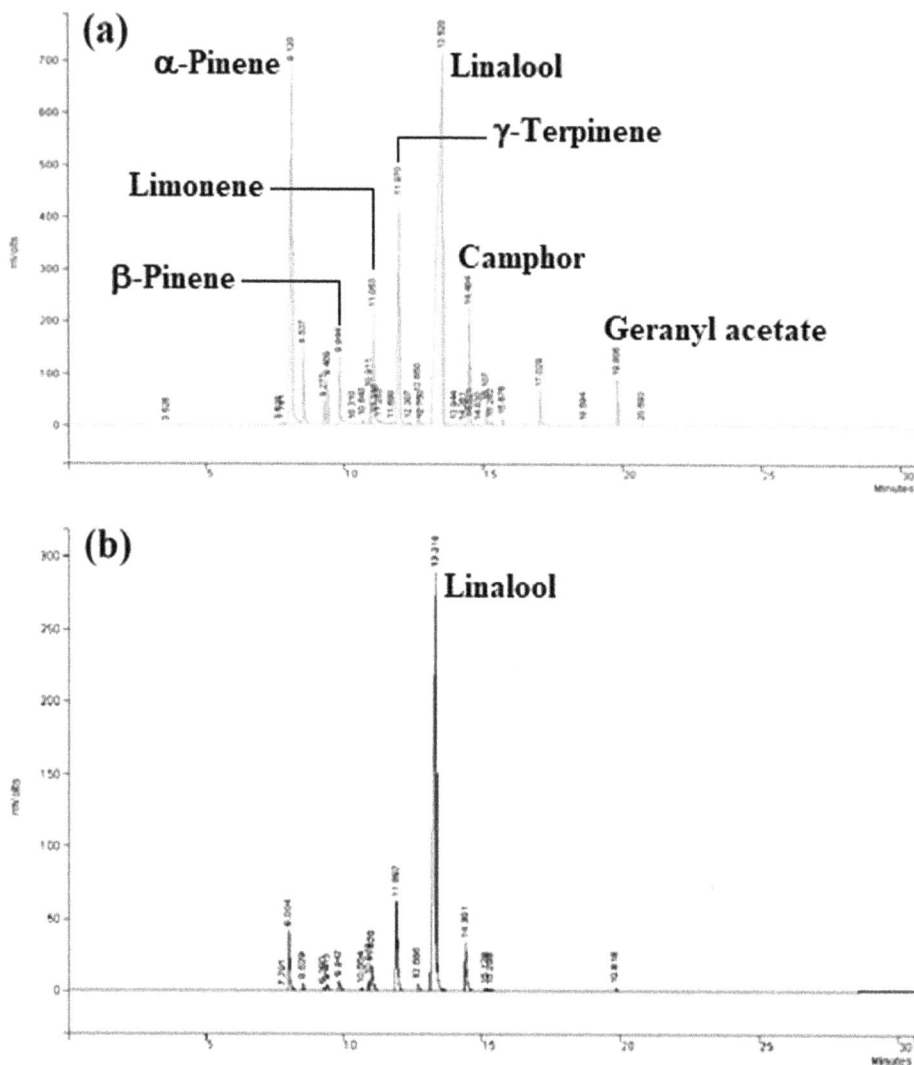

FIGURE 20.3   GC-FID analysis of the headspace of the odorous vegetable oil (a) and the condensate (aromatic water) (b) after their SPME sampling (Evon, 2019).

seems judicious to consider the coriander press cake valorization in this context of plant refinery (or biorefinery).

The vegetable oil is not entirely extracted during the twin-screw treatment of coriander seeds. The fraction of lipids that could not be pressed constitutes the residual oil of the cake. Organic solvent extraction will allow, if necessary, to extract these residual lipids during an additional extraction stage. Among the other chemical constituents present in the coriander cake, it is possible to mention minerals, water-soluble compounds, proteins, and plant fibers, which are themselves divided into three families of compounds: cellulose, hemicelluloses, and lignins. Table 20.2 presents the chemical composition of the press cake that was specifically used to manufacture bio-sourced materials through thermopressing and thermomolding (Uitterhaegen et al., 2016). This press cake was obtained using the operating conditions of trial number 4 presented in Table 20.1. However, its residual oil content was slightly higher as it was not from the same batch of coriander seeds. The chemical composition after the cake deoiling in a Soxhlet extractor is also mentioned in Table 20.2.

TABLE 20.2

Chemical Composition (% of Dry Matter) of a Press Cake Generated by Mechanical Pressing of Coriander Seeds in a Twin-Screw Extruder, and the Chemical Composition after the Cake's Deoiling in a Soxhlet Extractor

Deoiling	No	Yes
Minerals	6.2	6.7
Lipids	17.2	0.9
Proteins	18.2	26.7
Cellulose	26.6	34.7
Hemicelluloses	22.2	36.9
Lignins	5.0	1.0
Water-soluble compounds	16.0	15.6

Source: Uitterhaegen et al., 2016, 2017a; Evon, 2019.

Although the residual oil in the press cake collected at the twin-screw extruder outlet could be a disadvantage for its direct use, different valorizations are still possible concerning its chemical composition. In addition to obtaining bio-sourced materials, which will be discussed later, other uses are also possible. These are listed in the following paragraph.

20.3.2 Possible Valorizations Other than in the Field of Bio-Sourced Materials

The possible valorizations other than in the field of bio-sourced materials are listed under:

- The production of energy by combustion in pellet boilers, gasification, or pyrolysis is the first possible use for the cake from thermo-mechanical pressing (Yorgun et al., 2001; Gerçel, 2002).
- The press cake is rich in lignocellulosic fibers and proteins (Table 20.2). Moreover, it has not been in contact with any organic solvent during its production. This is why an important axis of valorization could be that of animal feed.
- Depending on the thermo-mechanical pressing conditions used in the twin-screw extruder, the pentosan content of the press cake is between 8.9% and 11.5% of its dry mass, which is far from negligible (Uitterhaegen, 2018). This is why the cake also appears as an adjuvant of choice in bread making. For example, the milling industry and other cookie manufacturers could use it in the coming years to develop gluten-free products.
- Various natural antioxidants are contained in coriander cake. A methanolic extraction would allow them to be isolated for future use due to their beneficial health impact (Sriti et al., 2014; Uitterhaegen et al., 2015).
- Moreover, last but not least, the coriander press cake is a source of essential oil (0.3% of its dry mass) corresponding to the fraction of essential oil initially present in the seed that was not extracted during the twin-screw extrusion fractionation (Uitterhaegen et al., 2015; Uitterhaegen, 2018). Therefore, hydrodistillation of the cake will allow the recovery of this essential oil, characterized by a high linalool content (up to 77%) and a fresh and flowery fragrance. Therefore, for these two reasons, this fragrant fraction could be used:
- As a mosquito repellent (Müller et al., 2009).
- In the cosmetic industry, for various applications, *i.e.*, the manufacture of perfumes, shampoos, soaps, household detergents, etc. (Lapczynski et al., 2008).

Nevertheless, since the essential oil in the press cake is present in small quantities, another strategy could be not to try to extract it necessarily. Instead, leaving it in other finished products such as bio-sourced materials would give them added value. This axis of valorization of the cake for the manufacture of bio-sourced materials will be discussed mainly in the following paragraph.

20.3.3 Bio-Sourced Materials through Thermopressing and Thermomolding

The high lignocellulosic fiber content of the coriander cake and its significant protein content (Table 20.2) make this agro-industrial co-product a kind of natural composite. Indeed, the thermomechanical behavior of the proteins in the press cake was studied by DSC analysis, and it was found that the fractionation by twin-screw extrusion had rendered the cake's proteins partially denatured (Uitterhaegen et al., 2016). This same phenomenon was previously observed for sunflower proteins (Evon et al., 2007, 2009, 2010b; Evon, 2008). Thus denatured, the proteins of coriander cake can be used as a natural binder under thermo-mechanical stress. In addition to the first uses envisaged in the previous paragraph, coriander cake can be shaped into bio-sourced materials, mainly by thermopressing, constituting an original way of valorization.

During thermopressing, also called hot pressing, the raw material is positioned between the two plates of a hydraulic press. It is then molded under temperature and pressure applied for a known period. In this way, particleboards or panels can be generated from the coriander cake. Furniture and construction are the two main sectors interested in such materials. However, the storage and transportation industries could also use them.

It should be noted here that thermopressing could eventually be carried out in a mold: the object obtained will then take the shape of the cavity. In this case, we speak rather of thermomolding. Even if no thermomolding test has been carried out to date on coriander cake, this technique would make it possible to transform it into containers of simple forms such as pet bowls or litter bins.

The valorization in the field of bio-sourced materials of the coriander cake generated at the end of the twin-screw treatment was mainly done with this thermocompression technology. More details about the results obtained for this use are mentioned below. Nevertheless, before presenting them, it is also important to point out that the press cake also has a high content of vegetable fibers. Its use in biocomposites could thus be considered in the future, even if such works are not mentioned in the recent literature. Indeed, the fibrous fraction of the cake could play a function of reinforcing filler inside various thermoplastic polymers, including compostable polymers such as PLA (Gamon et al., 2013), polyhydroxyalkanoates (PHA), and polycaprolactone (PCL) (Diebel et al., 2012). For this, the coriander cake will have to be incorporated into these plastic matrices *via* the first step of twin-screw compounding. Then, the composite granules thus formulated could be molded by thermoplastic injection or extruded, which would make it possible to obtain various parts such as ephemeral use parts (disposable cutlery, coffee capsules, etc.), food trays, films, profiles (*e.g.*, decking), etc.

When coriander cake is hot pressed (Uitterhaegen et al., 2016), the effect of temperature allows the initially amorphous proteins to reach the glassy transition, which then changes into a "molten" or rubbery form. This change of state of the proteins favors an efficient wetting of the fibers inside the cake. In the study, different thermopressing conditions (*i.e.*, mold temperature, applied pressure, and pressing time) were investigated, and the resulting particleboards were all cohesive, with the proteins and fibers acting as a natural binder and reinforcing filler, respectively. The operating conditions of thermopressing strongly influence the use properties of the boards. The optimized conditions are a temperature of 200°C and a pressure of 36.8 MPa, applied for 180 s. This results in particleboard with flexural strength, modulus of elasticity, and surface hardness of 11.3 MPa, 2.6 GPa, and 71° Shore D. And, although the water sensitivity of this board is not excellent (51% and 33%, respectively, for thickness swelling and water absorption after 24 h of immersion in water), it could be valued as a pallet interlayer sheeting or for the manufacture of containers in the packaging industry.

At the moment of hot pressing, the applied pressure causes some residual oil in the coriander cake to flow out. This oil can be quickly recovered by using a mold with vents located on the side walls of the mold (Uitterhaegen et al., 2016). Furthermore, an increase in the thermopressing conditions contributes to the increase in the amount of oil evacuated from the mold during hot pressing. The residual oil content in the particleboard is thus decreased, and the total pressed oil yield (*i.e.*, twin-screw extrusion plus thermopressing) is significantly increased, *i.e.*, up to 81%.

Once recovered, the oil pressed during the hot pressing process is filtered to remove any small solid particles it may contain and refined if necessary. However, the high thermo-mechanical stresses applied during the hot pressing process make this oil a much lower quality oil than oil pressed continuously by twin-screw extrusion. It is therefore not recommended to use it directly for human consumption. Various uses in the non-food field are, on the other hand, perfectly feasible:

- Its use in lipochemistry for synthesizing many new reaction intermediates because of its high petroselinic acid content (73%).
- Its use as a biolubricant.
- Its transformation into biodiesel by transesterification of triglycerides.

20.3.4 IMPROVEMENT OF PARTICLEBOARD PROPERTIES THANKS TO THE EXTRACTION OF RESIDUAL OIL FROM THE CAKE BEFORE HOT PRESSING

Although cohesive, the particleboards obtained by direct hot pressing of coriander cake still show relatively poor mechanical strength. This is because defects appear within the material when some residual lipids are expressed during molding. The creation of drain channels inside the materials then creates an excessive brittleness, which will make it break too quickly for specific uses at the time of mechanical stress. It was therefore imagined to optimize the thermocompression process. For this, the use of a cake previously deoiled with an organic solvent should be preferred to the cake still containing the residual lipids not extracted during the mechanical pressing of seeds in a twin-screw extruder. Hexane is the organic solvent of reference for such an extraction, which is very well mastered, including industrial-scale applications. At the end of the solvent extraction, the oil content in the cake becomes very low, *i.e.*, between 0.5% and 1.5%, at the exit of the extractor (Campbell, 1983). For coriander press cake, delipidation was performed on a laboratory scale using a Soxhlet extractor, and cyclohexane was preferred over hexane (Uitterhaegen et al., 2017a; Evon, 2019). The lipid content of the cake at the end of extraction was less than 1% of its dry mass (Table 20.2). This deoiled cake was used in a second step for hot pressing.

Using this deoiled cake, self-bonded particleboards are still obtained after hot pressing. Again, and as already observed in Uitterhaegen et al. (2016), the properties of the obtained panels, all of which are cohesive, vary depending on the conditions implemented during hot pressing (Uitterhaegen et al., 2017a), in particular the bending properties, *i.e.*, flexural strength (Figure 20.4) and elastic modulus (Figure 20.5). A material with flexural strength, elastic modulus, and surface hardness of 23.1 MPa, 4.4 GPa, and 80° Shore D, respectively, is obtained under process conditions (205°C, 21.6 MPa, 300 s) considered the optimized ones. Its swelling in thickness is equal to 31% and can even be reduced to only 20% by applying a heat treatment after thermopressing. This self-bonded panel is undoubtedly an environmentally friendly alternative to commercial wood-based panels such as oriented strand board (OSB) and particleboard.

In a logic of increasing the industrial forming rates, a further improvement of the hot pressing conditions can consist in a decrease of the molding time (from 300 s to 180 s) simultaneously with an increase of the applied pressure (from 21.6 MPa to 38.8 MPa) (Evon, 2019). When hot pressed under these perfectly controlled operating conditions, the particleboard is even more resistant to bending and water than in Uitterhaegen et al. (2017a). Its characteristics were compared to those of a panel obtained under the same thermopressing conditions but from a cake from which the residual oil had not been extracted with organic solvent beforehand (Table 20.3). The results show a

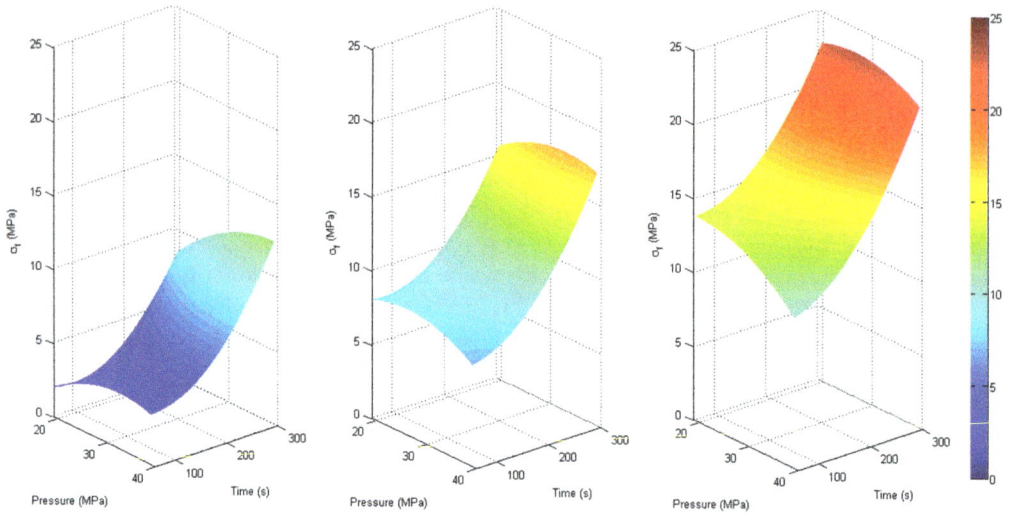

FIGURE 20.4 Surface plots for flexural strength (σ_f) for a mold temperature of 160°C, 180°C, and 200°C, respectively (from left to right) (reproduced from Uitterhaegen et al. (2017a) with the permission of the publisher).

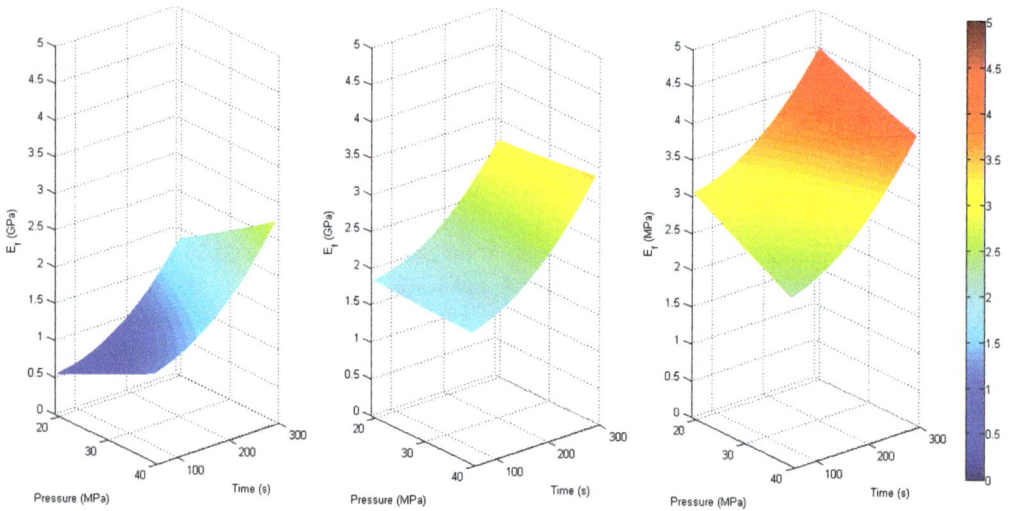

FIGURE 20.5 Surface plots for modulus of elasticity (E_f) for a mold temperature of 160°C, 180°C, and 200°C, respectively (from left to right) (reproduced from Uitterhaegen et al. (2017a) with the permission of the publisher).

noticeable improvement in the mechanical properties (bending strength and surface hardness) and water-resistance of the panel from the deoiled cake. Thus, it is reasonable to assume that when some of the residual oil in the cake is expressed during thermocompression, the internal defects generated contribute to the material's embrittlement.

In conclusion, the extraction of the residual oil from the coriander press cake with an organic solvent before the hot pressing process favors the production of particleboards that are much more resistant, both mechanically and to water. Although this process is well mastered on an industrial scale, it is not cost-neutral. According to the French standard NF EN 312 concerning particleboard requirements, the coriander board made from deoiled cake and whose characteristics are presented in Table 20.3 can already be classified as a board working in a dry environment (P4 type board). The

TABLE 20.3

Characteristics of Particleboards Resulting from the Hot Pressing (200°C Mold Temperature, 38.8 MPa Applied Pressure, and 180 s Molding Time) of a Coriander Press Cake before and after Its Deoiling in a Soxhlet Extractor

Cake Pre-Treatment	Non-Deoiled	Deoiled
Bending properties (NF EN 310 French standard)		
d (kg/m³)	1293 ± 7	1295 ± 30
σ_f (MPa)	4.7 ± 0.4	26.9 ± 2.8
E_f (MPa)	1548 ± 61	4773 ± 377
Surface hardness (NF EN ISO 868 French standard)		
Shore D (°)	64.1 ± 4.4	82.2 ± 1.6
Water-resistance (NF EN 317 French standard)		
TS (%)	76.7 ± 5.2	20.7 ± 1.8
WA (%)	66.8 ± 4.0	17.7 ± 0.8

Source: Evon, 2019.

d is the board's density. σ_f is the flexural strength. E_f is the elastic modulus. TS and WA are the thickness swelling and water absorption, respectively, after 24 h immersion in water.

standard's recommendations concerning the flexural properties (16 MPa for the flexural strength and 2.2 GPa for the modulus of elasticity, in the case of panels with a thickness between 4 mm and 6 mm) are primarily reached. The required minimum internal cohesion (0.40 MPa) is also achieved (0.68 MPa) (Uitterhaegen et al., 2017a). The same is true for the water-resistance (21% max for thickness swelling after 24 h of immersion). Various post-treatments could even further improve this dimensional stability criterion, which is of paramount importance for possible prolonged use of the materials in wet environments. In particular, thickness swelling could be reduced by cooking or through treatment of the panel surface by chemical or steam action (Widyorini et al., 2005; Okuda et al., 2006a; Evon, 2008; Halvarsson et al., 2009; Saadaoui et al., 2013; Uitterhaegen et al., 2017a).

This self-bonded coriander particleboard is an environmentally friendly alternative to commercially available products such as plywood, particleboard, OSB, medium-density fiberboard (MDF), high-density fiberboard (HDF), or hardwood particleboards and fiberboards (*e.g.*, Isorel or Unalit). The use of chemical adhesives such as urea-formaldehyde (UF) or phenol-formaldehyde (PF) thermosetting resins is not necessary to obtain it.

For the investigation of the emission of volatile organic compounds (VOC) from this optimal particleboard, a VOC concentration (adsorption) sampling method, followed by thermodesorption and then GC-FID/MS analysis, was developed at the laboratory scale and then validated (Uitterhaegen et al., 2018a). Volatile compounds emitted during the early life of the panel were identified. For this purpose, a comparison can be made of experimental mass spectra with those from spectra libraries and retention indices. It appears that it is mainly the presence of residual essential oil inside the deoiled press cake that is responsible for the VOC emissions from the coriander particleboard. Again, linalool was identified as the most abundant volatile compound. The VOC emissions were quantified after their active sampling on Tenax TA® adsorbent tubes, then thermal desorption, and finally by GC-FID analysis. The specific emission rate of linalool was first measured at 25°C and 50% relative humidity. The value obtained was 125 µg/m².h, increasing by +200% at 35°C. In parallel, linalool has high volatility, resulting in a progressive decrease in its specific emission rate over time (*e.g.*, only 12 µg/m².h after one year).

Conversely, no formaldehyde emission was found from this optimal particleboard even though formaldehyde can be naturally emitted from some plant species (Kesselmeier and Staudt, 1999;

Böhm et al., 2012). This confirms the more sustainable nature of such material compared to wood-based panels (FCBA, 2013).

The self-bonded particleboard based on coriander cake naturally emits linalool; it will be of real interest, whether for construction or agricultural uses. Indeed, linalool is a monoterpene with a characteristic odor that presents fascinating intrinsic characteristics. Besides its relaxing effect, it has proven bioactivity, being at the same time antioxidant, antimicrobial, and insect repellent. On the one hand, it will improve the quality of indoor air. On the other hand, it will promote a longer shelf life of agricultural or food products. The emission of linalool by this coriander panel can therefore be considered advantageous.

20.4 BIO-SOURCED MATERIALS FROM CORIANDER STRAW

20.4.1 Valorization Scheme for Coriander Straw and Its Chemical Composition

The straw constitutes the co-product of coriander cultivation. It can represent up to 85% (w/w) of the plant aerial part in the most favorable climatic conditions. Moreover, this co-product is particularly rich in lignocellulose: up to 65% of its dry mass (Uitterhaegen et al., 2016). As such, coriander straw can thus be considered as a promising source of reinforcement for bio-sourced materials. Figure 20.6 proposes a valorization scheme for coriander straw in this field, including applications used in a mixture with the press cake or other additives. The following paragraphs will present in more detail these different ways of valorization. As a reminder, the uses of the press cake alone in the field of bio-sourced materials presented in the previous chapter's section also appear in Figure 20.6.

20.4.2 Hardboards (i.e., Fiberboards with High Density)

First, whether used alone or mixed with the coriander cake from the thermo-mechanical pressing of the seeds in a twin-screw extruder, coriander straw can manufacture fiberboards by hot pressing. These materials are in the form of rigid panels of high density and are therefore hardboards. It is mainly the building industry that can use them. Various uses have already been identified, *e.g.*, floor underlays, interior partitions, ceiling tiles, etc. First of all, the thermopressing of coriander straw alone, simply crushed to 7.5 mm before shaping, makes it possible to generate cohesive panels by hot pressing. The cohesion of such materials is made possible by the presence of a fraction of hemi-celluloses (15.0%) but also a fraction of lignins (13.6%) inside the coriander straw, both families of biopolymers being able to be mobilized as natural binders under the influence of temperature (Uitterhaegen et al., 2016).

Nevertheless, the flexural properties of the fiberboard can be significantly increased by add-ing the deoiled coriander press cake to this straw. For example, in the case of a 100% coriander fiberboard containing 25% (w/w) deoiled cake and 75% crushed coriander straw, and produced at 200°C, 36.8 MPa, and 180 s, the flexural strength is increased by 89% (from 7.9 to 14.9 MPa), and the elastic modulus is increased by 73% (from 1.9 to 3.3 GPa) (Uitterhaegen, 2018). The proteins in the deoiled cake then act as a natural adhesive within the material, with the presence of this protein binder leading to improved wetting of the straw fibers (Uitterhaegen et al., 2016, 2017a, 2017b). This fiberboard also has substantially improved water-resistance, even though the densities of the two fiber materials (*i.e.*, with and without the addition of coriander cake) are perfectly comparable (about 1145 kg/m^3).

In order to optimize this fiberboard, it is possible to replace simply ground straw with a coriander straw fraction that has previously undergone thermo-mechanical defibration treatment with water in a twin-screw extruder. In Uitterhaegen et al. (2017b), several liquid-to-solid ratios (0.4 to 1.0) were tested during straw defibration by twin-screw extrusion. A ratio equal to 0.4 was found to be the most effective concerning the performance of the fiberboard, in particular its mechanical strength. Indeed, the fiberboard generated by thermopressing (205°C for the mold temperature and 21.6 MPa

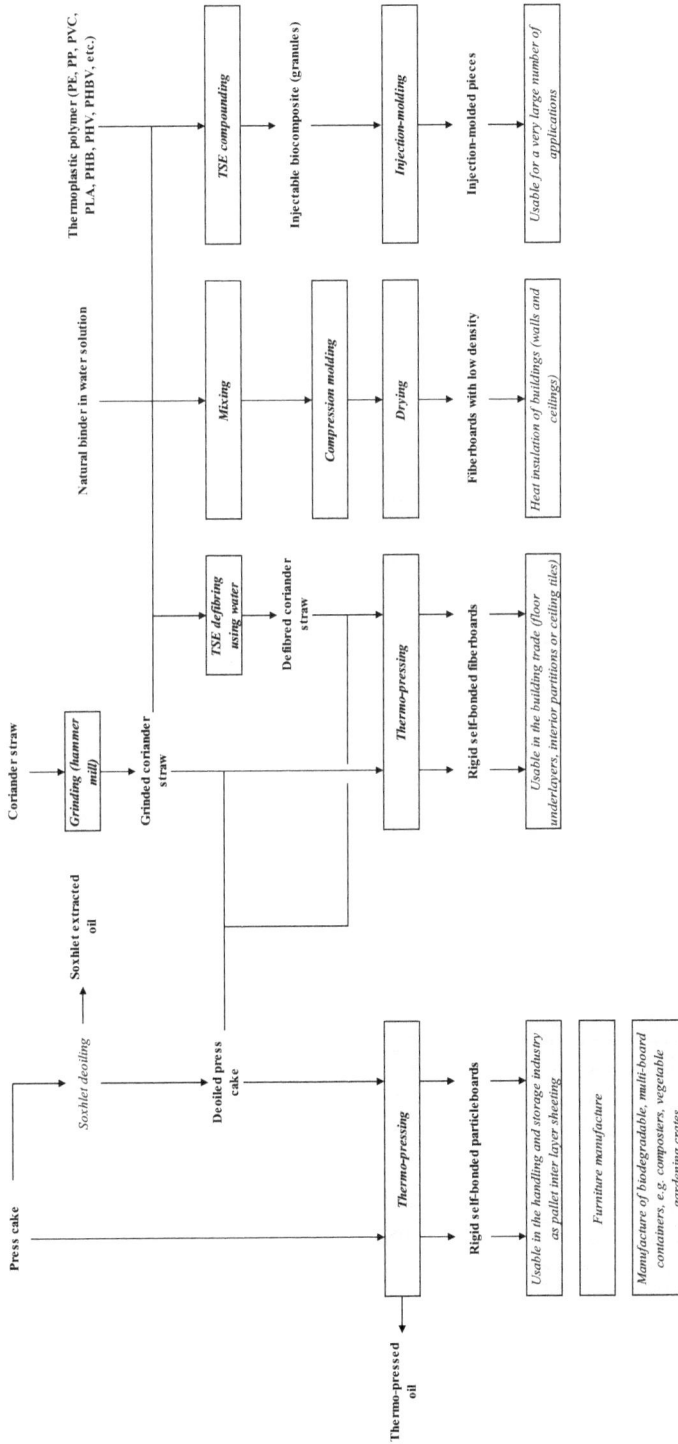

FIGURE 20.6 Valorization scheme of coriander press cake and straw in the field of bio-sourced materials (Evon, 2019). *TSE, twin-screw extrusion.*

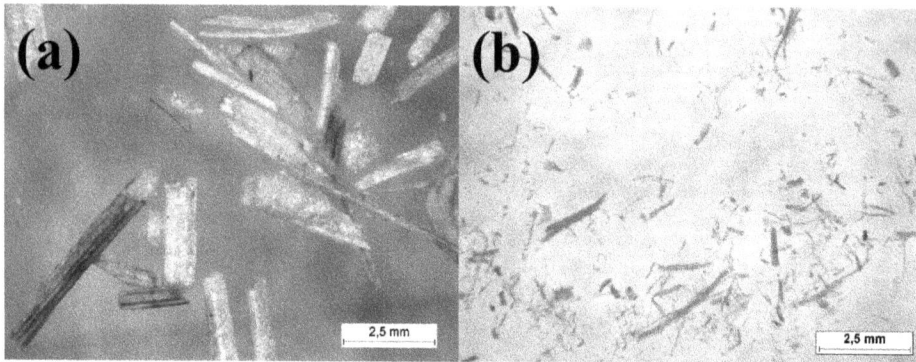

FIGURE 20.7   Macrographs of the crushed coriander straw (a) and the extrusion-refined straw using a liquid-to-solid ratio of 0.4 (b) (reproduced from Uitterhaegen et al. (2017b) with the permission of the publisher).

for the pressure, applied for 300 s) after mixing this extrusion-refined straw with 40% (w/w) deoiled press cake, followed by heat treatment, has significantly improved mechanical properties: 29.1 MPa for flexural strength, 3.9 GPa for elastic modulus, and 24% for thickness swelling, for a density of 1195 kg/m^3 (Uitterhaegen et al., 2017b). The difference between using crushed straw and extrusion-refined straw in hardboards comes from the morphological characteristics of the coriander fibers. These characteristics are significantly improved thanks to the thermo-mechanical defibration treatment in the twin-screw extruder. Indeed, an intense mechanical shear is applied to the coriander straw during its thermo-mechanical treatment. This operation favors fibrillation of the fibers, which generates fibers whose diameter is reduced compared to a simple grinding action (21 μm instead of 838 μm). Therefore, the lignocellulosic fibers obtained by extrusion-refining have a higher average aspect ratio of 22.9 instead of 4.5 in the ground material (Figure 20.7). As a result, the fibers adhere better to each other during the hot pressing process, and their entanglement is greatly enhanced, which explains the significantly improved mechanical properties of the fiberboard.

As previously for the particleboard produced from deoiled coriander cake only (Uitterhaegen et al., 2018a), this 100% coriander fiberboard was also investigated concerning its emissions of carbonyl compounds (aldehydes and ketones), in particular formaldehyde and acetaldehyde (Simon et al., 2020). Again, no formaldehyde emissions were observed. Indeed, whatever the day of exposure (until D+28) and the material temperature (from 23°C to 36°C, 50% relative humidity), the formaldehyde emissions from this material were systematically lower than 0.8 μg/m^2.h, corresponding to the detection threshold of the analytical method used. In Simon et al. (2020), formaldehyde emissions were studied for a particleboard and an MDF panel, two widely used commercial wood panels. They were found to be extremely high: up to 78 μg/m^2.h and 42 μg/m^2.h, respectively, at the beginning of exposure at 23°C.

In the wood panel industry, UF and PF resins are widely used, which is why most commercial wood-based panels emit formaldehyde during their use (FCBA, 2013). However, the regulatory context is becoming more and more demanding. Indeed, construction products will have to emit less and less VOC for the years to come, including formaldehyde. Formaldehyde is a carcinogenic, mutagenic, and reprotoxic (CMR) product of Group 1, classified by the International Agency for Research on Cancer (IARC). It can cause cancers of the nasopharynx and is often responsible for allergies and irritations of the eyes and respiratory tract. In addition, the weekly average concentration of pollutants in the indoor air of closed spaces will have to be vastly reduced in the coming years. It is currently 30 μg/m^3 in the vast majority of establishments receiving the public in France (*e.g.*, education, reception of children under six years old, health and social establishments with accommodation capacity, covered sports facilities, etc.). However, this value will have to be reduced to 10 µg/m³ as of January 1, 2023, following the French decree number 2011-1727 of December 2, 2011 (decree relating to the guide values in the indoor air for formaldehyde and benzene).

Using the coriander protein binder limits the health risk associated with these emissions. In this way, the 100% coriander hardboard presents an excellent environmental profile, and its commercialization should therefore be favored because of its harmlessness (Simon et al., 2020). It should be noted that this panel is also fully bio-based and biodegradable, which is not the case for commercial glued wood panels using UF- or PF-based thermosetting resins.

20.4.3   INSULATION BLOCKS (I.E., FIBERBOARDS WITH LOW DENSITY)

After being supplemented with a natural binder with physical curing in water solution (*e.g.*, a starch-based binder), the extrusion-refined coriander straw can also be compression-molded to make low-density insulation blocks (Uitterhaegen et al., 2020). Depending on the aspect ratio of the coriander fibers, the amount of starchy binder added, and the level of filling of the mold prior to compression, a wide range of densities (from 127 kg/m³ to 292 kg/m³) can be obtained for the blocks. Furthermore, a relatively clear correlation exists between the density of the blocks and their flexural and thermal insulation properties (Figure 20.8). Indeed, suppose a higher density contributes to more mechanically resistant blocks. In that case, it is preferable to reach low-density values to promote a better

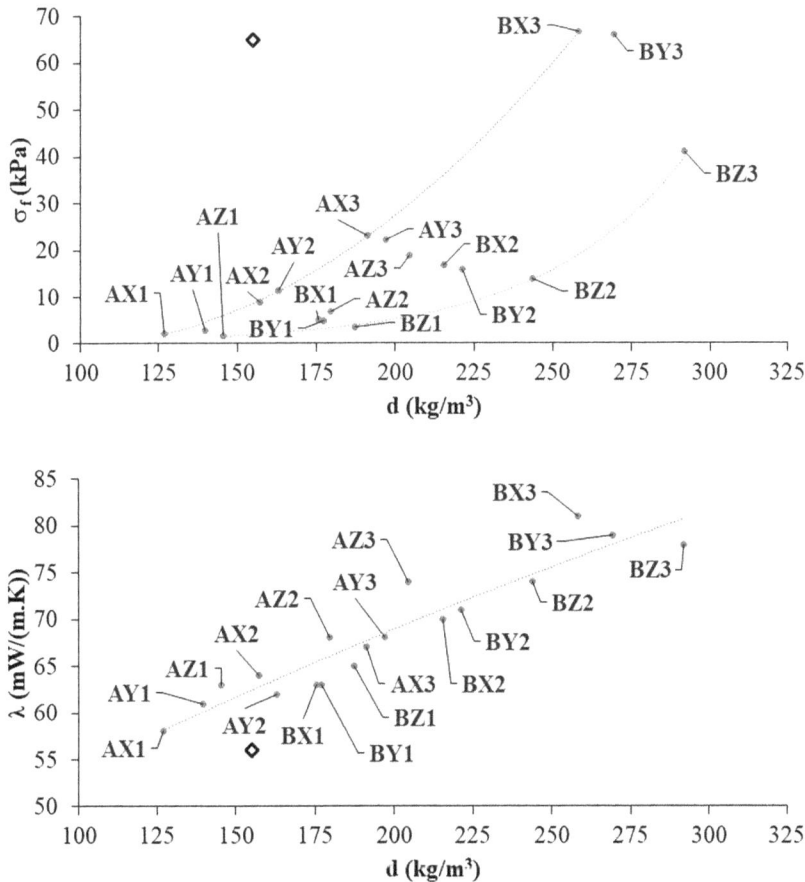

FIGURE 20.8   Flexural strength (σ_f) and thermal conductivity (λ) at 25°C of the insulation blocks as a function of their density (the open diamond corresponds to the insulation block made from the milled straw) (reproduced from Uitterhaegen et al. (2020) with the permission of the publisher).

FIGURE 20.9 Coriander straw in the form of a 7.5 mm crushed material (bottom left), and insulating block molded from this ground straw using a compression procedure at ambient temperature and a starch-based binder (top right) (reproduced from Uitterhaegen et al. (2020) with the permission of the publisher).

thermal insulation capacity: the lower the density of the insulating blocks, the lower their thermal conductivity, which then has a better insulating capacity. Such insulating materials can be used for thermal insulation of buildings, *e.g.*, when positioned in walls and ceilings. In Uitterhaegen et al. (2020), the insulating material made of a coriander straw refined by twin-screw extrusion using a water-to-straw ratio of 1.0 during the thermo-mechanical defibration pre-treatment, and 15% (w/w) of the starch binder appears as a good compromise. With a density of 163 kg/m³, this insulating block (referenced as AY2 in Figure 20.8) combines good mechanical properties and an essential thermal insulation capacity (thermal conductivity of only 62 mW/m.K at 25°C). The extrusion-refined straw (only 45 kg/m³ for its apparent density) has a thermal conductivity in bulk even reduced (47 mW/m.K). It can also be used in the attics of houses as loose insulation.

   A remarkable thermal insulating ability has also been observed with straw crushed to 7.5 mm. This is not only the case when used in bulk (thermal conductivity of only 51 mW/m.K at 25°C), due to its low apparent density value (95 kg/m³), but also in the form of a low-density insulating block (Uitterhaegen et al., 2020). The insulating block made of crushed straw and 15% (w/w) starchy binder, compression-molded under standard conditions (*i.e.*, 87 kPa for 30 s at ambient temperature), actually has a thermal conductivity only slightly higher, equal to 56 mW/m.K, for a density equal to 155 kg/m³ (Figure 20.9). For comparison, conventional thermal insulation materials have better thermal insulation capacities with thermal conductivity values for expanded polystyrene, rock wool, and glass wool panels of 37 mW/m.K, 36 mW/m.K, and 35 mW/m.K, respectively, for panel densities of 50 kg/m³, 115 kg/m³, and 26 kg/m³. Thus, for the same service of thermal insulation rendered as for these three commercial materials, the insulating block made of crushed coriander straw will have to be thicker: +51%, +56%, and +60%, respectively.

20.4.4 INJECTED OR EXTRUDED COMPOSITE PARTS

Injected or extruded composite parts can also be made from coriander straw (CS) after twin-screw compounding in a thermoplastic matrix, *e.g.*, PP or biopolyethylene (BioPE), and coupling of the matrix-to-fiber interface thanks to the addition of PP-g-MA (*i.e.*, PP grafted with maleic

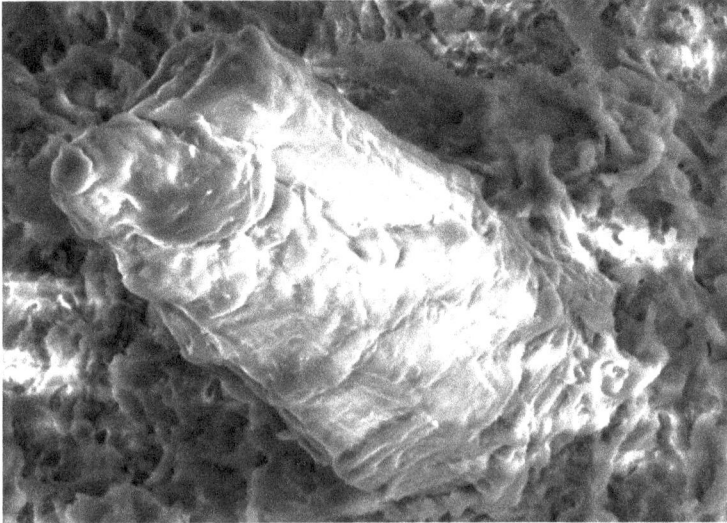

FIGURE 20.10 SEM image of the fracture surface of a PP/CS/PP-g-MA compound (72.5/25.0/2.5) at the end of a bending test (×1,500 for magnification) (Evon, 2019).

anhydride) (Figure 20.10) or PE-g-MA (*i.e.*, PE grafted with maleic anhydride) additives, respectively (Uitterhaegen et al., 2018b). Inside the obtained pieces, the coriander straw acts as a mechanical reinforcement of the matrix. These can be used in a wide variety of fields, such as:

- The building sector (*e.g.*, profiles used as technical coverings, floors, terraces, and door or window frames).
- The automotive sector (*e.g.*, for the interior design of cars).
- The sports equipment industry.

Coriander straw can be incorporated in PP up to 40% (w/w) and in BioPE up to 50%. Furthermore, the increase of the incorporation rate of the straw in the thermoplastic matrix results in a progressive increase of the strengths and the elastic moduli, both in flexion and tension, for both matrices used (Figure 20.11). In particular, a 40% filling rate highlights a particularly significant reinforcement effect of coriander straw in thermoplastic biocomposites, which then present adequate mechanical properties. For example, for polypropylene composites, an increase of about 50% in flexural and tensile strengths is obtained compared to the values of the native polymer: from 19 MPa to 28 MPa and from 12 MPa to 17 MPa, respectively. Moreover, the durability of these new biocomposites is particularly promising, as no alteration of their mechanical performance is observed after accelerated aging, whether UV aging and/or hygrothermal aging. Their recycling capacity is also real, and at least five recycling cycles can be applied without altering their mechanical performance. Finally, coriander straw exhibits the same ability to mechanically strengthen PP as two commercial wood flours, one from hardwood and the other from softwood (*i.e.*, spruce and fir) (Uitterhaegen et al., 2018b). These two wood flours had been chosen due to their proximity to coriander straw, namely equivalent particle size (200–400 μm) and similar chemical compositions.

Due to the expected progressive increase in the coming years in the need for coriander fragrant vegetable oil for the nutraceutical market, the availability of straw will increase simultaneously. Furthermore, this crop residue has the advantage of being very cheap (around €100 per ton including harvesting, bunching, and transportation) compared to commercial wood flours for which the price is rather around €600 per ton (based on large purchase volumes, *i.e.*, 6 tons and more), and especially synthetic reinforcements such as E-glass fibers whose cost is around €2.65 per kg (Bogoeva-Gaceva et al., 2007). This results in a significant economic advantage in using coriander

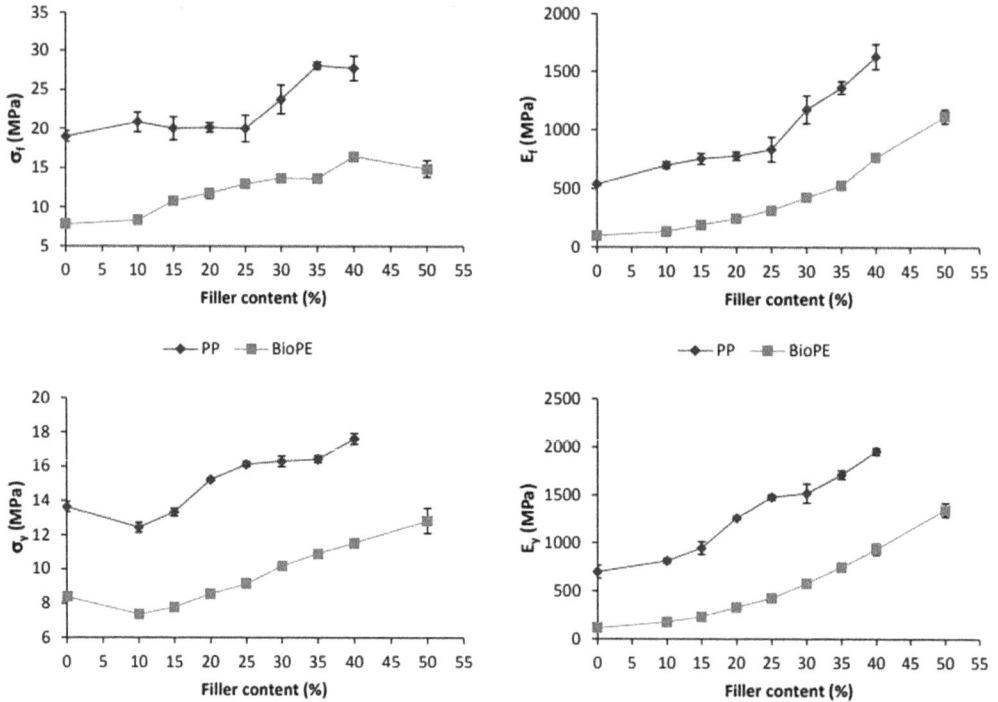

FIGURE 20.11 Flexural strength (σ_f), modulus of elasticity in bending (E_f), tensile strength (σ_y), and Young's modulus (E_y) of the PP and BioPE thermoplastic biocomposites reinforced with coriander straw as a function of the filler content (reproduced from Uitterhaegen et al. (2018b) with the permission of the publisher).

straw. For example, for the 30% filled thermoplastic composites, the cost reduction of composite pellets is estimated to be 28% for those containing coriander straw rather than commercial wood fibers (Uitterhaegen et al., 2018b).

When used in flour form, the industrial potential of coriander straw in thermoplastic composites as a replacement reinforcement for the synthetic or wood-based ones is therefore significant. In addition to its environmental and economic advantages, straw also offers significant recycling possibilities. For future work, it could also be incorporated into various bioplastics already available on the market:

- Poly(lactic acid) produces PLA/CS biocomposites that could be composted in industrial composting units.
- Other polyesters from the family of polyhydroxyalkanoates such as polyhydroxybutyrate (PHB), polyhydroxyvalerate (PHV), poly(3-hydroxybutyrate-co-3-hydroxyvalerate) (PHBV), etc.; in this way, the biocomposites produced will be even more favorable to domestic composting at their end of life.

20.5 CONCLUSIONS

In this chapter, various bio-sourced materials obtained from two co-products from the cultivation and first agro-industrial processing of coriander, namely the straw and the press cake from the seeds, respectively, have been presented. When the vegetable oil from coriander seeds is extracted by mechanical pressing *via* twin-screw extrusion technology, a solid raffinate, *i.e.*, the press cake, is effectively generated. Self-bonded particleboards can then be produced by hot pressing the cake, a material rich in proteins with thermoplastic and adhesive properties and lignocellulose.

The usability of the panels can even be significantly improved by adding a fraction of coriander straw, the aerial part of the plant, to the cake. Indeed, the straw contributes to the mechanical reinforcement of the panel, this aptitude being even exacerbated if the straw undergoes a preliminary stage of thermo-mechanical defibration in the presence of water by twin-screw extrusion. The optimized 100% coriander-based panels will constitute a viable and sustainable alternative to the wood-based materials found on the market today. They do not emit formaldehyde and are therefore more respectful of the environment and human health.

The richness in lignocellulose of straw also makes this co-product of coriander cultivation a raw material of choice for obtaining insulating materials. They can be used for the thermal insulation of buildings and can be presented in different forms: in bulk (straw simply crushed or extrusion-refined) or in the form of insulating blocks of low density, which are both cohesive and easily machinable.

Lastly, thermoplastic composites that can be molded by injection or extrusion can also be obtained from micronized straw (*i.e.*, in the form of flour), which has a real aptitude for mechanical reinforcement for commercial plastic matrices like polypropylene or biopolyethylene. This mechanical reinforcement ability is perfectly comparable to those of commercial wood flours. These coriander straw biocomposites also have excellent durability and recyclability and can be used in the automotive, construction (*e.g.*, door and window frames, decking, etc.), and sports equipment industries.

ACKNOWLEDGMENTS

The authors would like to express their sincere gratitude to Dr. Evelien Uitterhaegen for the importance and the high quality of her experimental work carried out during her PhD thesis conducted at the Laboratoire de Chimie Agro-industrielle (Toulouse, France) between 2014 and 2018 under the supervision of Dr. Philippe Evon and Dr. Othmane Merah, and whose results allowed the writing of this chapter. They also warmly thank Ovalie Innovation (Auch, France) for supplying coriander seeds and straw.

REFERENCES

Amalia Kartika, I., Pontalier, P.Y., & Rigal, L. (2006). Extraction of sunflower oil by twin-screw extruder: Screw configuration and operating condition effects. *Bioresource Technology*, 97(18), 2302–2310. https://doi.org/10.1016/j.biortech.2005.10.034.

Anglès, M.N., Ferrando, F., Farriol, X., & Salvadó, J. (2001). Suitability of steam exploded residual softwood for the production of binderless panels. Effect of the pre-treatment severity and lignin addition. *Biomass and Bioenergy*, 21(3), 211–224. https://doi.org/10.1016/S0961-9534(01)00031-9.

Ardente, F., Beccali, M., Cellura, M., & Mistretta, M. (2008). Building energy performance: A LCA case study of kenaf fibres insulation board. *Energy and Buildings*, 40(1), 1–10. https://doi.org/10.1016/j.enbuild.2006.12.009.

Benfratello, S., Capitano, C., Peri, G., Rizzo, G., Scaccianoce, G., & Sorrentino, G. (2013). Thermal and structural properties of a hemp lime biocomposite. *Construction and Building Materials*, 48, 745–754. https://doi.org/10.1016/j.conbuildmat.2013.07.096.

Bledzki, A.K., Letman, M., Viksne, A., & Rence, L. (2005). A comparison of compounding processes and wood type for wood fibre-PP composites. *Composites Part A: Applied Science and Manufacturing*, 36(6), 789–797. https://doi.org/10.1016/j.compositesa.2004.10.029.

Bogoeva-Gaceva, G., Avella, M., Malinconico, M., Buzarovska, A., Grozdanov, A., Gentile, G., & Errico, M.E. (2007). Natural fiber eco-composites. *Polymer Composites*, 28(1), 98–107. https://doi.org/10.1002/pc.20270.

Böhm, M., Salem, M.Z.M., & Srba, J. (2012). Formaldehyde emission monitoring from a variety of solid wood, plywood, blockboard and flooring products manufactured for building and furnishing materials. *Journal of Hazardous Materials*, 221–222, 68–79. https://doi.org/10.1016/j.jhazmat.2012.04.013.

Bouajila, J., Limare, A., Joly, C., & Dole, P. (2005). Lignin plasticization to improve binderless fiberboard mechanical properties. *Polymer Engineering and Science*, 45(6), 809–816. https://doi.org/10.1002/pen.20342.

Bouvier, J.M., & Guyomard, P. (1997). *Method and installation for continuous extraction of a liquid contained in a raw material* [French Patent FR97/00696].

Campbell, E.J. (1983). Sunflower oil. *Journal of the American Oil Chemists' Society*, 60, 387–392.

Crowe, T.W., Johnson, L.A., & Wang, T. (2001). Characterization of extruded-expelled soybean flours. *Journal of the American Oil Chemists' Society*, 78(8), 775–779. https://doi.org/10.1007/s11746-001-0341-9.

Diebel, W., Reddy, M.M., Misra, M., & Mohanty, A. (2012). Material property characterization of co-products from biofuel industries: Potential uses in value added biocomposites. *Biomass and Bioenergy*, 37, 88–96. https://doi.org/10.1016/j.biombioe.2011.12.028.

European Food Safety Authority (EFSA). (2013). Scientific opinion on the safety of "coriander seed oil" as a Novel Food Ingredient. *EFSA Journal*, 11(10), 3422. https://doi.org/10.2903/j.efsa.2013.3422.

Evon, Ph. (2008). New process for the biorefinery of sunflower whole plant by thermo-mechano-chemical fractionation in twin-screw extruder: Study of the aqueous extraction of lipids and manufacturing of the raffinate into agromaterials by compression moulding [Ph.D., University of Toulouse]. http://ethesis.inp-toulouse.fr/archive/00000668/.

Evon, Ph. (2019). *Compact processes for integrated biorefinery of whole plants: Multi-product recovery with added value* [HDR (empowerment to lead research), University of Toulouse]. https://oatao.univ-toulouse.fr/25201/.

Evon, Ph., Vandenbossche, V., Pontalier, P.Y., & Rigal, L. (2007). Direct extraction of oil from sunflower seeds by twin-screw extruder according to an aqueous extraction process: Feasibility study and influence of operating conditions. *Industrial Crops and Products*, 26(3), 351–359. https://doi.org/10.1016/j.indcrop.2007.05.001.

Evon, Ph., Vandenbossche, V., Pontalier, P.Y., & Rigal, L. (2009). Aqueous extraction of residual oil from sunflower press cake using a twin-screw extruder: Feasibility study. *Industrial Crops and Products*, 29(2–3), 455–465. https://doi.org/10.1016/j.indcrop.2008.09.001.

Evon, Ph., Vandenbossche, V., Pontalier, P.Y., & Rigal, L. (2010a). The twin-screw extrusion technology an original and powerful solution for the biorefinery of sunflower whole plant. *Oilseeds and Fats, Crops and Lipids*, 17(6), 404–417. https://doi.org/10.1051/ocl.2010.0339.

Evon, Ph., Vandenbossche, V., Pontalier, P.Y., & Rigal, L. (2010b). Thermo-mechanical behaviour of the raffinate resulting from the aqueous extraction of sunflower whole plant in twin-screw extruder: Manufacturing of biodegradable agromaterials by thermopressing. *Advanced Materials Research*, 112, 63–72. https://doi.org/10.4028/www.scientific.net/AMR.112.63.

Evon, Ph., Vandenbossche, V., & Rigal, L. (2012). Manufacturing of renewable and biodegradable fiberboards from cake generated during biorefinery of sunflower whole plant in twin-screw extruder: Influence of thermo-pressing conditions. *Polymer Degradation and Stability*, 97(10), 1940–1947. https://doi.org/10.1016/j.polymdegradstab.2012.01.025.

Evon, Ph., Amalia Kartika, I., Cerny, M., & Rigal, L. (2013). Extraction of oil from jatropha seeds using a twin-screw extruder: Feasibility study. *Industrial Crops and Products*, 47, 33–42. https://doi.org/10.1016/j.indcrop.2013.02.034.

Evon, Ph., Vandenbossche, V., Pontalier, P.Y., & Rigal, L. (2014a). New thermal insulation fiberboards from cake generated during biorefinery of sunflower whole plant in a twin-screw extruder. *Industrial Crops and Products*, 52, 354–362. https://doi.org/10.1016/j.indcrop.2013.10.049.

Evon, Ph., Amalia Kartika, I., & Rigal, L. (2014b). New renewable and biodegradable particleboards from jatropha press cakes. *Journal of Renewable Materials*, 2(1), 52–65. https://doi.org/10.7569/JRM.2013.634131.

Evon, Ph., Vinet, J., Labonne, L., & Rigal, L. (2015a). Influence of thermo-pressing conditions on mechanical properties of biodegradable fiberboards made from a deoiled sunflower cake. *Industrial Crops and Products*, 65, 117–126. https://doi.org/10.1016/j.indcrop.2014.11.036.

Evon, Ph., Vinet, J., Rigal, M., Labonne, L., Vandenbossche, V., & Rigal, L. (2015b). New insulation fiberboards from sunflower cake with improved thermal and mechanical properties. *Journal of Agricultural Studies*, 3(2), 194–211. https://doi.org/10.5296/jas.v3i2.7738.

Evon, Ph., Vandenbossche, V., Candy, L., Pontalier, P.Y., & Rouilly, A. (2018). Twin-screw extrusion: A key technology for the biorefinery. In Ali Ayoub & Lucian Lucia (Eds.), *Biomass extrusion and reaction technologies: Principles to practices and future potential*, 1304 (2) (pp. 25–44), American Chemical Society, ACS Symposium Series, eBooks, Washington, DC. https://doi.org/10.1021/bk-2018-1304.ch002.

Faruk, O., Bledzki, A.K., Fink, H.P., & Sain, M. (2014). Progress report on natural fiber reinforced composites. *Macromolecular Materials and Engineering*, 299(1), 9–26. https://doi.org/10.1002/mame.201300008.

FCBA. (2013). *Étude bibliographique sur les émissions de formaldéhyde des bois collés*. Institut technologique Forêt Cellulose Bois-construction Ameublement, Bordeaux, France.

Gamon, G., Evon, Ph., & Rigal, L. (2013). Twin-screw extrusion impact on natural fibre morphology and material properties in poly(lactic acid) based biocomposites. *Industrial Crops and Products*, 46, 173–185. https://doi.org/10.1016/j.indcrop.2013.01.026.

Gerçel, H.F. (2002). The production and evaluation of bio-oils from the pyrolysis of sunflower-oil cake. *Biomass and Bioenergy*, 23(4), 307–314. https://doi.org/10.1016/S0961-9534(02)00053-3.

Gunstone, F.D., Harwood, J.L., & Dijkstra, A.J. (Eds.). (2007). *The lipid handbook with CD-ROM* (3rd edition). CRC Press, Boca Raton. https://doi.org/10.1201/9781420009675.

Halvarsson, S., Edlund, H., & Norgren, M. (2009). Manufacture of non-resin wheat straw fiberboards. *Industrial Crops and Products*, 29(2–3), 437–445. https://doi.org/10.1016/j.indcrop.2008.08.007.

Isobe, S., Zuber, F., Uemura, K., & Noguchi, A. (1992). A new twin-screw press design for oil extraction of dehulled sunflower seeds. *Journal of the American Oil Chemists' Society*, 69(9), 884–889. https://doi.org/10.1007/BF02636338.

John, M.J., & Thomas, S. (2008). Biofibres and biocomposites. *Carbohydrate Polymers*, 71(3), 343–64. https://doi.org/10.1016/j.carbpol.2007.05.040.

Kamrozzaman, M.M., Ahmed, S., & Quddus, A.F.M.R. (2016). Effect of fertilizer on coriander seed production. *Bangladesh Journal of Agricultural Research*, 41(2), 345–352. https://doi.org/10.3329/bjar.v41i2.28236.

Kesselmeier, J., & Staudt, M. (1999). Biogenic Volatile Organic Compounds (VOC): An overview on emission, physiology and ecology. *Journal of Atmospheric Chemistry*, 33, 23–88. https://doi.org/10.1023/A:1006127516791.

Korjenic, A., Petránek, V., Zach, J., & Hroudová, J. (2011). Development and performance evaluation of natural thermal insulation materials composed of renewable resources. *Energy and Buildings*, 43(9), 2518–2523. https://doi.org/10.1016/j.enbuild.2011.06.012.

Kurniati, M., Fahma, F., Amalia Kartika, I., Sunarti, T.C., Syamsu, K., Hermawan, D., Saito, Y., & Sato, M. (2015). Binderless particleboard from castor seed cake: Effect of pressing temperature on physical and mechanical properties. *Asian Journal of Agricultural Research*, 9(4), 180–188. https://doi.org/10.3923/ajar.2015.180.188.

Kymäläinen, H.R., & Sjöberg, A.M. (2008). Flax and hemp fibres as raw materials for thermal insulations. *Building and Environment*, 43(7), 1261–1269. https://doi.org/10.1016/j.buildenv.2007.03.006.

Lapczynski, A., Letizia, C.S., & Api, A.M. (2008). Addendum to fragrance material review on linalool. *Food and Chemical Toxicology*, 46(11), S190–S192. https://doi.org/10.1016/j.fct.2008.06.087.

Msaada, K., Hosni, K., Ben Taarit, M., Hammami, M., & Marzouk, B. (2009). Effects of growing region and maturity stages on oil yield and fatty acid composition of coriander (*Coriandrum sativum* L.) fruit. *Scientia Horticulturae*, 120(4), 525–531. https://doi.org/10.1016/j.scienta.2008.11.033.

Müller, G.C., Junnila, A., Butler, J., Kravchenko, V.D., Revay, E.E., Weiss, R.W., & Schlein, Y. (2009). Efficacy of the botanical repellents geraniol, linalool, and citronella against mosquitoes. *Journal of Vector Ecology*, 34(1), 2–8. https://doi.org/10.1111/j.1948-7134.2009.00002.x.

Nonaka, S., Umemura, K., & Kawai, S. (2013). Characterization of bagasse binderless particleboard manufactured in high-temperature range. *Journal of Wood Science*, 59, 50–56. https://doi.org/10.1007/s10086-012-1302-6.

Okuda, N., & Sato, M. (2004). Manufacture and mechanical properties of binderless boards from kenaf core. *Journal of Wood Science*, 50, 53–61. https://doi.org/10.1007/s10086-003-0528-8.

Okuda, N., Hori, K., & Sato, M. (2006a). Chemical changes of kenaf core binderless boards during hot pressing. I: Influence of the pressing temperature condition. *Journal of Wood Science*, 52, 244–248. https://doi.org/10.1007/s10086-005-0761-4.

Okuda, N., Hori, K., & Sato, M. (2006b). Chemical changes of kenaf core binderless boards during hot pressing. II: Effects on the binderless board properties. *Journal of Wood Science*, 52, 249–254. https://doi.org/10.1007/s10086-005-0744-5.

Pickering, K.L., Efendy, M.G.A., & Le, T.M. (2016). A review of recent developments in natural fibre composites and their mechanical performance. *Composites Part A: Applied Science and Manufacturing*, 83, 98–112. https://doi.org/10.1016/j.compositesa.2015.08.038.

Saadaoui, N., Rouilly, A., Fares, K., & Rigal, L. (2013). Characterization of date palm lignocellulosic by-products and self-bonded composite materials obtained thereof. *Materials & Design*, 50, 302–308. https://doi.org/10.1016/j.matdes.2013.03.011.

Sabathier, V., Ahmed Maaloum, A., Magniont, C., Evon, Ph., & Labonne, L. (2017). *Contribution to the design and the characterization of a fully bio based insulating panel including sunflower pith.* Congrès National de la Recherche en IUT (CNRIUT2017), Auxerre, France.

Saiah, R., Perrin, B., & Rigal, L. (2010). Improvement of thermal properties of fired clays by introduction of vegetable matter. *Journal of Building Physics*, 34(2), 124–142. https://doi.org/10.1177/1744259109360059.

Salthammer, T., Mentese, S., & Marutzky, R. (2010). Formaldehyde in the indoor environment. *Chemical Reviews*, 110, 2536–2572. https://doi.org/10.1021/cr800399g.

Sharma, R.P., Singh, R.S., Verma, T.P., Tailor, B.L., Sharma, S.S., Singh, S.K. (2014). Coriander the taste of vegetables: Present and future prospectus for coriander seed production in southeast Rajasthan. *Economic Affairs*, 59(3), 345–354. https://doi.org/10.5958/0976-4666.2014.00003.5.

Simon, V., Uitterhaegen, E., Robillard, A., Ballas, S., Véronèse, T., Vilarem, G., Merah, O., Talou, T., & Evon, Ph. (2020). VOC and carbonyl compound emissions of a fiberboard resulting from a coriander biorefinery: Comparison with two commercial wood-based building materials. *Environmental Science and Pollution Research*, 27, 16121–16133. https://doi.org/10.1007/s11356-020-08101-y.

Singh, K.K., Wiesenborn, D.P., Tostenson, K., & Kangas, N. (2002). Influence of moisture content and cooking on screw pressing of crambe seed. *Journal of the American Oil Chemists' Society*, 79, 165–170. https://doi.org/10.1007/s11746-002-0452-3.

Sriti, J., Bettaieb, I., Bachrouch, O., Talou, T., & Marzouk, B. (2014). Chemical composition and antioxidant activity of the coriander cake obtained by extrusion. *Arabian Journal of Chemistry*, 12(7), 1765–1773. https://doi.org/10.1016/j.arabjc.2014.11.043.

Tajuddin, M., Ahmad, Z., & Ismail, H. (2016). A review of natural fibers and processing operations for the production of binderless boards. *BioResources*, 11(2), 5600–5617. https://doi.org/10.15376/biores.11.2.Tajuddin.

Uitterhaegen, E. (2018). *Study of the integrated biorefinery of vegetable and essential oil in Apiaceae seeds.* Ph.D., University of Toulouse, confidential document.

Uitterhaegen, E., Nguyen, Q.H., Sampaio, K.A., Stevens, C.V., Merah, O., Talou, T., Rigal, L., & Evon, Ph. (2015). Extraction of coriander oil using twin-screw extrusion: Feasibility study and potential press cake applications. *Journal of the American Oil Chemists' Society*, 92(8), 1219–1233. https://doi.org/10.1007/s11746-015-2678-4.

Uitterhaegen, E., Nguyen, Q.H., Merah, O., Stevens, C.V., Talou, T., Rigal, L., & Evon, Ph. (2016). New renewable and biodegradable fiberboards from a coriander (*Coriandrum sativum* L.) press cake. *Journal of Renewable Materials*, 4(3), 225–238. https://doi.org/10.7569/JRM.2015.634120.

Uitterhaegen, E., & Evon, Ph. (2017). Twin-screw extrusion technology for vegetable oil extraction: A review. *Journal of Food Engineering*, 212, 190–200. https://doi.org/10.1016/j.jfoodeng.2017.06.006.

Uitterhaegen, E., Labonne, L., Merah, O., Talou, T., Ballas, S., Véronèse, T., & Evon, Ph. (2017a). Optimization of thermopressing conditions for the production of binderless boards from a coriander twin-screw extrusion cake. *Journal of Applied Polymer Science*, 134(13), 44650. https://doi.org/10.1002/app.44650.

Uitterhaegen, E., Labonne, L., Merah, O., Talou, T., Ballas, S., Véronèse, T., & Evon, Ph. (2017b). Impact of a thermomechanical fiber pre-treatment using twin-screw extrusion on the production and properties of renewable binderless coriander fiberboards. *International Journal of Molecular Sciences*, 18(7), 1539. https://doi.org/10.3390/ijms18071539.

Uitterhaegen, E., Burianová, K., Ballas, S., Véronèse, T., Merah, O., Talou, T., Stevens, C.V., Evon, Ph., & Simon, V. (2018a). Characterization of volatile organic compound emissions from self-bonded boards resulting from a coriander biorefinery. *Industrial Crops and Products*, 122, 57–65. https://doi.org/10.1016/j.indcrop.2018.05.050.

Uitterhaegen, E., Parinet, J., Mérian, T., Ballas, S., Véronèse, T., Merah, O., Talou, T., Labonne, L., Stevens, C.V., Chabert, F., & Evon, Ph. (2018b). Performance, durability and recycling of thermoplastic biocomposites reinforced with coriander straw. *Composites Part A: Applied Science and Manufacturing*, 113, 254–263. https://doi.org/10.1016/j.compositesa.2018.07.038.

Uitterhaegen, E., Labonne, L., Merah, O., Talou, T., Ballas, S., Véronèse, T., & Evon, Ph. (2020). Innovative insulating materials from coriander (*Coriandrum sativum* L.) straw for building applications. *Journal of Agricultural Studies*, 8(4). https://doi.org/10.5296/jas.v8i4.17077.

Van Dam, J.E.G., Van den Oever, M.J.A., & Keijsers, E.R.P. (2004a). Production process for high density high performance binderless boards from whole coconut husk. *Industrial Crops and Products*, 20(1), 97–101. https://doi.org/10.1016/j.indcrop.2003.12.017.

Van Dam, J.E.G., Van den Oever, M.J.A., Teunissen, W., Keijsers, E.R.P., & Peralta, A.G. (2004b). Process for production of high density/high performance binderless boards from whole coconut husk: Part 1: Lignin as intrinsic thermosetting binder resin. *Industrial Crops and Products*, 19(3), 207–216. https://doi.org/10.1016/j.indcrop.2003.10.003.

Velásquez, J.A., Ferrando, F., Farriol, X., & Salvadó, J. (2003). Binderless fiberboard from steam exploded *Miscanthus sinensis*. *Wood Science and Technology*, 37, 269–278. https://doi.org/10.1007/s00226-003-0182-8.

Verdier, T., Balthazard, L., Montibus, M., Magniont, C., Evon, Ph., & Bertron, A. (2021). Using glycerol esters to prevent microbial growth on sunflower-based insulation panels. *Construction Materials*, 174(3), 140–149. https://doi.org/10.1680/jcoma.20.00002.

Widyorini, R., Xu, J., Umemura, K., & Kawai, S. (2005). Manufacture and properties of binderless particleboard from bagasse. I: Effect of raw material type, storage methods, and manufacturing process. *Journal of Wood Science*, 51, 648–654. https://doi.org/10.1007/s10086-005-0713-z.

Xu, J., Widyorini, R., Yamauchi, H., & Kawai, S. (2006). Development of binderless fiberboard from kenaf core. *Journal of Wood Science*, 52, 236–243. https://doi.org/10.1007/s10086-005-0770-3.

Yang, H.S., Wolcott, M.P., Kim, H.S., Kim, S., & Kim, H.J. (2006). Properties of lignocellulosic material filled polypropylene bio-composites made with different manufacturing processes. *Polymer Testing*, 25(5), 668–676. https://doi.org/10.1016/j.polymertesting.2006.03.013.

Yorgun, S., Şensöz, S., & Koçkar, Ö.M. (2001). Flash pyrolysis of sunflower oil cake for production of liquid fuels. *Journal of Analytical and Applied Pyrolysis*, 60(1), 1–12. https://doi.org/10.1016/S0165-2370(00)00102-9.

Zheng, Y., Wiesenborn, D.P., Tostenson, K., & Kangas, N. (2003). Screw pressing of whole and dehulled flaxseed for organic oil. *Journal of the American Oil Chemists' Society*, 80(10), 1039–1045. https://doi.org/10.1007/s11746-003-0817-7.

Zhou, X.Y., Zheng, F., Li, H.G., & Lu, C.L. (2010). An environment-friendly thermal insulation material from cotton stalk fibers. *Energy and Buildings*, 42(7), 1070–1074. https://doi.org/10.1016/j.enbuild.2010.01.020.

21 Composition and Functionality of Coriander Root

Monika Choudhary, Amarjeet Kaur, and Arashdeep Singh

CONTENTS

21.1 INTRODUCTION

Medicinal and aromatic plants have always been popular for culinary, medicinal, and other uses. Plants and their phytochemicals, also known as secondary metabolites, play an immensely important role in diets, health, and nutritional products. Essential oils are the most widely utilized secondary metabolites of plants, having been used for thousands of years (over 5000 years) for a wide range of purposes, mostly for their health advantages. Coriander is a medicinal and aromatic plant with many purposes (Sharifi-Rad et al., 2017). Coriander, also known as *cilantrillo*, cilantro, Chinese parsley, Arab parsley, Mexican parsley, Yuen sai, and Dhania, is a member of the Apiaceae family (formerly Umbelliferae) (Davis, 1970; Laribi et al., 2015). It is a perennial herb popular in Latin American, Middle Eastern, Indian, African, Mediterranean, and Southeast Asian cuisines (Aissaoui et al., 2008). Although the terms cilantro and coriander appear interchangeable, they have distinct meanings. The green leaves of plants are called cilantro when these are freshly gathered; the herb is called coriander when the dried fruits are consumed (Abascal & Yarnell, 2012).

21.2 AGRO-CLIMATIC CONDITIONS FOR CORIANDER CULTIVATION AND PLANT MORPHOLOGY

Coriander is native to the Mediterranean region (Randall et al., 2013) and is cultivated primarily in locations such as Argentina, Russia, Ukraine, Romania, India, Morocco, and Mexico (Priyadarshi et al., 2016). The ideal temperature for coriander germination and early growth is 20–25°C, especially in mild and reasonably dry frost. It preferably grows in arid regions, but it may also flourish in various soil types, including well-drained, light, moist, loamy soil, and light to heavy black soil

DOI: 10.1201/9781003204626-22

(Bhat et al., 2014). Coriander has small, pinkish-white blooms, thin, spindle-shaped roots, an erect stalk, alternating leaves, and short, spindle-shaped roots. Coriander is a glabrous aromatic herbaceous erect annual plant with a prominent taproot and slender branching stems reaching about 20–70 cm in height. The leaves of this plant are lanceolate in shape, green or dark green, both surfaces glabrous and lobed. Small umbels of light pink or white blooms appear asymmetrically, with petals pointing (directing) away from the center. Fruits of coriander are nearly ovoid globular dry schizocarps with two mericarps (Yeung & Bowra, 2011) and several longitudinal ridges on the surface. They have a sweet, somewhat spicy, citrus-like flavor similar to sage (Coşkuner & Karababa, 2007). As per Flora of Turkey, the genus *Coriandrum* L. is represented as *C. sativum* L. and *C. tordylium* (Fenzl) Bornm. Coriander (*C. sativum* L.) grows to a height of 30–100 cm and has pungent-smelling leaves. It is a plant that's grown in the home. Furthermore, coriander comes in two varieties: *C. sativum* L. var. *microcarpum* DC, which is a small-fruited species also known as oil-rich Russian coriander, and C. *sativum* L. var. *vulgare Alef.*, a larger-fruited species having low oil content recognized as Indian, Moroccan, and other Asiatic types (Rajeshwari & Andallu, 2011; Mandal & Mandal, 2015). Although all sections (seeds, leaves, and roots) of this plant are edible, they have quite different flavors and applications (Bhat et al., 2014). Nonetheless, coriander seeds are a popular spice throughout the Mediterranean, and finely powdered coriander seeds are available. Curry powder is one of the essential ingredients in Indian cuisine (Gil et al., 2002). Furthermore, the fresh green leaves have a distinct scent and are commonly used as a key element in Vietnamese and Thai cuisine. Coriander stems are chopped in a variety of Asian dishes, as well as in soups and stews. The coriander roots have a richer, stronger flavor than the leaves, and they are utilized in an array of Asian dishes (Rajeshwari & Andallu, 2011).

Various novel phytochemical, pharmacological, and pharmacognostical studies have been carried out on coriander and have been well documented. However, most of the studies on coriander are always focused and based on the aerial parts of this herb (Verma et al., 2011; Tang et al., 2013). Research on the medicinal properties of coriander roots in the scientific literature is quite limited and scanty to virtually almost none. Despite this limited data, the roots of this plant are frequently used in cooking and traditional medicine systems and are considered to contribute to health and protection against the onset of various diseases. Coriander root has medicinal importance concerning anticancer and antioxidant properties present in this herb, thus, preventing oxidative stress-related diseases. The essential oils in coriander and its extracts possess potential antifungal, antibacterial and antioxidative actions due to various chemical components in different parts of the plant.

21.3   COMPOSITION OF CORIANDER ROOTS

21.3.1   FATTY ACID COMPOSITION

Linoleic, oleic, and palmitic acid represent 90% of coriander total fatty acids (Neffati & Marzouk, 2009). Linoleic acid represents the largest share with 43%, followed by 26% oleic acid (Flagella et al., 2004). On the other hand, the proportion of palmitic acid is 22% (Table 21.1). In contrast, ripe coriander fruits mainly comprise petroselinic acid (68.8%) and linoleic acid (16.6%). Also, high amounts of oleic acid (7.5%) and palmitic acid (3.8%) are present in the fruit (Momin et al., 2012).

21.3.2   ESSENTIAL OILS

The total yield of essential oils in coriander roots at the vegetative stage has been reported as 0.06%. Essential oil compounds representing 95% of the total essential oil are presented in Table 21.2. (*E*)-2-dodecenal has been analyzed as the chief component (167.5 lg/g DW) in the composition, followed by decanal (67.28 lg/g DW), (*E*)-2-decenal (46.76 lg/g DW), (*E*)-2-tridecenal (43.46 lg/g DW), and α-thujene (41.55 lg/g DW). Other important compounds are 2-dodecenol, (*Z*)- myroxide, dodecanal, and tetradecanal. Other researchers have also identified similar components (Msaada et al.,

TABLE 21.1

Fatty Acid Profile of the Coriander Roots

Fatty Acid	%
Pentadecylic acid	2.74
Palmitic acid	21.6
Palmitoleic acid	0.41
Stearic acid	2.30
Oleic acid	25.4
Linoleic acid	43.0
α-Linolenic acid	3.33
Stearidonic acid	0.67

2007). They also documented (*E*)-2-dodecenal as the most abundant constituent in the different parts of Tunisian coriander. Alteration in the total yield of essential oil and composition of coriander depends on the climatic conditions (Gil et al., 2002).

One study has detected the influence of salt concentrations on both parameters. It was reported that an increase of 27% was noticed in the concentration of (*E*)-2-dodecenal at 25 mM sodium

TABLE 21.2

Essential Oils Present in Various Parts of Coriander

Roots	%	Seeds		Flowers	%	Leaves	%
α-Thujene	7.31	Linalool	58.0–80.3	Benzofuran,2,3-dihydro	15.4	Decanal	19.09
α-Pinene	0.81	g-Terpinene	0.3–11.2	Hexadecanoic acid, methyl ester	10.3	*Trans*-2-decenal	17.54
β-Pinene	0.70	α-Pinene	0.2–10.9	2,4a-epioxy-3,4,5,6, 7,8,-hexahydro-2,5,5 ,8a-tetramethyl-2h-1 -benzofuran	9.35	2-decen-1-ol	12.33
Octanal	0.26	p-Cymene	0.1–8.1	2-methyoxy-4-vinylphe-nol (8.8%)2,3,5,6-tet-rafluroanisole	8.62	Cyclodecane	12.15
Nonanal	0.30	Camphor	3.0–5.1	2,6-dimethyl-3- amino-benzoquinone (6.81%) dodecanoic acid	5	*Cis*-2-dodecen	10.72
(Z)-myroxide	4.50	Geranyl acetate	0.2–5.4			Dodecanal	4.1
Camphor	0.24					Dodecan-1-ol	3.13
(*E*)-2-decenol	0.51						
Decanal	11.96						
Linalyl acetate	0.14						
(*E*)-2-decenal	8.74						
a-Terpineol	0.42						
Borneol	3.85						
Eugenol	0.17						
(*E*)-2-undecenal	1.13						
Dodecanal	4.28						
(*E*)-2-dodecenal	30.81						
(*E*)-2-tridecenal	8.36						
Tetradecanal	2.85						

chloride concentrations. Similarly, the concentration of decanal (14%) was found at 50 mM salinity which further increased by 3.6 times. (*E*)-2-decenal increased by 57 and 97% at 25 and 75 mM NaCl, respectively. In contrast, (*E*)-2-tridecenal decreased by 18% with 25 mM and increased by 44% at 50 mM and was found three times higher under higher concentrations. These components were identified in leaves and possessed flowery, sharp, citronellol, and fruity odors (Msaada et al., 2007; Neffati & Marzouk, 2008). The concentration of α-thujene was significantly decreased at all salinity levels.

Further, the essential oil compounds in roots have been categorized due to the occurrence of aldehydes (Msaada et al., 2007), with a fraction of 70% represented by (*E*)-2-dodecenal (30.8%). The other main compounds are monoterpene alcohols (11.5%) and monoterpene hydrocarbons (8.82%). In comparison, linalool and other oxidized monoterpenes and monoterpene hydrocarbons are the main constituents of essential oil extracted from fully ripened coriander fruits (Bhuiyan et al., 2009). Coriander fruit has about 0.2–1.5% of volatile oil and 13–20% of oil (Olle & Bender, 2010); though, it has been reported that some cultivars have up to 2.6% of essential oil (Momin et al., 2012). The essential oil obtained from dried seeds with linalool as the main compound is colorless or pale yellow with a typical aroma and gentle and sweet flavor (Burdock & Carabin, 2009). Moreover, aliphatic aldehydes, with their obnoxious smell, are the major constituents of the fresh herb oil (Potter & Fagerson, 1990).

21.3.3 ANTIOXIDANTS

The total phenolics in coriander root extract have ranged between 1.73 and 31.3 mg GAE/g. The ethyl acetate extracts have shown the highest value of total phenols in the root, such as 31.3 mg GAE/g, compared to the combination of leaf and stem (24.5 mg GAE/g) (Tang et al., 2013). In comparison, the seed extract of coriander is comprised of various antioxidants such as total polyphenol (18.7 mg/g), tocopherol (0.181 mg/g), total ascorbate (0.287 mg/g), oxidized ascorbate (0.15 mg/g), reduced ascorbate (0.136 mg/g), riboflavin (0.0046 mg/g), caffeic acid (0.08 mg/g), gallic acid (0.173 mg/g), ellagic acid (0.162 mg/g), quercetin (0.608 mg/g), and kaempferol (0.233 mg/g) (Anita et al., 2014). Moreover, coriander leaves are rich sources of phytochemicals (carotenoids and phenolics) and essential oils (linalool). These compounds are responsible for the antioxidant activities of this part of the plant (Mandal & Mandal, 2015). Also, coriander leaf juice having flavonoids has high antioxidant activities due to its scavenging free radicals (hydroxyl- and superoxide-radicals) and strengthens the defense system against macromolecular oxidative damage by enhancing the natural antioxidant system of the human body (Panjwani et al., 2010).

21.4   FUNCTIONALITIES OF CORIANDER

21.4.1 ANTIOXIDANT ACTIVITY

Phenolic compounds in coriander root extracts can reduce the oxidized agents and give antioxidant protection to natural biological systems through electron transfer to free radicals (Conde et al., 2008). Some *in-vitro* experiments have been conducted to determine the antioxidant activity of coriander roots (Moirangthem et al., 2012). It has been documented that ethyl acetate extract of coriander root and dichloromethane extract of leaf and stem have the highest ferric reducing antioxidant power (FRAP) value at 0.129 mmol/g and 0.136 mmol/g, respectively (Table 21.3). Similarly, ethyl acetate extract of coriander exhibited the highest DPPH· radical scavenging activity (IC$_{50}$ = 2348 μg/mL) among all extracts. The aqueous extract of leaf and stem showed the lowest IC$_{50}$ value (335.0 μg/mL) (Tang et al., 2013). Similarly, another study has reported that all parts of coriander have antioxidant properties that protect body cells from the adverse effects of oxidative stress caused by reactive oxygen species (de Almeida et al., 2005; Tang et al., 2013).

TABLE 21.3
Antioxidant Activity of Coriander Root

Antioxidant Activity	Root	Seed	Leaves	Leaf and Stem
FRAP	0.129 mmol/g	-	-	0.136 mmol/g
DPPH·	2348 µg/mL	-	-	1335.0 µg/mL
RSA	-	66.4%	56.7%	-

FRAP, ferric reducing antioxidant power; DPPH·, 1,1-diphenyl-2-picrylhydrazyl; RSA, radical scavenging activity.

21.4.2 ANTI-PROLIFERATIVE ACTIVITY

Coriander root has anti-cancerous and antitumor activities due to essential oil compounds and antioxidants. In this context, one research study was conducted to evaluate anti-proliferative activity using breast adenocarcinoma cell line MCF-7 (Jänicke, 2009; Yang et al., 2006). MTT assay was used for assessing the activity in terms of inhibitory concentration (IC) at 20 and 50%. The ethyl acetate extract of coriander root exhibited the maximum antiproliferative activity having the lowest IC_{50} value (200 µg/mL). At the same time, the hexane extracts of leaf and stem exhibited the lowest IC_{50} value, i.e., 432 µg/mL. The ethyl acetate extract of roots exhibited the best antiproliferative activity on MCF-7 cells and showed less toxicity on non-malignant human breast epithelial cell line, 184B5, with an IC_{50} value (317 µg/mL) compared to MCF-7 cells (Tang et al., 2013).

21.4.3 INHIBITORY EFFECTS ON HYDROGEN PEROXIDE-INDUCED MCF-7 CELL MIGRATION AND DNA PROTECTION

Hydrogen peroxide (H_2O_2) is converted into reactive hydroxyl radicals linked with genetic instability, mutations, and ultimately DNA damage (Jayakumar & Kanthimathi, 2012). These processes lead to cancer development (Imlay & Linn, 1988). H_2O_2 can augment proliferation in cancer cells and migration, resulting in metastasis. These are the foremost causes of deaths due to cancer and the inefficiency of chemotherapeutic drugs (Nelson et al., 2003). Studies have reported that plant bio-actives may act as natural chemopreventive agents in healthy cells by maintaining H_2O_2 levels within adequate limits, inhibiting DNA damage. Concurrently, plant extracts can also perform as chemotherapeutic agents by growing H_2O_2 in dividing cancer cells at a swift rate to levels that cannot be compensated by the natural defense mechanisms, causing apoptotic cell death (López-Lázaro, 2007). Natural plant products are considered inducers of apoptosis in cancer cells of human origin (Taraphdar et al., 2001). Subsequently, substantial consideration focuses on natural agents as the source of novel chemo-therapeutic and -preventive agents. In this regard, one study used the scratch motility assay to exhibit the capacity of coriander root extract to restrain H_2O_2-induced migration of MCF-7 cells. The extract hinders cellular migration brought about by H_2O_2 following a dose-dependent pattern. A 60% inhibition of MCF-7 migration was observed at 150 µg/mL. The inhibition increased by 91 and 94% with increased concentrations such as at 250 and 300 µg/mL, respectively (Tang et al., 2013).

The DNA protection activity of coriander root extract was determined in one research trial using a comet assay on 3 T3-L1 cells. Fibroblasts pre-treated with the extract at 100–400 µg/mL exhibited a significant dose-dependent increase in DNA protection compared to the control. DNA protection activity was 21.5% at a 400 µg/mL concentration of coriander extract (Tang et al., 2013).

21.4.4 Antibacterial Activity

Some researchers have investigated the antibacterial activity of essential oils and fractions extracted from coriander seeds and roots against Gram-positive and Gram-negative bacteria. For instance, Delaquis et al. (2002) assessed the antibacterial activity of coriander seeds and cilantro leaves against some food spoilage bacteria such as *Salmonella typhimurium*, *Staphylococcus aureus*, and *Listeria monocytogenes*, etc. The potential antibacterial effect of coriander parts has also been reported on *Escherichia coli*, *Escherichia cloaca*, and *Enterococcus faecalis* spoilage bacteria (Keskin & Toroglu, 2011). The essential oil obtained for coriander fruit showed an inhibitory effect against *P. aeruginosa*, *E. coli*, and *Salmonella typhi* with a zone diameter of inhibition 10, 25, and 18 mm, respectively (Teshale et al., 2013). It could be due to the chief essential oil component linalool, which is considered to exhibit antibacterial effect against several bacterial strains (Ates & Erdogrul, 2003).

Further, extract obtained from coriander contains an antimicrobial peptide with 26 amino acids which exhibit antimicrobial activity for Gram-negative bacteria (*K. pneumonia*, *P. aeruginosa*). The minimum inhibitory concentration (MIC) values for *K. pneumoniae* and *P. aeruginosa* were 71.5 and 86.4 mg/mL, respectively. The antimicrobial effect for Gram-positive bacteria (*S. aureus*) was also reported with MIC 35.2 mg/mL (Zare-Shehneh et al., 2014). One study illustrated the inhibitory effect of coriander essential oils against 25 bacterial strains such as *Citrobacter freundii*, *Serratia marcescens*, and *Enterobacter aerogenes* with MIC of 4.2 mL/mL except for *Salmonella enterica* serotype, which depicted resistance to essential oils at MIC more than 62.5 mL/mL. Innocent and co-researchers (2011) determined the immune-stimulant potential of coriander in fish and recognized it as a diet supplement to induce resistance against diseases.

21.4.5 Antifungal Activity

Mono- and sesquiterpene hydrocarbons among essential oils indicate a synergistic effect on antifungal activity. The fractions containing these compounds exhibit antifungal activity against *Candida* spp. With minimum fungicidal concentrations (MFCs) such as 125 mg/mL to 1000 mg/mL. Therefore, the essential oil extracted from different parts of the coriander can be utilized as an effective ingredient in treating oral diseases, for instance, denture-related *candidiasis* (Freires et al., 2014). The antifungal activity of coriander has also been studied in plant crops such as paddy against seed-borne pathogens. These include *Fusarium moniliforme*, *Fusarium oxysprorum*, *Pyricularia oryzae*, *Alternaria alternate*, *Bipolaris oryzae*, *Tricoconis padwickii*, *Drechslera tetramera*, *Drechslera halodes*, and *Curvularia lunata* (Lalitha et al., 2011). Zare-Shehneh and co-researchers (2014) depicted the fungicidal activity of coriander leaf extract against *Penicillium lilacinum* (MIC 67.8 mg/mL) and *Asperjilus niger* (62.1 mg/mL), respectively.

21.5 CONCLUSIONS

Coriander is a well-accepted culinary herb with several health-promoting properties and has been documented in several pharmacological and pharmacognostical studies. However, most scientists have focused their attention on the aerial parts of this medicinal herb. Moreover, scientific studies on the root are scanty. So, investigation must be carried out to determine the unexploited potential of coriander roots. As a whole, this chapter has concluded that coriander root has therapeutic value in oxidative stress-related diseases such as cancer. Coriander may also be used in combination with traditional drugs to improve the treatment of foodborne diseases. Scientific studies in the literature support coriander use in the conventional medicine system.

REFERENCES

Abascal, K., & Yarnell, E. (2012). Cilantro-culinary herb or miracle medicinal plant? *Alternative and Complementary Therapies*, 18(5), 259–264.

Aissaoui, A., El-Hilaly, J., Israili, Z.H., & Lyoussi, B. (2008). Acute diuretic effect of continuous intravenous infusion of an aqueous extract of *Coriandrum sativum* L. in anesthetized rats. *Journal of Ethnopharmacology*, 115(1), 89–95.

Anita, D., Sharad, A., Amanjot, K., & Ritu, M. (2014). Antioxidant profile of *Coriandrum sativum* methanolic extract. *International Research Journal of Pharmacy*, 5, 220–224.

Ates, D.A., & Erdogrul, O.T. (2003). Antimicrobial activities of various medicinal and commercial plant extracts. *Turkish Journal of Biology*, 27, 157–162.

Bhat, S., Kaushal, P., Kaur, M., & Sharma, H.K. (2014). Coriander (*Coriandrum sativum* L.): Processing, nutritional and functional aspects. *African Journal of Plant Science*, 8(1), 25–33.

Bhuiyan, M.N.I., Begum, J., & Sultana, M. (2009). Chemical composition of leaf and seed essential oil of *Coriandrum sativum* L. from Bangladesh. *Bangladesh Journal of Pharmacology*, 4, 150–153.

Burdock, G.A., & Carabin, I.G. (2009). Safety assessment of coriander (*Coriandrum sativum* L.) essential oil as a food ingredient. *Food and Chemical Toxicology*, 47, 22–34.

Conde, E., Moure, A., Domínguez, H., & Parajó, J.C. (2008). Fractionation of antioxidants from autohydrolysis of barley husks. *Journal of Agricultural and Food Chemistry*, 56, 10651–10659.

Coşkuner, Y. & Karababa, E. (2007). Physical properties of coriander seeds (*Coriandrum sativum* L.). *Journal of Food Engineering*, 80(2), 408–416.

Davis, P.H. (1970). *Flora of Turkey and the East Aegean Islands*. Vol. 3.

de Almeida, M.E., Mancini Filho, J., & Barbosa, G.N. (2005). Characterization of antioxidant compounds in aqueous coriander extract (*Coriandrum sativum* L.). *LWT-Food Science and Technology*, 38, 15–19.

Delaquis, R.J., Stanich, K., Girard, B., & Massa, G. (2002). Antimicrobial activity of individual and mixed fractions of dill, cilantro, coriander, and eucalyptus essential oils. *International Journal of Food Microbiology*, 74, 101–109.

Flagella, Z., Giuliani, M.M., Rotunno, T., Di Caterina, R., & De Caro, A. (2004). Effect of saline water on oil yield and quality of a high oleic sunflower (*Helianthus annuus* L.) hybrid. *European Journal of Agronomy*, 21, 267–272.

Freires, I.D.A., Murata, R.M., Furletti, V.F., Sartoratto, A., de Alencar, S.M.D., & Figueira G.M. (2014). *Coriandrum sativum* L. (Coriander) essential oil: Antifungal activity and mode of action on *Candida* spp. and molecular targets affected in human whole genome expression. *PLoS One*, 9, e99086.

Gil, A., De La Fuente, E.B., Lenardis, A.E., López Pereira, M., Suárez, S.A., Bandoni, A., Van Baren, C., Di Leo Lira, P., & Ghersa, C.M. (2002). Coriander essential oil composition from two genotypes grown in different environmental conditions. *Journal of Agricultural and Food Chemistry*, 50(10), 2870–2877.

Imlay, J.A., & Linn, S. (1988). DNA damage and oxygen radical toxicity. *Science*, 240, 1302.

Innocent, B.X., Fathima, M.S.A., & Dhanalakshmi. (2011). Studies on the immouostimulant activity of *Coriandrum sativum* and resistance to *Aeromonas hydrophila* in *Catla catla*. *Journal of Applied Pharmaceutical Science*, 1, 132–135.

Jänicke, R.U. (2009). MCF-7 breast carcinoma cells do not express caspase-3. *Breast Cancer Research and Treatment*, 117, 219–221.

Jayakumar, R., & Kanthimathi, M.S. (2012). Dietary spices protect against hydrogen peroxide-induced DNA damage and inhibit nicotine-induced cancer cell migration. *Food Chemistry*, 134, 1580–1584.

Keskin, D., & Toroglu, S. (2011). Studies on antimicrobial activities of solvent extracts of different spices. *Journal of Environmental Biology*, 32, 251–256.

Lalitha, V., Kiran, B., & Raveesha, K.A. (2011). Antifungal and antibacterial potentiality of six essential oils extracted from plant source. *International Journal of Engineering, Science and Technology*, 3, 3029–3038.

Laribi, B., Kouki, K., M'Hamdi, M., & Bettaieb, T. (2015). Coriander (*Coriandrum sativum* L.) and its bioactive constituents. *Fitoterapia*, 103, 9–26.

López-Lázaro, M. (2007). Dual role of hydrogen peroxide in cancer: Possible relevance to cancer chemoprevention and therapy. *Cancer Letters*, 252, 1–8.

Mandal, S., & Mandal, M. (2015). Coriander (*Coriandrum sativum* L.) essential oil: Chemistry and biological activity. *Asian Pacific Journal of Tropical Biomedicine*, 5(6), 421–428.

Moirangthem, D.S., Talukdar, N.C., Kasoju, N., & Bora, U. (2012). Antioxidant, antibacterial, cytotoxic, and apoptotic activity of stem bark extracts of *Cephalotaxus griffithii* Hook. f. *BMC Complementary and Alternative Medicine*, 12, 30.

Momin, A.H., Acharya, S.S., & Gajjar, A.V. (2012). *Coriandrum sativum*-Review of advances in phytopharmacology. *International Journal of Pharmaceutical Sciences and Research*, 3, 1233–1239.

Msaada, K., Hosni, K., Ben Taarit, M., Chahed, T. & Marzouk, B. (2007). Variations in the essential oil composition from different parts of *Coriandrum sativum* L. cultivated in Tunisia. *Italian Journal of Biochemistry*, 56, 47–52.

Neffati, M., & Marzouk, B. (2008). Changes in essential oil and fatty acid composition in coriander (*Coriandrum sativum* L.) leaves under saline conditions. *Industrial Crops and Products*, 28, 137–142.

Neffati, M., & Marzouk, B. (2009). Roots volatiles and fatty acids of coriander (*Coriandrum sativum* L.) grown in saline medium. *Acta Physiologiae Plantarum*, 31, 455–461.

Nelson, K.K., Ranganathan, A.C., Mansouri, J., Rodriguez, A.M., Providence, K.M., Rutter, J.L., Pumiglia, K., Bennett, J.A., & Melendez, J.A. (2003). Elevated sod2 activity augments matrix metalloproteinase expression evidence for the involvement of endogenous hydrogen peroxide in regulating metastasis. *Clinical Cancer Research*, 9, 424–432.

Olle, M., & Bender, I. (2010). The content of oils in umbelliferous crops and its formation. *Agronomy Research*, 8, S687–696.

Panjwani, D., Mishra, B., & Banji, D. (2010). Time dependent antioxidant activity of fresh juice of leaves of *Coriandrum sativum. International Journal of Pharmaceutical Sciences and Drug Research*, 2, 63–66.

Potter, T.L., & Fagerson, I.S. (1990). Composition of coriander leaf volatiles. *Journal of Agricultural and Food Chemistry*, 38, 2054–2056.

Priyadarshi, S., Khanum, H., Ravi, R., Borse, B.B., & Naidu, M.M. (2016). Flavour characterisation and free radical scavenging activity of coriander (*Coriandrum sativum* L.) foliage. *Journal of Food Science and Technology*, 53(3):1670–1678.

Rajeshwari, U., & Andallu, B. (2011). Medicinal benefits of coriander (*Coriandrum sativum* L). *Spatula DD*, 1(1), 51–58.

Randall, K.M., Drew, M.D., Øverland, M., Østbye, T.K., Bjerke, M., Vogt, G., & Ruyter, B. (2013). Effects of dietary supplementation of coriander oil, in canola oil diets, on the metabolism of [1–14C] 18: 3n-3 and [1–14C] 18:2n-6 in rainbow trout hepatocytes. *Comparative Biochemistry and Physiology Part B: Biochemistry and Molecular Biology*, 166(1), 65–72.

Sharifi-Rad, J., Sureda, A., Tenore, G.C., Daglia, M., Sharifi-Rad, M., Valussi, M., Tundis, R., Sharifi-Rad, M., Loizzo, M.R., Ademiluyi, A.O., & Sharifi-Rad, R. (2017). Biological activities of essential oils: From plant chemoecology to traditional healing systems. *Molecules*, 22(1), 70.

Tang, E.L., Rajarajeswaran, J., Fung, S.Y., & Kanthimathi, M.S. (2013). Antioxidant activity of *Coriandrum sativum* and protection against DNA damage and cancer cell migration. *BMC Complementary and Alternative Medicine*, 13(1), 1–13.

Taraphdar, A.K., Roy, M., & Bhattacharya, R. (2001). Natural products as inducers of apoptosis: Implication for cancer therapy and prevention. *Current Science India*, 80, 1387–1396.

Teshale, C., Hussien, J., & Jemal, A. (2013). Antimicrobial activity of the extracts of selected Ethiopian aromatic medicinal plants. *Spatula DD*, 3, 175–180.

Yang, H.L., Chen, C.S., Chang, W.H., Lu, F.J., Lai, Y.C., Chen, C.C., Hseu, T.H., Kuo, C.T., & Hseu, Y.C. (2006). Growth inhibition and induction of apoptosis in MCF-7 breast cancer cells by *Antrodia camphorata. Cancer Letters*, 231, 215–227.

Yeung, E.C., & Bowra, S. (2011). Embryo and endosperm development in coriander (*Coriandrum sativum*). *Botany*, 89(4), 263–273.

Zare-Shehneh, M., Askarfarashah, M., Ebrahimi, L., Kor, N.M., Zare- Zardini, H., & Soltaninejad, H. (2014). Biological activities of a new antimicrobial peptide from *Coriandrum sativum. International Journal of Biosciences*, 4, 89–99.

22 *Coriandrum sativum* L. Root Phytochemicals

*Idowu Jonas Sagbo, Natascha Cheikhyoussef,
Ahmad Cheikhyoussef, and Ahmed A. Hussein*

CONTENTS

22.1 INTRODUCTION

Medicinal plants have been used for centuries, and many people continue to depend on them for their primary health care needs, cooking, and many other purposes, including beverages such as tea, dyes, fragrances, repellents, charms, smoking, cosmetics, and industrial uses. Over the years, they have evolved as fundamental parts of African civilization and are broadly accepted today, representing their rich cultural heritage (Dzoyem et al., 2013). The World Health Organization (WHO) stated that between 65% and 80% of the population in developing countries now use medicinal plants for several therapeutic purposes (WHO, 2019). Due to the widespread usage of medicinal plants, the WHO has encouraged developing countries, especially African member states, to recommend and incorporate medicinal plants or traditional medical practices in their health systems (WHO, 2008; Mahomoodally, 2013). In addition, several reports have indicated that treatment with medicinal plants is considered very safe and devoid of side effects (Maroyi, 2017; Sagbo & Mbeng-Otang, 2021). This is why medicinal plants are growing in popularity across the globe.

In general, medicinal plants contain mixtures of various phytochemical compounds, also called secondary metabolites, which can act individually, additively, or synergistically to enhance health (Mahomoodally, 2013; Sen & Samanta, 2015; Kasilo et al., 2019; Abubakar & Haque, 2020). Moreover, these phytochemical compounds also serve as a crucial herbal cure for numerous therapeutic purposes. Hence, we must elaborate more about these medicinal plants and their usefulness. Among numerous medicinal plants worldwide, *Coriandrum sativum*, also known as coriander or cilantro or Chinese parsley, is one of the most widespread species currently planted worldwide. This chapter discusses the composition and functionality of coriander root as an essential food ingredient for promoting well-being.

DOI: 10.1201/9781003204626-23

22.2 BOTANICAL DESCRIPTION AND TRADITIONAL USAGE OF CORIANDER

Coriander (*Coriandrum sativum* L.) is an aromatic annual plant that belongs to the family of Apiaceae (Laribi et al., 2015). It is commonly known as coriander, Arab parsley, Mexican parsley, and Chinese parsley. It is a soft plant with a height of 50 cm (20 in). The plant's leaves (Figure 22.1) are variable in form, widely lobed at the base, and thinner and feathery on the flowering stems. Coriander grows worldwide but is native to the Mediterranean regions. It is usually cultivated in Mexico, India, Romania, Argentina, Morocco, and Russia (Priyadarshi et al., 2016). The plant grows more efficiently in dry climates, and it can also grow in any soil (Bhat et al., 2014). Coriander is widely used in cooking and folkloric remedies (Marouti et al., 2010). Traditionally, the plant's fresh leaves are used as an herbal seasoning in numerous indigenous dishes. The ground seeds of coriander are also used to prepare poultry or meat (Singletary, 2016). The plant is also used as a functional food as a lipid-lowering, glucose-lowering, heavy metal detoxification, and antimicrobial agent (Omura et al., 1996; Chaudhry & Tariq, 2006; Prachayasittikul et al., 2018).

22.3 CORIANDER ROOTS

22.3.1 WHY ARE CORIANDER ROOTS SO IMPORTANT?

Coriander roots have a pure white central taproot covered by small hairy rootlets, usually darker (Figure 22.2). They are obtainable year-round and very aromatic (more than the leaves), with a slightly spicy, peppery flavor (Special Produce, 2021). The roots are used as a kitchen staple in various Asian cuisines, particularly in Vietnamese and Thai dishes such as soups or curry pastes (Iqbal et al., 2018; Allen, 2021). They are also used to pair with food ingredients such as carrots, tomato paste, galangal, lemongrass, chilli, and peppers (Allen, 2021). Coriander root is the foundation upon which many dishes are built. In Thailand, a report has indicated that numerous dishes are severely compromised without coriander roots, and the understanding of the use of the food culture is completely lost (Allen, 2021).

22.4 PHYTOCHEMICAL COMPONENTS OF CORIANDER ROOTS

There is a clear gap in the reports of phytochemical components of coriander root in the literature. Only a few articles have mentioned or reported some components. Kumar et al. (2014) revealed

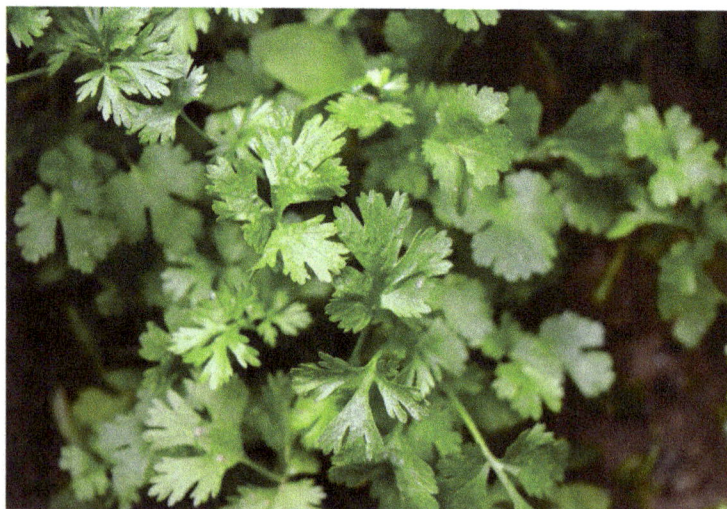

FIGURE 22.1 Coriander. Photo: Unsplash.

FIGURE 22.2   Coriander root. Photo: Freepik Company.

that the fresh roots of coriander contain alkaloids, carbohydrates, flavonoids, phenolic compounds, saponins, sterols, and terpenoids. Because of some phytoconstituents identified in the coriander root, we discuss essential phytoconstituents reported in the order.

22.4.1 Terpenoids and Volatile Constituents

The terpenoids (isoprenoids) are vital metabolites in coriander roots because of their broad spectrum of biological activities. Terpenoids are derived from the 5-carbon (5C) compound isoprene, and the isoprene polymers are known as terpenes. Terpenoids such as thujene, terpinene, pinene, and camphor (Figure 22.3) have been detected in the essential oil of the coriander root (Neffati & Marzouk, 2009). Thujene detected in the coriander root has been reported to contribute pungency to the flavor and antioxidant and antimicrobial activity of some medicinal plants such as black cumin (*Nigella sativa* L.) (Hassanien et al., 2015), while terpinene is also detected in the coriander root used as a fragrance compound (Rudbäck et al., 2012). On the other hand, pinene, the most encountered terpenoid in nature, is found in the coriander root essential oil reported to be often used in perfume and serves as the main constituent of aromatherapy essential oils (Tadtong et al., 2015). Camphor is a volatile aromatic terpenoid also detected in the coriander root. Camphor has been reported to be used in skin lotions and many ayurvedic medicines (Rahnama-Moghadam et al., 2015). In addition to these detected terpenoids, Sui et al. (2016) reported some major volatile components of coriander roots from China. In the study, the authors found that the major components were 2-dodecenal (41.8%) and 1-ethenyl-cyclododecanol (42.9%). A study conducted by Lu et al. (2007) also revealed that aldehydes comprise 49.2%, alcohols (18.8%), esters (6.48%), and hydrocarbons (5.25%) of the essential oil from Chinese Heilongjiang coriander roots. They further reported that furfural (23.0%) was the principal constituent, followed by 2-methylene-cyclopentanol (14.3%) and, lastly, 2-dodecenal (12.9%). This finding is almost similar to that of Msaada et al. (2007) in that coriander root essential oil had a higher level of aldehydes. It is imperative to suggest that these terpenoids, together with the volatile constituents found in the coriander root, may play a major role in its function such as an antidiabetic agent (Abou El-Soud et al., 2012; Andrade et al., 2020), and phototherapy (Bitterling et al., 2022).

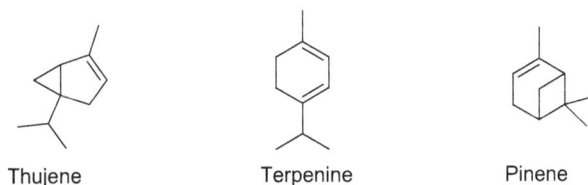

| Thujene | Terpenine | Pinene |

FIGURE 22.3   Structure of some terpenoids identified in coriander root.

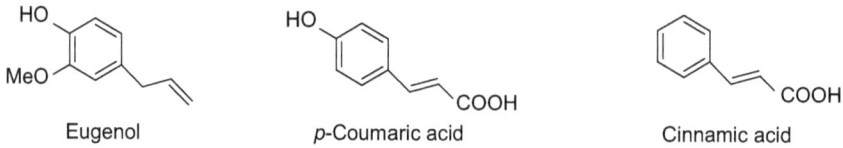

FIGURE 22.4 Structure of phenolics identified in coriander root.

22.4.2 PHENOLICS

Phenolics are also known as phenols, which generally are crucial secondary metabolites in plants. This chemical compound class comprises one or more hydroxyl groups bonded or attached directly to an aromatic hydrocarbon group. In coriander root, phenolic compounds such as eugenol, *p*-coumaric acid, and cinnamic acid (Figure 22.4) have been detected (Neffati & Marzouk, 2009; Tang et al., 2013). These phenolic compounds have been reported in the literature to possess potent antioxidants and several other biological properties (Rajarathnam et al., 2013; Ahmad et al., 2021). For example, eugenol is used as an additive for food (Cortés-Rojas et al., 2014), chronic diseases (Fujisawa & Murakami, 2016), antimicrobial (Marchese et al., 2017), and anti-inflammatory (Barboza et al., 2018) agents, while *p*-coumaric acid and cinnamic acid are also used to protect biologically important molecules from oxidation (Abramovič, 2015; Nejad et al., 2017; Kumar & Goel, 2019; Godlewska-Żyłkiewicz et al., 2020). Apart from these detected phenolic compounds, no chemical compounds in the phenolics class have been reported in the coriander root.

22.4.3 FATTY ACIDS

The essential oils of coriander root have been found to contain some amount of oleic (26%), palmitic (22%), and linoleic acids (43%) (Figure 22.5) (Neffati & Marzouk, 2009). Oleic acid (C18:1n-9) is a mono-unsaturated *omega*-9 fatty acid found naturally in many animal and vegetable fats and oils. It is considered highly stable against oxidation and stimulates antioxidant activity (Okeefe et al., 1993; Su et al., 2014). Palmitic acid (C16:0) is another fatty acid found in the essential oil of coriander roots. Palmitic acid is produced by adding an acetyl group to multiple malonyl groups linked by single bonds between carbons and, thereby, serving as emulsifiers and surfactants (Tabor et al., 1990). On the other hand, linoleic acid (C18:2n-6), an essential polyunsaturated fatty acid (PUFA) was detected in the coriander root. However, several reports have indicated that consumption or intake of linoleic acid plays a significant role in proper health, as it is an essential fatty acid (Whelan, 2013; den Hartigh, 2019; Marangoni et al., 2020). Therefore, in this study, it is imperative to suggest that these fatty acids reported in coriander root may play a significant role in its function and application.

22.5 BIOLOGICAL ACTIVITY OF CORIANDER ROOTS

The biological properties of coriander roots in the scientific literature are very limited. However, the plant's roots are regularly used in cooking and traditional medicine and are believed to play an essential role in improving health and protection against diseases. A study conducted by Tang et al.

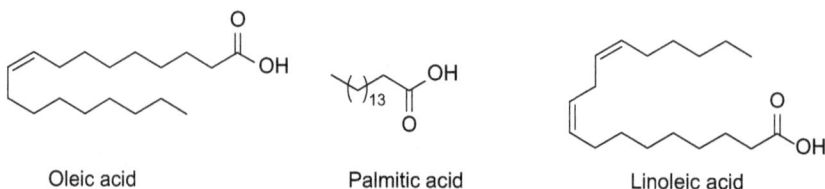

FIGURE 22.5 Structure of some fatty acids identified in the coriander root.

(2013) revealed the antioxidant potencies from coriander root extract against DPPH· and FRAP radical scavenging activities. It was indicated that the ethyl acetate extract of coriander root exhibited strong DPPH· and FRAP radical scavenging activities among other extracts of the root investigated in the study. In the same study, the author also indicated the important anticancer activity of coriander root against the MCF-7 cell line, exhibiting induction of apoptotic cell death. Another study conducted by Kumar et al. (2014) also revealed the antibacterial activity of ethanol extract from coriander root using the agar well diffusion method. The study reported that the root ethanol extract exhibited intense antibacterial activity against some bacteria pathogens (*Salmonella typhi*, *Staphylococcus aureus*, *Bacillus cereus*, and *Klebsiella* sp.) at the tested concentration (50 μg/mL). These biological properties of coriander root have been exceptionally linked with some bioactive components. Since not much work has been done on the biological properties of coriander roots, it is imperative to suggest that further biological studies be conducted.

22.6 CONCLUSION

The importance of coriander root and its composition based on the scientific literature have been discussed in this study. Its potential as a food ingredient and as a nutraceutical is significant along with a tremendous future, but information on its composition and biological activity remains scanty. Therefore, it is strongly recommended that the coriander root and its components be scientifically explored.

REFERENCES

Abou El-Soud, N. H., El-Lithy, N. A., El-Saeed, G. S., Wahby, M. S., Khalil, M. Y., Abou El-Kassem, L. T., Morsy, F., Shaffie, N. (2012). Efficacy of *Coriandrum sativum* L. essential oil as antidiabetic. *Journal of Applied Sciences Research*, 8(7), 3646–3655.

Abramovič, H. (2015). Antioxidant properties of hydroxycinnamic acid derivatives: A focus on biochemistry, physicochemical parameters, reactive species, and biomolecular interactions. In *Coffee in health and disease prevention*. Elsevier, 843–852.

Abubakar, A. R., Haque, M. (2020). Preparation of medicinal plants: Basic extraction and fractionation procedures for experimental purposes. *Journal of Pharmacy & Bioallied Sciences*, 12(1), 1–10. https://doi.org/10.4103/jpbs.JPBS_175_19

Ahmad, A., Elisha, I. L., van Vuuren, S., Viljoen, A. (2021). Volatile phenolics: A comprehensive review of the anti-infective properties of an important class of essential oil constituents. *Phytochemistry*, 190, 112864. https://doi.org/10.1016/j.phytochem.2021.112864

Andrade, C., Gomes, N., Duangsrisai, S., Andrade, P. B., Pereira, D. M., Valentão, P. (2020). Medicinal plants utilized in Thai Traditional Medicine for diabetes treatment: Ethnobotanical surveys, scientific evidence and phytochemicals. *Journal of Ethnopharmacology*, 263, 113177. https://doi.org/10.1016/j.jep.2020.113177

Allen, P. (2021). Coriander root - Demystifying this exquisite ingredient. Available at: https://onthegas.org/food/coriander-root/ (assessed on 10/10/2021).

Barboza, J. N., da Silva Maia Bezerra Filho, C., Silva, R. O., Medeiros, J., de Sousa, D. P. (2018). An overview on the anti-inflammatory potential and antioxidant profile of eugenol. *Oxidative Medicine and Cellular Longevity*, 3957262. https://doi.org/10.1155/2018/3957262

Bhat, S., Kaushal, P., Kaur, M., Sharma, H. K. (2014). Coriander (*Coriandrum sativum* L.): Processing, nutritional and functional aspects. *African Journal of Plant Science*, 8(1), 25–33.

Bitterling, H., Lorenz, P., Vetter, W., Kammerer, D. R., Stintzing, F. C. (2022). Photo-protective effects of selected furocoumarins on β-pinene, R-(+)-limonene and γ-terpinene upon UV-A irradiation. *Journal of Photochemistry and Photobiology A: Chemistry*, 113623, https://doi.org/10.1016/j.jphotochem.2021.113623

Chaudhry, N. M., Tariq, P. (2006). Bactericidal activity of black pepper, bay leaf, aniseed and coriander against oral isolates. *Pakistan Journal of Pharmaceutical Sciences*, 19, 214–218.

Cortés-Rojas, D. F., de Souza, C. R., Oliveira, W. P. (2014). Clove (*Syzygium aromaticum*): A precious spice. *Asian Pacific Journal of Tropical Biomedicine*, 4(2), 90–96. https://doi.org/10.1016/S2221-1691(14)60215-X

den Hartigh, L. J. (2019). Conjugated linoleic acid effects on cancer, obesity, and atherosclerosis: A review of pre-clinical and human trials with current perspectives. *Nutrients*, 11(2), 370. https://doi.org/10.3390/nu11020370

Dzoyem, J. P., Tshikalange, E., Kuete, V. (2013). Medicinal plants market and industry in Africa. In Victor, K. (ed.), *Medicinal plant research in Africa*. Amsterdam, Netherlands: Elsevier, 859–890. https://doi.org/10.1016/B978-0-12-405927-6.00024-2

Fujisawa, S., Murakami, Y. (2016). Eugenol and its role in chronic diseases. *Advances in Experimental Medicine and Biology*, 929, 45–66. https://doi.org/10.1007/978-3-319-41342-6_3

Godlewska-Żyłkiewicz, B., Świsłocka, R., Kalinowska, M., Golonko, A., Świderski, G., Arciszewska, Ż., Nalewajko-Sieliwoniuk, E., Naumowicz, M., Lewandowski, W. (2020). Biologically active compounds of plants: Structure-related antioxidant, microbiological and cytotoxic activity of selected carboxylic acids. *Materials (Basel, Switzerland)*, 13(19), 4454. https://doi.org/10.3390/ma13194454

Iqbal, M. J., Butt, M. S., Suleria, H. A. R. (2018). Coriander (*Coriandrum sativum* L.): bioactive molecules and health effects. In Mérillon, J. M., Ramawat, K. (eds), *Bioactive molecules in food. Reference series in phytochemistry*. Cham: Springer. https://doi.org/10.1007/978-3-319-54528-8_44-1

Kasilo, O., Wambebe, C., Nikiema, J. B., Nabyonga-Orem, J. (2019). Towards universal health coverage: Advancing the development and use of traditional medicines in Africa. *BMJ Global Health*, 4(Suppl 9), e001517. https://doi.org/10.1136/bmjgh-2019-001517

Kumar, N., Goel, N. (2019). Phenolic acids: Natural versatile molecules with promising therapeutic applications. *Biotechnology Reports (Amsterdam, Netherlands)*, 24, e00370. https://doi.org/10.1016/j.btre.2019.e00370

Kumar, S. R., Balasubramanian, P., Govindaraj, P., Krishnaven, T. (2014). Preliminary studies on phytochemicals and antimicrobial activity of solvent extracts of *Coriandrum sativum* L. roots (Coriander). *Journal of Pharmacognosy and Phytochemistry*, 2(6), 74–78.

Laribi, B., Kouki, K., M'Hamdi, M., Bettaieb, T. (2015). Coriander (*Coriandrum sativum* L.) and its bioactive constituents. *Fitoterapia*, 103, 9–26. https://doi.org/10.1016/j.fitote.2015.03.012

Lu, Z. G., Guo, H. Z., Li, W. (2007). Study on aromatic components of coriander root. *Chemistry and Adhesion*, 29(2), 79–81.

Mahomoodally, F. M. (2013). Traditional medicines in Africa: An appraisal of ten potent African medicinal plants. *Evidence-Based Complementary and Alternative Medicne*, 2013, 617459. https://doi.org/10.1155/2013/617459

Marangoni, F., Agostoni, C., Borghi, C., Catapano, A. L., Cena, H., Ghiselli, A., La Vecchia, C., Lercker, G., Manzato, E., Pirillo, A., Riccardi, G., Risé, P., Visioli, F., Poli, A. (2020). Dietary linoleic acid and human health: Focus on cardiovascular and cardiometabolic effects. *Atherosclerosis*, 292, 90–98. https://doi.org/10.1016/j.atherosclerosis.2019.11.018

Marchese, A., Barbieri, R., Coppo, E., Orhan, I. E., Daglia, M., Nabavi, S. F., Izadi, M., Abdollahi, M., Nabavi, S. M., Ajami, M. (2017). Antimicrobial activity of eugenol and essential oils containing eugenol: A mechanistic viewpoint. *Critical Reviews in Microbiology*, 43(6), 668–689. https://doi.org/10.1080/1040841X.2017.1295225

Marouti, K., Farahani, H. A., Darvishi, H. H. (2010). Importance of coriander (*Coriandrum sativum* L.) between the medicinal and aromatic plants. *Advances in Environmental Biology*, 4, 433–436.

Maroyi, A. (2017). Diversity of use and local knowledge of wild and cultivated plants in the Eastern Cape province, South Africa. *Journal of Ethnobiolology Ethnomededicine*, 13, 43.

Msaada. K., Hosni, K., Ben Taarit, M., Chahed, T., Marzouk, B. (2007). Variations in the essential oil composition from different parts of *Coriandrum sativum* L. cultivated in Tunisia. *Italian Journal of Biochemistry*, 56, 47–52.

Neffati, N.; Marzouk, M. (2009). Roots volatiles and fatty acids of coriander (*Coriandrum sativum* L.) grown in saline medium. *Acta Physiology Plantarum*, 31, 455–461.

Nejad, S. M., Ozgunes, H., Basarani, N. (2017). Pharmacological and toxicological properties of eugenol. *Turkish Journal of Pharmaceutical Science*, 14(2), 201–206.

Okeefe, S. F., Wiley, V. A., Knauft, D. A. (1993). Comparison of oxidative stability of high-oleic and normal-oleic peanut oils. *Journal of the American Oil Chemists' Society*, 20(70), 489–492.

Omura, Y., Shimotsuura, Y., Fukuoka, A., Fukuoka, H., Nomoto, T. (1996). Significant mercury deposits in internal organs following the removal of dental amalgam, & development of pre-cancer on the gingiva and the sides of the tongue and their represented organs as a result of inadvertent exposure to strong curing light (used to solidify synthetic dental filling material) & effective treatment: a clinical case report, along with organ representation areas for each tooth. *Acupuncture & Electro-Therapeutics Research*, 21(2), 133–160. https://doi.org/10.3727/036012996816356915

Prachayasittikul, V., Prachayasittikul, S., Ruchirawat, S., Prachayasittikul, V. (2018). Coriander (*Coriandrum sativum*): A promising functional food toward the well-being. *Food Research International*, 105, 305–323.

Priyadarshi, S., Khanum, H., Ravi, R., Borse, B. B., Naidu, M. (2016). Flavour characterization and free radical scavenging activity of coriander (*Coriandrum sativum* L.) foliage. *Journal of Food Science and Technology*, 53(3), 1670–1678.

Rahnama-Moghadam, S., Hillis, D. L., Lange, R. A. (2015). Environmental toxins and the heart. In *Heart and toxins*: Elsevier, 75–132. https://doi.org/10.1016/C2012-0-07917-3

Rajarathnam, S., Shashirakha, M. N., Mallikarjuna, S. E. (2013). Status of bioactive compounds in foods, with focus on fruits and vegetables. *Critical Reviews in Food Science and Nutrition*, 55, 1324–1339.

Rudbäck, J., Bergström, M. A., Börje, A., Nilsson, U., Karlberg, A.-T. (2012). α-Terpinene, an antioxidant in tea tree oil, autoxidizes rapidly to skin allergens on air exposure. *Chemical Research in Toxicology*, 25(3), 713–721.

Sagbo, I. J., Otang-Mbeng, W. (2021). Plants used for the traditional management of cancer in the Eastern Cape Province of South Africa: A review of ethnobotanical surveys, ethnopharmacological studies and active phytochemicals. *Molecules (Basel, Switzerland)*, 26(15), 4639. https://doi.org/10.3390/molecules26154639

Sen, T., Samanta, S. K. (2015). Medicinal plants, human health and biodiversity: A broad review. *Advances in Biochemical Engineering/Biotechnology*, 147, 59–110. https://doi.org/10.1007/10_2014_273

Singletary, K. (2016). Coriander: Overview of potential health benefits. *Nutrition Today*, 51, 151–161.

Special Produce (2021). *Cilantro Roots*. https://www.specialtyproduce.com/produce/Cilantro_Roots_6745.php (accessed 16-10-2021).

Su, M. H., Shih, M. C., Lin, K. H. (2014). Chemical composition of seed oils in native Taiwanese Camellia species. *Food Chemistry*, 156, 369–373.

Sui, H. S., Zou, Y., Zhou, W. Z., Dong, S. Q., Sun, K., Tang, Z. X., Hou, Y. (2016). Changes of volatile components in *Eryngium foetidum* L. and Yunnan *Coriandrum sativum* L. with different plant organs. *Food Research and Development*, 37(3), 161–165.

Tabor, B., Ikegami, M., Yamada, T., Jobe, A. (1990). Rapid clearance of surfactant-associated palmitic acid from the lungs of developing and adult animals. *Pediatric Research*, 27(3), 268–273.

Tadtong, S., Kamkaen, N., Watthanachaiyingcharoen, R., Ruangrungsi, N. (2015). Chemical components of four essential oils in aromatherapy recipe. *Natural Product Communication*, 10, 1091–1092.

Tang, E., Rajarajeswaran, J., Fung, S. Y., Kanthimathi, M. (2013). Antioxidant activity of *Coriandrum sativum* and protection against DNA damage and cancer cell migration. *BMC Complementary and Alternative Medicine*, 13, 347.

Whelan, J. (2013). Linoleic acid. *Advances in Nutrition*, 4(3), 311–312.

WHO (World Health Organization) (2008). *Fact sheet N4*. http://www.who.int/mediacentre/factsheets/2003/fs134/en/.

WHO (World Health Organization) (2019). *WHO organization global report on traditional and complementary medicine*. https://www.who.int/traditional-complementary-integrative- medicine/WhoGlobalReportOnTraditionalAndComplementaryMedicine2019.pdf. (assessed 28/10/2021)

23 Composition and Functionality of Agriculture Waste from Coriander Plant

Muhammad Imran, Muhammad Kamran Khan,
Muhammad Haseeb Ahmad, Rabia Shabir Ahmad,
Haseeb Anwar, Muhammad Nadeem,
and Muhammad Abdul Rahim

CONTENTS

23.1 INTRODUCTION

The coriander plant is predominantly photosynthetic eukaryotic of the kingdom Plantae, belonging to the family Apiaceae and its genus *Coriandrum sativum*. Coriander is one of the most widely cultivated culinary herbs in the world. Mainly, it is cultivated in southwest Asian countries such as India, Pakistan, Iran, Iraq, Armenia, Bangladesh, Oman, Palestine, and many other countries. However, all over the world, it has been used as food and medicine for multiple purposes. All parts of the coriander plant are consumable, but its seeds and leaves are widely used as a spice and flavoring agent in many dishes. Moreover, the coriander plant is also rich in micronutrients and nutrients, which is very useful to reduce the risk of chronic diseases in humans (Diederichsen, 1996; Shukla & Gupta, 2009; Nadeem et al., 2013).

The coriander seeds, leaves, and their oil have many health benefits. Coriander is most commonly used in the flavoring and spice industries for the production of many products, including bakery products such as bread, cookies, cakes, and biscuits, in confectionery like toffee, candy, and lollipops, cosmetics products, and nutraceutical products, as well as traditional medicine in research and development laboratories. These industries have produced a significant amount of coriander waste such as seeds, leaves, plucks, roots, and cake or meal (residues after extracting oil from coriander seeds). Coriander seeds are one of the major wastes of the food flavoring industries. Moreover, this waste of coriander seeds is the best source of essential oil (Mohamed et al., 2018). In addition, research work by Wahba et al. (2020) has indicated that coriander waste contains more essential oil than oil extracted from dried seeds. The essential oil was extracted from coriander waste and reduced environmental pollution. This essential oil accumulates more than 85% of industrially applicable fatty acids and petroselinic acid (Bhat et al., 2014).

In recent decades, agriculture waste management and researchers have been more interested in extracting oil from coriander plant waste. As a result, coriander waste oil (CWO) is gaining more popularity because it works as a natural remedy and has no side effects. In addition, the essential oil from coriander waste can reduce the risk of many health problems and improve body health. The major constituents of CWO are *trans*-anethole and linalool. Therefore, CWO has been used as a therapeutic agent, pharmaceutical agent, and household cleaning product.

Furthermore, CWO is also an essential source of phenolic compounds recognized as bioactive ingredients. The benefits of CWO come from their antioxidant, antidiabetic, anti-inflammatory, anxiolytic, antimicrobial, and anthelmintic properties. In addition, the potential benefits of CWO include losing weight, reducing pain, stimulating hormones, reducing bad breath, improving digestion, and improving erectile dysfunctions (Gray & Flatt, 1999; Rajeshwari & Andallu, 2011; Sahib et al., 2013; Mandal & Mandal, 2015).

23.2 HISTORY OF CORIANDER

Coriander is an ancient herb grown in large areas of Slovenia, Albania, Iraq, Monaco, Iran, Portugal, Qatar, the United Arab Emirates, Malta, Sudan, Ethiopia, Oman, Spain, Saudi Arabia, Croatia, Turkey, Serbia, and Bosnia. Coriander has a low germination rate and is a small plant (Ivanova & Stoletova, 1990). A study by Prance and Nesbitt (2012) reported that coriander has been used as traditional medicine and food to improve health since 1550 BC. Furthermore, coriander has been cultivated in southwestern European countries since around 2000 BC. In Germany, it was used as traditional medicine until about 900 AD. Since then, coriander seeds have been used as a spice, and their leaves are used as a flavoring ingredient in many foods. In addition, coriander is also used in many perfumes and dish recipes as a fragrant agent. In Greece, coriander is known as "Korannon" (Burdock & Carabin, 2009). Furthermore, it is also called the "spice of happiness" in Southwest Asia because of its natural effect on improving sexual behavior and improving sexual performance (Abascal & Yarnell, 2012). In the Roman Empire, coriander was used as a traditional herbal medicine to treat various diseases such as skin, brain, and heart diseases. In the district of "Voronezh", Russia, coriander oil was first extracted from the coriander seeds using the steam distillation process in 1885 (Eikani et al., 2007).

23.3 INDUSTRIAL WASTE OF CORIANDER

Figure 23.1 presents the main extract and bioactive chemicals from coriander wastes. Coriander seeds are most commonly used in the spice processing, food processing, pharmaceutical, nutraceutical, and flavoring industries. It is grown in different parts of the world and varies considerably in extrinsic quality. Therefore, coriander seeds contain damaged and rotten seeds that are the coriander's industrial waste (Parthasarathy & Zachariah, 2008; Sharma & Sharma, 2012). The wind-based separator method removes wrinkled and shrunken seeds from good-quality seeds in the spice processing industries. Wrinkled and withered seeds are low in weight and dry out faster than fresh and good quality seeds. Therefore, these seeds are easily separated by this method. These seeds are wasted in large quantities (Sharma & Sharma, 2012; Bhat et al., 2014). In other processing industries, these seeds are separated from the good quality seeds. These seeds are a waste of the coriander processing industry.

Furthermore, coriander waste contains coriander essential oil (CEO) in large amounts (Kassahun, 2020). In the oil processing industry, CEO is extracted from coriander seeds using various modern extraction methods. After oil extraction, industrial residues are discarded. The oil is not completely extracted from coriander seeds; the residues of coriander seeds contain essential oil and biological compounds. The results showed that the total phenolics in coriander waste ranged from 41 to 57 mg GAE/g (Aćimović et al., 2016).

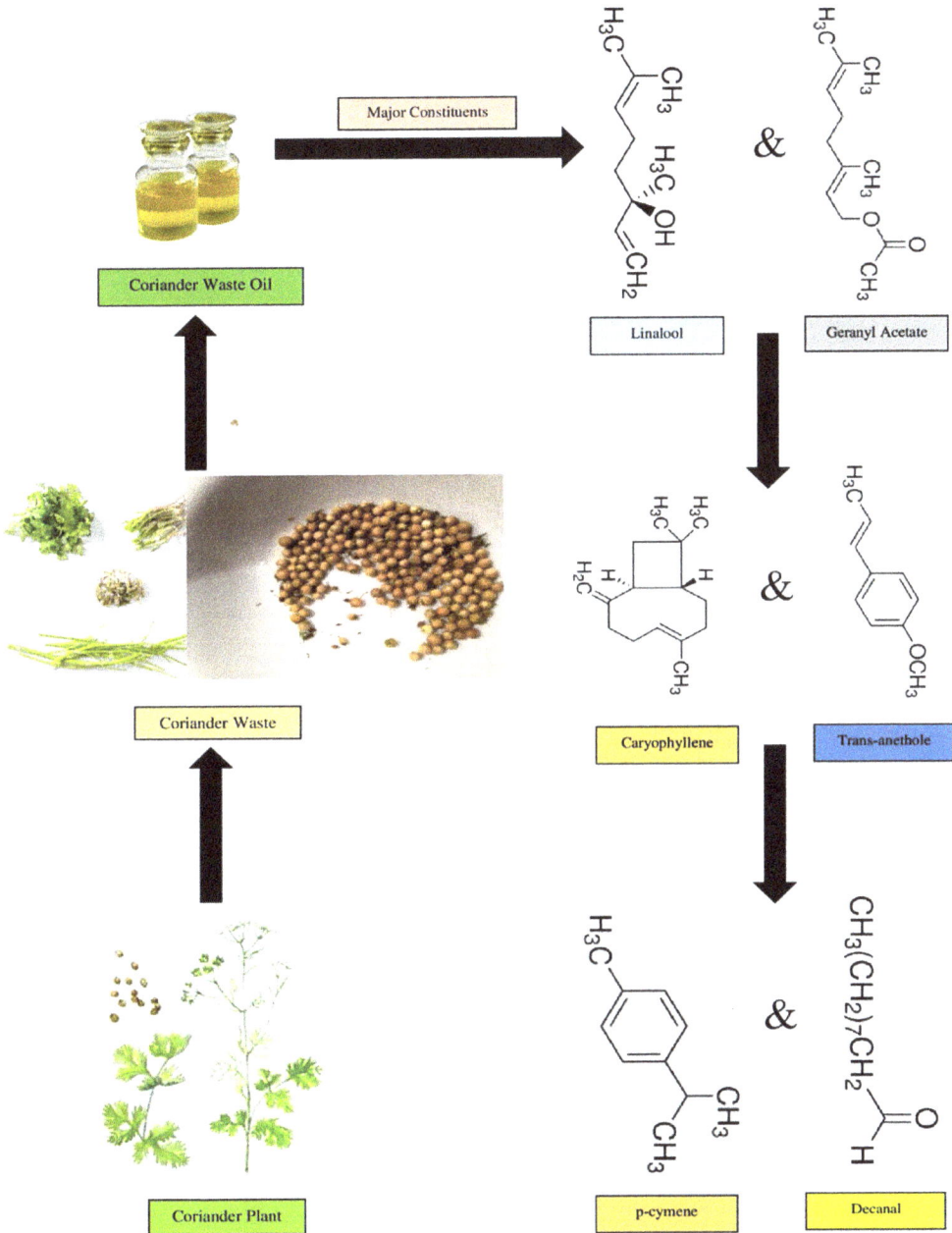

FIGURE 23.1 Main extracts and bioactive chemicals from coriander waste.

23.4 BIOLOGICAL ACTIVITY OF CORIANDER WASTE

Bioactive substances with antioxidant, anti-carcinogenic, antibiotic, antithrombotic, antimicrobial, anti-nutritional, anti-allergic, anti-inflammatory, antiviral, and anti-mutagenic activities have been used for years in the form of pure compounds. These bioactive substances are naturally found in a minor amount of plant waste. Furthermore, they provide many health benefits beyond the rudimentary nutritive value of the various products. These bioactive substances play a vital role against many chronic diseases; they have beneficial immunological, behavioral, and physiological properties. Moreover, bioactive substances can modulate the chemical reaction in living organisms and reveal beneficial properties (Rai & Carpinella, 2006; Chadwick & Marsh, 2008; Bernhoft, 2010; da Silva & Silva, 2016). In addition, fruits and vegetables are a good source of bioactive compounds such as carotenoids and phenolics. In vegetable sources, coriander waste is one of the best sources of bioactive substances. In recent research, the total phenolic compounds in coriander waste production after the extraction of CEO from coriander seeds were estimated. Coriander waste contains more total polyphenols (Aćimović et al., 2016). The effective use of industrial waste is essential for finding a way to use food and prevent consequent environmental pollution. Waste in the spice industry is an important source of biological compounds and active ingredients (Sowbhagya, 2019).

23.5   ISOLATION OF CORIANDER ESSENTIAL OIL (CEO) FROM WASTE

In the Wahba et al. (2020) study, the CEO was extracted from the damaged and shriveled seeds. The results indicated that the coriander seeds contained 0.31% of oil. The major components of waste oil were linalool and *trans*-anethole. The *trans*-anethole components found in the waste oil were 29.3%, while the linalool was 20.1%, respectively. In addition, coriander plant waste, such as dried leaves, stems, and broken seeds, is an excellent source of CEO. In a comparative study, coriander essential was extracted from the dry seeds and dry coriander plant waste using the hydro-distillation method. The results showed that the maximum CEO amount in dry seeds was 0.31%. Furthermore, the percentage of CEO in dried coriander plant waste was 1.09%, respectively. In addition, gas chromatography was used to identify the main constituents or components in the extracted oils. The high percentage of linalool found in dry seed oil was 59.6%, and a high content of *trans*-anethole was found in waste oil (Mohamed et al., 2018).

In general, the essential oil is extracted from the dry waste of coriander. For extraction, the hydro-distillation method is commonly used to extract oil from the dry residue of coriander. This method is easy to use, low in cost, has a low processing time, minimum loss of volatile compounds, and maximum efficiency. Therefore, coriander waste oil (CWO) was extracted from the dry coriander waste and dry coriander seeds using this method (hydro-distillation). Dry coriander waste provided the highest percentage of CWO compared to the dry seed oil. The maximum yield of CWO was 0.81%. Furthermore, the highest percentage of coriander seed oil was 0.44%, respectively. The results indicated that the main constituent in coriander seed oil was linalool, while *trans*-anethole was found as the main component of CWO. The results concluded that CWO is similar to coriander seed oil (El-Rokiek et al., 2021).

23.6   COMPOSITION OF CORIANDER WASTE OIL (CWO)

In recent research, the composition of oil and its biological activities are gaining more importance and increasingly attracting the interest of researchers; a scientific determination of the plants that have been used as herbal medicine is required to enhance the quality of health practices. The amount of major constituents in the oil depends on the species type. Oil constituents and composition are significantly affected by various factors such as cultural conditions, post-harvest processing methods, and environmental conditions. Therefore, it is necessary to analyze the composition of the oil. In a research work by Mohamed et al. (2018), the composition of CWO was identified using gas chromatography. The 17 essential components in CWO of different groups were detected with this method, such as carvacrol, α-pinene, linalool, *trans*-anethole, butanoic acid, germacrene-D, camphene, 2-methyl-, spathulenol, limonene, 2-methoxy-4-(2-propenyl) phenyl ester, zingiberene, β-pinene, estragole, longifolene, β-bisabolene, geranyl acetate, carvacrol, longifolene, and *p*-cymene. According to this analysis, CWO contained high levels of *trans*-anethole. A similar study was conducted by Wahba et al. (2020) on the isolation of CEO from dry seeds and waste. The CEO and CWO were extracted from the prepared samples and stored at various temperatures for one year. After the storage interval, the main components in the CEO and CWO were assessed using the mass spectrometer and gas chromatography method under optimal conditions. The obtained data indicate that the amount of main constituents in stored oil at room temperature was similar to that of CWO stored at a low temperature, while the concentration of stored CWO was slightly different. The results from all treatment groups indicated that the CWO contains the major components such as carvacrol, linalool, butanoic acid, 2-methoxy-4-(2-ropenyl) phenyl ester, estragole, 2-methyl-, and *trans*-anethole. The amount of *trans*-anethole and butanoic acid in CWO was higher under low temperatures than at room temperature. Moreover, the concentration of major constituents (*trans*-anethole and butanoic acid) was changed at low and room temperature. On the other hand, the amount of main components (linalool, estragole, and longifolene) in CWO was reduced throughout the treatment of groups.

23.7 FUTURE PROSPECTS OF CORIANDER WASTE

Food waste is rich in bioactive compounds such as essential oil, polyphenols, flavonoids, antioxidants, antimicrobials, flavoring, and essential constituents. It is also an excellent source of organic matter, so traditional land-filling and incineration methods produce toxic gases that can severely damage the environment and human health. Therefore, it can be used in various food processing and pharmaceutical industries, and using biological methods to treat such waste is proposed as a sustainable process for valorization (Alexander et al., 2013). In addition, fruit waste has been recognized as an excellent source of phytochemicals and vitamins (Imran et al., 2022). Similarly, a study by Jimenez-Lopez et al. (2020) reported that agricultural waste is rich in bioactive compounds that can be used for novel purposes. These bioactive compounds are extracted using innovative techniques and reutilized in many potential food applications but also represent an environmentally friendly measure while preparing functional products. Furthermore, there is growing interest in research work on producing innovative products based on the use of the waste extract. An important research approach to phenolic compounds is to clarify the ultimate effect of functional foods and nutraceuticals enriched by them, as the consumption and use of isolated phenolic compounds as part of nutrition lead to differences and many controversies (Cory et al., 2018).

23.8 CONCLUSION

Research activities in agricultural waste and environmental protection are more interested in recycling coriander waste generated after the food processing and harvest. Coriander waste contains essential nutrients that can be used as organic fertilizer for the normal growth and development of the plant. In addition, it could be used as a novel source of CWO and reduce environmental pollution. Therefore, we suggest that coriander waste be recycled by extracting the CEO and other essential nutrients to reduce pollution. Furthermore, extracting essential oil from coriander waste has substantial economic benefits.

REFERENCES

Abascal, K., & Yarnell, E. (2012). Cilantro-culinary herb or miracle medicinal plant? *Alternative and Complementary Therapies, 18*(5), 259–264.

Aćimović, M., Mara, D., Tešević, V., Stanković, J., Cvetković, M., Urošević, M., & Filipović, V. (2016). Analysis of total polyphenols from postdestillation waste material of different coriander accessions grown in Serbia. In *VII International Scientific Agriculture Symposium*, "Agrosym 2016", 6–9 October 2016, Jahorina, Bosnia and Herzegovina. *Proceedings* (pp. 796–802). University of East Sarajevo, Faculty of Agriculture.

Alexander, C., Gregson, N., & Gille, Z. (2013). Food waste. In *The handbook of food research, 1* (pp. 471–483). London, UK: Bloomsbury Academic. 471–484.

Bernhoft, A. (2010). A brief review on bioactive compounds in plants. In *Bioactive compounds in plants-benefits and risks for man and animals, 50* (pp. 11–17).

Bhat, S., Kaushal, P., Kaur, M., & Sharma, H. K. (2014). Coriander (*Coriandrum sativum* L.): Processing, nutritional and functional aspects. *African Journal of Plant Science*, 8(1), 25–33. Oslo, Norway: The Norwegian Academy of Science and Letters.

Burdock, G. A., & Carabin, I. G. (2009). Safety assessment of coriander (*Coriandrum sativum* L.) essential oil as a food ingredient. *Food and Chemical Toxicology*, 47(1), 22–34.

Chadwick, D. J., & Marsh, J. (Eds.). (2008). *Bioactive compounds from plants* (Vol. 154). John Wiley & Sons.

Cory, H., Passarelli, S., Szeto, J., Tamez, M., & Mattei, J. (2018). The role of polyphenols in human health and food systems: A mini-review. *Frontiers in Nutrition*, 5, 87.

da Silva, L. R., & Silva, B. (Eds.). (2016). *Natural bioactive compounds from fruits and vegetables as health promoters part I*. Bentham Science Publishers.

Diederichsen, A. (1996). *Coriander: Coriandrum sativum L.* (Vol. 3). Bioversity International.

Eikani, M. H., Golmohammad, F., & Rowshanzamir, S. (2007). Subcritical water extraction of essential oils from coriander seeds (*Coriandrum sativum* L.). *Journal of Food Engineering*, 80(2), 735–740.

El-Rokiek, K. G., Ibrahim, M. E., El-Din, S. A. S., & El-Sawi, S. A. (2021). Bioactivity of essential oil isolated from *Coriandrum sativum* plant against the weed *Avena fatua* associated wheat plants. *Journal of Materials and Environmental Sciences*, 12, 899–911.

Gray, A. M., & Flatt, P. R. (1999). Insulin-releasing and insulin-like activity of the traditional anti-diabetic plant *Coriandrum sativum* (coriander). *British Journal of Nutrition*, 81(3), 203–209.

Imran, M., Khan, M. K., Ahmad, M. H., Ahmad, R. S., Javed, M. R., Nisa, M. U., & Rahim, M. A. (2022). Valorization of peach (*Prunus persica*) fruit waste. In *Mediterranean fruits bio-wastes* (pp. 589–604). Springer.

Ivanova, K. V., & Stoletova, E. A. (1990). History of cultivation and infraspecific classification of coriander (*Coriandrum sativum* L.). *Sbornik Nauchnykh Trudov po Prikladnoĭ Botanike, Genetike i Selektsii*, 133, 26–40.

Jimenez-Lopez, C., Fraga-Corral, M., Carpena, M., García-Oliveira, P., Echave, J., Pereira, A. G., & Simal-Gandara, J. (2020). Agriculture waste valorisation as a source of antioxidant phenolic compounds within a circular and sustainable bioeconomy. *Food & Function*, 11(6), 4853–4877.

Kassahun, B. M. (2020). Unleashing the exploitation of coriander (*Coriander sativum* L.) for biological, industrial and pharmaceutical applications. *Academic Research Journal of Agricultural Science and Research*, 8(6), 552–564.

Mandal, S., & Mandal, M. (2015). Coriander (*Coriandrum sativum* L.) essential oil: Chemistry and biological activity. *Asian Pacific Journal of Tropical Biomedicine*, 5(6), 421–428.

Mohamed, M. A., Ibrahim, M. E., & Wahba, H. E. (2018). Flavoring compounds of essential oil isolated from agriculture waste of coriander (*Coriandrum sativum*) plant. *Journal of Materials and Environmental Science*, 9(1), 77–82.

Nadeem, M., Anjum, F. M., Khan, M. I., Tehseen, S., El-Ghorab, A., & Sultan, J. I. (2013). Nutritional and medicinal aspects of coriander (*Coriandrum sativum* L.): A review. *British Food Journal*. 115, 743–755.

Parthasarathy, V. A., & Zachariah, T. J. (2008). Coriander. In *Chemistry of spices* (p. 190). CAB International, Cambridge, MA.

Prance, G., & Nesbitt, M. (Eds.). (2012). *The cultural history of plants*. Routledge.

Rai, M., & Carpinella, M. C. (2006). *Naturally occurring bioactive compounds*. Elsevier.

Rajeshwari, U., & Andallu, B. (2011). Medicinal benefits of coriander (*Coriandrum sativum* L). *Spatula DD*, 1(1), 51–58.

Sahib, N. G., Anwar, F., Gilani, A. H., Hamid, A. A., Saari, N., & Alkharfy, K. M. (2013). Coriander (*Coriandrum sativum* L.): A potential source of high-value components for functional foods and nutraceuticals-A review. *Phytotherapy Research*, 27(10), 1439–1456.

Sharma, M. M., & Sharma, R. K. (2012). Coriander. In *Handbook of herbs and spices* (pp. 216–249). Woodhead Publishing.

Shukla, S., & Gupta, S. (2009). Coriander. In *Molecular targets and therapeutic uses of spices: Modern uses for ancient medicine* (pp. 149–171). World Scientific, Hackensack, NJ.

Sowbhagya, H. B. (2019). Value-added processing of by-products from spice industry. *Food Quality and Safety*, 3(2), 73–80.

Wahba, H. E., Abd Rabbu, H. S., & Ibrahim, M. E. (2020). Evaluation of essential oil isolated from dry coriander seeds and recycling of the plant waste under different storage conditions. *Bulletin of the National Research Centre*, 44(1), 1–7.

Section II

Coriander Leaves

Chemistry, Technology, Functionality, and Applications

24 Bioactive Compounds of Coriander Leaves

Chin Xuan Tan, Seok Shin Tan, and Seok Tyug Tan

CONTENTS

24.1 INTRODUCTION

Coriander (*Coriandrum sativum* L.) is an annual herbaceous plant that belongs to the Apiaceae family. It is widely cultivated in Asia, North Africa, and Central Europe regions nowadays, although the plant originated from the Mediterranean (Laribi, Kouki, M'Hamdi, & Bettaieb, 2015). All segments (leaves, fruits, stems, and roots) of the plant are edible and have been widely utilized in folk medicine and for culinary purposes.

Coriander leaves, also known as Chinese parsley and cilantro, are lanceolate and green in color. They are approved by the Flavour and Extract Manufacturers Association, the US Food and Drug Administration, and the Council of Europe as natural flavoring and preservation agents for food use (Mandal & Mandal, 2015). In Asian countries, coriander leaves are commonly used to prepare soup, salad, sauce, and cocktail decorations. In addition, the consumption of coriander leaves could stimulate appetite (Mani & Parle, 2009). Sometimes, the leaves mask particular food's unpleasant smell and taste (Laribi et al., 2015).

The physical appearance of coriander leaves highly resembles parsley leaves. An easy way to distinguish both leaves is by using the olfactory system. Coriander leaves have a strong scent, whereas parsley leaves have a mild one (Bhat, Kaushal, Kaur, & Sharma, 2014). Fresh coriander leaves have L^*, a^*, and b^* values of 44.2, −7.10, and 28.38, respectively (Yilmaz & Alibas, 2017). Proximate analysis revealed that 86.7% of the coriander leaves were moisture content, and crude fiber (5.24%), crude protein (4.05%), ash (1.9%), and crude fat (0.95%) contributed the rest (Shahwar et al., 2012). This chapter aims to summarize the micronutrients and bioactive compounds of coriander leaves. In addition, the application of coriander leaves is also compiled and discussed.

DOI: 10.1201/9781003204626-26

24.2 MICRONUTRIENT COMPOSITION

Micronutrients are nutrients required by the human body in small quantities for general well-being and disease prevention. As can be seen in Table 24.1, coriander leaves are rich in potassium, calcium, vitamin A, and ascorbic acid. The extraction of fresh leaves into juice concentrates the nutrients, a recommended therapeutic beverage for patients with vitamin A, B, and C deficiencies (Mani & Parle, 2009). However, it was reported that the loss of ascorbic acid in coriander leaves was high when the leaves were blanched in boiling water and dried for an extended period (Singh, Samsher, Sengar, & Kumar, 2020). This might be due to the increased oxidizing enzymes associated with heating, which leads to the destruction and leaching of ascorbic acid in the water.

24.2.1 CAROTENOIDS

Carotenoids are natural pigments responsible for the orange, yellow, and red colors of vegetables and fruits. In green foliage vegetables, carotenoids are mainly deposited in the leaves (Wei et al., 2019). Divya et al. (2012) measured the carotenoid content of ten commercial coriander varieties in India. Their study found that the total carotenoid content of coriander leaves at the mature stage (197.7–217.5 mg/100g) was higher than at the young stage (152.7–169.1 mg/100 g). In addition, a high level of β-carotene (39.0–73.6 mg/100g) was detected in coriander leaves harvested at a mature stage.

Aruna and Baskaran (2010) investigated the carotenoid profile of selected spices purchased from the local markets in India using high-performance liquid chromatography (HPLC). Among the 26 spices investigated, the β-carotene (67.5 mg/100g) and violaxanthin (83.4 mg/100g) of coriander leaves were the highest. Other carotenoid constituents detected in coriander leaves were neoxanthin (5.47 mg/100g) and lutein (9.92 mg/100g). Commercially, β-carotene is used as a coloring agent and antioxidant (Tan, Tan, & Tan, 2020). Therefore, the high level of β-carotene in coriander leaves suggests the potential of its use as a natural coloring or antioxidant ingredient in food products.

24.2.2 POLYPHENOLS

Polyphenols are a large group of secondary metabolites consisting of phenolic acids, flavonoids, stilbenes, and lignans. These polyphenols differ by the number of phenol rings and the structural compounds that bind to those rings. Phenolic acids and flavonoids have been widely extracted from

TABLE 24.1
Micronutrients Content (per 100 g of Leaves)

Mineral	Amount	Vitamin	Amount
Potassium	521.0 mg	Ascorbic acid	27.0 mg
Calcium	67.0 mg	Vitamin E	2.5 mg
Phosphorus	48.0 mg	Niacin	1.1 mg
Sodium	46.0 mg	Pantothenic acid	0.6 mg
Magnesium	26.0 mg	Riboflavin	0.2 mg
Iron	1.8 mg	Vitamin B6	0.2 mg
Zinc	0.5 mg	Thiamine	0.1 mg
Manganese	0.4 mg	Vitamin A	337 µg
Copper	0.2 mg	Vitamin K	310 µg
Selenium	0.9 µg	Folate	62 µg
		Vitamin D	0 µg
		Vitamin B12	0 µg

Source: USDA (2019)

TABLE 24.2

Total Phenolics and Flavonoids Content

Polyphenol	Extraction Solvent	Amount
TPC	Distilled water	2.85–4.48 mg GAE/g
	DMSO	5.63–10.0 mg GAE/g
	Hexane	5.87–6.08 mg GAE/g
	Methanol	4.91–6.16 mg GAE/g
TFC	Distilled water	6.97–10.1 mg QE/g
	DMSO	18.3–22.4 mg QE/g
	Hexane	59.3–65.5 mg QE/g
	Methanol	18.3–21.3 mg QE/g

Source: Agrawal, Sharma, Shirsat, & Saxena (2016)

a variety of plant sources as they are known to be rich in antioxidant activity associated with human health (Tan & Azrina, 2016).

Previous studies on the quantification of polyphenol constituents were mainly on the whole vegetative (i.e., leaves, stems, and roots) or the above-ground (i.e., leaves and stems) portions of the coriander plant (Barros et al., 2012; Oganesyan, Nersesyan, & Parkhomenko, 2007). Limited information on the polyphenol composition of coriander leaves is available in the literature. Three flavonoids (acacetin, quercetin, and kaempferol) and three phenolic acids (ferulic acid, *p*-coumaric acid, and vanillic acid) were observed in coriander leaves after calculating the relative retention factors in a thin layer chromatography (Nambiar, Daniel, & Guin, 2010).

The effects of different extraction solvents on the phenolics and flavonoids content of five different varieties of coriander leaves were summarized in Table 24.2. Results indicated that distilled water, dimethyl sulfoxide (DMSO), hexane, and methanol effectively extracted the polyphenols from coriander leaves. The average amount of total phenolic content [TPC, mg gallic acid equivalents (GAE)/g] isolated from coriander leaves was in the order below:

DMSO (8.69) > hexane (5.97) > methanol (5.37) > distilled water (3.73)

Meanwhile, the average amount of total flavonoid content [TFC; mg quercetin equivalents (QE)/g] isolated from coriander leaves was in the order below:

Hexane (61.6) > methanol (20.4) > DMSO (18.7) > distilled water (10.1)

The results above show that the ideal solvents to extract TPC and TFC from coriander leaves are DMSO and hexane, respectively. Further study on quantification of the individual constituents of polyphenols in coriander leaves should be carried out.

24.2.3 Chlorophyll

Chlorophyll is a natural green pigment widely distributed in plants. It is used in the medical field for diagnostics and remedies (Mishra, Bacheti, & Husen, 2011). Color analysis of fresh coriander leaves obtained an $a*$ value of –7.10 (Yilmaz & Alibas, 2017). A negative $a*$ reading indicates the presence of green pigment such as chlorophyll (Tan, Tan, Ghazali, & Tan, 2021). Spectrophotometry revealed that the chlorophyll level of raw coriander leaves was 1.69 mg/g, and the levels were affected by post-harvesting processing such as drying and blanching (Singh et al., 2020). Higher losses of chlorophyll content were observed when the leaves were blanched in boiling water (Singh et al., 2020). To conserve the chlorophyll content, it was recommended to blanch the coriander leaves at 80°C for 3 min (Ahmed, Shivhare, & Singh, 2001).

TABLE 24.3
Chlorophyll Content at Different Drying Conditions

Method	Value
Microwave drying (1000 W)	65.28
Microwave drying (500 W)	75.53
Microwave drying (100 W)	68.33
Convective drying (50°C)	54.30
Natural drying (25 ± 1°C)	61.22

Source: Yilmaz & Alibas (2017)

Fresh coriander leaves are highly perishable and have a short shelf-life due to high moisture content. Thus, immediate processing or preservation is required after harvest (Singh et al., 2020). Drying is one of the conventional methods to preserve coriander leaves. It works by reducing ethylene production and enzyme activity in the leaves (Yilmaz & Alibas, 2017). The effects of different drying methods and conditions on the chlorophyll content were analyzed (Yilmaz & Alibas, 2017). The leaves were dried by microwave, convective dryer, or natural drying until a moisture content of 0.1 was obtained. As measured using a SPAD meter, the chlorophyll content is shown in Table 24.3. Results indicated that microwave drying was the best method to retain the chlorophyll content, followed by natural and convective drying. Microwave drying at 500 W was the recommended method for drying the coriander leaves (Yilmaz & Alibas, 2017). Further study is needed to identify the individual chlorophyll constituent in coriander leaves.

24.2.4 FATTY ACIDS

Neffati and Marzouk (2008) investigated the chemical composition difference of the leaves collected from the base of the stem (basal leaves) and the leaves collected from the top of the stem (upper leaves). Their findings showed that the basal leaves' total fatty acid content (61.2 mg/g) was higher than the upper leaves (41.8 mg/g). Polyunsaturated fatty acids such as α-linolenic and linoleic acids were the main fatty acids detected in coriander leaves (Table 24.4).

TABLE 24.4
Fatty Acid Composition

	Leaves (mg/g Dried Weight)	
Content	Basal	Upper
Palmitic acid	7.8	5.73
Cis-palmitoleic acid	1.83	1.06
Trans-palmitoleic acid	0.39	0.23
Heptadecenoic acid	9.77	6.05
Stearic acid	0.49	0.31
Oleic acid	2.85	1.44
Linoleic acid	9.85	8.22
α-Linolenic acid	24.14	17.09
Stearidonic acid	4.09	4.67

Source: Neffati and Marzouk (2008)

TABLE 24.5

Main Volatile Compounds of Coriander Leaf Essential Oil

Potter & Fagerson (1990)	Eyres et al. (2005)	Matasyoh et al. (2009)	Padalia et al. (2011)	Shahwar et al. (2012)
E-2-decenal (46.10%)	*E*-2-decen-1-ol (26.00%)	*E*-2-decenal (15.90%)	*E*-2-decenal (18.02%)	*E*-2-decenal (32.23%)
E-2-dodecenal (10.30%)	1-decanol (19.64%)	Decanal (14.30%)	Decanal (14.36%)	Linalool (13.97%)
2-decen-1-ol (9.20%)	*E*-2-decenal (9.12%)	*E*-2-decen-1-ol (14.20%)	Dec-9-en-1-ol (11.66%)	*E*-2-dodecenal (7.51%)
E-2-tetradecenal (5.80%)	*E*-2-tetradecenal (7.03%)	*E*-2-tridececen-1-al (7.31%)	*E*-2-dodecenal (8.72%)	*E*-2-tetradecenal (6.56%)
E-2-undecenal (5.60%)	Decanal (6.56%)	*E*-2-dodecenal (6.23%)	*n*-Tetradecanol (6.09%)	2-decen-1-ol (5.45%)
Decanal (4.40%)	*E*-2-dodecenal (5.37%)	Dodecanal (4.36%)	Dodecanal (5.81%)	*E*-2-undecenal (4.31%)

24.2.5 BIOACTIVE VOLATILE COMPOUNDS

The utilization of coriander essential oil as an ingredient in cosmetics and perfume can be dated back to 2000 BCE (Prachayasittikul, Prachayasittikul, Ruchirawat, & Prachayasittikul, 2018). Table 24.5 summarizes the main volatile compounds detected in coriander leaf essential oil. Earlier, Potter and Fagerson (1990) performed a study that reported 41 volatile compounds in coriander leaf essential oil. These compounds were mainly aldehydes (82.6%) and alcohols (16.6%). Using gas chromatography-olfactometry and comprehensive two-dimensional gas chromatography-time-of-flight mass spectrometry, Eyres et al. (2005) detected 81 volatile compounds in coriander leaf essential oil, and more than 55% of the volatiles were contributed by alcohols.

Meanwhile, 24 volatile compounds were detected in the coriander leaf essential oil when using gas chromatography-mass spectroscopy (GC-MS) (Matasyoh et al., 2009). The majority of the compounds detected were aldehydes (55.5%), followed by alcohols (36.3%), alkanes (1.46%), and monoterpenes (0.36%). Padalia et al. (2011) reported that 90% of the volatile compounds detected in coriander leaf essential oil were aliphatic compounds such as aldehydes and alcohols when analyzed using GC-MS. These aliphatic compounds were responsible for the peculiar, fetid-like aroma of the oil. In agreement with Potter and Fagerson (1990), Matasyoh et al. (2009), and Padalia et al. (2011), Shahwar et al. (2012) also reported the most abundant volatile in coriander leaf essential oil was *E*-2-decenal, an aldehyde responsible for the fatty and floral taste.

The volatile composition of coriander leaf essential oil is greatly influenced by the geographical origins, varieties, post-harvest storage conditions, and analysis methods used by the researchers. Generally, alcohols and aldehydes are the main volatile constituents of coriander leaf essential oil.

24.3 APPLICATION OF CORIANDER LEAVES

24.3.1 ESSENTIAL OIL

Raw coriander leaves are characterized by a distinctive pungent, fatty, and aldehydic aroma (Eyres et al., 2005). These unique properties promote its use as a raw material for essential oil production. Hydro-distillation, also known as water distillation, is a conventional method to extract essential oils from various parts of plants. It is based on the principle that water is added in an adequate amount to the compartment containing plant materials and then brought to its boiling point. The

TABLE 24.6
Yield of Oil

Origin of Coriander Leaves	Yield	Reference
Tunisia	0.12%	Neffati and Marzouk (2008)
Bangladesh	0.10%	Bhuiyan, Begum, & Sultana (2009)
Pakistan	0.10%	Shahwar et al. (2012)

essential oil has been used in food preservatives, complementary and alternative medicine, and cosmetics (Mandal & Mandal, 2015).

Table 24.6 summarizes the yield of coriander leaf oil obtained by hydro-distillation. The coriander leaf essential oil yield is fairly consistent and ranges from 0.10 to 0.12%. This contrasts to literature data on the yield of essential oil obtained from the seed, which varied considerably and were influenced by factors such as geographical locations and varieties (Mandal & Mandal, 2015). It was reported that coriander seed essential oil yields isolated from different geographical regions were in the range of 0.31–0.82%, higher than the coriander leaf essential oil.

24.3.2 ANTIOXIDANT ACTIVITY

In the food industry, antioxidants are used to enhance the shelf-life of food products. For example, coriander is widely used as a spice in many cuisines as it possesses antioxidants that could prevent or delay the spoilage of food seasoned with this spice (Mandal & Mandal, 2015). A variety of assays with different mechanisms have been used to monitor the antioxidant activity of plants (Mishra, Ojha, & Chaudhury, 2012). Among these, 2,2-diphenyl-1-picrylhydrazyl (DPPH·) radical scavenging assay is commonly used to evaluate the antioxidant content of vegetables, herbs, edible seed oil, wheat grain, and bran (Mishra et al., 2012). This assay is based on the principle that DPPH· produces a violet color in ethanol solution and fades to shades of yellow in the presence of antioxidants.

Butylated hydroxyl toluene (BHT) is a synthetic antioxidant commonly added to commercial food products to retard lipid oxidation and prevent rancidity development (Frankel, 2012). Therefore, it is utilized as a reference standard in the comparative evaluation of the antioxidant activity of natural antioxidants isolated from plant products (Mishra et al., 2012). Figure 24.1 shows the DPPH· radical scavenging activity of ethanolic coriander leaf extract and BHT at different concentrations. At 250 and 500 µg/mL, the scavenging effect of coriander leaf extract was better than the BHT, suggesting

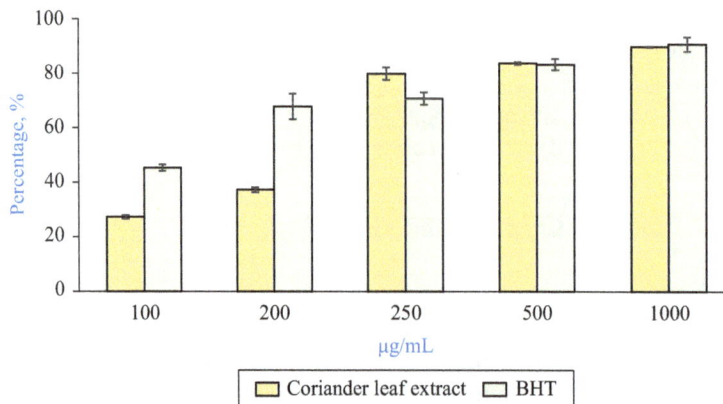

FIGURE 24.1 DPPH· Radical Scavenging Activity. Source: Farah, Elbadrawy & Al-Atoom (2015).

the potential of incorporating coriander leaf extract as a natural antioxidant in food formulation to improve shelf-life and nutritional value.

24.3.3 Antimicrobial Activity

Food poisoning is a condition caused by the intake of food or water contaminated with harmful germs. The antimicrobial properties of coriander leaf extracts and coriander leaf essential oil were investigated by Reddy et al. (2012). Coriander leaf extracts (methanol, chloroform, and ethyl acetate) and coriander leaf essential oil exhibited pronounced effects against Gram-positive (*Bacillus cereus*, *Staphylococcus aureus*, and *Enterobacter faecalis*) and Gram-negative (*Salmonella paratyphi*, *Serratia marcescens*, *Proteus vulgaris*, *Escherichia coli*, *Pseudomonas aeruginosa*, and *Klebsiella pneumoniae*) bacteria. The exerted antimicrobial effect was comparable with the conventional antibiotics such as ofloxacin, tobramycin, ciprofloxacin, and gentamicin sulphate screened under the same conditions. Their study highlights the potential of incorporating coriander leaf essential oil or extract into the formulation of antibacterial drugs.

24.3.4 Arthritis Management

Osteoarthritis is a degenerative joint disorder commonly experienced by the elderly. Oxidative stress is thought to be a crucial factor in developing this disorder (Rajeshwari, Siri, & Andallu, 2012). The study conducted by Rajeshwari et al. (2012) showed that daily consumption of 5 g coriander leaf powder for 60 days controlled the oxidative stress production in osteoarthritis patients. This was evidenced by improved levels of erythrocyte antioxidant enzyme (reduced glutathione, glutathione-S-transferase, and alkaline phosphate), erythrocyte sedimentation rate, serum non-enzymatic antioxidants (vitamin C and β-carotene), and serum calcium of osteoarthritis patients. This study suggests the potential role of coriander leaf powder in alleviating oxidative stress-associated disorders.

24.3.5 Memory-Enhancing

Memory is the capability to record information, events, and stimuli over some time. A study was conducted by Mani and Parle (2009) to determine the effect of fresh coriander leaves intake at different doses (5%, 10%, and 15%) on the cognitive functions of male Wistar rats. After 45 days, the memory scores of rats were improved. Furthermore, the memory deficits induced by scopolamine and diazepam were reversed successfully. The researchers postulated that the memory-enhancing effect of coriander leaves is due to high antioxidant compounds, which could reduce brain damage and enhance neuronal function. The promising results of this study indicate the potential of the leaves in the management of cognitive-related disorders such as Alzheimer's and dementia.

24.4 CONCLUSION

Coriander leaves are lanceolate and broadly lobed at the base of the plant. The leaves are rich in micronutrients such as vitamin A, vitamin C, potassium, and calcium. In addition, high levels of β-carotene, violaxanthin, polyphenols, α-linolenic, and linoleic acids are detected in the leaves. Alcohols and aldehydes are the main volatile constituents of coriander leaf essential oil. Consumption of coriander leaves could enhance memory, reduce oxidative stress-related disorder, and prevent the growth of harmful microorganisms. Further investigations are required to understand the effects of different drying methods on the bioactive compounds of coriander leaves.

REFERENCES

Agrawal, D., Sharma, L. K., Shirsat, M. K., & Saxena, S. N. (2016). Genetic variation in phenolics and antioxidant content in seed and leaf extracts of coriander (*Coriandrum sativum* L.) genotypes. *International Journal of Seed Spices*, 6(2), 7–12.

Ahmed, J., Shivhare, U. S., & Singh, G. (2001). Drying characteristics and product quality of coriander leaves. *Food and Bioproducts Processing, 79*(2), 103–106.

Aruna, G., & Baskaran, V. (2010). Comparative study on the levels of carotenoids lutein, zeaxanthin and β-carotene in Indian spices of nutritional and medicinal importance. *Food Chemistry, 123*(2), 404–409.

Barros, L., Dueñas, M., Dias, M. I., Sousa, M. J., Santos-Buelga, C., & Ferreira, I. C. F. R. (2012). Phenolic profiles of *in vivo* and *in vitro* grown *Coriandrum sativum* L. *Food Chemistry, 132*(2), 841–848.

Bhat, S., Kaushal, P., Kaur, M., & Sharma, H. K. (2014). Coriander (*Coriandrum sativum* L.): Processing, nutritional and functional aspects. *African Journal of Plant Science, 8*(1), 25–33.

Bhuiyan, M. N. I., Begum, J., & Sultana, M. (2009). Chemical composition of leaf and seed essential oil of *Coriandrum sativum* L. from Bangladesh. *Bangladesh Journal of Pharmacology, 4*(2), 150–153.

Divya, P., Puthusseri, B., & Neelwarne, B. (2012). Carotenoid content, its stability during drying and the antioxidant activity of commercial coriander (*Coriandrum sativum* L.) varieties. *Food Research International, 45*(1), 342–350.

Eyres, G., Dufour, J. P., Hallifax, G., Sotheeswaran, S., & Marriott, P. J. (2005). Identification of character-impact odorants in coriander and wild coriander leaves using gas chromatography-olfactometry (GCO) and comprehensive two-dimensional gas chromatography-time-of-flight mass spectometry (GC x GC-TOFMS). *Journal of Separation Science, 28*(9–10), 1061–1074.

Farah, H., Elbadrawy, E., & Al-Atoom, A. A. (2015). Evaluation of antioxidant and antimicrobial activities of ethanolic extracts of parsley (*Petroselinum erispum*) and coriander (*Coriandrum sativum*) plants grown in Saudi Arabia. *International Journal of Advanced Research, 3*(4), 1244–1255.

Frankel, E. N. (2012). Antioxidants. In E. N. Frankel (Ed.), *Lipid oxidation* (2nd ed., pp. 209–258). Cambridge, UK: Woodhead Publishing.

Laribi, B., Kouki, K., M'Hamdi, M., & Bettaieb, T. (2015). Coriander (*Coriandrum sativum* L.) and its bioactive constituents. *Fitoterapia, 103*, 9–26.

Mandal, S., & Mandal, M. (2015). Coriander (*Coriandrum sativum* L.) essential oil: Chemistry and biological activity. *Asian Pacific Journal of Tropical Biomedicine, 5*(6), 421–428.

Mani, V., & Parle, M. (2009). Memory-enhancing activity of *Coriandrum sativum* in rats. *Pharmacologyonline, 2*, 827–839.

Matasyoh, J. C., Maiyo, Z. C., Ngure, R. M., & Chepkorir, R. (2009). Chemical composition and antimicrobial activity of the essential oil of *Coriandrum sativum*. *Food Chemistry, 113*(2), 526–529.

Mishra, K., Ojha, H., & Chaudhury, N. K. (2012). Estimation of antiradical properties of antioxidants using DPPH– assay: A critical review and results. *Food Chemistry, 130*(4), 1036–1043.

Mishra, V. K., Bacheti, R. K., & Husen, A. (2011). Medicinal uses of chlorophyll: A critical overview. In *Chlorophyll: Structure, function and medicinal uses* (pp. 177–196). Hauppauge, NY: Nova Science Publishers.

Nambiar, V. S., Daniel, M., & Guin, P. (2010). Characterization of polyphenols from coriander leaves (*Coriandrum sativum*), red amaranthus (*A. paniculatus*) and green amaranthus (*A. frumentaceus*) using paper chromatography and their health implications. *Journal of Herbal Medicine and Toxicology, 4*(1), 173–177.

Neffati, M., & Marzouk, B. (2008). Changes in essential oil and fatty acid composition in coriander (*Coriandrum sativum* L.) leaves under saline conditions. *Industrial Crops and Products, 28*(2), 137–142.

Oganesyan, E. T., Nersesyan, Z. M., & Parkhomenko, A. Y. (2007). Chemical composition of the above-ground part of *Coriandrum sativum*. *Pharmaceutical Chemistry Journal, 41*(3), 149–153.

Padalia, R. C., Karki, N., Sah, A. N., & Verma, R. S. (2011). Volatile constituents of leaf and seed essential oil of *Coriandrum sativum* L. *Journal of Essential Oil-Bearing Plants, 14*(5), 610–616.

Potter, T. L., & Fagerson, I. S. (1990). Composition of coriander leaf volatiles. *Journal of Agricultural and Food Chemistry, 38*(11), 2054–2056.

Prachayasittikul, V., Prachayasittikul, S., Ruchirawat, S., & Prachayasittikul, V. (2018). Coriander (*Coriandrum sativum*): A promising functional food toward the well-being. *Food Research International, 105*, 305–323.

Rajeshwari, C. U., Siri, S., & Andallu, B. (2012). Antioxidant and antiarthritic potential of coriander (*Coriandrum sativum* L.) leaves. *E-SPEN Journal, 7*(6), e223–e228. https://doi.org/10.1016/j.clnme.2012.09.005

Reddy, L. J., Reshma, D. J., Jose, B., & Gopu, S. (2012). Evaluation of antibacterial and DPPH radical scavenging activities of the leaf extracts and leaf essential oil of *Coriandrum sativum* Linn. *World Journal of Pharmaceutical Research, 1*(3), 705–716.

Shahwar, M. K., El-Ghorab, A. H., Anjum, F. M., Butt, M. S., Hussain, S., & Nadeem, M. (2012). Characterization of coriander (*Coriandrum sativum* L.) seeds and leaves: Volatile and non volatile extracts. *International Journal of Food Properties*, *15*(4), 736–747.

Singh, S. K., Samsher, B. R., Sengar, R. S., & Kumar, P. (2020). Study on biochemical properties of dehydrated coriander leaves at different drying conditions. *International Journal of Chemical Studies*, *8*(4), 2348–2352.

Tan, C., & Azrina, A. (2016). Nutritional, phytochemical and pharmacological properties of *Canarium odontophyllum* Miq. (Dabai) Fruit. *PJSRR Pertanika Journal of Scholarly Research Reviews*, *2*(1), 80–94.

Tan, C. X., Tan, S. S., Ghazali, H. M., & Tan, S. T. (2021). Physical properties and proximate composition of Thompson red avocado fruit. *British Food Journal*, *124*(5), 1421–1429.

Tan, C. X., Tan, S. T., & Tan, S. S. (2020). An overview of papaya seed oil extraction methods. *International Journal of Food Science and Technology*, *55*(4), 1506–1514.

USDA. (2019). Coriander (cilantro) leaves, raw. Retrieved November 2, 2021, from https://fdc.nal.usda.gov/fdc -app.html#/food-details/169997/nutrients

Wei, J. N., Liu, Z. H., Zhao, Y. P., Zhao, L. L., Xue, T. K., & Lan, Q. K. (2019). Phytochemical and bioactive profile of *Coriandrum sativum* L. *Food Chemistry*, *286*, 260–267.

Yilmaz, A., & Alibas, I. (2017). Determination of microwave and convective drying characteristics of coriander leaves. *Journal of Biological and Environmental Sciences*, *11*(32), 75–85.

25 A Comprehensive Review on Chemistry, Nutritional Relevance, and Potential Applications of *Coriandrum sativum* L. Waste/By-Products

*Yassine M'Rabet, Hamza Ben Abdallah,
Mariem Ziedi, and Karim Hosni*

CONTENTS

25.1 INTRODUCTION

Coriander (*Coriandrum sativum* L.) is a member of the Apiaceae (Umbelliferae) family, which encompasses 3780 species distributed in 434 genera (Ahmad et al., 2017). It is a coveted multipurpose herb with potential health benefits associated with its antioxidant and antimicrobial (Duarte et al., 2012; 2016), antidiabetic and anti-inflammatory (Mechchate et al., 2021), hepatoprotective (Moustafa et al., 2012), immunomodulatory (Ahmed et al., 2020), anticancer (Tang et al., 2013), hypotensive (Hussain et al., 2018), stimulant, carminative, and diuretic (Kumar et al., 2016) properties (Figure 25.1). In addition, anxiolytic, sedative, anti-epileptic, anti-depressant, anti-mutagenic, anti-dyslipidemic, antiulcer, antiaging, anthelmintic, diuretic, cardioprotective, neuroprotective, and cognitive activities have also been reported (Sahib et al., 2013; Laribi et al., 2015; Önder, 2018).

In addition to their culinary use, all parts of coriander have been used in diverse traditional medicinal systems to manage different disorders (Table 25.1). Therefore, coriander curry powder and extracts or supplements in diet were used as folk remedies to treat headache, swelling,

DOI: 10.1201/9781003204626-27

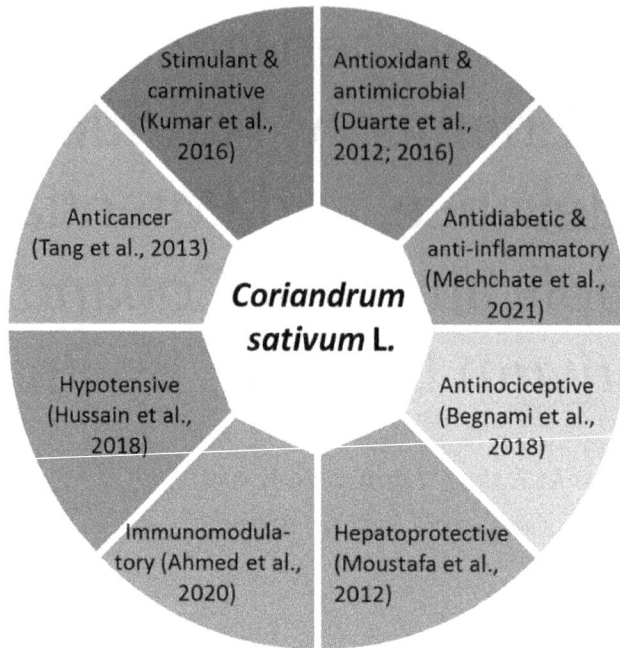

FIGURE 25.1   Biological activities of coriander.

stomatitis, dysentery, burning sensation, flatulence, eczema, indigestion, and vertigo (Bhat et al., 2014; Farzaei et al., 2017).

These biological activities have been associated with the presence of different classes of bioactive ingredients, including minerals, proteins, fibers, carbohydrates (Iqbal et al., 2018), fatty acids (Ramadan and Mörsel, 2002; Msaada et al., 2009a), sterols, tocopherols (Ramadan et al., 2003), essential oils (Msaada et al., 2007, 2009b), and phenolics (Msaada et al., 2017). The recovery of these bioactive compounds by different extraction processes, mainly using aqueous-organic solvent extraction or extrusion, usually generates residual matrices or underused by-products that could be valued as a consolidated source of chemically active and/or bioactive ingredients. Although the chemistry and bioactivity of coriander have been reported (Sahib et al., 2013; Laribi et al., 2015; Önder, 2018), no review papers updating the main chemical features and potential uses of coriander by-products have been published. This chapter summarizes the main coriander by-products, focusing on their chemical composition, nutritional and health benefits, and their main and prospective applications.

25.2   PHYTOCHEMICALS FROM CORIANDER WASTES/BY-PRODUCTS

25.2.1   Production of Coriander Wastes/By-Products

Coriander wastes (straw) and by-products (press cake or spent residue and hydrodistilled residue) are produced during agricultural processing and the transformation process of the raw materials (Figure 25.2). Traditionally, straws representing 60–80% of the aerial part are left in the field and can be used as animal feed, while the process residue by-products (press cake and hydrodistilled residue) are inadequately deposited, a disposal problem (Sowbhagya, 2019).

25.2.2   Chemistry of Coriander Wastes/By-Products

The available data about the chemistry of coriander wastes/by-products revealed that all matrixes (straw, press cake, and hydrodistilled residues) contain a plethora of phytochemicals with high added

TABLE 25.1

Medicinal Uses of Coriander in Different Folk Medicine Systems

Traditional Medicine	Form	Uses	References
Ayurvedic	• Curry powder • Aqueous extract of seeds	Digestive stimulant, bile acid secretion, stimulation of digestive enzymes (trypsin, chymotrypsin, amylase, lipase, disaccharidases, sucrase, lactase, maltase, alkaline phosphatase, aminopeptidases), treatment of local swelling, headache, lymphadenopathy, stomatitis, dysentery, burning sensation, conjunctivitis, vertigo, syncope, cough, bleeding disorder	• Platel and Srinivasan, 2004 • Gupta, 2010 • Bhat et al., 2014
Persian	• Aqueous/ethanol extract of seeds • Fruits as a supplement in diet and drinking water	Treatment of diabetes, indigestion, worms, flatulence, rheumatism, headache, pain in the joints, eczema, scabies, insomnia, scrofula, infected wounds, bronchitis, vomiting	• Asgarpanah and Kazemivach, 2014 • Farzaei et al., 2017
Greco-Arabic	• Leaf/fruit powder • Infusion • With diet or drinks	Stimulant, antispasmodic, diuretic, antirheumatic, treatment of flatulence, indigestion, diabetes, insomnia, dyslipidemia, renal, cardiovascular, and neurological disorders	• Alqethami et al., 2017 • Tahraoui et al., 2007 • Aissaoui et al., 2008
Chinese	• Aqueous extract of seeds	Treatment of indigestion, loss of appetite, influenza, bad breath	• Sahib et al., 2013
German	• Infusion or decoction of seeds	Carminative, laxative, treatment of dyspeptic complaints, abdominal discomforts, gastrointestinal upsets	• Pieroni and Gray, 2008
Turkish	• Brewed seeds	Carminative treatment of anorexia, appetite enhancer	• Ugulu et al., 2009
Peru	• Ethanol/water extracts from roots and stems	Antimalaria	• Ruiz et al., 2011

values (Figure 25.3). Consequently, the chemical investigation showed that coriander straw consists mainly of lignocellulosic materials with cellulose (51.5–52.5%), hemicellulose (15–21%), and lignin (9.8–13%) being the main constituents. In addition to hot-water extractives (10.4–10.7%), it also contains minerals (4.2–4.7%), lipids (0.8–1.6%), and proteins (2.8–3.7%) (Uitterhaegen et al., 2016; 2019). The volatile fraction of coriander straw was investigated (Wahba et al., 2020). In that study, the essential oil was obtained with a relatively high yield (>1%) and consisted predominantly of *trans*-anethole, linalool, eugenyl isovalerate, estragole, longifolene, carvacrol, and germacrene-D.

Some of the compounds mentioned above have been described in the thermo-mechanical-generated press cake (Uitterhaegen et al., 2016), but with a higher proportion being rich in minerals (6.3%), lipids (17.2%), proteins (18.2%), cellulose (26.6%), hemicellulose (22.2%), lignin (5%), and hot-water extractives (16%). The biochemical profile of lipids extracted from coriander cake yields 15 to 17% depending on the nozzle diameter (Sriti et al., 2013). Results showed that fatty acids (FA) were predominantly monounsaturated (MUFA) (82% of total FA) with petroselinic (C18:1(n-12), 75.5–77.3%), oleic (C18:1(n-9), 5.15–5.9%), and palmitoleic (C16:1(n-7), 0.23–0.25%) acids as the main representative of the MUFA fraction (Figure 25.4). Polyunsaturated FA (PUFA, 12.5–13.7%) was found to be dominated by linoleic (C18:2(n-6), 12.5–13.6%) acid, while linolenic acid (C18:3(n-3)) was detected with trace amounts (0.12–0.14%). Palmitic (C16:0), stearic (C18:0), and

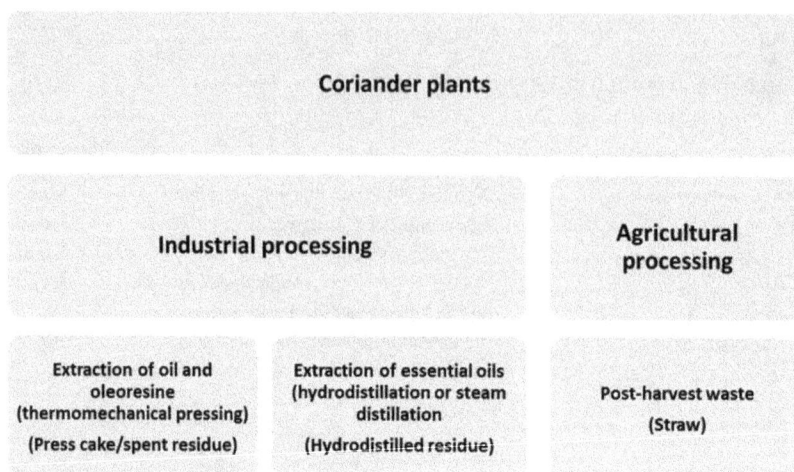

FIGURE 25.2   The main process behind the generation of coriander waste/by-products.

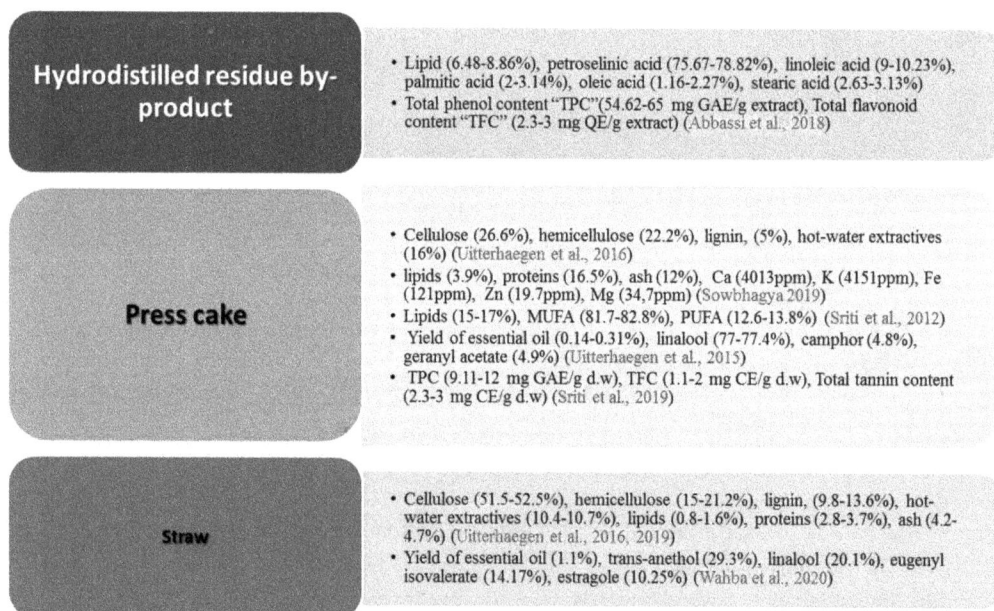

FIGURE 25.3   The main phytochemicals of coriander waste/by-products.

arachidic (C20:0) acids were the main saturated fatty acid (SFA) with proportions of 3.66–3.81%, 0.72–0.89%, and 0.08–0.09%, respectively. Lipids of the same raw materials also contain sterols and β-sitosterol (33.2–35.7 mg/100 g oil), stigmasterol (24.1–25.1 mg/100 g oil), Δ [7]-stigmasterol (14.4–15.5 mg/100 g oil), Δ [5],24-stigmastadienol (9.48–12.2 mg/100 g oil), campesterol (8.14–8.76 mg/100 g oil), Δ [7]-avenasterol (3.95–4.75 mg/100 g oil), and Δ [5]-avenasterol (2.32–2.79 mg/100 g oil). It also contains tocotrienols (93.2–94.2 mg/100 g oil) and tocopherols (5.73–6.71mg/100 g oil) (Sriti et al., 2013). In a more recent study, these authors have analyzed the volatile profile and determined the main phytochemicals of coriander cake (Sriti et al., 2019). Essentials oils were obtained with yields ranging from 0.04 to 0.11%, and the volatile bouquet was dominated by linalool, gearyl acetate, γ-terpinene, camphor, and linalyl acetate (Figure 25.4). The corresponding study obtained appreciable amounts of total phenol content, flavonoid content, and tannin content from the cake.

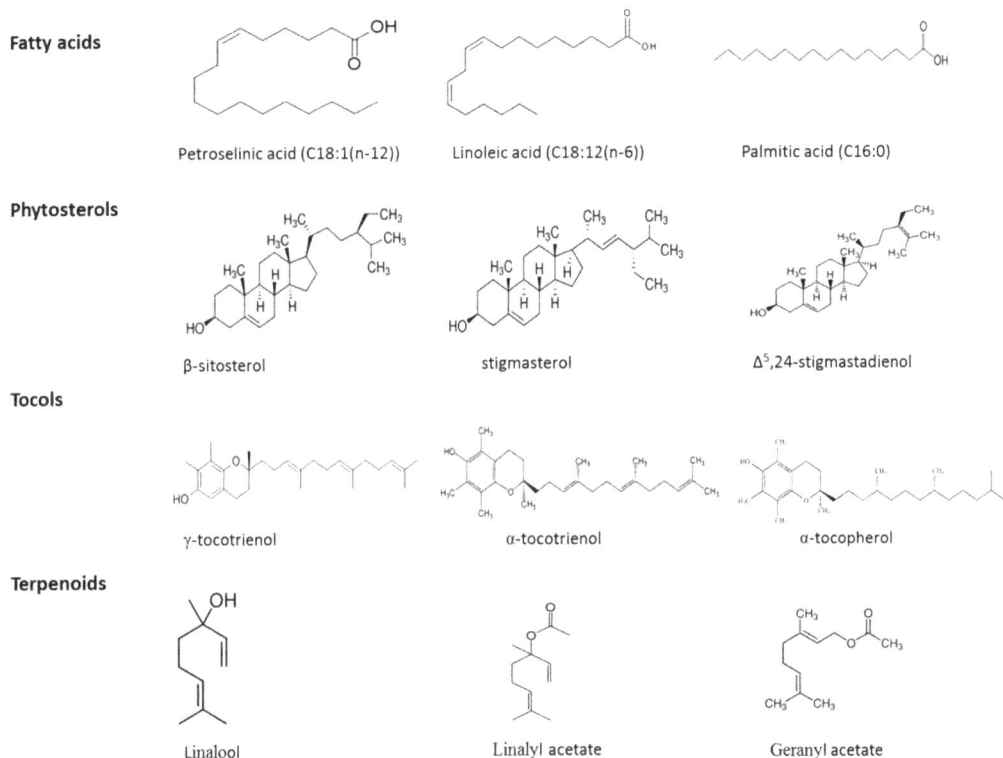

FIGURE 25.4 Structure of the main phytochemicals of coriander waste/by-products.

Another report from India showed that press cake contains high contents of dietary fiber (54%), proteins (16.5%), lipids (3.9%), and ash (12%), comprising appreciable amounts of the essential minerals like calcium (4013 ppm), potassium (4151 ppm), iron (124 ppm), zinc (19.7 ppm), and manganese (34.7 ppm) (Sowbhagya, 2019).

Regarding hydrodistilled residue by-products, the only published paper points to their richness in lipids (6.78–8.86%) with high amounts of UFA (91.7–92.5%) and bioactive phytochemicals including TPC, TFC, and TTC (Abbassi et al., 2018). However, they also showed significant variability in their contents depending on extracting solvent and treatment of the raw material.

In summary, the chemical composition of straw, press cake, and hydrodistilled residue by-products showed a dearth of energetic compounds and low caloric content with an appreciable percentage of dietary fibers, making them suitable for inclusion in obese/diabetic diets. Moreover, the excellent taste and smell of coriander wastes/by-products could improve their acceptance and compliance in such patients.

25.3 NUTRITIONAL VALUE AND HEALTH BENEFITS OF CORIANDER WASTE/BY-PRODUCTS: THE NEGLECTED SOURCE

Coriander wastes and by-products hold the potential to be used as rich sources of nutritionally active and bioactive ingredients.

25.3.1 DIETARY FIBERS

Although the paucity of data regarding dietary fibers in coriander wastes/by-products, coriander straw, and press cake could represent the uncommon dietary fibers because of their richness in cellulose, hemicellulose, lignin, and hot-water extractives (Uitterhaegen et al., 2016), compared with other Apiaceae species, coriander press cake is particularly rich in fiber. Hence, it could help achieve the daily recommended amount of dietary fibers. Furthermore, the possible use of coriander press cake as a potential substitute for fiber-rich foods such as whole grains, fruits, vegetables, legumes, nuts, and seeds could be envisaged to develop functional foods with enhanced nutritional and health attributes. For example, dietary rich coriander (10%) has been used to prepare functional bread without modification of loaf volume, flavor, crumb grain, and shape (Sowbhagya 2019).

Dietary fibers are nutritionally relevant because of their vast array of biological activities (Anderson et al., 2009). Dietary fiber intake has been associated with reduced risk of cardiovascular disease by decreasing serum lipids, lowering blood pressure, improving glucose metabolism, and reducing body mass index, LDL-cholesterol, and triglycerides values (Anderson 1990). Dietary fiber intake was also associated with a significant improvement in the prevention of diabetes and its associated metabolic disorders, including insulin resistance, dyslipidemia, visceral adiposity, and hypertension (Anderson et al., 2009). Moreover, epidemiologic studies indicate that higher-level fiber consumption protects against the development of obesity, enhances satiety, and reduces the prevalence of hiatal hernias, peptic ulcer disease, appendicitis, diverticular disease, gallbladder disease, colorectal cancer, and hemorrhoid, and improves immune function through gut health and fiber-microbiota interactions (Dong et al., 2016). The beneficial effects of dietary fibers on the management of depression and chronic inflammation have also been reported (Barber et al., 2020).

Because of the abundance of insoluble fibers (cellulose, hemicellulose, and lignin), coriander straw and press cake can be recommended to improve laxation, shorten bowel transit time, improve insulin sensitivity, and enhance the growth of probiotic intestinal microflora (Olive Li and Komarek, 2017).

25.3.2 Ash and Minerals

Coriander straw and press cake are a consolidated source of ash and minerals. Their amounts were, however, below that detected in many edible Apiaceae species (Tunçtürk and Özgökçe 2015). In the corresponding study, ash content was reported to be between 7% and 18% suggesting the richness of Apiaceae species in minerals. Potassium was the main component, with a concentration ranging from 488 ppm to 33,900 ppm in *Pimpinella aurea* and *Heracleum persicum*. Calcium was concentrated (20,600 ppm) in *Anthriscus nemorosa*, while magnesium was particularly abundant in *Heracleum persicum* (6020 ppm). Concerning iron, its content ranged from 81 ppm in *Prango ferulacea* to 324 ppm in *Pimpinella aurea*, while zinc concentrations were 6.78 ppm to 46.9 ppm, with *Anethum graveolens* being the richest Apiaceae species. In another report from the same country, Turan et al. (2003) have analyzed the mineral profile of 26 wild edible leaves and found that the Apiaceae *Eryngium billardieri* was distinguished by its high potassium (3453 g/kg) content confirming the consensus that the Apiaceae family is the richest source of potassium (Golubkina et al., 2021). Support for this assumption has already been reported by Paul et al. (2013), showing the high amounts of potassium (56.9–64.2 g/kg) in *Carum roxburghianum*. It also contains elevated amounts of macro-elements such calcium (15.9–17.5 g/kg), phosphorus (4.89–5.77 g/kg), and magnesium (2.24–2.49 g/kg). Micro-elements in that species were represented mainly by iron (2690–3162 mg/kg), zinc (74.3–87.9 mg/kg), and manganese (98.1–123 mg/kg), among others (Paul et al., 2013).

In the South African Apiaceae species *Heteromorpha arborescens*, considerable amounts of calcium (15.6 g/kg), potassium (16.8 g/kg), magnesium (4 g/kg), copper (7.5 mg/kg), zinc (21 mg/kg), iron (125 mg/kg), phosphorus (550 mg/kg), and sodium (750 mg/kg) have been noticed (Abifarin et al., 2020). Based on these values, the study authors have proposed *H. arborescens* for calcium and potassium supplements. The abundance of potassium, calcium, and, to a lesser extent, iron and magnesium, was expected due to their structural and functional roles in plants (Hänsch and Mendel, 2009).

From a nutritional standpoint, coriander press cake could be used as a supplement to ensure an adequate supply of calcium, magnesium, and iron for gluten-free diet consumers. Medicinally, the presence of some essential minerals which have pivotal physiological and biochemical roles in the human body (i.e., enzymatic reactions, energy production, and transmission of nerve impulses, among others) can support the use of coriander waste/by-products as a supplement. For example, potassium in elevated amounts makes coriander press cake suitable for managing hypertension, maintaining normal cell function, and regular muscle contraction (Abifarin et al., 2020). The abundance of calcium makes coriander press cake relevant for preventing osteopenia and osteoporosis in the celiac subject (Nascimento et al., 2014). Its simultaneous presence with zinc significantly improved the lipid-lowering and anti-obesity properties of the diet (Mills et al., 2011). It is also required for bones and muscle formation, blood coagulation, and synaptic nerve transmission (Abifarin et al., 2020). Additionally, the presence of magnesium and zinc in appreciable amounts was associated with better insulin release, high glucose uptake by fat cells, and reduced arterial pressure, with the effects being modulated by the presence of potassium and calcium (Ahsan and Neugut, 1998).

In addition to their deep involvement in the normal development and maintenance of the human body, potassium, calcium, iron, copper, zinc, and magnesium are crucial for maintaining bones and teeth and improving the muscle, heart, and brain functions (Jéquier and Constant, 2010). Apart from its involvement in glycolysis, oxidative phosphorylation, and protein synthesis, magnesium is required for maintaining osmotic equilibrium (Gröber et al., 2015). As part of many enzymes, zinc is required for normal body development, protein synthesis, and wound healing (Ogundola et al., 2018). Iron and copper, which are involved in several enzymatic reactions, are necessary for hemoglobin synthesis, energy metabolism, and oxygen transport (DiNicolantonio et al., 2018).

25.3.3 Fatty Acids, Phytosterols, and Tocols

As expected, lipid content was low in coriander straw (0.8–1.6%), but it was relatively higher in press cake (3.9%) and hydrodistilled residue (6.7–8.8%). Lipid contents in coriander wastes/by-products were within the range reported for other Apiaceae species, including *A. cerefolium* (4%) (Vyas et al., 2012) and *H. arborescens* (4.92%) (Abifarin et al., 2020). However, they were markedly lower than that observed for *Carum roxburghianum* (15.3–20.3%) (Paul et al., 2013).

Analytical data of the lipid fraction of coriander by-products showed the predominance of monounsaturated fatty acids (MUFA) followed by polyunsaturated fatty acids (PUFA). Saturated fatty acids (SFA) were the least representative fraction (Sriti et al., 2013). Petroselinic (C12:1n-12, 75.5–77.2%) and oleic acids (5.15–5.86%) were the main MUFA in the press cake (Sriti et al., 2013). In the hydrodistilled residue by-product, similar contents of petroselinic acid (75.7–78.8%) versus low contents of oleic acid (1.16–2.93%) were described (Abbassi et al., 2018). Linoleic acid (C18:2 n-6) was the major PUFA, with a proportion ranging from 12.4 to 13.6% in coriander press cake and 9.07 to 10.5% in the hydrodistilled residue by-product. Linolenic acid (C18:3n-3) was the less representative PUFA in the fatty profiles of coriander by-products. Palmitic acid (C16:0) and stearic acid (C18:0) were the main SFA in both press cake and hydrodistilled residue by-products, and their contents were found to be below 4% (Sriti et al., 2013; Abbassi et al., 2018).

In general, the reported profile of fatty acid was typical of the Apiaceae family, with the presence of petroselinic acid as the main MUFA and a small amount of SFA. In this context, the recent compositional analysis showed the dominance of the following fatty acid profile C18:1n-12 > C18:2 n-6 > C18:1n-9 in dill (*Anethum graveolens*), celery (*Apium graveolens*), caraway (*Carum carvi*), cumin (*Cuminum cyminum*), and fennel (*Foeniculum vulgare*) (Saini et al., 2021).

In order to evaluate the lipid quality index of coriander press cake and hydroditilled residue by-products, the following parameters were calculated based on their fatty acid profiles and compared with other Apiaceae species (Saini et al., 2021). These parameters include the sum of SFA, MUFA, PUFA, UFA (MUFA + PUFA), UFA/SFA ratio, *omega-3*/*omega*-6 PUFA ratio, atherogenic index (AI), thrombogenic index (TI) (Ulbicht and Southgate 1991), hypocholesterolemic fatty acids/hypercholesterolemic fatty acids ratio (h/H) (Santos et al., 2002), calculated oxidizability (Cox) (Fatemi and Hammond, 1980), and oxidative susceptibility (OS) (Cecchi et al., 2011):

$$AI = \frac{\left[(4 \times C14:0) + C16:0 + C18:0\right]}{£UFA + £É6\,PUFA + £É3\,PUFA]}$$

$$TI = \frac{(C14:0 + C16:0 + C18:0)}{\left[(0.5 \times MUFA) + 0.5 \times \omega6\,PUFA + 3 \times \omega3\,PUFA + \dfrac{\omega3}{\omega6}\,PUFA\right]}$$

$$\frac{h}{H} = \frac{\left[(C18:1n-9 + C18:2n-6 + C20n-6 + C18:3n-3 + C20:5n-3 + C22:5n-3 + C22:6n-6)\right]}{C14:0 + C16:0}$$

$$Cox = \frac{C18:1 + 10.3 \times C18:2 + 21.6\,C18:3}{100}$$

$$OS = MUFA + 45 \times C18:2 + 100 \times C18:3$$

Regarding the lipid quality index, Table 25.2 revealed that press cake and the hydrodistilled residue by-products have low SFA content (4.66–7.63%), giving a high PUFA/SFA ratio (1.3–2.7). These

---

TABLE 25.2

Lipid Nutritional Quality of Coriander Press Cake and Hydrodistilled By-Products of Edible Apiaceae Species

	Press Cake (Sriti et al., 2013)	Hydrodistilled Residue By-Product (Abbassi et al., 2018)	Coriander Seeds (Saini et al., 2021)	Dill Seeds (Saini et al., 2021)	Celery Seeds (Saini et al., 2021)	Caraway Fruits (Saini et al., 2021)	Parsley Leaves (Saini et al., 2021)
SFA	4.66	7.63	9.3	12.7	14.02	19.9	22.23
MUFA	82.75	81.87	69.49	65.97	57.11	45.93	2.14
PUFA	12.59	10.5	21.21	21.34	28.88	34.17	75.63
UFA	95.34	92.37	90.7	87.31	85.99	80.1	77.77
UFA/SFA	20.46	12.11	9.75	6.87	6.13	4.02	3.49
PUFA/SFA	2.7	1.37	2.28	1.68	2.08	1.71	3.4
ω3 PUFA	0.12	1.43	0.35	0.85	1.32	0.47	51.14
ω6 PUFA	12.47	9.07	20.85	20.49	27.55	33.7	24.49
ω3/ω6	0.009	0.157	0.016	0.041	0.047	0.014	2.08
AI	0.042	0.08	0.1	0.15	0.15	0.6	0.78
TI	0.465	0.12	0.19	0.24	0.29	0.46	0.11
h/H	4.83	3.666	12.12	8.22	7.45	4.94	4.61
Cox	2.13	2.04	2.91	2.95	3.691	4.03	9.75
Os	655.9	634.18	1108.09	1072.72	1428.86	1609.43	4444.19

Abbreviations: AI: atherogeic index, TI: thrombogenic index, h/H: hypoholesterolemic fatty acids/hypercholesterolemic fatty acids ratio, Cox: calculated oxidizability, OS: oxidative susceptibility, SFA: saturated fatty acids, MUFA: monounsaturated fatty acids, PUFA: polyunsaturated fatty acids.

TABLE 25.3
Dietary Fiber and Energy Values of Some Apiaceae Species

Specie	Dietary Fiber (%)	Energy (kcal/100 g)	References
Coriandrum sativum L.	5.53	390.3	Javid et al. (2009)
Anthriscus cerefolium L.	11	-	Vyas et al. (2012)
Carum roxburghianum Benth.	20.9–23.9	317–336	Paul et al. (2013)
Smyrnium olusatrum L.	2.4–10.4	43–330	Caprioli et al. (2014)
Daucus carota L.	9.07	358.9	Ayeni et al. (2018)
Eryngium caeruleum M. Bieb.	22.6	236.7	Ghajarieh Sepanlo et al. (2019)
Heteromorpha arborescens (Spring.)	21.4	271	Abifarin et al. (2020)

values were higher than the recommended ratio (0.45) for the diet (Briggs et al., 2017), suggesting the suitability of press cake and the hydrodistilled residue by-product to be included in the diet (Saini et al., 2021). Furthermore, the reduced SFA and the high UFA (MUFA + PUFA) contents resulted in reduced AI (0.04–0.08) and TI (0.12–0.46) versus a high h/H ratio (3.66–4.83). Such a trend reflects a reduced content of atherogenic fatty acids (i.e., myristic and palmitic) *vs.* high content of MUFA (i.e., petroselinic and oleic acids). Interestingly, the fatty acid profiles of press cake and hydrodistilled residue by-products exhibited the highest nutritional quality compared to common commercial Apiaceae species (Table 25.3). Both fatty acids profiles were also distinguished by their low Cox (2.04–2.13) and Os (634–656) because of their lox PUFA contents, namely C18:3n-3. These characteristics make oils from press cake and hydrodistilled residue by-products relatively stable to oxidative deterioration. In contrast, oils from the common edible Apiaceae were prone to oxidation, as reflected in their Cox (2.91–9.75) and Os (1072–4444) values (Saini et al., 2021).

Conversely to linoleic acid, considered a pro-inflammatory agent (as a precursor of arachidonic acid and its derivatives prostaglandins and leukotrienes) and cancer promoter (Whelan and Fritsche, 2013; Jandack, 2017), petroselinic acid can counteract arachidonic acid-induced vasoconstriction by inhibiting the synthesis of arachidonic acid (Shukla and Gupta, 2009). It is also considered low-fat owing to its low lipolysis by pancreatic lipase (Delbeke et al., 2016). Additionally, petroselinic acid has been shown to inhibit the growth of the opportunistic pathogen *Burkholderia cenocepacia* K56-2 which is responsible for the multiresistant lung infection in a patient with cystic fibrosis (Mil-Homens et al., 2012). It also exhibits an antibacterial effect against some *Borrelia* sp., which could be associated with its surfactant-type property that increases membrane fluidity (Goc et al., 2019).

Petroselinic acid is also recognized for its inhibitory effects on topoisomerase I and II reflecting its potential in treating cancer, improving the ability of epidermal cell differentiation, and reducing skin inflammation (Avato and Tava, 2021).

Additional phytochemical investigations on active lipophilic compounds in coriander press cake have been performed (Sriti et al., 2019). These authors successfully detected seven sterols in the unsaponifiable fraction with β-sitosterol (33.2–35.7 mg/100 g oil), stigmasterol (24.1–25.1 mg/100 g oil), Δ [7]-stigmasterol (14.4–15.5 mg/100 g oil), and Δ [5], 24-stigmastadienol (9.48–10.5 mg/100 g oil) being the main sterols. Regarding tocopherols and tocotrienols, the isoforms γ-, and α-tocotrienols were dominant with 71.1–76.4 mg/100 g oil and 12.2–19.7 mg/100 g oil. In addition, α-tocopherol, γ-, δ-tocopherol and δ-tocotrienol were also found in relatively small amounts (< 5 mg/100 g oil) (Sriti et al., 2019).

Phytosterols have received particular attention because of their beneficial health effects, especially their cholesterol-lowering properties, thereby reducing the risk of cardiovascular diseases (Moreau et al., 2002). The mechanism behind lowering total plasma cholesterol and low-density lipoprotein (LDL) properties of phytosterols was mediated through their inhibitory effect on intestinal absorption of cholesterol (Ryan et al., 2007). In addition, increasing evidence suggests that phytosterols are endowed with antioxidant and anticarcinogenic effects (de Jong et al., 2003). Furthermore, they have shown anti-inflammatory effects as revealed by their ability to reduce plasma levels of C-reactive protein, IL-6, TNF-α, phospholipase A1, and fibrinogen with concomitant disturbance of membrane fluidity, sensitivity, and signaling pathways (Othman and Moghadasian, 2011). Furthermore, there is consistent evidence to support the *in vitro* and *in vivo* anti-inflammatory effect of phytosterols (i.e., β-sitosterol, stigmasterol, etc.) by inhibiting cytokine-mediated cytotoxicity nitric acid and prostaglandin production (Shishodia and Aggarwal, 2004). Moreover, β-sitosterol has been shown to exhibit anti-neoplastic, anti-pyretic, and immune-modulating activities in *in vitro* and *in vivo* experiments. By targeting T-helper (Th) lymphocytes, β-sitosterol improved the activity of T-lymphocytes and natural killer cells (Frail et al., 2012).

Tocols (tocopherols and tocotrienols), commonly known as vitamin E, are the essential lipid-soluble antioxidants and the most potent scavengers of lipid peroxyl radicals. Their intake is inversely associated with cardiovascular diseases and effectively contributes to preventing Alzheimer's disease and cancer (Ryan et al., 2007). Potential health benefits of tocopherols and tocotrienols include preventing cardiovascular diseases, cancers, obesity, and diabetes (Shahidi and de Camargo, 2016).

Given these nutritional and biological properties of the main bioactive phytochemicals, it can be inferred that the oil from coriander by-products (press cake and hydrodistilled residue) has excellent nutritional quality making it suitable for the production of commercial and healthy vegetable oil.

25.3.4 Proteins

With 16.5% proteins, coriander press cake could be considered a good source of protein and a potential protein supplement in the diet (Uitterhaegen et al., 2016). The reported value was higher than that found in other Apiaceae such as *Anethum graveolens* (7.4%), *Anthriscus nemorosa* (6.0%), *Chaerophyllum macropodum* (8.0%), *Ferula rigidula* (8.8%), *Heracleum persicum* (8.3%), *Hippomarathrum microcarpum* (5.1%), *Pimpinella aurea* (6.0%), *Parangos ferulacea* (8.4%) (Tunçtürk and Özgökçe 2015), and *H. arborescens* (15.7%) (Abifarin et al., 2020). In contrast, it was lower than that of *Anthriscus cerefolium* (23.2%) (Vyas et al., 2012) and *Carum roxburghianum* (18.5–22.3%) (Paul et al., 2013).

25.3.5 Secondary Metabolites

In addition to dietary fibers, ash, minerals, proteins, and lipids, press cake and hydrodistilled residue by-products contain considerable amounts of phenolics, bioactive metabolites with well-recognized food-related biological activities like antioxidant, antimicrobial, anti-diabetic, anticancer, and immune-modulatory properties, among others (Sahib et al., 2013). In press cake, the total phenol content (TPC) and total flavonoid content (TFC) were estimated to be 9–12 mg EGA/g d.w. and 1.1–2.0 mg CE/g d.w., respectively (Sriti et al., 2019). However, these bioactive metabolites were more abundant in hydrodistilled residue by-products (Abbassi et al., 2018) (Figure 25.2). The corresponding study mentioned TPC and TFC values ranging from 54.6 to 64.9 mg EGA/g extract and 8.56 to 13.8 mg EQ/g extract, respectively (Figure 25.2). The highest releases of TPC and TFC were attributed to the thermal treatment and the subsequent acidification of the bulk medium during distillation (Abbassi et al., 2018). The previously reported data stated TPC values of (23.8 mg GAE/g d.w.) for parsley (Śledź et al., 2013), (0.94–1.09 mg GAE/g d.w.) for coriander (Msaada et al., 2017), and (3.7–7 mg GAE/g d.w.) for cumin (Demir and Korukluoglu, 2020) suggesting that coriander press cake and hydrodistilled residue by-product could serve as a potential source of phenolic compounds. Phenolic extracts are routinely assayed for their antioxidant properties, given their recognized antioxidant activity.

The antioxidant property of coriander press cake and hydrodistilled has been evaluated using different *in vitro* tests, including 2,2-diphenyl-1-picrylhydrazyl (DPPH), 2,2-azino-bis-(3-ethylben zothiozoline-6-sulfonic acid di-ammonium salt) (ABTS); reducing power, β-carotene bleaching, ferric reducing antioxidant power (FRAP), and cellular antioxidant assay (CAA) (Abbassi et al., 2018; Sriti et al., 2019). The methanol extract of coriander cake was evaluated for its DPPH· radical scavenging activity, reducing power, and its β-carotene bleaching ability and compared with the reference antioxidant BHT. The methanol extract exhibited intense scavenging activity against the radical DPPH· with IC_{50} values of 55–88 µg/mL versus 25 µg/mL for BHT. The methanol extract inhibited lipid peroxyl radicals (IC_{50} 543–610 µg/mL) as determined in the β-carotene bleaching assay. Its ability to reduce Fe^{3+} was 14 to 18 times lower than that of the standard ascorbic acid (Sriti et al., 2019). The antioxidant activity of coriander press cake has been attributed to its total phenol and flavonoid contents.

Evidence for the *in vitro* antioxidant activity emphasizes that coriander hydrodistilled residue by-products effectively scavenged the DPPH· (EC_{50} 118–149 µg/mL), but it showed a weak ability to reduce the Fe^{3+} in the FRAP assay. Using splenocytes as cellular models, coriander hydrodistilled residue by-product showed vigorous cellular antioxidant activity with IC_{50} values ranging from 13.4 to 14.4 µg/mL (Abbassi et al., 2018). These authors attributed the antioxidant activity to putative phenolic antioxidant compounds such as quercetin, kaempferol, isorhamnetin, rutin, gallic acid, and coumaric acid. They also proposed coriander hydrodistilled residue by-products as a natural raw material for treating oxidant-related diseases.

In addition to the well-known antioxidant activity, phenolic compounds are endowed with a plethora of intriguing biological properties, including the prevention of diabetes, cardiovascular diseases, neurodegenerative diseases, liver diseases, and cancer (Arora et al., 2019). Epidemiological studies and associated meta-analyses supported the anti-platelet, anti-inflammatory, improved endothelial function, and atheroma plaque stability and reduced the oxidation of LDL particles minimizing thus the risk of coronary heart disease and myocardial infarctions (Pandy and Rizvi, 2009). In their comprehensive study, these authors indicated that polyphenols had shown anti-aging activity by inhibiting lipid peroxidation and the inflammatory mediators' cyclooxygenase (COX)-1 and 2. They are also beneficial in ameliorating the adverse effect of aging on the nervous system and brain. Moreover, polyphenols have effectively treated ulcers and asthma and protected against UV radiation (Pandy and Rizvi, 2009; Gutiérrez-Grijalva et al., 2018).

Essential oils responsible for the pleasant smell of coriander were extracted from coriander press cake (Uitterhaegen et al., 2015; Sriti et al., 2019) and the whole coriander plant waste (Whaba et al., 2020). The former raw material essential oils were obtained with yields ranging from 0.04 to 0.31% (w/w). Chromatographic analysis showed the dominance of oxygenated monoterpenes (88.9%) with linalool (52.1–76.1%), geranyl acetate (3.43–7.45%), camphor (1.26–3.1%), linalyl acetate (1.01–4.04%), and α-terpineol (0.33–2.26%) as the main components. Except for γ-terpinene (1.1–7.6%), the remaining volatile components were detected in very low proportions (Uitterhaegen et al., 2015; Sriti et al., 2019). In another report from Egypt, the hydrodistillation of whole plant waste yielded 1.1% of essential oil. *Trans*-anethol (29.2%), linalool (20.0%), eugenyl isovalerate (14.1%), estragole (10.2%), and carvacrol (5.1%) were the main oxygenated monoterpenes. Longifolene (6.8%) and germacrene-D (2.31%) were found as the main sesquiterpene hydrocarbons (Whaba et al., 2020). These studies show that coriander waste/by-products can be considered an alternative source of linalool-, linalyl acetate-, or *trans*-anethol-rich essential oils.

Linalool-rich essential oils of coriander are unequivocally the most studied bioactive ingredient of coriander. The anti-inflammatory, antihyperalgesic, and antinociceptive effects of linalool and its ester linalyl acetate have been established (Kamatou and Viljoen, 2008). Earlier *in vitro* experiments showed that linalool exhibited potent anticancer activity against a broad range of cancer cells, such as cervix (IC_{50} 0.37 µg/mL), stomach (IC_{50} 14.1 µg/mL), skin (IC_{50} 14.9 µg/mL), lung (IC_{50} 21.5 µg/mL), and bone (IC_{50} 21.7 µg/mL) (Cherng et al., 2007). The ability of linalool to inhibit large bowel and duodenal tumor formation has been confirmed in azoxymethane-induced neoplasia in the rat (Gould et al., 1990). Moreover, linalool- and linalyl acetate-producing species have been reported to possess antibacterial, antifungal, antioxidant, and sedative activities (Mandal and Mandal, 2015; Gastón et al., 2016). They are also used to alleviate spasms, gastric complaints, bronchitis, and gout and treat gastrointestinal disorders such as anorexia and diarrhea (Ahmad et al., 2017).

Given these data, it could be concluded that coriander waste/by-products hold the potential to be used as a renewable, sustainable, economical source of nutritionally, medicinally, and industrially relevant ingredients.

25.4 CURRENT AND PROSPECTIVE APPLICATIONS

Owing to their rich chemical profile and manifest biological properties, many potential applications have been proposed, with considerable attention to valorizing lignocellulosic raw material, the petroselinic-rich oil, and linalool-rich essential oils (Figure 25.5). At this point, the pioneering works of Uitterghaegen and co-workers have successfully employed the binder properties of coriander press cake protein and its reinforcing fibers in the fabrication of biodegradable fiberboard with improved physicochemical (thinness swelling, water absorption, thermal properties) and mechanical properties (strength, surface hardness, thermoplastic behaviors).

The authors proposed the application of their innovative material as pallet interlayers sheeting for the manufacture of containers or furniture or in the building trade (Uitterhagaen et al., 2016). Using the same starting raw material (coriander straw and press cake) without any chemical adhesives, these authors have also manufactured binderless fiberboards to replace wood-based materials (Uitterhagaen et al., 2017). Furthermore, additional improvements in their bio-based materials have replaced formaldehyde emission with a linalool-rich volatile bouquet, enhancing indoor air quality and extending the possible application of the self-bonded coriander boards for the storage of food and agricultural products (Uitterhagaen et al., 2019).

The petroselinic acid-rich oil from coriander by-products could also find potential applications in different industrial sectors (Figure 25.6). Petroselinic acid (C18:1n-12), the main fatty acid of the coriander press cake and hydrodistilled residue, is a double bond isomer of oleic acid (C18:1n-9). In addition to the increase in melting temperature (30°C), the unique double bond confers better

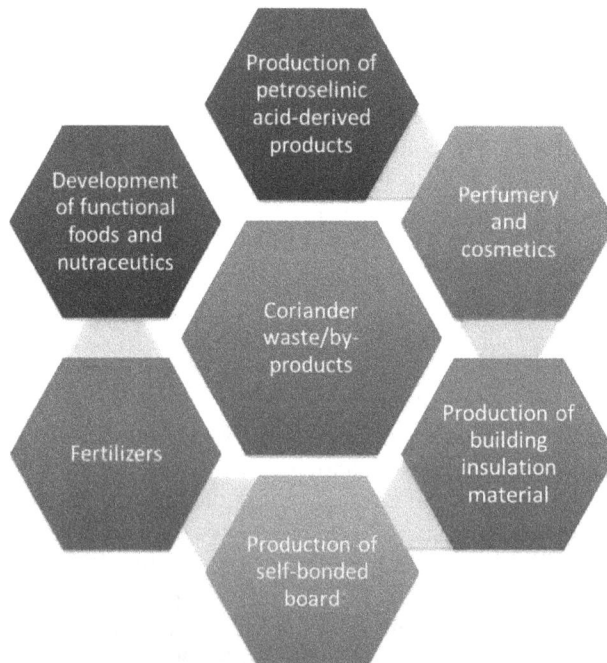

FIGURE 25.5 Main applications of coriander waste/by-products.

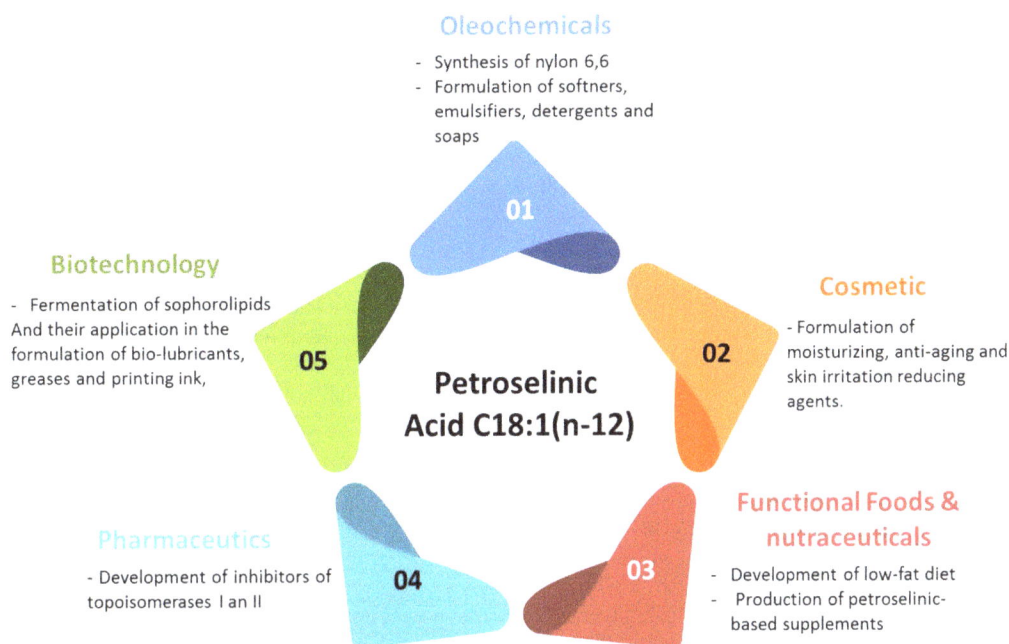

Oleochemicals
- Synthesis of nylon 6,6
- Formulation of softners, emulsifiers, detergents and soaps

01

Biotechnology
- Fermentation of sophorolipids And their application in the formulation of bio-lubricants, greases and printing ink,

05

Petroselinic Acid C18:1(n-12)

Cosmetic
- Formulation of moisturizing, anti-aging and skin irritation reducing agents.

02

Pharmaceutics
- Development of inhibitors of topoisomerases I an II

04

Functional Foods & nutraceuticals
- Development of low-fat diet
- Production of petroselinic-based supplements

03

FIGURE 25.6 Industrial applications of petroselinic acid.

oxidative stability to petroselinic acid-rich oil, enlarging its spectrum of application (Moser, 2009). Petroselinic acid's oxidative cleavage yielded adipic acid and lauric acid in the chemical industry. The former acid can synthesize nylon 6.6, while the latter was used in softeners, emulsifiers, detergents, and soap formulations.

Their intriguing biological activities (i.e., anti-aging and anti-inflammatory) make this unusual fatty acid suitable for the cosmetic (formulation of moisturizing, anti-aging skin irritation reducing agent), pharmaceutical (inhibitors of topoisomerases), and functional food industry (Avato and Tava, 2021). Petroselinic acid exhibited various biological activities (Delbeke et al., 2016). Furthermore, petroselinic-based estolide esters have been synthesized with good physical properties as bio-based lubricants, greases, and printing ink (Cermak et al., 2011). Further applications include the development of a petroselinic-based low-fat diet and petroselinic-based supplements (Avato and Tava, 2021).

Alternatively, coriander by-products (press cake) could represent an additional linalool and linalyl acetate source, representing one of the most used fragrance ingredients. Linalool and its esters linalyl acetate are used in cosmetics and may be found in decorative cosmetics, fine fragrances, shampoos, soaps, deodorants, and other toiletries. It is also used in household cleaners and detergents (Letizia et al., 2003a, b). It is also used as a food additive and in the formulation of pesticides (Pereira et al., 2018).

In food, the incorporation of deoiled coriander (10%) into bread to produce a dietary fiber-rich bread has been successfully achieved without altering the rheological and structural properties of the product (Sowbhagya, 2019). Furthermore, the European Commission approved the use of petroselininc acid as a supplement (EFSA, 2013).

25.5 CONCLUSION

This review has identified a line of evidence in the scientific literature that supports the valorization of coriander waste/by-products as a potential candidate for the recovery and the application of putative bioactive and/or valuable ingredients that can find numerous industrial applications. The few available data point to the richness of these uncommon raw materials in active phytochemicals, including essential minerals, dietary fibers, lignocellulosic materials, petroselinic acid-rich oil, linalool-, linalyl-rich essential oil, fat-soluble antioxidants, and antioxidant phenolic compounds. Despite these encouraging results about the valorization of coriander waste/by-products, the detailed chemical characterization, the *in vitro* and *in vivo* biological properties, and the process optimization for the extraction of valuable ingredients, as well as their toxicity safety features remain unexplored and may prove to be a rewarding challenge for future research.

REFERENCES

Abbassi, A., Mahmoudi, H., Zaouali, W., M'Rabt, Y., Casabianca, H., Hosni, K. (2018). Enzyme-aided release of bioactive compounds from coriander (*Coriandrum sativum* L.) seeds and their residue by-products and evaluation of their antioxidant activity. *Journal of Food Science and Technology*, 55:3065–3076.

Abifarin, T.O., Otunola, G.A., Afolayan, A.J. (2020). Nutritional composition and antinutrient content of *Heteromorpha arborescens* (Spring.) Cham. & Shletdl. leaves: An underutilized wild vegetable. *Food Science & Nutrition*, 19:172–179.

Ahmad, B.S., Talou, T., Saad, Z., Hijazi, A., Merah, O. (2017). The Apiaceae: Ethnomedicinal family as source of medicinal uses. *Industrial Crops and Products*, 109:661–671.

Ahmed, A.A.S., Reda, R.M., ElHady, M. (2020). Immunomodulation by *Coriandrum sativum* seeds (coriander) and its ameliorative effect on lead-induced immunotoxicity in Nile tilapia (*Oreochromis niloticus* L.). *Aquaculture Research*, 51:1077–1088.

Ahsan, H., Neugut, A.I. (1998). Radiation therapy for breast cancer and increased risk for esophageal carcinoma *Annals of Internal Medicine*, 128, 114–117.

Aissaoui, A., El-Hilaly, J., Israili, Z.H., Lyoussi, B. (2008). Acute diuretic effect of continuous intravenous infusion of an aqueous extracts of *Coriandrum sativum* L. in anesthetized rats. *Journal of Ethnopharmacology*, 115:89–85.

Alqethami, A., Hawkins, J.A., Teixidor-Toneu, I. (2017). Medicinal plants used by women in Mecca: Urban, Muslim and gendered knowledge. *Journal of Ethnobiology and Ethnomedicine*, 13:62.

Anderson, J.W. (1990). Dietary fiber and human health. *HortScience*, 25(12):1488–1495.

Anderson, J.W., Baird, P., Dais Jr, R.H., Ferreri, S., Knudtson, M., Koraym, A., Waters, V., Williams, C.L. (2009). Health benefits of dietary fibers. *Nutrition Reviews*, 67(4):188–205.

Arora, I., Sharma, M., Tollefsbol, T.O. (2019). Combinatorial epigenetics impact of polyphenols and phytochemicals in cancer prevention and therapy. *International Journal of Molecular Sciences*, 20:4567. https://doi.org/10.3390/ijms20184567.

Asgarpanah, J., Kazemivach, N. (2014). Phytochemistry, pharmacology and medicinal properties of *Coriandrum sativum* L. *African Journal of Pharmacy and Pharmacology*, 6(31):2340–2345.

Avato, P., Tava, A. (2021). Rare fatty acids and lipids in plant oilseeds: Occurrence and bioactivities. *Phytochemistry Review*. https://doi.org/10.1007/s11101-021-09770-4.

Ayeni, E.A., Abubakar, A., Ibrahim, G., Atinga, V., Muhammed, Z. (2018). Phytochemical, nutraceutical and antioxidant studies of the aerial parts of *Daucus carota* (Apiaceae). *Journal of Herbmed Pharmacology*, 7(2):68–73.

Barber, T.M., Kabisch, S., Pfeiffer, A.F.H., Weickert, M.O. (2020). The health benefits of dietary fibre. *Nutrients*, 12:3209. https://doi.org/10.3390/nu12103209.

Bhat, S.P., Rizvi, W., Kumar, A. (2014). Dose-dependent effect of *Coriandrum sativum* Linn. Seeds on thermal pain stimulus. *The Journal of Phytopharmacology*, 3(4):254–258.

Briggs, M., Petersen, K., Kris-Etherton, P. (2017). Saturated fatty acids and cardiovascular disease: Replacements for saturated fat to reduce cardiovascular risk. *Healthcare*, 5:29.

Cherng, J.M., Shieh, D.-E., Chiang, W., Chang, M.Y., Chaing, L.C. (2007). Chemopreventive effects of minor dietary constituents in common foods on human cancer cells. *Bioscience, Biotechnology, and Biochemistry*, 71:1500–1504.

De Jong, N., Plat, J., Mensink, R.P. (2003). Metabolic effects of plant sterols and stanols. *Journal of Nutritional Biochemistry*, 4:362–369.

Delbeke, E.I.P., Everaert, J., Uitterhaegen, E., Verweire, S., Verlee, A., Talou, T., Soetaert, W., Van Bogaert, I.N.A., Stevens, C.V. (2016). Petroselinic acid purification and its use for the fermentation of new sophoroplipids. *AMB Express*, 6:28. https://doi.org/10.1186/s13568-016-0199-7.

DiNicolantonio, J.J., mangan, D., O'Keefe, J.H. (2018). Copper deficiency may be a healing cause of ischaemic heart disease. *Open Heart*, 5:748. https://doi.org/10.1136/openhrt-2018-000784.

Dong, H., Sargent, L.J., Chatzidiako, Y., Saunders, C., Harkness, L., Bordenave, N., Rowland, I., Spencer, J.P.E., Lovegrove, J.A. (2016). Orange pomace fibre increases a composite scoring of subjective ratings of hunger and fullness in healthy adults. *Appetite*, 107:478–485.

Duarte, A., Ferreira, S., Silva, F., Domingues, F.C. (2012). Synergistic activity of coriander oil and conventional antibiotics against *Acinetobacter baumannii*. *Phytomedicine*, 19:236–238.

Duarte, A., Luís, Â., Oleatro, M., Domingues, F.C. (2016). Antioxidant properties of coriander essential oil and linalool and their potential to control *Campylobacter* spp. *Food Control*, 61:115–122.

Caprioli, G., Fiorini, D., Maggi, F., Marangoni, M., Papa, F., Vittori, S., Sagratini, G. (2014). Ascorbic acid content, fatty acid composition and nutritional value of the neglected vegetable Alexanders (*Smyrnium olusatrum* L., Apiaceae). *Journal of Food Composition and Analysis*, 35:30–36.

Cecchi, T., Passamonti, P., Alfei, B., Cecchi, P. (2011). Monovarietal extra virgin olive oil from the Marche region, Italy: Analytical and sensory characterization. *International Journal of Food Properties*, 14:483–495.

Cermak, S.C., Isbell, T.A., Evangelista, R.L., Johnson, B.L. (2011). Synthesis and physical properties of petroselinic acid based estolide esters. *Industrial Crops and Products*, 33:132–139.

Demir, S., Korukluoglu, M. (2020). A comparative study about antioxidant activity and phenolic composition of cumin (*Cuminum cyminum* L.) and coriander (*Coriandrum sativum* L.). *Indian Journal of Traditional Knowledge*, 19(2):383–393.

EFSA. (2013). Scientific opinion on the safety of "coriander seed oil" as a novel food ingredient. *EFSA Journal*, 11:3422–3442.

Farzaei, F., Morovati, M.R., Farjadmand, F., Farzaei, M.H. (2017). A mechanistic review on medicinal plants used for diabetes mellitus in Traditional Persian medicine. *Journal of evidence-Based Complementary & Alternative medicine*, 22(4):944–955.

Fatemi, S.H., Hammond, E.G. (1980). Analysis of oleate, linoleate and linolenate hydroperoxides in oxidized esters mixture. *Lipids*, 15:179–385.

Frail, L., Crisci, E., Córdoba, L., Navarro, M.A., Osada, J., Montoya, M. (2012). Immunomodulatory properties of Beta-sitosterol in pig immune responses. *International Immunopharmacology*, 13:316–321.

Gastón, M.S., Cid, M.P., Vásquez, A.M., Decarlini, M.F., Demmel, G.I., Rossi, L.I., Aimar, M.L., Salvatierra, N.A. (2016). Sedative effect of central administration of *Coriandrum sativum* essential oil and its major components linalool in neonatal chicks. *Pharmaceutical Biology*, 10:1954–1961.

Ghajarieh Sepanlou, M., Salami, F., Mirabzadeh Ardakani, M., Sadati Lamardi, S.N., Sadrai, S., Amin, G.R., Sadeghi, N. (2019). The proximate, mineral and amino acid composition of spring, autumn leaves and roots of *Eryngium caeruleum* M. Bieb. *Research Journal of Pharmacognosy*, 6(3):1–7.

Goc, A., Niedzwiecki, A., Rath, M. (2019). Anti-Borreliae efficacy of selected organic oils and fatty acids. *BMC Complementary and Alternative Medicine*, 19:40. https://doi.org/10.1186/s12906-019-2450-7.

Golubkina, N., Kharchenko, V., Moldovan, A., Zayachkovsky, V., Stepanov, V., Pivovarov, V., Sekara, A., Tallarita, A., Caruso, G. (2021). Nutritional value of Apiaceae seeds as affected by 11 species and 43 cultivars. *Horticulturae*, 7:57. https://doi.org/10.3390/hoticulturae7030057.

Gould, M.N., Wacker, W.D., Maltzman, T.H. (1990). Chemoprevention and chemotherapy of mammary tumors by monoterpenoids. Progress in clinical and biological research. *Mutagens and Carcinogens in the Diet*, 347:255–268.

Gröber, U., Schmidt, J., Kisters, K. (2015). Magnesium in prevention and therapy. *Nutrients*, 7:8199–8226.

Gupta, M. (2010). Pharmacological properties and traditional therapeutic uses of important Indian spices: A review. *International Journal of Food Properties*, 13:1092–1116.

Gutiérrez-Grijalva, E.P., Picos-Salas, M.A., Leyva-López, N., Criollo-Mendoza, M.S., Vazquez-Olivo, G., Heredia, J.B. (2018). Flavonoids and phenolic acids from oregano: Occurrence, biological activity and health benefits. *Plants*, 7:2. https://doi.org/10.3390/plants7010002.

Hänsch, R., Mendel, R.R. (2009). Physiological functions of mineral micronutrients (Cu, Zn, Mn, Fe, Ni, Mo, B, Cl). *Current Opinion in Plant Biology*, 12:259–266.

Hussain, F., Jahan, N., ur-Rahman, K., Sultana, B., Jamil, S. (2018). Identification of hypotensive biofunctional compounds of *Coriandrum sativum* and evaluation of their angiotensin-converting enzyme (ACE) inhibition potential. *Oxidative Medicine and Cellular Longevity*. ID46473736. https://doi.org/10.1155/2018/4647376.

Iqbal, M.J., Butt, M.S., Suleria, H.A.R. (2018). Coriander (*Coriandrum sativum* L.): Bioactive molecules and health effects. In Mérillon, J.-M. and Ramawat, K.G. (eds), *Bioactive molecules in food*, Springer, Berlin, Germany, 1–37. https://doi.org/10.1007/978-3-319-54528-8_44-1.

Jandacek, R.J. (2017). Linoleic acid: A nutritional quandary. *Healthcare*, 5:25. https://doi.org/10.3390/healthcare5020025.

Javid, H., Khan, A.L., Farmanullah, K., Hussain, S.T., Shinwari, Z.K. (2009). Proximate and nutrient investigation of selected medicinal plants species of Pakistan. *Pakistan Journal of Nutrition*, 8(5):620–624.

Jéquier, E., Constant, F. (2010). Water as an essential nutrient: The physiological basis of hydration. *European Journal of Clinical Nutrition*, 64:615. https://doi.org/10.1038/sjcn.2009.111.

Kamatou, G.P.P., Viljoen, A.M. (2008). Linalool- A review of a biologically active compounds of commercial importance. *Natural Product Communications*, 3(7):1183–1192.

Kumar, V., Marković, T., Emerald, M., Dey, A. (2016). *Herbs: Composition and dietary importance. Encyclopedia of food and health*, 332–337. https://doi.org/10.1016/b978-0-12-384947-2.00376-7.

Laribi, B., Kouki, K., M'Hamdi, M., Bettaieb, T. (2015). Coriander (*Coriandrum sativum* L.) and its bioactive constituents. *Fitoterapia*, 103:9–26.

Letizia, C.S., Cocchiara, J., Lalko, J., Api, A.M. (2003a). Fragrance material review on linalool. *Food and Chemical Toxicology*, 41:943–964.

Letizia, C.S., Cocchiara, J., Lalko, J., Api, A.M. (2003b). Fragrance material review on linalyl acetate. *Food and Chemical Toxicology*, 41:965–976.

Mandal, S., Mandal, M. (2015). Coriander (*Coriandrum sativum* L.) essential oil: Chemistry and biological activity. *Asian Pacific Journal of Tropical Biomedicine*, 5(6):421–428. https://doi.org./10.1016/j.apjtb.2015.04.001.

Mechchate, H., Es-Safi, I., Amaghnouje, A., Boukhira, S., Alotaibi, A.A., Al-Zharani, M., Nasr, F.A., Noman, O.M., Conte, R., Amal, E.H.E.Y., Bekkari, H., Bousta, D. (2021). Antioxidant, anti-inflammatory, and antidiabetic properties of LC-MS/MS identified polyphenols from coriander seeds. *Molecules*, 26:487.

Mil-Homens, D., Bernardes, N., Fialho, A.M. (2012). The antibacterial properties of docosahexaenoic omega-3 fatty acid against the cystis fibrosis multiresistant pathogen *Burkholderia cenocepacia*. *FEMS Microbiology Letters*, 328:61–69.

Mills, S., Ross, R.P., Hill, C., Fitzgerald, G.F., Stanton, C. (2011). Milk intelligence: Mining milk for bioactive substances associated with human health. *International Dairy Journal*, 21:377–401.

Moreau, R.A., Whitaker, B.D., Hicks, K.B. (2002). Phytosterols, phytostanols, and their conjugates in foods: Structural diversity, quantitative analysis, and health-promoting uses. *Progress in Lipid Research*, 41(6):457–500.

Moser, B.R. (2009). Comparative oxidative stability of fatty acid alkyl esters by accelerated methods. *Journal of the American Oil Chemists' Society*, 86:699–706.

Moustafa, A.H.A., Ali, E.M.M., Moselhy, S.S., Tousson, E., El-Said, K.S. (2012). Effect of coriander on thioacetamide-induced hepatotoxicity in rats. *Toxicology and Industrial Health*, 30(7):621–629.

Msaada, K., Hosni, K., Taarit, M.B., Chahed, T., Kchouk, M.E., Marzouk, B. (2007). Changes on essential oil composition of coriander (*Coriandrum sativum* L.) during three stages of maturity. *Food Chemistry* 102:1131–1134.

Msaada, K., Hosni, K., Taarit, M.B., Hammami, M., Marzouk, B. (2009a). Effect of growing region and maturity stages on oil yield and fatty acid composition of coriander (*Coriandrum sativum* L.) fruits. *Scientia Horticulturae*, 120:525–531.

Msaada, K., Hosni, K., Taarit, M.B., Hammami, M., Marzouk, B. (2009b). Regional and maturational effects on essential oils yields and composition of coriander (*Coriandrum sativum* L.) fruits. *Scientia Horticulturae*, 122:116–124.

Msaada, K., Jemia, M.B., Salem, N., Bachrouch, O., Sriti, J., Tammar, S., Bettaieb, I., Jabri, I., Kefi, S., Limam, F., Marzouk, B. (2017). Antioxidant activity of methanolic extracts from three coriander (*Coriandrum sativum* L.) fruit varieties. *Arabian Journal of Chemistry* 10, S3176–S3183.

Nascimento, A.C., Mota, C., Coelho, I., Gueifão, S., Santos, M., Matos, A.S., Gimenez, A., Lobo, M., Samman, N., Castanheira, I. (2014). Characterization of nutrient profile of quinoa (*Chenopodium quinoa*), amaranth (*Amaranthus caudatus*), and purple corn (*Zea mays* L.) consumed in the North of Argentina: Proximates, minerals and trace elements. *Food Chemistry*, 148:420–426.

Ogundola, A.F., Wintola, O.A., Afolayan, A.J. (2018). Nutrient, antinutrient composition and heavy metal uptake and accumulation in *S. nigrum* cultivated in different soil types. *Scientific World Journal*, 1–20.

Olive Li, Y., Komarek, A.R. (2017). Dietary fibers basics: Health, nutrition, analysis, and applications. *Food Quality and Safety*, 1:47–59.

Önder, A. (2018). Coriander and its phytoconstituents for the beneficial effects. In El-Shemy H.A. (ed.), *Potential of essential oils*, IntechOpen, 165–185. https://doi.org/10.5572/intechopen.78656.

Othman, R.A., Moghadasian, M.H. (2011). Beyond cholesterol-lowering effects of plant sterols: Clinical and experimental evidence of the anti-inflammatory properties. *Nutrition Reviews*, 69(7):371–382.

Pandey, K.B., Rizvi, S.I. (2009). Plant polyphenols as dietary antioxidant in human health and disease. *Oxidative Medicine and Cellular Longevity*, 2(5):270–278.

Paul, B.K., Saleh-e-In, M.M., Ara, A., Roy, S.K. (2013). Mineral and nutritional composition of radhuni (*Carum roxburghianum* Benth.) seeds. *International Food Research Journal*, 20(4):1731–1737.

Platel, K., Srinivasan, K. (2004). Digestive stimulant action of spices: A myth or reality? *Indian Journal of Medicinal Research*, 119:167–179.

Pereira, I., Severino, P., Santos, A.C., Silva, A.M., Souto, E.B. (2018). Linalool bioactive properties and potential applicability in drug delivery systems. *Colloids and Surfaces B: Biointerfaces*, 171:566–578.

Pieroni, A., Gray, C. (2008). Herbal and folk medicines of the Russlanldeutschen living in Kuzelsau/Talacker, Southwestern Germany. *Phytotherapy Research*, 22:889–901.

Ramadan, M.F., Mörsel, J.-T. (2002). Oil composition of coriander (*Coriandrum sativum* L.) fruit-seeds. *European Food Research and Technology*, 215:204–209.

Ramadan, M.F., Kroh, L.W., Mörsel, J.-T. (2003). Radical scavenging activity of black cumin (*Nigella sativa* L.), coriander (*Coriander sativum* L.), and niger (*Guizotia abyssinica* Cass.) crude seed oils and oil fractions. *Journal of Agricultural and Food Chemistry*, 51:6961–6969.

Ruiz, L., Ruiz, L., Maco, M., Cobos, M., Gutierrez-Choquevilca, A.-L., Roumy, V. (2011). Plant used by native Amazonian groups from the Nanay river (Peru) for the treatment of malaria. *Journal of Ethnopharmacology*, 133:917–921.

Ryan, E., Galvin, K., O'Connor, T.P., Maguire, A.R., O'Brien, N.M. (2007). Phytosterol, squalene, tocopherol content and fatty acid profile of selected seeds, grains, and legumes. *Plant Foods for Human Nutrition*, 62:85–91.

Sahib, N.G., Anwar, F., Gilani, A.-H., Hamid, A.A., Saari, N., Alkharfy, K.M. (2013). Coriander (*Coriandrum sativum* L.): A potential source of high-added value components for functional foods and nutraceuticals-A review. *Phytotherapy Research*, 27:1439–146.

Saini, R.K., Assefa, A.D., Keum, Y.-S. (2021). Spices in the Apiaceae family represent the healthiest fatty acid profile: A systematic comparison of 34 widely used spices and herbs. *Foods*, 10:854. https://doi.org/10.3390/foods10040854.

Santos-Silva, J., Bessa, R.J.B., Santos-Silva, F. (2002). Effect of genotype, feeding system and slaughter weight on the quality of light lambs II: Fatty acid composition of meat. *Livestock Production Science*, 77:187–194.

Shahidi, F., de Camargo, A.C. (2016). Tocopherols and tocotrienols in common and emerging dietary sources: Occurrence, applications, and health benefits. *International Journal of Molecular Sciences*, 17:1745. https://doi.org/10.3390/ijms17101745.

Shishodia, S., Aggarwal, B.B. (2004). Guggulsterone inhibits NF-kappaB and IkappaBalpha kinase activation, suppresses expression of anti-apoptotic gene products, and enhances apoptosis. *Journal of Biological Chemistry*, 279:47148–47158.

Shukla, S., Gupta, S. (2009). Coriander. In Aggarwal, B.B., Kunnumakkara, A.B. (eds), *Molecular targets and therapeutic uses of spices: Modern uses for ancient medicine*, World Scientific, Singapore, 149–171.

Śledź, M., Nowacka, M., Wiktor, A., Witowa-Rajchert, D. (2013). Selected chemical and physic-chemical properties of microwave-connective dried herbs. *Food and Bioproducts Processing*, 91(4):421–428.

Sowbhagya, H.B. (2019). Value-added processing of by-products from spice industry. *Food Quality and Safety*, 3:73–80.

Sriti, J., Neffati, M., Msaada, K., Talou, T., Marzouk, B. (2013). Biochemical characterization of coriander cakes obtained by extrusion. *Journal of Chemistry*, ID871631. https://doi.org/10.1155/2013/871631.

Sriti, J., Bettaieb, I., Bachrouch, O., Talou, T., Marzouk, B. (2019). Chemical composition and antioxidant activity of the coriander cake obtained by extrusion. *Arabian Journal of Chemistry*, 12:1765–1773.

Tahraoui, A., El-Hilaly, J., Israili, Z.H., Lyoussi, B. (2007). Ethnopharmacological survey of plants used in the traditional treatment of hypertension and diabetes in south-eastern Morocco (*Errachidia province*). *Journal of Ethnopharmacology*, 110:105–117.

Tang, E.L.H., Rajarajeswaran, J., Fung, S.Y., Kanthimathi, M.S. (2013). Antioxidant activity of *Coriandrum sativum* and protection against DNA damage and cancer cell migration. *BMC Complementary Medicine and Therapies*, 13:347.

Tunçtürk, M., Özgökçe, F. (2015). Chemical compostion of some Apiaceae plants commonly used in herby cheese in eastern Anatolia. *Turkish Journal of Agriculture and Forestry*, 39:55–62.

Turan, M., Kordali, S., Zengin, H., Dursun, A., Sezen, Y. (2003). Macro and micro mineral content of some wild edible leaves consumed in Eastern Anatolia. *Acta Agriculturae Scandinavica, Section B-Plant Soil Science*, 53(3):129–137.

Ugulu, I., Baslar, S., Yorek, N., Dogan, Y. (2009). The investigation and quantitative ethnobotanical evaluation of medicinal plants used around Izmir province, Turkey. *Journal of Medicinal Plant Research*, 119:167–179.

Uitterhagaen, E., Nguyen, Q.H., Sampaio, K., Stevens, C., Merah, O., Talou, T., Rigal, L., Evon, Ph. (2015). Extraction of coriander oil by using twin-screw extrusion: Feasibility study and potential press cake applications. *Journal of the American Oil Chemists' Society*, 82(8):1219–1233.

Uitterhagaen, E., Nguyen, Q.H., Merah, O., Stevens, C., Talou, T., Rigal, L., Evon, Ph. (2016). New renewable and biodegradable fiberboards from a coriander press cake. *Journal of Renewable Materials*, 4(3):225–238.

Uitterhagaen, E., Labonne, L., Merah, O., Talou, T., Ballas, S., Véronèse, T., Evon, Ph. (2017). Impact of thermo-mechanical fiber pre-treatment using twin-screw extrusion on the production and properties of renewable binderless coriander fiberboards. *International Journal of Molecular Sciences*, 18:139. https://doi.org/10.3390/ijms18071539.

Uitterhagaen, E., Labonne, L., Ballas, S., Véronèse, T., Evon, Ph. (2019). The coriander straw an original agricultural by-product for the production of building insulation materials. *Academic Journal of Civil Engineering*, 37(2):627–633.

Ulbricht, T.L.V., Southgate, D.A.T. (1991). Coronary heart disease: Seven dietary factors. *The Lancet*, 338(8773):985–992.

Vyas, A., Shukla, S.S., Pandey, R., Jain, V., Joshi, V., Gidwani, B. (2012). Chervil: A multifunctional miraculous Nutritional herb. *Asian Journal of Plant Sciences*, 11(4):163–171.

Wahba, H.E., Abd Rabbu, H.S., Ibrahim, M.E. (2020). Evaluation of essential oil isolated from dry coriander seeds and recycling of the plant waste under different storage conditions. *Bulletin of the National Research Center*, 44:192. https://doi.org/10.1186/s42269-020-00448-z.

Whelan, J., Fritsche, K. (2013). Linoleic acid. *Advances in Nutrition*, 4:311–312.

26 Food Applications of Coriander (*Coriandrum sativum*) Leaves

Muhammad Kamran Khan, Muhammad Imran,
Muhammad Haseeb Ahmad, Rabia Shabir Ahmad,
and Saira Sattar

CONTENTS

26.1 INTRODUCTION

The green leaves of coriander play an essential role in its nutrition as green vegetables are a rich source of iron, minerals, and vitamins. In addition, coriander leaves contain 84–89% water, are low in cholesterol and saturated fat, and contain a high amount of vitamin C and vitamin A, and they are also rich in dietary fiber, zinc, and thiamine (Bhat, Kaushal, Kaur, & Sharma, 2014). The complete nutritional composition of coriander leaves is mentioned in Table 26.1.

People worldwide have used coriander leaves for ages in different forms like sauces and dried leaves as a tea for treating gastrointestinal problems like diarrhea, colic, and flatulence (Kumar, 2020). Coriander stimulates the secretion of bile salts in the liver and activates the enzymes that help in digestion, leading to an overall acceleration in the digestive process. Carotenoid is one of the essential phenolic compounds present in coriander and deposited in its leaves in the form of β-carotene (11.2 mg/g), which is an excellent source of antioxidants and can also be used as a coloring agent (Nadihira, Ramar, Jegadeeswari, & Srinivasan, 2021). Being a good source of β-carotene, coriander leaves are used for patients deficient in vitamin A (Ashokkumar et al., 2020). Coriander leaves also contain a high amount of anthocyanin, which, along with its antioxidant activity, serves as a preventive agent for different chronic diseases, hence playing an essential role in maintaining body health (Kozłowska & Dzierżanowski, 2021). These phenolic compounds in coriander leaves and other phenols such as rutin help control diabetes by lowering the blood and urinary glucose

DOI: 10.1201/9781003204626-28

TABLE 26.1

Chemical Composition of Fresh Coriander Leaves

Component	Amount
Moisture	87.9%
Protein	3.3%
Carbohydrate	6.5%
Total ash	1.7%
Calcium	0.14%
Phosphorus	0.06%
Iron	0.01%
Vitamin B_2	60 mg/100 g
Niacin	0.8 mg/100 g
Vitamin C	135 mg/100 g
Vitamin A	10,460 IU/100 g

and postprandial glycemia by inhibiting α-glucosidase activity in the gastrointestinal tract (Lee et al., 2021). According to a recent review by Hosseini and Boskabady (2021), coriander leaves are also beneficial in the proper functioning of the brain; they prevent histological changes in the brain such as lymphocytic, cellular edema, gliosis, and infiltration. Moreover, the leaves of the coriander plant were also observed to alleviate anxiety and insomnia; the lower anxiety levels were seen in rodents after giving intraperitoneal doses of 50, 100, and 200 mg/kg of aqueous leaves extracts (Latha, Rammohan, Sunanda, Maheswari, & Mohan, 2015), while the insomniac relief was confirmed when hydroalcoholic extracts (400 and 600 mg/kg) and aqueous extracts (200, 400, and 600 mg/kg) were administered to mice resulting in increasing the sleep duration to 130–180 min and 160–220 min, respectively (Emam & Heydari, 2006).

26.2 PHYTOCHEMICAL CONSTITUENTS IN CORIANDER LEAVES

26.2.1 Lipids

Polyunsaturated fatty acids are among the major nutrient components; they also function as precursors of different signal molecules and are important constituents of cell membranes (Djuricic & Calder, 2021). The determination of fatty acids in the coriander plants' basal and upper leaves was first done by Neffati and Marzouk (2008). The researchers reported that fatty acid content in upper leaves was 41.8 mg/g DW whereas the highest content of fatty acids was found in basal leaves (61.2 mg/g DW). Among all fatty acids, α-linolenic acid was predominant in coriander leaves with 41.1%. The other fatty acids such as palmitic acid, heptadecenoic acid, linoleic acid, and some traces of stearic, palmitoleic, and oleic acids were also detected in the upper and basal leaves of the coriander plant.

26.2.2 Polyphenols

Phenolic compounds are an important component because they exhibit antioxidant activity and have a beneficial physiological effect (Albuquerque, Heleno, Oliveira, Barros, & Ferreira, 2021). Considering their functional activity, these phenolic compounds are isolated from various plants, including *Coriandrum sativum*. The total phenolic content and flavonoid content in coriander leaves were 1013 mg/kg and 5259 mg/kg, respectively, as determined by Barros et al. (2012). The characterization of different bioactive compounds from coriander leaves is represented in Figure 26.1.

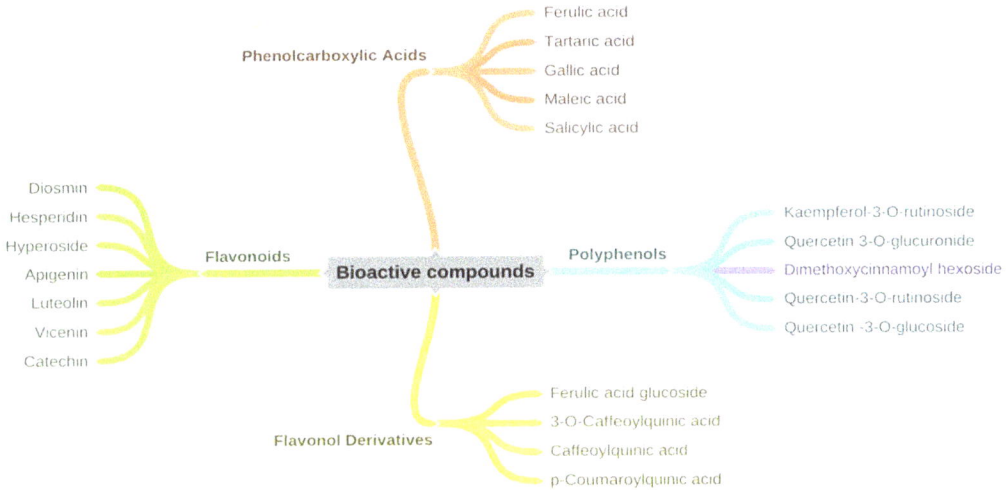

FIGURE 26.1 Characterization of bioactive compounds from coriander leaves.

26.2.3 ESSENTIAL OIL

Various researchers worldwide have investigated essential oil from coriander leaf during its blooming stage. It was found that coriander leaves mainly comprise alcohols and aldehydes (Laribi, Kouki, M'Hamdi, & Bettaieb, 2015). The major constituents of aromatic acids in leaf oil are tridecanoic acid (5.5%), undecyl alcohol (6.4%), capric acid (12.7%), *E*-11-tetradecenoic acid (13.4%), and 2-decenoic acid (30.8%). Other major constituents are decamethylene glycol (1.15%), 2-dodecanal (1.32%), decanal (1.35%), undecanoic acid (2.13%), cyclododecane (2.45%), dodecanoic acid (2.63%), and 2-undecenal (3.87%) respectively (Zeb, 2016). The high concentration of 2-deconic acid makes coriander leaf useful as a flavoring agent and perfume. The essential oils in coriander leaves make them useful in manufacturing other products such as soap, lotion, shampoo, and detergent (Bhuiyan, Begum, & Sultana, 2009).

26.3 PROCESSING OF CORIANDER LEAVES FOR FOOD APPLICATIONS

Coriander leaves are often consumed in an unprocessed form, but processing is beneficial to improve their quality, trading purpose, and shelf life (Thamkaew, Sjöholm, & Galindo, 2021). Harvesting is considered the first step in processing in which a crop is collected at a specific time and stage. After harvesting, the crop is exposed to vacuum cooling in which free moisture is vaporized from the product, thus enhancing the shelf life of leaves (Bhat et al., 2014). Drying is another process to eliminate moisture from coriander leaves after harvesting, which can be done by freeze-drying, microwave drying, or sun drying, etc. Air temperature plays an important role in accelerating the drying process; the higher the air temperature the faster the drying process is accomplished (Alibas, Yilmaz, Asik, & Erdoğan, 2021). However, the drying process causes the loss of chlorophyll from the green plants (particularly sun drying); therefore, blanching is mainly done prior to drying to minimize the loss of chlorophyll, and using alternative drying methods like microwave drying also helps in retaining the chlorophyll content in coriander leaves as it is a faster process than traditional drying (Shaw, Meda, Tabil Jr, & Opoku Jr, 2006). Coriander leaves may also be exposed to irradiation such as gamma rays or microwaves for the elimination of microbial load before consumption. The effect of different processing conditions on components of coriander leaves is summarized in Table 26.2.

The coriander leaves undergo different processing in food industries to formulate various food products in fast food, traditional cuisine, or different beverages. These processing

TABLE 26.2

Effect of Different Processing Techniques on Coriander Leaf Components

Processing Technique	Treatment Condition	Effect	References
Drying	**Convective drying**: 70°C for 120 min **Microwave drying**: 360 W	Preserved aromatic sensory attribute of coriander leaves.	(Łyczko et al., 2021)
High power ultrasound and supercritical CO_2 drying	**HPU**: 40 W, 40°C **scCO$_2$**: 10 MPa, up to 90 min	Yeast and molds not detected; mesophilic spores reduced to 1 log; mesophilic bacteria reduced up to 4 log.	(Michelino, Zambon, Vizzotto, Cozzi, & Spilimbergo, 2018)
Freeze drying	**Temp**: –32°C **Time**: 72 h	Maximum retention of total volatile contents (only 1–3% reduction compared to non-treated sample).	(Pirbalouti, Salehi, & Craker, 2017)
Blanching	**Temp**: 100°C **Time**: 2 min	Preserved the phytochemical content to 84%.	(Meena, Agrawal, & Agrawal, 2016)
Refrigeration	**Temp**: 4°C **Time**: 5 days	Minimal effect on phenolic content and antioxidant activity.	(Chawla & Thakur, 2014)
Shade drying	**Temp**: 35°C **Time**: 2 days	Total polyphenols were retained to up to 78%.	(Kamel, Thabet, & Algadi, 2013)
Microwave drying	**Power**: 850 W 600 W, 450 W, 300 W, and 180 W **Time**: 30 s, 60 s, 90 s, 120 s, and 150 s	850W and 90 s was the best-determined MW drying condition to enhance bioactive contents.	(Divya, Puthusseri, & Neelwarne, 2012)
Irradiation (gamma rays)	**Dose**: 0.5 kGy	Aerobic mesophilic count reduced by 99.9% Total coliform bacteria reduced from 871,000 cfu/g to 100 cfu/g.	(Cruz-Zaragoza, Ruiz-Gurrola, Wacher, Flores Espinosa, & Barboza-Flores, 2011)
Blanching	**Type**: water blanching **Temp**: 80°C	Higher rehydration and better chlorophyll retention.	(Ahmed, Randhawa, Yusuf, & Khalid, 2011)

techniques also protect the components of fresh coriander leaves from degradation (Bhat et al., 2014). Thermal processing, like boiling, pasteurization, microwave heating, etc., is used to make a puree, paste, and sauces from coriander leaves, and it also increases the final product's shelf life (Qiu, Zhang, Mujumdar, & Liu, 2020). Blanching (both steam and water blanching) is used in making powder from coriander leaves as it inactivates the peroxidase enzyme, which catalyzes hydrogen peroxide into a water molecule after blanched leaves are dried, ground uniformly, and preserved for further use in salads and cooking purposes (Kaur, Kaur, Aggarwal, & Grover, 2021). Blanching is also used in making products such as salsa, and it helps retain levels of antioxidants and other phenolic compounds present in coriander leaves (Kaiser, Kammerer, & Carle, 2013). Another major processing technique used for coriander leaves is extraction; various extraction methods such as solvent extraction, cold press extraction, CO_2 extraction, water distillation, and steam distillation are used for extracting the oil from plant leaves (Hanif, Nisar, Khan, Mushtaq, & Zubair, 2019). The oil is an important component in determining food quality, and coriander leaf oil is majorly utilized in flavoring and seasoning food products like curries, salads, rice, etc. Moreover, coriander oil is also used in flavoring meat, candies, tobacco, sauces, and alcoholic beverages (Sharma & Sharma, 2012).

26.4 PRESERVATION OF CORIANDER LEAVES FOR FOOD APPLICATION

The demand for coriander leaves is high due to their multiple uses in cooking, pharmacology, food industry, etc., but coriander leaves have a very short shelf life under ambient conditions due to their high perishability. Therefore, the need for innovative post-harvesting technologies and study on microbial stability is gaining attention among processors and growers, and they are finding new ways to improve the shelf life of coriander leaves to resist shipping and storage conditions (Pragalyaashree & Thirupathi, 2016). Researchers have applied various processing techniques and packaging styles to fresh green leafy vegetables, which help destroy microorganisms and enzymes, improve the chlorophyll content in leaves, and control the formation of unwanted secondary metabolites (Pavani & Aduri, 2018). The role of different processing techniques has already been discussed in the previous section; in this section, different packaging styles and the use of chemical preservatives to enhance the shelf life of fresh coriander leaves will be discussed.

The fresh coriander leaves are very delicate and have a high content of moisture which, when handled poorly, can cause damage, bruising, microbial infection, and excessive loss of water which results in oxidation and withering (Gao et al., 2017). Dehydration significantly affects the sensory attributes of coriander leaves like color, aroma, texture, and overall appearance. Hence, a suitable packaging material is required to minimize the loss of moisture content, microbial contamination, and insect infestation and to reduce chilling injuries during transportation (Ranjitha et al., 2017). The selection of packaging material is an important step as it should be economical and functional to perform all needful. For years, plastic bags have been used for the packaging of leafy vegetables due to their easy availability, compatibility, selective barrier properties, light weight, heat sealability, and printability; these bags minimize moisture loss and can be easily stored in refrigeration (Bhide, 2006). A study on the effect of perforated and non-perforated packaging on coriander leaves was done in which researchers found that non-perforated packets were more acceptable in enhancing the shelf life of products instead of perforated packets throughout the storage (Guha, Sharangi, & Debnath, 2013). In another study by Sharangi, Guha, and Chakrabarty (2015), the effects of different packaging materials, including newspaper packets, brown paper packets, laminated paper packets, perforated polyethylene packets, and non-perforated polythene packets, were observed on fresh coriander leaves kept in storage. According to the result, the leaves packed in non-perforated polythene bags showed minimal color changes and physiological weight loss. The sealed polythene bags acted as a barrier against water loss as they created high relative humidity in the package which retards the water loss through the process of transpiration.

Modified atmospheric packaging (MAP) is another technique adopted by processors to enhance the shelf life of coriander leaves. MAP is commonly used for perishable foods as it uses polymeric packaging films and acts as selectively permeable for water vapor, O_2, and CO_2. This permeability of the package and the respiration process of products create a modified atmosphere around the packaged product (Haque et al., 2021). The increased CO_2 and decreased O_2 levels around the fresh coriander leaves help in decreasing the texture losses, respiration rate, and ethylene production; delays in ripening improves the pigment retention and reduces microbial spoilage (Kumara & Beneragama, 2020). In one of the studies, the effect of MAP was observed along with aqueous chlorine dioxide washing on the microbial growth and quality of fresh coriander leaves. The researchers observed that the microbial loads were initially reduced by chlorine dioxide washing, and MAP packaging with $15\%O_2 + 3\%CO_2 + 82\%N_2$ inhibits the microbial growth completely and enhances coriander's sensory and nutritional attributes compared to aerobic packaging (Fang et al., 2016). In another study, the combined effect of modified atmospheric packaging (5% O_2, 10% CO_2, and 85% N_2) with sodium hypochlorite dipping (100 ppm) on the quality and shelf life of coriander leaves for 25 days at 5°C was observed by Waghmare & Annapure (2015). The results showed that the combined effect of sodium hypochlorite dipping with MAP treatment decreased the microbial count and retained the sensory attributes of packed coriander leaves better compared to their individual effect.

26.5   USE OF FRESH CORIANDER LEAVES IN DIFFERENT REGIONS OF THE WORLD

The coriander plant as a whole is used as a flavor ingredient in food industries. It is one of the most important herbs globally with great significance in international trade. It is approved as generally recognized as safe (GRAS) by the FDA to be used as a flavoring agent (Fukushima et al., 2020). In the food industry of Asia, Central and South America, and the Middle East, coriander leaves are mainly used for flavoring purposes (Karakol & Kapi, 2021). The essential oil from coriander leaves is used to flavor all types of food like pickles, tobacco, candy, alcoholic beverages, seasoning, meat sauce, baked goods, frozen dairy, chewing gum, hard candy, and soft candy (Burdock & Carabin, 2009).

Coriander leaves are used to prepare various types of cuisine worldwide (Bower, Marquez, & de Mejia, 2016). Chopped fresh coriander leaves are used to garnish Thai, Chinese, and Indian dishes. They are also used to flavor pickles, salads, soups, curries, salsas, liqueurs, cakes, bread, and chutneys. In addition, the juice of coriander leaves is often mixed with smoothies, tomatoes, apples, and celery to make energy and medicinal drinks (Devi, Gupta, Sahu, & Mishra, 2020). Table 26.3 represents the culinary uses of coriander leaves in different world regions.

Coriander leaves have wide application in the pharmaceutical industry; they are used in many traditional medicines to treat arthritis, sore muscles, and indigestion, improve appetite and diabetes, and relieve anxiety (Slimestad, Fossen, & Brede, 2020). Coriander helps in the smooth functioning of the digestive system as it stimulates the secretion of gastric juices, which eases flatulence and colic. The chemical substances extracted from coriander leaves have an antibacterial effect mainly used against *Salmonella choleraesuis*. Coriander is also used to treat toothache, joint pains, hemorrhoids, and measles. The presence of fiber content, potassium, and magnesium in coriander leaves makes it useful in eliminating intestinal irritations, boosting immunity, and purifying blood (Al-Juhaimi & Ghafoor, 2011).

Coriander leaves are also a valuable ingredient in the cosmetics, perfume, detergent, and soap industries (Önder, 2018). In addition, the essential oil extracted from leaves is used in aromatherapy (Paul, Sen, Karim, & Faruque, 2013). Coriander is also used in emulsifiers and surfactants, and the residues left after extraction are also used as feedstuff for ruminants (Tajodini, Moghbeli, Saeedi, & Effati, 2014).

26.6   CORIANDER LEAVES AS A FUNCTIONAL FOOD

Coriander leaves possessing a high amount of phytochemicals, and various pharmacological activities are ascribed to the essential oil, such as antioxidant, antimicrobial, anti-inflammatory,

---

TABLE 26.3

Culinary Uses of Coriander Leaves around the World

Country	Uses	References
Vietnam	Aromatic herb	(Ron Wolf & Petersen)
Mexico	Salads	(Sachan, Kumar, Kumari, & Singh, 2018)
Indonesia	Flavoring agent	(Sahib et al., 2013)
Thailand	Flavoring agent	(Gil et al., 2002)
Pakistan	Garnishing and sauces	(Chaudhry & Tariq, 2006)
India	Sauces, flavoring, and garnishing	(Verma & Jheeba, 2016)
France	Dressing, flavoring sausages, and pickling	(Bower et al., 2016)
United States	Garnish meat dishes and eaten with duck eggs	(Kuebel & Tucker, 1988)
Germany and South Africa	Used for making sausages	(J.-H. Lee, Won, & Song, 2015)

anti-diabetic, antihypertensive, etc. These biological activities of the coriander plant make it a valuable herb as a functional food for the nutraceutical and food industries (Prachayasittikul, Prachayasittikul, Ruchirawat, & Prachayasittikul, 2018).

26.6.1 ANTIOXIDANT ACTIVITY

The damage to tissues and biomolecules caused by oxidative stress mainly occurs due to reactive oxygen species (Barros et al., 2012). The antioxidants, therefore, act as protective agents against this biological damage. The coriander leaves enriched in polyphenols (particularly carotenoids) possess a high antioxidant activity (Divya et al., 2012). The essential oil extracted from coriander leaves is considered a natural antioxidant containing terpenoid components that inhibit the primary and secondary oxidation products. This essential oil prevents oxidative rancidity caused by high lipid products such as cakes in the food industry. In a study, the effect of the antioxidant activity of coriander essential oil on ghee was studied by Patel, Shende, Arora, and Singh (2013); the authors concluded that the addition of essential oil during frying significantly prevents the deterioration of ghee due to its higher antioxidant activity. Another study (Marangoni & Moura, 2011) revealed that when coriander oil was used as a natural antioxidant instead of a synthetic one in preserving Italian salami, it improved its shelf life without affecting its sensory attributes. Thus, this essential oil is useful in stabilizing and enhancing the shelf life of many food products (Silva, Caldera, Trotta, Nerín, & Domingues, 2019).

26.6.2 ANTIMICROBIAL ACTIVITY

The aqueous extract and essential oil from coriander leaves are found to have an inhibitory effect against different Gram-positive bacteria (e.g., *Bacillus* spp. and *Staphylococcus aureus*) and Gram-negative bacteria (e.g., *Salmonella typhi, E. coli, Pseudomonas aeruginosa*, etc.), yeast species, and Candida species (Begnami, Duarte, Furletti, & Rehder, 2010). The effect of coriander oil in preventing microbial growth by direct incorporation of it into minced beef film was studied by Michalczyk, Macura, Tesarowicz, and Banaś (2012); it was deduced that 0.02% of coriander oil inhibits *Enterobacteriacae* counts and also inhibits meat spoilage which causes undesirable sensory changes in the product. In another study, 0.5% coriander essential oil successfully eliminated *C. jejuni* from beef and chicken meat, and 0.1 and 0.25% essential oil reduced the bacterial count to 2- and 4-log from the products, respectively (Rattanachaikunsopon & Phumkhachorn, 2010). This antimicrobial effect of coriander leaves makes them a suitable food preservative in the food industry which extends the shelf life of food by preventing the growth of microbes (Silva & Domingues, 2017).

26.6.3 ANTIFUNGAL ACTIVITY

High moisture foods like cakes and other baked goods deteriorate significantly due to fungi, mainly *Aspergillus niger, Penicillium expansum, Rhizopus stolonifer, Monila sitophila*, etc. (Jayan, Priyadharsini, Ramya, & Rajkumar, 2020). Coriander essential oil works effectively in controlling the mold growth in such food products. In one of the studies by Darughe, Barzegar, and Sahari (2012) 0.05% of coriander essential oil (CEO) acts similarly to 0.01% BHA in controlling the mold growth in cakes, and increasing the dose of CEO up to 0.15% inhibits the mold growth more efficiently as compared to BHA samples.

26.7 CONCLUSION

Coriander leaves have been used for food products as fresh leaves in culinary support and in dried form in the food and pharmaceutical industries. The use of fresh leaves varies in different regions of the world, but they are mainly used as a taste improver in culinary dishes. Due to their short shelf life, the fresh leaves are usually dried using different conventional and innovative technologies. The dried forms are generally used in fast-food chains, beverage formulations, and functional food. Coriander leaves' antioxidant, antimicrobial, and esthetic properties may potentially replace synthetic ingredients for different food products and need to be explored.

REFERENCES

Ahmed, A., Randhawa, M. A., Yusuf, M. J., & Khalid, N. (2011). Effect of processing on pesticide residues in food crops: A review. *Journal of Agricultural Research*, *49*(3), 379–390.

Al-Juhaimi, F., & Ghafoor, K. (2011). Total phenols and antioxidant activities of leaf and stem extracts from coriander, mint and parsley grown in Saudi Arabia. *Pakistan Journal of Botany*, *43*(4), 2235–2237.

Albuquerque, B. R., Heleno, S. A., Oliveira, M. B. P., Barros, L., & Ferreira, I. C. (2021). Phenolic compounds: Current industrial applications, limitations and future challenges. *Food & Function*, *12*(1), 14–29.

Alibas, I., Yilmaz, A., Asik, B. B., & Erdoğan, H. (2021). Influence of drying methods on the nutrients, protein content and vitamin profile of basil leaves. *Journal of Food Composition and Analysis*, *96*, 103758.

Ashokkumar, K., Pandian, A., Murugan, M., Dhanya, M., Sathyan, T., Sivakumar, P., ... Warkentin, T. D. (2020). Profiling bioactive flavonoids and carotenoids in select south Indian spices and nuts. *Natural Product Research*, *34*(9), 1306–1310.

Barros, L., Duenas, M., Dias, M. I., Sousa, M. J., Santos-Buelga, C., & Ferreira, I. C. (2012). Phenolic profiles of in vivo and in vitro grown *Coriandrum sativum* L. *Food Chemistry*, *132*(2), 841–848.

Begnami, A., Duarte, M., Furletti, V., & Rehder, V. (2010). Antimicrobial potential of *Coriandrum sativum* L. against different *Candida* species in vitro. *Food Chemistry*, *118*(1), 74–77.

Bhat, S., Kaushal, P., Kaur, M., & Sharma, H. (2014). Coriander (*Coriandrum sativum* L.): Processing, nutritional and functional aspects. *African Journal of Plant Science*, *8*(1), 25–33.

Bhide, M. (2006). *Buying, using and storing their favorite spices and herbs*. Cookbook.

Bhuiyan, M. N. I., Begum, J., & Sultana, M. (2009). Chemical composition of leaf and seed essential oil of *Coriandrum sativum* L. from Bangladesh. *Bangladesh Journal of Pharmacology*, *4*(2), 150–153.

Bower, A., Marquez, S., & de Mejia, E. G. (2016). The health benefits of selected culinary herbs and spices found in the traditional Mediterranean diet. *Critical Reviews in Food Science and Nutrition*, *56*(16), 2728–2746.

Burdock, G. A., & Carabin, I. G. (2009). Safety assessment of coriander (*Coriandrum sativum* L.) essential oil as a food ingredient. *Food and Chemical Toxicology*, *47*(1), 22–34.

Chaudhry, N., & Tariq, P. (2006). Bactericidal activity of black pepper, bay leaf, aniseed and coriander against oral isolates. *Pakistan Journal of Pharmaceutical Sciences*, *19*(3), 214–218.

Chawla, S., & Thakur, M. (2014). Effect of thermal processing on total phenolic content and antioxidant activity of *Coriandrum sativum* L. leaves. *Asian Journal of Bio Science*, *9*(1), 58–62.

Cruz-Zaragoza, E., Ruiz-Gurrola, B., Wacher, C., Flores Espinosa, T., & Barboza-Flores, M. (2011). Gamma radiation effects in coriander (*Coriandrum sativum* L) for consumption in Mexico. *Revista Mexicana de Física*, *57*, 80–86.

Darughe, F., Barzegar, M., & Sahari, M. (2012). Antioxidant and antifungal activity of coriander (*Coriandrum sativum* L.) essential oil in cake. *International Food Research Journal*, *19*(3), 1253–1260.

Devi, S., Gupta, E., Sahu, M., & Mishra, P. (2020). Proven health benefits and uses of coriander (*Coriandrum sativum* L.). In *Ethnopharmacological investigation of Indian spices* (pp. 197–204). IGI Global.

Divya, P., Puthusseri, B., & Neelwarne, B. (2012). Carotenoid content, its stability during drying and the antioxidant activity of commercial coriander (*Coriandrum sativum* L.) varieties. *Food Research International*, *45*(1), 342–350.

Djuricic, I., & Calder, P. (2021). Beneficial outcomes of Omega-6 and Omega-3 polyunsaturated fatty acids on human health: An update for 2021. *Nutrients* 13, 2421.

Emam, G. M., & Heydari, H. G. (2006). Sedative-hypnotic activity of extracts and essential oil of coriander seeds. *Iranian Journal of Medical Sciences*, *31*(1), 22–27.

Fang, X., Chen, H., Gao, H., Yang, H., Li, Y., Mao, P., & Jin, T. Z. (2016). Effect of modified atmosphere packaging on microbial growth, quality and enzymatic defence of sanitiser washed fresh coriander. *International Journal of Food Science & Technology*, *51*(12), 2654–2662.

Fukushima, S., Cohen, S. M., Eisenbrand, G., Gooderham, N. J., Guengerich, F. P., Hecht, S. S., ... Harman, C. L. (2020). FEMA GRAS assessment of natural flavor complexes: Lavender, Guaiac Coriander-derived and related flavoring ingredients. *Food and Chemical Toxicology*, *145*, 111584.

Gao, H., Fang, X., Li, Y., Chen, H., Zhao, Q. F., & Jin, T. Z. (2017). Effect of alternatives to chlorine washing for sanitizing fresh coriander. *Journal of Food Science and Technology*, *54*(1), 260–266.

Gil, A., De La Fuente, E. B., Lenardis, A. E., López Pereira, M., Suárez, S. A., Bandoni, A., ... Ghersa, C. M. (2002). Coriander essential oil composition from two genotypes grown in different environmental conditions. *Journal of Agricultural and Food Chemistry, 50*(10), 2870–2877.

Guha, S., Sharangi, A., & Debnath, S. (2013). Effect of different sowing times and cutting management on phenology and yield of off season coriander under protected cultivation. *Trends in Horticultural Research, 3*(1), 27–32.

Hanif, M. A., Nisar, S., Khan, G. S., Mushtaq, Z., & Zubair, M. (2019). Essential oils. In *Essential oil research* (pp. 3–17). Springer.

Haque, M. A., Asaduzzaman, M., Mahomud, M. S., Alam, M. R., Khaliduzzaman, A., Pattadar, S. N., & Ahmmed, R. (2021). High carbon-di-oxide modified atmospheric packaging on quality of ready-to-eat minimally processed fresh-cut iceberg lettuce. *Food Science and Biotechnology, 30*(3), 413–421.

Hosseini, M., & Boskabady, M. H. (2021). Neuroprotective effects of *Coriandrum sativum* and its constituent, linalool: A review. *Avicenna Journal of Phytomedicine, 11*(5), 436–450.

Jayan, L., Priyadharsini, N., Ramya, R., & Rajkumar, K. (2020). Evaluation of antifungal activity of mint, pomegranate and coriander on fluconazole-resistant *Candida glabrata*. *Journal of Oral and Maxillofacial Pathology: JOMFP, 24*(3), 517.

Kaiser, A., Kammerer, D. R., & Carle, R. (2013). Impact of blanching on polyphenol stability and antioxidant capacity of innovative coriander (*Coriandrum sativum* L.) pastes. *Food Chemistry, 140*(1–2), 332–339.

Kamel, S. M., Thabet, H. A., & Algadi, E. A. (2013). Influence of drying process on the functional properties of some plants. *Chemistry and Materials Research, 3*(7), 1–8.

Karakol, P., & Kapi, E. (2021). Use of selected antioxidant-rich spices and herbs in foods. *Antioxidants*. In (Ed.), Antioxidants - Benefits, Sources, Mechanisms of Action edited by, Viduranga Waisundara. IntechOpen.

Kaur, S., Kaur, N., Aggarwal, P., & Grover, K. (2021). Bioactive compounds, antioxidant activity, and color retention of beetroot (*Beta vulgaris* L.) powder: Effect of steam blanching with refrigeration and storage. *Journal of Food Processing and Preservation, 45*(3), e15247.

Kozłowska, A., & Dzierżanowski, T. (2021). Targeting inflammation by anthocyanins as the novel therapeutic potential for chronic diseases: An update. *Molecules, 26*(14), 4380.

Kuebel, K., & Tucker, A. O. (1988). Vietnamese culinary herbs in the United States. *Economic Botany, 42*(3), 413–419.

Kumar, V. (2020). Seven spices of India-from kitchen to clinic. *Journal of Ethnic Foods, 7*(1), 1–16.

Kumara, B., & Beneragama, C. (2020). Modified atmospheric packaging extends the postharvest shelf life of Mukunuwenna (*Alternanthera sessilis* L.). *Tropical Agricultural Research, 31*(2), 87–96.

Laribi, B., Kouki, K., M'Hamdi, M., & Bettaieb, T. (2015). Coriander (*Coriandrum sativum* L.) and its bioactive constituents. *Fitoterapia, 103*, 9–26.

Latha, K., Rammohan, B., Sunanda, B., Maheswari, M. U., & Mohan, S. K. (2015). Evaluation of anxiolytic activity of aqueous extract of *Coriandrum sativum* Linn. in mice: A preliminary experimental study. *Pharmacognosy Research, 7*(Suppl 1), S47.

Lee, J.-H., Won, M., & Song, K. B. (2015). Physical properties and antimicrobial activities of porcine meat and bone meal protein films containing coriander oil. *LWT-Food Science and Technology, 63*(1), 700–705.

Lee, L.-C., Hou, Y.-C., Hsieh, Y.-Y., Chen, Y.-H., Shen, Y.-C., Lee, I.-J., ... Liu, H.-K. (2021). Dietary supplementation of rutin and rutin-rich buckwheat elevates endogenous glucagon-like peptide 1 levels to facilitate glycemic control in type 2 diabetic mice. *Journal of Functional Foods, 85*, 104653.

Łyczko, J., Masztalerz, K., Lipan, L., Iwiński, H., Lech, K., Carbonell-Barrachina, Á. A., & Szumny, A. (2021). *Coriandrum sativum* L.-Effect of multiple drying techniques on volatile and sensory profile. *Foods, 10*(2), 403.

Marangoni, C., & Moura, N. F. d. (2011). Sensory profile of Italian salami with coriander (*Coriandrum sativum* L.) essential oil. *Food Science and Technology, 31*, 119–123.

Meena, S., Agrawal, M., & Agrawal, K. (2016). Effect of blanching and drying on antioxidants and antioxidant activity of selected green leafy vegetables. *International Journal of Science and Research, 5*(10), 1811–1814.

Michalczyk, M., Macura, R., Tesarowicz, I., & Banaś, J. (2012). Effect of adding essential oils of coriander (*Coriandrum sativum* L.) and hyssop (*Hyssopus officinalis* L.) on the shelf life of ground beef. *Meat Science, 90*(3), 842–850.

Michelino, F., Zambon, A., Vizzotto, M. T., Cozzi, S., & Spilimbergo, S. (2018). High power ultrasound combined with supercritical carbon dioxide for the drying and microbial inactivation of coriander. *Journal of CO2 Utilization, 24*, 516–521.

Nadihira, S., Ramar, A., Jegadeeswari, V., & Srinivasan, S. (2021). Microgreen production in herbal spices. *Journal of Pharmacognosy and Phytochemistry, 10*, 168–170.

Neffati, M., & Marzouk, B. (2008). Changes in essential oil and fatty acid composition in coriander (*Coriandrum sativum* L.) leaves under saline conditions. *Industrial Crops and Products, 28*(2), 137–142.

Önder, A. (2018). Coriander and its phytoconstituents for the beneficial effects. In *Potential of essential oils* (pp. 165–185).

Patel, S., Shende, S., Arora, S., & Singh, A. K. (2013). An assessment of the antioxidant potential of coriander extracts in ghee when stored at high temperature and during deep fat frying. *International Journal of Dairy Technology, 66*(2), 207–213.

Paul, N., Sen, S. K., Karim, H. N., & Faruque, A. (2013). Phytochemical and biological investigations of *Coriandrum sativum* (Cilantro) leaves. *International Journal of Innovative Pharmaceutical Sciences and Research, 1*(1), 170–184.

Pavani, K., & Aduri, P. (2018). Effect of packaging materials on retention of quality characteristics of dehydrated green leafy vegetables during storage. *International Journal of Environment, Agriculture and Biotechnology, 3*(1), 239062.

Pirbalouti, A. G., Salehi, S., & Craker, L. (2017). Effect of drying methods on qualitative and quantitative properties of essential oil from the aerial parts of coriander. *Journal of Applied Research on Medicinal and Aromatic Plants, 4*, 35–40.

Prachayasittikul, V., Prachayasittikul, S., Ruchirawat, S., & Prachayasittikul, V. (2018). Coriander (*Coriandrum sativum*): A promising functional food toward the well-being. *Food Research International, 105*, 305–323.

Pragalyaashree, M. M., & Thirupathi, V. (2016). Coriander. In *Leafy medicinal herbs* (p. 116).

Qiu, L., Zhang, M., Mujumdar, A. S., & Liu, Y. (2020). Recent developments in key processing techniques for oriental spices/herbs and condiments: A review. *Food Reviews International*, 1–21.

Ranjitha, K., Shivashankara, K., Rao, D. S., Oberoi, H. S., Roy, T., & Bharathamma, H. (2017). Improvement in shelf life of minimally processed cilantro leaves through integration of kinetin pretreatment and packaging interventions: Studies on microbial population dynamics, biochemical characteristics and flavour retention. *Food Chemistry, 221*, 844–854.

Rattanachaikunsopon, P., & Phumkhachorn, P. (2010). Potential of coriander (*Coriandrum sativum*) oil as a natural antimicrobial compound in controlling *Campylobacter jejuni* in raw meat. *Bioscience, Biotechnology, and Biochemistry, 74*(1), 31–35.

Ron Wolf, C., & Petersen, M. (2007). *The cuisine of Southeast Asia culinary focus: Vietnam*. Center for the Advancement of Foodservice Education, Vietnam.

Sachan, A., Kumar, S., Kumari, K., & Singh, D. (2018). Medicinal uses of spices used in our traditional culture: Worldwide. *Journal of Medicinal Plants Studies, 6*(3), 116–122.

Sahib, N. G., Anwar, F., Gilani, A. H., Hamid, A. A., Saari, N., & Alkharfy, K. M. (2013). Coriander (*Coriandrum sativum* L.): A potential source of high-value components for functional foods and nutraceuticals-A review. *Phytotherapy Research, 27*(10), 1439–1456.

Sharangi, A. B., Guha, S., & Chakrabarty, I. (2015). Effect of different packaging materials on storage life of fresh coriander (*Coriandrum sativum* L) leaves. *Nature and Science, 13*(6), 100–108.

Sharma, M., & Sharma, R. (2012). Coriander. In *Handbook of herbs and spices* (pp. 216–249). Elsevier.

Shaw, M., Meda, V., Tabil Jr, L., & Opoku Jr, A. (2006). Drying and color characteristics of coriander foliage using convective thin-layer and microwave drying. *Journal of Microwave Power and Electromagnetic Energy, 41*(2), 56–65.

Silva, F., Caldera, F., Trotta, F., Nerín, C., & Domingues, F. C. (2019). Encapsulation of coriander essential oil in cyclodextrin nanosponges: A new strategy to promote its use in controlled-release active packaging. *Innovative Food Science & Emerging Technologies, 56*, 102177.

Silva, F., & Domingues, F. C. (2017). Antimicrobial activity of coriander oil and its effectiveness as food preservative. *Critical Reviews in Food Science and Nutrition, 57*(1), 35–47.

Slimestad, R., Fossen, T., & Brede, C. (2020). Flavonoids and other phenolics in herbs commonly used in Norwegian commercial kitchens. *Food Chemistry, 309*, 125678.

Tajodini, M., Moghbeli, P., Saeedi, H., & Effati, M. (2014). The effect of medicinal plants as a feed additive in ruminant nutrition. *Iranian Journal of Applied Animal Science, 4*(4), 681–686.

Thamkaew, G., Sjöholm, I., & Galindo, F. G. (2021). A review of drying methods for improving the quality of dried herbs. *Critical Reviews in Food Science and Nutrition, 61*(11), 1763–1786.

Verma, V., & Jheeba, S. (2016). Marketing cost and prices spread of coriander in Kota District of Rajasthan. *International Journal Seed Spices, 6*(1), 59–65.

Waghmare, R. B., & Annapure, U. S. (2015). Integrated effect of sodium hypochlorite and modified atmosphere packaging on quality and shelf life of fresh-cut cilantro. *Food Packaging and Shelf Life, 3*, 62–69.

Zeb, A. (2016). Coriander (*Coriandrum sativum*) oils. In *Essential oils in food preservation, flavor and safety* (pp. 359–364). Elsevier.

27 Green Synthesis of Nanoparticles from Coriander Extract

Sarfaraz Ahmed Mahesar, Muhammad Saqaf Jagirnai,
Abdul Rauf Khaskheli, Aamna Balouch,
and Syed Tufail Hussain Sherazi

CONTENTS

27.1 INTRODUCTION

Nanotechnology is an emerging field of material science that manipulates the materials on the nanoscale (Fathi et al., 2012). Nanoparticles (NPs) are the main class of nanomaterials that contain particulate substances, which have different dimensions such as 0D, 1D, 2D, and 3D less than 100 nm (Tiwari et al., 2012). NPs exhibit unique physical and chemical properties based on actual size, shape, distribution, and high surface-to-volume ratio (Singh et al., 2017b). Several NPs have been developed, including silver (Ag) (Guilger-Casagrande & Lima, 2019), gold (Au) (Hamelian et al., 2018), palladium (Pd) (Favier et al., 2019), platinum (Pt) (Jeyaraj et al., 2019), iron (Fe) (Pan et al., 2020), copper (Cu) (Vasantharaj et al., 2019), zinc oxide (ZnO) (Emamhadi et al., 2020), and carbon NPs such as graphene oxide (GO) (Menazea & Ahmed, 2020), carbon nanotubes (CNTs) (Zhang et al., 2020), and diamond NPs (Li et al., 2020). Due to their extraordinary characteristics, NPs have a wide range of applications such as sensing (Karimi-Maleh et al., 2020), drug delivery (Mitchell et al., 2021), catalysis (Gao et al., 2020), data transmission and storage (Harada et al., 2020), biotechnology (Salem & Fouda, 2021), electronics (He et al., 2020), water treatment (Punia et al., 2021), conversion of solar energy (Ismael, 2020), and antimicrobials properties (Garibo et al., 2020).

Several techniques are available to prepare NPs, such as photochemical, chemical reduction, reverse micelles, thermal decomposition, electrochemical, radiation assisted, and microwave-assisted

DOI: 10.1201/9781003204626-29

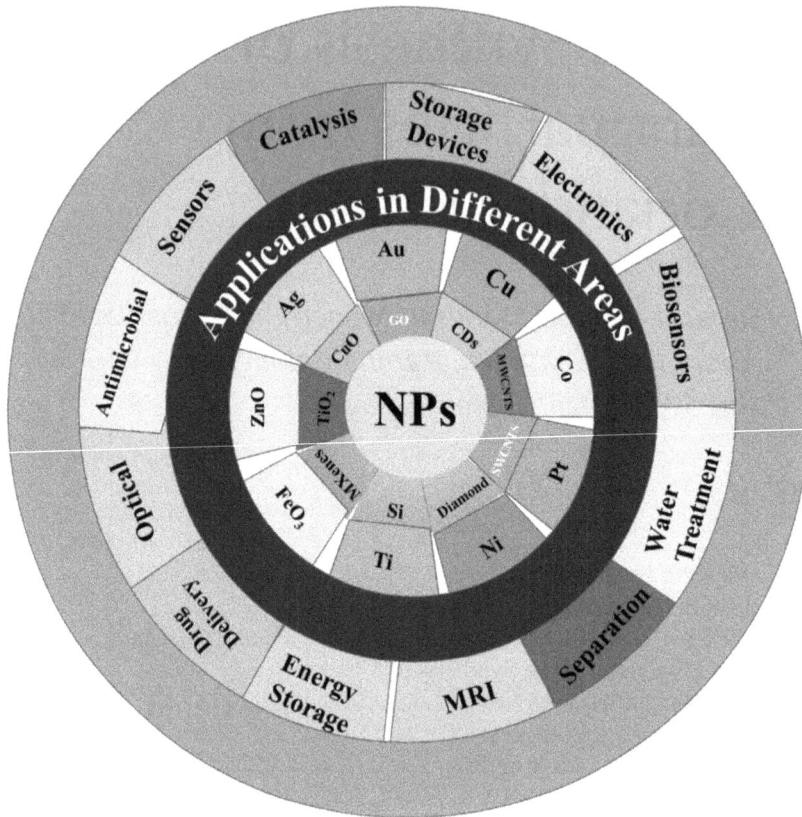

FIGURE 27.1   Applications of NPs in various fields.

(Jamkhande et al., 2019). These techniques have some limitations; they use toxic chemicals to synthesize NPs, they are costly, there is high consumption of organic solvents, and they need high safety protocols. Nanotechnology researchers have developed a green route to synthesize nanoparticles to overcome these limitations. The green synthesis method is environmentally friendly and economically beneficial (Jadoun et al., 2021). Due to environmental pollution and its toxic effects on living organisms during the last 15 years, the researchers focused on developing a sustainable method and following the green chemistry protocols. Figure 27.1 represents the types and applications of NPs.

27.2   GREEN SYNTHESIS OF NANOPARTICLES

Green chemistry deals with the sustainability of the materials and minimizes the use of hazardous chemicals (Clark & Macquarrie, 2008). Green chemistry procedures are essential for the prospect of NPs. This field should come to a climax in developing nanoscience; the eco-friendly and safe synthesis of NPs should have acceptance in nanotechnology (Varma, 2012). The nano-revolution has emerged due to the current development and implementation of advanced technologies, which offer tools and technology as platforms for assessing biological entities that serve as inspirational models for bio-assembled materials in the production of NPs. The biosynthesis of NPs is a bottom-up approach that applies biological ordination for the production of NPs. Biomaterials play an essential role in synthesizing NPs from raw material through the reduction process (Li et al., 2007). Biological species may vary from prokaryotic to eukaryotic bacteria such as fungi and plants during the formation of NPs. The plants have wide applications compared to microorganisms because plant-based preparation is a single-step process with no need for extra protocols during the synthesizing of NPs.

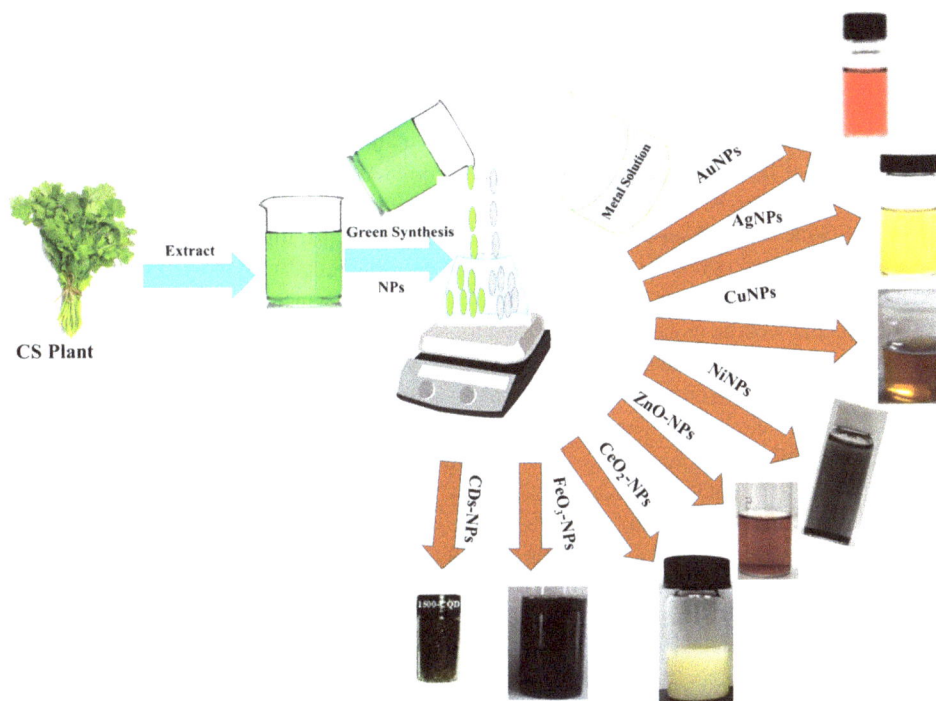

FIGURE 27.2   The synthesis process of NPs using CS.

In contrast, the microorganisms have limitations, loss of mutation during the NPs preparation, need extra multiple steps, and need extra protocols and time-consuming processes. Due to the straightforward, safe, and one-step synthesis of NPs using plant material, plant-based synthesis expanded rapidly (Mandal et al., 2006). The plant material is a unique and sustainable alternative for the preparation of NPs compared to conventional methods. In the current research scenario, the plant material-based preparation of NPs with controlled size and shape has seen outstanding progress. The plants are a well-known natural product with metal tolerance which has resulted from phytoremediation and phytomining. The plant material-based preparation of NPs has been developed due to active biomaterials, including amino acids, proteins, vitamins, polyphenols, polysaccharides, terpenoids, and organic acids. These molecules play a crucial role during the development of NPs in reduction and capping. Due to their excellent capping efficiency, they form the desired shape and size (Iravani, 2011). Figure 27.2 represents the green synthesis process of NPs.

27.2.1   ROLE OF PLANTS IN GREEN SYNTHESIS OF NANOPARTICLES

The green synthesis of NPs is classified into different classes, such as biosynthetic and plant extract routes (Nath & Banerjee, 2013). The plant extract acts as a reducing and capping agent utilized for the preparation of NPs and has a significant effect on the size, shape, morphology, and physicochemical properties. The biosynthesis method is a sustainable, biocompatible, greener, and environmentally friendly way to synthesize the NPs using microorganisms (algae, fungi, and bacteria) and plants (seeds, roots, stems, leaves, and fruits) (Razavi et al., 2015). The plant's substances are an excellent green chemistry route for the synthesis of NPs due to the presence of phytochemicals, and simultaneously they work as reducing and capping agents (Vickers, 2017). Green procedures of preparation are significantly attractive due to their excellent efficiency in reducing the toxicity of NPs. Therefore, amino acids, vitamins, and plant extracts are prevalent nowadays (Baruwati et al., 2009). Conventional approaches have been used for the last years. Still, scientists have proved that the greener techniques are more efficient for synthesizing NPs with vast advantages, low chances of failure, cost-effectiveness, and ease of characterization (Abdelghany et al., 2018). Chemical and physical methods of preparation of NPs have some limitations due to the consumption of toxic solvents and chemicals that affect the environment. Plant-based green synthesis of NPs is a straightforward procedure due to the rich amount of natural substances present in the plants, which are economically beneficial and need little care. Plant material has unique characteristics of accumulating environmental pollution and metal detoxification; metals are very toxic on the trace level (Shahid et al., 2017). Compared to other green materials, the plant material-based synthesis of nanomaterials are highly efficient, and there is no need to use different protocols. It is an easy route and environmentally friendly, and there is no need for toxic solvents during the extraction and synthesis process of nanomaterials (Hulkoti & Taranath, 2014). Different parts of the plants have been used for the green synthesis of NPs, including root, stem, leaf, and fruit. The phytochemicals such as terpenoids, polyols, and polyphenols are responsible for reducing and capping NPs (Ovais et al., 2018). Take a selective part of the plant, appropriately wash it with distilled water, take a certain amount of water and boil. After that extract is filtered and used for the preparation of NPs. Take certain amount of extract to prepare NPs after some time, the color of the solution will start to change, indicating the formation of plant-based NPs. The prepared NPs can be separated using a centrifuge, and the NPs can be collected for further study. Synthesis via plant extract is economically beneficial and environmentally friendly, and can be used to prevent the use of any intermediate base group (Carolin et al., 2017). Several plant materials have been reported for the synthesis of NPs, such as *Emblica officinalis* (Muthuchudarkodi & Ashli, 2020), *Aloe vera* (Rahmani et al., 2020), alfalfa (Baraka et al., 2017), *Ocimum basilicum* (Altikatoglu et al., 2017), *Tamarindus indica* (Sheikh & Bramhecha, 2019), and *Coriandrum sativum* (CS) (Ashraf et al., 2020). Table 27.1 represents different plant materials used for the synthesis of NPs.

27.3   CHEMISTRY OF CS

The origination of CS is in Mediterranean countries; it is also used as a traditional herb, and belongs to the Apiaceae family. CS is commercially cultivated in Asia, Europe, and Africa. It is one of the amazing plants with two functions used in herbal medicine and as a spice (Bhat et al., 2014). The CS fruit is antipyretic, diuretic, stomachic, stimulant, aphrodisiac, anathematic, laxative, and can treat bronchitis, vomiting, biliousness, and gland tuberculosis. The seed of CS exhibited important hypoglycemic action on rats (Dash, 2011). All parts of CS are edible, but the fresh leaves and dried seeds are commonly used for cooking. CS is also used in the Indian traditional medicines to treat different illnesses such as digestive problems and respiratory and urinary systems. It has diuretic, diaphoretic, stimulant, and carminative activity. CS is also used in traditional Iranian medicines to treat many diseases, including loss of appetite, dyspeptic complaints, anxiety, convulsion, and insomnia (Pathak et al., 2011). The chief sources of different constituents of CS are fatty oil and essential oils. A variety of macromolecules are mainly responsible for bio-activity. The aldehydes are a major component of CS leaves. The plant's phenolic compounds (flavonoids and lignins) can be used as signaling and antioxidants, compounds, and tannins as defense response chemicals. Furthermore, it has protective properties as an antioxidant, anti-aging, anti-proliferative, and anti-inflammatory due to the presence of phenolic compounds. CS is used for digestive issues such as appetite loss, stomach discomfort, vomiting, hernia, nausea, bowel spasms, and bowel gas. It is used to treat hemorrhoids, measles, joint pain, and toothaches, and CS has excellent anti-bacterial and antifungal properties (da Silva et al., 2015; Sharma et al., 2010). The quantity of essential oil contained is up to 0.03% to 2.6% in the dried seed of CS. The fatty oil content is 9.9% to 27.7%, respectively. According to the reported studies, the chemical composition of CS varies with the region and age of the plant. The volatile oil and different constituent contents in the seed are about 8%, α-pinene (1.5%), α-terpinyl acetate (31.3%), α-phellandrene (0.2%), α-terpineol (2.6%), β-pinene (0.2%), limonene (11.6%), γ-terpinene (0.7%), linalool (3%), 1,8-cineole (36.3%), linalyl acetate (2.5%), citronellol (0.3%),

TABLE 27.1

Plant Materials Used for the Preparation of NPs

Plant Origin	NPs	Size (nm)	Morphology	Applications	References
Aloe barbadensis Miller (*Aloe vera*)	Au and Ag	10–30	Spherical, triangular	Cancer hyperthermia, optical coatings	(Fair & Tor, 2014)
Acalypha indica	Ag	20–30	Spherical	Antibacterial activity against waterborne pathogens	(Krishnaraj et al., 2010)
Camellia sinensis (black tea leaf extracts)	Au and Ag	20	Spherical, prism	Catalysts, sensors	(Mondal et al., 2011)
Citrus limon (lemon)	Ag	<50	Spherical, spheroidal	-	(Prathna et al., 2011)
Eucalyptus citriodora (neelagiri)	Ag	20	Spherical	Antibacterial	(Ravindra et al., 2010)
Ludwigia adscendens (ludwigia)	Ag	100–400	Spherical	-	(Mochochoko et al., 2013)
Morus (mulberry)	Ag	15–20	Spherical	Antimicrobial activity against *E. coli, B. subtilis*	(Singh et al., 2017a)
Nelumbo nucifera (lotus)	Ag	25–80	Spherical, triangular, truncated triangular, decahedral	Larvicidal activity against malaria and flariasis vectors	(Santhoshkumar et al., 2011)
Ocimum sanctum (tulsi; root extract)	Ag	10 and 5	Spherical	Catalytic reduction	(Singh et al., 2018)
Ocimum sanctum (tulsi; leaf extract)	Au and Ag	30 and 10–20	Crystalline, hexagonal, triangular, and spherical	Biolabeling, biosensor	(Philip & Unni, 2011)
Acacia nilotica (Fabaceae) babul	Ag	20–30	Distorted spherical	Culex quinquefasciatus and Culex gelidus	(Velayutham et al., 2013)
Coccinia grandis (Cucurbitaceae) – ivy gourd	Ag	20–30	Spherical	-	(Dayem et al., 2017)
Orange peel	Ag	9.5	Spherical	-	(Kahrilas et al., 2014)
Kiwifruit juice	Ag	31	Spherical	Bioanalytical	(Gao et al., 2014)
Holarrhena antidysenterica (L.)	Ag	550	Spherical	Larvicidal activity against dengue and filariasis vectors	(Kumar et al., 2018)
Coleus aromaticus (Lamiaceae) – borage	Ag	40–50	Spherical	Antibacterial and cytotoxic activities	(Gogoi et al., 2015)

TABLE 27.2

CS Leaves-Based Preparation of NPs

NPs	Size (nm)	Morphology	Applications	References
CeO$_2$	13	Spherical Fluorite	Antibacterial activity	(Rajeshkumar & Naik, 2017)
FeO$_3$	161	Spherical	Antioxidant	(Singh et al., 2020)
Ag	86	Spherical	Biological evaluation	(Alsubki et al., 2021)
Au	70		Optical studies	(Rao, 2011)
Ag	37	Spherical	Anti-acne, anti-dandruff, and anti-breast cancer efficacy	(Sathishkumar et al., 2016)
Au	57	-	Biomedical and biotechnological	(Narayanan & Sakthivel, 2008)
Ag	6	Face-centered-cubic	-	(Khan et al., 2018)
Ag		-	Antimicrobial activity	(Mandal & Mandal, 2015)
Cu	13	-	Biocatalyst	(Kulkarni et al., 2015)
CDs	3	Spherical	Antioxidants, sensors and bioimaging agents	(Chan et al., 2018)
Ag	11	Globular shape	Antibacterial	(Ashraf et al., 2019)
Ni	30	-	Catalytic	(Vasudeo & Pramod, 2016)
Ag	5	-	Antibacterial	(Luna et al., 2016)
ZnO	128	-	Free radical scavenging	(Siregar et al., 2017)
Ag	35	Cubic	Biological characteristics	(Senthilkumar et al., 2018)
Ag	26	-	Nonlinear optics	(Sathyavathi et al., 2010)
Ag	60	-	Energy and waste	(Kannan Bhagyanathan & Ernest Thoppil, 2018)
ZnO	-	-	Photocatalytic degradation	(Hassan et al., 2015)
FeO$_3$	90	Spherical	Antimicrobial activity	(Sathya et al., 2017)

myrcene (1.6%), nerd (0.5%), sabinene (2.8%), terpinene (0.9%), terpinolene (0.5%), methyl eugenol (0.2%) geraniol (0.5%), and trans-nerolidol (2.7%).

The α-terpinyl acetate and 1,8-cineole are responsible for the aroma in CS. The CS oil is used in perfumery products and food, and as a carminative and flavor. CS is also used in pharmaceuticals. It is a powerful antiseptic, aromatic, carminative, expectorant, stimulant, antispasmodic, stomachic, and diuretic. Several plant extract materials can reduce the metal ions and make the production of NPs cost-effective (Alsubki et al., 2021). CS is one of the common plant spices and a major source of valuable ingredients, with hypolipidemic, antimicrobial, insecticidal effect, and hypoglycemic action (Mandal & Mandal, 2015). Due to its unique properties, CS has also been used to prepare NPs. Table 27.2 shows the reported methods for the CS leaves-based preparation of NPs.

27.4 APPLICATION OF CS EXTRACT IN THE SYNTHESIS OF METAL-NPs

CS contains a rich amount of phytochemicals; due to this unique natural property, it is an extraordinary material for synthesizing highly stable NPs. Several papers have reported the CS-based synthesis of NPs and their wide applications in different fields such as gold nanoparticles (AuNPs), silver nanoparticles (AgNPs), copper nanoparticles (CuNPs), etc. Figure 27.3 shows the CS-based NPs applications.

27.4.1 SILVER NANOPARTICLES (AGNPS)

AgNPs have attracted intensive research interest due to their unique properties. These include high-efficiency, cost-effective electronic properties, and thermal, unique optical, and physiochemical properties.

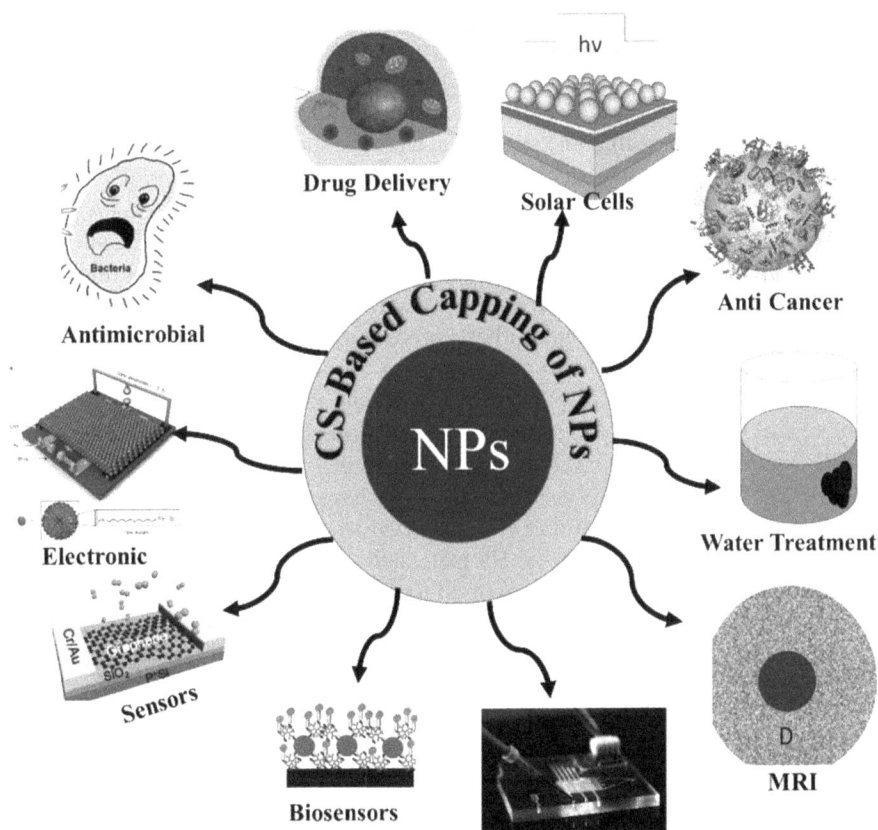

FIGURE 27.3 Applications of CS-based NPs.

These lead to various applications in different fields such as antimicrobial agents, catalysts, sensors, drug delivery, pharmaceutical industry, cosmetics, food industry, orthopedics, diagnostics, and anti-cancer agents (Chernousova & Epple, 2013). AgNPs have been communally used in different wound dressings, textiles, keyboards, and biomedical devices (Li et al., 2014). Nano-size AgNPs have unique characteristics and can substantially vary in their biological and physio-chemical properties due to their high surface-to-volume ratio; therefore, AgNPs have been explored for different purposes. Different traditional methods have been used to prepare AgNPs, such as physical and chemical methods (Mahdavi et al., 2013), but these methods are expensive, need extra operational protocols, and use hazardous chemicals. To highlight these limitations, the researchers pay attention to the biological method. Subsequently, biologically fabricated AgNPs show excellent stability and high yield with well-defined morphology and size under optimal conditions (Gurunathan et al., 2015). The biological procedure is easy, non-toxic, rapid, dependable, and follows the green chemistry protocols. The biosynthesis of AgNPs using different types of microorganisms, including bacteria and fungi, and plants, is the biomaterial for the preparation of AgNPs. Medicinal plants like CS are the best material for the preparation of AgNPs. Yakout et al. proposed a easy and eco-friendly procedure for synthesizing AgNPs using CS leaf extract as a bio-reducing and capping agent to stabilize the AgNPs. The effect of time and preparation mechanisms has been checked during the synthesis process. The synthesized AgNPs were confirmed by changing color from colorless to dark brown. The silver nitrate was used as a precursor of AgNPs and extract of CS. The reaction was completed in one hour without using any other substances. The prepared material has been characterized by UV/Vis spectroscopy (Yakout et al., 2014).

Luna et al. developed a new method for synthesizing AgNPs using CS leaf extracts using the wet chemical method. The resultant AgNPs were ultrafine nanoparticles and nanoclusters with tree-like morphology. Prepared AgNPs have been used for antibacterial activity (Gram-positive and Gram-negative) (Luna et al., 2016). Senthilkumar et al. reported the CS-based preparation of AgNPs. The formation of AgNPs was confirmed by UV/Vis. The maximum absorbance was shown at 400–450 nm. The average size obtained was 35–53 nm. The X-ray diffraction (XRD) results confirmed that the AgNPs were crystalline with cubic structure. After the successful synthesis and characterization, the AgNPs were used as antibacterial against the *Klebsiella pneumonia* (Gram -ve) bacteria (Senthilkumar et al., 2018). Khan et al. proposed a biological method for preparing AgNPs using CS leaf extract. The average size of AgNPs obtained was up to 6.45 nm. A Fourier transform infrared spectroscopy (FTIR) study shows different components capped and stabilized the AgNPs present in the CS extract, such as protein, amine, hydroxyl, and carbonyl functional groups (Khan et al., 2018). Ashraf et al. developed a biosynthesis procedure for synthesizing AgNPs using CS leaf extract. After optimizing different parameters such as the amount of salt, temperature, pH, and stabilizing agent associated with CS leaf extract, the average size obtained was up to 11.9 nm. This study aimed to check the anti-microbial activity in Gram-negative and Gram-positive bacteria such as *Pasteurella multocida*, *Bacillus subtilis*, *Staphylococcus aureus*, and *Enterobacter aerogenes*. The fabricated AgNPs were more efficient for the *P. multocida* and *B. subtilis* than *S. aureus* and *E. aerogenes* (Ashraf et al., 2019). Abbas et al. reported an easy AgNPs-based bio-remediation of naphthalene (PAHs) from the wastewater. He fabricated AgNPs using CS leaf extract. After successful synthesis, the CS-based AgNPs were used for the removal of PAHs; the removal efficiency obtained was up to 95% (Abbas et al., 2020).

27.4.2 GOLD NANOPARTICLES (AUNPS)

AuNPs are a promising material with unique tunable electronic and optical and surface plasmon resonance (SPR) properties. Due to their unique properties, AuNPs attract more attention from the research community. AuNPs have a wide range of applications in different fields such as sensing, photothermal therapeutic, imaging, diagnostic, detecting, catalysis, drug delivery, and anti-bacterial activity. Several traditional methods, such as physical, chemical, and laser ablation, are available to synthesize AuNPs (Jagirani et al., 2021). Currently, biosynthesis is receiving more attention due to its being simple, easy, economically beneficial, and environmentally friendly.

Along with the development in biosynthesis, the plant's material, especially of medicinal plants, plays a significant role in reducing and capping AuNPs due to highly active ingredients, including alkaloids, phenols, terpenes, and different essential oils. Narayanan and Sakthivel first proposed a new biosynthesis-based method for preparing AuNPs using CS leaf extract as a reducing and capping agent. The resultant AuNPs were highly stable with spherical morphology with truncated, triangle, and decahedral shapes and 6.75–57.91 nm in size. The proposed synthesis has been completed in 12 h, a rapid synthesis method compared to the reported biosynthesis process (Narayanan & Sakthivel, 2008). The prepared AuNPs have been characterized by different modern analytical techniques such as UV-Vis spectroscopy (UV/Vis), FTIR, XRD, and transmission electron microscopy (TEM). CS-based AuNPs have been applied in the biomedical, non-linear optics, and biotechnological fields (Narayanan & Sakthivel, 2008). Yang Jiao et al. developed a CS leaf extract-based biosynthesis of AuNPs. The absorption peak was observed in the UV/Vis range up to 540–550 nm. According to the results, the AuNPs were spherical and crystalline, with a tiny size of 5.25 nm. According to the FTIR study, the observed peaks were closely linked to flavonoids, anthocyanins, phenols, and benzophenones, and the prepared AuNPs have been used as analgesic agents and antioxidants (Jiao et al., 2021).

27.4.3 COPPER NANOPARTICLES (CUNPS)

CuNPs have broad scope due to the high natural abundance of Cu with cost-effective and straightforward various methods to prepare Cu-based nanomaterials (Huang et al., 2015). CuNPs have been widely applied in different fields such as drug delivery, catalysts in organic synthesis, food preservation, agriculture, water treatment, paint, semiconducting, antimicrobial products, sensors, capacitor materials, sintering additives, construction materials, metal-metal bonding process, and nanometal lubricant additives (Huang et al., 2015). Bulk Cu has a strong background with numerous applications, e.g., optics, and Cu's essential instability controls electronics. The use of CuNPs under atmospheric conditions makes them prone to oxidation. The research community has developed a novel material that increases the CuNPs' stability without losing sensitivity to water, oxygen, and other chemicals. These valuable efforts encourage the investigation into alternative CuNPs with a highly complex structure like copper oxide and core-shell CuNPs. The CuNPs have wide applications in different fields such as organic chemistry, chemical engineering, inorganic chemistry, nanotechnology, and materials chemistry. CuNPs have been prepared with different methods such as reducing microemulsions, thermal decomposition, laser irradiation, vapor deposition, reverse micelles, sonoelectrochemical methods, chemical reduction, and flame spray (Gawande et al., 2015). However, these reported methods have some disadvantages, such as expensive, time-consuming reaction processes requiring different protocols. Besides these reported methods, the biosynthesis of CuNPs has been reported from different ingredients such as milk, and herbal extract, lemon juice and tamarind, *Syzygium aromaticum*, *Magnolia kobus*, and *Ocimum sanctum* (Burda et al., 2005). From these plants, CS has been considered an efficient plant extract material for the preparation of CuNPs, due to the extraordinary properties and rich amount of natural products. CS was used as a reducing and capping agent during the fabrication of CuNPs. Different researchers have reported CS-based synthesis of CuNPs and their applications. Vasudeo Kulkarni et al. developed a new CS plant leaf extract-based biosynthesis of CuNPs. During synthesis, copper sulfate was used as a starting precursor, and CS leaf extract was used as a reducing and capping agent. The resultant material was obtained up to 13 nm in the range according to the results. The CS extract is highly efficient towards the Cu compared to the reported materials; the prepared material has been used for catalytic applications (Kulkarni et al., 2015).

27.4.4 Nickel Nanoparticles (NiNPs)

NiNPs have remarkable operational simplicity, high reactivity, bacterial resistance, excellent biocompatibility, economic benefits, anti-inflammatory activities, abundance, and environmental compatibilities. NiNPs attracted tremendous interest due to their catalyst, photo-catalyst, electro-catalyst, heat-exchanger, and biosensor properties (Bibi et al., 2017; Seo et al., 2018). Several methods have been reported to prepare NiNPs, such as photocatalytic reduction, discharge route, chemical reduction processes, etc. (El-Khatib et al., 2018). Most of these synthesis processes have disadvantages; for example they are time-consuming and require complex reaction processes and high temperatures. However, the biosynthesis process is highly efficient in NiNPs preparation. The plant-based extract plays a significant role in the preparation of NPs. Vasudeo and Pramod reported CS leaf extract-based preparation of NiNPs using the biosynthesis route. Nickel chloride has been used as a precursor and CS leaf extract as a reducing and capping agent for synthesizing NiNPs at room temperature under normal conditions. The resultant NiNPs were smaller in size up to 30.7 nm than the reported methods and were successfully applied as biocatalysts (Vasudeo & Pramod, 2016).

27.4.5 Zinc Oxide Nanoparticles (ZnO-NPs)

ZnO-NPs are n-type semiconducting materials that have drawn interest in past years due to the wide range of applications in the field of optical, electronics, and biomedical systems (Arthur & Uzairu, 2019). ZnO-NPs are easy to prepare and economically beneficial. ZnO-NPs exhibits remarkable semiconducting properties due to large bandgap (3.37 eV) and higher excitation bonding energy (60 meV) with high catalytic activity, UV filtering, optical properties, wound healing, and anti-inflammatory activity (Mirzaei & Darroudi, 2017). Due to its UV filtering characteristics, it has been significantly used in cosmetics products (sunscreen lotions) (Wodka et al., 2010). ZnO-NPs have also been used in biomedical applications, due to their anti-cancer, drug delivery, antibacterial, anti-diabetic, agricultural, and antifungal properties (Hameed et al., 2016). ZnO-NPs have strong microbial activity at a very low concentration towards Gram-negative and Gram-positive bacteria, as reported (Venkatachalam et al., 2017). They have been used to manufacture rubber, paints, and as efficient adsorbent material for toxic metals such as arsenic, sulfur, and proteins and used in dental materials. ZnO-NPs have piezoelectric and pyroelectric characteristics. They are also used to subtract aquatic weeds, which resist all eradication methods like physiochemical and mechanical means (Rajeshkumar, 2016). The biosynthesis of ZnO-NPs is a unique route due to the simple, easy, and environmentally friendly plant extract used to prepare ZnO-NPs. The resultant ZnO-NPs have been found in a spherical shape, highly stable, and easy to prepare. The bio-reduction comprises the reduction of metal oxides to 0 valence NPs. Due to the rich amount of phytochemicals in plants, they play an essential role in preparing ZnO-NPs. Plants contain various natural products (phytochemicals) such as polyphenolic, polysaccharides, amino acids, vitamins, terpenoids, and alkaloids (Qu et al., 2011). CS is one of the plants that contain a rich amount of phytochemicals, also called medicinal plants. It also has excellent reducing and capping properties during the preparation of ZnO-NPs.

Different researchers also reported CS leaf extract for the synthesis of ZnO-NPs. Gnanasangeetha and Thambavani proposed a new green and chemical method to prepare ZnO-NPs using CS leaf extract as a precursor with zinc acetate solutions for the pH adjustment using sodium hydroxide. The prepared ZnO-NPs were up to 66 nm with the biosynthesis method, while with the chemical method, the average size was up to 81 nm. According to the results, the biogenic green synthesis was best as it is eco-friendly and due to the good average size of particles. The CS-based ZnO-NPs have been used in different fields such as sensors, catalysis, medical, biotechnology, DNA labeling, optical devices, water remediation, and drug delivery (Gnanasangeetha & SaralaThambavani, 2013). Siregar et al. proposed a new study on the synthesis and free radical scavenging activity of ZnO-NPs. The rich phytochemical plant (CS) has been used to reduce zinc metal into ZnO-NPs. This study aims to check the effect of different solvent polarities used on the CS leaf reducing power and free radical scavenging capability of prepared ZnO-NPs and calcination effect at 300 and 500°C. After calcination, the average size of particles was up to 128.1 nm to 552.3 nm (Siregar et al., 2017).

27.4.6 CERIUM OXIDE NANOPARTICLES (CeO$_2$-NPs)

CeO$_2$-NPs exhibit unique antioxidant characteristics *in vivo* and *in vitro* with novel self-regeneration of surface properties. At the nanoscale, the CeO$_2$-NPs have been used in several fields such as chemical mechanical planarization/polishing (Reed et al., 2014), fuel oxidation catalysis, corrosion protection, and automotive exhaust treatment, oxidase, peroxidase, and phosphatase, solar cells, catalase, nitric oxide radicals, peroxynitrite, and scavenging hydroxyl radicals. Different synthesis protocols have been used to prepare CeO$_2$-NPs, such as chemical, mechanochemical, sonochemical, hydrothermal, semi-batch reactor, sol-gel, spray-pyrolysis, and micro-emulsion protocols (Farahmandjou et al., 2016). The biological method is simple, easy, eco-friendly, and cheap and follows the green analytical chemistry protocols. Along with the development in the biological method, plant-based materials have significant importance. The metals (cerium) produce CeO$_2$-NPs with high stability and very small particles using the CS extract. Rajeshkumar et al. developed the CS-based biosynthesis of CeO$_2$-NPs. Prepared CeO$_2$-NPs were confirmed from UV/Vis; the surface plasmon resonance (SPR) was found at 276 nm. The FTIR spectra confirmed the functionality present in the biomolecules. According to the XRD, the particles had a cubic fluorite structure and were crystalline in nature, with an average size of 13.56 nm. After the successful synthesis and characterization, CeO$_2$-NPs have been used for their antibacterial activity (Rajeshkumar & Naik, 2017).

27.4.7 IRON OXIDE NANOPARTICLES (FeO$_3$-NPs)

FeO$_3$-NPs are a unique type of metal oxide nanoparticles due to their superparamagnetic properties, low toxicity, high surface area to volume ratio, ease of the preparation process, and separation from the solution using an external magnetic field (no need to centrifuge). FeO_3-NPs are widely used in different fields such as magnetic resonance imaging (MRI), biomedical, protein immobilization, drug delivery, sensing, catalysis, thermal therapy and macroscopic applications, detoxification, tissue repair, and hyperthermia (Abegunde et al., 2020). Prepared FeO_3-NPs have unique shapes: porous spheres, nanorods, nanocubes, nanohusks, self-oriented flowers, and distorted cubes. The surface modification is a key step during the synthesis of FeO_3-NPs, as it plays a significant role in controlling the shape, size, and morphology. In addition, the proper functionalization and molecular conjugation result in high stability with high biocompatibility (Sheng-Nan et al., 2014). FeO_3-NPs have been prepared using pyrolysis, electrodeposition, sol-gel synthesis, hydrothermal, reverse micelle, and co-precipitation. These reported methods have multiple steps with limitations, requiring different protocols with expensive instruments, poor size control property, and high temperature and pressure.

Furthermore, the biosynthesis process has unique characteristics; it is easy, environmentally friendly, and has low economic constraints. CS biomaterials have novel properties to control the size and shape and minimize the extra protocols during FeO_3-NPs preparation. The presence of bio-reductant and capping agents in the plant materials controls the size of the prepared NPs (Hoag et al., 2009). Singh et al. developed a new biosynthesis method to prepare FeO_3-NPs using CS leaf extract under the optimized conditions (metal salt amount and CS leaf extract volume). The resultant FeO_3-NPs were highly stable and smaller in size, up to 161.5 nm. The prepared NPs have been used in biomedical applications (Singh et al., 2020).

27.4.8 CARBON DOT NANOPARTICLES (CDS-NPS)

Carbonaceous materials are used to prepare carbon-based NPs. Xu et al. discovered the CDs-NPs during the separation of the single-walled carbon nanotubes from the carbon soot using the discharge process (Xu et al., 2004). CDs are spherical nanocarriers with a diameter of <10 nm. CDs have a carbon core with different functionalities such as (–OH, –COOH, –NH$_2$), which provides high stability, higher biological activity, and excellent efficiency in conjugating with organic and inorganic compounds. This is due to CDs' novel properties, such as smaller size, ease of preparation and modification, high degree of oxidation, low toxicity, and excellent water retention property (Lim et al., 2015). Carbon quantum dots (CQDs) have been significantly used in different fields such as targeted drug delivery, cancer treatment, fluorescent labeling, imaging, heavy metal ions determination, and bacterial labeling. Still, different types of CDs have been reported, like carbon nanodots (CDs), CQDs, graphene quantum dots (GQDs), and carbonized polymer dots (CPDs) depending upon morphology, carbon core, characteristics, and surface group. Several methods have been used to synthesize CDs, including polymerization, dehydration, arc-discharge, laser ablation carbonization, electrochemical oxidation, hydrothermal, and microwave pyrolysis (Chan et al., 2018). These reported methods have some limitations, such as requiring sophisticated and energy-consuming, expensive instruments, and also toxic chemicals. Still, the green synthesis of CDs-NPs is an attractive research area that exploits renewable natural carbon compounds. However, it always exists to discover greener sources for CDs-NPs due to simplicity, clean, inexpensive, non-toxic, and easily available. Accordingly, CS can functionalize the CDs-NPs. Chen et al. reported CS-based synthesis fluorescent CDs-NPs with superior optical properties and established excitation, solvent, and pH-dependent emission performance. CS-based CDs-NPs had excellent antioxidant properties with good selectivity during the detection of selective Fe^{3+} ions. Due to low toxicity, it may be used for bio-imaging under *in vitro* conditions (Chan et al., 2018).

27.5 CONCLUSION

The traditional process for the synthesis of NPs is costly, uses toxic chemicals, produces by-products, and violates green chemistry rules. The alternative approach found for the synthesis of NPs is green synthesis. Plant-based materials have unique potential due to the rich amount of phytochemicals, and they are a potential candidate for the preparation of NPs. As long as development in plant-based extraction, the CS received intensive attention from the research community to prepare NPs. The CS plant material-based preparation of NPs is simple, easy, cost-effective, convenient, eco-friendly, and follows the green chemistry protocols to minimize toxic chemical substances. In addition, the CS-based NPs can provide high stability in terms of size, shape, and structure and obtain a high yield of synthesized NPs compared to the traditional synthesis methods and have been applied in several fields.

REFERENCES

Abbas, S., Nasreen, S., Haroon, A., & Ashraf, M.A. (2020) Synthesis of silver and copper nanoparticles from plants and application as adsorbents for naphthalene decontamination. *Saudi Journal of Biological Sciences* 27:1016–1023. doi:10.1016/j.sjbs.2020.02.011.

Abdelghany, T., Al-Rajhi, A.M., Al Abboud, M.A., Alawlaqi, M., Magdah, A.G., Helmy, E.A., & Mabrouk, A.S. (2018) Recent advances in green synthesis of silver nanoparticles and their applications: About future directions. A review. *BioNanoScience* 8:5–16. doi:10.1007/s12668-017-0413-3.

Abegunde, S.M., Idowu, K.S., & Sulaimon, A.O. (2020) Plant-mediated iron nanoparticles and their applications as adsorbents for water treatment-a review. *Journal of Chemical Reviews* 2:103–113. doi:10.33945/SAMI/JCR.2020.2.3.

Alsubki, R., Tabassum, H., Abudawood, M., Rabaan, A.A., Alsobaie, S.F., & Ansar, S. (2021) Green synthesis, characterization, enhanced functionality and biological evaluation of silver nanoparticles based on *Coriander sativum. Saudi Journal of Biological Sciences* 28:2102–2108. doi:10.1016/j.sjbs.2020.12.055.

Altikatoglu, M., Attar, A., Erci, F., Cristache, C.M., & Isildak, I. (2017) Green synthesis of copper oxide nanoparticles using *Ocimum basilicum* extract and their antibacterial activity. *Fresenius Environmental Bulletin* 25:7832–7837.

Arthur, D.E., & Uzairu, A. (2019) Molecular docking studies on the interaction of NCI anticancer analogues with human Phosphatidylinositol 4, 5-bisphosphate 3-kinase catalytic subunit. *Journal of King Saud University-Science* 31:1151–1166. doi:10.1016/j.jksus.2019.01.011.

Ashraf, A., Zafar, S., Zahid, K., Shah, M.S., Al-Ghanim, K.A., Al-Misned, F., & Mahboob, S. (2019) Synthesis, characterization, and antibacterial potential of silver nanoparticles synthesized from *Coriandrum sativum* L. *Journal of Infection and Public Health* 12:275–281. doi:10.1016/j.jiph.2018.11.002.

Ashraf, R., Ghufran, S., Akram, S., Mushtaq, M., & Sultana, B. (2020) Cold pressed coriander (*Coriandrum sativum* L.) seed oil. In *Cold pressed oils.* Elsevier, pp. 345–356.

Baraka, A., et al. (2017) Synthesis of silver nanoparticles using natural pigments extracted from Alfalfa leaves and its use for antimicrobial activity. *Chemical Papers* 71:2271–2281. doi:10.1007/s11696-017-0221-9.

Baruwati, B., Polshettiwar, V., & Varma, R.S. (2009) Glutathione promoted expeditious green synthesis of silver nanoparticles in water using microwaves. *Green Chemistry* 11:926–930. doi:10.1039/B902184A.

Bhat, S., Kaushal, P., Kaur, M., & Sharma, H. (2014) Coriander (*Coriandrum sativum* L.): Processing, nutritional and functional aspects. *African Journal of Plant Science* 8:25–33. doi:10.5897/AJPS2013.1118.

Bibi, I., et al. (2017) Nickel nanoparticle synthesis using *Camellia sinensis* as reducing and capping agent: Growth mechanism and photo-catalytic activity evaluation. *International Journal of Biological Macromolecules* 103:783–790. doi:10.1016/j.ijbiomac.2017.05.023.

Burda, C., Chen, X., Narayanan, R., & El-Sayed, M.A. (2005) Chemistry and properties of nanocrystals of different shapes. *Chemical Reviews* 105:1025–1102. doi:10.1021/cr030063a.

Carolin, C.F., Kumar, P.S., Saravanan, A., Joshiba, G.J., & Naushad, M. (2017) Efficient techniques for the removal of toxic heavy metals from aquatic environment: A review. *Journal of Environmental Chemical Engineering* 5:2782–2799. doi:10.1016/j.jece.2017.05.029.

Chan, K.K., Yap, S.H.K., & Yong, K.-T. (2018) Biogreen synthesis of carbon dots for biotechnology and nanomedicine applications. *Nano-Micro Letters* 10:1–46. doi:10.1007/s40820-018-0223-3.

Chernousova, S., & Epple, M. (2013) Silver as antibacterial agent: Ion, nanoparticle, and metal. *Angewandte Chemie International Edition* 52:1636–1653. doi:10.1002/anie.201205923.

Clark, J.H., & Macquarrie, D.J. (2008) *Handbook of green chemistry and technology.* John Wiley & Sons.

da Silva, S.B., Amorim, M., Fonte, P., Madureira, R., Ferreira, D., Pintado, M., & Sarmento, B. (2015) Natural extracts into chitosan nanocarriers for rosmarinic acid drug delivery *Pharmaceutical Biology* 53:642–652. doi:10.3109/13880209.2014.935949.

Dash, B.K. (2011) Antibacterial activities of methanol and acetone extracts of fenugreek (*Trigonella foenum*) and coriander (*Coriandrum sativum*). *Life Sciences and Medicine Research.* 2011: LSMR-27.

Dayem, A.A., et al. (2017) The role of reactive oxygen species (ROS) in the biological activities of metallic nanoparticles. *International Journal of Molecular Sciences* 18. doi:10.3390/ijms18010120.

El-Khatib, A.M., Badawi, M.S., Roston, G.D., Moussa, R.M., & Mohamed, M.M. (2018) Structural and magnetic properties of nickel nanoparticles prepared by arc discharge method using an ultrasonic nebulizer. *Journal of Cluster Science* 29:1321–1327. doi:10.1007/s10876-018-1451-x.

Emamhadi, M.A., Sarafraz, M., Akbari, M., Fakhri, Y., Linh, N.T.T., & Khaneghah, A.M. (2020) Nanomaterials for food packaging applications: A systematic review. *Food and Chemical Toxicology*:111825. doi:10.1016/j.fct.2020.111825.

Fair, R.J., & Tor, Y. (2014) Antibiotics and bacterial resistance in the 21st century. *Perspectives in Medicinal Chemistry* 6:S14459.

Farahmandjou, M., Zarinkamar, M., & Firoozabadi, T. (2016) Synthesis of cerium oxide (CeO_2) nanoparticles using simple CO-precipitation method. *Revista Mexicana de Física* 62:496–499.

Fathi, M., Mozafari, M.R., & Mohebbi, M. (2012) Nanoencapsulation of food ingredients using lipid based delivery systems. *Trends in Food Science & Technology* 23:13–27. doi:10.1016/j.tifs.2011.08.003.

Favier, I., Pla, D., & Gómez, M. (2019) Palladium nanoparticles in polyols: Synthesis, catalytic couplings, and hydrogenations. *Chemical Reviews* 120:1146–1183. doi:10.1021/acs.chemrev.9b00204.

Gao, C., Lyu, F., & Yin, Y. (2020) Encapsulated metal nanoparticles for catalysis. *Chemical Reviews* 121:834–881. doi:10.1021/acs.chemrev.0c00237.

Gao, Y., Huang, Q., Su, Q., & Liu, R. (2014) Green synthesis of silver nanoparticles at room temperature using kiwifruit juice. *Spectroscopy Letters* 47:790–795. doi:10.1080/00387010.2013.848898.

Garibo, D. et al. (2020) Green synthesis of silver nanoparticles using *Lysiloma acapulcensis* exhibit high-antimicrobial activity. *Scientific Reports* 10:1–11. doi:10.1038/s41598-020-69606-7.

Gawande, M.B., Zboril, R., Malgras, V., & Yamauchi, Y. (2015) Integrated nanocatalysts: A unique class of heterogeneous catalysts. *Journal of Materials Chemistry A* 3:8241–8245. doi:10.1039/C5TA00119F.

Gnanasangeetha, D., & SaralaThambavani, D. (2013) One pot synthesis of zinc oxide nanoparticles via chemical and green method. *Research Journal of Material Sciences* 2320:6055.

Gogoi, N., Babu, P.J., Mahanta, C., & Bora, U. (2015) Green synthesis and characterization of silver nanoparticles using alcoholic flower extract of *Nyctanthes arbortristis* and in vitro investigation of their antibacterial and cytotoxic activities. *Materials Science and Engineering: C* 46:463–469. doi:10.1016/j.msec.2014.10.069.

Guilger-Casagrande, M., & Lima, R.d. (2019) Synthesis of silver nanoparticles mediated by fungi: A review. *Frontiers in Bioengineering and Biotechnology* 7:287. doi:10.3389/fbioe.2019.00287.

Gurunathan, S., Park, J.H., Han, J.W., & Kim, J.-H. (2015) Comparative assessment of the apoptotic potential of silver nanoparticles synthesized by *Bacillus tequilensis* and Calocybe indica in MDA-MB-231 human breast cancer cells: Targeting p53 for anticancer therapy. *International Journal of Nanomedicine* 10:4203. doi:10.2147/IJN.S83953.

Hameed, A.S.H., Karthikeyan, C., Ahamed, A.P., Thajuddin, N., Alharbi, N.S., Alharbi, S.A., & Ravi, G. (2016) In vitro antibacterial activity of ZnO and Nd doped ZnO nanoparticles against ESBL producing *Escherichia coli* and *Klebsiella pneumoniae*. *Scientific Reports* 6:1–11. doi:10.1038/srep24312.

Hamelian, M., Hemmati, S., Varmira, K., & Veisi, H. (2018) Green synthesis, antibacterial, antioxidant and cytotoxic effect of gold nanoparticles using *Pistacia atlantica* extract. *Journal of the Taiwan Institute of Chemical Engineers* 93:21–30. doi:10.1016/j.jtice.2018.07.018.

Harada, M., Kuwa, M., Sato, R., Teranishi, T., Takahashi, M., & Maenosono, S. (2020) Cation distribution in monodispersed MFe2O4 (M= Mn, Fe, Co, Ni, and Zn) nanoparticles investigated by X-ray absorption fine structure spectroscopy: Implications for magnetic data storage, catalysts, sensors, and ferrofluids. *ACS Applied Nano Materials* 3:8389–8402. doi:10.1021/acsanm.0c01810.

Hassan, S.S., El Azab, W.I., Ali, H.R., & Mansour, M.S. (2015) Green synthesis and characterization of ZnO nanoparticles for photocatalytic degradation of anthracene. *Advances in Natural Sciences: Nanoscience and Nanotechnology* 6:045012. doi:10.1088/2043-6262/6/4/045012.

He, Z., Zhang, Z., & Bi, S. (2020) Nanoparticles for organic electronics applications *Materials Research Express* 7:012004.

Hoag, G.E., Collins, J.B., Holcomb, J.L., Hoag, J.R., Nadagouda, M.N., & Varma, R.S. (2009) Degradation of bromothymol blue by 'greener' nano-scale zero-valent iron synthesized using tea polyphenols. *Journal of Materials Chemistry* 19:8671–8677. doi:10.1039/B909148C.

Huang, H. et al. (2015) Catalytic oxidation of gaseous benzene with ozone over zeolite-supported metal oxide nanoparticles at room temperature. *Catalysis Today* 258:627–633. doi:10.1016/j.cattod.2015.01.006.

Hulkoti, N.I., & Taranath, T. (2014) Biosynthesis of nanoparticles using microbes-a review. *Colloids and Surfaces B: Biointerfaces* 121:474–483. doi:10.1016/j.colsurfb.2014.05.027.

Iravani, S. (2011) Green synthesis of metal nanoparticles using plants. *Green Chemistry* 13:2638–2650. doi:10.1039/C1GC15386B.

Ismael, M. (2020) A review and recent advances in solar-to-hydrogen energy conversion based on photocatalytic water splitting over doped-TiO$_2$ nanoparticles. *Solar Energy* 211:522–546. doi:10.1016/j.solener.2020.09.073.

Jadoun, S., Arif, R., Jangid, N.K., & Meena, R.K. (2021) Green synthesis of nanoparticles using plant extracts: A review. *Environmental Chemistry Letters* 19:355–374. doi:10.1007/s10311-020-01074-x.

Jagirani, M.S. et al. (2021) Functionalized gold nanoparticles based optical, surface plasmon resonance-based sensor for the direct determination of mitoxantrone anti-cancer agent from real samples. *Journal of Cluster Science*:1–7. doi:10.1007/s10876-020-01948-8.

Jamkhande, P.G., Ghule, N.W., Bamer, A.H., & Kalaskar, M.G. (2019) Metal nanoparticles synthesis: An overview on methods of preparation, advantages and disadvantages, and applications. *Journal of Drug Delivery Science and Technology* 53:101174. doi:10.1016/j.jddst.2019.101174.

Jeyaraj, M., Gurunathan, S., Qasim, M., Kang, M.-H., & Kim, J.-H. (2019) A comprehensive review on the synthesis, characterization, and biomedical application of platinum nanoparticles. *Nanomaterials* 9:1719. doi:10.3390/nano9121719.

Jiao, Y., Wang, X., & Chen, J.-h. (2021) Biofabrication of AuNPs using *Coriandrum sativum* leaf extract and their antioxidant, analgesic activity. *Science of the Total Environment* 767:144914. doi:10.1016/j.scitotenv.2020.144914.

Kahrilas, G.A., Wally, L.M., Fredrick, S.J., Hiskey, M., Prieto, A.L., & Owens, J.E. (2014) Microwave-assisted green synthesis of silver nanoparticles using orange peel extract. *ACS Sustainable Chemistry & Engineering* 2:367–376. doi:10.1021/sc4003664.

Kannan Bhagyanathan, N., & Ernest Thoppil, J. (2018) Plant-mediated synthesis of silver nanoparticles by two species of *Cynanchum* L. (Apocynaceae): A comparative approach on its physical characteristics. *International Journal of Nano Dimension* 9:104–111. doi:20.1001.1.20088868.2018.9.2.1.7.

Karimi-Maleh, H., Cellat, K., Arıkan, K., Savk, A., Karimi, F., & Şen, F. (2020) Palladium-Nickel nanoparticles decorated on Functionalized-MWCNT for high precision non-enzymatic glucose sensing. *Materials Chemistry and Physics* 250:123042. doi:10.1016/j.matchemphys.2020.123042.

Khan, M., Tareq, F., Hossen, M., & Roki, M. (2018) Green synthesis and characterization of silver nanoparticles using *Coriandrum* sativum leaf extract. *Journal of Engineering Science and Technology* 13:158–166.

Krishnaraj, C., Jagan, E.G., Rajasekar, S., Selvakumar, P., Kalaichelvan, P.T., & Mohan, N. (2010) Synthesis of silver nanoparticles using *Acalypha indica* leaf extracts and its antibacterial activity against water borne pathogens. *Colloids and Surfaces B: Biointerfaces* 76:50–56. doi:10.1016/j.colsurfb.2009.10.008.

Kulkarni, V., Kale, N., Kute, N., & Kulkarni, P. (2015) Coriander leaf extract is efficient biocatalyst for synthesis of copper nanoparticles. *ChemXpress* 8:127–132.

Kumar, D., Kumar, G., & Agrawal, V. (2018) Green synthesis of silver nanoparticles using *Holarrhena antidysenterica* (L.) Wall. bark extract and their larvicidal activity against dengue and filariasis vectors. *Parasitology Research* 117:377–389. doi:10.1007/s00436-017-5711-8.

Li, C. et al. (2014) In vivo real-time visualization of tissue blood flow and angiogenesis using Ag$_2$S quantum dots in the NIR-II window. *Biomaterials* 35:393–400. doi:10.1016/j.biomaterials.2013.10.010.

Li, S., Shen, Y., Xie, A., Yu, X., Qiu, L., Zhang, L., & Zhang, Q. (2007) Green synthesis of silver nanoparticles using *Capsicum annuum* L. extract. *Green Chemistry* 9:852–858. doi:10.1039/B615357G.

Li, W. et al. (2020) High pyroelectric effect in poly (vinylidene fluoride) composites cooperated with diamond nanoparticles. *Materials Letters* 267:127514. doi:10.1016/j.matlet.2020.127514.

Lim, S.Y., Shen, W., & Gao, Z. (2015) Carbon quantum dots and their applications. *Chemical Society Reviews* 44:362–381. doi:10.1039/C4CS00269E.

Luna, C., Barriga-Castro, E.D., Gómez-Treviño, A., Núñez, N.O., & Mendoza-Reséndez, R. (2016) Microstructural, spectroscopic, and antibacterial properties of silver-based hybrid nanostructures biosynthesized using extracts of coriander leaves and seeds. *International Journal of Nanomedicine* 11:4787. doi:10.2147/IJN.S105166.

Mahdavi, M., Ahmad, M.B., Haron, M.J., Namvar, F., Nadi, B., Rahman, M.Z.A., & Amin, J. (2013) Synthesis, surface modification and characterisation of biocompatible magnetic iron oxide nanoparticles for biomedical applications. *Molecules* 18:7533–7548. doi:10.3390/molecules18077533

Mandal, D., Bolander, M.E., Mukhopadhyay, D., Sarkar, G., & Mukherjee, P. (2006) The use of microorganisms for the formation of metal nanoparticles and their application. *Applied Microbiology and Biotechnology* 69:485–492. doi:10.1007/s00253-005-0179-3.

Mandal, S., & Mandal, M. (2015) Coriander (*Coriandrum sativum* L.) essential oil: Chemistry and biological activity. *Asian Pacific Journal of Tropical Biomedicine* 5:421–428. doi:10.1016/j.apjtb.2015.04.001.

Menazea, A., & Ahmed, M. (2020) Synthesis and antibacterial activity of graphene oxide decorated by silver and copper oxide nanoparticles. *Journal of Molecular Structure* 1218:128536. doi:10.1016/j.molstruc.2020.128536.

Mirzaei, H., & Darroudi, M. (2017) Zinc oxide nanoparticles: Biological synthesis and biomedical applications. *Ceramics International* 43:907–914. doi:10.1016/j.ceramint.2016.10.051.

Mitchell, M.J., Billingsley, M.M., Haley, R.M., Wechsler, M.E., Peppas, N.A., & Langer, R. (2021) Engineering precision nanoparticles for drug delivery. *Nature Reviews Drug Discovery* 20:101–124. doi:10.1038/s41573-020-0090-8.

Mochochoko, T., Oluwafemi, O.S., Jumbam, D.N., & Songca, S.P. (2013) Green synthesis of silver nanoparticles using cellulose extracted from an aquatic weed; water hyacinth *Carbohydrate Polymers* 98:290–294. doi:10.1016/j.carbpol.2013.05.038.

Mondal, S., Roy, N., Laskar, R.A., Sk, I., Basu, S., Mandal, D., & Begum, N.A. (2011) Biogenic synthesis of Ag, Au and bimetallic Au/Ag alloy nanoparticles using aqueous extract of mahogany (*Swietenia mahogani* JACQ.) leaves. *Colloids and Surfaces B: Biointerfaces* 82:497–504. doi:10.1016/j.colsurfb.2010.10.007.

Muthuchudarkodi R.R., & Ashli A. (2020) Green synthesis of undoped and Ti doped CeO_2 nanoparticles by the fruit extract of emblica officinalis and its photocatalytic activity. *International Journal of Chemical Concepts* 6: 43–49.

Narayanan, K.B., & Sakthivel, N. (2008) Coriander leaf mediated biosynthesis of gold nanoparticles. *Materials Letters* 62:4588–4590. doi:10.1016/j.matlet.2008.08.044.

Nath, D., & Banerjee, P. (2013) Green nanotechnology-a new hope for medical biology. *Environmental Toxicology and Pharmacology* 36:997–1014. doi:10.1016/j.etap.2013.09.002.

Ovais, M. et al. (2018) Role of plant phytochemicals and microbial enzymes in biosynthesis of metallic nanoparticles. *Applied Microbiology and Biotechnology* 102:6799–6814. doi:10.1007/s00253-018-9146-7.

Pan, Z., Huang, Y., Qian, H., Du, X., Qin, W., & Liu, T. (2020) Superparamagnetic iron oxide nanoparticles drive miR-485-5p inhibition in glioma stem cells by silencing Tie1 expression. *International Journal of Biological Sciences* 16:1274. doi:10.7150/ijbs.42887.

Pathak, N., Kasture, S., & Bhatt, N. (2011) Phytochemical screening of *Coriander sativum* Linn. *International Journal of Pharmaceutical Sciences Review and Research* 9.

Philip, D., & Unni, C. (2011) Extracellular biosynthesis of gold and silver nanoparticles using Krishna tulsi (*Ocimum sanctum*) leaf. *Physica E: Low-dimensional Systems and Nanostructures* 43:1318–1322. doi:10.1016/j.physe.2010.10.006.

Prathna, T., Chandrasekaran, N., Raichur, A.M., & Mukherjee, A. (2011) Biomimetic synthesis of silver nanoparticles by *Citrus limon* (lemon) aqueous extract and theoretical prediction of particle size. *Colloids and Surfaces B: Biointerfaces* 82:152–159. doi:10.1016/j.colsurfb.2010.08.036.

Punia, P., Bharti, M.K., Chalia, S., Dhar, R., Ravelo, B., Thakur, P., & Thakur, A. (2021) Recent advances in synthesis, characterization, and applications of nanoparticles for contaminated water treatment-a review. *Ceramics International* 47:1526–1550. doi:10.1016/j.ceramint.2020.09.050.

Qu, J., Yuan, X., Wang, X., & Shao, P. (2011) Zinc accumulation and synthesis of ZnO nanoparticles using *Physalis alkekengi* L. *Environmental Pollution* 159:1783–1788. doi:10.1016/j.envpol.2011.04.016.

Rahmani, R., Gharanfoli, M., Gholamin, M., Darroudi, M., Chamani, J., Sadri, K., & Hashemzadeh, A. (2020) Plant-mediated synthesis of superparamagnetic iron oxide nanoparticles (SPIONs) using aloe vera and flaxseed extracts and evaluation of their cellular toxicities. *Ceramics International* 46:3051–3058. doi:10.1016/j.ceramint.2019.10.005.

Rajeshkumar, S. (2016) Synthesis of silver nanoparticles using fresh bark of *Pongamia pinnata* and characterization of its antibacterial activity against gram positive and gram negative pathogens *Resource-Efficient Technologies* 2:30–35. doi:10.1016/j.reffit.2016.06.003.

Rajeshkumar, S., & Naik, P. (2017) Synthesis and biomedical applications of cerium oxide nanoparticles - A review. *Biotechnology Reports* 17:1–5. doi:10.1016/j.reffit.2016.06.003.

Rao, S.V. (2011) Picosecond nonlinear optical studies of gold nanoparticles synthesised using coriander leaves (*Coriandrum sativum*) *Journal of Modern Optics* 58:1024–1029. doi:10.1080/09500340.2011.590903.

Ravindra, S., Mohan, Y.M., Reddy, N.N., & Raju, K.M. (2010) Fabrication of antibacterial cotton fibres loaded with silver nanoparticles via "Green Approach". *Colloids and Surfaces A: Physicochemical and Engineering Aspects* 367.31–40. doi:10.1016/j.colsurfa.2010.06.013.

Razavi, M., Salahinejad, E., Fahmy, M., Yazdimamaghani, M., Vashaee, D., & Tayebi, L. (2015) Green chemical and biological synthesis of nanoparticles and their biomedical applications. In *Green processes for nanotechnology*, pp. 207–235. doi:10.1007/978-3-319-15461-9_7.

Reed, K., Cormack, A., Kulkarni, A., Mayton, M., Sayle, D., Klaessig, F., & Stadler, B. (2014) Exploring the properties and applications of nanoceria: Is there still plenty of room at the bottom? *Environmental Science: Nano* 1:390–405. doi:10.1039/C4EN00079J.

Salem, S.S., & Fouda, A. (2021) Green synthesis of metallic nanoparticles and their prospective biotechnological applications: An overview. *Biological Trace Element Research* 199:344–370. doi:10.1007/s12011-020-02138-3.

Santhoshkumar, T. et al. (2011) Synthesis of silver nanoparticles using Nelumbo nucifera leaf extract and its larvicidal activity against malaria and filariasis vectors. *Parasitology Research* 108:693–702. doi:10.1007/s00436-010-2115-4.

Sathishkumar, P. et al. (2016) Anti-acne, anti-dandruff and anti-breast cancer efficacy of green synthesised silver nanoparticles using *Coriandrum sativum* leaf extract. *Journal of Photochemistry and Photobiology B: Biology* 163:69–76. doi:10.1016/j.jphotobiol.2016.08.005.

Sathya, K., Saravanathamizhan, R., & Baskar, G. (2017) Ultrasound assisted phytosynthesis of iron oxide nanoparticle. *Ultrasonics Sonochemistry* 39:446–451. doi:10.1016/j.ultsonch.2017.05.017.

Sathyavathi, R., Krishna, M.B., Rao, S.V., Saritha, R., & Rao, D.N. (2010) Biosynthesis of silver nanoparticles using *Coriandrum sativum* leaf extract and their application in nonlinear optics. *Advanced Science Letters* 3:138–143. doi:10.1166/asl.2010.1099.

Senthilkumar, N., Aravindhan, V., Ruckmani, K., & Potheher, I.V. (2018) *Coriandrum sativum* mediated synthesis of silver nanoparticles and evaluation of their biological characteristics. *Materials Research Express* 5:055032.

Seo, S., Perez, G.A., Tewari, K., Comas, X., & Kim, M. (2018) Catalytic activity of nickel nanoparticles stabilized by adsorbing polymers for enhanced carbon sequestration. *Scientific Reports* 8:1–11. doi:10.1038/s41598-018-29605-1.

Shahid, M., Dumat, C., Khalid, S., Schreck, E., Xiong, T., & Niazi, N.K. (2017) Foliar heavy metal uptake, toxicity and detoxification in plants: A comparison of foliar and root metal uptake. *Journal of Hazardous Materials* 325:36–58. doi:10.1016/j.jhazmat.2016.11.063.

Sharma, V., Kansal, L., & Sharma, A. (2010) Prophylactic efficacy of *Coriandrum sativum* (Coriander) on testis of lead-exposed mice. *Biological Trace Element Research* 136:337–354. doi:10.1007/s12011-009-8553-0.

Sheikh, J., & Bramhecha, I. (2019) Multi-functionalization of linen fabric using a combination of chitosan, silver nanoparticles and *Tamarindus indica* L. seed coat extract. *Cellulose* 26:8895–8905. doi:10.1007/s10570-019-02684-7.

Sheng-Nan, S., Chao, W., Zan-Zan, Z., Yang-Long, H., Venkatraman, S.S., & Zhi-Chuan, X. (2014) Magnetic iron oxide nanoparticles: Synthesis and surface coating techniques for biomedical applications. *Chinese Physics B* 23:037503.

Singh, J., Mehta, A., Rawat, M., & Basu, S. (2018) Green synthesis of silver nanoparticles using sun dried tulsi leaves and its catalytic application for 4-nitrophenol reduction. *Journal of Environmental Chemical Engineering* 6:1468–1474. doi:10.1016/j.jece.2018.01.054.

Singh, J., Singh, N., Rathi, A., Kukkar, D., & Rawat, M. (2017a) Facile approach to synthesize and characterization of silver nanoparticles by using mulberry leaves extract in aqueous medium and its application in antimicrobial activity. *Journal of Nanostructure* 7: 134–140. doi:10.22052/jns.2017.02.007.

Singh, K., Chopra, D.S., Singh, D., & Singh, N. (2020) Optimization and ecofriendly synthesis of iron oxide nanoparticles as potential antioxidant. *Arabian Journal of Chemistry* 13:9034–9046. doi:10.1016/j.arabjc.2020.10.025.

Singh, T., Shukla, S., Kumar, P., Wahla, V., Bajpai, V.K., & Rather, I.A. (2017b) Application of nanotechnology in food science: Perception and overview. *Frontiers in Microbiology* 8:1501. doi:10.3389/fmicb.2017.01501.

Siregar, T.M., Cahyana, A.H., & Gunawan, R.J. (2017) Characteristics and free radical scavenging activity of zinc oxide (ZnO) Nanoparticles derived from extract of coriander (*Coriandrum sativum* L.). *Reaktor* 17:145–150. doi:10.14710/reaktor.17.3.145-151.

Tiwari, J., Tiwari, R., & Kim, K. (2012) Progress in materials science three-dimensional nanostructured materials for advanced electrochemical energy devices. *Progress in Materials Science* 57:724–803. doi:10.1016/j.pmatsci.2011.08.003.

Varma, R.S. (2012) Greener approach to nanomaterials and their sustainable applications. *Current Opinion in Chemical Engineering* 1:123–128. doi:10.1016/j.coche.2011.12.002.

Vasantharaj, S. et al. (2019) Synthesis of ecofriendly copper oxide nanoparticles for fabrication over textile fabrics: Characterization of antibacterial activity and dye degradation potential. *Journal of Photochemistry and Photobiology B: Biology* 191:143–149. doi:10.1016/j.jphotobiol.2018.12.026.

Vasudeo, K., & Pramod, K. (2016) Biosynthesis of nickel nanoparticles using leaf extract of coriander. *BioTechnology: An Indian Journal* 12:1–6.

Velayutham, K. et al. (2013) Larvicidal activity of green synthesized silver nanoparticles using bark aqueous extract of *Ficus racemosa* against *Culex quinquefasciatus* and *Culex gelidus*. *Asian Pacific Journal of Tropical Medicine* 6:95–101. doi:10.1016/S1995-7645(13)60002-4.

Venkatachalam, P., Jayaraj, M., Manikandan, R., Geetha, N., Rene, E.R., Sharma, N., & Sahi, S. (2017) Zinc oxide nanoparticles (ZnONPs) alleviate heavy metal-induced toxicity in *Leucaena leucocephala* seedlings: A physiochemical analysis. *Plant Physiology and Biochemistry* 110:59–69. doi:10.1016/j. plaphy.2016.08.022.

Vickers, N.J. (2017) Animal communication: When i'm calling you, will you answer too? *Current Biology* 27:R713–R715. doi:10.1016/j.cub.2017.05.064.

Wodka, D. et al. (2010) Photocatalytic activity of titanium dioxide modified by silver nanoparticles. *ACS Applied Materials & Interfaces* 2:1945–1953. doi:10.1021/am1002684.

Xu, X., Ray, R., Gu, Y., Ploehn, H.J., Gearheart, L., Raker, K., & Scrivens, W.A. (2004) Electrophoretic analysis and purification of fluorescent single-walled carbon nanotube fragments. *Journal of the American Chemical Society* 126:12736–12737. doi:10.1021/ja040082h.

Yakout, S.M., Abdeltawab, A.A., Mostafa, A.A., & Sherif, H. (2014) Green and facile approach for synthesis and characterization of silver nanoparticles using coriander (*Coriandrum sativum*) leaf extract. *Journal of Pure and Applied Microbiology* 8:313–316.

Zhang, W., Yang, Q., Luo, Q., Shi, L., & Meng, S. (2020) Laccase-carbon nanotube nanocomposites for enhancing dyes removal. *Journal of Cleaner Production* 242:118425. doi:10.1016/j.jclepro.2019.118425.

28 Green Synthesis of Nanoparticles from Coriander Leaf Extract

Aiman K. H. Bashir, Natascha Cheikhyoussef,
Ahmad Cheikhyoussef, and Ahmed A. Hussein

CONTENTS

28.1 INTRODUCTION

Many remarkable improvements in medical and applied sciences have begun with the discovery of nanotechnology in 1959 (Feynman, 1960). Nanotechnology, often known as nanoscale technology, is concerned with scales smaller than 100 nm. At this size domain, materials display extraordinary features and unique properties attracting significant interest from researchers in biomedical and industrial fields (Schmid, 1992). Nanoparticles (NPs) have qualities that are dependent on their size and shape. As a result, nanoparticles with novel applications can be manufactured by manipulating their size and shape on the nanoscale scale.

The physical and chemical strategies of synthesis of NPs were widely used until several recent studies revealed some of their disadvantages (Iravani et al., 2014; Gebre & Sendeku, 2019; Ijaz et al., 2020; Makvandi et al., 2020; Ali et al., 2021; Ramanathan et al., 2021; Sahani & Sharma, 2021). However, NPs can be engineered to change their physical and chemical properties to influence their interactions in their biological environments and enable precision delivery to their target destinations (Poon et al., 2020). In most of these strategies, synthesis was carried out at high temperatures, necessitating the use of specialized instruments, along with hazardous materials such as hydrazine and sodium borohydride, among others. (Jamkhande et al., 2019; Kharissova et al., 2019; Soni et al., 2021). Quite recently, plant-based compounds have been used in green chemistry to synthesize nanoparticles that are physiologically safe (Mittal et al., 2013; Hussain et al., 2016; Ishak et al., 2019; Verma et al., 2019; Atanasov et al., 2021; Hernández-Díaz et al., 2021; Ettadili et al., 2022). Green synthesis, unlike the procedures outlined above, provides a number of advantages such as the use of plant materials that are particularly accessible, inexpensive, and usually eco-friendly, ensuring that the environment is free of hazardous residues (Hussain et al., 2016; Parveen et al., 2016; Ishak et al.,

2019; Soni et al., 2021). Plant extracts contain a wide range of phytochemicals that can cause the synthesis of nanoparticles by reducing metal salts due to their reducing and antioxidant properties (Saranya et al., 2017; Ifeanyichukwu et al., 2020; Zhang et al., 2020; Hernández-Díaz et al., 2021; Ettadili et al., 2022). Recently, coriander leaves and seeds have been used in the green synthesis of different nanoparticles (Luna et al., 2016; Senthilkumar et al., 2018; Alsubki et al., 2021). The origin of coriander is the Mediterranean region, but it is widely distributed around the world (Iqbal et al., 2018). It was reported to be used in the traditional medicine of different ethnic groups around the world and has a wide range of pharmacological properties (Laribi et al., 2015; Prachayasittikul et al., 2018).

28.2 OVERVIEW OF GREEN SYNTHESIS OF NPs

In a green chemistry method, the extract from a piece of a plant such as the bark, leaves, root, shoot, stems, or the entire plant is used to synthesize NPs (Figure 28.1). The bioactive constituents in a desired plant are extracted by soaking the air-dried plant part at room temperature or boiling it at a high temperature for a suitable extraction time. According to numerous reports on the synthesis of NPs using green chemistry, distilled water is commonly used as an extracting solvent to extract the bioactive components from plants (Hassan et al., 2015; Hussain et al., 2016; Ahmed et al., 2016; Goutam et al., 2017; Prabhakar et al., 2017; Khan et al., 2018; Gour, & Jain, 2019; Ishak et al., 2019; Ifeanyichukwu et al., 2020; Mondal et al., 2020; Kannaiyan et al., 2021; Ettadili et al., 2022). Hence, the successful extraction of bioactive phytochemicals in plant extracts is a key factor in the production of diverse types of NPs employing various salt precursors. These bioactive compounds, such as flavonoids, phenolic, and organic acids, among others, operate as reducing agents by reacting with the salt precursor to generate NPs, which are subsequently stabilized by

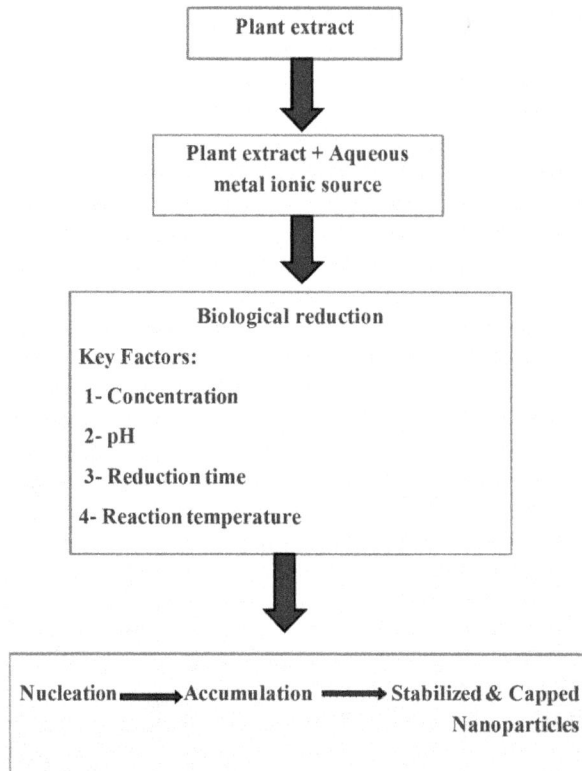

FIGURE 28.1 Flow chart demonstrating the steps of the green synthesis of nanoparticles (NPs).

the extract's chemical constituents (Castro et al., 2011; Huston et al., 2021). The quality, size, and morphology of the produced biosynthesized NPs are greatly influenced by properties of the plant extract, concentration, metal salt concentration, pH, time, and temperature of the reaction (Shah et al., 2015; Singh et al., 2021).

28.3 BIOACTIVE CONSTITUENTS IN CORIANDER LEAVES

Coriander (*Coriandrum sativum*) provides both significant therapeutic and nutritional benefits due to its abundant amounts of fiber, vitamin B, vitamin C, carotene, mineral elements, and more (Bhat et al., 2014; Iqbal et al., 2018; Zhao et al., 2021). On the other hand, coriander leaves, also called cilantro (Łyczko et al., 2021), contain a large number of bioactive constituents such as essential oil, phenolic acids (chlorogenic acid), and polyphenols such as flavonoids (quercetin), in addition to the other constituents previously reported in the leaves (Matasyoh et al., 2009; Nambiar et al., 2010). Several studies reported groups of bioactive compounds existing in the leaves of coriander, which are mainly classified into lipids, polyphenols, and essential oil (Laribi et al., 2015; Prachayasittikul et al., 2018). A study on the fatty acid concentration and composition of coriander basal and upper leaves revealed that the total fatty acid amount was found to be 61.21 mg/g dry weight (DW) in the basal leaves compared with a lower amount of the fatty acid in the upper ones (41.8 mg/g DW) (Neffati & Marzouk, 2008). The total phenolic content along with the flavonoid content of methanolic extracts of coriander vegetative parts were 1013.95 mg/kg and 5259.52 mg/kg respectively (Barros et al., 2012). According to the literature, coriander leaves have less essential oil than coriander seeds (Bhuiyan et al., 2009; Shahwar et al., 2012). A quantitative study reported by Nambiar et al. (2010) revealed that coriander leaves are a rich source of phenolics such as *p*-coumaric acid (4.39 ± 0.55%), *trans*-ferulic acid (11.07 ± 0.66%), vanillic acid (7.51 ± 0.55%), 4'-*O* methylquercetin (5.62 ± 0.66%), 3'-*O*-methylquercetin (9.19 ± 0.24%), kaempferol (3.49 ± 0.40%), and acacetin (3.91 ± 0.45%).

28.4 GREEN SYNTHESIS OF NPs USING CORIANDER LEAF EXTRACTS

Due to its environmentally friendly character, the utilization of coriander plant resources is considered a green approach and a trustworthy method for NPs (Iravani, 2011; Kalpana & Devi Rajeswari, 2018; Fierascu et al., 2020; Primo et al., 2020; Soltys et al., 2021). Coriander leaves have been successfully used to synthesize noble metal (MNPs) and metal oxide (MONPs) nanoparticles in a fast and extracellular manner (Alsubki et al., 2021). Coriander leaf extracts are rich with different bioactive constituents which could serve as both bio-reducing and stabilizing agents in the synthesis of nanoparticles (Elia et al., 2014; Alsubki et al., 2021). Table 28.1 provides an overview of recent publications on metallic and metal oxide nanoparticles biosynthesized using coriander leaf extract.

28.4.1 SYNTHESIS OF METAL NANOPARTICLES

Several studies have reported the green synthesis of metal nanoparticles (MNPs) using coriander leaf extract (Narayanan & Sakthivel, 2008; Kulkarni et al., 2015; Luna et al., 2016; Goutam et al., 2017; Senthilkumar et al., 2018; Alsubki et al., 2021). These MNPs include silver (Ag), gold (Au), copper (Cu), and nickel (Ni). Different types of morphology as well as particle sizes of MNPs were obtained. Yakout et al. (2014) reported on the formation of Ag NPs using coriander leaf water extract through the bio-reduction of $AgNO_3$. The shift in color from colorless to dark brown indicated the formation of Ag NPs. Also, the effect of reaction time and the mechanism of formation of Ag NPs were explored (Khan et al., 2018). It is observed that employing coriander extract for the synthesis of Ag NPs could reduce Ag^+ ions into Ag NPs approximately within an hour of a reaction without using any additional procedures (Figure 28.2).

Senthilkumar et al. (2018) prepared silver Ag NPs via a percolated green synthesis method using *C. sativum* leaf, root, seed, and stem extracts with different sizes (35 to 53 nm) and reported

TABLE 28.1

Metals and Metal Oxide NPs Synthesized Using Green Chemistry from Coriander Leaf Extract

Type of NPs	Precursor	Shape/Size	Application	Reference
Ag	AgNO$_3$	(2.9 nm)	Antibacterial	Luna et al. (2016)
Ag	AgNO$_3$	Spherical (37 nm)	Anti-acne, antidandruff, and cytotoxic	Sathishkumar et al. (2016)
Ag	AgNO$_3$	–	–	Yakout et al. (2014)
Ag	AgNO$_3$	Spherical (6.45 nm)	–	Khan (2018)
Ag	AgNO$_3$	(11.9 nm)	Antibacterial	Ashraf et al. (2019)
Ag	AgNO$_3$	Spherical	Antimicrobial	Alsubki et al. (2021)
Ag	AgNO$_3$	Spherical (26 nm)	Nonlinear optics	Sathyavathi et al. (2010)
Ag	AgNO$_3$	–	Antimicrobial and wound healing	Balamurugan & Sami Nathan (2014)
Ag[1]	AgNO$_3$	Poly-disperse spherical (13.5 nm)	Antimicrobial	Nazeruddin et al. (2014)
Ag[2]	AgNO$_3$	Spherical	Antimicrobial	Manisha et al. (2014)
Au	HAuCl$_4$	Spherical, triangle, truncated triangles, and decahedral morphologies (20.65 ± 7.09 nm)	–	Narayanan & Sakthivel (2008)
Au	HAuCl$_4$	Spherical (30 nm)	Nonlinear optics	Venugopal (2011)
Au	HAuCl$_4$.3H$_2$O	–	–	Sumit & Nayak (2012)
Cu	CuSO$_4$.5H$_2$O	(11.66–14.65 nm)	–	Kulkarni et al. (2015)
Ni	NiCl$_2$	(30.71 nm)	–	Vasudeo & Pramod (2016)
Se[3]	Na$_2$SeO$_3$	–	Antimicrobial	Jai Rexlin et al. (2020).
Au-Ag	AgNO$_3$ and HAuCl$_4$	Spherical (10– 12)		Alti et al. (2020)
Ni/Cr/Cu	CuSO$_4$.5H$_2$O, Ni(NO$_3$)$_2$.6H$_2$O, Cr (NO$_3$)$_3$.9H$_2$O	(17.72 nm)	Antimicrobial	Kannaiyan et al. (2021)
ZnO	Zn(NO$_3$)$_2$.6H$_2$O	Spherical (40 nm)	Antibacterial	Goutam et al. (2017)
ZnO	Zn(CH$_3$CO$_2$)$_2$·2H$_2$O	Cubic (100–190 nm)	–	Gnanasangeetha & Thambavani (2013)
ZnO	Zn(CH$_3$CO$_2$)$_2$.2H$_2$O	(9–18 nm)	Photocatalytic degradation	Hassan et al. (2015)
CeO$_2$	Ce(NO$_3$)$_3$.6H$_2$O	–	Antibacterial	Sathiyapriya et al. (2020)
TiO$_2$	Ti{(OCH(CH$_3$)$_2$}$_4$	Spherical (6.7–8.3 nm)	Photocatalytic application	Pushpamalini et al. (2020)
Fe$_3$O$_4$	FeCl$_3$	Spherical (20–90nm)	Antimicrobial	Sathya et al. (2017)

[1,2] Synthesized from seed extract. [3] Synthesized from coriander oleoresin.

excellent antimicrobial activity against *Klebsiella pneumoniae*. Another study by Ashraf et al. (2019) on green synthesis has revealed the ability of coriander leaf extract in the formation of Ag NPs. In this study, the synthesis of Ag NPs was performed by adding 20 mL of the leaf extract to a 1 mM solution of silver nitrate (80 mL). The particle sizes and polydispersity index of the formed Ag NPs, respectively, were found to be 11.9 nm and 0.452. Recently, Alsubki et al. (2021) reported

FIGURE 28.2 (a) Image of fresh coriander leaves, (b) photograph of synthesized Ag NPs solution over several reaction times (a) 10 min, (b) 20 min, (c) 70 min. The shift in color confirming the bio-reduction of Ag$^+$ to Ag with time. Modified from (Khan et al., 2018) with permission.

on the characterization and functionality of Ag NPs synthesized from an extract of coriander leaves. Their results showed that the reduction of Ag$^+$ ions was obtained within 10 minutes of the reaction time, confirmed by the color change to brownish yellow. The morphological studies revealed the spherical shape of Ag NPs, where the particles were distributed in a good manner without aggregation. Moreover, silver-based hybrid nanostructures were synthesized using an extract from coriander leaves (Luna et al., 2016) (Figure 28.3).

The precursor used for synthesizing these silver-based hybrid nanostructures was AgNO$_3$. It has been found that the nanostructures synthesized with a low concentration of the precursor Ag salt (AgNO$_3$ = 5 mM) have an ultrafine size and a narrow size distribution, whereas particles synthesized with a high concentration of the precursor Ag salt (AgNO$_3$ = 0.5 M) are polydisperse with supramolecular structures.

Besides Ag NPs, some others MNPs were synthesized using an extract from coriander leaves. The noble metallic gold nanoparticles (Au NPs) were biosynthesized by a simple method using an

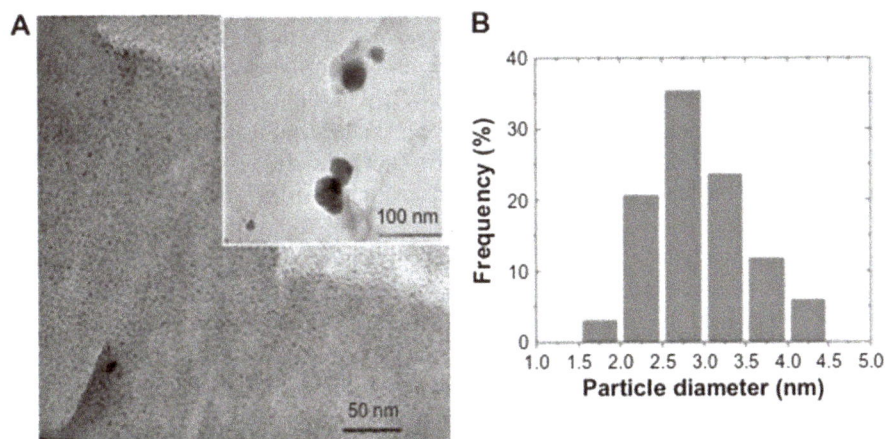

FIGURE 28.3 Green synthesis of silver-based hybrid nanostructures using an extract from coriander leaves. (a) Transmission electron microscopy (TEM) image, inset high-resolution TEM (HRTEM) image showing the formed Ag nanostructures. (b) Histogram displaying the particle size distribution. Reprinted from (Luna et al., 2016) with permission.

extract from coriander leaves as a bio-reducing agent (Narayanan & Sakthivel, 2008). The surface study showed the existence of different shapes such as spherical, triangle, truncated triangles, and decahedral morphologies with 20.65 ± 7.09 nm as an average size. Mittal et al. (2013) produced Ag and Au nanoparticles from the leaf extract of *C. sativum* with different sizes ranging from 7 to 58 nm.

Nickel nanoparticles (Ni NPs) were also biosynthesized using an extract from coriander leaves at room temperature under normal conditions utilizing nickel chloride as a precursor (Vasudeo and Pramod, 2016). The process of synthesis was performed by adding 10 mL of coriander leaf extract to 100 mL of 1 mm nickel chloride solution, and then the progress of the bio-reaction was observed by a shift in the color of the mixture indicating the bio-reduction of Ni ions. The particle size of the formed Ni NPs was obtained from the analysis of XRD data based on Debye-Scherrer's equation, and found to be 30 nm. On the other hand, the efficiency of coriander leaf extract as a biocatalyst in the synthesizing of copper nanoparticles (Cu NPs) was also investigated (Kulkarni et al., 2015). The starting material in the synthesizing of Cu NPs was copper sulphate solution at a pH value of 2.16. Hence, the biosynthesis of Cu NPs was confirmed by changing the pH value of the copper sulphate solution to 2.64 after adding a certain amount of coriander leaf extract. Selenium nanoparticles (Se NPs) were also fabricated by adding coriander extract to sodium selenite (Jai Rexlin et al., 2020).

The bimetallic Au/Ag NPs were successfully synthesized in one step using coriander leaf extract (Alti et al., 2020). In a typical synthesis process, 1.5 mL of coriander extract was devolved separately in 15 mL of 10^{-3} M of $HAuCl_4$ and $AgNO_3$, and then mixed in a ratio of (1:1). The mixture was heated to 80°C for half an hour, resulting in the colorless solution changing to a red color, indicating the synthesis of Au/Ag NPs. The synthesized Au/Ag NPs appeared without agglomerations in spherical shapes with diameters ranging from 10 to 12 nm. Also, the formation of trimetallic Ni/Cr/Cu NPs with a particle size of 17.72 nm was obtained using coriander leaf extract which acts as a bio-reducing agent (Kannaiyan et al., 2021). Multimetallic nanoparticles are fabricated from different metallic elements such as Au, Ag, Co, Cu, and Ni via several methods such as polymer nanoreactor-mediated synthesis and libraries with their nanostructures have been reported (Chen et al., 2016; Kang et al., 2019; Huang et al., 2020).

28.4.2 SYNTHESIS OF METAL OXIDE NANOPARTICLES

Similar to MNPs, metal oxide nanoparticles (MONPs) were biosynthesized with the green approach using several plant extracts (Ghafar et al., 2018; Singh et al., 2018; Patiño-Ruiz et al., 2021), among these plant materials being coriander leaves. Zinc oxide nanoparticles (ZnO NPs) were successfully biosynthesized and can be considered as one of the most important semiconductor materials that is widely used in diverse applications (Sajjad et al., 2021; Rajeshkumar & Jeevitha, 2021). Spherical ZnO NPs with sizes ranging from 30 to 150 nm were synthesized using an extract from coriander leaves (Goutam et al., 2017). For the biosynthesis of ZnO NPs, 100 mL of 0.1 M zinc acetate dihydrate was mixed with the leaf extract in a ratio 1:1, then 20 mL of 1.0 M NaOH solution was dropped under in order to increase the pH value of the mixture. During the stirring process, the color of the solution changed gradually from a solid brown to a yellowish white, indicating the creation of NPs. Then the precipitation was exposed to an annealing process at 400°C. A TEM image of the synthesized ZnO NPs confirmed the nanostructure with spherical form (Goutam et al., 2017).

Cerium oxide nanoparticles (CeO_2 NPs) are an indispensable rare-earth metal oxide used in most technological applications (Xu & Qu, 2014; Nadeem et al., 2020). The green synthesis of CeO_2 NPs was also achieved using an extract from coriander leaves (Sathiyapriya et al., 2020). In this process of synthesis, a gel solution was formed in 30 mL of 0.5 M of aqueous cerium nitrate hexahydrate solution, then NaOH solution was added drop by drop under continuous stirring for 2 h. This was followed by annealing the product in the open air at 700°C for two hours. The morphological studies on the biosynthesized Ce_2O NPs revealed that particle sizes increase upon increasing the concentration of the leaf extract.

Titanium oxide nanoparticles (TiO$_2$ NPs) were synthesized by adding 5 mL each of coriander extract and titanium tetra isopropoxide (TTIP) to 25 mL of double distilled water (Pushpamalini et al., 2020). The prepared solution was stirred well using a magnetic stirrer for 3 hours at 70°C, and then filtered and annealed at 400°C for 3 hours. The shapes of the synthesized TiO$_2$ NPs were determined to be approximately spherical with size distribution between 6.7 and 8.3 nm. A green synthesis of TiO$_2$ NPs using extracts of coriander was reported using a modified sol–gel method (Bhullar et al., 2021). The authors used the green approach to ensure the use of safe materials with no production of toxic by-products and for the rapid synthesis of the TiO$_2$ NPs. The size for the obtained NPs ranged from 5 to 20 nm. In addition to the above mentioned MNPs, iron oxide nanoparticles (Fe$_3$O$_4$ NPs) were also synthesized using coriander leaf extract in spherical shapes with sizes ranging from 20 to 90 nm (Sathya et al., 2017) and 164.5 nm (Singh et al., 2020).

28.4.3 CHARACTERIZATION TECHNIQUES OF NANOPARTICLES

Various techniques were independently used to characterize the NPs that were synthesized using an extract of coriander leaves. Generally, there are microscopy-based techniques such as transmission electron microscopy (TEM), high-resolution TEM (HRTEM), and atomic force microscopy (AFM), which are able to provide information on the size, morphology, and crystal structure of the nanomaterials. Other techniques which provide further information on elemental composition, optical properties, and structure include X-ray, spectroscopy, and scattering techniques (Mourdikoudis et al., 2018).

As a preliminarily investigation, X-ray diffraction (XRD) patterns were used to identify the crystalline phase of the biosynthesized metallic and metal oxide nanoparticles. Also, XRD data were used to estimate the particle sizes based on Debye-Scherrer's equation of Ag, Au, Ni, and Cu MNPs biosynthesized using coriander leaf extract (Narayanan & Sakthivel, 2008; Kulkarni et al., 2015; Vasudeo and Pramod, 2016; Ashraf et al., 2019). The detailed morphological characterization was performed using HRSEM and HRTEM techniques on the biosynthesized nanoparticles. Moreover, the particle sizes of some of the synthesized nanoparticles were determined using high-resolution scanning electron microscopy (HRSEM) and HRTEM.

UV–visible spectroscopy is a widely used technique for studying optical characteristics. The wavelengths of light in the 200–700 nm range are commonly employed to describe a variety of metallic and metal oxide nanoparticles. Hence, UV–visible spectra were used to detect the absorption peaks of various MNPs and MONPs synthesized using coriander leaf extract. Absorption peaks located around 430, 530, 272, and 276 nm were observed for Ag, Au, ZnO, and CeO$_2$ NPs, respectively (Narayanan & Sakthivel, 2008; Hassan et al., 2015; Khan et al., 2018; Sathiyapriya et al., 2020). Moreover, the monitoring of reaction time was performed using UV–Vis spectra for some Ag and Au NPs obtained from coriander extracts (Narayanan & Sakthivel, 2008; Venugopal Rao, 2011; Khan, et al. 2018).

The chemical interaction at the surface can be studied using Fourier-transform infrared (FTIR) spectroscopy. FTIR spectroscopy is used to detect organic functional groups (e.g., carbonyls, hydroxyls) adhering to the surface of nanoparticles as well as other surface chemical residues. However, the functional groups available in coriander leaf extract were explored using FTIR spectroscopy (Luna et al., 2016). The FTIR spectrum of coriander leaf extract revealed a broad peak centered at wave number (3300 cm^{-1}) that can be ascribed to N–H and O–H stretching modes of coriander proteins and phenols. Another peak also observed at 2961 cm^{-1} is attributed to methyl group stretching vibrations, while the absorbance bands at 1603, 1405, and 1119 cm^{-1} are attributed to C=O stretching vibrations, C–OH bending vibrations (from phenols, polysaccharides, and polyols), and –C–O and C–O–H stretching vibrations of alcohols, respectively. Also, C–N stretching vibrations of amine could contribute to the band at 1119 cm^{-1}. On the other hand, the capping and efficient stabilization of MNPs by biomolecules found in coriander leaf extract were identified using FTIR spectra (Luna et al., 2016).

28.4.4 MECHANISM OF FORMATION OF NANOPARTICLES USING CORIANDER LEAF EXTRACT

Understanding the mechanism of formation of nanoparticles synthesized from plant extracts is still one of the challenges facing researchers in the green chemistry field. This follows from the fact that these plant extracts contain large numbers of bioactive elements, some of which are responsible for the formation of NPs.

The presence of different types of bioactive compounds in the leaves of coriander plant may play a significant role in the formation of NPs. Among them are secondary metabolites such as phenolic compounds which are considered a big family containing groups of aromatic hydroxyl groups. Moreover, phenolic compounds were reported to possess hydroxyl and carbonyl groups which have the ability to bind with metals (Yakout et al., 2014). However, the chelating capacity of phenolic compounds can inactivate the ions, which is more likely due to nucleophilic properties of the aromatic instead of specific chelating groups available in molecules (Yakout et al., 2014).

Techniques such as FTIR were used to study and propose the possible mechanism of formation of NPs synthesized from plants extracts (Matinise et al., 2017). This can be achieved by comparing the different functional groups observed in the FTIR spectrum of the plant extract used in green synthesis with those observed in the biosynthesized NPs. For instance, Luna et al. (2016) have used FTIR spectra to propose the mechanism of formation of Ag-based hybrid nanostructure synthesized using an extract from coriander leaves. According to the analysis of the FTIR spectra of the coriander leaves extract and Ag-based nanostructure, they found that the functional groups assigned to coriander proteins and polyphenols are very effective in the reduction of Ag⁺ and stabilization of the synthesized NPs.

More precisely, the vegetative component of the coriander contains some polyphenols such as dimethoxycinnamoyl hexoside, kaempferol-3-*O*-rutinoside, quercetin-3-*O*-rutinoside, quercetin-3-*O*-glucoside, and quercetin 3-*O*-glucuronide (Nambiar et al., 2010; Laribi et al., 2015). In addition, the four flavonoids identified in the vegetative part of coriander are 3-*O*-caffeoylquinic acid, caffeoylquinic acid, ferulic acid glucoside, and *p*-coumaroylquinic acid (Laribi et al., 2015). Hence, it could be deduced that one of these compounds may be responsible for the formation of NPs upon using coriander leaf extract.

28.5 BIOLOGICAL EVALUATION OF NPS SYNTHESIZED FROM CORIANDER LEAVES

According to the therapeutic nature and antioxidant activity of coriander plant, most of the MNPs and MONPs synthesized using an extract from coriander leaves were biologically evaluated. Different microorganisms such as *Bacillus subtilis*, *Enterobacter aerogenes*, *Pasteurella multocida*, and *Staphylococcus aureus* were inhibition tested using NPs fabricated using coriander leaf extracts (Narmadha, 2013; Ashraf et al., 2019). Ashraf et al. (2019) have studied the antibacterial potential of Ag NPs synthesized from coriander leaf extract. Their tests revealed that the activity of Ag NPs destroyed large number of bacteria and destroyed their exterior membrane. Also, Ag-based hybrid nanostructures fabricated from coriander leaf extract showed enhanced bactericidal activities (Figure 28.4) against Gram-positive and Gram-negative bacteria (Luna et al., 2016). Moreover, Ag NPs synthesized from coriander extract have shown antimicrobial activity assisting in wound healing (Balamurugan & Sami Nathan, 2014).

On the other hand, metal oxide nanoparticles synthesized from coriander leaf extract also have shown strong potency against bacterial growth. The biosynthesized ZnO NPs exhibited effective and robust antibacterial action against *E. coli* (Goutam et al., 2017). In addition, the antibacterial tests on Gram-negative and Gram-positive bacteria were done using CeO_2 synthesized using coriander leaves. According to the findings, the biosynthesized CeO_2 NPs have the highest zone of inhibition against *Bacillus cereus*, with a 12 mm zone of inhibition (Sathiyapriya et al., 2020).

The anticancer activity was tested *in vitro* using Ag NPs synthesized from coriander leaf extract (Alsubki et al., 2021). The toxic effect of Ag NPs on cancerous cells was confirmed through tests

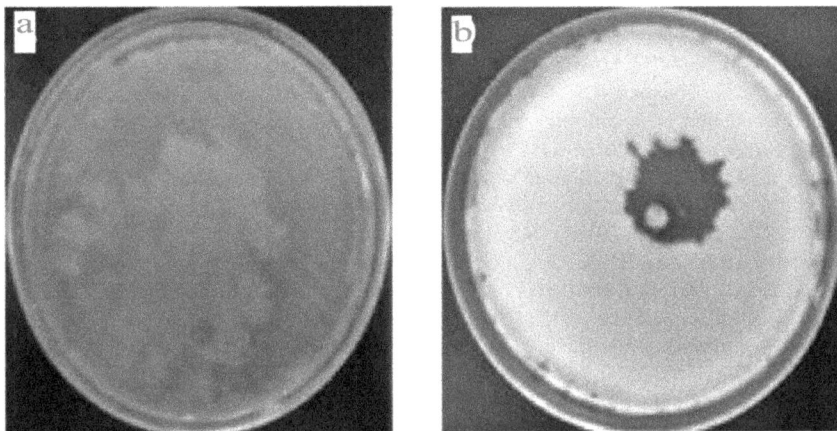

FIGURE 28.4 Antibacterial activity against *Staphylococcus aureus* bacteria grown on agar plates. (a) In the presence of an extract from coriander leaves. (b) In the presence of Ag nanostructures synthesized using an extract from coriander leaves. Reprinted from (Luna et al., 2016), with permission.

at different concentrations on the MCF-7 cell lines. This observed *in-vitro* anticancer activity indicates that Ag NPs synthesized from coriander extract could be effective therapeutic agents in treating cancer.

28.6 CONCLUSION

This chapter summarizes the green synthesis of different metallic and metal oxide nanoparticles using coriander leaf extract. The effectiveness of coriander leaves in the biosynthesis of these nanoparticles is attributed to some bioactive compounds present in the leaves. The mechanism of formation of these nanoparticles is proposed through bio-reduction effects of the secondary metabolites polyphenols and/or coriander proteins. In addition, the biological evaluation of these green synthesized NPs showed bactericidal potency, assisting in healing wounds and cytotoxic activity with potential product development for food packaging and the pharmaceutical industry.

REFERENCES

Ahmed, S., Annu, Ikram, S., & Yudha, S. S. (2016). Biosynthesis of gold nanoparticles: A green approach. *Journal of Photochemistry and Photobiology. B, Biology*, *161*, 141–153. https://doi.org/10.1016/j.jphotobiol.2016.04.034

Ali, S. S., Al-Tohamy, R., Koutra, E., Moawad, M. S., Kornaros, M., Mustafa, A. M., Mahmoud, Y. A., Badr, A., Osman, M., Elsamahy, T., Jiao, H., & Sun, J. (2021). Nanobiotechnological advancements in agriculture and food industry: Applications, nanotoxicity, and future perspectives. *The Science of the Total Environment*, *792*, 148359. https://doi.org/10.1016/j.scitotenv.2021.148359

Alsubki, R., Tabassum, H., Abudawood, M., Rabaan, A. A., Alsobaie, S. F., & Ansa, S. (2021). Green synthesis, characterization, enhanced functionality and biological evaluation of silver nanoparticles based on *Coriander sativum*. *Saudi Journal of Biological Sciences*, *28*(4), 2102–2108. https://doi.org/10.1016/j.sjbs.2020.12.055

Alti, D., Veeramohan Rao, M., Rao, D. N., Maurya, R., & Kalangi, S. K. (2020). Gold-silver bimetallic nanoparticles reduced with herbal leaf extracts induce ROS-mediated death in both promastigote and amastigote stages of *Leishmania donovani*. *ACS Omega*, *5*(26), 16238–16245. https://doi.org/10.1021/acsomega.0c02032

Ashraf, A., Zafar, S., Zahid, K., Salahuddin Shah, M., Al-Ghanim, K. A., Al-Misned, F., & Mahboob, S. (2019). Synthesis, characterization, and antibacterial potential of silver nanoparticles synthesized from *Coriandrum sativum* L. *Journal of Infection and Public Health*, *12*(2), 275–281. https://doi.org/10.1016/j.jiph.2018.11.002

Atanasov, A. G., Zotchev, S. B., Dirsch, V. M., International Natural Product Sciences Taskforce, & Supuran, C. T. (2021). Natural products in drug discovery: Advances and opportunities. *Nature Reviews Drug Discovery, 20*(3), 200–216. https://doi.org/10.1038/s41573-020-00114-z

Balamurugan, M., & Sami Nathan, R. (2014). Characterization of *Coriander sativum* mediated silver nanoparticles and evaluation of its antimicrobial and wound healing activity. *IOSR Journal of Polymer and Textile Engineering (IOSR-JPTE), 1*, 22–25.

Barros, L., Dueňas, M., Dias, M. I., Sousa, M. J., Santos-Buelga, C., & Ferreira, I. C. F. R. (2012). Phenolic profiles of in vivo and in vitro grown *Coriandrum sativum* L. *Food Chemistry, 132*(2), 841–848. https://doi.org/10.1016/j.foodchem.2011.11.048

Bhat, S., Kaushal, P., & Sharma, H. K. (2014). Coriander (*Coriandrum sativum* L.): Processing, nutritional and functional aspects. *African Journal of Plant Science, 8*(1), 25–33. https://doi.org/10.5897/AJPS2013.1118

Bhuiyan, N. I., Begum, J., & Sultana, M. (2009). Chemical composition of leaf and seed essential oil of *Coriandrum sativum* L. from Bangladesh. *Bangladesh Journal of Pharmacology, 4*, 150–153. https://doi.org/10.3329/bjp.v4i2.2800

Bhullar, S., Goyal, N., & Gupta, S. (2021). Rapid green-synthesis of TiO2 nanoparticles for therapeutic applications. *RSC Advances, 11*, 30343–30352. https://doi.org/10.1039/D1RA05588G

Castro, L., Blazquez, M. L., Munoz, J. A., Gonzalez, F., Garcıa-Balboa, C., & Ballester, A. (2011). Biosynthesis of gold nanowires using sugar beet pulp. *Process Biochemistry, 46*(5), 1076–1082. https://doi.org/10.1016/j.procbio.2011.01.025

Chen, P. C., Liu, X., Hedrick, J. L., Xie, Z., Wang, S., Lin, Q. Y., Hersam, M. C., Dravid, V. P., & Mirkin, C. A. (2016). Polyelemental nanoparticle libraries. *Science, 352*(6293), 1565–1569. https://doi.org/10.1126/science.aaf8402

Elia, P., Zach, R., Hazan, S., Kolusheva, S., Porat, Z., & Zeiri, Y. (2014). Green synthesis of gold nanoparticles using plant extracts as reducing agents. *International Journal of Nanomedicine, 9*, 4007–4021. https://doi.org/10.2147/IJN.S57343

Ettadili, F. E., Aghris, S., Laghrib, F., Farahi, A., Saqrane, S., Bakasse, M., Lahrich, S., & El Mhammedi, M. A. (2022). Recent advances in the nanoparticles synthesis using plant extract: Applications and future recommendations. *Journal of Molecular Structure, 1248*, 131538. https://doi.org/10.1016/j.molstruc.2021.131538

Feynman, R. (1960). There's plenty of room at the bottom (reprint from speech given at annual meeting of the American Physical Society). *Engineering and Science, 23*(5), 22–36. https://resolver.caltech.edu/CaltechES:23.5.1960Bottom

Fierascu, I., Fierascu, I. C., Brazdis, R. I., Baroi, A. M., Fistos, T., & Fierascu, R. C. (2020). Phytosynthesized metallic nanoparticles-between nanomedicine and toxicology. A brief review of 2019's findings. *Materials, 13*(3), 574. https://doi.org/10.3390/ma13030574

Gebre, S. H., & Sendeku, M. G. (2019). New frontiers in the biosynthesis of metal oxide nanoparticles and their environmental applications: An overview. *SN Applied Sciences, 1*, 928. https://doi.org/10.1007/s42452-019-0931-4

Ghafar, A., Yang, J. P., Jung, W. K., & Sung, O. C. (2018). A green, general, and ultrafast route for the synthesis of diverse metal oxide nanoparticles with controllable sizes and enhanced catalytic activity. *ACS Applied Nano Materials, 1*(11), 6112–6122. https://doi.org/10.1021/acsanm.8b01220

Gnanasangeetha, D., & Thambavani, D. S. (2013). One pot synthesis of zinc oxide nanoparticles via chemical and green method. *Research Journal of Material Sciences, 1*(7), 1–8.

Gour, A., & Jain, N. K. (2019). Advances in green synthesis of nanoparticles. *Artificial Cells, Nanomedicine, and Biotechnology, 47*(1), 844–851. https://doi.org/10.1080/21691401.2019.1577878

Goutam, S. G., Yadav, A. K., & Das, A. J. (2017). Coriander extract mediated green synthesis of zinc oxide nanoparticles and their structural, optical and antibacterial properties. *Journal of Nanoscience and Technology, 3*(1), 249–252.

Hassan, S. S. M., El Azab, W. I. M., Ali, H. R., & Mansour, M. S. M. (2015). Green synthesis and characterization of ZnO nanoparticles for photocatalytic degradation of anthracene. *Advances in Natural Sciences: Nanoscience and Nanotechnology, 6*(4) 045012. https://doi.org/10.1088/2043-6262/6/4/045012

Hernández-Díaz, J. A., Garza-García, J. J., Zamudio-Ojeda, A., León-Morales, J. M., López-Velázquez, J. C., & García-Morales, S. (2021). Plant-mediated synthesis of nanoparticles and their antimicrobial activity against phytopathogens. *Journal of the Science of Food and Agriculture, 101*(4), 1270–1287. https://doi.org/10.1002/jsfa.10767

Huang, L., Lin, H., Zheng, C. Y., Kluender, E. J., Golnabi, R., Shen, B., & Mirkin, C. A. (2020). Multimetallic high-index faceted heterostructured nanoparticles. *Journal of the American Chemical Society, 142*(10), 4570–4575. https://doi.org/10.1021/jacs.0c00045

Hussain, I., Singh, N. B., Singh, A., Singh, H., & Singh, S. C. (2016). Green synthesis of nanoparticles and its potential application. *Biotechnology Letters*, *38*(4), 545–560. https://doi.org/10.1007/s10529 -015-2026-7

Huston, M., DeBella, M., DiBella, M., & Gupta, A. (2021). Green synthesis of nanomaterials. *Nanomaterials*, *11*(8), 2130. https://doi.org/10.3390/nano11082130

Ifeanyichukwu, U. L., Fayemi, O. E., & Ateba, C. N. (2020). Green synthesis of zinc oxide nanoparticles from pomegranate (*Punica granatum*) extracts and characterization of their antibacterial activity. *Molecules*, *25*(19), 4521. https://doi.org/10.3390/molecules25194521

Ijaz, I., Gilani, E., Nazir, A., & Bukhari, A. (2020). Detail review on chemical, physical and green synthesis, classification, characterizations and applications of nanoparticles. *Green Chemistry Letters and Reviews*, *13*(3), 223–245. https://doi.org/10.1080/17518253.2020.1802517

Iqbal, M. J., Butt, M. S., & Suleria, H. A. R. (2018). Coriander (*Coriandrum sativum* L.): Bioactive molecules and health effects. In Mérillon, J. M., & Ramawat, K. (Eds.), *Bioactive Molecules in Food. Reference Series in Phytochemistry*. Springer, Cham. https://doi.org/10.1007/978-3-319-54528-8_44-1

Iravani, S. (2011). Green synthesis of metal nanoparticles using plants. *Green Chemistry*, *13*(10), 2638–2650. https://doi.org/10.1039/C1GC15386B

Iravani, S., Korbekandi, H., Mirmohammadi, S. V., & Zolfaghari, B. (2014). Synthesis of silver nanoparticles: Chemical, physical and biological methods. *Research in Pharmaceutical Sciences*, *9*(6), 385–406.

Ishak, N. M., Kamarudin, S. K., & Timmiati, S. N. (2019). Green synthesis of metal and metal oxide nanoparticles via plant extracts: An overview. *Materials Research Express*, *6*(11), 112004. https://doi.org/10 .1088/2053-1591/ab4458

Jai Rexlin, P. E., Roy, A., Rajeshkumar, S., & Lakshmi, T. (2020). Antimicrobial activity of coriander oleoresin mediated selenium nanoparticles against oral pathogens. *Journal of Complementary Medicine Research*, *11*(5), 35–40.

Jamkhande, P. G., Ghule, N. W., Bamer, A. H., & Kalaskar, M. G. (2019). Metal nanoparticles synthesis: An overview on methods of preparation, advantages and disadvantages, and applications. *Journal of Drug Delivery Science and Technology*, *53*, 101174–101184. https://doi.org/10.1016/j.jddst.2019. 101174

Kalpana, V. N., & Devi Rajeswari, V. (2018). A review on green synthesis, biomedical applications, and toxicity studies of ZnO NPs. *Bioinorganic Chemistry and Applications*, *2018*, 3569758. https://doi.org/10 .1155/2018/3569758

Kang, X., Wei, X., Jin, S., Yuan, Q., Luan, X., Pei, Y., Wang, S., Zhu, M., & Jin, R. (2019). Rational construction of a library of M_{29} nanoclusters from monometallic to tetrametallic. *Proceedings of the National Academy of Sciences of the United States of America*, *116*(38), 18834–18840. https://doi.org/10.1073/ pnas.1912719116

Kannaiyan, S. K., Rengaraj, R., Gayathri, P. K., Lavanya, G., & Hemapriya, D. (2021). Antimicrobial activity of green synthesized tri-metallic oxide Ni/Cr/Cu nanoparticles. *Journal of the Nigerian Society of Physical Sciences*, *3*(3), 144–147. https://doi.org/10.46481/jnsps.2021.237

Khan, M. Z. H., Tareq, F. K., Hossen, M. A., & Roki, M. N. A. M. (2018). Green synthesis and characterization of silver nanoparticles using *Coriandrum sativum* leaf extract. *Journal of Engineering Science and Technology*, *1*, 158–166.

Kharissova, O. V., Kharisov, B. I., Oliva González, C. M., Méndez, Y. P., & López, I. (2019). Greener synthesis of chemical compounds and materials. *Royal Society Open Science*, *6*(11), 191378. https://doi.org/10 .1098/rsos.191378

Kulkarni, V., Kale, N., Kute, N., & Kulkarni, P. (2015). Coriander leaf extract is efficient biocatalyst for synthesis of copper nanoparticles. *Chem Xpress*, *8*(2), 127–132.

Laribi, B., Kouki, K., M'Hamdi, M., & Bettaieb, T. (2015). Coriander (*Coriandrum sativum* L.) and its bioactive constituents. *Fitoterapia*, *103*, 9–26. https://doi.org/10.1016/j.fitote.2015.03.012

Luna, C., Barriga-Castro, E. D., Gómez-Treviño, A., Núñez, N. O., & Mendoza-Reséndez, R. (2016). Microstructural, spectroscopic, and antibacterial properties of silver-based hybrid nanostructures biosynthesized using extracts of coriander leaves and seeds. *International Journal of Nanomedicine*, *11*, 4787–4798. https://doi.org/10.2147/IJN.S105166

Łyczko, J., Masztalerz, K., Lipan, L., Iwiński, H., Lech, K., Carbonell-Barrachina, Á. A., & Szumny, A. (2021). *Coriandrum sativum* L.-effect of multiple drying techniques on volatile and sensory profile. *Foods*, *10*(2), 403. https://doi.org/10.3390/foods10020403

Makvandi, P., Wang, C.-y., Zare, E. N., Borzacchiello, A., Niu, L.-n., & Tay, F. R. (2020). Metal-based nanomaterials in biomedical applications: Antimicrobial activity and cytotoxicity aspects. *Advanced Functional Materials*, *30*, 1910021. https://doi.org/10.1002/adfm.201910021

Manisha, R. D., Merugu, R., Vijay babu, A. R., & Pratap Rudra, M. P. (2014). Microwave assisted biogenic synthesis of silver nanoparticles using dried seed extract of *Coriandrum sativum*, characterization and antimicrobial activity. *International Journal of ChemTech Research*, 6(7), 3957–3961.

Matasyoh, S. C., Maiyo, Z. C., Ngure, R. M., & Chepkorir, R. (2009). Chemical composition and antimicrobial activity of the essential oil of *Coriandrum sativum*. *Food Chemistry*, 113(2), 526–529. https://doi.org/10.1016/j.foodchem.2008.07.097

Matinise, N., Fuku, X. G., Kaviyarasu, K., Mayedwa, N., & Maaza, M. (2017). ZnO nanoparticles via *Moringa oleifera* green synthesis: Physical properties & mechanism of formation. *Applied Surface Science*, 406, 339–347. https://doi.org/10.1016/j.apsusc.2017.01.219

Mittal, A. K., Chisti, Y., & Banerjee, U. C. (2013). Synthesis of metallic nanoparticles using plant extracts. *Biotechnology Advances*, 31(2), 346–356. https://doi.org/10.1016/j.biotechadv.2013.01.003

Mondal, P., Anweshan, A., & Purkait, M. K. (2020). Green synthesis and environmental application of iron-based nanomaterials and nanocomposite: A review. *Chemosphere*, 259, 127509. https://doi.org/10.1016/j.chemosphere.2020.127509

Mourdikoudis, S., Pallares, R. M., & Thanh, N. T. K. (2018). Characterization techniques for nanoparticles: Comparison and complementarity upon studying nanoparticle properties. *Nanoscale*, 10, 12871–12934. https://doi.org/10.1039/C8NR02278J

Nadeem, M., Khan, R., Afridi, K., Nadhman, A., Ullah, S., Faisal, S., Mabood, Z. U., Hano, C., & Abbasi, B. H. (2020). Green synthesis of cerium oxide nanoparticles (CeO₂ NPs) and their antimicrobial applications: A review. *International Journal of Nanomedicine*, 15, 5951–5961. https://doi.org/10.2147/IJN.S255784

Nambiar, V. S., Daniel, M., & Guin, P. (2010). Characterization of polyphenols from coriander leaves (*Coriandrum sativum*), red amaranthus (*A. paniculatus*) and green amaranthus (*A. frumentaceus*) using paper chromatography and their health implications. *Journal of Herbal Medicine & Toxicology*, 4(1), 173–177.

Narayanan, K. B., & Sakthivel, N. (2008). Coriander leaf mediated biosynthesis of gold nanoparticles. *Materials Letters*, 62(30), 4588–4590. https://doi.org/10.1016/j.matlet.2008.08.044

Narmadha, E. (2013). Characterization of *Coriander sativum* mediated silver nanoparticles and evaluation of its antimicrobial and wound healing activity. *Indo American Journal of Pharmaceutical Research*, 3(7), 5404–5410.

Nazeruddin, G. M., Prasad, N. R., Prasad, S. R., Shaikh, Y. I., Waghmare, S. R., & Adhyapak, P. (2014). *Coriandrum sativum* seed extract assisted *in situ* green synthesis of silver nanoparticle and its antimicrobial activity. *Industrial Crops and Products*, 60, 212–216. https://doi.org/10.1016/j.indcrop.2014.05.040

Neffati, M., & Marzouk, B. (2008). Changes in essential oil and fatty acid composition in coriander (*Coriandrum sativum* L.) leaves under saline conditions. *Industrial Crops and Products*, 28(2), 137–142. https://doi.org/10.1016/j.indcrop.2008.02.005

Parveen, K., Banse, V., & Ledwani, L. (2016). Green synthesis of nanoparticles: Their advantages and disadvantages. *AIP Conference Proceedings*, 1724, 020048. https://doi.org/10.1063/1.4945168

Patiño-Ruiz, D. A., Meramo-Hurtado, S. I., González-Delgado, Á. D., & Herrera, A. (2021). Environmental sustainability evaluation of iron oxide nanoparticles synthesized via green synthesis and the coprecipitation method: A comparative life cycle assessment study. *ACS Omega*, 6(19), 12410–12423. https://doi.org/10.1021/acsomega.0c05246

Prabhakar, R., Samadder, S. R., & Jyotsana. (2017). Aquatic and terrestrial weed mediated synthesis of iron nanoparticles for possible application in wastewater remediation. *Journal of Cleaner Production*, 168, 1201–1210. https://doi.org/10.1016/j.jclepro.2017.09.063

Prachayasittikul, V., Prachayasittikul, S., Ruchirawat, S., & Prachayasittikul, V. (2018). Coriander (*Coriandrum sativum*): A promising functional food toward the well-being. *Food Research International*, 105, 305–323. https://doi.org/10.1016/j.foodres.2017.11.019

Primo, J. O., Bittencourt, C., Acosta, S., Sierra-Castillo, A., Colomer, J. F., Jaerger, S., Teixeira, V. C., & Anaissi, F. J. (2020). Synthesis of zinc oxide nanoparticles by eco-friendly routes: Adsorbent for copper removal from wastewater. *Frontiers in Chemistry*, 8, 571790. https://doi.org/10.3389/fchem.2020.571790

Poon, W., Kingston, B. R., Ouyang, B., Ngo, W., & Chan, W. (2020). A framework for designing delivery systems. *Nature Nanotechnology*, 15(10), 819–829. https://doi.org/10.1038/s41565-020-0759-5

Pushpamalini, T., Keerthana, M., Sangavi, R., Nagaraj, A., & Kamaraj, P. (2020). Comparative analysis of green synthesis of TiO₂ nanoparticles using four different leaf extract. *Materials Today Proceedings*, 40(1), S180–S184.

Rajeshkumar, S., & Jeevitha, M. V. (2021). Plant-mediated biosynthesis and characterization of zinc oxide nanoparticles. In Kamel A. Abd-Elsalam (Ed.), *Nanobiotechnology for Plant Protection, Zinc-Based Nanostructures for Environmental and Agricultural Applications.* Elsevier, pp. 37–51. https://doi.org/10.1016/B978-0-12-822836-4.00023-9

Ramanathan, S., Gopinath, S. C. B., Md Arshad, M. K., Poopalan, P., & Perumal, V. (2021). Nanoparticle synthetic methods: Strength and limitations. In Gopinath, S. C. B., & Gang, F. (Eds.), *Nanoparticles in Analytical and Medical Devices.* Elsevier, pp. 31–43. https://doi.org/10.1016/B978-0-12-821163-2.00002-9

Sahani, S., & Sharma, Y. C. (2021). Advancements in applications of nanotechnology in global food industry. *Food Chemistry, 342,* 128318. https://doi.org/10.1016/j.foodchem.2020.128318

Sajjad, A., Bhatti, S. H., Ali, Z., Jaffari, G. H., Khan, N. A., Rizvi, Z. F., & Zia, M. (2021). Photoinduced fabrication of zinc oxide nanoparticles: Transformation of morphological and biological response on light irradiance. *ACS Omega, 6*(17), 11783–11793. https://doi.org/10.1021/acsomega.1c01512

Saranya, S., Eswari, A., Gayathri, E., Eswari, S., & Vijayarani, K. (2017). Green synthesis of metallic nanoparticles using aqueous plant extract and their antibacterial activity. *International Journal of Current Microbiology and Applied Sciences, 6*(6), 1834–1845. https://doi.org/10.20546/ijcmas.2017.606.214

Sathishkumar, P., Preethi, J., Vijayan, R., Mohd Yusoff, A. R., Ameen, F., Suresh, S., Balagurunathan, R., & Palvannan, T. (2016). Anti-acne, anti-dandruff and anti-breast cancer efficacy of green synthesised silver nanoparticles using *Coriandrum sativum* leaf extract. *Journal of Photochemistry and Photobiology. B, Biology, 163,* 69–76. https://doi.org/10.1016/j.jphotobiol.2016.08.005

Sathiyapriya, R., Balaji, M., & Rajesh, S. (2020). Bio-synthesis of cerium oxide nanoparticles from *Coriandrum sativum* L. leaf extract and their antibacterial activity. *International Journal of Advanced Science, Engineering, 6,* 1439–1444. https://doi.org/10.29294/IJASE.6.3.2020.1439-1444

Sathya, K., Saravanathamizhan, R., & Baskar, G. (2017). Ultrasound assisted phytosynthesis of iron oxide nanoparticle. *Ultrasonics Sonochemistry, 39,* 446–451. https://doi.org/10.1016/j.ultsonch.2017.05.017

Sathyavathi, R., Krishna, M. B., Rao, S. V., Saritha, R., & Rao, D. N. (2010). Biosynthesis of silver nanoparticles using *Coriandrum sativum* leaf extract and their application in nonlinear optics. *Advanced Science Letters, 3*(2), 1–6. https://doi.org/10.1166/asl.2010.1099

Senthilkumar, N., Aravindhan, V., Ruckman, K., & Potheher, I. V. (2018). *Coriandrum sativum* mediated synthesis of silver nanoparticles and evaluation of their biological characteristics. *Materials Research Express, 5*(5), 055032. https://doi.org/10.1088/2053-1591/aac312

Schmid, G. (1992). Large clusters and colloids. Metals in the embryonic state. *Chemical Reviews, 92*(8), 1709–1727. https://doi.org/10.1021/cr00016a002

Shah, M., Fawcett, D., Sharma, S., Tripathy, S. K., & Poinern, G. (2015). Green synthesis of metallic nanoparticles via biological entities. *Materials, 8*(11), 7278–7308. https://doi.org/10.3390/ma8115377

Shahwar, M. K., El-Ghorab, A. H., Anjum, F. M., Hussain, S., & Nadeem, M. (2012). Characterization of coriander (*Coriandrum sativum* L.) seeds and leaves: Volatile and non-volatile extracts. *International Journal of Food Properties, 15*(4), 736–747. https://doi.org/10.1080/10942912.2010.500068

Singh, J., Dutta, T., Kim, K. H., Rawat, M., Samddar, P., & Kumar, P. (2018). 'Green' synthesis of metals and their oxide nanoparticles: Applications for environmental remediation. *Journal of Nanobiotechnology, 16*(1), 84. https://doi.org/10.1186/s12951-018-0408-4

Singh, K., Chopra, D. S., Singh, D., & Singh, N. (2020). Optimization and ecofriendly synthesis of iron oxide nanoparticles as potential antioxidant. *Arabic Journal of Chemistry, 13*(12), 9034–9046. https://doi.org/10.1016/j.arabjc.2020.10.025

Singh, N. B., Jain, P., De, A., & Tomar, R. (2021). Green synthesis and applications of nanomaterials. *Current Pharmaceutical Biotechnology, 22*(13), 1705–1747. https://doi.org/10.2174/1389201022666210412142734

Soltys, L., Olkhovyy, O., Tatarchuk, T., & Naushad, M. (2021). Green synthesis of metal and metal oxide nanoparticles: Principles of green chemistry and raw materials. *Magnetochemistry, 7*(11), 145. https://doi.org/10.3390/magnetochemistry7110145

Soni, V., Raizada, P., Singh, P., Cuong, H. N., Rangabhashiyam, S., Saini, A., Saini, R. V., Le, Q. V., Nadda, A. K., Le, T. T., & Nguyen, V. H. (2021). Sustainable and green trends in using plant extracts for the synthesis of biogenic metal nanoparticles toward environmental and pharmaceutical advances: A review. *Environmental Research, 202,* 111622. https://doi.org/10.1016/j.envres.2021.111622

Sumit, S. L., & Nayak, P. L. (2012). Green synthesis of gold nanoparticles using various extract of plants and spices. *International Journal of Science Innovations and Discoveries, 2*(3), 325–350.

Vasudeo, K., & Pramod, K. (2016). Biosynthesis of nickel nanoparticles using leaf extract of coriander. *BioTechnology: An Indian Journal, 12*(7), 106–111.

Venugopal Rao, S. (2011). Picosecond nonlinear optical studies of gold nanoparticles synthesised using coriander leaves (*Coriandrum sativum*). *Journal of Modern Optics*, *58*(12), 1024–1029. https://doi.org/10.1080/09500340.2011.590903

Verma, A., Gautam, S. P., Bansal, K. K., Prabhakar, N., & Rosenholm, J. M. (2019). Green nanotechnology: Advancement in phytoformulation research. *Medicines (Basel, Switzerland)*, *6*(1), 39. https://doi.org/10.3390/medicines6010039

Xu, C., & Qu, X. (2014). Cerium oxide nanoparticle: A remarkably versatile rare earth nanomaterial for biological applications. *NPG Asia Materials*, *6*, e90. https://doi.org/10.1038/am.2013.88

Yakout, S., Abdeltawab, A., Mostafa, A., & Abd-Alrahman, S. (2014). Green and facile approach for synthesis and characterization of silver nanoparticles using coriander (*Coriandrum sativum*) leaf extract. *Journal of Pure and Applied Microbiology*, *8*, 313–316.

Zhang, D., Ma, X. L., Gu, Y., Huang, H., & Zhang, G. W. (2020). Green synthesis of metallic nanoparticles and their potential applications to treat cancer. *Frontiers in Chemistry*, *8*, 799. https://doi.org/10.3389/fchem.2020.00799

Zhao, L., Zhao, X., Xu, Y., Liu, X., Zhang, J., & He, Z. (2021). Simultaneous determination of 49 amino acids, B vitamins, flavonoids, and phenolic acids in commonly consumed vegetables by ultra-performance liquid chromatography–tandem mass spectrometry. *Food Chemistry*, *344*, 128712. https://doi.org/10.1016/j.foodchem.2020.128712

Section III

Coriander Fixed Oil

Chemistry, Technology, Functionality, and Applications

29 Composition and Functionality of Coriander Fixed Oil

Mustafa Kiralan and Onur Ketenoglu

CONTENTS

29.1 INTRODUCTION

Coriander (*Coriandrum sativum* L.) belongs to the family Umbelliferae (Apiaceae) and is an annual herb generally used for seasoning and medicinal purposes. Coriander is native to the Mediterranean and Western Asia regions. Coriander is extensively cultivated in Ukraine, Russia, Romania, Morocco, Mexico, India, and Argentina. The edible parts of this plant are the seeds, leaves, and roots which have distinct flavors such as light and fresh (Bhat, Kaushal, Kaur, & Sharma, 2014; Priyadarshi, Khanum, Ravi, Borse, & Naidu, 2016). The coriander seeds are common ingredients in fish, meat, bakery, and confectionery products. The fresh leaves are called cilantro or Chinese parsley and are used in food preparation in the Middle East and Southeast Asia as a flavoring or agent for masking the unpleasant aroma of some foods (Laribi, Kouki, M'Hamdi, & Bettaieb, 2015). In addition, the roots and seeds are utilized in Thai cuisine for different purposes (Prachayasittikul, Prachayasittikul, Ruchirawat, & Prachayasittikul, 2018). The main reason for using coriander in various food products is to enhance aroma and increase the shelf life by exhibiting antioxidant and antimicrobial effects.

The chemical composition of coriander is responsible for its antioxidant and antimicrobial activity. For example, 100 g of fresh coriander leaves contain 92.2 g moisture, 2.13 g protein, 0.52 g lipid, 1.47 g ash, 3.67 carbohydrates, and 2.8 g dietary fiber (USDA, 2021). Besides, coriander contains many bioactive compounds such as minerals, phenolics, fatty acids, and essential oils (Laribi et al., 2015).

The essential and fixed oils of coriander are valuable products. Essential oil is extracted using traditional methods such as hydrodistillation and solid-liquid extraction with methylene chloride, and modern methods such as microwave-assisted hydrodistillation, supercritical fluid extraction, and subcritical water extraction (Ghazanfari, Mortazavi, Yazdi, & Mohammadi, 2020; Pavlić,

DOI: 10.1201/9781003204626-32

Vidović, Vladić, Radosavljević, & Zeković, 2015). Linalool is the predominant constituent of coriander essential oil. Coriander essential oil is reported to exhibit potent antioxidant and antimicrobial activity (Ghazanfari et al., 2020; Silva & Domingues, 2017). Fixed oil is extracted using different extraction techniques such as cold pressing and Soxhlet extraction with various extraction solvents used (Ashraf, Ghufran, Akram, Mushtaq, & Sultana, 2020). The predominant fatty acid identified in the fixed oil is petroselinic acid (C18:1n-12), making up 73% of all fatty acids (Uitterhaegen et al., 2016). Besides, coriander fixed oil is rich in tocopherols. Among tocopherols, β-tocopherol is the main tocopherol isomer in coriander fixed oil (56% of total tocopherol content) (Ramadan, 2013b). Tocopherols and tocotrienols exhibit great antioxidant capacity and biological activities (Uitterhaegen et al., 2016). β-Sitosterol, campesterol, stigmasterol, and Δ ^{7}-stigmasterol are the major oil sterols from coriander fruit parts, including the seed, pericarp, and whole fruit (Sriti, Talou, Wannes, Cerny, & Marzouk, 2009).

29.2 COMPOSITION OF CORIANDER (*CORIANDRUM SATIVUM*) FIXED OIL

29.2.1 OIL CONTENT

The coriander seed contains different amounts of oil depending on the cultivar, geographic location, weather conditions, and extraction techniques. Nguyen, Talou, Evon, Cerny, and Merah (2020) evaluated the oil content of coriander seeds during different fruit development stages, and the oil yield varied between 5.8% and 24.9%. A rapid oil accumulation was observed gradually from blooming to maturity. Ghobadi and Ghobadi (2012) stated a slight increase in oil content with a delay at sowing. The oil yield also varied according to geographical regions (Table 29.1). The coriander from Tunisia and France has higher oil contents. The extraction technique can also influence the oil content of coriander. The oil content of coriander obtained using ultrasound-assisted solvent extraction increased up to 30.74% (Senrayan & Venkatachalam, 2019). Sriti, Msaada, Talou, Faye, Kartika, and Marzouk (2012) emphasized that a twin-screw extruder could enhance the oil content of coriander.

29.3 FATTY ACID AND ACYL LIPIDS PROFILE

29.3.1 TRIGLYCERIDES AND FATTY ACIDS

Neutral lipids (NL) in coriander oil represent 94.8% of total lipids and are mainly composed of triglycerides (95.5%), followed by free fatty acids (2.05%), diacylglycerols (1.8%), and monoacylglycerols

---

TABLE 29.1

Yields of Coriander Fixed Oils from Different Geographical Regions

Coriander Origin	Oil Yield (%)	References
France	23	(Uitterhaegen et al., 2016)
Tunisia	21.8–26.4	(Msaada, Hosni, Ben Taarit, Chahed, Hammami, & Marzouk, 2009; Sriti, Talou, Msaada, & Marzouk, 2011)
Canada	15.8	(Sriti, Talou, Msaada et al., 2011)
Turkey	10.6–23.0	(Kiralan, Calikoglu, Ipek, Bayrak, & Gurbuz, 2009; Matthäus, Özcan, & Al Juhaimi, 2015)
Germany	28.4	(Ramadan & Mörsel, 2002)
Iran	13.7–14.6	(Ghobadi and Ghobadi, 2012)
Ethiopia	11.1–16.5	(Mengesha & Getinetalemaw, 2010)
Vietnam	19.7	(Matthaus, Vosmann, Pham, & Aitzetmüller, 2003)
India	12.5	(Daga, Vaishnav, Dalmia, & Tumaney, 2021)

(0.5%) (Sriti, Wannes, Talou, Mhamdi, Cerny, & Marzouk, 2010). Ramadan and Morsel (2002) stated that the major neutral lipid subclass was triglycerides (91.1% of the total NL). The other NL subclasses were free fatty acids (5.87%), diacylglycerols (0.69%), sterol esters (0.33%), and mono-acylglycerols (0.29%) in coriander oil. In these NL subclasses, petroselinic acid and linoleic acid were the major fatty acids, and the sum of these two fatty acids constituted 85% of total fatty acids in all NL subclasses.

The distribution of NL, glycolipid (GL), and phospholipid (PL) changed according to the oils from different parts of coriander. Among the TL, the higher content in NL is in the seed (93.1%) and whole fruit (93.7%); GL was the predominant fraction in the pericarp (73.2%) (Sriti, Wannes, Talou, Mhamdi, Hamdaoui, & Marzouk, 2010). Six TAG species were detected in coriander oils, and the major triglycerides were tripetroselinin and/or dipetroselinoyloleoyl glycerol which comprised more than 50% of the total triglyceride content (Ramadan and Morsel, 2002).

The major fatty acids in coriander seeds are petroselinic acid (C18:1n-12), linoleic acid (C18:2n-6), palmitic acid (C16:0), and oleic acid (Table 29.2). Saturated fatty acids include mainly palmitic acid (C16:0) and stearic acid (C18:0); monounsaturated fatty acids (MUFA) are primarily petroselinic acid (C18:1n-12) and oleic acid (C18:1n-9); and polyunsaturated fatty acids (PUFA) consist largely of linoleic (C18:2n-6) and linolenic acid (C18:3n-3).

Different stages of maturation influence fatty acid composition (Table 29.3). Msaada, Hosni, Taarit, Hammami, and Marzouk (2009) reported that saturated and polyunsaturated fatty acids decreased significantly and monounsaturated fatty acids increased during the maturation of coriander fruit. In particular, an increase was observed in petroselinic acid, and a decrease in palmitic acid with the fruit development of coriander. Nguyen, Talou, Cerny, Evon, and Merah (2015) emphasized that saturated and polyunsaturated acids were major subclasses of fatty acids at earlier stages (2-12 DAF).

Genotype is another factor that affects the fatty acid composition of coriander. For example, Kiralan et al. (2009) reported that a small variety of coriander (*C. sativum* var. *microcarpum*) had a slightly higher content of petroselinic acid compared to other varieties (*C. sativum* var. *vulgare*).

Although seeds are generally used in oil production, different parts have different fatty acid compositions. In the study by Sriti et al. (2009), coriander oils obtained from different parts of the fruit were investigated. While the fatty acids composition of the oils obtained from the whole seed and fruit were similar, the composition of the fatty acids obtained from the pericarp part showed differences. Pericarp oil had higher saturated fatty acids than seed and whole fruit oil, while pericarp oil was poor in terms of polyunsaturated fatty acids. Regarding individual fatty acids, palmitic acid (18.4%), myristic acid (4.83%), and stearic acid (2.50%) content was higher in pulp oil, while petroselinic acid was lower in this oil than seed and whole fruit oils.

29.3.2 POLAR LIPIDS

Polar lipids comprise a wide range of molecules, including phospholipids and glycolipids. Some of these components exhibit biological activities such as antioxidant activity, reducing/preventing cardiovascular diseases, memory improvement, and immunological function (Alves, Simoes, & Domingues, 2021; Sun, Chen, Wang, & Lin, 2018).

Sriti, Wannes, Talou, Mhamdi, Cerny et al. (2010) isolated and identified five phospholipid fractions: phosphatidylinositol (PI) phosphatidylcholine (PC), phosphatidylglycerol (PG), phosphatidylethanolamine (PE), and phosphatidic acid (PA). In the study of Ramadan and Morsel (2002), seven subclasses were isolated and identified: PG, PE, PI, PC, lysophosphatidylethanolamine (LPE), phosphatidylserine (PS), and lysophosphatidylcholine (LPC). The fatty acid compositions in individual phospholipid fractions are exhibited in Table 29.4. In all fractions, petroselinic acid's major fatty acid comprises approximately 50% of total fatty acids, followed by linoleic and palmitic acids, respectively.

TABLE 29.2
Fatty Acids in Coriander Oils from Different Countries

	Vietnam	Hungary	Tunisia	Turkey	Canada	USA	Egypt
Saturated fatty acids (SFA)							
C12:0	-	-	-	-	-	-	0.33
C14:0	0.05	-	0.08	0.03–0.08	0.08	-	0.09
C16:0	2.91	5.54	3.50	3.74–5.18	3.97	5.3	7.82
C18:0	0.51	1.36	0.78	0.80–0.97	0.95	3.1	4.97
C20:0	0.07	-	0.10	0.07–0.11	0.10	-	-
C22:0	0.06	-	-	-	-	-	-
C24:0	0.07	-	-	-	-	-	-
Monounsaturated fatty acids (MUFA)							
C16:1 n-7	0.19	-	0.23	-	0.19	0.3	0.26
C18:1 n-12	78.7	67.0	76.6	72.3–77.1	73.2	68.5	47.9
C18:1 n-9	-	7.86	5.47	2.94–3.44	6.48	7.6	
C18:1 n-7	-	-	-	-	-	1.0	
C18:1 n-11	0.82	-	-	-	-	-	
C20:1	0.31	-	-	0.19–0.27	-	-	
C22:1 n-9	-	0.77	-	-	-	-	
Polyunsaturated fatty acid (PUFA)							
C18:2 n-6	14.2	15.9	13.0	14.7–17.3	14.8	13.0	36.6
C18:3 n-3	0.20	-	0.15	0.18–0.43	0.19	-	1.66
C18:3 n-6	-	1.00	-	-	-	-	-
C22:1	-	-	-	-	-	-	0.27
C22:2 n-6	0.06	-	-	-	-	-	-
C22:6 n-3	-	0.57	-	-	-	-	-
References	(Matthaus et al., 2003)	(Ramadan et al., 2003)	(Sriti, Wannes, Talou, Mhamdi, Cerny et al., 2010)	(Kiralan et al., 2009)	(Sriti, Talou, Msaada et al., 2011)	(Moser & Vaughn, 2010)	(Ramadan, 2013a)

TABLE 29.3

Major Fatty Acids and Subclass of Fatty Acids during Coriander Fruit Maturation Cultivated in Different Patterns and Regions

Sample	Days after Flowering (DAF)	Subclass of Fatty Acid (%)			Palmitic Acid C16:0	Petroselinic Acid (C18:1n-12)	Linoleic Acid (C18:2n-6)	Reference
		Saturated (SFA)	Monounsaturated (MUFA)	Polyunsaturated (PUFA)				
Tunisia, fruits	5	23.3	32.1	44.6	17.7	25.9	29.2	(Msaada, Hosni, Ben Taarit et al., 2009)
	9	16.5	44.3	39.2	14.4	42.4	30.9	
	13	11.0	61.4	27.4	9.4	60.3	22.9	
	18	5.4	67.8	26.8	4.5	53.1	20.8	
	23	6.8	74.2	18.9	5.5	73.6	16.5	
	27	9.4	79.5	11.1	3.5	74.6	10.5	
	32	0.9	84.9	14.2	0.1	84.8	13.6	
	37	1.1	84.4	14.5	0.1	84.2	13.7	
	55	4.5	81.2	14.3	3.6	80.9	13.6	
France, seeds (data from 2011a season)	3	42.1	3.6	54.3	27.6	1.0	35.6	(Nguyen et al., 2020)
	11	44.1	10.5	45.4	24.2	6.2	32.7	
	24	16.3	62.7	21.0	10.0	52.0	19.6	
	30	16.0	61.8	22.2	9.8	51.8	20.1	
	35	6.1	79.5	14.4	4.1	72.3	14.1	
	59	4.8	80.4	14.9	3.8	74.3	14.7	
France, seeds	2	32.6	5.6	61.8	20.5	2.8	44.2	(Nguyen et al., 2015)
	5	32.0	6.7	65.4	21.1	3.6	45.1	
	10	34.9	11.3	57.4	20.9	7.8	36.0	
	12	14.4	55.2	32.0	10.9	48.8	26.9	
	14	10.8	65.7	25.1	7.6	59.3	22.1	
	18	5.7	79.1	16.1	4.0	74.4	15.4	
	25	4.6	81.1	14.6	3.5	76.4	14.3	
	35	4.9	79.5	15.7	3.7	74.7	15.0	
	53	6.6	78.2	15.6	5.1	71.9	15.0	

TABLE 29.4
The Major Fatty Acids in Individual Phospholipid Fractions

PL Subclasses*	Unsaturated Fatty Acids		Saturated Fatty Acids	
	C18:1 n-12	C18:2 n-6	C16:0	C18:0
PG	53.2	25.7	10.2	4.45
PE	48.7	23.2	16.2	5.00
PI	49.2	22.9	21.5	1.60
LPE	57.0	28.6	9.50	0.50
PS	52.4	21.9	21.0	0.50
PC	57.7	23.7	9.10	3.74
LPC	56.6	24.6	10.4	1.89

Source: adapted from Ramadan et al., 2002.

* Phosphatidylserine (PS), phosphatidylethanolamine (PE), phosphatidylinositol (PI), phosphatidylcholine (PC), phosphatidylglycerol (PG), lysophosphatidylcholine (LPC), lysophosphatidylethanolamine (LPE).

29.4   MINOR BIOACTIVE LIPIDS

29.4.1   Tocopherols and Tocotrienols

Tocols (tocopherol and tocotrienols) are minor oil components and exhibit strong antioxidant activity towards lipid oxidation in foods and biological systems (Kamal-Eldin & Appelqvist, 1996; Shahidi & Zhong, 2010). Depending on location, the tocol composition of the oil is exhibited in Table 29.5. γ-Tocotrienol is the major tocol isomer in oils from France, Vietnam, Tunisia, and Canada. In contrast, δ-tocotrienol was the major isomer in Egyptian coriander oil (Ramadan, 2013a). In Hungary and the USA, coriander oil tocotrienols are not determined. β-Tocopherol was the major tocol found in Hungarian coriander oils, while only two isomers, γ- and δ-tocopherols, were detected in the coriander oils from the USA.

The tocol composition of the coriander oils shows differences depending on the different parts of the plant. For example, Sriti, Wannes, Talou, Mhamdi, Hamdaoui et al. (2010) determined that γ-tocotrienol was the major tocol isomer in oils from whole fruit and seed (19.5 and 19.5 mg/100 g, respectively), while α-tocopherol and γ-tocotrienol were the predominant tocols in pericarp oil. Besides, the highest tocol content was detected in whole fruit and seed oils (27.7 and 26.4 mg/100 g, respectively), while pericarp oil was poor for tocol content (5.36 mg/100 g).

Illés, Daood, Perneczki, Szokonya, and Then (2000) examined the effects of different extraction techniques (carbon dioxide and propane as solvents, under sub- and supercritical conditions) on the tocopherol content of coriander oil. Under CO_2 at subcritical conditions (100–200 bar and 25°C), the amount of α-tocopherol increased in oils. The average concentration of α-tocopherol increased up to 5.2 mg/g in oils under mild conditions (100 bar and 25°C). A similar trend was observed for β- and γ-tocopherol.

29.4.2   Phytosterols

Coriander oil is a rich source of phytosterols, wherein the total phytosterol content was determined as 6.70 g/kg (Uitterhaegen et al., 2016). The sterol compositions of coriander oils from different locations are presented in Table 29.6. The predominant compound is β-sitosterol, followed by stigmasterol. Other major compounds include Δ⁷-stigmastenol, Δ⁵,²⁴- stigmastadienol, and campesterol. Besides sterols, squalene was determined in the unsaponifiable matter with 14.5 mg/100 g oil (Daga et al., 2021).

TABLE 29.5
Tocol Composition of Coriander Oil

Tocol	France mg/kg	Vietnam mg/kg	Hungary mg/kg	Tunisia mg/100 g	Canada mg/100 g	USA mg/kg	Egypt mg/kg
α-Tocopherol	12.4	46	86	1.12	0.59	196	45.95
β-Tocopherol	-	-	672	0.05	-	-	38.76
γ-Tocopherol	10.1	31	162	0.33	0.75	18	3944
δ-Tocopherol	-	-	347	-	0.18	15	110.2
α-Tocotrienol	98.0	96	-	6.04	4.64	-	1327
β-Tocotrienol	-	-	-	-	-	-	40.10
γ-Tocotrienol	350	231	-	19.5	14.5	-	507.2
δ-Tocotrienol	25.7	41	-	0.69	0.70	-	1005
Reference	(Uitterhaegen et al., 2016)	(Matthaus et al., 2003)	(Ramadan et al., 2003)	(Sriti, Talou, Msaada et al., 2011)	(Sriti, Talou, Msaada et al., 2011)	(Moser et al., 2010)	(Ramadan, 2013a)

TABLE 29.6
Sterol Composition of Coriander Fixed Oil

Sterol	France mg/kg	Hungary g/kg	Tunisia mg/g oil	Canada mg/g oil	Egypt g/kg
Cholesterol	0.02	-	-	-	-
Campesterol	0.54	0.735	4.86	5.02	-
Stigmasterol	1.61	1.512	14.7	14.7	0.803
β-Sitosterol	2.31	1.553	21.3	21.9	2.644
Δ^5-Avenasterol	0.27	1.466	1.44	1.27	1.511
Δ^7-Stigmastenol	1.22	-	9.42	10.9	0.141
Δ^7-Avenasterol	0.40	0.365	2.70	2.46	0.192
$\Delta^{5,24}$- Stigmastadienol	-	-	6.58	6.61	-
Gramisterol	0.07	-	-	-	-
Citrostadienol	0.10	-	-	-	-
Cycloartenol	0.08	-	-	-	-
Methylene cycloartenol	0.06	-	-	-	-
Lanosterol	-	-	-	-	1.456
Sitostanol	-	-	-	-	0.112
References	(Uitterhaegen et al., 2016)	(Ramadan et al., 2003)	(Sriti, Talou, Msaada et al., 2011)	(Sriti, Talou, Msaada et al., 2011)	(Ramadan, Amer, & Awad, 2008)

The oil extraction method also affects the amount and type of phytosterol extracted coriander oil. Sriti et al. (2011) examined the effects of two extraction methods, screw press and Soxhlet methods, on the sterol composition of oils. The percentage of β-sitosterol detected in Soxhlet-extracted oil (34.9% of total sterols) was higher than that of screw pressed oil (28.83% of total sterols). However, the concentration of $\Delta^{5,24}$-stigmastadienol, Δ^7-stigmasterol, and Δ^7-avenasterol was higher in screw pressed oils than in Soxhlet-extracted oils. Another study by Sriti et al. (2012) evaluated the effects of screw speed and fruit flow rate in twin-screw extruders on individual sterols. The content of $\Delta^{5,24}$-stigmastadienol decreased with the increase of the screw speed from 50 to 100 rpm, while the levels of Δ^5-avenasterol increased at the same conditions for screw speed but increasing fruit flow rate from 2.27 to 3.4 kg/h.

29.4.3 PHENOLICS

Total phenolics of the extracts from oils are generally analyzed with the Folin-Ciocalteu reagent. The total phenolics contents were determined to be 11 ppm caffeic acid in the coriander oil from Germany (Ramadan & Mörsel, 2004), 4.3 mg gallic acid equivalents (GAE)/g for Egyptian oil (Ramadan, 2013a), and 157.26 mg GAE/100 g in Turkish oil (Dedebas, Ekici, & Sagdic, 2021). Daga et al. (2021) also examined the individual phenolic compounds of coriander oils. In addition, hydroxybenzoic acid (470 μg/100 g oil), *trans*-cinnamic acid (348 μg/100 g oil), *p*-coumaric acid (227 μg/100 g oil), and kaempferol (91 μg/100 g oil) were identified as phenolics in oil samples.

The polarity of solvents used in the oil extraction from the seeds also significantly affects the chemical composition and, thus, total phenolic content. Kozłowska, Gruczyńska, Ścibisz, and Rudzińska (2016) evaluated the effect of two different solvents (chloroform/methanol and hexane) used in oil extraction on total phenolics. For example, the level of total phenolics in chloroform/methanol extracted oils (0.20 mg GAE/g oil) was greater than in those extracted with *n*-hexane (0.17 mg GAE/g oil).

29.4.4 Volatiles

The vacuum-assisted sorbent extraction (VASE) method is an extraction system that reduces the sampling time, enriches detectable compounds, and improves reproducibility (Trujillo-Rodríguez, Anderson, Dunham, Noad, & Cardin, 2020). Jeleń, Marcinkowska, and Marek (2021) used the VASE method to extract terpenes from the cold-pressed coriander oil. A total of 17 terpenes were identified with the VASE. At 20°C, the predominant compound was linalool (23.0%), followed by α-pinene (17.9%), γ-terpinene (13.9%), sylvestrene (10.1%), β-pinene (9.70%), and o-cymene (8.92%). The characteristic flavor of coriander originates from the presence of terpenes, especially linalool which is the major volatile in coriander essential oils.

29.5 FUNCTIONALITY OF CORIANDER OIL

Coriander oil exhibits strong antioxidant activity due to its chemical compounds. Ramadan and Moersel (2006) evaluated the antiradical activity of various vegetable oils (coriander, black cumin, cottonseed, peanut, sunflower, walnut, hemp seed, linseed, olive, niger seed oils) using the DPPH· method. Coriander oil exhibited potent radical scavenging activity (26.7%). Ramadan, Kroh, and Mörsel (2003) evaluated the antiradical activity of crude black cumin, coriander, and niger seed oils. In the DPPH· test, coriander oil had the highest radical scavenging activity (35.0% of DPPH· radicals were quenched), followed by black cumin (25.1%) and niger seed oils (14.0%). In the same study, electron spin resonance (ESR) measurements showed similar results to DPPH·. The quenching percentages of galvinoxyl radical were 32.4%, 23.3%, and 12.8% for coriander, black cumin, and niger crude seed oils, respectively.

Kozłowska et al. (2016) reported a higher antiradical activity in chloroform/methanol extracted oils (4.96 μmol TEAC/g oil) than those of oils extracted with n-hexane (2.24 μmol TEAC/g oil). Daga et al. (2021) used three tests, including DPPH·, ABTS, and FRAP assays, to determine the antioxidant activity of coriander oils. Coriander oil exhibited moderate antioxidant activity among fixed oils. DPPH·, ABTS, and FRAP levels were found to be 8.15 mg ascorbic acid equivalent/100 g oil, 599 mM Trolox equivalent/100 g oil, and 956 mM Trolox equivalent/100 g oil, respectively.

Coriander oil is also used to increase the stability of vegetable oils having low oxidative stability. Ramadan (2013a) used blends (10% and 20%, w/w) of cold-pressed coriander oil with high linoleic sunflower oil. Under oxidative conditions (60°C) for 8 days, lower levels in peroxide value (PV), conjugated diene (CD), conjugated triene (CT), and p-anisidine were observed in blended oils compared to the control sample (sunflower oil). In addition to increasing oxidative stability, coriander oil also enhanced the tocopherol MUFA contents of sunflower oils while decreasing PUFA content. In another study by Ramadan and Wahdan (2012), the researchers utilized coriander oil in corn oil, and they stated that their PV, CD, and CT values were lower in blended oils than in corn oil. Blending with CO resulted in significant increases in the MUFA and tocopherol content, especially β- and δ-tocopherols which were the main tocopherols in coriander oil.

29.6 CONCLUSION

Coriander is a well-known plant and is commonly used in seasonings or for medical purposes due to its rich content in many functional components such as tocopherols, phenolic components, sterols, and fatty acids. The fixed oil of coriander is rich in petroselinic acid; moreover, the oil is a good source for such unsaturated fatty acids as oleic, linoleic, and linolenic. β-Sitosterol constitutes the major part of the phytosterols present in coriander oil, followed by stigmasterol. Linalool is the most abundant component in the volatile part of the plant, and many phenolic components with various functional properties exist in coriander. In addition, coriander oil is reported to have such functional activities as antioxidant and antiradical activities, which are beneficial in prolonging the oxidative stability of such oils as sunflower or corn oils. Several studies have been conducted on the composition and functionality of this oil, and it is evident that further research is needed to reveal other health-promoting effects of this valuable plant and its extracts.

REFERENCES

Alves, E., Simoes, A., & Domingues, M. R. (2021). Fruit seeds and their oils as promising sources of value-added lipids from agro-industrial byproducts: Oil content, lipid composition, lipid analysis, biological activity and potential biotechnological applications. *Critical Reviews in Food Science and Nutrition*, *61*(8), 1305–1339. https://doi.org/10.1080/10408398.2020.1757617.

Ashraf, R., Ghufran, S., Akram, S., Mushtaq, M., & Sultana, B. (2020). Chapter 31- Cold pressed coriander (*Coriandrum sativum* L.) seed oil. In M. F. Ramadan (Ed.), *Cold Pressed Oils* (pp. 345–356): Academic Press.

Bhat, S., Kaushal, P., Kaur, M., & Sharma, H. (2014). Coriander (*Coriandrum sativum* L.): Processing, nutritional and functional aspects. *African Journal of Plant Science*, *8*(1), 25–33. https://doi.org/10.5897/AJPS2013.1118.

Daga, P., Vaishnav, S. R., Dalmia, A., & Tumaney, A. W. (2021). Extraction, fatty acid profile, phytochemical composition and antioxidant activities of fixed oils from spices belonging to Apiaceae and Lamiaceae family. *Journal of Food Science and Technology*. https://doi.org/10.1007/s13197-021-05036-1.

Dedebas, T., Ekici, L., & Sagdic, O. (2021). Chemical characteristics and storage stabilities of different cold-pressed seed oils. *Journal of Food Processing and Preservation*, *45*(2), e15107. https://doi.org/10.1111/jfpp.15107.

Ghazanfari, N., Mortazavi, S. A., Yazdi, F. T., & Mohammadi, M. (2020). Microwave-assisted hydrodistillation extraction of essential oil from coriander seeds and evaluation of their composition, antioxidant and antimicrobial activity. *Heliyon*, *6*(9), e04893. https://doi.org/10.1016/j.heliyon.2020.e04893.

Ghobadi, M.-E., & Ghobadi, M. (2012). Effects of late sowing on quality of coriander (*Coriandrum sativum* L.). *International Journal of Agricultural and Biosystems Engineering*, *6*(7), 425–428. https://doi.org/10.5281/zenodo.1061808.

Illés, V., Daood, H. G., Perneczki, S., Szokonya, L., & Then, M. (2000). Extraction of coriander seed oil by CO₂ and propane at super- and subcritical conditions. *The Journal of Supercritical Fluids*, *17*(2), 177–186. https://doi.org/10.1016/S0896-8446(99)00049-2.

Jeleń, H. H., Marcinkowska, M. A., & Marek, M. (2021). Determination of volatile terpenes in coriander cold pressed oil by vacuum assisted sorbent extraction (VASE). *Molecules*, *26*(4), 884. https://doi.org/10.3390/molecules26040884.

Kamal-Eldin, A., & Appelqvist, L.-Å. (1996). The chemistry and antioxidant properties of tocopherols and tocotrienols. *Lipids*, *31*(7), 671–701. https://doi.org/10.1007/BF02522884.

Kiralan, M., Calikoglu, E., Ipek, A., Bayrak, A., & Gurbuz, B. (2009). Fatty acid and volatile oil composition of different coriander (*Coriandrum sativum*) registered varieties cultivated in Turkey. *Chemistry of Natural Compounds*, *45*(1), 100–102. https://doi.org/10.1007/s10600-009-9240-2.

Kozłowska, M., Gruczyńska, E., Ścibisz, I., & Rudzińska, M. (2016). Fatty acids and sterols composition, and antioxidant activity of oils extracted from plant seeds. *Food Chemistry*, *213*, 450–456. https://doi.org/10.1016/j.foodchem.2016.06.102.

Laribi, B., Kouki, K., M'Hamdi, M., & Bettaieb, T. (2015). Coriander (*Coriandrum sativum* L.) and its bioactive constituents. *Fitoterapia*, *103*, 9–26. https://doi.org/10.1016/j.fitote.2015.03.012.

Matthäus, B., Özcan, M., & Al Juhaimi, F. (2015). Variations in oil, fatty acid and tocopherol contents of some Labiateae and Umbelliferae seed oils. *Quality Assurance and Safety of Crops & Foods*, *7*(2), 103–107. https://doi.org/10.3920/QAS2013.0296.

Matthaus, B., Vosmann, K., Pham, L. Q., & Aitzetmüller, K. (2003). FA and tocopherol composition of Vietnamese oilseeds. *Journal of the American Oil Chemists' Society*, *80*(10), 1013–1020. https://doi.org/10.1007/s11746-003-0813-y.

Mengesha, B., & Getinetalemaw, G. (2010). Variability in Ethiopian coriander accessions for agronomic and quality traits. *African Crop Science Journal*, *18*(2), 43–49. https://doi.org/10.4314/acsj.v18i2.65795.

Moser, B. R., & Vaughn, S. F. (2010). Coriander seed oil methyl esters as biodiesel fuel: Unique fatty acid composition and excellent oxidative stability. *Biomass and Bioenergy*, *34*(4), 550–558. https://doi.org/10.1016/j.biombioe.2009.12.022.

Msaada, K., Hosni, K., Ben Taarit, M., Chahed, T., Hammami, M., & Marzouk, B. (2009). Changes in fatty acid composition of coriander (*Coriandrum sativum* L.) fruit during maturation. *Industrial Crops and Products*, *29*(2), 269–274. https://doi.org/10.1016/j.indcrop.2008.05.011.

Msaada, K., Hosni, K., Taarit, M. B., Hammami, M., & Marzouk, B. (2009). Effects of growing region and maturity stages on oil yield and fatty acid composition of coriander (*Coriandrum sativum* L.) fruit. *Scientia Horticulturae*, *120*(4), 525–531. https://doi.org/10.1016/j.scienta.2008.11.033.

Nguyen, Q.-H., Talou, T., Cerny, M., Evon, P., & Merah, O. (2015). Oil and fatty acid accumulation during coriander (*Coriandrum sativum* L.) fruit ripening under organic cultivation. *The Crop Journal*, *3*(4), 366–369. https://doi.org/10.1016/j.cj.2015.05.002.

Nguyen, Q.-H., Talou, T., Evon, P., Cerny, M., & Merah, O. (2020). Fatty acid composition and oil content during coriander fruit development. *Food Chemistry*, *326*, 127034. https://doi.org/10.1016/j.foodchem.2020.127034.

Pavlić, B., Vidović, S., Vladić, J., Radosavljević, R., & Zeković, Z. (2015). Isolation of coriander (*Coriandrum sativum* L.) essential oil by green extractions versus traditional techniques. *The Journal of Supercritical Fluids*, *99*, 23–28. https://doi.org/10.1016/j.supflu.2015.01.029.

Prachayasittikul, V., Prachayasittikul, S., Ruchirawat, S., & Prachayasittikul, V. (2018). Coriander (*Coriandrum sativum*): A promising functional food toward the well-being. *Food Research International*, *105*, 305–323. https://doi.org/10.1016/j.foodres.2017.11.019.

Priyadarshi, S., Khanum, H., Ravi, R., Borse, B. B., & Naidu, M. M. (2016). Flavour characterisation and free radical scavenging activity of coriander (*Coriandrum sativum* L.) foliage. *Journal of Food Science and Technology*, *53*(3), 1670–1678. https://doi.org/10.1007/s13197-015-2071-1.

Ramadan, M., & Mörsel, J.-T. (2002). Oil composition of coriander (*Coriandrum sativum* L.) fruit-seeds. *European Food Research and Technology*, *215*(3), 204–209. https://doi.org/10.1007/s00217-002-0537-7.

Ramadan, M. F. (2013a). Healthy blends of high linoleic sunflower oil with selected cold pressed oils: Functionality, stability and antioxidative characteristics. *Industrial Crops and Products*, *43*, 65–72. https://doi.org/10.1016/j.indcrop.2012.07.013.

Ramadan, M. F. (2013b). Improving the stability and radical-scavenging activity of sunflower oil upon blending with black cumin (*Nigella sativa*) and coriander (*Coriandrum sativum*) seed oils. *Journal of Food Biochemistry*, *37*(3), 286–295. https://doi.org/10.1111/j.1745-4514.2011.00630.x.

Ramadan, M. F., Amer, M. M. A., & Awad, A. E.-S. (2008). Coriander (*Coriandrum sativum* L.) seed oil improves plasma lipid profile in rats fed a diet containing cholesterol. *European Food Research and Technology*, *227*(4), 1173–1182. https://doi.org/10.1007/s00217-008-0833-y.

Ramadan, M. F., Kroh, L. W., & Mörsel, J.-T. (2003). Radical scavenging activity of black cumin (*Nigella sativa* L.), coriander (*Coriandrum sativum* L.), and niger (*Guizotia abyssinica* Cass.) crude seed oils and oil fractions. *Journal of Agricultural and Food Chemistry*, *51*(24), 6961–6969. https://doi.org/10.1021/jf0346713.

Ramadan, M. F., & Moersel, J.-T. (2006). Screening of the antiradical action of vegetable oils. *Journal of Food Composition and Analysis*, *19*(8), 838–842. https://doi.org/10.1016/j.jfca.2006.02.013.

Ramadan, M. F., & Mörsel, J.-T. (2004). Oxidative stability of black cumin (*Nigella sativa* L.), coriander (*Coriandrum sativum* L.) and niger (*Guizotia abyssinica* Cass.) crude seed oils upon stripping. *European Journal of Lipid Science and Technology*, *106*(1), 35–43. https://doi.org/10.1002/ejlt.200300895.

Ramadan, M. F., & Wahdan, K. M. M. (2012). Blending of corn oil with black cumin (*Nigella sativa*) and coriander (*Coriandrum sativum*) seed oils: Impact on functionality, stability and radical scavenging activity. *Food Chemistry*, *132*(2), 873–879. https://doi.org/10.1016/j.foodchem.2011.11.054.

Senrayan, J., & Venkatachalam, S. (2019). Optimization of ultrasound-assisted solvent extraction (UASE) based on oil yield, antioxidant activity and evaluation of fatty acid composition and thermal stability of *Coriandrum sativum* L. seed oil. *Food Science and Biotechnology*, *28*(2), 377–386. https://doi.org/10.1007/s10068-018-0467-1.

Shahidi, F., & Zhong, Y. (2010). Lipid oxidation and improving the oxidative stability. *Chemical Society Reviews*, *39*(11), 4067–4079. https://doi.org/10.1039/B922183M.

Silva, F., & Domingues, F. C. (2017). Antimicrobial activity of coriander oil and its effectiveness as food preservative. *Critical Reviews in Food Science and Nutrition*, *57*(1), 35–47. https://doi.org/10.1080/10408398.2013.847818.

Sriti, J., Msaada, K., Talou, T., Faye, M., Kartika, I. A., & Marzouk, B. (2012). Extraction of coriander oil by twin-screw extruder: Screw configuration and operating conditions effect. *Industrial Crops and Products*, *40*, 355–360. https://doi.org/10.1016/j.indcrop.2012.03.034.

Sriti, J., Talou, T., Faye, M., Vilarem, G., & Marzouk, B. (2011). Oil extraction from coriander fruits by extrusion and comparison with solvent extraction processes. *Industrial Crops and Products*, *33*(3), 659–664. https://doi.org/10.1016/j.indcrop.2011.01.005.

Sriti, J., Talou, T., Msaada, K., & Marzouk, B. (2011). Comparative analysis of fatty acid, sterol and tocol composition of Tunisian and Canadian coriander (*Coriandrum sativum* L.) fruit. *Analytical Chemistry Letters*, *1*(5–6), 375–383. https://doi.org/10.1080/22297928.2011.10648241.

Sriti, J., Talou, T., Wannes, W. A., Cerny, M., & Marzouk, B. (2009). Essential oil, fatty acid and sterol composition of Tunisian coriander fruit different parts. *Journal of the Science of Food and Agriculture*, *89*(10), 1659–1664. https://doi.org/10.1002/jsfa.3637.

Sriti, J., Wannes, W. A., Talou, T., Mhamdi, B., Cerny, M., & Marzouk, B. (2010). Lipid profiles of Tunisian coriander (*Coriandrum sativum*) seed. *Journal of the American Oil Chemists' Society*, *87*(4), 395–400. https://doi.org/10.1007/s11746-009-1505-1.

Sriti, J., Wannes, W. A., Talou, T., Mhamdi, B., Hamdaoui, G., & Marzouk, B. (2010). Lipid, fatty acid and tocol distribution of coriander fruit's different parts. *Industrial Crops and Products*, *31*(2), 294–300. https://doi.org/10.1016/j.indcrop.2009.11.006.

Sun, N., Chen, J., Wang, D., & Lin, S. (2018). Advance in food-derived phospholipids: Sources, molecular species and structure as well as their biological activities. *Trends in Food Science & Technology*, *80*, 199–211. https://doi.org/10.1016/j.tifs.2018.08.010.

Trujillo-Rodríguez, M. J., Anderson, J. L., Dunham, S. J. B., Noad, V. L., & Cardin, D. B. (2020). Vacuum-assisted sorbent extraction: An analytical methodology for the determination of ultraviolet filters in environmental samples. *Talanta*, *208*, 120390. https://doi.org/10.1016/j.talanta.2019.120390.

Uitterhaegen, E., Sampaio, K. A., Delbeke, E. I. P., De Greyt, W., Cerny, M., Evon, P., ... Stevens, C. V. (2016). Characterization of French coriander oil as source of petroselinic acid. *Molecules*, *21*(9), 1202. https://doi.org/10.3390/molecules21091202.

United States Department of Agriculture (USDA). Food Data Central: Coriander (cilantro) leaves, raw. FDC ID: 169997 NDB Number:11165. Released in April 2018. https://fdc.nal.usda.gov/fdc-app.html#/food-details/169997/nutrients (accessed on 5 April 2021).

30 Effect of Processing on the Composition and Quality of Coriander Fixed Oil

Rabia Shabir Ahmad, Ali Imran, Muhammad Imran,
Muhammad Kamran Khan, Muhammad Haseeb Ahmad,
Muhammad Sajid Arshad, Muhammad Abdul Rahim,
and Abdul Sami

CONTENTS

30.1 INTRODUCTION

Coriander (*Coriandrum sativum* L.) is a member of the Apiaceae family known as cilantro, parsley, or dhania. Its young plant is called cilantro, while the fresh fruit is coriander. Coriander originally came from the word "korannon," which means "koris" and "anon" in Greece (Önder, 2018). Coriander is a flavoring herb that grows year after year. The plant's leaves, seeds, and roots are edible, and their taste is pleasant and fresh (Bhat et al., 2014). Although coriander has a hollow, green stem and green leaves about 50 cm long, the herb may be purple or red when flowering. Coriander has heterophylly, which means that it has different shapes of leaves on the same plant. It belongs to the Umbelliferae family, which means its blooms are arranged in umbels and are white or pale pink. It is a brown globular fruit with many longitudinal ridges that grow up to 6 mm in diameter. It is split into two varieties based on the size of the fruit: *C. sativum* var. *Vulgare alef.* and *C. sativum* var. *vulgare alef.* Also available is *C. sativum* var. *microcarpum* DC, which contains smaller fruits (Uitterhaegen, 2014).

Using various extraction methods, coriander fixed oil (CFO) is extracted from the seeds. CFO is extracted from various species of coriander, and 20 major essential oils are available in the world

DOI: 10.1201/9781003204626-33

market. The commercial production of CFO depends on its flavoring properties, chemical composition, and physical description. The CFO is analyzed by gas chromatography and mass spectroscopy to determine its chemical composition. The CFO content of the dried seeds varies from 0.1 to 0.36% depending on the stage of plant development. CFO's fatty acid composition indicates that it contains the highest amount of linalool, phytol, and oleic acid. Linalool is the major volatile compound, more than 50% of essential oil. CFO is used as an antimicrobial agent because of its broad-spectrum antimicrobial activity (Politeo et al., 2007; Khani & Rahdari, 2012; Nurzynska-Wierdak, 2013; Mandal & Mandal, 2015).

Coriander and its essential oil have a fragrance component. Coriander possesses antifungal, antioxidant, and antibacterial properties that are effective in food preservation and preventing spoilage. It can also be used as an emulsifier and flavoring agent in the preparation of various foods (Politeo et al., 2007; Mandal & Mandal, 2015). It has many medicinal properties as well as therapeutic properties. In traditional medicine, its seeds and oil are used to treat gastrointestinal problems, rheumatism, and joint pains. Furthermore, it has been reported that they have a hypoglycemic effect and affect carbohydrate metabolism and that both leaves and seeds have antibacterial properties. Coriander seeds, leaves, and their oil are also used to flavor liqueurs, beverages, meat items, and pickles, among other things (Zeković et al., 2011).

30.2   EXTRACTION OF CORIANDER FIXED OIL (CFO)

Coriander fixed oil (CFO) is extracted from various plant components such as the leaves, shoots, seeds, roots, and bark, but the essential oil composition varies depending on the part of the plant. CFO is a colorless or pale yellow oil; this oil is collected from fully ripe and dried seeds. It has a warm and aromatic flavor with a distinct odor and a slight sweetness. The volatile oil in coriander fruit is 0.25–1.5% and fixed oil 14–20%. The CFO ranges from 1.77% to 2.43% (Bhat et al., 2014).

The CFO is extracted from ripe coriander seeds using steam distillation, supercritical CO_2 extraction, and other methods (Illés et al., 2000). Traditionally, hydro distillation has been used to extract CFO from seeds, but nowadays, supercritical fluid extraction with carbon dioxide is used instead. A partial separation process is added to the supercritical extract to obtain pure essential oil, preferably accelerating the non-volatile components. The advantages of this method include that no solvent is used. Furthermore, only the steam volatile compound can be extracted in hydro distillation (Uitterhaegen, 2014).

30.3   CHEMICAL COMPOSITION AND QUALITY OF CFO

The composition of fatty oil in the ripe fruit includes linoleic acid (18.6%), petroselinic acid (70.8%), oleic acid (8.5%), and palmitic acid (4.8%), as described in Table 30.1. There is a difference between

TABLE 30.1
Composition of Coriander Fixed Oil (CFO)

Compound	Quantity (%)
Linoleic acid	18.6
Petroselinic acid	70.8
Oleic acid	8.50
Palmitic acid	4.80
Linalool	55.4
Geranyl acetate	17.5
Caryophyllene	2.32
Camphor and *p*-cymene	4.12

the production of essential oils and the production of coriander seeds grown in different places. The content of CFO in fruits varies greatly and increases with the ripening of the fruit (Mandal & Mandal, 2015). The CFO is used to flavor perfumes and soaps. It is a rich source of beneficial phytonutrients. Moreover, antibacterial and antifungal activities are also present. Other vegetable oils are less stable and can retain their odor longer than coriander essential oil (Uitterhaegen, 2014).

The CFO contains the maximum content of aliphatic aldehyde compounds of which decanal, E-2-dodecanol, and E-2-decenol are found in the higher percentages. The amount of these essential compounds in the CFO varies due to various environmental factors. The essential components are identified, ranging from 96.6 to 99.7% of the total oil composition. The CFO of the composition contains linalool (55.4%), geranyl acetate (17.5%), caryophyllene (2.32%), camphor (4.12%), and p-cymene, respectively, as shown in Table 30.1. The contents of cis- and *trans*-linalool oxide are also found in a low range from 0.1 to 0.4%. The results conclude that linalool is a major component of CFO (Bandoni et al., 1998; Raal et al., 2004).

30.3.1 NUTRITIONAL ASPECTS

The green leaves and dried berries of coriander provide nutrients such as iron, vitamins, and minerals. For example, coriander leaves are rich in vitamin A and vitamin C. In addition, saturated fat, high-density lipoprotein, thiamine, zinc, dietary fiber, and water make up 83% of green coriander (Iwatani et al., 2003).

30.3.2 MAJOR CHEMICAL CONSTITUENTS

According to their origin, coriander seeds contain up to 2.1% volatile oil. The distilled oil contains 66 to 71% linalool, depending on its source (Anju & Banerjee, 2011).

30.3.3 MINOR CHEMICAL CONSTITUENTS

In coriander, minor chemical constituents include monoterpene hydrocarbons such as pinene, limonene, terpinene, borneol, clymene, geraniol, and geranyl acetate; heterocyclic compounds such as pyrazine, pyridine, thiazole, furan, and tetrahydrofuran derivatives; heterocyclic compounds such as pyridine, isocoumacins, dihyrocoriandrin, and coriandrones A-E, glazonoids; and phthalides such as neochidilide and Z-digustilide; phenolic acids, sterols, and flavonoids were reported (Wallis, 2005).

Carotenoid concentrations are higher in the reproductive organ. Carotenoids, particularly β-carotene, are found in the leaves of leafy green veggies. Carotenoids are also a rich source of antioxidants, and they can be processed to be used as a coloring agent. However, their primary role is to scavenge free radicals created by chlorophyll during photo-oxidation (Kandlakunta, Rajendran, & Thingnganing, 2008). Coriander foliage is utilized in various meals, primarily in the diets of those who are vitamin A deficient. Anthocyanin is found in green foliage. Bioactive flavonoid molecules protect against a variety of chronic illnesses. Coriander contains antioxidants, which are beneficial in improving nutrition and maintaining health (Omidbaigi, 2005).

Coriander leaves contain a variety of nutrients. Many essential oils and fatty acids contribute to the fragrant flavor of its seeds. Coriander contains a high concentration of essential oils needed for brain development and effective functioning. Linoleic and linolenic acids are the mail fatty acids present in coriander (USDA, 2013).

30.4 PROCESSING

Coriander is consumed without processing, but it can be treated to improve its palatability and profitability. Harvesting is the first step taken at the best possible time to ensure the production of plant material. Coriander seeds ripen two to three months after seeding. Harvesting depends on whether green or ripe coriander could be harvested till it turns brown (Bhat et al., 2014). Coriander seeds should be cut at the right ripening stage to have a full aroma. Immature coriander seeds have a pungent taste and lack a specific spicy aroma. On the other hand, overripe seeds are scattered and yields are reduced (Sue Azam-Ali, 2008).

The taste of green coriander is more delicate and more important than ripe seeds. Mature seeds can be ground into powder form. Plants can be uprooted from their roots or cut off the seed heads with a few inches of stalk and when the seeds become dehydrated, separate them from the stalk and store them in a cool, dry place. The coriander collected before shipment must be pre-cooled to remove heat from the crop and improve shelf life after harvest. For this purpose, vacuum cooling is used most commonly. It is accomplished by a portion of the coriander's liquid. It has a significant impact on the product's shelf life (Bhat et al., 2014).

After harvesting, the coriander leaves are washed and dried in the shade. Sun-drying, microwave drying, freeze-drying, and other processes are mostly used at the international level for drying the seeds. The drying time depends on the air temperature, while chlorophyll deficiency is due to drying or heat processing. The conversion of chlorophyll to pheophytins is the most common change in green plants during heat processing, resulting in a color shift from brilliant green to olive-brown, an unappealing hue for the customer. Chlorophyll loss can be reduced by bleaching before drying. Furthermore, microwave drying instead of sun-drying can prevent chlorophyll loss because microwave drying is faster than other traditional drying processes. Microwave drying also reduces the moisture level of coriander and retards the oxidation (Alibas, 2006).

Coriander seeds do not fully develop their specific aroma and taste until dried. Therefore, the whole plant is dried to less than 18% moisture level. After that, the seeds are removed and dried to a final moisture of 9%. It prevents the seeds from becoming overheated. The sun-drying time depends on the climate. An artificial dryer can also be used to dry the seeds, but it is important to make sure that the temperature does not exceed 100°C as it reduces volatile oil contents (Sue Azam-Ali, 2008).

The oil is usually encapsulated in alginate to retain its aromatic flavor and reduce nutritional loss. When calcium ions create intermolecular cross-connections with the carboxyl gluconate groups in an aqueous sodium alginate solution, the result is the well-known "egg-box" structure that preserves CFO. The CFO is frequently used in sauces and salsas, and its fruits are ground into powder form, which is used for flavoring various products like meat, fish, pickles, bakery, and curry recipes (Ravi at al., 2007).

Coriander leaves may also be used in cooking and processed paste and puree, among other things. In addition, coriander leaves are used for product formulations such as pasta and sauces to prevent degradation. If fresh leaf powder is blanched at 90°C for at least 2 min, it can be pulverized by inactivating the peroxidase enzyme (Ahmed et al., 2004). In contrast, short-term water blanching provides better value than long-term heat treatment. The results concluded that the coriander fruits and leaves could be bleached with short-term water treatment at the beginning of processing (Kaiser et al., 2013).

30.5   FUNCTIONAL ASPECTS

The oil's antibacterial activity was unaffected by the type of meat or the temperature. As a result, coriander oil acts as a natural antibacterial agent in food. Antioxidant activity is the most essential and well-known functional feature (Rattanachaikunsopon & Phumkhachorn, 2010).

30.6   BIOLOGICAL ACTIVITY OF CORIANDER FIXED OIL (CFO)

The CFO has promising biological activities like anti-bacterial, anti-microbial, anti-fungal, and anti-oxidative activities, which are useful for food preservation and prevent spoilage and foodborne illness.

30.6.1   ANTI-MICROBIAL ACTIVITY

Worldwide, foodborne illness is a significant problem and a severe public threat to human health. A new challenge has arisen for the pharmaceutical and food industries to develop strategies for reducing foodborne diseases from natural sources. Thus, CFO is an excellent natural food preservative against foodborne pathogens. The CFO reduces the activity of Gram-positive (+) and negative (–) environmental bacteria and is also effective against yeast (Silva & Domingues, 2017). In the research work of Silva et al. (2011), the results concluded that the tested Gram-positive bacteria exhibited lower susceptibility than the Gram-negative bacteria to CFO. CFO is very effective against Gram-positive bacteria such as *Stenotrophomonas maltophilia* and *Penicillium expansum* (blue mold). CFO is more effective against blue mold than *Stenotrophomonas maltophilia*. Furthermore, linalool is a major component of CFO which has been reported to have antimicrobial effects against many bacterial strains (Ateş & Turgay, 2003; Kačániová et al., 2020).

30.6.2   ANTIOXIDANT ACTIVITY

In recent years, CFO has been qualified as a natural antioxidant and suggested as a potential alternative to synthetic antioxidants in food preservation. Lipid oxidation is a major risk factor in the food industry. Therefore, various synthetic antioxidants are used in the food industry to prevent oxidation processes and improve the shelf life of food products. However, the safety of these antioxidants has been questioned due to their toxic effects. Numerous research studies have shown that high consumption of synthetic antioxidants increases the risk of liver damage and cancer. Therefore, the researchers and the food industry are more interested in developing a natural source of antioxidants. A variety of refined oils have been used for this purpose, but CFO has been shown to have effective antioxidant activity and good fragrance properties. The results conclude that CFO has no toxic effects on food products (Wangensteen et al., 2004; Burdock & Carabin, 2009; Kačániová et al., 2020).

The high lipid content of food products is a big issue, as it causes rancidity, which results in an unpleasant flavor change and nutrition loss, as well as changes in look and texture. Coriander essential oil helps to decrease oxidative rancidity. A terpenoid component in coriander essential oil contributes to its antioxidant activity. Essential oil's radical-scavenging activity is employed as a natural antioxidant to extend food's shelf life and stability (Divya et al., 2012). Coriander is a good source of phenolics and phytochemicals due to its potent antioxidant activity. It is found in both the leaves and the seeds; however, the leaves have a more significant concentration than the seeds. Its high pigment concentration, particularly carotenoid, is attributed to its antioxidant content. Carotenoids have a stronger hydroxyl radical scavenging capacity, protecting cells from oxidative damage. Phenolic compounds are the most significant group of secondary metabolites (Balasundram et al., 2006).

30.6.3   ANTIFUNGAL ACTIVITY

Coriander fixed oil (CFO) has a wide range of functional properties and is used as a preventative and medicinal herb. Due to a considerable drop in serum progesterone, it has antifertility properties. It is in many common drugs to treat things like seasonal fever, vomiting, stomach problems, nasty colds, indigestion, worms, and joint pain. It is called a bioactive chemical storage facility. The oil it contains can be used as an antibacterial agent. Coriander essential oil is effective against bacteria, both Gram-positive and Gram-negative, as well as dangerous fungi (Darughe et al., 2012).

30.6.4   ANTI-HYPERGLYCEMIC ACTIVITY

A coriander seed extract has long been used to treat diabetes. The addition of ground coriander seed extract to the diet resulted in a considerable reduction in blood glucose levels and a rise in insulin levels. The addition of its seed extract reactivated antioxidant enzymes and boosted antioxidant levels, in addition to preventing peroxidative damage (Deepa & Anuradha, 2011).

30.6.5 ANTI-ANXIETY EFFECT

Because anxiety can be annoying and overwhelming at times, coriander is used as a folk remedy for insomnia. Coriander extract can be utilized as a muscle relaxant and sedative (Mahendra & Bisht, 2011).

30.7 CONCLUSIONS

Herbs and spices are treated to extend the shelf life of food, and processed spices and herbs have been used to season dishes since the dawn of time. Coriander is a multipurpose plant that may be used as a spice and an herbal medication. Processed coriander improves flavor, profitability, and international traceability. The leaves and berries of coriander have a pleasant scent and are abundant in protein, vitamins, and minerals. Coriander is antibacterial and anti-cancer in nature. Coriander's usage as an antioxidant is one of its most essential and well-known features. As a result, the best technique to preserve coriander is to prepare its fruit and leaves.

REFERENCES

Ahmed, J., Shivhare, U., & Singh, P. (2004). Colour kinetics and rheology of coriander leaf puree and storage characteristics of the paste. *Food Chemistry*, 84(4), 605–611.

Alibas, I. (2006). Characteristics of chard leaves during microwave, convective, and combined microwave-convective drying. *Drying Technology*, 24(11), 1425–1435.

Anju, M., & Banerjee, D. (2011). Associations of cadmium, zinc, and lead in soils from a lead and zinc mining area as studied by single and sequential extractions. *Environmental Monitoring and Assessment*, 176(1), 67–85.

Ateş, D. A., & Turgay, Ö. (2003). Antimicrobial activities of various medicinal and commercial plant extracts. *Turkish Journal of Biology*, 27(3), 157–162.

Balasundram, N., Sundram, K., & Samman, S. (2006). Phenolic compounds in plants and agri-industrial by-products: Antioxidant activity, occurrence, and potential uses. *Food Chemistry*, 99(1), 191–203.

Bandoni, A. L., Mizrahi, I., & Juárez, M. A. (1998). Composition and quality of the essential oil of coriander (*Coriandrum sativum* L.) from Argentina. *Journal of Essential Oil Research*, 10(5), 581–584.

Bhat, S., Kaushal, P., Kaur, M., & Sharma, H. (2014). Coriander (*Coriandrum sativum* L.): Processing, nutritional and functional aspects. *African Journal of Plant Science*, 8(1), 25–33.

Burdock, G. A., & Carabin, I. G. (2009). Safety assessment of coriander (*Coriandrum sativum* L.) essential oil as a food ingredient. *Food and Chemical Toxicology*, 47(1), 22–34.

Darughe, F., Barzegar, M., & Sahari, M. (2012). Antioxidant and antifungal activity of coriander (*Coriandrum sativum* L.) essential oil in cake. *International Food Research Journal*, 19(3), 1253–1260.

Deepa, B., & Anuradha, C. (2011). Antioxidant potential of *Coriandrum sativum* L. seed extract. *Indian Journal Experimental Biology*, 49(1):30–8.

Divya, P., Puthusseri, B., & Neelwarne, B. (2012). Carotenoid content, its stability during drying and the antioxidant activity of commercial coriander (*Coriandrum sativum* L.) varieties. *Food Research International*, 45(1), 342–350.

Illés, V., Daood, H., Perneczki, S., Szokonya, L., & Then, M. (2000). Extraction of coriander seed oil by CO_2 and propane at super-and subcritical conditions. *The Journal of Supercritical Fluids*, 17(2), 177–186.

Iwatani, Y., Arcot, J., & Shrestha, A. K. (2003). Determination of folate contents in some Australian vegetables. *Journal of Food Composition and Analysis*, 16(1), 37–48.

Kačániová, M., Galovičová, L., Ivanišová, E., Vukovic, N. L., Štefániková, J., Valková, V., … & Tvrdá, E. (2020). Antioxidant, antimicrobial and antibiofilm activity of coriander (*Coriandrum sativum* L.) essential oil for its application in foods. *Foods*, 9(3), 282.

Kaiser, A., Kammerer, D. R., & Carle, R. (2013). Impact of blanching on polyphenol stability and antioxidant capacity of innovative coriander (*Coriandrum sativum* L.) pastes. *Food Chemistry*, 140(1–2), 332–339.

Kandlakunta, B., Rajendran, A., & Thingnganing, L. (2008). Carotene content of some common (cereals, pulses, vegetables, spices and condiments) and unconventional sources of plant origin. *Food Chemistry*, 106(1), 85–89.

Khani, A., & Rahdari, T. (2012). Chemical composition and insecticidal activity of essential oil from *Coriandrum sativum* seeds against *Tribolium confusum* and *Callosobruchus maculatus*. *International Scholarly Research Notices*, 2012.

Mahendra, P., & Bisht, S. (2011). Anti-anxiety activity of *Coriandrum sativum* assessed using different experimental anxiety models. *Indian Journal of Pharmacology*, 43(5), 574.

Mandal, S., & Mandal, M. (2015). Coriander (*Coriandrum sativum* L.) essential oil: Chemistry and biological activity. *Asian Pacific Journal of Tropical Biomedicine*, 5(6), 421–428.

Nurzynska-Wierdak, R. (2013). Essential oil composition of the coriander (*Coriandrum sativum* L.) herb depending on the development stage. *Acta Agrobotanica*, 66(1).

Omidbaigi, R. (2005). *Production and processing of medicinal plants*. Mashhad: Beh-Nashr, 210–225.

Önder, A. (2018). Coriander and its phytoconstituents for the beneficial effects. *Potential of Essential Oils*, 165–185.

Politeo, O., Jukic, M., & Milos, M. (2007). Chemical composition and antioxidant capacity of free volatile aglycones from basil (*Ocimum basilicum* L.) compared with its essential oil. *Food Chemistry*, 101(1), 379–385.

Raal, A., Arak, E., & Orav, A. (2004). Chemical composition of coriander seed essential oil and their conformity with EP standards. *Agraarteadus*, 15(4), 234–239.

Rattanachaikunsopon, P., & Phumkhachorn, P. (2010). Potential of coriander (*Coriandrum sativum*) oil as a natural antimicrobial compound in controlling *Campylobacter jejuni* in raw meat. *Bioscience, Biotechnology, and Biochemistry*, 74(1), 31–35.

Ravi, R., Prakash, M., & Bhat, K. K. (2007). Aroma characterization of coriander (*Coriandrum sativum* L.) oil samples. *European Food Research and Technology*, 225(3), 367–374.

Silva, F., & Domingues, F. C. (2017). Antimicrobial activity of coriander oil and its effectiveness as food preservative. *Critical Reviews in Food Science and Nutrition*, 57(1), 35–47.

Silva, F., Ferreira, S., Queiroz, J. A., & Domingues, F. C. (2011). Coriander (*Coriandrum sativum* L.) essential oil: Its antibacterial activity and mode of action evaluated by flow cytometry. *Journal of Medical Microbiology*, 60(10), 1479–1486.

Sue Azam-Ali, D. (2008). *Coriander processing*. The Schumacher Centre Bourton-on-Dunsmore Rugby, Warwickshire, United Kingdom.

Uitterhaegen, E. (2014). *Coriander oil–extraction, applications and biologically active molecules*. Master's dissertation, Faculty of Bioscience Engineering Universiteit Gent.

USDA. (2013). *National nutrient database for standard reference, release 28*. US Department of Agriculture, Agricultural Research Service, Nutrient Data Laboratory.

Wallis, T. E. (2005). *Textbook of pharmacognosy*. CBS Publishers and Distributors. New Delhi, India.

Wangensteen, H., Samuelsen, A. B., & Malterud, K. E. (2004). Antioxidant activity in extracts from coriander. *Food Chemistry*, 88(2), 293–297.

Zeković, Z., Adamović, D., Ćetković, G., Radojković, M., & Vidović, S. (2011). Essential oil and extract of coriander (*Coriandrum sativum* L.). *Acta Periodica Technologica*, 42, 281–288.

31 Food Applications of Coriander Fixed Oil

Mustafa Kiralan and Onur Ketenoglu

CONTENTS

31.1 INTRODUCTION

Coriander (*Coriandrum sativum* L.) is mainly grown in Mediterranean environments. It is an annual herb of the family Apiaceae. The plant is cultivated for green leaves and its seeds throughout the world. The seeds and leaves are used as seasonings in the food industry, such as liqueurs, teas, meat products, and pickles, due to their unique, strong aroma (Carrubba, 2009; Nguyen, Talou, Cerny, Evon, & Merah, 2015). Besides, coriander is also used in the aromatherapy, medicine, perfumery, and pharmaceutical industries (Nguyen et al., 2015). In addition to its specific flavor, the antioxidant and antimicrobial properties of coriander also make it preferable in food applications (Mandal & Mandal, 2015).

The seed of coriander contains water (11.3%), crude protein (11.5%), fat (19.1%), crude fiber (28.4%), starch (10.5%), pentosans (10.3%), sugar (1.92%), minerals (4.98%), and essential oil (0.84%) (Nadeem, Muhammad Anjum, Issa Khan, Tehseen, El-Ghorab, & Iqbal Sultan, 2013). Among these constituents, fatty acids, sterols, tocols, and volatile compounds are effective components that make coriander essential for health and nutrition (Laribi, Kouki, M'Hamdi, & Bettaieb, 2015).

Coriander fixed oil is extracted with solvent extraction or mechanical pressing. Oil yield differs depending on the ripening period, varying between 4.6% and 25.1% (Nguyen et al., 2015). Coriander fixed oil is a rich source of monounsaturated fatty acids (FA), especially petroselinic acid (C18:1n12), which represents 60 to 80% of total FA in coriander seed oil (Msaada, Hosni, Ben Taarit, Chahed, Hammami, & Marzouk, 2009; Saini, Assefa, & Keum, 2021; Sriti, Talou, Msaada, & Marzouk, 2011; Uitterhaegen et al., 2016). Petroselinic acid (18:1Δ ^{6}) is an uncommon fatty acid and a positional isomer of oleic acid (18:1Δ ^{9}) (Figure 31.1). The difference in its double bond position affects the melting point. The melting point of petroselinic acid is 33°C, while that of oleic acid is 12°C (Yang et al., 2020). On the aspect of health benefits, a study by Weber, Richter, Schulte, and Mukherjee (1995) emphasized that petroselinic acid from dietary triglycerides is adsorbed easily by rats and also reduces the amount of arachidonic acid in tissue lipids. This fatty acid could be used as a healthy substitute for margarine and shortening due to its solid form under room conditions (Yang et al., 2020).

Besides fatty acids, coriander fixed oil contains various phytochemicals, including sterols, phenols, and tocopherols, contributing to their antioxidant property (Daga, Vaishnav, Dalmia, & Tumaney, 2021). Δ5-Stigmasterol (1095 mg/kg oil, *n*-hexane extracted oil; 1180 mg/kg oil, chloroform/methanol extracted oil) and β-sitosterol (997 mg/kg oil, *n*-hexane extracted oil; 1009 mg/kg oil, chloroform/

DOI: 10.1201/9781003204626-34

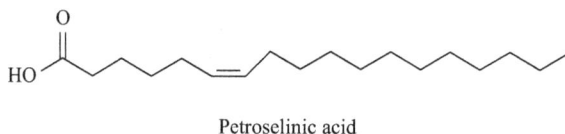

Petroselinic acid

FIGURE 31.1   Chemical structure of petroselinic acid.

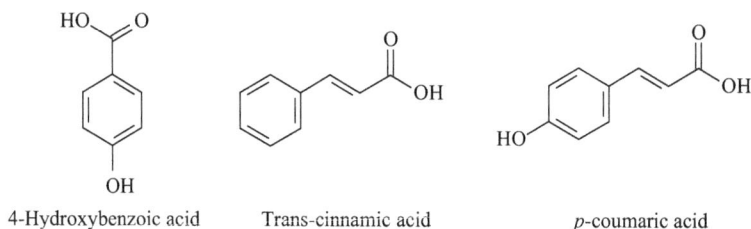

4-Hydroxybenzoic acid   Trans-cinnamic acid   *p*-coumaric acid

FIGURE 31.2   Phenolic acids in coriander seed oil.

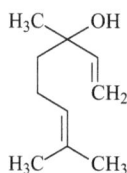

FIGURE 31.3   Chemical structure of linalool.

methanol extracted oil) are present as major sterols in coriander fixed oil (Kozłowska, Gruczyńska, Ścibisz, & Rudzińska, 2016). B-Tocopherol was the predominant tocopherol isomer in coriander oil with 0.672 g/kg, followed by δ-tocopherol (0.347 g/kg) (Ramadan & Mörsel, 2004). According to Daga et al. (2021), the total phenol content of fixed oil was 28.5 mg gallic acid equivalent/100 g oil. The most abundant phenolics (Figure 31.2) present in fixed oil are 4-hydroxybenzoic acid (470 µg/100 g oil), *trans*-cinnamic acid (348 µg/100 g oil), and *p*-coumaric acid (227 µg/100 g oil).

The coriander seeds contain both essential oil and fixed oil. Fixed oils contain different amounts of volatile compounds depending on extraction methods. For this reason, the food applications of coriander fixed oil and essential oil will be discussed together in this section. The main component of coriander essential oil (CEO) is linalool (Figure 31.3), comprising up to 66.0–91.7% of total volatiles (Beyzi, Karaman, Gunes, & Buyukkilic Beyzi, 2017; Kačániová et al., 2020; Kiralan, Calikoglu, Ipek, Bayrak, & Gurbuz, 2009). In addition, coriander essential oil has broad-spectrum antimicrobial activity and potent antioxidant activity. Due to these activities, coriander essential oil has been used in the production of various foods and also is preferred by consumers because it is a safe and natural preservative (Burdock & Carabin, 2009; Kačániová et al., 2020).

31.2   MEAT AND MEAT PRODUCTS

Randall, Reaney, and Drew (2013) added cold-pressed coriander oil to vegetable oils in the diet of rainbow trout. The increase of eicosapentaenoic acid (20:5n-3) and docosahexaenoic acid (22:6n-3) fatty acids was observed in the fillet-fed diet containing coriander oil. The addition of coriander oil to vegetable oil improved fatty acid composition in the fillets of rainbow trout, increasing the bioconversion of 18:3n-3 to 20:5n-3 and 22:6n-3.

CEO was utilized as a natural antimicrobial compound in ground chicken meat and beef at 4°C and 32°C. The antimicrobial activity of CEO against *Campylobacter jejuni*, which is the most commonly reported bacterial cause of foodborne infection in the United States, varied depending on concentration. At a concentration of 0.5% v/w, the CEO was more effective on this pathogen in ground chicken meat and beef samples compared to the usage at 0.1% (v/w) and 0.25%. At both 4 and 32°C, 0.5% concentration was more effective (Rattanachaikunsopon & Phumkhachorn, 2010).

The effects of CEO (0.02%) on the shelf life of ground veal stored at 0 and 4°C for two weeks were investigated. CEO improved the sensory properties of meat and reduced the spoiled meat odor during storage. Besides, the CEO exhibited an inhibition effect against Enterobacteriaceae and molds (Macura, Michalczyk, & Banaś, 2011).

The use of coriander essential oil in Italian salami (0.01%) exhibited lower undesirable sensory attributes such as rancid taste and rancid odor compared to the control sample (Marangoni & Fernandes de Moura, 2011b). Furthermore, in another study using the same concentration of essential oil in Italian salami, a decrease in peroxide value (PV) and Thiobarbituric acid reactive substances (TBARS) values was observed compared to Butylated hydroxytoluene (BHT) used as a food preservative and also control sample (Marangoni & Fernandes de Moura, 2011a).

The effects of CEO and hyssop (*Hyssopus officinalis* L.) essential oils on the shelf life of vacuum-packed ground beef during storage at 0.5 and 6°C for 15 days were evaluated. The addition of essential oil positively affected the smell and taste of beef samples stored at 0.5 and 6°C (extending acceptability by up to 3 days). Also, it inhibited the growth of Enterobacteriaceae (up to approximately 1–2 log cycles compared to the control samples) (Michalczyk, Macura, Tesarowicz, & Banaś, 2012).

Sodium alginate was incorporated with CEO as a bioactive edible coating on refrigerated chicken fillet. The alginate with CEO exhibited a stronger inhibition effect against a wide range of foodborne pathogens on coated fillets than uncoated fillets. Besides antimicrobial activity, the coating of fillet with Alg/CEO also inhibited lipid oxidation and lower values were observed in total volatile base nitrogen, TBARS, and peroxide formation in the samples (Kargozari, Hamedi, Amir Amirnia, Montazeri, & Abbaszadeh, 2018).

In pork sausages, coriander essential oil was used with nitrite to retard lipid oxidation and microbial spoilage. For example, at 0.12 µL/g essential oil of coriander with 60 mg/kg of nitrites retarded the thiobarbituric acid reactive substances (TBARS, approx. 0.12 mg MDA/kg) and total plate count (TPC, approx. 2.50 log CFU/g) (Šojić et al., 2019).

Zhang and Zhong (2019) evaluated the effect of CEO (2 mL/100 g), garlic essential oil (2 mL/100 g), and a mixture of CEO and garlic essential oil (1 mL/100 g) on the fishy smell and refrigeration quality of surimi products from silver carp (*Hypophthalmichthys molitrix*). A decrease in the fishy smell was observed in the surimi products with the addition of essential oils. Besides, the growth of microorganisms was inhibited by essential oils, and also protein oxidation and water loss were reduced in surumi. Furthermore, the essential oil mixture improved the refrigeration quality and prolonged the shelf life of these products.

Omidi-Mirzaei, Hojjati, Behbahani, and Noshad (2020) evaluated coriander essential oil against *Listeria innocua* in ground lamb. The coriander essential oil enhanced the shelf life of ground lamb by increasing this pathogen's growth rate lag phase.

Turan et al. (2017) examined the effects of different essential oils such as rosemary, coriander, laurel, and garlic on anchovy marinade's lipid oxidation and sensory attributes (*Engraulis encrasicolus* L.). All essential oils exhibited a strong reducing effect on rancidity compared to the control sample. The panelists preferred laurel essential oil more than the other essential oils, while the CEO was the least desired sample.

31.3   VEGETABLE OILS

Due to the antioxidant properties of coriander oil, it is used in vegetable oils to increase oxidation stability. Ramadan and Wahdan (2012) blended coriander oil (10% and 20%, w/w) with corn oil. Coriander oil improved the oxidative stability of corn oil under oxidative conditions (60°C) for 15 days. At the end of the storage period, the peroxide value of corn oil increased up to 18.5 meq/kg, while the PV of corn oil blended with coriander oil in a proportion of 9:1 and 8:2 (w/w) was 7.5 and 6.8 meq/kg, respectively. Moreover, coriander oil increased MUFA content in blended oils while decreasing PUFA content. The blended oils with coriander oil also enhanced the tocopherol content, in particular β-tocopherol. The researchers utilized hexane extracted coriander oil to improve the oxidative stability of sunflower oil. The PV, conjugated diene, and conjugated triene values of sunflower oil with coriander oil showed lower values than the control sample (sunflower oil with no addition) during the thermal oxidation experiment at 60°C. The amount of added coriander oil also had varying oxidative stability. Samples with 20% of coriander oil addition contributed more to the oxidative stability of sunflower oil than 10% coriander oil-containing samples. Researchers also evaluated the antiradical activity using DPPH· and galvinoxyl radicals. The antiradical activity results showed that sunflower oil blended with coriander oil showed more intense scavenging than sunflower oil alone. Besides, enrichment in the tocopherol isomers was achieved by mixing coriander oil with sunflower oil. β-Tocopherol was the main component in coriander oil (approx. 56% of total tocopherol content). β- and γ-tocopherol isomers increased with the addition of coriander oil, and δ-tocopherol was identified in blended oils, which is not found in sunflower oil. On the other hand, blended oils with coriander oil contained higher levels of MUFA and low levels of PUFA compared to sunflower oil.

Wang, Fan, Guan, Huang, Yi, and Ji (2018) applied CEO directly to the refined sunflower oil at 300, 600, and 1200 ppm. These flavored oils were subjected to thermal oxidation at 65°C. At 1200 ppm of CEO, intense activity was exhibited against lipid oxidation, comparable with TBHQ. Besides, there was no adverse effect on sensory evaluation in flavored oil with CEO (1200 ppm).

The CEO (0.2 mg/mL) obtained by hydrodistillation and microwave-assisted extraction was added to the purified soybean oil (2 g) and stored at 60°C to determine essential oils' effects on oxidative stability soybean oil. At the end of 8 days' storage, lower peroxide values were observed in soybean oil blended with CEO (5.34 meq O_2/kg oil and 5.28 meq O_2/kg oil for oils obtained by hydrodistillation and microwave-assisted extraction, respectively) compared to purified soybean oil (8.09 meq O_2/kg oil) (Ghazanfari, Mortazavi, Yazdi, & Mohammadi, 2020).

31.4   FRUIT AND VEGETABLES

Pellegrini et al. (2020) utilized coriander essential oil as a sustainable washing treatment to remove *Salmonella* spp. from fresh-cut carrots. Their results revealed that the application of essential oil (5 μL/mL) for a short time (2 min) had potent activity against *Salmonella* spp. in artificially inoculated stick carrots up to 1 day of storage time. Besides, no adverse effects were observed in terms of sensory evaluation of the stick carrots. Zamindar, Sadrarhami, and Doudi (2016) evaluated the CEO against *Byssochlamys fulva* inoculated in tomato sauce. Samples were stored at 30°C for 2 months. The samples with 800 ppm CEO inhibited up to 32% of fungi growth, while samples with a mixture of cinnamon (250 ppm) and coriander (800 ppm) essential oils had significant inhibition (90%) against fungi growth.

31.5   EDIBLE FILMS AND PACKAGING MATERIAL

In a study where essential oils were used together with hake proteins edible films, the edible films incorporated with CEO showed stronger DPPH· radical-scavenging capacity and also reducing power compared to control sample and edible films having various essential oils including citronella, tarragon, and thyme (Pires, Ramos, Teixeira, Batista, Nunes, & Marques, 2013).

Porcine meat and bone meal (MBM) protein films containing CEO were investigated against *E. coli* O157:H7 and *Listeria monocytogenes*. The films containing 1% CEO exhibited antimicrobial activity against *E. coli* O157:H7 and *L. monocytogenes*. The inhibition zones of the films against the food pathogens were 20.54 mm for *E. coli* O157:H7 and 25.56 mm for *L. monocytogenes* (Lee, Won, & Song, 2015).

In another study by Kačániová et al. (2020), the CEO's antioxidant, antimicrobial, and antibiofilm activities were investigated against *Stenotropomonas maltophilia* and *Bacillus subtilis*. The inhibition rate was 51.0% in the DPPH· method. When tested using the disc diffusion method, the strongest antibacterial activity was against *B. subtilis* (10.7 mm), followed by *S. maltophilia* (9.22 mm), and *Penicillium expansum* (8.99 mm). Furthermore, the CEO exhibited strong anti-biofilm activity against *S. maltophilia* and intense antifungal activity against *P. expansum* on the bread after 14 days with MID_{50} 367.19 and MID_{90} 445.92 μL/L of air.

Kostova et al. (2020) utilized CEO in packaging paper materials (bleached, unbleached, and recycled). They stored up to 5 days to determine the antimicrobial activity of essential oil against various microorganisms, including against *Staphylococcus aureus*, *Bacillus subtilis*, *Escherichia coli*, *Pseudomonas aeruginosa*, *Salmonella abony*, *Candida albicans*, and *Aspergillus brasiliensis*. The highest antimicrobial activity was observed in the bleached paper coated with CEO against Gram-positive bacteria during the 5-day storage period. Besides, CEO showed strong fungicidal activity against the tested molds and yeasts (inhibition zone diameter between 20.3 mm and 24.8 mm).

31.6 OTHER FOOD APPLICATIONS

The effect of CEO was investigated on the shelf life of cake samples stored at room temperature for 60 days. The cake samples with CEO at 0.05, 0.10, and 0.15% concentrations contained lower primary and secondary oxidation products than the control. Besides, CEO at 0.15% inhibited fungal growth in the cake (Darughe, Barzegar, & Sahari, 2012).

Shori (2020) utilized coriander leaves to enrich yogurt to assess such functional properties of coriander as proteolytic and antioxidant activities, total phenolic content, and α-amylase inhibition activity. The findings concluded that the addition of coriander to yogurt achieved higher total phenolic contents than cumin seeds. Besides, coriander exhibited a higher radical scavenging activity during the storage period. The addition of coriander also increased the α-amylase inhibition activity compared to the control group.

In a similar study by Mohite and Waghmare (2020), coriander was used to fortify biscuits. The results revealed that the addition of coriander to the biscuits achieved higher levels of macronutrients, antioxidant activity, phenolic concentration, and hardness. According to the researchers, the utilization of coriander also resulted in a reduction in the fat content of the biscuits.

31.7 CONCLUSION

Coriander has been proven to possess many health-promoting attributes due to various functional components such as fatty acids, sterols, tocols, and volatile compounds present in many parts of the plant, including fixed and essential oil. The fixed oil of coriander is rich in petroselinic acid, which can replace margarine and shortenings. Also, the high sterol, tocopherol, and phenolic components make coriander fixed oil preferable in the food industry regarding many positive health traits. In addition to fixed oil, CEO is also a valuable product due to its high content in many volatile components, in particular, linalool, as the major compound of CEO. Furthermore, due to the potent antimicrobial and antioxidant activities that CEO possesses, it has many usage areas as a preservative agent in such food products as meat, fruit and vegetables, edible films and packaging materials, and vegetable oils. In this chapter, the effects of coriander in the food industry have been discussed. Therefore, it can be concluded that coriander extracts will have increasing demand and will be utilized in a broader range of food products due to their desirable attributes.

REFERENCES

Beyzi, E., Karaman, K., Gunes, A., & Buyukkilic Beyzi, S. (2017). Change in some biochemical and bioactive properties and essential oil composition of coriander seed (*Coriandrum sativum* L.) varieties from Turkey. *Industrial Crops and Products*, *109*, 74–78. https://doi.org/10.1016/j.indcrop.2017.08.008.

Burdock, G. A., & Carabin, I. G. (2009). Safety assessment of coriander (*Coriandrum sativum* L.) essential oil as a food ingredient. *Food and Chemical Toxicology*, *47*(1), 22–34. https://doi.org/10.1016/j.fct.2008.11.006.

Carrubba, A. (2009). Nitrogen fertilisation in coriander (*Coriandrum sativum* L.): A review and meta-analysis. *Journal of the Science of Food and Agriculture, 89*(6), 921–926. https://doi.org/10.1002/jsfa.3535.

Daga, P., Vaishnav, S. R., Dalmia, A., & Tumaney, A. W. (2021). Extraction, fatty acid profile, phytochemical composition and antioxidant activities of fixed oils from spices belonging to Apiaceae and Lamiaceae family. *Journal of Food Science and Technology.* https://doi.org/10.1007/s13197-021-05036-1.

Darughe, F., Barzegar, M., & Sahari, M. (2012). Antioxidant and antifungal activity of coriander (*Coriandrum sativum* L.) essential oil in cake. *International Food Research Journal, 19*(3), 1253–1260.

Ghazanfari, N., Mortazavi, S. A., Yazdi, F. T., & Mohammadi, M. (2020). Microwave-assisted hydrodistillation extraction of essential oil from coriander seeds and evaluation of their composition, antioxidant and antimicrobial activity. *Heliyon, 6*(9), e04893. https://doi.org/10.1016/j.heliyon.2020.e04893.

Kačániová, M., Galovičová, L., Ivanišová, E., Vukovic, N. L., Štefániková, J., Valková, V., ... Tvrdá, E. (2020). Antioxidant, antimicrobial and antibiofilm activity of coriander (*Coriandrum sativum* L.) essential oil for its application in foods. *Foods, 9*(3), 282. https://doi.org/10.3390/foods9030282.

Kargozari, M., Hamedi, H., Amir Amirnia, S., Montazeri, A., & Abbaszadeh, S. (2018). Effect of bioactive edible coating based on sodium alginate and coriander (*Coriandrum sativum* L.) essential oil on the quality of refrigerated chicken fillet. *Food & Health, 1*(3), 30–38. https://fh.srbiau.ac.ir/article_12858_b2e35ca6fb7e67bbf52d4f80ad6d8572.pdf.

Kiralan, M., Calikoglu, E., Ipek, A., Bayrak, A., & Gurbuz, B. (2009). Fatty acid and volatile oil composition of different coriander (*Coriandrum sativum*) registered varieties cultivated in Turkey. *Chemistry of Natural Compounds, 45*(1), 100–102. https://doi.org/10.1007/s10600-009-9240-2.

Kostova, I., Lasheva, V., Georgieva, D., Damyanova, S., Fidan, H., Stoyanova, A., & Gubenia, O. (2020). Characterization of active paper packaging materials with coriander essential oil (*Coriandrum sativum* l.). *Journal of Chemical Technology & Metallurgy, 55*(6), 2085–2093. http://search.ebscohost.com/login.aspx?direct=true&db=asn&AN=146662254&lang=tr&site=eds-live&authtype=ip,uid.

Kozłowska, M., Gruczyńska, E., Ścibisz, I., & Rudzińska, M. (2016). Fatty acids and sterols composition, and antioxidant activity of oils extracted from plant seeds. *Food Chemistry, 213*, 450–456. https://doi.org/10.1016/j.foodchem.2016.06.102.

Laribi, B., Kouki, K., M'Hamdi, M., & Bettaieb, T. (2015). Coriander (*Coriandrum sativum* L.) and its bioactive constituents. *Fitoterapia, 103*, 9–26. https://doi.org/10.1016/j.fitote.2015.03.012.

Lee, J.-H., Won, M., & Song, K. B. (2015). Physical properties and antimicrobial activities of porcine meat and bone meal protein films containing coriander oil. *LWT-Food Science and Technology, 63*(1), 700–705. https://doi.org/10.1016/j.lwt.2015.03.043.

Macura, R., Michalczyk, M., & Banaś, J. (2011). Effect of essential oils of coriander (*Coriandrum sativum* L.) and lemon balm (*Melissa officinalis* L.) on quality of stored ground veal. *Zywnosc. Nauka. Technologia. Jakosc/Food. Science Technology. Quality, 4*(77), 127–137. https://doi.org/10.15193/zntj/2011/77/127-137.

Mandal, S., & Mandal, M. (2015). Coriander (*Coriandrum sativum* L.) essential oil: Chemistry and biological activity. *Asian Pacific Journal of Tropical Biomedicine, 5*(6), 421–428. https://doi.org/10.1016/j.apjtb.2015.04.001.

Marangoni, C., & Fernandes de Moura, N. (2011a). Antioxidant activity of essential oil from *Coriandrum sativum* L. in Italian salami. *Food Science and Technology (Campinas), 31*(1), 124–128. https://doi.org/10.1590/S0101-20612011000100017.

Marangoni, C., & Fernandes de Moura, N. (2011b). Sensory profile of Italian salami with coriander (*Coriandrum sativum* L.) essential oil. *Food Science and Technology, 31*(1), 119–123. https://doi.org/10.1590/S0101-20612011000100016.

Michalczyk, M., Macura, R., Tesarowicz, I., & Banaś, J. (2012). Effect of adding essential oils of coriander (*Coriandrum sativum* L.) and hyssop (*Hyssopus officinalis* L.) on the shelf life of ground beef. *Meat Science, 90*(3), 842–850. https://doi.org/10.1016/j.meatsci.2011.11.026.

Mohite, D., & Waghmare, R. (2020). The fortification of biscuits with coriander leaf powder and its effect on physico-chemical, antioxidant, nutritional and organoleptic characteristics. *International Journal of Food Studies, 9*(1), 225–237. https://doi.org/10.7455/ijfs/9.1.2020.a8.

Msaada, K., Hosni, K., Ben Taarit, M., Chahed, T., Hammami, M., & Marzouk, B. (2009). Changes in fatty acid composition of coriander (*Coriandrum sativum* L.) fruit during maturation. *Industrial Crops and Products, 29*(2), 269–274. https://doi.org/10.1016/j.indcrop.2008.05.011.

Nadeem, M., Muhammad Anjum, F., Issa Khan, M., Tehseen, S., El-Ghorab, A., & Iqbal Sultan, J. (2013). Nutritional and medicinal aspects of coriander (*Coriandrum sativum* L.). *British Food Journal, 115*(5), 743–755. https://doi.org/10.1108/00070701311331526.

Nguyen, Q.-H., Talou, T., Cerny, M., Evon, P., & Merah, O. (2015). Oil and fatty acid accumulation during coriander (*Coriandrum sativum* L.) fruit ripening under organic cultivation. *The Crop Journal*, *3*(4), 366–369. https://doi.org/10.1016/j.cj.2015.05.002.

Omidi-Mirzaei, M., Hojjati, M., Behbahani, B. A., & Noshad, M. (2020). Modeling the growth rate of *Listeria innocua* influenced by coriander seed essential oil and storage temperature in meat using FTIR. *Quality Assurance and Safety of Crops & Foods*, *12*(SP1), 1–8. https://doi.org/10.15586/qas.v12iSP1.776.

Pellegrini, M., Rossi, C., Palmieri, S., Maggio, F., Chaves-López, C., Lo Sterzo, C., ... Serio, A. (2020). *Salmonella enterica* control in stick carrots through incorporation of coriander seeds essential oil in sustainable washing treatments. *Frontiers in Sustainable Food Systems*, *4*(14). https://doi.org/10.3389/fsufs.2020.00014.

Pires, C., Ramos, C., Teixeira, B., Batista, I., Nunes, M. L., & Marques, A. (2013). Hake proteins edible films incorporated with essential oils: Physical, mechanical, antioxidant and antibacterial properties. *Food Hydrocolloids*, *30*(1), 224–231. https://doi.org/10.1016/j.foodhyd.2012.05.019.

Ramadan, M. F., & Mörsel, J.-T. (2004). Oxidative stability of black cumin (*Nigella sativa* L.), coriander (*Coriandrum sativum* L.) and niger (*Guizotia abyssinica* Cass.) crude seed oils upon stripping. *European Journal of Lipid Science and Technology*, *106*(1), 35–43. https://doi.org/10.1002/ejlt.200300895.

Ramadan, M. F., & Wahdan, K. M. M. (2012). Blending of corn oil with black cumin (*Nigella sativa*) and coriander (*Coriandrum sativum*) seed oils: Impact on functionality, stability and radical scavenging activity. *Food Chemistry*, *132*(2), 873–879. https://doi.org/10.1016/j.foodchem.2011.11.054.

Randall, K. M., Reaney, M. J. T., & Drew, M. D. (2013). Effect of dietary coriander oil and vegetable oil sources on fillet fatty acid composition of rainbow trout. *Canadian Journal of Animal Science*, *93*(3), 345–352. https://doi.org/10.4141/cjas2013-001.

Rattanachaikunsopon, P., & Phumkhachorn, P. (2010). Potential of coriander (*Coriandrum sativum*) oil as a natural antimicrobial compound in controlling *Campylobacter jejuni* in raw meat. *Bioscience, Biotechnology, and Biochemistry*, *74*(1), 31–35. https://doi.org/10.1271/bbb.90409.

Saini, R. K., Assefa, A. D., & Keum, Y.-S. (2021). Spices in the Apiaceae family represent the healthiest fatty acid profile: A systematic comparison of 34 widely used spices and herbs. *Foods*, *10*(4), 854. https://doi.org/10.3390/foods10040854.

Shori, A. B. (2020). Proteolytic activity, antioxidant, and α-Amylase inhibitory activity of yogurt enriched with coriander and cumin seeds. *LWT*, *133*, 109912. https://doi.org/10.1016/j.lwt.2020.109912.

Šojić, B., Pavlić, B., Ikonić, P., Tomović, V., Ikonić, B., Zeković, Z., ... Ivić, M. (2019). Coriander essential oil as natural food additive improves quality and safety of cooked pork sausages with different nitrite levels. *Meat Science*, *157*, 107879. https://doi.org/10.1016/j.meatsci.2019.107879.

Sriti, J., Talou, T., Msaada, K., & Marzouk, B. (2011). Comparative analysis of fatty acid, sterol and tocol composition of Tunisian and Canadian coriander (*Coriandrum sativum* L.) fruit. *Analytical Chemistry Letters*, *1*(5–6), 375–383. https://doi.org/10.1080/22297928.2011.10648241.

Turan, H., Kocatepe, D., Keskin, İ., Altan, C. O., Köstekli, B., Candan, C., & Ceylan, A. (2017). Interaction between rancidity and organoleptic parameters of anchovy marinade (*Engraulis encrasicolus* L. 1758) include essential oils. *Journal of Food Science and Technology*, *54*(10), 3036–3043. https://doi.org/10.1007/s13197-017-2605-9.

Uitterhaegen, E., Sampaio, K. A., Delbeke, E. I. P., De Greyt, W., Cerny, M., Evon, P., ... Stevens, C. V. (2016). Characterization of French coriander oil as source of petroselinic acid. *Molecules*, *21*(9), 1202. https://doi.org/10.3390/molecules21091202.

Wang, D., Fan, W., Guan, Y., Huang, H., Yi, T., & Ji, J. (2018). Oxidative stability of sunflower oil flavored by essential oil from *Coriandrum sativum* L. during accelerated storage. *LWT*, *98*, 268–275. https://doi.org/10.1016/j.lwt.2018.08.055.

Weber, N., Richter, K.-D., Schulte, E., & Mukherjee, K. D. (1995). Petroselinic acid from dietary triacylglycerols reduces the concentration of arachidonic acid in tissue lipids of rats. *The Journal of Nutrition*, *125*(6), 1563–1568. https://doi.org/10.1093/jn/125.6.1563.

Yang, Z., Li, C., Jia, Q., Zhao, C., Taylor, D. C., Li, D., & Zhang, M. (2020). Transcriptome analysis reveals candidate genes for petroselinic acid biosynthesis in fruits of *Coriandrum sativum* L. *Journal of Agricultural and Food Chemistry*, *68*(19), 5507–5520. https://doi.org/10.1021/acs.jafc.0c01487.

Zamindar, N., Sadrarhami, M., & Doudi, M. (2016). Antifungal activity of coriander (*Coriandrum sativum* L.) essential oil in tomato sauce. *Journal of Food Measurement and Characterization*, *10*(3), 589–594. https://doi.org/10.1007/s11694-016-9341-0.

Zhang, D., & Zhong, Y. (2019). Effects of coriander (*Coriandrum sativum* L.) and garlic (*Allium sativum* L.) essential oil on the quality and fishy smells during refrigerated storage of restructured product from Silver Carp (*Hypophthalmichthys molitrix*) Surimi. *Nanoscience and Nanotechnology Letters*, *11*(10), 1470–1476. https://doi.org/10.1166/nnl.2019.3022.

32 Non-Food Applications of Coriander Fixed Oil

Zeliha Ustun-Argon, Zinar Pinar Gumus, and Cigdem Yengin

CONTENTS

32.1 INTRODUCTION

Coriander (*Coriandrum sativum* L.) is one of the oldest known herbs, used for culinary and medicinal purposes for over 3000 years (dating back at least to the Ebers papyrus of 1550 BC) (Ishikawa, Kondo, & Kitajima, 2003). It is an annual, 30–100 cm high (Mahendra & Bisht, 2011), hairless, aromatic, herbaceous plant (Chahal, Singh, Kumar, & Bhardwaj, 2017), and belongs to the Apiaceae (also known as Umbelliferae) family. The plant seeds, leaves, and roots are edible, although they have different flavors and uses (Bhat, Kaushal, Kaur, & Sharma, 2014). The ripe fruit of *Coriandrum sativum*, with strongly divided, strongly scented leaves, has a light and fresh flavor and is widely used worldwide in the ground or volatile isolated form. The oval and spherical coriander seeds are the most consumed and popular ingredient globally as a local spice, traditional medicine, and flavoring agent (Gupta, 2010). Although the plant is native to the Mediterranean and the Middle East regions, *Coriandrum sativum* is best found worldwide between October and February in Central and Eastern Europe (Russia, Hungary, and the Netherlands), in Mediterranean regions (Morocco, Malta, Egypt), and in Africa and Asia (China, Pakistan, India, and Bangladesh) (Sahib et al., 2013; Weiss, 2002). Coriander is grown almost everywhere except Japan (Shivanand, 2010). It is cultivated in India, in Andhra Pradesh, Maharashtra, West Bengal, Uttar Pradesh, Rajasthan, Jammu, and Kashmir, Madhya Pradesh, and is widely cultivated in Tamil Nadu, Karnataka, and Bihar (Sriti et al., 2010). While coriander grows best in dry climates, it can grow in any soil, including light, well-drained, moist, loamy soil and light to heavy black soil (Verma, Pandeya, Sanjay, & Styawan, 2011).

The production amount is around 500,000 tons/year globally (Anwar et al., 2011; Uitterhaegen et al., 2017). The whole plant can be used for flavoring purposes, but it needs to be processed to make its seeds more palatable and because its leaves are perishable (Bhat et al., 2014). The coriander plant yields two main products that differ markedly in smell and taste, used for flavoring: the fresh green herb and the spice (ripe dried seed pod or fruit). Coriander berries, often called 'seeds', are aromatic with a slightly bitter-sweet, spicy flavor (Aluko, McIntosh, & Reaney, 2001; Baba, Xiao, Taniguchi, Ohishi, & Kozawa, 1991; Ceska et al., 1988; Msaada, Hosni, Taarit, Hammami, & Marzouk, 2009; Wangensteen, Samuelsen, & Malterud, 2004).

DOI: 10.1201/9781003204626-35

The coriander plant is rich in major and minor nutrients such as fixed and essential oils, proteins, sugars, vitamins, and minerals. Coriander fruits are also valuable for fixed and essential oil components. Therefore the seeds and both fractions of the oil and the different parts of the plant have a wide range of usages (Beyzi, Karaman, Gunes, & Buyukkilic Beyzi, 2017; Uitterhaegen et al., 2015). Fruit essential oils can be obtained in amounts ranging from 0.5% to 2.5%. Many compounds have been identified, such as geraniol, α-pinene, γ-terpinene, geranyl acetate, and limonene (Chahal et al., 2017). Coriander is very low in saturated lipids and contains α-tocopherol, linoleic acid, and vitamin K. The plant leaves are rich in vitamins, and the seeds are rich in phenolics and essential oils. The aroma of coriander belongs to its essential oil, which contains significant amounts of linoleic and furanocoumarin (coriander and dihydrochoridrin) (Potter & Fagerson, 1990).

The seeds of *C. sativum* are used in culinary applications for seasoning, developing organoleptic properties, in preparations of curry powder, pastries, buns, sausages, pickles, and chocolates, and for increasing the shelf life of products with their antioxidant and antimicrobiological effects (Ramadan, Kroh, & Mörsel, 2003; Ramadan & Wahdan, 2012; Silva, Domeno, & Domingues, 2020). The young leaves of coriander are used in making sauces and chutneys, and the green leaves are consumed as fresh herbs, salads, and garnishes due to their attractive green color and aroma (Begnami, Duarte, Furletti, & Rehder, 2010; Matasyoh, Maiyo, Ngure, & Chepkorir, 2009; Norman, 1990; Saeed & Tariq, 2007). In addition, it is used to sweeten various alcoholic beverages such as coriander and gin (Jansen, 1981). Coriander also has lead detoxifying potential (Sahib et al., 2013).

In traditional medicine, the seeds are used for convulsion, cough, insomnia, spasms, worms, dysentery, seasonal fever, gout, flatulent colic, anxiety, rheumatism, hypertension, interdigital tinea pedis, and nausea and also are preferred for urinary and cardiovascular disorders. Recent studies also showed that coriander has insulin-like effects for the treatment of diabetes and hypolipidemic activities against high cholesterol levels in the blood, and it also can be used successfully against bacteria and fungus (Abbassi et al., 2018; Beikert, Anastasiadou, Fritzen, Frank, & Augustin, 2013; Cortés-Eslava, Gómez-Arroyo, Villalobos-Pietrini, & Espinosa-Aguirre, 2004; Darughe, Barzegar, & Sahari, 2012; Dias, Barros, Sousa, & Ferreira, 2011; Duarte, Ferreira, Oliveira, & Domingues, 2013; Eidi et al., 2009; Ghazanfari, Mortazavi, Yazdi, & Mohammadi, 2020; Gray & Flatt, 1999; Jabeen, Bashir, Lyoussi, & Gilani, 2009; Mahendra & Bisht, 2011; Paarakh, 2009; Silva, Ferreira, Queiroz, & Domingues, 2011; Sourmaghi, Kiaee, Golfakhrabadi, Jamalifar, & Khanavi, 2015; Sriti, Bettaieb, Bachrouch, Talou, & Marzouk, 2019; Sunil, Agastian, Kumarappan, & Ignacimuthu, 2012; Uitterhaegen et al., 2015).

The composition of coriander plants is related to different factors such as growing conditions, maturity stages, and processing conditions. The fixed oil content is 20–28% of the dry fruit weight. The main fatty acid of coriander is petroselinic (6Z-octadecenoic) acid in the range of 31–75% of the total fatty acid composition. Petroselinic acid is an uncommon positional isomer of oleic acid, and its unique structure makes the fruit an exciting ingredient for pharmaceuticals, food, and cosmetics. Furthermore, this plant is accepted as a safe food supplement by the European Food Safety Authority, allowing up to 600 mg intake per day for healthy adults (Uitterhaegen et al., 2015; 2017). Although its essential oil is preferred in pharmaceuticals, perfumes, cosmeceuticals, and the food industry, the fixed oil also has many different applications, including as a mouth wash (Havale et al., 2021), as a feed (Abd El-Hack et al., 2020), and for biodiesel (Moser & Vaughn, 2010) production. Coriander vegetable oil has a long history in traditional medicine and also has had its monographs in the United States Pharmacopeia since 1965 and the Food Chemicals Codex since 1972, has had GRAS status by the Foreign Exchange Management Act (FEMA) since 1965, and was included in the list of substances, spices, and seasonings by the Council of Europe in 1970.

The coriander seed fixed oil has various applications in different industries. Food and pharmaceuticals are the main industries where the oil can be employed. As reported by Parthasarathy, Chempakam, and Zachariah (2008), the essential oil form of coriander oil is used instead of synthetic compounds (Chahal et al., 2017) to solve the problem with coriander's distinctive flavors, both in aromas and in perfume and soap production (Mahendra & Bisht, 2011). The commercial value of coriander essential oil varies depending on its physical properties, chemical composition, and aroma quality. This chapter will describe the non-food applications of coriander vegetable oil.

32.2 WORLD PRODUCTION AND CONSUMPTION

Certified for food use by the FDA, coriander essential oil is among the 20 most widely used essential oils in the world market. It has been given GRAS status by FEMA and the Council of Europe (Silva et al., 2011). Coriander seed is grown mainly for its essential oil (from 0.3% to 1.1%). In addition to India (Peter, 2004), which is one of the leading producers of coriander seeds and which has cultivated 5.25×105 hectares and a yield of 3.10×105 tons per year, commercial coriander is produced in Morocco, Canada, Pakistan, Romania, and the former Soviet Union, Iran, Turkey, and Egypt. India is the world's largest coriander producer, consumer, and exporter. The United States and Canada are the primary export markets for large seed coriander. Secondary export markets include Sri Lanka, Trinidad and Tobago, the United Kingdom, Mexico, and Guatemala. The leading coriander importers from India are Europe, the USA, Singapore, and Gulf countries. Ukraine and India mainly control the coriander oil market (British Pharmacopoeia, 2003).

32.3 CHEMICAL COMPOSITION OF CORIANDER

Steam distillation, organic solvent extraction (Soxhlet), and supercritical fluid extraction (Mhemdi, Rodier, Kechaou, & Fages, 2011) are the three most important techniques that are widely used for the extraction of vegetable oil from its fruits. Dried, ripe coriander fruit contains essential oil, fixed (fatty) oil, protein, cellulose, pentosans, tannins, calcium oxalate, and minerals. The main components of coriander were found to be fiber (23–36%), carbohydrates (about 20%), fatty oil (16–28%), and proteins (11–17%). The most important components of coriander seeds are essential oil and fixed oil (Coşkuner & Karababa, 2007). Light yellow fatty oil (physical oil/fixed oil) with a characteristic odor constitutes about 25% of the seed, while the essential oil content is less than 1%.

The other components of dried seeds are crude protein (11.5–21.3%), oil (17.8–19.1%), crude fiber (28.4–29.1%), and ash (4.9–6.0%) (Uitterhaegen et al., 2016). Selenium content has been reported to be higher in coriander than other herbs and herbal teas (Özcan, Ünver, Uçar, & Arslan, 2008). Mg and minerals such as Al, Si, P, Cl, K Ca, Ti, Mn, Fe, Cu, and Zn have also been reported (Al-Bataina, Maslat, & Al-Kofahi, 2003). Non-nutritive compounds such as glucosinolates (27.5 μmol/g), sinapine (4 mg/g), condensed tannins (1.1 mg/g), and inositol phosphates (17.4 mg/g) are also found in *C. sativum* seeds (Matthäus & Angelini, 2005).

Essential oil accumulation and chemical composition in plants are determined by many different factors such as cultivation, environmental factors (Rakic & Johnson, 2002; Sriti, Wannes, Talou, Vilarem, & Marzouk, 2011), genetics (Ebrahimi, Hadian, & Ranjbar, 2010; Zheljazkov et al., 2008), and ontogenetic factors (Mohammadi & Saharkhiz, 2013; Msaada et al., 2007). For example, fresh coriander contains essential oil (Oganesyan, Nersesyan, & Parkhomenko, 2007; Telci & Hişil, 2008), fatty acids (Neffati, Sriti, Hamdaoui, Kchouk, & Marzouk, 2011), flavonoids, carotenoids (Taniguchi, Yanai, Xiao, Kido, & Baba, 1996), and coumarin compounds (Bhuiyan, Begum, & Sultana, 2009; Raju, Varakumar, Lakshminarayana, Krishnakantha, & Baskaran, 2007).

Coriander seed essential oil, the yield of which varies between 0.03 and 2.6% depending on the species, growing region, and climatic conditions, was obtained as a result of essential oil extraction from coriander seeds and leaves by hydrodistillation (Shahwar et al., 2012). It has been reported that the main compound in coriander seed essential oil is linalool. Gil et al. (2002) reported the chemical composition of coriander as linalool 72.3% and 77.7%, α-pinene 5.9% and 4.4%, γ-terpinene 4.7% and 5.6%, camphor 4.6% and 2.4%, limonene 2.0% and 0.9% in Argentinian and European coriander, respectively. In Russian coriander seed essential oil, it has been reported that linalool constitutes approximately 68% of the oil (Misharina, 2001).

The studies determined that the compounds in the seeds and leaves of coriander fruits vary significantly depending on the different maturity stages, and it was reported that coriander seed oil contains 60–70% linalool and 20% hydrocarbons. This means that the vegetable oil composition is entirely different from the seed oil. The essential oil content of the ripe and dried fruit weight of coriander varies between 0.03% and 2.6%, and the fatty oil content varies between 9.9% and 27.7%. The aliphatic aldehydes (mainly C10-C16 aldehydes) are the main components of the herb's essential oil but are off-odor. Monoterpene hydrocarbons, linalool, and other oxidized monoterpenes are the predominant compounds in fruit oil (Bhuiyan et al., 2009).

According to another study, coriander seeds have zero cholesterol content (USDA, 2013) and are rich in vitamin C (21 mg/100 g). Seeds could be considered to be a source of vitamins, minerals, and lipids. Iwatani et al. reported mineral amounts, potassium (1267 mg/100 g), calcium (709 mg/100 g), phosphorus (409 mg/100 g), magnesium (330 mg/100 g), sodium (35 mg/100 g), and zinc (4.70 mg/100 g) (Iwatani, Arcot, & Shrestha, 2003).

Coriandrum sativum is a potential lipid source rich in petroselinic acid (C18:1n-12) and an essential oil isolated from seeds and aerial parts high in linalool (Sahib et al., 2013). Moser and Vaughn reported that the yield of *C. sativum* seeds containing 26–29% by weight vegetable oil is around 1 in 954 kg/da. In their studies, the primary fatty acid (FA) in *C. sativum* oil (CSO) was petroselinic (9Z-octadecenoic) acid and linoleic with lesser amounts of stearic and palmitic acids (Moser & Vaughn, 2010).

The ash content of the coriander fruit has been determined to be 5–7%, resin content was determined to be 13%, and the fruits contained coriandrol, vebriniol, jireniol, malic acid, and alkaloids (Rao, Ahmed, Ibrahim, & Ahmed, 2012). Coriander oil is a good source of tocol (327 mg/g), g tocopherol dominates (26.4 mg/g), and the only sugar detected was glucose (Sriti, Aidi, & Thierry, 2010; Sriti, Talou, Wannes, & Marzouk, 2009; Sriti, Wannes, et al., 2010).

32.4   FIXED OIL PROFILE OF CORIANDER

In addition to oilseeds, many plant species have been investigated as potential vegetable oil sources in recent years. This is due to nutritional, industrial, and pharmaceutical interests. This section gives information about the chemical properties of coriander, an annual herb that can contain a significant amount of oil and is widely used as a flavor and spice in temperate regions. Petroselinic acid (C18:1n-12) is the main fatty acid ranging from 65.7% to 76.6%, followed by linoleic acid with 13.0–16.7%. Being rich in petroselinic acid, coriander oil has many industrial applications. Petroselinic acid presents an uncommon positional isomer of oleic acid. While petroselinic acid could be used as a precursor of adipic acid, the monomeric component of nylon 66, it could also be used as a precursor to lauric acid, a component of detergents and surfactants. Thanks to the phospholipid structures it contains, it can be a multifunctional additive for food, pharmaceutical, and industrial applications. Its oil composition strengthens the value-added use of coriander (Neffati & Marzouk, 2008; Reiter, Lechner, & Lorbeer, 1998; Sahib et al., 2013; Sriti, Aidi, & Thierry, 2010).

The essential components of the fruits (seed and pericarp), which are the most widely used components of the coriander plant, are essential oil and fatty oil. Coriander fruits have an essential oil fraction of about 1% by weight of dry fruit and a fraction of vegetable oil in the range of 20–28% by weight of dried fruit (Sahib et al., 2013; Uitterhaegen et al., 2018). In the studies given in Table 32.1, different extraction techniques and different solvent or solvent ratios were applied to obtain oil from coriander seeds, fruit, and pericarp. Apart from oil extraction, differences in oil ratios are also affected by the geographical origin of coriander.

Nguyen et al. determined the oil content of coriander at different ripening stages from flowering to maturity in 2010 and 2011. The oil yield varied between 5.8% and 24.9% at different maturation stages. They reported a higher oil yield in 2010 when rainfall was higher. This study demonstrated the effect of climate change (Nguyen et al., 2020).

TABLE 32.1
Yield of Fixed Oil from Different Parts of *Coriandrum sativum*

Country	Part and Type of Oil	% Yield	Reference
Northwestern Tunisia	Fruits	22.6	(Sriti, Aidi, & Thierry, 2010)
Germany	Seed	28.4	(Ramadan & Mörsel, 2002)
Northwestern Tunisia	Fruits	19.2	(Sriti et al., 2010)
	Seed	22.6	
	Pericarp	9.30	
France	Fruits	22.9	(Uitterhaegen et al., 2016)
Vietnam	Seed	19.7	(Matthaus, Vosmann, Pham, & Aitzetmüller, 2003)
Egypt	Seed	16.8	(Nguyen, Aparicio, & Saleh, 2015)
Tunisia	Fruits	21.8	(Sriti, Talou, Msaada, & Marzouk, 2011)
Canada		15.8	
Poland	Seed	20.0	(Kozłowska, Gruczyńska, Ścibisz, & Rudzińska, 2016)
		22.1	
Canada	Seed	19.0	(Mhemdi et al., 2011)
England	Fruit	17.0	(Griffiths, Robertson, Millam, & Holmes, 1992)
Northeastern Tunisia	Fruit	21.2	(Sriti, Talou, Faye, Vilarem, & Marzouk, 2011)

Coriander fruits of the same genotype, sown on the same day in 2002 in Menzel Temime (Northeast Tunisia) and Oued Beja (Northwest Tunisia), were randomly collected at different ripening stages. In this study by Msaada et al., oil yields for the Oued Beja region between 5 and 22 days after flowering and the Menzel Temime region between 5 and 41 days were found in the range 2.70–25.80% and 3.30–25.90%, respectively. Thus, they showed the effect of a growing region and maturity stages on oil yield (Msaada et al., 2009). In another study by Msaada et al., coriander fruits were randomly collected from plants cultivated in the Charfine region (Northeast Tunisia) at different maturation periods (between the 5th and 55th days after flowering). While the lipid content was 9.6% on the 5th day, it gradually increased to 26.4% on the 55th day (Msaada et al., 2009). Finally, in the study of Sriti et al., oil extraction was conducted by Soxhlet extraction and mechanical pressing, without solvent extraction, for the cake obtained from the coriander fruits collected from the Korba area region in the Northeast of Tunisia (Sriti et al., 2011).

While the *n*-hexane solvent extraction of French origin coriander fruits had 22.9% oil yield, 21.8% and 15.8% oil yields were obtained for Tunisian and Canadian coriander fruits, respectively (Sriti et al., 2011; Uitterhaegen et al., 2016). Kozlowska et al. compared *n*-hexane and 2:1 chloroform/methanol as different extraction solvents for Polish coriander oil and found 20.0% and 22.1% oil yields, respectively (Kozłowska et al., 2016). Different English coriander fruits were extracted using the Welch method using 2% sulfuric acid in methanol as solvent, resulting in a mean of oil yield of 17% (Griffiths et al., 1992). A 2:1 mixture of chloroform and methanol was used to extract the oil from German coriander fruits, and a high oil yield of 28.4% was obtained (Ramadan & Mörsel, 2002). Mhemdi et al. implemented a sustainable process for Canadian coriander fruits using supercritical CO_2 extraction and separated essential and vegetable oil. They achieved an oil yield of 19% (Mhemdi et al., 2011). Vietnamese coriander fruits yielded 19.7% oil in *n*-hexane solvent extraction, while the new strain produced from breeding studies for Turkish coriander fruits was only 10.6% (Kiralan, Calikoglu, Ipek, Bayrak, & Gurbuz, 2009; Matthäus & Angelini, 2005). When these results are examined, they show that fruit oil content depends on fruit origin, variety, maturity, and growing conditions.

Apart from solvent extraction, some studies showed an increased oil yield by applying mechanical pressing. For example, for Tunisian coriander fruits, oil yields obtained by a single screw and twin-screw extrusion method reported 65% and 47%, respectively (Sriti et al., 2012; Sriti, Talou,

Faye, et al., 2011). A similar oil recovery of 47% was achieved with twin-screw extrusion of French coriander fruits (Uitterhaegen et al., 2015). Although it can be seen from Table 32.1 that the oil yields are generally close to each other, there are differences between the oil yields of coriander obtained from different regions. Factors such as which part of the plant oil is extracted, extraction techniques, geographical origins, and climate affect these differences.

Phospholipids (PL) are commonly found in foods and have pro- and antioxidant effects attributed to them. According to the results of Uitterhaegen et al.'s analysis of the phospholipid composition of coriander vegetable oil, the total phospholipid content of this oil is 0.31%. They found that phosphatidic acid (PA), constituting the phospholipid subclass, constitutes 32.5% of phospholipids. Phosphatidylinositol (PI) and phosphatidylethanolamine (PE) both represent 17% and 16.7%, respectively, while phosphatidylcholine (PC), another important class of phospholipids, accounts for 25.4% (Uitterhaegen et al., 2016). In their study, Ramadan and Morsel reported the lipids classes of coriander as NL, GL, and PL 93.0%, 4.14%, and 1.57%, respectively. The amounts of phospholipid subclasses that make up the 1.57% group are PG, PE, PI, LPE, PS, PC, and LPC 1.80%, 25.4%, 13.1%, 2.10%, 7.18%, 45.1%, and 3.02%, respectively (Ramadan & Mörsel, 2002). It was also shown by Sriti et al. that there are significant amounts of NL among the total lipid (TL). The seed and whole fruit lipid contents were NL 95.6%, GL 2.42%, and PL 1.94% (Sriti et al., 2010). In another study by Sriti et al., in which the lipid profiles of Tunisian coriander seeds were studied, PC, PG, PE, and PA were found to be 15.4%, 35.9%, 6.68%, 33.8%, and 8.11%, respectively (Sriti, Aidi, & Thierry, 2010). In the study of Ramadan et al. with Hungarian coriander seeds, neutral lipids were reported as 960 g/kg, glycolipids as 23.9 g/kg, and phospholipids as 8.50 g/kg (Ramadan et al., 2003). The amounts of glycerolipid content of oils as monoacylglycerols (MAGs), diacylglycerols (DAGs), and triacylglycerols (TAGs) are given in Table 32.2.

The TAGs amount is over 90% in seeds and fruits but very low in the pericarp. Chromatographic techniques are generally used to identify and determine the distribution of fatty acids and other oil components. Table 32.3 summarizes coriander seed oil's major fatty acid composition from different regions.

Ripe seeds of coriander were harvested in 2006 from the Korba region in northeastern Tunisia. Petroselinic acid constitutes 76.6% of total fatty acids, linoleic, oleic, and palmitic acids, constituting 13.0%, 5.4%, and 3.4% of total fatty acids, respectively (Sriti, Aidi, & Thierry, 2010). The fatty acid components of Ramadan and Morsel's study are given in Tables 32.2 and 32.4 (Ramadan & Mörsel, 2002). Sriti et al. determined the fatty acid compositions of different lipid classes in different parts of the fruit. Tables 32.2 and 32.3 show the results of the TAGs classes (Sriti et al., 2010).

The FAME compositions of seed, pericarp, and whole fruit lipids are shown in Tables 32.2 and 32.3. Fatty acids have been identified, where petroselinic acid (C18:1n-12) forms the main FA of different parts of the fruit. Its percentage is 76.3% in the seed, 42.2% in the pericarp, and 75.0% in the whole fruit. These results suggest that this fatty acid accumulation occurs in the seed, the main

TABLE 32.2

MAGs, DAGs, and TAGs Composition of Coriander Oil

Country	Part and Type of Oil	% TAGs	% DAGs	% MAGs	References
Northwestern Tunisia	Fruits	95.5	1.88	0.57	(Sriti, Aidi, & Thierry, 2010)
Germany	Seed	91.1	0.69	0.29	(Ramadan & Mörsel, 2002)
Northwestern Tunisia	Fruits	93.7	1.27	0.14	(Sriti, Wannes, et al., 2010)
	Seed	93.1			
	Pericarp	0.26			
France	Fruits	97.8	1.05	0.02	(Uitterhaegen et al., 2016)

TABLE 32.3
Fatty Acid Composition of Coriander Seed Oil

Fatty Acid	Northeastern Tunisia	Northwestern Tunisia	Germany	Vietnam	Northwestern Tunisia	Turkey	USA	Poland*	Hungary
Parts	Seed	Seed	Seed	Seed	Seed	Seed	Seed	Seed	Seed
C14:0 (Myristic A.)	0.08	0.08	-	0.07	-	0.03	-	-	-
C16:0 (Palmitic A.)	3.50	3.48	1.49	3.64	1.34	3.74	5.3	3.49 / 3.01	5.54
C16:1n-7 (Palmitoleic A.)	0.23	0.23	0.37	0.21	0.23	-	0.30	0.28 / 0.33	-
C18:0 (Stearic A.)	0.78	0.77	2.79	0.87	1.96	0.97	3.10	0.84 / 0.70	1.36
C18:1n-12 (Petroselinic A.)	76.6	76.3	71.5	78.7	78.2	77.1	68.5	73.4 / 73.8	67.0
C18:1n-9 (Oleic A.)	5.47	5.45	8.32	7.79	6.46	2.94	7.60	6.03 / 5.97	7.86
C18:2n-6 (Linoleic A.)	13.0	13.0	13.1	5.51	11.5	14.7	13.0	15.3 / 15.6	15.9
C18:3n-3 (Linolenic A.)	0.15	0.15	0.16	0.34	0.14	0.18	-	0.22 / 0.18	1.0
C20:0 (Arachidic A.)	0.10	0.15	0.15	0.12	0.04	0.08	-	0.09 / 0.10	-
Reference	(Sriti, Aidi, & Thierry, 2010)	(Sriti et al., 2009)	(Ramadan & Mörsel, 2002)	(Matthaus et al., 2003)	(Sriti, Wannes, et al., 2010)	(Kiralan et al., 2009)	(Moser & Vaughn, 2010)	(Kozłowska et al., 2016)	(Ramadan et al., 2003)

Handbook of Coriander (*Coriandrum sativum*)

TABLE 32.4
Fatty Acid Composition of Fruit and Pericarp Parts of Coriander

Fatty Acid	Northwestern Tunisia	Northwestern Tunisia	Germany	Northwestern Tunisia	Northwestern Tunisia	Northeastern Tunisia	Tunisia	Canada	France	England*
Parts	Pericarp	Fruit	Fruit	Fruit	Pericarp	Fruit	Fruit	Fruit	Fruit	Fruit
C14:0 (Myristic A.)	4.83	0.23	-	-	-	-	0.08	0.08	-	-
C16:0 (Palmitic A.)	18.4	3.96	3.96	1.76	8.73	0.10	3.82	3.97	2.90	4.80
C16:1n-7 (Palmitoleic A.)	0.58	0.24	0.41	0.25	-	-	0.22	0.19	0.40	-
C18:0 (Stearic A.)	2.50	0.81	2.91	1.02	5.17	0.60	0.82	0.95	<0.1	2.00
C18:1n-12 (Petroselinic A.)	42.2	75.0	65.7	77.5	55.6	84.8	75.6	73.23	72.6	71.8
C18:1n-9 (Oleic A.)	9.88	5.91	7.85	6.42	11.9	0.10	6.38	6.48	6.00	7.60
C18:2n-6 (Linoleic A.)	18.0	13.4	16.7	12.7	18.5	13.60	12.8	14.8	13.7	14.3
C18:3n-3 (Linolenic A.)	1.35	0.20	0.20	-	-	0.30	0.14	0.19	0.10	-
C20:0 (Arachidic A.)	2.19	0.18	0.25	0.05	-	0.2	0.09	0.10	0.10	-
Reference	(Sriti et al., 2009)	(Sriti et al., 2009)	(Ramadan & Mörsel, 2002)	(Sriti et al., 2010)	(Sriti et al., 2010)	(Msaada et al., 2009)	(Sriti et al., 2011)	(Sriti et al., 2011)	(Uitterhaegen et al., (2016)	(Griffiths et al., 1992)

organ of biosynthesis (Sriti et al., 2009). On the other hand, this acid is used as a raw material for fine chemicals and broken down by oxidative ozonolysis to produce adipic and lauric acids (Sriti et al., 2009). Adipic acid is used in the production of plasticizers and nylon. Lauric acid (C12:0) is used as a raw material for softeners, emulsifiers, detergents, and soaps. The changes in fatty acids during the ripening of coriander fruits grown in the northeast of Tunisia (Charfine) were investigated by Msaada et al. Since the fruits ripened 55 days after flowering and the synthesis of oil and petroselinic acid proceeded at a constant rate until 32 days after flowering, Table 32.2 shows the composition of fatty acids for the 32 days. After 32 days, while palmitic acid increased, petroselinic acid slightly decreased (Msaada et al., 2009). During fruit ripening, palmitic and petroselinic acids follow evolutionary changes in opposite directions suggesting metabolic linkage. However, Cahoon et al. reported that palmitic acid is the precursor of petroselinic acid in coriander metabolism (Cahoon, Shanklin, & Ohlrogge, 1992).

Although the differences in the fatty acid composition are not significant, the differences may be due to the geography of the plant (Gumus, Ertas, Yasar & Gumus, 2018). Based on current research results, the fruits and mainly coriander seeds are valuable sources of essential unsaturated fatty acids. The nutritional value of polyunsaturated fatty acids (PUFA) and their pharmaceutical and industrial uses have made coriander an alternative source of unsaturated fat.

Tocol composition (Table 32.5) could be affected by genetic variation, region, and growing conditions such as climate and soil.

Sterols are structures that make up the bulk of the unsaponifiable matter in many oils. These 'minor' lipids, essential in nutrition, are of great interest to food chemists because they are also important in food labeling. The sterol compositions found in studies with coriander are given in Table 32.6.

The sterol contents of different parts of the coriander are given in Table 32.6. Coriander oils with a content of 6.29 g kg^{-1} fruit oil, 4.31 g kg^{-1} pericarp oil, and 36.92 g kg^{-1} seed oil appear to have high sterol levels. The sterol composition of fruit and pericarp oils consists of high levels of β-sitosterol with 36.7% and 49.4%, respectively. While β-sitosterol content was high in fruit and pericarp oils, stigmasterol (29.5%) was found to be a sterol marker in seed oils (Sriti et al., 2009). This difference may be mainly due to geographical conditions, climate, and seed maturation. It can also be caused by other agricultural and technological factors such as variety, soil type, and extraction and preservation procedures (Gumus, Yasar, Gumus, & Ertas, 2020). In the sterol results of the study by Sriti et al., the upper results were obtained by screw extruder, and the lower values were obtained by Soxhlet extraction (Sriti et al., 2011). Studies to determine the sterol composition of coriander seem to have high β-sitosterol content.

In the study of Uitterhaegen et al. with French coriander vegetable oil, both β-carotene and chlorophyll content were measured using UV spectrophotometry. The content of β-carotene was 10.1 mg/kg, while the amount of chlorophyll was reported to be 11.1 mg/kg (Uitterhaegen et al., 2016). Ramadan et al. reported a β-carotene content of 890 mg/kg for Hungarian coriander vegetable oil (Ramadan et al., 2003).

According to the study of Sriti et al., the main component of fruit, seed, and pericarp essential oils is linalool with 86.1%, 91.1%, and 24.6%, respectively. The essential oil of coriander fruit contains camphor, γ-terpinene, α-pinene, and geraniol with 2.57%, 2.15%, 1.65%, and 1.63%, respectively (Sriti et al., 2009). This change in the ratios of the main components may be due to geographical and environmental factors and the effect of fertilization (Gil et al., 2002). This compound, which has a floral and pleasant odor, can also be used in the perfumery industries (Bandoni, Mizrahi, & Juárez, 1998).

C. sativum seed is a promising oil plant for both food and non-food use due to the properties of its seed oil. Although coriander fruit lipid has positive effects on health in terms of fatty acids, tocopherols, and tocotrienols, the composition of fatty acids is essential in non-food use. In addition, due to the high content of petroselinic acid, a general feature of coriander seed oil, it has a wide range of applications in the industry. The fraction of PC, the marker component in lecithin production,

TABLE 32.5

Tocol Composition of Coriander Oil (T: Tocopherol, T$_3$: Tocotrienol)

Country	Part of Coriander	αT	αT$_3$	βT	γT	γT$_3$	δT	δT$_3$	Reference
Tunisia (mg/g)	Fruits	1.12	6.04	0.05	0.33	19.5	-	0.69	(Sriti et al., 2011)
Canada (mg/g)	Fruits	0.59	4.64	-	0.75	14.5	0.18	0.70	(Sriti et al., 2011)
Northwestern Tunisia (mg/100 g)	Fruits	1.12	6.04	0.05	0.33	19.5	-	0.69	(Sriti et al., 2010)
	Seed	0.34	5.68	-	0.16	19.5	0.09	0.61	
	Pericarp	1.82	-	0.48	0.67	1.71	0.68	-	
France (mg/kg)	Fruits	12.4	98.0		10.1	350	-	25.7	(Uitterhaegen et al., 2016)
Vietnam (mg/kg)	Seed	46	96		31	231	-	41	(Matthaus et al., 2003)
Belgium (μg/g)	Seed	5	6	1.1	4.4	36.5	-	-	(Horvath et al., 2006)
Hungary (g/kg)	Seed	0.086	-	0.672	0.162	-	0.347	-	(Ramadan et al., 2003)

TABLE 32.6
Sterol Composition of Coriander Oil

Part of coriander sterols	Tunisia Fruit (%)	Canada Fruit (%)	Northwestern Tunisia Seed (%) Pericarp (%) Fruit (%)	France Fruit (g/kg)	Germany Fruit (µg/g)	Poland Seed (mg/kg)	Northeastern Tunisia Fruit (%)	Hungary Seed (g/kg)
Campesterol	4.86	5.02	8.82 4.92 7.06	0.54	508	385	8.29 7.95	0.735
Stigmasterol	14.75	14.71	29.5 4.02 21.7	1.61	1548	1095	24.8 24.1	1.512
β-Sitosterol	21.33	21.97	24.8 49.4 36.7	2.31	1464	997	28.8 34.9	1.553
Δ5-Avenasterol	1.44	1.27	4.81 1.96 3.34	0.27	1235	516	2.73 2.36	1.466
Δ5,24-Stigmastadienol	6.58	6.61	9.24 27.7 9.38	-	-	-	13.9 10.7	-
Δ7-Stigmastenol	9.42	10.98	16.3 5.29 16.9	1.22	-	-	16.3 15.4	-
Δ7-Avenasterol	2.70	2.46	5.44 1.87 4.75	0.40	244	142	5.07 4.41	0.365
Totals sterols	61.09%	63.03%	-	-	5186	3474	-	-
Reference	(Sriti et al., 2011)	(Sriti et al., 2011)	(Sriti et al., 2009)	(Uiterhaegen et al., 2016)	(Ramadan & Mörsel, 2002)	(Kozłowska et al., 2016)	(Sriti et al., 2011)	(Ramadan et al., 2003)

indicates that it is suitable for use in the production of lecithin, which is currently the commercial source of its production. Furthermore, the GLs found in the coriander pericarp indicate that coriander is a good source of nutritionally essential fatty acids, which may have many applications in the food, pharmaceutical, and cosmetic industries. In conclusion, the fruit of *C. sativum* is a promising oil plant due to the specific traits of the seed oil. According to the results obtained from the studies, it is seen that the chemical composition of coriander fruits depends on the origin. Petroselinic acid, an unusual fatty acid in the coriander fruit lipid, is a marker for coriander. In addition, it is rich in other health-promoting substances such as β-sitosterol and γ-tocotrienol, which increases coriander use in food and non-food applications. Therefore, it is seen that the effect of fruit origin on the vegetable oil composition is important, especially in terms of phospholipids, sterols, and tocol composition.

32.5   CORIANDER OIL AND BIODIESEL PRODUCTION

Diesel engines are widely used in many important sectors all over the world. These leading sectors are transportation, agriculture, manufacturing, and energy production facilities. While the need for diesel fuel increases, the decrease in diesel reserves and the current consumption rate of petroleum and diesel fuels are causing a critical environmental problem due to global warming, and this has led to the need for alternative fuels that can be used instead of diesel. The best example of this is the alternative fuel, biodiesel (ASTM, 2008; Babu, Kumar, Sathiyaraj, & Senthilkumar, 2020). Biodiesel is made of monoalkyl esters of long-chain fatty acids derived from vegetable oils, animal fats, or spent cooking oils containing triglycerides. Its advantages over conventional petroleum-based fuels (petrodiesel) are that it is derived from renewable raw materials, has good heating power, displaces imported oil, offers superior lubrication and biodegradability, has lower toxicity, has almost no sulfur and no aromatic polycyclic compound content, and it has a higher flash point and offers a reduction in most exhaust emissions (Babu et al., 2020; Moser & Vaughn, 2010). Its disadvantages include low oxidation and storage stability, low volumetric energy content, low-temperature operability, and higher nitrogen exhaust emission. Biodiesel must meet the requirements of accepted fuel standards such as those of the Committee for Standardization (CEN) standard EN 14214 (ASTM, 2008; Knothe, Krahl, & Gerpen, 2010; Moser & Vaughn, 2010).

Vegetable oils have been used as an energy source for lighting and heating worldwide since ancient times. Due to the increase in fuel prices, today, the technology for converting vegetable oil to biodiesel, which is not new, has made biodiesel economically attractive as an alternative fuel. The most important example of this situation is the fair in Paris in the 1900s when a diesel-cycle engine was shown to run entirely on peanut oil (Babu et al., 2020). In terms of impact upon environmental conditions, using cars with a conventional fuel source leads to many toxic gaseous substances (carbon monoxide, carbon dioxide, nitrogen oxides, sulfur oxide, hydrocarbons, etc.) (Babu et al., 2020).

The supply of raw materials for biodiesel production varies according to geography and climate. The most commonly used raw materials are rapeseed/canola oil in Europe, palm oil in tropical countries, and soybean oil and animal fats in the United States. However, these fats and oils are not sufficient to replace the current use of petrodiesel. Therefore, as an alternative to biodiesel production, jatropha (*Jatropha curcas* L.), wild mustard (*Brassica juncea* L.), field dried (*Thlaspi arvense* L.), moringa (*Moringa oleifera* L.), and camellia (*Camelina sativa* L.) oils are of great interest (Moser & Vaughn, 2010).

The fact that the fuel properties of biodiesel depend on the fatty acid (FA) composition of the lipid from which it is prepared (Knothe, 2009; Moser, 2009) results in different fuel properties of biodiesel fuels with different FA compositions. This could serve as a model for other oils with similar FA profiles, leading to genetic modification of existing oilseed crops for optimum biodiesel fuel properties (Kinney & Clemente, 2005). The best example of this is coriander, which contains a vegetable oil with a different FA profile.

Moser and Vaughn evaluated coriander seed oil methyl esters as alternative biodiesel fuel and revealed that the transesterification procedure with methanol and sodium methoxide catalyst yielded 94% wt. They also determined that biodiesel fuels contain petroselinic (6Z-octadecenoic; 68.5% by weight) acid as the main component, and the remaining fatty acid profile is made up of linoleic (9Z, 12Z-octadeca-dienoic; 13% by weight), oleic (9Z-octadecanoic; 7.6% by weight), and stearic (octadecanoic; 3.1% by weight) acids. Using standard methods in their studies, low-temperature properties, oxidative stability, cetane number, sulfur content, free and total glycerol content, and kinematics were used. In addition, content fuel properties such as viscosity, acid value (AV), phosphorus content, lubricity, combustion heat, Gardner color, iodine value (IV), FA profile, and tocopherol were determined (Moser & Vaughn, 2010).

The biodiesel production process produces methyl ester (biodiesel) and glycerol. Most biodiesel production processes use excess methanol to achieve high yields (Babu et al., 2020). Moser and Vaughn used a standard transesterification procedure with methanol and sodium methoxide catalyst to provide *C. sativum* oily methyl esters (CSME) and obtained 94% wt yield. Acid-catalyzed pretreatment was required to reduce the AV of the oil from 2.66 to 0.47 mg KOH g^{-1}. Cold filter plugging (CFPP) and pour points (PP) at 40°C and 14.6 h (110°C) were found to be 15°C and 19°C, respectively. Experimental results including acid value (AV), free and total glycerol content, iodine value, and other properties such as sulfur and phosphorus content were reported meeting the requirements of biodiesel fuel standards ASTM D6751 and EN 14214. In conclusion, CSME's unique oil content indicated attractive fuel properties due to its fatty acid (FA) composition.

Stating that coriander seed oil can be used as an alternative fuel to diesel fuel, Mahesh et al. (2020) revealed that coriander seed oil methyl ester was prepared using transesterification using a single-cylinder VCR engine. In another study, Illes et al. (2000) reported that coriander seed oil could be used both as an alternative to diesel fuel and as an additive for diesel yachts and revealed that the oil is extracted in CO_2 and propane and solvents under subcritical conditions. The reaction temperature was 60°C, and the yield was 94% over 1.5 h. Tamilselvan et al. (2017) used coriander seed oil as methanol and CH_3Ona with a 6:1 alcohol/oil ratio as a catalyst.

Nguyen et al. (2015) reported that the European Union had shown more interest in coriander seed oil, that coriander seed oil was used as a new food ingredient in the European Union, and that this oil was authorized as a food supplement. He noticed that coriander seed oil is particularly rich in the less common monounsaturated oleic acid isomer, petroselinic acid. Eidi et al. (2009) summarized that it contains coriander seed oil and that its main component, linalool, has a mild acute toxic potential.

Considering the studies of Babu et al., a 3 to 5% improvement of biodiesel in brake thermal efficiency analysis and an increase of about 7% in stated thermal efficiency are due to a 10% higher oxygen content. The parameter is mechanical efficiency, and the experimental results showed that B80 biodiesel had the highest mechanical efficiency. The results showed that B80, B60, B50, B40, and B20 biodiesels were efficient. It was stated that B80 fuel with the highest biodiesel content had the lowest EGT, followed by B60, B50, B40, and B20 biodiesels, respectively. The combination of oilseed oil and biodiesel was evaluated according to experimental analysis such as mechanical efficiency and specified thermal efficiency calculation. After the first experiment using conventional diesel fuel, the engine was started using different diesel and biodiesel blend combinations, and performance characteristics were observed for varying loads of different biodiesel blend combinations. At the end of the experimental parameters, it was reported that the biodiesel consumption was lower at higher loads than diesel. Apart from B20 (60-10-10) biodiesel, it was observed that the brake thermal efficiency regularly decreased with the decrease in the amount of coriander seed biodiesel (Babu et al., 2020).

32.6 CORIANDER AND FIBER

Today, commercially used materials cause a loss of sustainability due to the rapid depletion of our resources. The use of toxic and dangerous synthetic resins such as urea-formaldehyde in the production process and other harmful substances can be detrimental to the environment and human health; the alternative that is being proposed here in this chapter results in new materials that are more environmentally friendly, cost-effective, and highly available. In this case, the production of renewable non-binding fiberboards and degreased coriander press cakes from coriander straw emerges as an alternative. Another important result of this process is the evaluation of crop residues and process by-products and making them applicable to some existing wood-based resin-bonded products (Uitterhaegen et al., 2017). Panels bonded with formaldehyde resins used indoors have led to changes in air quality regulations (Salthammer, Mentese, & Marutzky, 2010). Relatively more expensive synthetic adhesives in particleboard or fiberboard often represent more than 30% of the total production cost (Van Dam, Van Den Oever, & Keijsers, 2004).

The press cake acts as a natural binder inside the sheets due to the thermoplastic behavior of the protein fraction during thermos-pressing. By evaluating the effect of different fiber refining methods, it has been shown that the twin-screw extrusion process effectively improves fiber morphology and results in fiberboards with better performance compared to a conventional milling process. The best fiberboard was produced with extruded refined straw using a 0.4 liquid/solid (L/S) ratio and the addition of 40% press cake. In addition, the water sensitivity of the boards was effectively reduced by 63% by adding the extruded feedstock premix and heat-treating the panels at 200°C, resulting in well-performing materials showing a flexural strength (Salthammer et al., 2010) MPa and a thickness swelling of 24%. Manufactured without chemical adhesives, these chipboards can offer viable, sustainable alternatives to existing commercial wood-based materials such as oriented particleboard, particleboard, and medium-density fiberboard with high-cost effectiveness.

Adhesive bonded sheets are produced by thermos-pressing lignocellulosic fiber material. This was based on the self-binding capacity of lignins, and the presence of less hemicellulose on wheat straw, *Miscanthus sinensis*, sugarcane pulp, kenaf kernel, palm oil, and banana stem. In addition, press cake is formed as a by-product of the mechanical pressing of oilseeds and is used to produce self-adhesive sheets (Uitterhaegen et al., 2017).

Coriandrum sativum, whose production results in approximately 500,000 tons/year of ready-made coriander straw, has been approved as a new foodstuff (NFI), and coriander vegetable oil is derived from fruits (Kamrozzaman, Ahmed, & Quddus, 2016; Sharma et al., 2014). It could be extracted efficiently by mechanical pressing using a high-quality twin-screw extruder (Uitterhaegen et al., 2015). It was reported in one study that coriander press cake, a by-product of the oil extraction process, was used in the production of self-adhesive boards (as particleboard and oriented particle board (OSB)) with mechanical properties comparable to commercial wood-based panels (Uitterhaegen et al., 2017). The produced plates have two different raw materials: coriander straw and coriander press cake. The coriander press cake is obtained from the mechanical pressing of coriander fruits for vegetable oil extraction, while the lignocellulosic material, coriander straw, is formed from the vegetative stem parts of the coriander plant. Like jute fiber and wheat straw, coriander straw can be used as a raw material for producing renewable panels with high cellulose content and good mechanical performance.

Extrudates and ground coriander straw were used to produce binder-free fiberboards; a natural gasket was used to obtain adhesive panels. Different amounts of degreased coriander press cake (10, 25, and 40% by weight of the total raw material) were added to the raw material as a binder. It was reported that the best mechanical and water resistance chipboard performance was obtained with the addition of extruded refined coriander straw and the L/S ratio of 0.4% and 40% coriander degreased press cake. In addition, twin-screw extrusion with strong mixing capacity achieved high micro-mixing and good product homogeneity. Coriander binder-free boards produced for use in dry conditions have been reported to comply with international ISO requirements for general purpose medium density fiberboard (MDF) and furniture grade and load-bearing MDF (Bouvier & Campanella, 2014; Faruk, Bledzki, Fink, & Sain, 2012; ISO, 2010; Letizia, Cocchiara, Lalko, & Api, 2003; Li, Tabil, & Panigrahi, 2007).

32.7 CORIANDER OIL PRESS CAKE

Organic by-products or residues are becoming necessary for different applications such as vitamins, antibiotics, industrial enzymes, and bio-pesticide production. The composition of coriander oil and its cake is mainly based on the characteristics of coriander fruits, which are affected by various factors such as geographical conditions and genetic impacts. In addition, the oil extraction process conditions and storage parameters are also influential in the nutritional availability of the seedcakes (Sriti et al., 2019; Uitterhaegen et al., 2015). Petroselinic acid, an isomer of oleic acid, is found as the main fatty acid constituent of *Coriandrum sativum*. The amount of this component was determined to be between 31 and 75% of the fatty acid profile, and it is considered to be a safe supplement to be consumed for adults at the highest level of 8.6 mg/kg bw per day (Sriti et al., 2019; E. Uitterhaegen et al., 2015).

The oil content of the seeds and the chemical composition of the press cake were affected by inlet flow rate, screw rotation speed, and nozzle diameter. Uitterhaegen et al. (2015) found the highest oil yield at 2.3 kg/h inlet flow rate and 50 rpm screw rotation speed, and the oil rate was found at a minimum of 16.6% of the dry matter in the press cake. The solid particles content in the filtrates was found between 47.5–66.0%. The acid value was 2.90–3.58 mg of KOH/g oil; acidity changed from 1.41–1.80% FFA as petroselinic acid for the oil was extracted with the Soxhlet method and screw press. Based on the acid value and acidity criteria examined, it was clear that all analyzed pressed oils were high-quality vegetable oils.

The potential industrial applications of process by-products help increase the extrusion's economic efficacy. Studies showed that particle size distribution in coriander press cake is not related to the applied screw profile. The coriander press cakes were found to contain 15–18% oil. The oil level can be considered an advantage for energy applications such as gasification, pyrolysis, or combustion. Thermo-processing could also improve the usability of the press cakes of coriander in producing renewable novel agro-materials. Combining the press cake with biodegradable polymer materials, including polylactic and polycaprolactone, can be used to develop an alternative material for the biocomposite industry.

Natural antioxidants were also found in the press cake. The isolation of these antioxidants using methanol extraction could help to protect their benefits and health impacts. However, it should be considered that the temperature level of the isolation process could affect the antioxidant level of the press cake inversely or increase the migration of some water-soluble antioxidants to the hydrodistillation's water phase, resulting in fewer antioxidants in the final product.

The most abundant essential oil components of coriander fruits were determined as linalool (71.7%), α-pinene (5.6%), γ-terpinene (5.0%), camphor (4.2%), geranyl acetate (3.0%), and linalyl acetate (2.9%). The amounts of essential oil components changed drastically in press cake except for linalool. The essential oil content in the cake was at the levels of 0.14–0.31% of the dry matter, and the residue amount found was inversely correlated with the applied temperature. Applying heat at 65°C allowed 40% of the essential oil to be kept in the press cake, of which the composition is similar to the essential oil profile of the coriander fruits. The linalool is found most abundant among other components, and it is preferred in various applications as a fragrance material for soaps, perfumes, and cosmetics. The residual essential oil might be regained by the hydrodistillation method. Furthermore, the essential oil residues in the coriander press cake could be evaluated in the production of mosquito repellents or value-added agro-materials (Uitterhaegen et al., 2015).

Coriander cake extract is evaluated to determine the DPPH· radical scavenging activities, and antioxidant capacity is positively related to increased nozzle diameter. These results also showed that the total phenolic content decreases with the increased nozzle diameter. The research about the antioxidant activity with different methods also showed that coriander oil cake has a lower antioxidant level than synthetic antioxidants BHA and BHT. However, the coriander oil cake still can be a safe natural antioxidants source. The total phenolics and flavonoid content found was decreased by the thermal process or polymerization, while the amount of total tannin increased with the increased nozzle diameter (Sriti et al., 2019).

Different applications of coriander press cake in various industries are also possible. The formulation of a fiberboard produced with extrusion-refined coriander straw and 40% press cake was evaluated for a binderless fiberboard industry. The boards were found stronger against water effects by 63%. The other parameters of the panels, such as thickness swelling (24%) and flexural strength (29 MPa), were determined after the 200°C thermal treatment. The results show that the fiberboard material is environmentally friendly, bio-based, does not need any chemical adhesives, and could be a cost-effective, sustainable alternative for wood-based materials in the market. However, since the current raw commercial materials are based on forest resources and fossils, and the process ingredients have synthetic resins and harmful volatile organic compounds, they are significantly dangerous for the environment and human health (Uitterhaegen et al., 2017).

After the oil extraction, the residue press cake characteristics make the material an attractive source for feed, food, and pharmaceutical applications. The high amount of bioactive compounds such as flavonoids, tannins, and phenolics and the antioxidant potential could be preferred for formulating wellness products in the supplement industry or specific formulations in the food or feed industry. The fiber content of the different parts of the coriander plant makes the press cake a valuable additive for different applications such as pasta products. Considering the press cake's sustainability, bioavailability, and cost-effectiveness, it can be helpful in green biofuel applications (Chakraborty, Bhattacharya, Bhattacharyya, Bandyopadhyay, & Ghosh, 2016; Sriti et al., 2019).

32.8   APPLICATIONS OF CORIANDER OIL IN PACKAGING

Various coriander oil applications have been employed in different areas in the food industry. A study evaluated the effectiveness of the addition of coriander oil for porcine meat and bone meal (MBM) film packaging materials during storage. The extraction of the protein content was used to create a base for a film, and coriander oil (CO) was added to increase the effectiveness of the packaging material. Compared to the control groups, MBM-CO packaging material helped reduce *E. coli* population by 1.36 log CFU/g after 3 days of storage and the overall microbial population by absorbing excess water of the packaged patties. Since the MBM-CO packaging material has a better oxygen barrier property, its peroxide value (POV) and thiobarbituric acid reactive substances (TBARS) were lower than the control group. The increase was lower in MetMB values in CO added materials compared to the control and MBM film group. The antimicrobial effect of the packaging material with 1% CO inclusion increased against *Listeria monocytogenes* and *Escherichia coli* O157:H7. Therefore, adding coriander oil into the packaging film material could increase the effectiveness of preservation during storage (Lee & Song, 2015; Lee, Won, & Song, 2015). Furthermore, combining the coriander seed oil with hyssop oil (0.02% v/w) was also found to increase the shelf life of stored ground beef. In addition, a delay in undesirable biochemical changes due to the additives and antimicrobial properties against Enterobacteriaceae were reported (Michalczyk, Macura, Tesarowicz, & Banaś, 2012).

32.9   CONCLUSION

Coriander, which functions both as a spice and herbal medicine, is an important and valuable bioactive resource for improving human health, thanks to its edible, safe, essential and fixed oils. Considering its bioactivity, it is also inspiring further research to standardize its commercial applications, especially existing food and medical applications. In addition, although it is a plant that can be grown throughout the year, the processing of the fruits and leaves of coriander is the best way to protect the herb, and processing the plant and using it in different ways is important for both increasing its profitability and its place in international trade.

When the literature was examined, the toxicity of coriander oil and its components was related to the dose and gender at different doses of coriander oil (Fujii, Furukawa, & Suzuki, 1972). All results showed that coriander essential oil in appropriate dosages could be used as a food ingredient and for human consumption in cosmetics.

A biodegradable (without sulfur content), non-toxic, and natural lubricant obtained by the transesterification process from vegetable oils, biodiesel is a safe alternative fuel to diesel due to its flash point above 130°C. Moreover, considering the limited fossil fuel reserves in today's world, it is a preferable alternative because it is renewable and environmentally friendly. Moreover, the realization of biodiesel production, which could be done both on a batch and continuous scale, could be made economical by growing the right crop.

The hygroscopic nature of the plant material causes the formation of sheets with insufficient water resistance, resulting in detrimental results in potential industrial applications. These studies show that there is still a need for optimization in the bare fiberboard production process and improvement of the dimensional stability of the boards. In order to better protect the fiber structure and reduce the production cost, it is necessary to optimize and improve the techniques in the raw material preparation process. Coriander sheets are suitable for commercialization due to their biobased structure, environmental friendliness, and affordable cost. In addition, the evaluation of coriander straw, a ready-made product residue, and the press cakes, which are a by-product of the process, is critical in terms of sustainability and helps establish a coriander bio-refinery.

REFERENCES

Abbassi, A., Mahmoudi, H., Zaouali, W., M'Rabet, Y., Casabianca, H., & Hosni, K. (2018). Enzyme-aided release of bioactive compounds from coriander (*Coriandrum sativum* L.) seeds and their residue by-products and evaluation of their antioxidant activity. *Journal of Food Science and Technology*, 55(8), 3065–3076. https://doi.org/10.1007/s13197-018-3229-4

Abd El-Hack, M. E., Abdelnour, S. A., Taha, A. E., Khafaga, A. F., Arif, M., Ayasan, T., … Abdel-Daim, M. M. (2020). Herbs as thermoregulatory agents in poultry: An overview. *Science of the Total Environment*, 703, 134399. https://doi.org/10.1016/j.scitotenv.2019.134399

Al-Bataina, B. A., Maslat, A. O., & Al-Kofahi, M. M. (2003). Element analysis and biological studies on ten oriental spices using XRF and Ames test. *Journal of Trace Elements in Medicine and Biology*, 17(2), 85–90. https://doi.org/10.1016/S0946-672X(03)80003-2

Aluko, R. E., McIntosh, T., & Reaney, M. (2001). Comparative study of the emulsifying and foaming properties of defatted coriander (*Coriandrum sativum*) seed flour and protein concentrate. *Food Research International*, 34(8), 733–738. https://doi.org/10.1016/S0963-9969(01)00095-3

Anwar, F., Sulman, M., Hussain, A. I., Saari, N., Iqbal, S., & Rashid, U. (2011). Physicochemical composition of hydro-distilled essential oil from coriander (*Coriandrum sativum* L.) seeds cultivated in Pakistan. *Journal of Medicinal Plants Research*, 5(15), 3537–3544. https://doi.org/10.5897/JMPR.9000978

ASTM. (2008). *Standard Specification for Biodiesel Fuel Blend Stock (B100) for Middle Distillate Fuels*. West Conshohocken, PA: ASTM International. https://doi.org/10.1520/D6751-08

Baba, K., Xiao, Y. Q., Taniguchi, M., Ohishi, H., & Kozawa, M. (1991). Isocoumarins from *Coriandrum sativum*. *Phytochemistry*, 30(12), 4143–4146. https://doi.org/10.1016/0031-9422(91)83482-Z

Babu, B. S., Kumar, D. B., Sathiyaraj, S., & Senthilkumar, A. (2020). Experimental investigation and performance of diesel engine using biodiesel coriander seed oil. *Materials Today: Proceedings*, 33, 1044–1048. https://doi.org/10.1016/j.matpr.2020.07.055

Bandoni, A. L., Mizrahi, I., & Juárez, M. A. (1998). Composition and quality of the essential oil of coriander (*Coriandrum sativum* L.) from Argentina. *Journal of Essential Oil Research*, 10(5), 581–584. https://doi.org/10.1080/10412905.1998.9700977

Begnami, A. F., Duarte, M. C. T., Furletti, V., & Rehder, V. L. G. (2010). Antimicrobial potential of *Coriandrum sativum* L. against different Candida species *in vitro*. *Food Chemistry*, 118(1), 74–77. https://doi.org/10.1016/j.foodchem.2009.04.089

Beikert, F. C., Anastasiadou, Z., Fritzen, B., Frank, U., & Augustin, M. (2013). Topical treatment of tinea pedis using 6% coriander oil in unguentum leniens: A randomized, controlled, comparative pilot study. *Dermatology*, 226(1), 47–51. https://doi.org/10.1159/000346641

Beyzi, E., Karaman, K., Gunes, A., & Buyukkilic Beyzi, S. (2017). Change in some biochemical and bioactive properties and essential oil composition of coriander seed (*Coriandrum sativum* L.) varieties from Turkey. *Industrial Crops and Products*, 109(April), 74–78. https://doi.org/10.1016/j.indcrop.2017.08.008

Bhat, S., Kaushal, P., Kaur, M., & Sharma, H. K. (2014). Coriander (*Coriandrum sativum* L.): Processing, nutritional and functional aspects. *African Journal of Plant Science*, 8(1), 25–33. https://doi.org/10.5897/AJPS2013.1118

Bhuiyan, N. I., Begum, J., & Sultana, M. (2009). Chemical composition of leaf and seed essential oil of *Coriandrum sativum* L. from Bangladesh. *Bangladesh Journal of Pharmacology*, 4, 150–153.

Bouvier, J. M., & Campanella, O. H. (2014). Extrusion processing technology: Food and non-food biomaterials. *Extrusion Processing Technology: Food and Non-Food Biomaterials*, *9781444338*, 1–518. https://doi.org/10.1002/9781118541685

British Pharmacopoeia. (2003). *Introduction General Notices Monographmedicinal and Pharmaceutical*. London: British Pharmacopeia Commission.

Cahoon, E. B., Shanklin, J., & Ohlrogge, J. B. (1992). Expression of a coriander desaturase results in petroselinic acid production in transgenic tobacco. *Proceedings of the National Academy of Sciences of the United States of America*, *89*(23), 11184–11188. https://doi.org/10.1073/pnas.89.23.11184

Ceska, O., Chaudhary, S. K., Warrington, P., Ashwood-Smith, M. J., Bushnell, G. W., & Poultont, G. A. (1988). Coriandrin, a novel highly photoactive compound isolated from *Coriandrum sativum*. *Phytochemistry*, *27*(7), 2083–2087. https://doi.org/10.1016/0031-9422(88)80101-8

Chahal, K. K., Singh, R., Kumar, A., & Bhardwaj, U. (2017). Chemical composition and biological activity of *Coriandrum sativum* L.: A review. *Indian Journal of Natural Products and Resources*, *8*(3), 193–203.

Chakraborty, P., Bhattacharya, A., Bhattacharyya, D. K., Bandyopadhyay, N. R., & Ghosh, M. (2016). Studies of nutrient rich edible leaf blend and its incorporation in extruded food and pasta products. *Materials Today: Proceedings*, *3*(10), 3473–3483. https://doi.org/10.1016/j.matpr.2016.10.030

Cortés-Eslava, J., Gómez-Arroyo, S., Villalobos-Pietrini, R., & Espinosa-Aguirre, J. J. (2004). Antimutagenicity of coriander (*Coriandrum sativum*) juice on the mutagenesis produced by plant metabolites of aromatic amines. *Toxicology Letters*, *153*(2), 283–292. https://doi.org/10.1016/j.toxlet.2004.05.011

Coşkuner, Y., & Karababa, E. (2007). Physical properties of coriander seeds (*Coriandrum sativum* L.). *Journal of Food Engineering*, *80*(2), 408–416. https://doi.org/10.1016/j.jfoodeng.2006.02.042

Darughe, F., Barzegar, M., & Sahari, M. A. (2012). Antioxidant and antifungal activity of Coriander (*Coriandrum sativum* L.) essential oil in cake. *International Food Research Journal*, *19*(3), 1253–1260.

Dias, M. I., Barros, L., Sousa, M. J., & Ferreira, I. C. F. R. (2011). Comparative study of lipophilic and hydrophilic antioxidants from in vivo and in vitro grown *Coriandrum sativum*. *Plant Foods for Human Nutrition*, *66*(2), 181–186. https://doi.org/10.1007/s11130-011-0227-3

Duarte, A. F., Ferreira, S., Oliveira, R., & Domingues, F. C. (2013). Effect of coriander oil (*Coriandrum sativum*) on planktonic and biofilm cells of *Acinetobacter baumannii*. *Natural Product Communications*, *8*(5), 673–678. https://doi.org/10.1177/1934578x1300800532

Ebrahimi, S. N., Hadian, J., & Ranjbar, H. (2010). Essential oil compositions of different accessions of *Coriandrum sativum* L. from Iran. *Natural Product Research, 6419*. https://doi.org/10.1080/14786410903132316

Eidi, M., Eidi, A., Saeidi, A., Molanaei, S., Sadeghipour, A., Bahar, M., & Bahar, K. (2009). Effect of coriander seed (*Coriandrum sativum* L.) ethanol extract on insulin release from pancreatic beta cells in streptozotocin-induced diabetic rats. *Phytotherapy Research*, *23*(3), 404–406. https://doi.org/10.1002/ptr.2642

Faruk, O., Bledzki, A. K., Fink, H., & Sain, M. (2012). Progress in polymer science biocomposites reinforced with natural fibers: 2000–2010. *Progress in Polymer Science*, *37*(11), 1552–1596. https://doi.org/10.1016/j.progpolymsci.2012.04.003

Fujii, T., Furukawa, S., & Suzuki, S. (1972). Studies on compounded perfumes for toilet goods. *Journal of Japan Oil Chemists Society*, *21*, 904–908.

Ghazanfari, N., Mortazavi, S. A., Yazdi, F. T., & Mohammadi, M. (2020). Microwave-assisted hydrodistillation extraction of essential oil from coriander seeds and evaluation of their composition, antioxidant and antimicrobial activity. *Heliyon*, *6*(9), e04893. https://doi.org/10.1016/j.heliyon.2020.e04893

Gil, A., De la Fuente, E. B., Lenardis, A. E., López Pereira, M., Suárez, S. A., Bandoni, A., … Ghersa, C. M. (2002). Coriander essential oil composition from two genotypes grown in different environmental conditions. *Journal of Agricultural and Food Chemistry*, *50*(10), 2870–2877. https://doi.org/10.1021/jf011128i

Gray, A. M., & Flatt, P. R. (1999). Insulin-releasing and insulin-like activity of the traditional anti-diabetic plant *Coriandrum sativum* (coriander). *British Journal of Nutrition*, *81*(3), 203–209. https://doi.org/10.1017/S0007114599000392

Griffiths, D. W., Robertson, G. W., Millam, S., & Holmes, A. C. (1992). The determination of the petroselinic acid content of coriander (*Coriandrum sativum*) oil by capillary gas chromatography. *Phytochemical Analysis*, *3*(6), 250–253. https://doi.org/10.1002/pca.2800030603

Gumus, O., Yasar, E., Gumus, Z. P., & Ertas, H. (2020). Comparison of different classification algorithms to identify geographic origins of olive oils. *Journal of Food Science and Technology*, *57*(4), 1535–1543. https://doi.org/10.1007/s13197-019-04189-4

Gumus, Z. P., Ertas, H., Yasar, E., & Gumus, O. (2018). Classification of olive oils using chromatography, principal component analysis and artificial neural network modelling. *Journal of Food Measurement and Characterization*, *12*(2), 1325–1333. https://doi.org/10.1007/s11694-018-9746-z

Gupta, M. (2010). Pharmacological properties and traditional therapeutic uses of important indian spices: A review. *International Journal of Food Properties*, *13*(5), 1092–1116. https://doi.org/10.1080/10942910902963271

Havale, R., Rao, D. G., Shrutha, S. P., Fatima, B. O., Sara, S. S., & Bemalgi, N. (2021). A comperative evaluation of kidodent moutwash and *Coriandrum sativum* oil mouthwash in reducing *Streptococcus mutans* count- A parallel double blinded randomized control trial. *Annals of R.S.C.B.*, *25*(1), 2638–2650.

Horvath, G., Wessjohann, L., Bigirimana, J., Jansen, M., Guisez, Y., Caubergs, R., & Horemans, N. (2006). Differential distribution of tocopherols and tocotrienols in photosynthetic and non-photosynthetic tissues. *Phytochemistry*, *67*(12), 1185–1195. https://doi.org/10.1016/j.phytochem.2006.04.004

Illés, V., Daood, H. G., Perneczki, S., Szokonya, L., & Then, M. (2000). Extraction of coriander seed oil by CO_2 and propane at super- and subcritical conditions. *Journal of Supercritical Fluids*, *17*(2), 177–186. https://doi.org/10.1016/S0896-8446(99)00049-2

Ishikawa, T., Kondo, K., & Kitajima, J. (2003). Water-soluble constituents of coriander. *Chemical and Pharmaceutical Bulletin*, *51*(1), 32–39. https://doi.org/10.1248/cpb.51.32

ISO. (2010). *ISO 16895-2:2010 Wood-Based Panels - Dry Process Fibreboard*. ISO.

Iwatani, Y., Arcot, J., & Shrestha, A. K. (2003). Determination of folate contents in some Australian vegetables. *Journal of Food Composition and Analysis*, *16*(1), 37–48. https://doi.org/10.1016/S0889-1575(02)00159-X

Jabeen, Q., Bashir, S., Lyoussi, B., & Gilani, A. H. (2009). Coriander fruit exhibits gut modulatory, blood pressure lowering and diuretic activities. *Journal of Ethnopharmacology*, *122*(1), 123–130. https://doi.org/10.1016/j.jep.2008.12.016

Jansen, P. C. M. (1981). Spices, condiments and medicinal plants in Ethiopia, their taxonomy and agricultural significance. *Agricultural Research Reports*, *56–67*.

Kamrozzaman, M., Ahmed, S., & Quddus, A. (2016). Effect of fertilizer on coriander seed production. *Bangladesh Journal of Agricultural Research*, *41*(2), 345–352. https://doi.org/10.3329/bjar.v41i2.28236

Kinney, A. J., & Clemente, T. E. (2005). Modifying soybean oil for enhanced performance in biodiesel blends. *Fuel Processing Technology*, *86*(10), 1137–1147. https://doi.org/10.1016/j.fuproc.2004.11.008

Kiralan, M., Calikoglu, E., Ipek, A., Bayrak, A., & Gurbuz, B. (2009). Fatty acid and volatile oil composition of different coriander (*Coriandrum sativum*) registered varieties cultivated in Turkey. *Chemistry of Natural Compounds*, *45*(1), 100–102. https://doi.org/10.1007/s10600-009-9240-2

Knothe, G. (2009). Improving biodiesel fuel properties by modifying fatty ester composition. *Energy & Environmental Science*, 759–766. https://doi.org/10.1039/b903941d

Knothe, G., Krahl, J., & Gerpen, J. Van. (2010). *The Biodiesel Handbook*. G. Knothe, J. Krahl, & J. Van Gerpen, Eds. (2nd ed.). AOCS Press. https://doi.org/10.1016/C2015-0-02453-4

Kozłowska, M., Gruczyńska, E., Ścibisz, I., & Rudzińska, M. (2016). Fatty acids and sterols composition, and antioxidant activity of oils extracted from plant seeds. *Food Chemistry*, *213*, 450–456. https://doi.org/10.1016/j.foodchem.2016.06.102

Lee, J. H., & Song, K. B. (2015). Application of an antimicrobial protein film in beef patties packaging. *Korean Journal for Food Science of Animal Resources*, *35*(5), 611–614. https://doi.org/10.5851/kosfa.2015.35.5.611

Lee, J. H., Won, M., & Song, K. B. (2015). Physical properties and antimicrobial activities of porcine meat and bone meal protein films containing coriander oil. *LWT*, *63*(1), 700–705. https://doi.org/10.1016/j.lwt.2015.03.043

Letizia, C. S., Cocchiara, J., Lalko, J., & Api, A. M. (2003). Fragrance material review on linalool. *Food and Chemical Toxicology*. https://doi.org/10.1016/S0278-6915(03)00015-2

Li, X., Tabil, L. G., & Panigrahi, S. (2007). Chemical treatments of natural fiber for use in natural fiber-reinforced composites: A review. *Journal of Polymers and the Environment*. https://doi.org/10.1007/s10924-006-0042-3

Mahendra, P., & Bisht, S. (2011). *Coriandrum sativum*: A daily use spice with great medicinal effect. *Pharmacognosy Journal*, *3*(21), 84–88. https://doi.org/10.5530/pj.2011.21.16

Mahesh, R., Kumar, K. V., Ali, S. S., & Prasanth, S. (2020). Emission characteristics of VCR engine with coriander seed oil mixed diesel. *International Journal of Recent Technology and Engineering*, *8*(5), 5550–5553. https://doi.org/10.35940/ijrte.e3118.018520

Matasyoh, J. C., Maiyo, Z. C., Ngure, R. M., & Chepkorir, R. (2009). Chemical composition and antimicrobial activity of the essential oil of *Coriandrum sativum*. *Food Chemistry*, *113*(2), 526–529. https://doi.org/10 .1016/j.foodchem.2008.07.097

Matthäus, B., & Angelini, L. G. (2005). Anti-nutritive constituents in oilseed crops from Italy. *Industrial Crops and Products*, *21*(1), 89–99. https://doi.org/10.1016/j.indcrop.2003.12.021

Matthaus, B., Vosmann, K., Pham, L. Q., & Aitzetmüller, K. (2003). FA and tocopherol composition of vietnamese oilseeds. *JAOCS, Journal of the American Oil Chemists' Society*, *80*(10), 1013–1020. https://doi .org/10.1007/s11746-003-0813-y

Mhemdi, H., Rodier, E., Kechaou, N., & Fages, J. (2011). A supercritical tuneable process for the selective extraction of fats and essential oil from coriander seeds. *Journal of Food Engineering*, *105*(4), 609–616. https://doi.org/10.1016/j.jfoodeng.2011.03.030

Michalczyk, M., Macura, R., Tesarowicz, I., & Banaś, J. (2012). Effect of adding essential oils of coriander (*Coriandrum sativum* L.) and hyssop (*Hyssopus officinalis* L.) on the shelf life of ground beef. *Meat Science*, *90*(3), 842–850. https://doi.org/10.1016/j.meatsci.2011.11.026

Misharina, T. A. (2001). Effect of conditions and duration of storage on composition of essential oil from coriander seeds. *Prikladnaia Biokhimiia i Mikrobiologiia*, *37*, 726–732. http://www.helpa-prometheus .gr/διαγνωστικές-εξετάσεις-για-τον-καρκί/

Mohammadi, S., & Saharkhiz, M. J. (2013). Changes in essential oil content and composition of catnip (*Nepeta cataria* L.) during different developmental stages. *Journal of Essential Oil Bearing Plants*, *5026*. https://doi.org/10.1080/0972060X.2011.10643592

Moser, B. R. (2009). Biodiesel production, properties, and feedstocks. *In Vitro Cellular & Developmental Biology - Plant*, *45*, 229–266. https://doi.org/10.1007/s11627-009-9204-z

Moser, B. R., & Vaughn, S. F. (2010). Coriander seed oil methyl esters as biodiesel fuel: Unique fatty acid composition and excellent oxidative stability. *Biomass and Bioenergy*, *34*(4), 550–558. https://doi.org/10 .1016/j.biombioe.2009.12.022

Msaada, K., Hosni, K., Ben Taarit, M., Chahed, T., Hammami, M., & Marzouk, B. (2009). Changes in fatty acid composition of coriander (*Coriandrum sativum* L.) fruit during maturation. *Industrial Crops and Products*, *29*(2–3), 269–274. https://doi.org/10.1016/j.indcrop.2008.05.011

Msaada, K., Hosni, K., Taarit, M. B., Chahed, T., Kchouk, M. E., & Marzouk, B. (2007). Changes on essential oil composition of coriander (*Coriandrum sativum* L.) fruits during three stages of maturity. *Food Chemistry*, *102*(4), 1131–1134. https://doi.org/10.1016/j.foodchem.2006.06.046

Msaada, K., Hosni, K., Taarit, M. B., Hammami, M., & Marzouk, B. (2009). Effects of growing region and maturity stages on oil yield and fatty acid composition of coriander (*Coriandrum sativum* L.) fruit. *Scientia Horticulturae*, *120*(4), 525–531. https://doi.org/10.1016/j.scienta.2008.11.033

Musa Özcan, M., Ünver, A., Uçar, T., & Arslan, D. (2008). Mineral content of some herbs and herbal teas by infusion and decoction. *Food Chemistry*, *106*(3), 1120–1127. https://doi.org/10.1016/j.foodchem.2007 .07.042

Neffati, M., & Marzouk, B. (2008). Changes in essential oil and fatty acid composition in coriander (*Coriandrum sativum* L.) leaves under saline conditions. *Industrial Crops and Products*, *28*(2), 137–142. https://doi.org/10.1016/j.indcrop.2008.02.005

Neffati, M., Sriti, J., Hamdaoui, G., Kchouk, M. E., & Marzouk, B. (2011). Salinity impact on fruit yield, essential oil composition and antioxidant activities of *Coriandrum sativum* fruit extracts. *Food Chemistry*, *124*(1), 221–225. https://doi.org/10.1016/j.foodchem.2010.06.022

Nguyen, Q. H., Talou, T., Cerny, M., Evon, P., & Merah, O. (2015). Oil and fatty acid accumulation during coriander (*Coriandrum sativum* L.) fruit ripening under organic cultivation. *Crop Journal*, *3*(4), 366–369. https://doi.org/10.1016/j.cj.2015.05.002

Nguyen, Q. H., Talou, T., Evon, P., Cerny, M., & Merah, O. (2020). Fatty acid composition and oil content during coriander fruit development. *Food Chemistry*, *326*(May), 127034. https://doi.org/10.1016/j.foodchem .2020.127034

Nguyen, T., Aparicio, M., & Saleh, M. (2015). Accurate mass GC/LC-quadrupole time of flight mass spectrometry analysis of fatty acids and triacylglycerols of spicy fruits from the Apiaceae family. *Molecules*, *20*(12), 21421–21432. https://doi.org/10.3390/molecules201219779

Norman, J. (1990). *The Complete Book of Spices*. London, UK: Dorling Kindersley Limited.

Oganesyan, E. T., Nersesyan, Z. M., & Parkhomenko, A. Y. (2007). Chemical composition of the aboveground part of *Coriandrum sativum*. *Pharmaceutical Chemistry Journal*, *41*(3), 149–153.

Paarakh, P. M. (2009). *Coriandrum sativum* Linn. *Pharmacologyonline*, *3*(3), 561–573. https://doi.org/10.1007 /978-0-387-70638-2_403

Parthasarathy, V. A., Chempakam, B., & Zachariah, T. J. (2008). Coriander: Chemistry of spices. *CAB International*, *11*(75), 190–206. https://doi.org/10.1079/9781845934057.0190

Peter, K. V. (2004). *Handbook of Herbs and Spices. Handbook of Herbs and Spices*. Cambridge, UK: Woodhead Publishing.

Potter, T., & Fagerson, I. (1990). Composition of coriander leaf volatiles. *Journal of Agriculture and Food Chemistry*, *38*, 2054–2056.

Raju, M., Varakumar, S., Lakshminarayana, R., Krishnakantha, T. P., & Baskaran, V. (2007). Carotenoid composition and vitamin A activity of medicinally important green leafy vegetables. *Food Chemistry*, *101*(4), 1598–1605. https://doi.org/10.1016/j.foodchem.2006.04.015

Rakic, Z., & Johnson, C. B. (2002). Influence of environmental factors (including UV-B radiation) on the composition of the essential oil of *Ocimum basilicum* - Sweet basil. *Journal of Herbs, Spices and Medicinal Plants*, *9*, 157–162. https://doi.org/10.1300/J044v09n02_22

Ramadan, M. F., Kroh, L. W., & Mörsel, J. T. (2003). Radical scavenging activity of black cumin (*Nigella sativa* L.), coriander (*Coriandrum sativum* L.), and niger (*Guizotia abyssinica* Cass.) crude seed oils and oil fractions. *Journal of Agricultural and Food Chemistry*, *51*(24), 6961–6969. https://doi.org/10.1021/jf0346713

Ramadan, M. F., & Mörsel, J. T. (2002). Oil composition of coriander (*Coriandrum sativum* L.) fruit-seeds. *European Food Research and Technology*, *215*(3), 204–209. https://doi.org/10.1007/s00217-002-0537-7

Ramadan, M. F., & Wahdan, K. M. M. (2012). Blending of corn oil with black cumin (*Nigella sativa*) and coriander (*Coriandrum sativum*) seed oils : Impact on functionality, stability and radical scavenging activity. *Food Chemistry*, *132*(2), 873–879. https://doi.org/10.1016/j.foodchem.2011.11.054

Rao, A. S., Ahmed, M. F., Ibrahim, M., & Ahmed, M. F. (2012). Hepatoprotective activity of *Melia azedarach* leaf extract against simvastatin induced Hepatotoxicity in rats. *Journal of Applied Pharmaceutical Science*, *02*(July), 144–148. https://doi.org/10.7324/JAPS.2012.2721

Reiter, B., Lechner, M., & Lorbeer, E. (1998). The fatty acid profiles -including petroselinic and cis-vaccenic acid-of different Umbelliferae seed oils. *Lipid-Fett*, *100*(11), 498–502. https://doi.org/10.1002/(sici)1521-4133(199811)100:11<498::aid-lipi498>3.3.co;2-z

Saeed, S., & Tariq, P. (2007). Antimicrobial activities of *Emblica officinalis* and *Coriandrum sativum* against gram positive bacteria and *Candida albicans*. *Pakistan Journal of Botany*, *39*(3), 913–917.

Sahib, N. G., Anwar, F., Gilani, A. H., Hamid, A. A., Saari, N., & Alkharfy, K. M. (2013). Coriander (*Coriandrum sativum* L.): A potential source of high-value components for functional foods and nutraceuticals - A review. *Phytotherapy Research*, *27*(10), 1439–1456. https://doi.org/10.1002/ptr.4897

Salthammer, T., Mentese, S., & Marutzky, R. (2010). Formaldehyde in the indoor environment. *Chemical Reviews*, *110*(4), 2536–2572. https://doi.org/10.1021/cr800399g

Shahwar, M. K., El-ghorab, A. H., Anjum, M., Butt, M. S., Hussain, S., Butt, M. S., … Nadeem, M. (2012). Characterization of coriander (*Coriandrum sativum* L.) seeds and leaves: Volatile and non volatile extracts. *International Journal of Food Properties*, *15*, 736–747. https://doi.org/10.1080/10942912.2010.500068

Sharma, R. P., Singh, R. S., Verma, T. P., Tailor, B. L., Sharma, S. S., & Singh, S. K. (2014). Coriander the taste of vegetables: Present and future prospectus for coriander seed production in Southeast Rajasthan. *Economic Affairs*, *59*(3), 345. https://doi.org/10.5958/0976-4666.2014.00003.5

Shivanand, P. (2010). Coriandrum sativum : A biological description and its uses in the treatment of various diseases. *International Journal of Pharmacy and Life Sciences*, *1*(3), 119–126.

Silva, F., Domeno, C., & Domingues, F. C. (2020). *Coriandrum sativum* L.: Characterization, biological activities, and applications. In V. R. Preedy & R. R. Watson (Eds.), *Nuts & Seeds in Health and Disease Prevention* (2nd ed., pp. 497–520). London, UK: Academic Press, Elsevier.

Silva, F., Ferreira, S., Queiroz, J. A., & Domingues, F. C. (2011). Coriander (*Coriandrum sativum* L.) essential oil: Its antibacterial activity and mode of action evaluated by flow cytometry. *Journal of Medical Microbiology*, *60*(10), 1479–1486. https://doi.org/10.1099/jmm.0.034157-0

Sourmaghi, M. H. S., Kiaee, G., Golfakhrabadi, F., Jamalifar, H., & Khanavi, M. (2015). Comparison of essential oil composition and antimicrobial activity of *Coriandrum sativum* L. extracted by hydrodistillation and microwave-assisted hydrodistillation. *Journal of Food Science and Technology*, *52*(4), 2452–2457. https://doi.org/10.1007/s13197-014-1286-x

Sriti, J., Aidi, W., Talou, T., Mhamdi, B., Hamdaoui, G., & Marzouk, B. (2010). Lipid, fatty acid and tocol distribution of coriander fruit's different parts, *31*, 294–300. https://doi.org/10.1016/j.indcrop.2009.11.006

Sriti, J., Aidi, W., & Thierry, W. (2010). Lipid profiles of Tunisian coriander (*Coriandrum sativum*) seed. *Journal of the American Oil Chemists' Society*, *87*, 395–400. https://doi.org/10.1007/s11746-009-1505-1

Sriti, J., Bettaieb, I., Bachrouch, O., Talou, T., & Marzouk, B. (2019). Chemical composition and antioxidant activity of the coriander cake obtained by extrusion. *Arabian Journal of Chemistry*, *12*(7), 1765–1773. https://doi.org/10.1016/j.arabjc.2014.11.043

Sriti, J., Msaada, K., Talou, T., Faye, M., Kartika, I. A., & Marzouk, B. (2012). Extraction of coriander oil by twin-screw extruder: Screw configuration and operating conditions effect. *Industrial Crops and Products*, *40*(1), 355–360. https://doi.org/10.1016/j.indcrop.2012.03.034

Sriti, J., Talou, T., Faye, M., Vilarem, G., & Marzouk, B. (2011). Oil extraction from coriander fruits by extrusion and comparison with solvent extraction processes. *Industrial Crops and Products*, *33*(3), 659–664. https://doi.org/10.1016/j.indcrop.2011.01.005

Sriti, J., Talou, T., Msaada, K., & Marzouk, B. (2011). Comparative analysis of fatty acid, sterol and tocol composition of Tunisian and Canadian coriander (*Coriandrum sativum* L.) fruit. *Analytical Chemistry Letters*, *1*(5–6), 375–383. https://doi.org/10.1080/22297928.2011.10648241

Sriti, J., Talou, T., Wannes, A., & Marzouk, B. (2009). Essential oil, fatty acid and sterol composition of Tunisian coriander fruit different parts. *Journal of the Science of Food and Agriculture*, *89*, 1659–1664. https://doi.org/10.1002/jsfa.3637

Sriti, J., Wannes, W. A., Talou, T., Mhamdi, B., Hamdaoui, G., & Marzouk, B. (2010). Lipid, fatty acid and tocol distribution of coriander fruit's different parts. *Industrial Crops and Products*, *31*(2), 294–300. https://doi.org/10.1016/j.indcrop.2009.11.006

Sriti, J., Wannes, W. A., Talou, T., Vilarem, G., & Marzouk, B. (2011). Chemical composition and antioxidant activities of tunisian and Canadian coriander (*Coriandrum sativum* L.) Fruit. *Journal of Essential Oil Research*, *23*(4), 7–15. https://doi.org/10.1080/10412905.2011.9700462

Sunil, C., Agastian, P., Kumarappan, C., & Ignacimuthu, S. (2012). In vitro antioxidant, antidiabetic and antilipidemic activities of *Symplocos cochinchinensis* (Lour.) S. Moore bark. *Food and Chemical Toxicology*, *50*(5), 1547–1553. https://doi.org/10.1016/j.fct.2012.01.029

Tamilselvan, P., Nallusamy, N., & Rajkumar, S. (2017). A comprehensive review on performance, combustion and emission characteristics of biodiesel fuelled diesel engines. *Renewable and Sustainable Energy Reviews*, *79*, 1134–1159. https://doi.org/10.1016/j.rser.2017.05.176

Taniguchi, M., Yanai, M., Xiao, Y. Q., Kido, T., & Baba, K. (1996). Three isocoumarins from *Coriandrum sativum*. *Phytochemistry*, *42*(3), 843–846. https://doi.org/10.1016/0031-9422(95)00930-2

Telci, I., & Hişil, Y. (2008). Biomass yield and herb essential oil characters at different harvest stages of spring and autumn sown *Coriandrum sativum*. *European Journal of Horticultural Science*, *73*(6), 267–272.

Uitterhaegen, E., Burianová, K., Ballas, S., Véronèse, T., Merah, O., Talou, T., … Simon, V. (2018). Characterization of volatile organic compound emissions from self-bonded boards resulting from a coriander biorefinery. *Industrial Crops and Products*, *122*(May), 57–65. https://doi.org/10.1016/j.indcrop.2018.05.050

Uitterhaegen, E., Labonne, L., Merah, O., Talou, T., Ballas, S., Véronèse, T., & Evon, P. (2017). Optimization of thermopressing conditions for the production of binderless boards from a coriander twin-screw extrusion cake. *Journal of Applied Polymer Science*, *134*(13). https://doi.org/10.1002/app.44650

Uitterhaegen, E., Nguyen, Q. H., Sampaio, K. A., Stevens, C. V., Merah, O., Talou, T., … Evon, P. (2015). Extraction of coriander oil using twin-screw extrusion: Feasibility study and potential press cake applications. *JAOCS, Journal of the American Oil Chemists' Society*, *92*(8), 1219–1233. https://doi.org/10.1007/s11746-015-2678-4

Uitterhaegen, E., Labonne, L., Merah, O., Talou, T., Ballas, S., Véronèse, T., & Evon, P. (2017). Impact of thermomechanical fiber pre-treatment using twin-screw extrusion on the production and properties of renewable binderless coriander fiberboards. *International Journal of Molecular Sciences*, *18*(7). https://doi.org/10.3390/ijms18071539

Uitterhaegen, E., Sampaio, K. A., Delbeke, E. I. P., De Greyt, W., Cerny, M., Evon, P., … Stevens, C. V. (2016). Characterization of French coriander oil as source of petroselinic acid. *Molecules*, *21*(9), 1–13. https://doi.org/10.3390/molecules21091202

USDA. (2013). Coriander seed. *National Nutrient Database for Standard Reference Release 26 Full Report (All Nutrients) Nutrient Data for Spices*. U.S. Department of Agriculture.

Van Dam, J. E. G., Van Den Oever, M. J. A., & Keijsers, E. R. P. (2004). Production process for high density high performance binderless boards from whole coconut husk. *Industrial Crops and Products*, *20*, 97–101. https://doi.org/10.1016/j.indcrop.2003.12.017

Verma, A., Pandeya, S. N., Sanjay, K. Y., & Styawan, S. (2011). A review on *Coriandrum sativum* (Linn.): An ayurvedic medicinal herb of happiness. *Journal of Advanced Pharmaceutical Healthcare Research*, *1*(3), 28–48.

Wangensteen, H., Samuelsen, A. B., & Malterud, K. E. (2004). Antioxidant activity in extracts from coriander. *Food Chemistry*, *88*(2), 293–297. https://doi.org/10.1016/j.foodchem.2004.01.047

Weiss, E. A. (2002). *Spice Crops* (E. A. Weiss, Ed.). Wallingford, UK: CABI Publishing.

Zheljazkov, V. D., Pickett, K. M., Caldwell, C. D., Pincock, J. A., Roberts, J. C., & Mapplebeck, L. (2008). Cultivar and sowing date effects on seed yield and oil composition of coriander in Atlantic Canada. *Industrial Crops and Products*, *28*(1), 88–94. https://doi.org/10.1016/j.indcrop.2008.01.011

33 Health-Promoting Activities of Coriander Fixed Oil

Seok Shin Tan, Chin Xuan Tan, and Seok Tyug Tan

CONTENTS

33.1 INTRODUCTION

Coriander (*Coriandrum sativum* L.), in the Apiaceae family, originated from the Mediterranean and spread widely in Asia, North Africa, and Central Europe (Laribi, Kouki, M'Hamdi & Bettaieb, 2015). It is widely employed for culinary and folk medicine purposes, particularly the leaves and seeds. All segments of coriander, including the leaves, seeds, fruits, stems, and roots of the plant, are edible. Therefore, it is widely adopted for medicinal uses as an antioxidant, anti-microbial, anti-diabetic, anti-xiolytic, anti-epileptic, anti-depressant, anti-mutagenic, anti-inflammatory, anti-dyslipidemic, anti-hypertensive, and neuroprotective as well as a diuretic (Deeksha et al., 2015).

Research has focused on determining the antioxidants with pharmacological potential and at the same time with low or no side effects for preventive medicine. Coriander is one of the examples that fit perfectly into the criteria and interests of the researchers. Caffeic acid and chlorogenic acid are the phenolic acids available in coriander, while quercetin, kaempferol, rhamnetin, and apigenin are the flavonoids found in coriander. The bioactive compounds identified can be obtained through dietary intake and are well known for their potential to inhibit free radicals (Iqbal, Butt & Suleria, 2019).

The seed of coriander is the most crucial constituent contributing to the fixed oil and essential oil, whereby the fixed oil is about 25% of the seed, and the essential oil is less than 1%. The fixed oil is light yellow and has a high amount of petroselinic acid (C18:1n-12), with a characteristic smell (Sahib et al., 2012). The proximate analysis of coriander seeds and the yield of fixed oil are listed in Table 33.1, and the fatty acids composition of the coriander seed oil is tabulated in Table 33.2. This chapter will explore and discuss the health-promoting activities of coriander fixed oil.

33.2 ANTIOXIDANT PROPERTIES OF CORIANDER FIXED OIL

The antioxidant property of coriander fixed oil was evaluated using DPPH radicals and galvinoxyl radicals assay. Results showed that coriander seed oil has reached 35% and 32.4% for DPPH radicals and galvinoxyl radicals. Further investigation found that the radical scavenging activity of coriander oil contributed by the high composition of unsaponifiables and phospholipids content in the coriander seed oil (Ramadan et al., 2003).

Another study by Ramadan and Wahdan (2012) investigated the functionality, stability, and radical scavenging activity of blended corn oil with coriander seed oil at 10% to 20%. In a dose-dependent manner, the blended oil was found to enhance oxidative stability and DPPH free radical

DOI: 10.1201/9781003204626-36

TABLE 33.1

Proximate Analysis of Coriander Seed

Proximate Analysis	%
Moisture	6.3–8.0
Carbohydrates	16.4–31.1
Protein	11.5–21.3
Ash	4.9–6.0
Fat	17.8–19.2
Fiber	28.4–29.1
Essential oil	0.3–2.1
Fixed oil	13.0–22.7

Reference: Ramadan and Morsel, 2002; Sriti et al., 2010

TABLE 33.2

Fatty Acid Composition of Coriander Seed Oil

Fatty Acid	%
Petroselinic acid (C18:1n-12)	65.7–84.2
Linoleic acid (C18:2n-6)	13.1–16.7
Oleic acid (C18:1n-9)	5.45–7.50
Palmitic acid (C16:0)	3.27–3.96
Stearic acid (C18:0)	0.49–2.91
Palmitoleic acid (C16:1n-7)	0.23–1.10
α-Linolenic acid (C18:3n-3)	0.15–0.50
Arachidic acid (C20:0)	0.10–0.25

Reference: Ramadan and Morsel, 2002; Sriti et al. 2009; Sriti et al., 2010; Msaada et al., 2009

scavenging capacity compared to the pure corn oil. The main contributors to improving oxidation parameters and antioxidant properties were the fatty acid compositions, tocopherol profiles, and the bioactive compounds available in the coriander seed oil. In addition, the oxidative stability of sunflower oil was improved as it was blended with coriander seed oil due to the decreased linoleic acid and increased tocols levels of the blended oil (Ramadan and Wahdan, 2012).

33.3   HEALTH-PROMOTING ACTIVITIES OF CORIANDER FIXED OIL

Kharade et al. (2011) investigated an *in vivo* model for anti-depressant activity with the "Immobility Time" test, whereby a longer immobility time showed a weak anti-depressant activity. Furthermore, the forced swimming test and tail suspension test were carried out under the anti-depressant study, and the outcome showed the anti-depressant potential of coriander seed fixed oil. On the other hand, coriander has been used as an anti-diabetic plant in parts of Europe. In addition, India has widely used coriander for its anti-inflammatory properties, while the United States adopted coriander for its cholesterol-lowering effects (Msaada et al., 2007). Ramadan et al. (2008) reported that *C. sativum* fixed seed oil improved the plasma lipid profile in rats fed with a high cholesterol diet.

33.4 CONCLUSION

The rich content of bioactive compounds in coriander has made it a potential candidate for therapeutic purposes in promoting human health. However, to date, a limited number of studies have focused on the fixed oil of coriander. This has rendered a research gap for researchers to explore further the functionality of coriander fixed oil in health-promoting activities. Therefore, more research is needed to fill this knowledge gap and, subsequently, potential nutraceutical development for the health and well-being of humans.

REFERENCES

Deeksha, S., Asmita, T., Parul, A. (2015). An overview of coriander. *Journal of Biomedical and Pharmaceutical Research*, 4(2), 67–70.

Iqbal, M. J., Butt, M. S., Suleria, H. A. R. (2019). Coriander (*Coriandrum sativum* L.): Bioactive molecules and health effects. In Mérillon, J.M., Ramawat, K. (eds.), *Bioactive Molecules in Food, Reference Series in Phytochemistry*. https://doi.org/10.1007/978-3-319-78030-6_44.

Kharade, S. M., Gumate, D. S., Patil, V. M., Kokane, S. P., Narikwade, N. S. (2011). Behavioral and biochemical studies of seeds of *Coriandrum sativum* in various stress models of depression. *International Journal of Current Research and Review*, 1003, 4–8.

Laribi, B., Kouki, K., M'Hamdi, M., Bettaieb, T. (2015). Coriander (*Coriandrum* sativum L.) and its bioactive constituents. *Fitoterapia*, 103, 9–26.

Msaada, K., Hosni, K., Taarit, M. B., Chahed, T., Hammami, M., Marzouk, B. (2009). Changes in fatty acid composition of coriander (*Coriandrum sativum* L.) fruits during maturation. *Industrial Crops and Products*, 29(2–3), 269–274.

Msaada, K., Hosni, K., Taarit, M. B., Chahed, T., Kchouk, M. E., Marzouk, B. (2007). Changes on essential oil composition of coriander (*Coriandrum sativum* L.) fruits during three stages of maturity. *Food Chemistry*, 102, 1131–1134.

Ramadan, M. F., Amer, M. M. A., Awad, A. E. S. (2008). Coriander (*Coriandrum sativum* L.) seed oil improves plasma lipid profile in rats fed a diet containing cholesterol. *European Food Research and Technology*, 227, 1173–1182. https://doi.org/10.1007/s00217-008-0833-y.

Ramadan, M. F., Kroh, L. W., Morsel, J. T. (2003). Radical scavenging activity of black cumin (*Nigella sativa* L.), coriander (*Coriandrum sativum* L.) and niger (*Guitoza abyssinica* Cass.) crude seed oils and oil fractions. *Journal of Agricultural and Food Chemistry*, 51, 6961–6969.

Ramadan, M. F., Morsel, J. T. (2002). Oil composition of coriander (*Coriandrum sativum* L.) fruitseeds. *European Food Research and Technology*, 215, 204–209.

Ramadan, M. F., Wahdan, K. M. M. (2012). Blending of corn oil with black cumin (*Nigella sativa*) and coriander (*Coriandrum sativum*) seeds oil: Impact on functionality, stabililty and radical scavenging activity. *Food Chemistry*, 132(2), 873–879.

Sahib, N. G., Anwar, F., Gilani, A. H., Hamid, A. A., Saari, N., Alkharfy, K. M. (2012). Coriander (*Coriandrum sativum* L.): A potential source of high-value components for functional foods and nutraceuticals- A review. *Phytotherapy Research*, 27(10), 1439–1456.

Sriti, J., Talou, T., Wannes, W. A., Cerny, M., Marzouk, B. (2009). Essential oil, fatty acid and sterol composition of Tunisian coriander fruit different parts. *Journal of the Science of Food and Agriculture*, 89, 1659–1664.

Sriti, J., Wannes, W. A., Talou, T., Mhamdi, B., Handaoui, G., Marzouk, B. (2010). Lipid fatty acid and tocol distribution of coriander fruits' different parts. *Industrial Crops and Products*, 31, 294–300.

Section IV

Coriander Essential Oil

Chemistry, Technology, Functionality, and Applications

34 Composition and Functionality of Coriander Essential Oil

Saša Đurović and Stevan Blagojević

CONTENTS

34.1 INTRODUCTION

Coriander (*Coriandrum sativum* L.) is a famous plant from the Apiaceae botanical family and is widely cultivated and distributed all around the world, but mainly in Mediterranean countries for its seed (Bhuiyan, Begum, and Sultana 2009; Eikani, Golmohammad, and Rowshanzamir 2007; Grosso et al. 2008; Pavlić et al. 2015; Wangensteen, Samuelsen, and Malterud 2004). The seeds are spherical and longitudinally straited with slight pointing at one end. Their length is 3–5 mm, and they are brown in most cases (Coşkuner and Karababa 2007). They may be found commercially in ground powder and/or whole. They are characterized by a sweet aromatic smell and a slightly spicy bittersweet taste. This plant is traditionally used to prepare dishes in many regions, e.g., South Asia, North Africa, Latin America, and Europe (Dima et al. 2016).

It is well-known that the plant synthesizes many compounds that are part of the primary and secondary metabolism. It has been shown that these compounds often possess many different biological activities. Because of such behavior, they attract quite a lot of attention from the scientific community these days. Extracts made from plant materials often find application in the food, pharmaceutical, or cosmetic industry. Therefore, different extraction techniques for the isolation of these molecules have been developed and are still being developed by many different research groups all around the globe (Azmir et al. 2013; Wang and Weller 2006).

The first developed extraction techniques were conventional ones. These techniques are liquid-liquid, solid-liquid, maceration, Soxhlet extraction, and hydrodistillation. They have been used for a long time and still have an important place in the laboratory and industrial practice. However, they usually require the application of elevated temperatures and organic solvents, mainly toxic. They are also time-consuming techniques, i.e., requiring several hours or days for completion, or, in some cases, just to achieve satisfactory yield (Azmir et al. 2013; da Silva, Rocha-Santos, and Duarte 2016; Radojković et al. 2016; Wang and Weller 2006). To overcome these issues, nonconventional extraction techniques have been developed. Maceration and Soxhlet techniques are developed to ultrasound-assisted and microwave-assisted techniques. On the other hand, hydrodistillation is successfully developed to supercritical fluid extraction for the isolation of volatile compounds such as terpenes which are the main constituents of the

441

essential oils (Mašković et al. 2018; Mašković et al. 2018; Mašković et al. 2017; Cvetanović et al. 2017; Veličković et al. 2017).

Essential oils are mainly the primary aroma of plants. The term "essential" derives from the word *essence*, as it carries the distinctive scent or essence of the plant (Riabov et al. 2020). Because of their evaporative nature, they are also known as volatile oils (Attokaran 2017). They consist mainly of terpenes, but other compounds may be present in lower amounts. Chemical structure dictates their nature, which allows them to dissolve in moderately polar and nonpolar solvents, such as methylene chloride and hexane. The essential oil may be isolated from different parts of the plant, e.g., flower, seed, leaves, twigs, and root, while compounds that are essential oils' constituents usually accumulate in the secretory glands and trichomes of plants (Micić et al. 2019). It is well-known that essential oils possess many biological activities, such as antioxidant and antimicrobial. This allows them to be used as a natural antioxidant agent. Recently, there have been many studies about their application as an antioxidant agent for the preservation of meat (Šojić et al. 2019; Šojić, Tomović, et al. 2020; Šojić, Pavlić, et al. 2020) and sunflower oil (Micić et al. 2021). The quality and biological activity are strongly dependent on the chemical composition (Riabov et al. 2020; Micić et al. 2021) while the chemical composition correlates with the applied extraction techniques (Pavlić et al. 2021; Radivojac et al. 2020). Therefore, it is essential to carefully choose an extraction technique to isolate the oil from the plant material.

The main goal of this chapter is to provide an insight into the isolation of the essential oil from the coriander seed using different extraction techniques. Information about the chemical composition of the isolated essential oil will be given together with the extraction process's optimization analysis using response surface methodology (RSM) and artificial neural network (ANN). At the end of the chapter, the biological activity of the essential oils will be given and discussed.

34.2   EXTRACTION TECHNIQUES

Generally, extraction techniques may be divided into the two major groups:

- Conventional techniques (maceration, Soxhlet extraction, and hydrodistillation)
- Nonconventional techniques (ultrasound-assisted, microwave-assisted, supercritical fluid, and subcritical water extraction)

34.2.1   CONVENTIONAL EXTRACTION TECHNIQUES

Conventional extraction techniques are mainly based on applying an elevated temperature and organic solvent for the extraction and the agitation (Wang and Weller 2006). Hydrodistillation is the most common process for isolating the essential oils (terpenes). For this purpose, the Clevenger apparatus has been developed (Figure 34.1, left), where the volatile compounds are to be isolated using water vapor. The round flask (1) is filled with water and plant material. Vapor is created by heating the flask. It carries volatile compounds with it straight to the condenser where it is condensed. The volatile compounds are dissolved into the nonpolar organic solvent (e.g., hexane), evaporating after the process. Condensed water returns to the flask through the recirculating pipe and evaporates again (Đurović, 2019).

On the other hand, the Soxhlet extraction technique is one of the most frequently used techniques and represents an etalon to evaluate other techniques' performance (Cvetanović et al. 2017; Wang and Weller 2006). In most cases, this extraction technique was proven to be more efficient. However, it is limited by natural compounds' thermal stability (or lability). Typical apparatus for the Soxhlet extraction is given in Figure 34.1 (right). First, the round flask is filled with the solvent (usually ethanol or hexane). Then, the shell with the plant material is embedded into a Soxhlet extractor equipped with the condenser on the top. When condensed solvent fills the Soxhlet extractor up to the

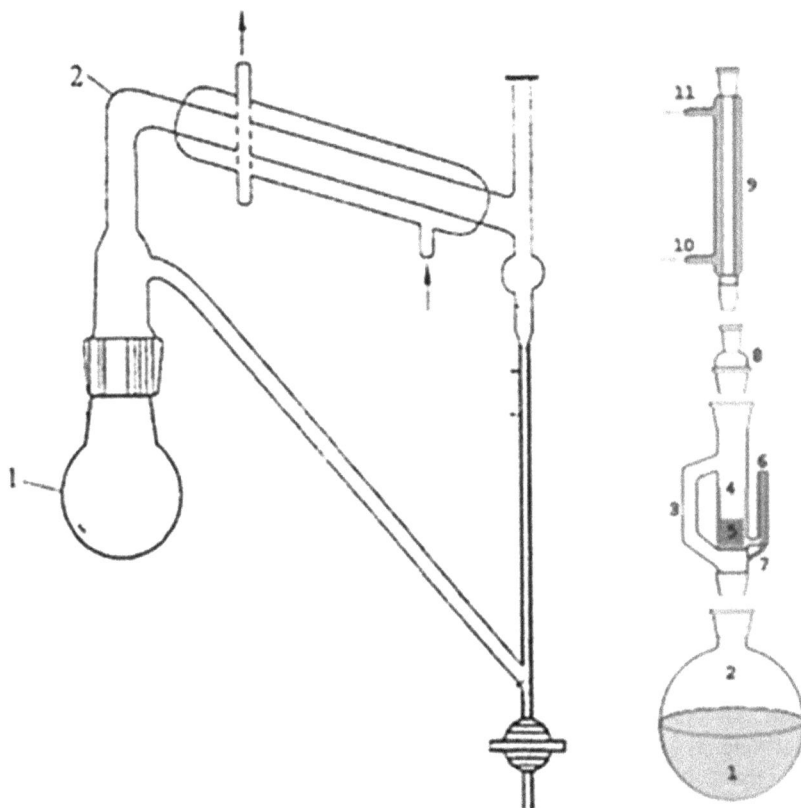

FIGURE 34.1 Apparatus for the isolation of volatile compounds and Soxhlet extraction. Clevenger apparatus (left): (1) round flask, (2) Clevenger apparatus with condenser. Soxhlet extraction (right): (1) solvent, (2) round flask, (3) side tube, (4) extractor, (5) shell for plant material, (6) syphon, (7) recirculation tube, (8) extension, (9) condenser, (10) coolant in, (11) coolant out.

overflow level, it flows back to the flask through the syphon. This is one cycle of the process, usually comprised of 15 cycles (Wang and Weller 2006).

Maceration is one more conventional technique widely used to prepare homemade tonics. The process itself may be divided into several steps: grinding the plant material, immersion of the prepared material into the desired solvent, separating the solvent, and pressing the mark for recovery of the occluded solution. Maceration is usually done in closed vessels in combination with occasional mixing. Mixing is essential because it increases the diffusion of the compounds from the plant material into the solvent and removes the concentrated solution from the surface of the immersed plant material. The latter process influences the extraction yield positively, i.e., the yield increases (Azmir et al. 2013). However, besides the wide application of these techniques, new ones have been developed to overcome many disadvantages of the conventional ones. Some of these disadvantages are the application of elevated temperatures, toxic solvents for extraction, time-consuming processes, and possible degradation of the desired compounds.

34.2.2 NONCONVENTIONAL TECHNIQUES

This group of extraction techniques consists of ultrasound-assisted, microwave-assisted, supercritical fluid, and subcritical water extraction. Each technique uses a different effect and consequently shows different effectiveness and/or selectivity.

The ultrasound-assisted (UAE) technique applies ultrasonic waves with frequencies of 20 kHz to 100 MHz. Ultrasonic waves go directly through the medium, causing compression and expansion, known as cavitation, the main phenomenon in this technique. In addition, it includes production, growth, and explosion of the bubbles (Azmir et al. 2013; Chemat, Tomao, and Virot 2008; Rostagno, Palma, and Barroso 2003; Zeković, Đurović, and Pavlić 2016). Mechanical effects are also included, and they are responsible for enhanced penetration of the solvent into the cellular material (Rostagno, Palma, and Barroso 2003; Zeković, Đurović, and Pavlić 2016). The advantages of this technique are its cheapness, high economic efficiency, low equipment requirements, high reproducibility, simplified manipulation, and low consumption of both solvent and energy (Zeković, Đurović, and Pavlić 2016). However, there are also some disadvantages, such as a possible change in chemical composition and degradation of the compounds due to the creation of free radical species inside the gas bubbles (Zeković, Đurović, and Pavlić 2016; Paniwnyk et al. 2001).

The microwave-assisted extraction (MAE) technique uses a microwave with a frequency in the range of 300 MHz to 300 GHz. This technique offers rapid delivery of the energy, which causes homogenous and efficient heating of the system (Wang and Weller 2006). Ion conduction and dipole rotation are responsible for the heating effect. During the exposure to the microwaves, the resistance of the medium to ion flow occurs, which causes the heating of the system (Azmir et al. 2013; Jain et al. 2009). There are three steps in the MAE process: separation of the solutes from the active site under increased pressure and temperature, diffusion of the solvent across the matrix, and release of the solutes from the sample matrix to the solvent (Azmir et al. 2013). Advantages of this process include quicker heating, reduced thermal gradient, smaller equipment size, and significantly increased yield (Cravotto et al. 2008). Moreover, this technique allows the application of green solvents for the extraction, such as water.

Subcritical water extraction (SWE) is a modern extraction technique whose application relies on the changes in the dielectric constant of the water. It is known that water is a polar medium with a dielectric constant of about 80 at the ambient conditions. However, the dielectric constant depends on the temperature and decreases with increasing temperature. Thus, at 110°C the dielectric constant of water is 53 and at 190°C it is 36.5 (similar to the constant of methanol). In order to maintain water in a liquid state at those temperatures, pressure up to 10 MPa is to be applied (Ko, Cheigh, and Chung 2014; Cvetanović et al. 2016). The selectivity of this process may be directed by changing the operational temperature and consequently changing the polarity of water, which influences the solubility of the compounds according to their polarity (nature). Investigation of the effects of the operational parameters, i.e., temperature and pressure, on the solubility of natural compounds in SWE showed that both parameters have a significant influence, but the temperature was the most influential parameter (Cvetanović et al. 2018; Cvetanović et al. 2019).

Supercritical fluid extraction (SFE) is also one of the modern nonconventional techniques, and is considered a green one. It is usually applied as an alternative technique to hydrodistillation and solid-liquid extraction, mainly for the isolation of essential oils, as well as for the isolation of other compounds, such as carotenoids, chlorophyll, fatty acids, etc. (da Silva, Rocha-Santos, and Duarte 2016; Đurović et al. 2022; Đurović et al. 2018; Zeković et al. 2017; Zeković, Pavlić, et al. 2016). The early steps of this technique were in the 1980s, while the first industrial application was for the decaffeination of coffee beans and black tea leaves. Application of this technique continues in different directions, such as the isolation of essential oils, oleoresins, fractionation of edible oils, and removing pesticides from plants (Brunner 2005; Brunner 1994; Đurović et al. 2018; Jesus and Meireles 2014). The main principle of this technique is to bring the fluids to their supercritical state by increasing the pressure and temperature above the critical values for the given fluid. In such a state, fluid shows properties of both gas and liquid state, i.e., the density is almost like the density of the liquid, while the viscosity is similar to the gas. This ensures that diffusivity ranges between the gaseous and liquid state, consequently increasing the transport properties (Herrero, Cifuentes, and Ibanez 2006; Meullemiestre et al. 2015). The selectivity of the extraction process is quickly changed by simple manipulation of the pressure and temperature. These changes modulate the density of the

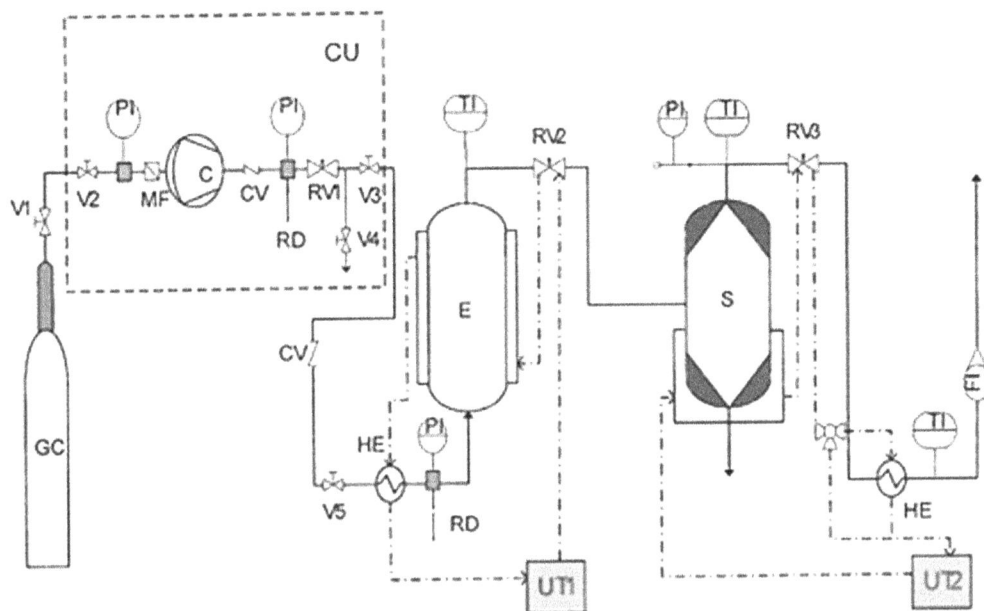

FIGURE 34.2 Laboratory-scale SFE plant. (GC) gas cylinder, (CU) compressor unit, (C) compressor with diaphragm, (E) extractor, (S) separator, (HE) heat exchanger, (UT) ultra-thermostat, (RV) regulation valve, (V) on/off valve, (MF) microfilter, (CV) cut-off valve, (RD) rupture disc, (PI) pressure indicator, (TI) temperature indicator, (FI) flow indicator (modified after Filip et al. 2016).

supercritical fluid and, consequently, the solubility of the target compounds in the fluid (de Melo, Silvestre, and Silva 2014; Filip et al. 2016).

The most common gas used in this process is carbon dioxide. The reasons for such wide application are its cheapness, nontoxicity, nonexplosive nature, nonflammable, inertness, and moderate critical parameters ($p_c = 73.8$ bar and $t_c = 31.1°C$) (de Melo, Silvestre, and Silva 2014; Đurović et al. 2018; Radojković et al. 2016; Zeković et al. 2017). Typical laboratory-scaled equipment for supercritical fluid extraction is given in Figure 34.2.

34.3 EXTRACTION OF THE CORIANDER SEEDS AND OPTIMIZATION OF THE PROCESS

There are different studies of the bioactive compounds' extraction from coriander seeds with both conventional and nonconventional techniques (Grosso et al. 2008; Pavlić et al. 2015; Zeković et al. 2017; Zeković et al. 2015; Zeković et al. 2014; Zeković, Pavlić, et al. 2016; Zeković, Vladić, et al. 2016; Zeković, Đurović, and Pavlić 2016).

Pavlić et al. (2015) and Zeković et al. (2015) proved that the Soxhlet technique is the most efficient for the total extraction yield. It was also shown that particle size significantly impacted the extraction (Zeković et al. 2015). Obtained results showed that the extraction process was more efficient when using material of smaller particle size (Table 34.1). Results also showed that the Soxhlet technique gave higher extraction yield when compared to the SWE technique. On the other hand, SWE was more efficient than hydrodistillation (Eikani, Golmohammad, and Rowshanzamir 2007).

Moreover, nonconventional techniques were also applied to isolate bioactive compounds from the coriander seeds. The UAE and MAE techniques were used for the isolation of moderately polar compounds such as phenolic compounds (Gallo et al. 2010; Zeković, Vladić, et al. 2016; Zeković, Đurović, and Pavlić 2016). For optimization of the extraction processes, RSM was used. It is quite important to optimize the process, ensuring its maximal response accompanying minimal losses.

TABLE 34.1

Total Extraction Yield of Hydrodistillation and Soxhlet Extraction of Coriander Seeds

Technique	Conditions	Yield (%, w/w)
Hydrodistillation*	t < 100°C, 2 h	0.600
Hydrodistillation**	t < 100°C, 2 h, d_1	0.350
	t < 100°C, 2 h, d_2	0.575
	t < 100°C, 2 h, d_3	0.600
Soxhlet extraction*	Boiling point, 15 exchanges	14.450
Soxhlet extraction**	Boiling point, 15 exchanges, d_1	2.540
	Boiling point, 15 exchanges, d_2	8.100
	Boiling point, 15 exchanges, d_3	16.770

* Result from Pavlić et al. (2015), **Results published by Zeković et al. (2015), $d_1 = 1.368$ mm, $d_2 = 0.775$ mm, $d_3 = 0.631$ mm.

In our cases, UAE and MAE processes were optimized. For the UAE, ultrasonic power, extraction time, and temperature were the investigated parameters (Zeković, Đurović, and Pavlić 2016). In the case of the MAE, those parameters were irradiation power, the concentration of ethanol, and temperature (Zeković, Vladić, et al. 2016). The most common model in RSM is the Box–Behnken design (BBD), consisting of several numerical factors, 3 levels, and a minimum of 15 randomized experiments, with a minimum of 3 replicates in the central point. Investigated process parameters are independent variables ranging from +1 to –1 (normalization). Normalization ensures more influence on the response parameters and makes the parameters' units irrelevant (Zeković, Vladić, et al. 2016). Table 34.2 shows natural and coded values for independent variables in UAE and MAE.

There are generally two approaches for choosing the operational parameters for optimization. The first is to conduct a series of preliminary extractions to evaluate the influence of the investigated parameters prior to the optimization. The second approach is consulting the available literature and gathering available data. Both approaches are beneficial for the initial evaluation of the parameters and their range for optimization. The UAE and MAE optimization results are given in Tables 34.3 and 34.4.

TABLE 34.2

Natural and Coded Levels of Independent Variables in RSM

Extraction Technique	Variable	Coded Levels		
		–1	0	1
		Natural Levels		
UAE*	Temperature (°C)	40	60	80
	Extraction time (min)	40	60	80
	Ultrasonic power	96	156	216
MAE**	Extraction time (min)	15	25	35
	Ethanol concentration (%)	50	70	90
	Irradiation power (W)	400	600	800

* Parameters from Zeković, Đurović, and Pavlić (2016), **Parameters from Zeković, Vladić, et al. (2016).

TABLE 34.3

Experimental Conditions for the BBD Design of UAE of Coriander

Independent Variable			Measured Response*			
X_1	X_2	X_3	TPC (mg GAE/100 g)	TFC (mg CE/100 g)	IC_{50} (µg/mL)	EC_{50} (µg/mL)
0	0	0	287.76	150.40	43.44	152.30
0	−1	1	307.28	203.40	45.16	158.20
1	−1	0	350.82	203.90	53.98	165.10
0	0	0	282.67	160.40	47.98	148.30
0	1	1	296.61	162.90	51.05	157.70
−1	1	0	222.32	124.60	53.92	163.40
1	0	1	364.74	192.60	35.69	144.80
−1	−1	0	221.53	126.90	50.94	154.50
0	0	0	288.17	164.70	49.90	156.60
−1	0	1	260.05	145.70	48.20	155.20
0	0	0	265.60	155.40	48.89	158.40
0	−1	−1	326.05	190.50	52.12	165.00
1	0	−1	374.25	198.70	48.62	147.40
0	1	−1	310.64	165.01	52.52	158.30
−1	0	−1	240.85	138.90	48.81	148.30
1	1	0	372.10	199.50	48.68	143.60
0	0	0	287.46	149.30	48.86	14.64

* Results from Zeković, Đurović, and Pavlić (2016).

TABLE 34.4

Experimental Conditions for the BBD Design of MAE of Coriander Seed

Independent Variable			Measured Response*			
X_1	X_2	X_3	TPC (mg GAE/100 g)	TFC (mg CE/100 g)	IC_{50} (µg/mL)	EC_{50} (µg/mL)
0	0	0	294.13	213.62	35.40	135.00
0	−1	1	346.35	201.79	66.50	153.20
1	−1	0	384.54	210.39	54.10	175.30
0	0	0	287.96	210.93	35.30	117.90
0	1	1	145.90	104.31	60.30	182.40
−1	1	0	136.92	94.50	49.90	169.00
1	0	1	291.33	211.83	46.00	130.20
−1	−1	0	358.15	201.26	38.10	161.70
0	0	0	284.03	198.57	35.20	115.30
−1	0	1	268.31	189.08	35.00	133.60
0	0	0	284.03	208.06	38.50	126.50
0	−1	−1	372.24	204.66	41.80	178.40
1	0	−1	299.19	214.69	36.40	1476.30
0	1	−1	154.42	109.91	55.30	157.70
−1	0	−1	250.90	202.15	30.20	143.70
1	1	0	163.31	120.45	53.90	159.50
0	0	0	279.54	202.69	35.60	127.40

* Results from Zeković, Vladić, et al. (2016).

The results of the optimization process showed that the MAE technique is superior to UAE. Furthermore, the MAE technique gave higher extraction yield and extracts with higher antioxidant activity. These results are consistent with other studies which also showed that MAE provided an extract of higher quality (Veličković et al. 2017; P. Z. Mašković et al. 2018; Mašković et al. 2017). Finally, obtained results from the conducted experiments were fitted into a second-order polynomial model, which could describe the relationship between the investigated and independent variables:

$$Y = \beta_0 + \sum_{i=1}^{3} \beta_i X_i + \sum_{i=1}^{3} \beta_{ii} X_i^2 + \Sigma \sum_{i<j=1}^{3} \beta_{ij} X_i X_j$$

where Y is a measured response, β_0 is constant, and *bj*, *bjj*, *bij* are the linear, quadratic, and interactive coefficients of the model, respectively; X_i and X_j are the levels of the independent variables. All calculations were done in Design-Expert v.7 Trial (State-Eset, Minneapolis, Minnesota, USA). The significance level was 0.05, and adequacy was evaluated by the coefficient of multiple determination (R^2), coefficient of variance (*CV*), *p*-values, and lack of fit. All these parameters were obtained from the analysis of variance (ANOVA) in the above-mentioned software (Zeković, Vladić, et al. 2016; Zeković, Đurović, and Pavlić 2016).

A report of the ANOVA analysis for optimization of the UAE extraction is given in Figure 34.3. In this case, the applied model for total phenolic content (TPC) was significant, while the lack of fit was insignificant. Analysis of the parameters showed that *A* and C^2 showed significant influence (temperature and quadratic term of ultrasonic power). In addition, the coefficient of multiple determination was higher than 0.90 ($R^2 = 0.9709$), which indicated that the model represents a good approximation of the experimental results in the case of TPC.

Figure 34.4 shows estimated coefficients for TPC. According to the results shown in Figure 34.3, only temperature and the quadratic term of ultrasonic power significantly influenced TPC. The positive influence of the temperature was somewhat expected because temperature increases diffusion and consequently influences the mass transfer process. Moreover, temperature elevation may cause degradation of the plant material and improve the solvent properties (Ramić et al. 2015). On the other hand, results also showed the negative influence of the linear term of ultrasonic power and the positive influence of its quadratic term. This means TPC will decrease with increasing ultrasonic power up to specific values, after which TPC will slightly increase. In the end, the predicted model may be expressed with the following equation:

$$TPC = 282.24 + 64.64X_1 - 0.50X_2 - 2.89X_3 + 5.12X_1X_2 - 7.18X_1X_3 + 1.19X_2X_3$$

$$+ 4.64X_1^2 + 4.81X_2^2 + 23.09X_3^2$$

Results showed that all quadratic terms showed positive influence, contrary to the linear terms of extraction time and ultrasonic power, which showed a negative influence. Interactive coefficients are mostly positive, except for the interaction of temperature and ultrasonic power. Generated 3D plots (Figure 34.5) showed the influence of independent variables on all investigated parameters given in Table 34.3.

Response surface methodology was also used to optimize the SFE extraction of the coriander seeds. In this case, the measured response was extraction yield (Zeković, Pavlić, et al. 2016). There were 15 randomized experiments with 3 replicates in the central point. The independent variables were pressure, temperature, and CO_2 flow rate. Results are presented in Table 34.5.

Results showed that Y ranges from 0.59 to 7.00 g/100 g. The highest Y was achieved at 200 bar, 55°C, and 0.4 kg/h CO_2, while the lowest Y was noticed at 100 bar, 70°C, and 0.3 kg/h. Experimental results were processed in the Design-Expert v.7 Trial software, and reports of ANOVA were generated. Reports are presented in Figures 34.6 and 34.7. The pressure ranged from 100 to 200 bar, temperature from 40 to 70°C, and gas flow between 0.2 and 0.4 kg/h.

Response   2   TP

ANOVA for Response Surface Quadratic Model

Analysis of variance table [Partial sum of squares - Type III]

Source	Sum of Squares	df	Mean Square	F Value	p-value Prob > F	
Model	36378.25	9	4042.03	25.94	0.0001	significant
A-Temperature	33431.81	1	33431.81	214.55	< 0.0001	
B-Time	2.01	1	2.01	0.013	0.9128	
C-Ultrasonic power	66.76	1	66.76	0.43	0.5337	
AB	104.96	1	104.96	0.67	0.4389	
AC	206.07	1	206.07	1.32	0.2879	
BC	5.62	1	5.62	0.036	0.8548	
A^2	90.65	1	90.65	0.58	0.4705	
B^2	97.52	1	97.52	0.63	0.4549	
C^2	2245.32	1	2245.32	14.41	0.0068	
Residual	1090.77	7	155.82			
Lack of Fit	725.40	3	241.80	2.65	0.1852	not significant
Pure Error	365.37	4	91.34			
Cor Total	37469.02	16				

Std. Dev.	12.48	R-Squared	0.9709
Mean	297.56	Adj R-Squared	0.9335
C.V. %	4.20	Pred R-Squared	0.6750
PRESS	12177.25	Adeq Precision	17.052

FIGURE 34.3 Generated and printed report regarding the ANOVA for total phenolic content. The report has been generated in Design-Expert v.7 Trial.

The results of the ANOVA test showed that the proposed model was significant, while lack of fit was again insignificant. In this case, the coefficient of multiple regression is particularly high ($R^2 = 0.9847$), while the coefficient of variance is slightly higher than in the case of UAE ($CV = 10.6\%$). Linear terms significantly influenced the model, which was expected, especially in the pressure and temperature. The pressure effect was strong and positive, while the quadratic term was low and negative. The linear term of the temperature showed a negative effect, while the quadratic term was positive. This means that yield would increase with increasing pressure and decrease in temperature. The pressure's positive influence was expected because pressure directly influences the density of supercritical carbon dioxide (Paixao Coelho and Figueiredo Palavra 2015; Pourmortazavi and Hajimirsadeghi 2007).

On the other hand, the temperature had a more complex influence. Namely, temperature also influences the density of the supercritical fluid. However, it also influences the vapor pressure of the compounds. Thus, an increase in the temperature causes a decrease in the supercritical fluid density, but vapor pressure increases. Therefore, lower density means lower solubility. However, an increase in the vapor pressure leads to increased solubility (Paixao Coelho and Figueiredo Palavra 2015;

Factor	Coefficient Estimate	df	Standard Error	95% CI Low	95% CI High	VIF
Intercept	282.24	1	5.58	269.04	295.44	
A-Temperature	64.64	1	4.41	54.21	75.08	1.00
B-Time	-0.50	1	4.41	-10.94	9.93	1.00
C-Ultrasonic power	-2.89	1	4.41	-13.32	7.55	1.00
AB	5.12	1	6.24	-9.64	19.88	1.00
AC	-7.18	1	6.24	-21.94	7.58	1.00
BC	1.19	1	6.24	-13.57	15.94	1.00
A2	4.64	1	6.08	-9.75	19.03	1.01
B2	4.81	1	6.08	-9.57	19.20	1.01
C2	23.09	1	6.08	8.71	37.48	1.01

Final Equation in Terms of Coded Factors: **Final Equation in Terms of Actual Factors:**

TP =	TP =
+282.24	+282.24000
+64.64 * A	+64.64500 * Temperature
-0.50 * B	-0.50125 * Time
-2.89 * C	-2.88875 * Ultrasonic power
+5.12 * A * B	+5.12250 * Temperature * Time
-7.18 * A * C	-7.17750 * Temperature * Ultrasonic power
+1.19 * B * C	+1.18500 * Time * Ultrasonic power
+4.64 * A2	+4.64000 * Temperature2
+4.81 * B2	+4.81250 * Time2
+23.09 * C2	+23.09250 * Ultrasonic power2

FIGURE 34.4 Continuation of ANOVA report for total phenolic content generated in Design-Expert v.7 Trial.

Jesus and Meireles 2014). In this case, the influence of the temperature on the density was the main phenomenon, so yield consequently decreased.

The influence of the linear flow's linear term was also positive. This indicated that an increase in the supercritical carbon dioxide flow would increase the extraction yield. An explanation for such behavior may be reducing the film thickness around the solid particles. This causes decreased resistance and increased mass transfer (Döker et al. 2004). A polynomial model for the extraction yield was generated:

$$Y = 3.93 + 2.22X_1 - 0.89X_2 + 0.64X_3 + 0.38X_1X_2 + 0.46X_1X_3$$

$$+0.032X_2X_3 - 0.32X_1^2 + 0.038X_2^2 - 0.097X_3^2$$

D models of the combined influence of described parameters on total extraction yield are given in Figure 34.8. The presented plots showed a positive influence of pressure and carbon dioxide flow on Y and a negative influence of temperature on the same parameter.

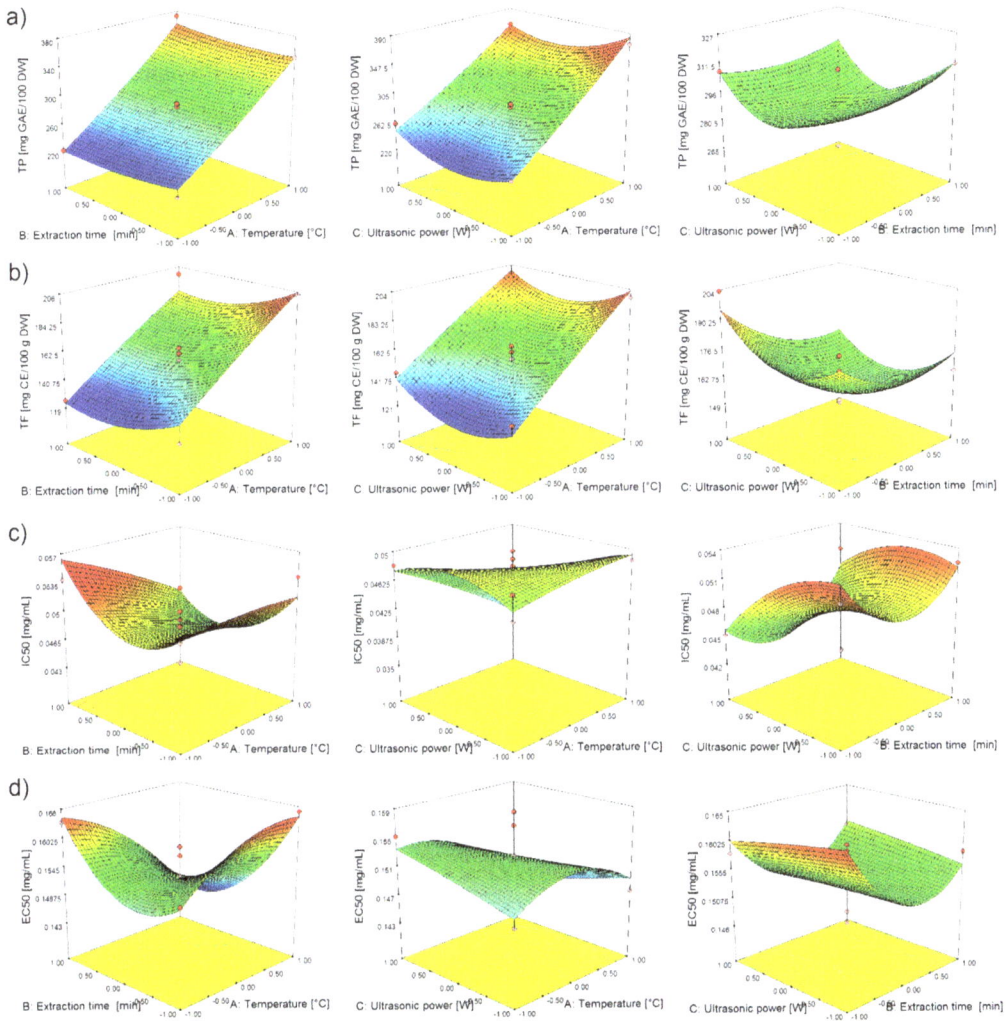

FIGURE 34.5 RSM plots showing the combined influence of UAE parameters on TPC (a), TFC (b), IC_{50} (c), and EC_{50} (d) (modified after Zeković, Đurović, and Pavlić, 2016).

Another possibility of optimization is an artificial neural network (ANN). This system is similar to the natural neural system. This approach's advantages are non-linearity, adaptiveness, generalization, model independence, ease of use, and high accuracy. The relative importance of the various inputs is determined by their connection weight (Gevrey, Dimopoulos, and Lek 2003; Olden and Jackson 2002; Tchaban, Taylor, and Griffin 1998; Yoon, Swales, and Margavio 1993). ANN has been successfully used for the prediction of the extraction kinetics and optimization of the SFE process (Azmir et al. 2014; Khajeh, Moghaddam, and Shakeri 2012; Kuvendziev et al. 2014; Shokri, Hatami, and Khamforoush 2011; Sodeifian, Sajadian, and Saadati Ardestani 2016; Zahedi and Azarpour 2011; Zeković et al. 2017).

The optimization model for the SFE process was developed in MATLAB® software using the MATLAB Neural Network Toolbox. There were three inputs (pressure, temperature, and carbon dioxide flow) and one output (initial slope obtained from the kinetics model). The model is schematically given in Figure 34.9. This model was used to optimize the SFE, like in the case of RSM. Calculation of the kinetics parameters was combined with optimization to achieve maximal yield as quickly as possible. The initial slope was calculated and used as a representative parameter of the

TABLE 34.5

Experimental Results for Optimization of SFE Using BBD Design

Independent Variable			Measured Response*
X_1	X_2	X_3	Y (g/100 g)
0	0	0	3.77
0	−1	1	5.64
1	−1	0	5.95
0	0	0	4.00
0	1	1	3.50
−1	1	0	0.59
1	0	1	7.00
−1	−1	0	2.69
−1	0	1	1.20
0	0	0	4.02
0	−1	−1	4.31
1	0	−1	4.90
0	1	−1	2.05
−1	0	−1	0.95
1	1	0	5.36

* Results from Zeković, Pavlić, et al. (2016).

process, i.e., the solubility control phase. Because the extraction time was 4 h, the initial slope was selected as the response variable (Zeković et al. 2017).

ANN results, including weights values, varied with the changes in initial values used for ANN construction and fitting. Different numbers of hidden neurons may also give different results. To avoid these impacts on the final results, the number of neurons in the hidden layer was varied from 1 to 20, while the training process of each network was repeated 10 times using random initial values of weights and biases. As the final results, 200 ANNs were created, of which 194 were used for further analysis ($R^2 > 0.8$). The mean value for R^2 was 0.932, while the best fit was achieved with 5 hidden neurons ($R^2 = 0.979$) (Zeković et al. 2017). The influence of hidden neurons on fitting values after ten repeated trainings is shown in Figure 34.10.

Figure 34.10 demonstrates the positive influence of the neurons on the R^2 mean value, i.e., increasing neurons in the hidden layer induced an increase in R^2 mean value, which was always higher than 0.9. The best of 200 ANNs with 5 hidden neurons are used for further optimization calculations to avoid "overfitting" with a high number of hidden neurons. Investigation of the hidden neurons' influence on relative importance (RI) mean values after ten trainings is presented in Figure 34.11a, while mean values of all calculated RI with standard deviation are given in Figure 34.11b.

Results presented in Figure 34.11 indicated that pressure was the most influential parameter with a relative importance of 50%. Temperature and flow showed the relative importance of about 18% and 32%, respectively. Pressure and flow rate had a positive influence, while the temperature had a negative influence, following the results obtained by RSM. In the end, optimized parameters were obtained for both approaches and are summarized in Table 34.6.

34.4 CHEMICAL COMPOSITION OF CORIANDER ESSENTIAL OIL

Studies regarding coriander and its chemical composition showed that this plant is exciting for the scientific community. Many of these studies are about the essential oil's isolation, characterization, and biological activity. Generally, the essential oil was obtained by hydrodistillation and SFE in

Response   1   Yield

ANOVA for Response Surface Quadratic Model

Analysis of variance table [Partial sum of squares - Type III]

Source	Sum of Squares	df	Mean Square	F Value	p-value Prob > F	
Model	51.00	9	5.67	35.70	0.0005	significant
A-Pressure	39.56	1	39.56	249.25	< 0.0001	
B-Temperature	6.27	1	6.27	39.53	0.0015	
C-CO2 flow	3.31	1	3.31	20.83	0.0060	
AB	0.57	1	0.57	3.61	0.1159	
AC	0.86	1	0.86	5.41	0.0675	
BC	3.982E-003	1	3.982E-003	0.025	0.8804	
A^2	0.38	1	0.38	2.42	0.1803	
B^2	5.326E-003	1	5.326E-003	0.034	0.8619	
C^2	0.034	1	0.034	0.22	0.6607	
Residual	0.79	5	0.16			
Lack of Fit	0.76	3	0.25	13.38	0.0703	not significant
Pure Error	0.038	2	0.019			
Cor Total	51.80	14				

Std. Dev.	0.40	R-Squared	0.9847	
Mean	3.73	Adj R-Squared	0.9571	
C.V. %	10.68	Pred R-Squared	0.7648	
PRESS	12.18	Adeq Precision	20.545	

FIGURE 34.6 Generated report regarding the ANOVA for total extraction yield. The report has been generated in Design-Expert v.7 Trial.

most cases. However, despite the differences in the applied techniques, all studies agree that monoterpene linalool is the principal compound in coriander essential oil, with content of ≥50% (Alves-Silva et al. 2013; Baratta et al. 1998; Gil et al. 2002; Mandal and Mandal 2015; Micić et al. 2019; Mhemdi et al. 2011; Msaada et al. 2007; Msaada et al. 2009; Ravi, Prakash, and Bhat 2007; Sahib et al. 2012; Shahwar et al. 2012; Singh et al. 2006; Sourmaghi et al. 2015; Sriti et al. 2009; Sriti et al. 2011; Teixeira et al. 2013; Zoubiri and Baaliouamer 2010). Besides linalool, other compounds were also found in significant amounts, e.g., geraniol, camphor, and limonene (Zeković et al. 2015; Pavlić et al. 2015).

Terpenes are a class of natural compounds and consist of isoprene units. Depending on the number of isoprene units in the terpene molecule, they may be divided into hemiterpenes, monoterpenes, sesquiterpenes, diterpenes, sesterterpenes, triterpenes, tetraterpenes, and polyterpenes. Monoterpenes may be further classified according to their structure into acyclic monoterpenes, cyclic monoterpenes, acyclic oxygenated monoterpenes, and cyclic and aromatic oxygenated monoterpenes. The presence of several of these mentioned groups was previously reported by Zeković, Pavlić, et al. (2016). Their results are summarized in Table 34.7.

Results showed that chemical profiles significantly differ. These differences in the composition of the selected samples are the consequence of the influence of the operational parameters, i.e.,

Factor	Coefficient Estimate	df	Standard Error	95% CI Low	95% CI High	VIF
Intercept	3.93	1	0.23	3.34	4.52	
A-Pressure	2.22	1	0.14	1.86	2.59	1.00
B-Temperature	-0.89	1	0.14	-1.25	-0.52	1.00
C-CO2 flow	0.64	1	0.14	0.28	1.00	1.00
AB	0.38	1	0.20	-0.13	0.89	1.00
AC	0.46	1	0.20	-0.049	0.98	1.00
BC	0.032	1	0.20	-0.48	0.54	1.00
A²	-0.32	1	0.21	-0.86	0.21	1.01
B²	0.038	1	0.21	-0.50	0.57	1.01
C²	-0.097	1	0.21	-0.63	0.44	1.01

Final Equation in Terms of Coded Factors:

$$
\begin{aligned}
\text{Yield} =\ & +3.93 \\
& +2.22 \ * A \\
& -0.89 \ * B \\
& +0.64 \ * C \\
& +0.38 \ * A * B \\
& +0.46 \ * A * C \\
& +0.032 \ * B * C \\
& -0.32 \ * A^2 \\
& +0.038 \ * B^2 \\
& -0.097 \ * C^2
\end{aligned}
$$

Final Equation in Terms of Actual Factors:

$$
\begin{aligned}
\text{Yield} =\ & +3.93279 \\
& +2.22387 \ * \text{Pressure} \\
& -0.88564 \ * \text{Temperature} \\
& +0.64289 \ * CO_2 \text{ flow} \\
& +0.37841 \ * \text{Pressure} * \text{Temperature} \\
& +0.46337 \ * \text{Pressure} * CO_2 \text{ flow} \\
& +0.031550 \ * \text{Temperature} * CO_2 \text{ flow} \\
& -0.32276 \ * \text{Pressure}^2 \\
& +0.037981 \ * \text{Temperature}^2 \\
& -0.096644 \ * CO_2 \text{ flow}^2
\end{aligned}
$$

FIGURE 34.7 Continuation of ANOVA report for total extraction yield generated in Design-Expert v.7 Trial.

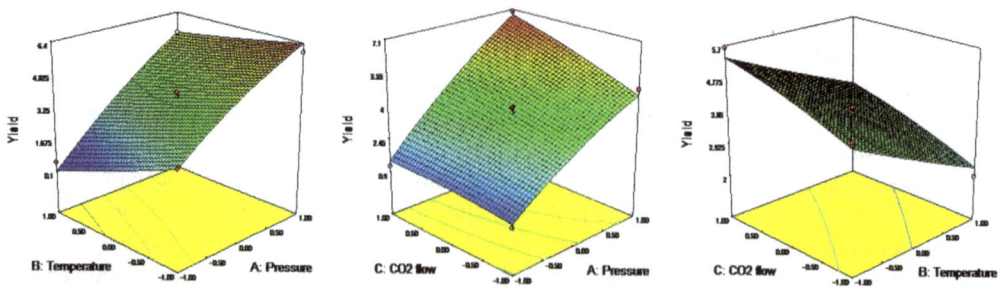

FIGURE 34.8 RSM plots show the combined influence of SFE parameters on Y (Zeković, Pavlić, et al. 2016).

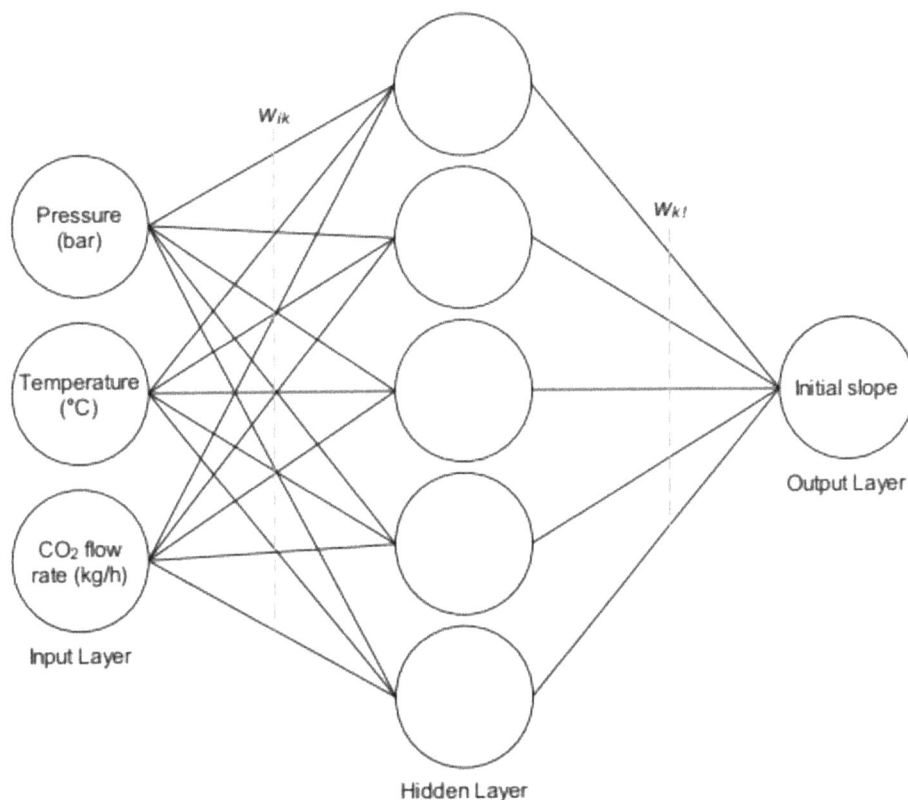

FIGURE 34.9 Schematic representation of ANN architecture (modified after Zeković et al. 2017).

pressure, temperature, and flow, on the SFE process. These differences may be easily noticed in the chromatograms (Figure 34.12).

Besides the operational parameters, the structure of the molecules also dictates the occurrence in the extract. Chemical structure directly influences the solubility of the molecule into the solvent. Therefore, to analyze the structure-solubility relationship, the structure of the main terpenes in coriander essential oil is given in Figure 34.13.

Limonene and γ-terpinene have the same molecular formula ($C_{10}H_{16}$) but different structures (position of the double bonds). Moreover, geraniol, α-terpineol, eucalyptol, and linalool have the same formula ($C_{10}H_{18}O$). However, their structures show significant differences, where geraniol and

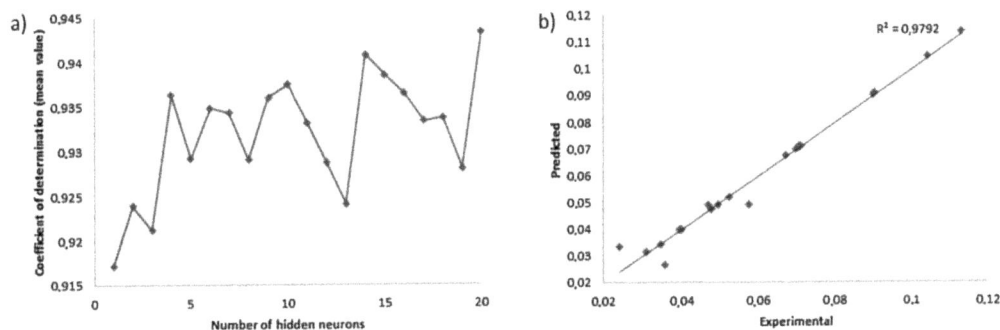

FIGURE 34.10 Influence of hidden neurons on the coefficient of determination (a) and regression plot for ANN with five hidden neurons (b) (modified after Zeković et al. 2017).

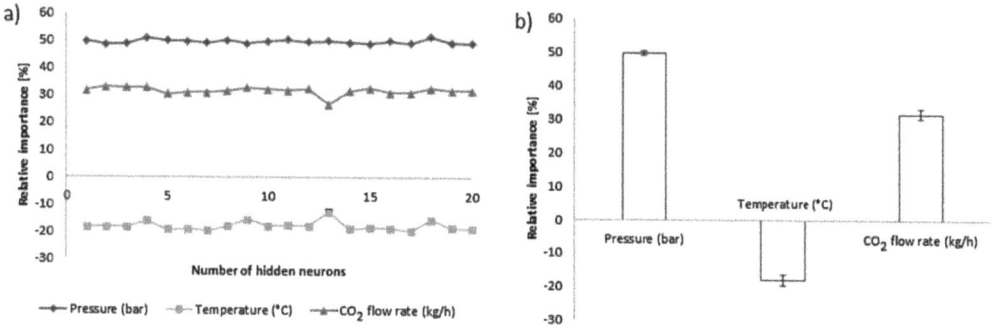

FIGURE 34.11 Influence of the number of hidden neurons on RI mean values (a) and mean values of all calculated RIs with standard deviation (b) (modified after Zeković et al. 2017).

TABLE 34.6

Optimized SFE Parameters Using Different Optimization Approaches

Optimization Process	Pressure (bar)	Temperature (°C)	Flow Rate (kg/h)
RSM*	199.50	40.15	0.396
ANN**	200	40	0.4

* Results from Zeković, Đurović, and Pavlić (2016), **Results from Zeković et al. (2017).

linalool have a hydroxyl group in a different position. Geraniol is a primary alcohol, while linalool is a tertiary alcohol. Results presented by Zeković et al. (2016) showed that geraniol was better extracted at a higher temperature, while linalool was more soluble at a lower temperature. Such behavior may be explained by the occurrence of self-association among molecules (Tufeu, Subra, and Plateaux 1993). Alcohols with an increased number of branches showed lower self-associative affinity, hence increasing solubility (Friedrich and Schneider 1989). Consequently, linalool is more soluble at a lower temperature, while a higher temperature is required to extract geraniol. The same explanation goes for α-terpineol, also tertiary alcohol with a cyclic system and a double bond.

Camphor is a ketone, while eucalyptol possesses an etheric functional group; both are cyclic systems. Aldehyde and ketones are generally soluble in supercritical fluid. On the other hand, eucalyptol is a saturated cyclic system, hance its solubility is weaker in supercritical carbon dioxide (Dandge, Heller, and Wilson 1985). Eugenol and methyl chavicol are aromatic systems. Both compounds have propenyl and methoxy groups. However, eugenol has an additional hydroxy functional group. The presence of methoxy and hydrocarbon groups increases the solubility (Dandge, Heller, and Wilson 1985). An additional hydroxy group modifies the solubility of eugenol, where higher temperatures are required for its isolation. This may be explained by the above-mentioned self-association of the molecules (Friedrich and Schneider 1989; Tufeu, Subra, and Plateaux 1993).

The most common technique is gas chromatography coupled with a flame ionization detector (FID) and/or a mass spectrometer (MS) when analyzing the essential oils. When the analysis is done with GC-FID, identification of the compounds has to be done by comparing the retention times and retention indices (RI). On the other hand, GC/MS ensures identification with higher certainty without using the RIs. In this case, identification was made by comparing the mass spectrum of the unknown compounds with the mass spectrum available in the library (usually NIST), while applying the RIs is unnecessary (Đurović 2021). Furthermore, the most common column used to analyze the essential oil is non-polar HP 5-MS or equivalent. However, it has been shown that polar column TR WAX-MS ensures better separation of the compounds in this case (Đurović 2021; Micić et al. 2021;

TABLE 34.7

Chemical Composition of SFE Extracts of Coriander Seeds

Compound	Sample*		
	3	11	15
Cyclic monoterpenes			
(+)-Limonene	2.0	1.0	<0.1
α-Pinene	<0.1	<0.1	<0.1
β-Pinene	<0.1	<0.1	<0.1
γ-Terpinene	3.0	<0.1	2.1
p-Cymene	D**	D	D
Acyclic oxygenated monoterpenes			
Geraniol	3.0	5.2	10.7
Geranyl acetate	D	ND***	ND
Linalool	717.0	642.0	608.2
Nerol	D	ND	ND
Cyclic oxygenated monoterpenes			
Camphor	21.0	10.9	9.6
Eucalyptol	<0.1	<0.1	1.2
Terpinen-4-ol	D**	ND***	ND
α-Terpineol	2.0	6.1	4.2
Aromatic oxygenated monoterpenes			
Carvacrol	<0.1	<0.1	<0.1
Eugenol	<0.1	2.0	20.7
Methyl cavicol	1.0	18.2	13.2
Furanoids			
cis-Linalool oxide	D	D	D
trans-Linalool oxide	D	D	D

* Results from Zeković, Pavlić, et al. (2016), **D – detected, ***ND – not detected.

Micić et al. 2019). Therefore, Micić et al. (2019) analyzed the coriander essential oils obtained by hydrodistillation using polar column (TR WAX-MS), and the results are summarized in Table 34.8, while the chromatogram is given in Figure 34.14.

Results showed that linalool was the most abundant compound. Also, separation of the *cis* and *trans* isomers was done completely. Compounds were separated during the 41 min of the GC/MS program. Although linalool was abundant at 64.0%, there were several more compounds found in significant amounts: geranyl acetate (5.76%), γ-terpinene (5.59%), camphor (4.24%), α -pinene (7.31%), p-cymene (3.83%), geraniol (2.15%), and limonene (1.60%). All other compounds were found with content <1%.

Although the seed is the primary source of the essential oil, leaves were also used. Different aldehydes and alcohols, e.g., 2E-decenal (15.9%), decanal (14.3%), 2E-decen-1-ol (14.2%), and n-decanol (13.6%), have been reported as major compounds in leaf essential oil. On the other hand, 2E-tridecen-1-al, 2E-dodecenal, dodecanal, undecanol, and undecanal were reported as minor compounds, and alkanes (1.46%) were reported as minor compounds (Matasyoh et al. 2009). The essential oil derived from the coriander leaves from Bangladesh showed a different profile. The principal compounds were 2-decenoic acid (30.8%), E-11-tetradecenoic acid (13.4%), capric acid (12.7%), undecyl alcohol (6.4%), tridecanoic acid (5.5%), and undecanoic acid (7.1%) (Bhuiyan, Begum, and Sultana 2009). In conclusion, the oil composition varies from region to region, but the main

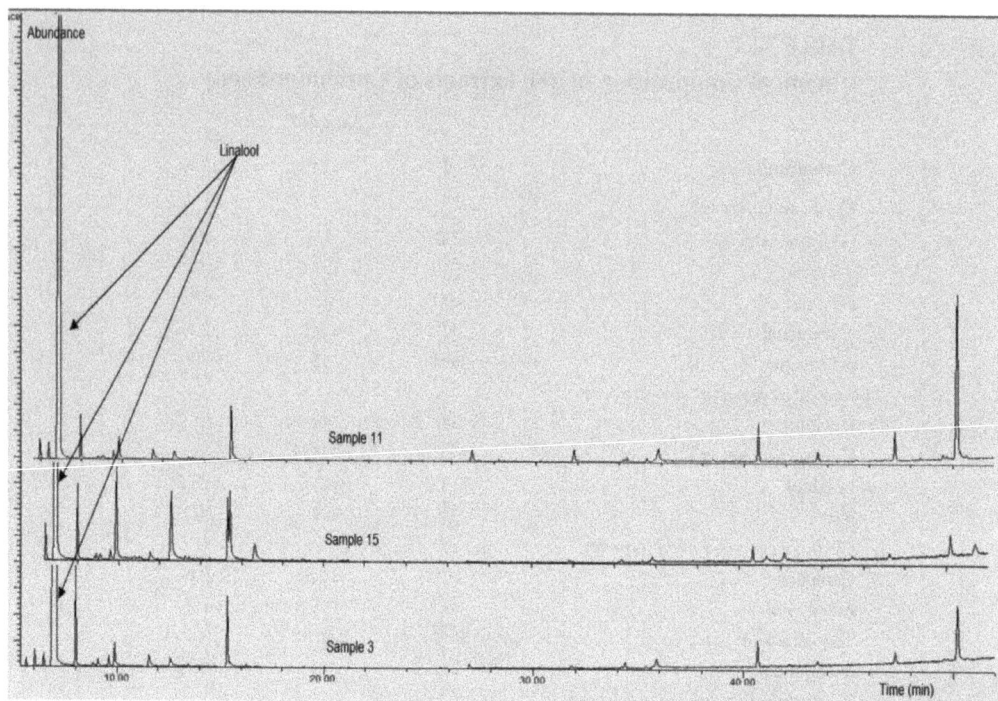

FIGURE 34.12 Chromatograms of samples no. 3, 11, and 15 obtained after analysis on nonpolar HP 5-MS column (modified after Zeković, Pavlić, et al. 2016).

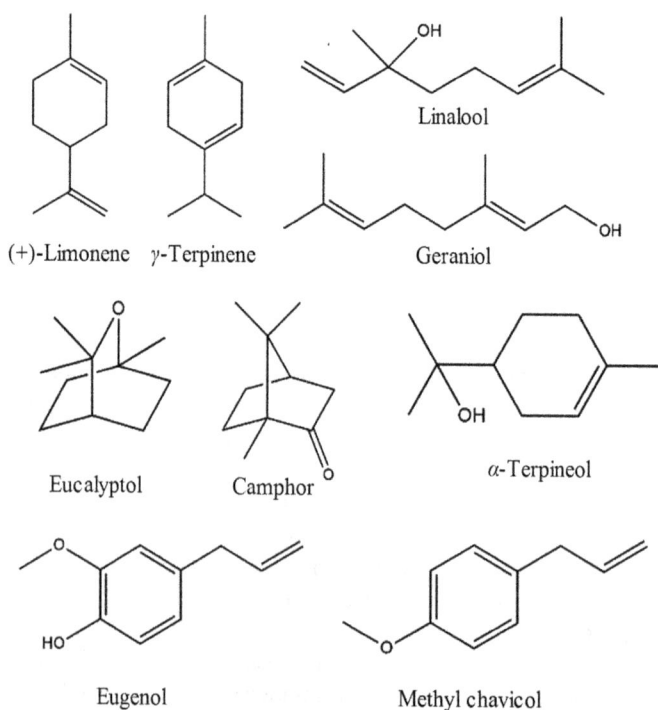

FIGURE 34.13 Structures of most common monoterpenes in coriander seed extracts.

TABLE 34.8
Composition of the Coriander Essential Oil Obtained by Hydrodistillation

Compound	Content (%)	Content (mg/g)
α-Pinene	7.31	19.85
Camphene	0.77	-
β-Pinene	0.65	-
3-Carene	0.30	-
Myrcene	0.70	3.25
Limonene	1.60	2.90
Eucalyptol	0.14	0.15
p-Cymene	3.83	6.10
Terpinolene	0.49	-
γ-Terpinene	5.59	12.08
4-Methyl-2-pentanol	0.35	-
Citronellal	0.05	0.77
Sabinene	0.10	1.15
cis-Linalool oxide	0.08	-
trans-Linalool oxide	0.06	-
Camphor	4.24	8.51
Linalool	64.06	256.65
trans-Caryophyllene	0.22	0.06
α-Caryophyllene	0.02	-
Terpinen-4-ol	0.28	-
Borneol	0.13	0.76
Thymol	0.03	<0.01
Carvacrol	0.08	<0.01
cis-Verbenol	0.01	-
Neral	0.02	0.28
Myrthenyl acetate	0.15	-
α-Terpineol	0.38	<0.01
Neryl acetate	0.07	0.31
Geranial	0.04	0.60
Geranyl acetate	5.76	15.76
Carane	0.09	-
Methyl salicylate	0.02	-
Myrthenol	0.03	-
Nerol	0.02	<0.01
Geraniol	2.15	5.78
Isosaromadendrene epoxide	0.02	-
γ-Elemene	0.02	-
Viridiflorol	0.14	-

compounds in the leaves' essential oil are aldehydes and alcohols, while linalool dominates in the seeds' essential oil (Sahib et al. 2012). Besides the regional differences, the chemical composition of plant extracts also depends on the plant's maturity (Msaada et al. 2007), environmental conditions (Gil et al. 2002), and the extraction technique applied (Micić et al. 2019; Pavlić et al. 2015; Zeković, Pavlić, et al. 2016; Sourmaghi et al. 2015).

Besides the terpenes, aldehyde, and alcohols, fatty acids and sterols have also been found in plant material (Sriti et al. 2009; Singh et al. 2006; Mhemdi et al. 2011). Singh et al. (2006) reported oleic

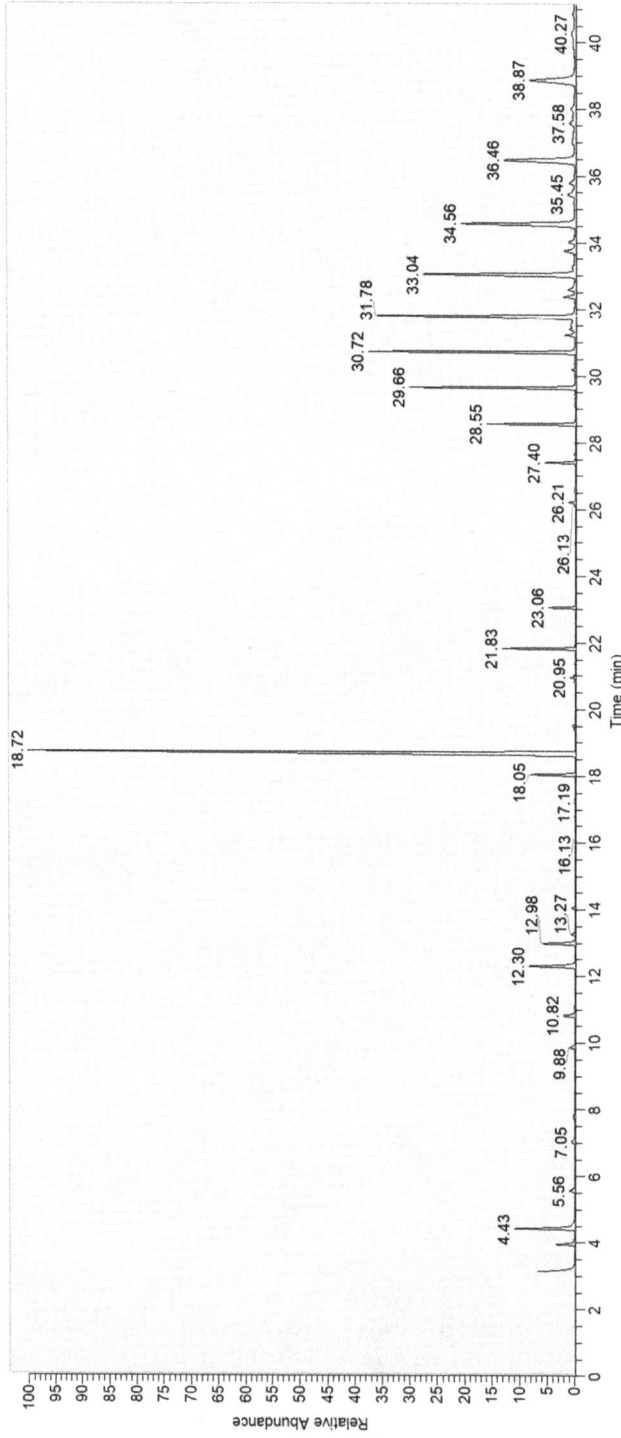

FIGURE 34.14 Chromatogram of the coriander essential oil obtained by hydrodistillation after analysis on polar TR WAX-MS column (modified after Micić et al. 2019).

acid as the principal fatty acid (36.5%), followed by linoleic acid (33.2%). Sriti et al. (2009) have analyzed fatty acids and sterols in the seed, pericarp, and whole fruit of coriander. The authors have reported petroselinic acid as a principal component, with contents of 76.3%, 42.2%, and 75.0% in the seed, pericarp, and whole fruit of coriander, respectively. Linoleic, oleic, and palmitic acids achieved higher content. In total, saturated fatty acids were present in 4.44%, 27.9%, and 5.17%, while unsaturated acid achieved 95.2%, 72.0%, and 94.8% in the seed, pericarp, and whole fruit of coriander, respectively. When summarizing, seeds and fruits are richer in unsaturated acids than pericarp. On the other side, pericarp has a higher content of saturated acids. Mhemdi et al. (2011) investigated monounsaturated and polyunsaturated fatty acids separately. Authors have confirmed that unsaturated acids are more prevalent than saturated, but monounsaturated showed higher content than polyunsaturated acids. The content of saturated acids varied from 4.59% to 5.64%, monounsaturated from 78.8% to 81.4%, and polyunsaturated varied from 14.0% to 15.5%. The most abundant saturated acid is palmitic acid (3.45–4.24%), the most abundant monounsaturated acid is petroselinic acid (72.3–75.3%), and the most abundant polyunsaturated acid is linoleic acid (13.7–15.3%) (Mhemdi et al. 2011).

As far as sterols are concerned, Sriti et al. (2009) have reported the presence of cholesterol, campesterol, stigmasterol, β-sitosterol, Δ5-avenasterol, Δ5, 24-stigmastadienol, Δ7-stigmasterol, and Δ7-avenasterol. These compounds have been detected in all analyzed parts of the plant, but stigmasterol is the most abundant in seed, and β-sitosterol in pericarp and whole fruit. It should be mentioned that Δ5, 24-stigmastadienol is detected in pericarp in higher amounts, while stigma sterol and β-sitosterol dominate in seed and fruit, followed by Δ7-stigmasterol (Sriti et al. 2009).

34.5 BIOLOGICAL ACTIVITY OF CORIANDER

Coriander and its parts and products are well-known and used worldwide in traditional medicine. The most commonly applied parts and products are leaves, seeds, and essential oils. Leaves are known for their antispasmodic, dyspeptic, and appetizer effects. Coriander is also used to treat abdominal discomfort. Preparations made from leaves have been used to treat coughs, chest pain, and bladder complaints, and as an aphrodisiac. Coriander fruit has been used to treat inflammation, indigestion, cough, bronchitis, vomiting, dysentery, diarrhea, gout, rheumatism, etc. (Sahib et al. 2012).

Coriander, its extracts, and its essential oil have shown a wide range of biological activity. One of the most investigated ones is antimicrobial activity. Previously conducted studies showed that essential oil and extracts had shown activity against many bacteria and yeast species (Begnami et al. 2010; Kubo et al. 2004; Matasyoh et al. 2009). The essential oil showed activity against both Gram-positive and Gram-negative bacteria and antifungal activity against *Candida albicans*. The essential oil derived from the leaves also inhibits numerous *Candida* species in the concentration range of 125–500 μg/mL (Sahib et al. 2012). Antifungal and sprout suppressant activities have also been studied and proved. In this case, essential oils showed higher activity than oleoresin (Singh et al. 2006). Investigations have shown that aqueous and methanol extracts of the leaves induced cell damage and consequently inhibited the growth of the microbes. Methanol extracts showed higher activity, while stem extracts had higher TPC and higher activity than leaf extracts indicating a correlation between the TPC and antimicrobial activity (Wong and Kitts 2006). The antibacterial activity of the essential oils has been attributed to the presence of alcohols, aldehydes, and other compounds such as α-pinene, camphene, and linalool. These compounds are active against a wide range of both Gram-positive and Gram-negative bacteria (Delaquis et al. 2002).

The antioxidant activity of the coriander essential oils and other extracts was also extensively investigated (Baratta et al. 1998; Gallo et al. 2010; Wangensteen, Samuelsen, and Malterud 2004; Zeković, Pavlić, et al. 2016; Zeković, Đurović, and Pavlić 2016; Zeković et al. 2014). Wangensteen et al. (2004) investigated the antioxidant activity of the extracts with different polarities and oil itself. Extracts obtained using ethyl acetate (medium polarity) showed intense activity in the case of the DPPH· test. They concluded that leaf extracts have higher activity when compared to seed extract. Another study showed that methanol and aqueous extract of coriander leaves showed higher activity against free radicals than stem extract obtained with the same solvents (Wong and Kitts 2006). The activity of coriander extract on H_2O_2-induced stress in human lymphocytes was also studied. Treatment with phenolic fraction in 50 μg/mL concentration efficiently protected cells from oxidative stress by restoring it to normal levels (Hashim et al. 2005). There is evidence that coriander has a hepatoprotective effect against CCl_4-induced toxicity (Pandey et al. 2011). These effects of the extracts have been attributed to the high content of iso-quercetin and quercetin. Antioxidant activity was also the primary target of the optimization (Zeković, Đurović, and Pavlić 2016; Zeković, Pavlić, et al. 2016; Zeković, Vladić, et al. 2016; Zeković et al. 2014). RSM was used to obtain the extract with the highest possible antioxidant activity prepared with different extraction techniques such as UAE (Zeković, Đurović, and Pavlić 2016), MAE (Zeković, Vladić, et al. 2016), SWE (Zeković et al. 2014), and SFE (Zeković, Pavlić, et al. 2016). Although all these techniques used different solvents of different polarities, they showed significant antioxidant activity.

As previously mentioned, coriander has been used as an antidiabetic agent. Application of coriander induced an increase in glucose uptake, glucose oxidation, glycogenesis, and dose-dependent effect on insulin secretion (Gray and Flatt 1999). Decreasing serum glucose concentration and an increase in the activity of betta cells were observed after the addition of 200–250 mg/kg of ethanol extract (Eidi et al. 2009). Aqueous extract of the coriander seeds suppressed the hyperglycemia, while a normal glucose level was reached after 4 h dosing (Sahib et al. 2012). It has been shown that pretreatment with coriander seed powder causes specific changes in the metabolism of carbohydrates. Increases in the concentrations of hepatic glycogen and glycogen synthase have been noticed (Chithra and Leelamma 1999). Besides the hyperglycemic effect, coriander showed antidyslipidemic activity. In addition, it shows a lipid-lowering effect. Decreases in triglyceride level, LDL, VLDL, and increased HDL have been observed (Chithra and Leelamma 1997).

Moreover, coriander and its extracts have also been tested for other activities such as anticonvulsant, anxiolytic, sedative, antidepressant, cognitive, antimutagenic, diuretic, antihypertensive, anti-inflammatory, and miscellaneous effects (Sahib et al. 2012), while studies regarding this plant are still going on.

34.6 CONCLUSION

Coriander has been used in various civilizations and nations for an extended period as food and medicine. Various parts of the plant have been investigated to determine chemical composition. In order to do so, different extraction approaches have been applied. The most investigated part of the plant is its seed, mainly for essential oil, widely used worldwide. The investigation has shown that monoterpene linalool is the main compound, followed by geraniol, camphor, pinene, and limonene. Different solvents have been used to extract compounds from coriander ranging from polar to nonpolar. Besides chemical composition, the biological activity of extracts has been assessed. The obtained results showed that plants extract expressed a wide range of activity, which justifies its application in folk medicine in many different countries. Studies have also shown that chemical composition and activity are in strong correlation. Response surface methodology and artificial neural networks have been applied to optimize extraction processes to obtain extracts with a maximal concentration of biologically active compounds and maximal activity. Overall, results have shown that this plant justifiably possesses one of the most significant places in the pharmaceutical and food industries these days.

REFERENCES

Alves-Silva, Jorge M., Sandra M. Dias dos Santos, Manuela E. Pintado, José A. Pérez-Álvarez, Juana Fernández-López, and Manuel Viuda-Martos. 2013. "Chemical Composition and in Vitro Antimicrobial, Antifungal and Antioxidant Properties of Essential Oils Obtained from Some Herbs Widely Used in Portugal." *Food Control* 32 (2): 371–378. doi:10.1016/j.foodcont.2012.12.022.

Attokaran, Mathew. 2017. *Natural Food Flavors and Colorants*. 2nd ed. Chichester, UK: John Wiley & Sons, Ltd. doi:10.1002/9781119114796.

Azmir, J., I.S.M. Zaidul, M.M. Rahman, K.M. Sharif, A. Mohamed, F. Sahena, M.H.A. Jahurul, K. Ghafoor, N.A.N. Norulaini, and A.K.M. Omar. 2013. "Techniques for Extraction of Bioactive Compounds from Plant Materials: A Review." *Journal of Food Engineering* 117 (4): 426–436. doi:10.1016/j.jfoodeng.2013.01.014.

Azmir, J., I.S.M. Zaidul, K.M. Sharif, M.S. Uddin, M.H.A. Jahurul, S. Jinap, Parvaneh Hajeb, and A. Mohamed. 2014. "Supercritical Carbon Dioxide Extraction of Highly Unsaturated Oil from Phaleria Macrocarpa Seed." *Food Research International* 65 (November): 394–400. doi:10.1016/j.foodres.2014.06.049.

Baratta, M.T., H. J. Damien Dorman, Stanley G. Deans, Daniela M. Biondi, and Giuseppe Ruberto. 1998. "Chemical Composition, Antimicrobial and Antioxidative Activity of Laurel, Sage, Rosemary, Oregano and Coriander Essential Oils." *Journal of Essential Oil Research* 10 (6): 618–627. doi:10.1080/10412905.1998.9700989.

Begnami, A.F., M.C.T. Duarte, V. Furletti, and V.L.G. Rehder. 2010. "Antimicrobial Potential of *Coriandrum sativum* L. against Different Candida Species in Vitro." *Food Chemistry* 118 (1): 74–77. doi:10.1016/j.foodchem.2009.04.089.

Bhuiyan, Mohammad Nazrul Islam, Jaripa Begum, and Mahbuba Sultana. 2009. "Chemical Composition of Leaf and Seed Essential Oil of *Coriandrum sativum* L. from Bangladesh." *Bangladesh Journal of Pharmacology* 4 (2). doi:10.3329/bjp.v4i2.2800.

Brunner, Gerd. 1994. *Gas Extraction: An Introduction to Fundamentals of Supercritical Fluids and the Application to Separation Processes, Topics in Physical Chemistry*. Vol. 4. *Topics in Physical Chemistry*. Heidelberg: Steinkopff. doi:10.1007/978-3-662-07380-3.

Brunner, Gerd. 2005. "Supercritical Fluids: Technology and Application to Food Processing." *Journal of Food Engineering* 67 (1–2): 21–33. doi:10.1016/j.jfoodeng.2004.05.060.

Chemat, Farid, Valérie Tomao, and Matthieu Virot. 2008. "Ultrasound-Assisted Extraction in Food Analysis." In *Handbook of Food Analysis Instruments*, edited by Semih Ötles, 85–94. Boca Raton, FL: CRC Press.

Chithra, V., and S. Leelamma. 1997. "Hypolipidemic Effect of Coriander Seeds (*Coriandrum sativum*): Mechanism of Action." *Plant Foods for Human Nutrition (Dordrecht, Netherlands)* 51 (2): 167–172. http://www.ncbi.nlm.nih.gov/pubmed/9527351.

Chithra, V., and S. Leelamma. 1999. "*Coriandrum sativum*-Mechanism of Hypoglycemic Action." *Food Chemistry* 67 (3): 229–231. doi:10.1016/S0308-8146(99)00113-2.

Coşkuner, Yalçın, and Erşan Karababa. 2007. "Physical Properties of Coriander Seeds (*Coriandrum sativum* L.)." *Journal of Food Engineering* 80 (2): 408–416. doi:10.1016/j.jfoodeng.2006.02.042.

Cravotto, Giancarlo, Luisa Boffa, Stefano Mantegna, Patrizia Perego, Milvio Avogadro, and Pedro Cintas. 2008. "Improved Extraction of Vegetable Oils under High-Intensity Ultrasound and/or Microwaves." *Ultrasonics Sonochemistry* 15 (5): 898–902. doi:10.1016/j.ultsonch.2007.10.009.

Cvetanović, Aleksandra, Jaroslava Švarc-Gajić, Uroš Gašić, Živoslav Tešić, Gökhan Zengin, Zoran Zeković, and Saša Đurović. 2016. "Isolation of Apigenin from Subcritical Water Extracts: Optimization of the Process." *The Journal of Supercritical Fluids* 120 (October): 32–42. doi:10.1016/j.supflu.2016.10.012.

Cvetanović, Aleksandra, Jaroslava Švarc-Gajić, Zoran Zeković, Uroš Gašić, Živoslav Tešić, Gokhan Zengin, Pavle Mašković, Mohamad Fawzi Mahomoodally, and Saša Đurović. 2018. "Subcritical Water Extraction as a Cutting Edge Technology for the Extraction of Bioactive Compounds from Chamomile: Influence of Pressure on Chemical Composition and Bioactivity of Extracts." *Food Chemistry* 266 (November): 389–396. doi:10.1016/j.foodchem.2018.06.037.

Cvetanović, Aleksandra, Jaroslava Švarc-Gajić, Zoran Zeković, Jelena Jerković, Gokhan Zengin, Uroš Gašić, Živoslav Tešić, et al. 2019. "The Influence of the Extraction Temperature on Polyphenolic Profiles and Bioactivity of Chamomile (*Matricaria chamomilla* L.) Subcritical Water Extracts." *Food Chemistry* 271 (January): 328–337. doi:10.1016/j.foodchem.2018.07.154.

Cvetanović, Aleksandra, Jaroslava Švarc-Gajić, Zoran Zeković, Pavle Mašković, Saša Đurović, Gökhan Zengin, Cristina Delerue-Matos, Jesus Lozano-Sánchez, and Aleksandra Jakšić. 2017. "Chemical and Biological Insights on Aronia Stems Extracts Obtained by Different Extraction Techniques: From Wastes to Functional Products." *The Journal of Supercritical Fluids* 128 (October): 173–181. doi:10.1016/j.supflu.2017.05.023.

da Silva, Rui P.F.F., Teresa A.P. Rocha-Santos, and Armando C. Duarte. 2016. "Supercritical Fluid Extraction of Bioactive Compounds." *TrAC Trends in Analytical Chemistry* 76: 40–51. doi:10.1016/j.trac.2015.11.013.

Dandge, Dileep K., John P. Heller, and Kennard V. Wilson. 1985. "Structure Solubility Correlations: Organic Compounds and Dense Carbon Dioxide Binary Systems." *Industrial & Engineering Chemistry Product Research and Development* 24 (1): 162–166. doi:10.1021/i300017a030.

de Melo, M.M.R., A.J.D. Silvestre, and C.M. Silva. 2014. "Supercritical Fluid Extraction of Vegetable Matrices: Applications, Trends and Future Perspectives of a Convincing Green Technology." *The Journal of Supercritical Fluids* 92: 115–176. doi:10.1016/j.supflu.2014.04.007.

Delaquis, Pascal J., Kareen Stanich, Benoit Girard, and G. Mazza. 2002. "Antimicrobial Activity of Individual and Mixed Fractions of Dill, Cilantro, Coriander and Eucalyptus Essential Oils." *International Journal of Food Microbiology* 74 (1–2): 101–109. doi:10.1016/S0168-1605(01)00734-6.

Dima, Cristian, George Adrian Ifrim, Gigi Coman, Petru Alexe, and Ştefan Dima. 2016. "Supercritical CO2 Extraction and Characterization of *Coriandrum sativum* L. Essential Oil." *Journal of Food Process Engineering* 39 (2): 204–211. doi:10.1111/jfpe.12218.

Döker, O., U. Salgın, İ. Şanal, Ü. Mehmetoğlu, and A. Çalımlı. 2004. "Modeling of Extraction of β-Carotene from Apricot Bagasse Using Supercritical CO2 in Packed Bed Extractor." *The Journal of Supercritical Fluids* 28 (1): 11–19. doi:10.1016/S0896-8446(03)00006-8.

Đurović, Saša. 2019. "Urtica Dioica, Contemporary Extraction Techniques, Chemical Profile, Biological Activity, Formulation of Food Product." University of Novi Sad, Faculty of Technology.

Đurović, Saša. 2021. "Application of the GC/MS Technique in Environmental Analytics: Case of the Essential Oils." In *Green Sustainable Process for Chemical and Environmental Engineering and Science*, edited by Innamudin, R. Boddula, and A. Asiri, 197–208. Elsevier. doi:10.1016/B978-0-12-821883-9.00005-9.

Đurović, Saša, Saša Šorgić, Saša Popov, Lato Pezo, Pavle Mašković, Stevan Blagojević, and Zoran Zeković. 2022. "Recovery of Biologically Active Compounds from Stinging Nettle Leaves Part I: Supercritical Carbon Dioxide Extraction." *Food Chemistry* 373 (March): 131724. doi:10.1016/j.foodchem.2021.131724.

Đurović, Saša, Saša Šorgić, Saša Popov, Marija Radojković, and Zoran Zeković. 2018. "Isolation and GC Analysis of Fatty Acids: Study Case of Stinging Nettle Leaves." In *Carboxylic Acid - Key Role in Life Sciences*. InTech. doi:10.5772/intechopen.73533.

Eidi, Maryam, Akram Eidi, Ali Saeidi, Saadat Molanaei, Alireza Sadeghipour, Massih Bahar, and Kamal Bahar. 2009. "Effect of Coriander Seed (*Coriandrum sativum* L.) Ethanol Extract on Insulin Release from Pancreatic Beta Cells in Streptozotocin-Induced Diabetic Rats." *Phytotherapy Research* 23 (3): 404–406. doi:10.1002/ptr.2642.

Eikani, Mohammad H., Fereshteh Golmohammad, and Soosan Rowshanzamir. 2007. "Subcritical Water Extraction of Essential Oils from Coriander Seeds (*Coriandrum sativum* L.)." *Journal of Food Engineering* 80 (2): 735–740. doi:10.1016/j.jfoodeng.2006.05.015.

Filip, Snežana, Senka Vidović, Jelena Vladić, Branimir Pavlić, Dušan Adamović, and Zoran Zeković. 2016. "Chemical Composition and Antioxidant Properties of *Ocimum basilicum* L. Extracts Obtained by Supercritical Carbon Dioxide Extraction: Drug Exhausting Method." *The Journal of Supercritical Fluids* 109 (March): 20–25. doi:10.1016/j.supflu.2015.11.006.

Friedrich, Jörg, and Gerhard M. Schneider. 1989. "Near-Infrared Spectroscopic Investigations on Phase Behaviour and Association of 1-Octadecanol, 3-Hexanol, and 3-Methyl-3-Pentanol in Carbon Dioxide and Chlorotrifluoromethane." *The Journal of Chemical Thermodynamics* 21 (3): 307–319. doi:10.1016/0021-9614(89)90020-7.

Gallo, Monica, Rosalia Ferracane, Giulia Graziani, Alberto Ritieni, and Vincenzo Fogliano. 2010. "Microwave Assisted Extraction of Phenolic Compounds from Four Different Spices." *Molecules* 15 (9): 6365–6374. doi:10.3390/molecules15096365.

Gevrey, Muriel, Ioannis Dimopoulos, and Sovan Lek. 2003. "Review and Comparison of Methods to Study the Contribution of Variables in Artificial Neural Network Models." *Ecological Modelling* 160 (3): 249–264. doi:10.1016/S0304-3800(02)00257-0.

Gil, Alejandra, Elba B. de la Fuente, Adriana E. Lenardis, Mónica López Pereira, Susana A. Suárez, Arnaldo Bandoni, Catalina van Baren, Paola Di Leo Lira, and Claudio M. Ghersa. 2002. "Coriander Essential Oil Composition from Two Genotypes Grown in Different Environmental Conditions." *Journal of Agricultural and Food Chemistry* 50 (10): 2870–2877. doi:10.1021/jf011128i.

Gray, A.M., and P.R. Flatt. 1999. "Insulin-Releasing and Insulin-like Activity of the Traditional Anti-Diabetic Plant *Coriandrum sativum* (Coriander)." *The British Journal of Nutrition* 81 (3): 203–209. http://www.ncbi.nlm.nih.gov/pubmed/10434846.

Grosso, C., V. Ferraro, A.C. Figueiredo, J.G. Barroso, J.A. Coelho, and A. M. Palavra. 2008. "Supercritical Carbon Dioxide Extraction of Volatile Oil from Italian Coriander Seeds." *Food Chemistry* 111 (1): 197–203. doi:10.1016/j.foodchem.2008.03.031.

Hashim, M., S. Lincy, V. Remya, M. Teena, and L. Anila. 2005. "Effect of Polyphenolic Compounds from on HO-Induced Oxidative Stress in Human Lymphocytes." *Food Chemistry* 92 (4): 653–660. doi:10.1016/j.foodchem.2004.08.027.

Herrero, M., A. Cifuentes, and E. Ibanez. 2006. "Sub- and Supercritical Fluid Extraction of Functional Ingredients from Different Natural Sources: Plants, Food-by-Products, Algae and Microalgae: A Review." *Food Chemistry* 98 (1): 136–148. doi:10.1016/j.foodchem.2005.05.058.

Jain, Tripti, V. Jain, R. Pandey, A. Vyas, and S. S. Shukla. 2009. "Microwave Assisted Extraction for Phytoconstituents-An Overview." *Asian Journal of Research in Chemistry* 2 (1): 19–25.

Jesus, S. P., and M. A. M. Meireles. 2014. "Supercritical Fluid Extraction: A Global Perspective of the Fundamental Concepts of This Eco-Friendly Extraction Technique." In *Alternative Solvents for Natural Products Extraction*, edited by F. Chemat and M. A. Vian, 39–72. Berlin: Springer-Verlag.

Khajeh, Mostafa, Mansour Ghaffari Moghaddam, and Mohammad Shakeri. 2012. "Application of Artificial Neural Network in Predicting the Extraction Yield of Essential Oils of *Diplotaenia cachrydifolia* by Supercritical Fluid Extraction." *The Journal of Supercritical Fluids* 69: 91–96. doi:10.1016/j.supflu.2012.05.006.

Ko, Min-Jung, Chan-Ick Cheigh, and Myong-Soo Chung. 2014. "Relationship Analysis between Flavonoids Structure and Subcritical Water Extraction (SWE)." *Food Chemistry* 143: 147–155. doi:10.1016/j.foodchem.2013.07.104.

Kubo, Isao, Ken-ichi Fujita, Aya Kubo, Ken-ichi Nihei, and Tetsuya Ogura. 2004. "Antibacterial Activity of Coriander Volatile Compounds against *Salmonella choleraesuis*." *Journal of Agricultural and Food Chemistry* 52 (11): 3329–3332. doi:10.1021/jf0354186.

Kuvendziev, Stefan, Kiril Lisichkov, Zoran Zeković, and Mirko Marinkovski. 2014. "Artificial Neural Network Modelling of Supercritical Fluid CO2 Extraction of Polyunsaturated Fatty Acids from Common Carp (*Cyprinus carpio* L.) Viscera." *The Journal of Supercritical Fluids* 92 (August): 242–248. doi:10.1016/j.supflu.2014.06.007.

Mandal, Shyamapada, and Manisha Mandal. 2015. "Coriander (*Coriandrum sativum* L.) Essential Oil: Chemistry and Biological Activity." *Asian Pacific Journal of Tropical Biomedicine* 5 (6): 421–428. doi:10.1016/j.apjtb.2015.04.001.

Mašković, Pavle, Marija Radojković, Aleksandra Cvetanović, Milan Mitić, Zoran Zeković, and Saša Đurović. 2018. "Chemical Profile and Biological Activity of Tart Cherry Twigs : Possibilities of Plant Waste Utilization." *Journal of Food and Nutrition Research* 57 (3): 222–230.

Mašković, Pavle, Vesna Veličković, Milan Mitić, Saša Đurović, Zoran Zeković, Marija Radojković, Aleksandra Cvetanović, Jaroslava Švarc-Gajić, and Jelena Vujić. 2017. "Summer Savory Extracts Prepared by Novel Extraction Methods Resulted in Enhanced Biological Activity." *Industrial Crops and Products* 109: 875–881. doi:10.1016/j.indcrop.2017.09.063.

Mašković, Pavle Z., Vesna Veličković, Saša Đurović, Zoran Zeković, Marija Radojković, Aleksandra Cvetanović, Jaroslava Švarc-Gajić, Milan Mitić, and Jelena Vujić. 2018. "Biological Activity and Chemical Profile of *Lavatera thuringiaca* L. Extracts Obtained by Different Extraction Approaches." *Phytomedicine* 38: 118–124. doi:10.1016/j.phymed.2017.11.010.

Matasyoh, J.C., Z.C. Maiyo, R.M. Ngure, and R. Chepkorir. 2009. "Chemical Composition and Antimicrobial Activity of the Essential Oil of *Coriandrum sativum*." *Food Chemistry* 113 (2): 526–529. doi:10.1016/j.foodchem.2008.07.097.

Meullemiestre, Alice, Cassandra Breil, Maryline Abert-Vian, and Farid Chemat. 2015. *Modern Techniques and Solvents for the Extraction of Microbial Oils. SpringerBriefs in Molecular Science*. Cham: Springer International Publishing. doi:10.1007/978-3-319-22717-7.

Mhemdi, Houcine, Elisabeth Rodier, Nabil Kechaou, and Jacques Fages. 2011. "A Supercritical Tuneable Process for the Selective Extraction of Fats and Essential Oil from Coriander Seeds." *Journal of Food Engineering* 105 (4): 609–616. doi:10.1016/j.jfoodeng.2011.03.030.

Micić, Darko, Saša Đurović, Pavel Riabov, Ana Tomić, Olja Šovljanski, Snežana Filip, Tomislav Tosti, et al. 2021. "Rosemary Essential Oils as a Promising Source of Bioactive Compounds: Chemical Composition, Thermal Properties, Biological Activity, and Gastronomical Perspectives." *Foods* 10 (11): 2734. doi:10.3390/foods10112734.

Micić, Darko, Sanja Ostojić, Lato Pezo, Stevan Blagojević, Branimir Pavlić, Zoran Zeković, and Saša Đurović. 2019. "Essential Oils of Coriander and Sage: Investigation of Chemical Profile, Thermal Properties and QSRR Analysis." *Industrial Crops and Products* 138 (October): 111438. doi:10.1016/j.indcrop.2019.06.001.

Msaada, Kamel, Karim Hosni, Mouna Ben Taarit, Thouraya Chahed, Mohamed Elyes Kchouk, and Brahim Marzouk. 2007. "Changes on Essential Oil Composition of Coriander (*Coriandrum sativum* L.) Fruits during Three Stages of Maturity." *Food Chemistry* 102 (4): 1131–1134. doi:10.1016/j.foodchem.2006.06.046.

Msaada, Kamel, Mouna Ben Taarit, Karim Hosni, Mohamed Hammami, and Brahim Marzouk. 2009. "Regional and Maturational Effects on Essential Oils Yields and Composition of Coriander (*Coriandrum sativum* L.) Fruits." *Scientia Horticulturae* 122 (1): 116–124. doi:10.1016/j.scienta.2009.04.008.

Olden, Julian D., and Donald A. Jackson. 2002. "Illuminating the "Black Box": A Randomization Approach for Understanding Variable Contributions in Artificial Neural Networks." *Ecological Modelling* 154 (1–2): 135–150. doi:10.1016/S0304-3800(02)00064-9.

Paixao Coelho, J. A., and A. M. Figueiredo Palavra. 2015. "Supercritical Fluid Extraction of Compounds from Spices and Herbs." In *High Pressure Fluid Technology for Green Food Processing*, edited by T. Fornari and R. P. Stateva, 357–396. New York: Springer.

Pandey, A., P. Bigoniya, V. Raj, and K. K. Patel. 2011. "Pharmacological Screening of *Coriandrum sativum* Linn. for Hepatoprotective Activity." *Journal of Pharmacy & Bioallied Sciences* 3 (3): 435–441. doi:10.4103/0975-7406.84462.

Paniwnyk, L., E. Beaufoy, J. P. Lorimer, and T. J. Mason. 2001. "The Extraction of Rutin from Flower Buds of *Sophora japonica*." *Ultrasonics Sonochemistry* 8 (3): 299–301. doi:10.1016/S1350-4177(00)00075-4.

Pavlić, Branimir, Nemanja Teslić, Gokhan Zengin, Saša Đurović, Dušan Rakić, Aleksandra Cvetanović, A.K. Gunes, and Zoran Zeković. 2021. "Antioxidant and Enzyme-Inhibitory Activity of Peppermint Extracts and Essential Oils Obtained by Conventional and Emerging Extraction Techniques." *Food Chemistry* 338 (February): 127724. doi:10.1016/j.foodchem.2020.127724.

Pavlić, Branimir, Senka Vidović, Jelena Vladić, Robert Radosavljević, and Zoran Zeković. 2015. "Isolation of Coriander (*Coriandrum sativum* L.) Essential Oil by Green Extractions versus Traditional Techniques." *The Journal of Supercritical Fluids* 99 (April): 23–28. doi:10.1016/j.supflu.2015.01.029.

Pourmortazavi, Seied Mahdi, and Seiedeh Somayyeh Hajimirsadeghi. 2007. "Supercritical Fluid Extraction in Plant Essential and Volatile Oil Analysis." *Journal of Chromatography A* 1163 (1–2): 2–24. doi:10.1016/j.chroma.2007.06.021.

Radivojac, Aleksandar, Oskar Bera, Darko Micić, Saša Đurović, Zoran Zeković, Stevan Blagojević, and Branimir Pavlić. 2020. "Conventional versus Microwave-Assisted Hydrodistillation of Sage Herbal Dust: Kinetics Modeling and Physico-Chemical Properties of Essential Oil." *Food and Bioproducts Processing* 123 (September): 90–101. doi:10.1016/j.fbp.2020.06.015.

Radojković, Marija, Zoran Zeković, Pavle Mašković, Senka Vidović, Anamarija Mandić, Aleksandra Mišan, and Saša Đurović. 2016. "Biological Activities and Chemical Composition of Morus Leaves Extracts Obtained by Maceration and Supercritical Fluid Extraction." *The Journal of Supercritical Fluids* 117 (November): 50–58. doi:10.1016/j.supflu.2016.05.004.

Ramić, Milica, Senka Vidović, Zoran Zeković, Jelena Vladić, Aleksandra Cvejin, and Branimir Pavlić. 2015. "Modeling and Optimization of Ultrasound-Assisted Extraction of Polyphenolic Compounds from Aronia Melanocarpa By-Products from Filter-Tea Factory." *Ultrasonics Sonochemistry* 23: 360–368. doi:10.1016/j.ultsonch.2014.10.002.

Ravi, Ramasamy, Maya Prakash, and K. Keshava Bhat. 2007. "Aroma Characterization of Coriander (*Coriandrum sativum* L.) Oil Samples." *European Food Research and Technology* 225 (3–4): 367–374. doi:10.1007/s00217-006-0425-7.

Riabov, Pavel A., Darko Micić, Rade B. Božović, Dušan V. Jovanović, Ana Tomić, Olja Šovljanski, Snežana Filip, et al. 2020. "The Chemical, Biological and Thermal Characteristics and Gastronomical Perspectives of Laurus Nobilis Essential Oil from Different Geographical Origin." *Industrial Crops and Products* 151: 112498. doi:10.1016/j.indcrop.2020.112498.

Rostagno, Mauricio A., Miguel Palma, and Carmelo G. Barroso. 2003. "Ultrasound-Assisted Extraction of Soy Isoflavones." *Journal of Chromatography A* 1012 (2): 119–128. doi:10.1016/S0021-9673(03)01184-1.

Sahib, Najla Gooda, Farooq Anwar, Anwarul-Hassan Gilani, Azizah Abdul Hamid, Nazamid Saari, and Khalid M. Alkharfy. 2012. "Coriander (*Coriandrum sativum* L.): A Potential Source of High-Value Components for Functional Foods and Nutraceuticals - A Review." *Phytotherapy Research* 1456 (November 2012). doi:10.1002/ptr.4897.

Shahwar, Muhammad Khuram, Ahmed Hassan El-Ghorab, Faqir Muhammad Anjum, Masood Sadiq Butt, Shahzad Hussain, and Muhammad Nadeem. 2012. "Characterization of Coriander (*Coriandrum sativum* L.) Seeds and Leaves: Volatile and Non Volatile Extracts." *International Journal of Food Properties* 15 (4): 736–747. doi:10.1080/10942912.2010.500068.

Shokri, A., T. Hatami, and M. Khamforoush. 2011. "Near Critical Carbon Dioxide Extraction of Anise (*Pimpinella anisum* L.) Seed: Mathematical and Artificial Neural Network Modeling." *The Journal of Supercritical Fluids* 58 (1): 49–57. doi:10.1016/j.supflu.2011.04.011.

Singh, Gurdip, Sumitra Maurya, M. P. de Lampasona, and Cesar A. N. Catalan. 2006. "Studies on Essential Oils, Part 41. Chemical Composition, Antifungal, Antioxidant and Sprout Suppressant Activities of Coriander (*Coriandrum sativum*) Essential Oil and Its Oleoresin." *Flavour and Fragrance Journal* 21 (3): 472–479. doi:10.1002/ffj.1608.

Sodeifian, Gholamhossein, Seyed Ali Sajadian, and Nedasadat Saadati Ardestani. 2016. "Optimization of Essential Oil Extraction from Launaea Acanthodes Boiss: Utilization of Supercritical Carbon Dioxide and Cosolvent." *The Journal of Supercritical Fluids* 116: 46–56. doi:10.1016/j.supflu.2016.05.015.

Šojić, Branislav, Branimir Pavlić, Vladimir Tomović, Predrag Ikonić, Zoran Zeković, Sunčica Kocić-Tanackov, Saša Đurović, Snežana Škaljac, Marija Jokanović, and Maja Ivić. 2019. "Essential Oil versus Supercritical Fluid Extracts of Winter Savory (*Satureja montana* L.) - Assessment of the Oxidative, Microbiological and Sensory Quality of Fresh Pork Sausages." *Food Chemistry* 287 (July): 280–286. doi:10.1016/j.foodchem.2018.12.137.

Šojić, Branislav, Branimir Pavlić, Vladimir Tomović, Sunčica Kocić-Tanackov, Saša Đurović, Zoran Zeković, Miona Belović, et al. 2020. "Tomato Pomace Extract and Organic Peppermint Essential Oil as Effective Sodium Nitrite Replacement in Cooked Pork Sausages." *Food Chemistry* 330 (November): 127202. doi:10.1016/j.foodchem.2020.127202.

Šojić, Branislav, Vladimir Tomović, Sunčica Kocić-Tanackov, Danijela Bursać Kovačević, Predrag Putnik, Živan Mrkonjić, Saša Đurović, et al. 2020. "Supercritical Extracts of Wild Thyme (*Thymus serpyllum* L.) By-Product as Natural Antioxidants in Ground Pork Patties." *LWT* 130 (August): 109661. doi:10.1016/j.lwt.2020.109661.

Sourmaghi, Mohammad Hossein Salehi, Gita Kiaee, Fereshteh Golfakhrabadi, Hossein Jamalifar, and Mahnaz Khanavi. 2015. "Comparison of Essential Oil Composition and Antimicrobial Activity of *Coriandrum sativum* L. Extracted by Hydrodistillation and Microwave-Assisted Hydrodistillation." *Journal of Food Science and Technology* 52 (4): 2452–2457. doi:10.1007/s13197-014-1286-x.

Sriti, Jazia, Thierry Talou, Wissem Aidi Wannes, Muriel Cerny, and Brahim Marzouk. 2009. "Essential Oil, Fatty Acid and Sterol Composition of Tunisian Coriander Fruit Different Parts." *Journal of the Science of Food and Agriculture* 89 (10): 1659–1664. doi:10.1002/jsfa.3637.

Sriti, Jazia, Wissem Aidi Wannes, Thierry Talou, Gerard Vilarem, and Brahim Marzouk. 2011. "Chemical Composition and Antioxidant Activities of Tunisian and Canadian Coriander (*Coriandrum sativum* L.) Fruit." *Journal of Essential Oil Research* 23 (4): 7–15. doi:10.1080/10412905.2011.9700462.

Tchaban, T., M. J. Taylor, and J. P. Griffin. 1998. "Establishing Impacts of the Inputs in a Feedforward Neural Network." *Neural Computing & Applications* 7 (4): 309–317. doi:10.1007/BF01428122.

Teixeira, Bárbara, António Marques, Cristina Ramos, Nuno R. Neng, José M.F. Nogueira, Jorge Alexandre Saraiva, and Maria Leonor Nunes. 2013. "Chemical Composition and Antibacterial and Antioxidant Properties of Commercial Essential Oils." *Industrial Crops and Products* 43 (1). Elsevier B.V.: 587–595. doi:10.1016/j.indcrop.2012.07.069.

Tufeu, Roland, Pascale Subra, and Christine Plateaux. 1993. "Contribution to the Experimental Determination of the Phase Diagrams of Some (Carbon Dioxide+a Terpene) Mixtures." *The Journal of Chemical Thermodynamics* 25 (10): 1219–1228. doi:10.1006/jcht.1993.1120.

Veličković, Vesna, Saša Đurović, Marija Radojković, Aleksandra Cvetanović, Jaroslava Švarc-Gajić, Jelena Vujić, Srećko Trifunović, and Pavle Z. Mašković. 2017. "Application of Conventional and Non-Conventional Extraction Approaches for Extraction of *Erica carnea* L.: Chemical Profile and Biological Activity of Obtained Extracts." *The Journal of Supercritical Fluids* 128 (October): 331–337. doi:10.1016/j.supflu.2017.03.023.

Wang, Lijun, and Curtis L. Weller. 2006. "Recent Advances in Extraction of Nutraceuticals from Plants." *Trends in Food Science & Technology* 17 (6): 300–312. doi:10.1016/j.tifs.2005.12.004.

Wangensteen, Helle, Anne Berit Samuelsen, and Karl Egil Malterud. 2004. "Antioxidant Activity in Extracts from Coriander." *Food Chemistry* 88 (2): 293–297. doi:10.1016/j.foodchem.2004.01.047.

Wong, P., and D. Kitts. 2006. "Studies on the Dual Antioxidant and Antibacterial Properties of Parsley (*Petroselinum crispum*) and Cilantro (*Coriandrum sativum*) Extracts." *Food Chemistry* 97 (3): 505–515. doi:10.1016/j.foodchem.2005.05.031.

Yoon, Younguhc, George Swales, and Thomas M. Margavio. 1993. "A Comparison of Discriminant Analysis versus Artificial Neural Networks." *Journal of the Operational Research Society* 44 (1): 51–60. doi:10.1057/jors.1993.6.

Zahedi, Gholamreza, and Abbas Azarpour. 2011. "Optimization of Supercritical Carbon Dioxide Extraction of Passiflora Seed Oil." *The Journal of Supercritical Fluids* 58 (1): 40–48. doi:10.1016/j.supflu.2011.04.013.

Zeković, Zoran, Oskar Bera, Saša Đurović, and Branimir Pavlić. 2017. "Supercritical Fluid Extraction of Coriander Seeds: Kinetics Modelling and ANN Optimization." *The Journal of Supercritical Fluids* 125: 88–95. doi:10.1016/j.supflu.2017.02.006.

Zeković, Zoran, Arijana Bušić, Draženka Komes, Jelena Vladić, Dušan Adamović, and Branimir Pavlić. 2015. "Coriander Seeds Processing: Sequential Extraction of Non-Polar and Polar Fractions Using Supercritical Carbon Dioxide Extraction and Ultrasound-Assisted Extraction." *Food and Bioproducts Processing* 95 (July): 218–227. doi:10.1016/j.fbp.2015.05.012.

Zeković, Zoran, Sasa Đurović, and Branimir Pavlić. 2016. "Optimization of Ultrasound-Assisted Extraction of Polyphenolic Compounds from Coriander Seeds Using Response Surface Methodology." *Acta Periodica Technologica* 47: 249–263. doi:10.2298/APT1647249Z.

Zeković, Zoran, Branimir Pavlić, Aleksandra Cvetanović, and Saša Đurović. 2016. "Supercritical Fluid Extraction of Coriander Seeds: Process Optimization, Chemical Profile and Antioxidant Activity of Lipid Extracts." *Industrial Crops and Products* 94: 353–362. doi:10.1016/j.indcrop.2016.09.008.

Zeković, Zoran, Senka Vidović, Jelena Vladić, Robert Radosavljević, Aleksandra Cvejin, Mohamed A Elgndi, and Branimir Pavlić. 2014. "Optimization of Subcritical Water Extraction of Antioxidants from *Coriandrum sativum* Seeds by Response Surface Methodology." *The Journal of Supercritical Fluids* 95: 560–566. doi:10.1016/j.supflu.2014.09.004.

Zeković, Zoran, Jelena Vladić, Senka Vidović, Dušan Adamović, and Branimir Pavlić. 2016. "Optimization of Microwave-Assisted Extraction (MAE) of Coriander Phenolic Antioxidants - Response Surface Methodology Approach." *Journal of the Science of Food and Agriculture* 96 (13): 4613–4622. doi:10.1002/jsfa.7679.

Zoubiri, Safia, and Aoumeur Baaliouamer. 2010. "Essential Oil Composition of *Coriandrum sativum* Seed Cultivated in Algeria as Food Grains Protectant." *Food Chemistry* 122 (4): 1226–1228. doi:10.1016/j.foodchem.2010.03.119.

35 Effect of Agricultural Processes and Practices on the Composition and Quality of Coriander (*Coriandrum sativum* L.) Essential Oil

Kevser Karaman, Erman Beyzi, Sibel Turan Sirke, and Adem Gunes

CONTENTS

35.1 INTRODUCTION

Coriander (*Coriandrum sativum* L. 2n=2x=22) is among the oldest plant species known to human-kind, and it has been cultivated for about 8000 years. Archaeological excavations reveal some information about the origin of coriander and provide evidence that it was grown in ancient times (Diederichsen, 1996). Coriander is named after a list of medicinal plants found in an Egyptian papyrus dating from 1550 BC (Harten, 1974). It is an annual herbaceous dicotyledonous plant that belongs to the family of Umbelliferae/Apiaceae (Singh et al., 2017; Arif et al., 2014; Song et al., 2020), and it has been used for various purposes, including as spice, in folk medicine, fragrances, dyes, jewelry, repellents, and for aroma and flavoring in the food industry (Nadeem et al., 2013). The plant's fruit contains different components such as essential oil, essential oil components, crude oil, fatty acids, protein, minerals, fiber, sugar, and starch (Diederichsen, 1996). Among these components, especially essential oil, crude oil, protein, and mineral substances play a decisive role in the quality of the plant.

In different studies, the essential oil ratio was reported as 0.83–0.90% by Argañosa et al. (1998), 0.18–0.60% by Ayanoğlu et al. (2002), and 0.14–0.37% by Ramezani et al. (2009). In terms of essential oil composition, linalool is found in coriander stems, leaves, and fruits as the major volatile compound constituting about 60–70% of the essential oils of coriander (Teneva et al., 2016; Beyzi et al., 2017). The other compounds include geraniol, pinene, limonene, geranyl acetate, terpinene, and borneol (Nadeem et al., 2013). The essential oil can be used to manufacture food, perfumery, soap, cosmetics, medicine, and aroma (Carrubba et al., 2002; Neffati and Marzouk, 2008). Besides, coriander essential oil possesses different biological activities such as antioxidant, anti-inflammatory, analgesic, antibacterial, antifungal, and insecticidal properties (Kiralan et al., 2009; Lo Cantore et al., 2004). The essential oil yield and composition of coriander depend on many factors such

DOI: 10.1201/9781003204626-39

as plant genetics, growth development, appropriate fertilizer application, plant and soil nutrient contents, geographical location, climatic conditions, stress factors, and post-harvest storage (Lubbe and Verpoorte, 2011; Gil et al., 2002; Telci et al., 2006; Figueiredo et al., 2008). In this chapter, agricultural-postharvest processes and genetic factors that cause essential oil composition and quality changes have been explained.

35.2 AGRICULTURAL PROCESSES

Many agricultural factors are influential in the change of essential oil components and the quality of coriander plants. These factors can be listed as different planting times, harvest time, plant density, fertilization, water and salt stress applications, and the effects of climate and soil characteristics. Studies have shown that different planting times significantly affect the change of essential oil components of the coriander plant (Zheljazkov et al., 2008; Nowak and SempliĚski, 2014; Delibaltova, 2020). Delibaltova (2020) reported that different sowing periods affected coriander essential oil yield and components, and also, 12 essential oil components were determined while linalool was the main component. Apart from linalool, camphor, α-pinene, γ-terpinene, ρ-cymene, limonene, α-terpineol, geranyl acetate, geraniol, camphene, sabinene, and myrcene components were also determined. In another study, Zheljazkov et al. (2008) reported that different planting dates affected the essential oil composition of coriander and stated that the linalool ratio was in the range of 64.0–84.6%. In a different study examining the effects of different years and times on plant planting, Nowak and SempliĚski (2014) reported that the essential oil ratio in the plant differed between years and the essential oil ratio increased slightly when the planting date was delayed. In addition, they stated that the essential oil composition differed by the treatment years rather than the planting date.

The essential oil components of the coriander plant differ according to the development period. Especially in different maturity periods, the chemical composition of the essential oil of coriander fruits shows differences. For example, an increase was observed in linalool content from 36.7% to 72.3% during to initial and final maturity stage. In comparison, a decrease was seen in geranyl acetate from 35.2% to 1.49%. These differences occurring at different maturation levels were attributed to changes in secondary metabolism in plants (Msaada et al., 2009). El-Zaeddi et al. (2020) reported that the coriander plant should be harvested three weeks after planting in their study, which determined the optimization of the harvest date according to the volatile composition of some aromatic plants at different vegetative stages. The highest total concentrations of volatile compounds were 2279 mg kg^{-1} in that harvest time. In their study, Katar and Katar (2016) stated that the optimum plant density for two different coriander cultivars (Arslan and Gürbüz) was 40 and 50 plant/m^2, respectively. In another study, Kızıl and İpek (2004) reported that the essential oil ratio varied between 0.29% and 0.32%, and higher essential oil yield was obtained from lower row spacing.

The amount of plant nutrients that the plant can take from the soil affects the yield and quality of the coriander plant and the amount of essential oil. Mineral nutrition is an essential factor in the composition and yield of essential oils in plants, and they could be affected by genetic, physiological, and climatic factors (El Gendy et al., 2015; Chrysargyris et al., 2017a, 2017b). In studies on phosphorus, which is one of these nutrients, a significant positive effect on the essential oil ratio of the coriander plant occurred. At increasing levels, α-pinene, geranyl acetate, and camphor ratios were positively affected by the amount of phosphorus. However, components such as linalool and γ-terpinene were not significantly affected by phosphorus doses (İzgi, 2020). In the other study, vermicompost and NPK-containing fertilizer applications increased the coriander plant's total biomass and essential oil content (Singh, 2012). In another study in which organic fertilizers such as Vermicompost were applied, the highest essential oil yield was obtained compared to the control group, depending on the fertilizer application (Rajesh et al., 2015). In soils where irrigation conditions are suitable, the essential oil rate of coriander increases depending on the added phosphorus fertilizers (Hani et al., 2015). Potassium is also one of the essential nutrients for the coriander plant. It was reported that depending on the increase in the amount of potassium, fruit yield and essential oil content increased (Freitas et al., 2020). Sulfur is a component of acetyl-CoA molecule involved in terpene synthesis and, being a protein component, and it has an essential effect on essential oil production (Dubey et al., 2003). With the application of Zn, the yield amount of the coriander plant and the essential oil yield increased. Optimum essential oil yield was obtained from 1.33 kg da⁻¹ Zn application, especially in soils with low Zn content (Özyazıcı, 2020).

Instead of chemical fertilizers, arbuscular mycorrhizal fungi were evaluated in coriander, and an increase in the contents of macro- and micro-nutrients was observed. Mycorrhiza inoculation caused a significant change in the α-pinene, α-phellandrene, and p-cymene of essential oil of coriander shoots both in sole cropping and in intercropping systems (Weisany et al., 2021). Additionally, the use of biological control agents based on *Trichoderma* species in coriander cultivation was studied, and linalool content was positively affected by *Trichoderma* treatments compared to non-treatment (Gebarowska et al., 2019).

Depending on the temperature and precipitation factors, the accumulation of active substances such as linalool could be positively or negatively affected (Sangwan et al., 2001). A positive relationship was determined between the temperature increase and linalool (Telci et al., 2010). In a different study, in which the important effect of the climate on the coriander plant was revealed, it was determined that there were significant differences in essential oil components between years due to the changes in the amount of precipitation (Özyazıcı, 2021). Although some plants vary in yield levels according to the stress tolerance level, there may not be a difference in essential oil ratios. Unlukara et al. (2016) observed a decrease in the coriander plant's biomass, seed yield, and essential oil yield under different water stress conditions but determined that there was no statistically significant change in the essential oil ratio. In addition, in studies where a significant decrease was observed in parameters such as the seed and oil yield of coriander with the increase of water stress, it was stated that the coriander plant is sensitive to water stress (Ghamarnia and Daichin 2013; Hani et al., 2015). However, in another study, it was reported that water deficiency increased the percentage of essential oil in the coriander plant but decreased the essential oil yield (Amir et al., 2019). The level of salt stress could cause different effects on the essential oil. When the essential oil yield of coriander leaves was examined under different levels of salt stress, essential oil yield was stimulated only under low and medium stress, while essential oil yield decreased at high salinity levels (Neffati and Marzouk 2008). The same authors studied the effect of salt stress on vegetative growth, essential oil content, and composition of Tunisian coriander grown in hydroponic culture. They monitored that salt stress increased the contents of (E)-2-decenal, (E)2-dodecenal, and dodecanal under 25 mM and 50 mM NaCl treatments in the leaf oils; however, the content of these compounds decreased significantly under 75 mM NaCl (Neffati and Marzouk 2010). In the study of Amiripour et al. (2021), salt-stressed coriander plants had higher essential oil yield and 2E-decanal content while having lower n-decanal. The authors investigated the effects of cadmium and lead levels in coriander cultivation under greenhouse conditions as a different stress condition. The lowest essential oil content was obtained from control plants (0.18%), while the highest (0.31%) was 20 mg kg⁻¹ Cd treatment. The authors also recommended coriander cultivation in soils contaminated with Cd or Pb because no heavy metal residual was observed in the extracted essential oil (Fattahi et al., 2021).

35.3 POSTHARVEST PROCESSES

For the edible seeds, postharvest processing is required to prepare the material for consumption. The seeds are generally exposed to several processing stages between harvest and consumption. The first step of the postharvest process is storing the material after harvesting. And then, the samples are subjected to cleaning, including removing the husk or other undesired parts of the seed. Moreover, the seeds are subjected to final processing such as drying, grinding, tempering, parboiling, soaking, etc. All of these steps applied as postharvest processing steps could affect the quality characteristics of the final samples (FAO, https://www.fao.org/3/au104e/au104e.pdf).

Drying is the most often applied process for all seeds to remove the excess water. Many investigations have been conducted to determine the effects of the drying process on the food samples' quality attributes. Also, coriander was characterized under different drying conditions regarding essential oil yield and quality characteristics. Pirbalouti et al. (2017) investigated the effects of different drying methods on the qualitative and quantitative properties of essential oil from the aerial parts of coriander. They applied different drying methods such as sunlight, shade, mechanical ovens (40 and 60°C), microwave oven (500 and 700 W), and freeze-drying. They concluded that the highest total essential oil yield was observed for the freeze-dried tissue (0.18 mL/100 g dry matter) followed by shade-dried samples (0.13 mL/100 g dry matter). The authors attributed the decrease in the oil yield with the temperature increase to the biological structure of the oil glands of coriander that decreased essential oil yield at high temperatures. For the fresh and dried coriander samples, 39 aromatic compounds were identified, and the major components were decanal (0–37.5%), *cis*-phytol (1.0–34.1%), 1-tetradecanol (trace–31.7%), 2*E*-dodecenal (8.3–17.2%), dode-canal (0.5–14.8%), *n*-decanol (0.5–14.8%), *trans*-2-undecen-1-ol (trace–12.9%), 2*E*-decanal (0–11.3%), 1-eicosanol (0–6.4%), and methyl chavicol (0–6.0%). It was seen clearly that the drying of the samples in the oven at 60°C or microwave caused a significant decrease in the levels of decanal and *n*-decanol percentages. They recommended that for the high-oil yield, freeze-drying was the most suitable drying type. They demonstrated that the drying process could change the oil composition because the chemical profiles of the essential oils from the leaves and stems of coriander are affected by the drying methods. Pirbalouti et al. (2013) showed that the changes such as increase or losses of the compounds occur in the aromatic profile of the essential oil caused by the formation of new compounds by oxidation, glycol-side hydrolysis, esterification, and/or other processes. In other research conducted by Sellami et al. (2011) and Rahimmalek and Goli (2013), the essential oil profile of dried samples was determined to be quite different compared to that of fresh samples. Łyczko et al. (2021) performed detailed research to reveal the effects of different drying processes and their combinations on the volatile profile of *C. sativum* cilantro, a leaf that is the valuable part of coriander. The authors dried the cilantro samples by convective drying, vacuum microwave drying, and a combination of convective pre-drying and vacuum-microwave finish-drying. They reported that applying of various drying methods caused a significant shift in all odor active compound contributions in the cilantro volatile compound profile. Application of drying regardless of the method used increased the contribution of all cilantro odor active compounds except for (Z)-hex-3-en-1-ol. Additionally, drying methods including only vacuum-microwave drying caused a significant increase in α-pinene and terpinolene contribution, which are responsible for moldy, earthy, and mushroom aroma notes of dried products.

Storage is one of the most important postharvest processes applied for plant food. Many factors such as temperature, light, humidity, and air can significantly affect the volatile oils of spice and herbs (Njoroge et al., 1996). Because of these reasons, storage is an important factor in identifying the most suitable way to get a high-quality product. Wahba et al. (2020) performed comparative research to reveal the best storage conditions for *C. sativum* essential oil, whether to store it in the form of essential oils or to keep it inside the dry seeds until extracted by hydrodistillation. They recorded the linalool level as 59.6%, 59.2%, and 47.9% for the time zero treatment, cool conditions treatment, and for the conditions of stored dry seeds at room temperature, respectively. Additionally, the levels of the second major compound, named ethyl hexanoic acid, and sabinene hydrate decreased dramatically from 4.92 and 4.36% to 0.56 and 0.0% after one year of storage at room temperature, respectively. They also reported that the monoterpenes hydrocarbons levels decreased from 8.6 to 4.9%, while oxygenated monoterpenes decreased from 66.7 to 53.7% during storage for one year at room temperature.

In another study, Misharina (2001) studied the effect of storage conditions on the volatile composition of coriander essential oil. In this study, coriander essential oil was stored in daylight and dark conditions, and it was reported that the linalool was the dominant aromatic compound in the samples. For the essential oil stored in dark conditions, the linalool level was 69.2, 68.8, and

68.7% for the stored sample for 1, 6, and 12 months, respectively. It was revealed that there was no significant change in the major compounds of the sample during storage in dark conditions. On the contrary, the linalool levels of this sample were determined as 67.8, 45.1, and 23.5% for the samples stored for 1, 6, and 12 months in daylight conditions, respectively.

35.4 GENETIC FACTORS

The variation in essential oil compositions may occur due to genetic characteristics (Sifola and Barbieri, 2006; Gharib et al., 2008). In this context, molecular characterization with markers plays an essential role in germplasm conservation, identifying genetic relationships among individuals, and coriander breeding. Omidbaigi et al. (2009) have investigated for the first time the molecular and essential oil variation of 20 local species of Iranian coriander in order to contribute to the development of germplasm management and breeding programs. It was determined that the genetic similarity calculated from the data of 15 selected RAPD primers ranged from 0.294 to 0.740, with a mean of 0.541. According to the results obtained, they reported the suitability of RAPD markers to identify local coriander species. In line with the genetic similarity data, it was observed that there was a high level of variation in essential oil yield among local species for both grass and seed parts.

Developing highly efficient new-generation sequencing technologies makes it easier to access genetic information for breeding programs. At the same time, transcriptome analysis is an essential technology for identifying unknown genes and determining the functionality of existing genes. For example, coriander mericarps produce C10 terpenoid, a rich essential oil. To investigate essential oil metabolism, the transcriptome of coriander mericarps was sequenced at three growth stages (early, middle, late), and a transcript library was constructed. The resulting transcriptome sequences were validated using two terpene synthase genes named CsγTRPS and CsLINS, which encode 558 and 562 amino acid proteins. It was reported that by developing a transcript library for coriander, two mTPS genes that make up most of the monoterpene content of total essential oils would be identified, facilitating future studies on essential oil production in coriander (Galata et al., 2014).

Song et al. (2020) sequenced and assembled the complete genome of the coriander using Pacific Biosciences, Illumina, 10X Genomics, and HiC next-generation sequence technologies. They determined that the sequenced genome was 2118.3 Mb, the length of the contig N50 was determined to be 604.13 kb, and scaffold N50 was obtained as 160.99 Mb. The sequenced genome was reported to the Coriander Genomics Database (CGDB, http://http://cgdb.bio2db.com) as a reference genome.

A study related to gene expression and annotation of coriander determined that 1249 specific gene families control the taste and aroma of coriander. Furthermore, the expression, metabolomics, and genomic analyses of the terpene synthase (TPS) gene family in the terpenoid biosynthesis pathway were essential for expanding the coriander TPS gene family. The gene, which was highly expressed in all four tissues and three developmental stages, encoded linalool synthase and myrcene synthase. In addition, it was concluded that two tetraploid events likely provided a high level of crossbreeding and enhanced the rapid divergence of ancestral plants, eventually producing the Apiaceae family (Song et al., 2020).

Plant tissue culture methods are also one of the methods used to improve many herbal compounds, including essential oils. Ali et al. (2019) investigated the effect of methyl jasmonate (MeJA) on essential oil yield in various coriander tissues cultured *in vitro*. It was determined that there were significant changes in the essential oil level due to the stress state caused by MeJA in the tissues. In the study, five different MeJA levels (50, 100, 150, 200 μM) and a control were used, and the essential oil yield of coriander in different tissues cultured *in vitro* was estimated using GC-MS. It was determined that MeJA positively increased the synthesis of essential oil in coriander tissues grown *in vitro*. Furthermore, the essential oil began to accumulate with the differentiation of the callus tissue, and the yield was highest in the maturation stage of the somatic embryo. Moreover, modification of MeJA as an elicitor was reported to significantly increase essential oil yield at growth stages of somatic embryos.

35.5 CONCLUSION

Many studies have been conducted regarding the compositional change of coriander essential oil in recent years. Different parameters could be the reason for these changes, such as agricultural practices (fertilizer, irrigation, and harvest time), stress conditions, post-harvest practices, and genetic-environmental factors. Obtaining standard essential oil composition is very difficult because of these parameters and their interaction. The essential oil content of the coriander plant may vary with the nutrient level of the plant. The nutrient content of the coriander plant varies depending on agricultural practices, climatic conditions, soil characteristics, and the genetic characteristics of the plant. However, if the amount of plant nutrients increases or decreases, it should be considered that the plant essential oil components change. Therefore, it can be concluded that more study is crucial in this area.

REFERENCES

Ali, M., Mujib, A., Gulzar, B., & Zafar, N. (2019). Essential oil yield estimation by gas chromatography-mass spectrometry (GC-MS) after methyl jasmonate (MeJA) elicitation in *in vitro* cultivated tissues of *Coriandrum sativum* L. *3 Biotech, 9*(11), 1–16.

Amir, G., Hamid, D., & Mostafa, K. (2019). The effect of different levels of drought stress on some morphological, physiological and phytochemical characteristics of different endemic coriander (*Coriandrum sativum* L.) genotypes. *Environmental Stresses in Crop Sciences, 12*(2), 459–470.

Amiripour, A., Jahromi, M. G., Soori, M. K., & Mohammadi Torkashvand, A. (2021). Changes in essential oil composition and fatty acid profile of coriander (*Coriandrum sativum* L.) leaves under salinity and foliar-applied silicon. *Industrial Crops and Products, 168*, 113599.

Arganosa, G. C., Sosulski, F. W., & Slikard, A. E. (1998). Seed yields and essential oil of northern-grown coriander (*Coriandrum sativum* L.). *Journal of Herbs, Spices & Medicinal Plants, 6*(2), 23–32.

Arif, M., Khurshid, H., & Khan, S. (2014). Genetic structure and green leaf performance evaluation of geographically diverse population of coriander (*Coriandrum sativum* L.). *European Academic Research, 2*(3), 3269–3285.

Ayanoğlu, F., Mert, A., Aslan, N., & Gürbüz, B. (2002). Seed yields, yield components and essential oil of selected coriander (*Coriandrum sativum* L.) lines. *Journal of Herbs, Spices & Medicinal Plants, 9*(2–3), 71–76.

Beyzi, E., Karaman, K., Gunes, A., & Beyzi, S. B. (2017). Change in some biochemical and bioactive properties and essential oil composition of coriander seed (*Coriandrum sativum* L.) varieties from Turkey. *Industrial Crops and Products, 109*, 74–78.

Carrubba, A., la Torre, R., Prima, A. D., Saiano, F., & Alonzo, G. (2002). Statistical analyses on the essential oil of Italian coriander (*Coriandrum sativum* L.) fruits of different ages and origins. *Journal of Essential Oil Research, 14*(6), 389–396.

Chrysargyris, A., Drouza, C., & Tzortzakis, N. (2017a). Optimization of potassium fertilization/nutrition for growth, physiological development, essential oil composition and antioxidant activity of *Lavandula angustifolia* Mill. *Journal of Soil Science and Plant Nutrition, 17*(2), 291–306.

Chrysargyris, A., Xylia, P., Botsaris, G., & Tzortzakis, N. (2017b). Antioxidant and antibacterial activities, mineral and essential oil composition of spearmint (*Mentha spicata* L.) affected by the potassium levels. *Industrial Crops and Products, 103*, 202–212.

Delibaltova, V. (2020). Effect of sowing period on seed yield and essential oil composition of coriander (*Coriandrum sativum* L.) in south-east Bulgaria condition. *Series A. Agronomy, LXIII*(1), 233–240.

Diederichsen, A. (1996). Promoting the conservation and use of under utilized and neglected crops 3. Coriander. *Institute of Plant Genetics and Crop Plant Research, 3*, 45–48.

Dubey, V. S., Bhalla, R., & Luthra, R. (2003). An overview of the non-mevalonate pathway for terpenoid biosynthesis in plants. *Journal Bioscience, 28*, 637–646.

El Gendy, A. G., El Gohary, A. E., Omer, E. A., Hendawy, S. F., Hussein, M. S, Petrova, V., & Stancheva, I. (2015). Effect of nitrogen and potassium fertilizer on herbage and oil yield of chervil plant (*Anthriscus cerefolium* L.). *Industrial Crop and Products, 69*, 167–174.

El-Zaeddi, H., Calín-Sánchez, Á., Noguera-Artiaga, L., Martínez-Tomé, J., & Carbonell-Barrachina, Á. A. (2020). Optimization of harvest date according to the volatile composition of Mediterranean aromatic herbs at different vegetative stages. *Scientia Horticulturae, 267*, 109336.

Fattahi, B., Arzani, K., Souri, M. K., & Barzegar, M. (2021). Morphophysiological and phytochemical responses to cadmium and lead stress in coriander (*Coriandrum sativum* L.). *Industrial Crops and Products, 171,* 113979.

Figueiredo, A. C., Barroso, J. G., Pedro, L. G., & Scheffer, J. J. (2008). Factors affecting secondary metabolite production in plants: Volatile components and essential oils. *Flavour and Fragrance Journal, 23*(4), 213–226.

Freitas, M. S. M., Gonçalves, Y. S., Lima, T. C., dos Santos, P. C., Peçanha, D. A., Vieira, M. E., Carvalho, A. J. C., & Vieira, I. J. C. (2020). Potassium sources and doses in coriander fruit production and essential oil content. *Horticultura Brasileira, 38,* 268–273.

Galata, M., Sarker, L. S., & Mahmoud, S. S. (2014). Transcriptome profiling, and cloning and characterization of the main monoterpene synthases of *Coriandrum sativum* L. *Phytochemistry, 102,* 64–73.

Gębarowska, E., Pytlarz-Kozicka, M., Nöfer, J., Łyczko, J., Adamski, M., & Szumny, A. (2019). The effect of *Trichoderma* spp. on the composition of volatile secondary metabolites and biometric parameters of coriander (*Coriandrum sativum* L.). *Journal of Food Quality, 2019,* 5687032.

Ghamarnia, H., & Daichin, S. (2013). Effect of different water stress regimes on different Coriander (*Coriander sativum* L.) parameters in a semi-arid climate. *International Journal of Agronomy and Plant Production, 4*(4), 822–832.

Gharib, F. A., Moussa, L. A., & Massoud, O. N. (2008). Effect of compost and bio-fertilizers on growth, yield and essential oil of sweet marjoram (*Majorana hortensis*) plant. *International Journal of Agriculture and Biology, 10*(4), 381–387.

Gil, A., De La Fuente, E. B., Lenardis, A. E., López Pereira, M., Suárez, S. A., Bandoni, A., ... & Ghersa, C. M. (2002). Coriander essential oil composition from two genotypes grown in different environmental conditions. *Journal of Agricultural and Food Chemistry, 50*(10), 2870–2877.

Hani, M. M., Hussein, S. H. A., Mursy, M. H., Ngezimana, W., & Mudau, F. N. (2015). Yield and essential oil response in coriander to water stress and phosphorus fertilizer application. *Journal of Essential Oil Bearing Plants, 18*(1), 82–92.

Harten, A. M. Van. (1974). Coriander: The history of an old crop [in Dutch]. *Landbouwkd. Tijdschr, 86,* 58–64.

İzgi, N. M. (2020). The effect of different phosphorus doses on agronomic and quality characteristics of coriander (*Coriandrum sativum* L.). *Applied Ecology and Environmental Research, 18*(6), 8205–8216.

Katar, D., & Katar, N. (2016). Determination of the effect of plant density on yield and yield components for two different coriander cultivars (*Coriandrum sativum* L.). *International Journal of Agricultural and Wildlife Sciences, 2*(1), 33–42.

Kızıl, S., & İpek, A. (2004). The effects of different row spacing on yield, yield components and essential oil content of some coriander (*Coriandrum sativum* L.) lines. *Journal of Agricultural Sciences, 10*(3), 237–244.

Kiralan, M., Calikoglu, E., Ipek, A., Bayrak, A., & Gurbuz, B. (2009). Fatty acid and volatile oil composition of different coriander (*Coriandrum sativum*) registered varieties cultivated in Turkey. *Chemistry of Natural Compounds, 45*(1), 100–102.

Lo Cantore, P., Iacobellis, N. S., De Marco, A., Capasso, F., & Senatore, F. (2004). Antibacterial activity of *Coriandrum sativum* L. and *Foeniculum vulgare* Miller var. vulgare (Miller) essential oils. *Journal of Agricultural and Food Chemistry, 52*(26), 7862–7866.

Lubbe, A., & Verpoorte, R. (2011). Cultivation of medicinal and aromatic plants for specialty industrial materials. *Industrial Crops and Products, 34,* 785–801.

Łyczko, J., Masztalerz, K., Lipan, L., Iwiński, H., Lech, K., Carbonell-Barrachina, Á. A., & Szumny, A. (2021). *Coriandrum sativum* L.-Effect of multiple drying techniques on volatile and sensory profile. *Foods, 10*(2), 403.

Misharina, T. A. (2001). Influence of the duration and conditions of storage on the composition of the essential oil from coriander seeds. *Applied Biochemistry and Microbiology, 37*(6), 622–628.

Msaada, K., Hosni, K., Taarit, M. B., Ouchikh, O., & Marzouk, B. (2009). Variations in essential oil composition during maturation of coriander (*Coriandrum sativum* L.) fruits. *Journal of Food Biochemistry, 33*(5), 603–612.

Nadeem, M., Anjum, F. M., Khan, M. I., Tehseen, S., El-Ghorab, A., & Sultan, J. I. (2013). Nutritional and medicinal aspects of coriander (*Coriandrum sativum* L.): A review. *British Food Journal.* 115(5), 743 755.

Neffati, M., & Marzouk, B. (2010). Salinity impact on growth, essential oil content and composition of coriander (*Coriandrum sativum* L.) stems and leaves. *Journal of Essential Oil Research, 22*(1), 29–34.

Neffati, M., & Marzouk, B. (2008). Changes in essential oil and fatty acid composition in coriander (*Coriandrum sativum* L.) leaves under saline conditions. *Industrial Crops and Products, 288*(2), 137–142.

Njoroge, S. M., Ukeda, H., & Sawamura, M. (1996). Changes in the volatile composition of yuzu (Citrus junos Tanaka) cold-pressed oil during storage. *Journal of Agricultural and Food Chemistry, 44*(2), 550–556.

Nowak, J., & Szempliëski, W. (2014). Influence of sowing date on yield and fruit quality of coriander (*Coriandrum sativum* L.). *Acta Scientiarum Polonorum Hortorum Cultus*, *13*(2), 83–96.

Omidbaigi, R., Rahimi, S., & Naghavi, M. R. (2009). Evaluation of molecular and essential oil diversity of coriander (*Coriandrum sativum* L.) landraces from Iran. *Journal of Essential Oil Bearing Plants*, *12*(1), 46–54.

Özyazıcı, G. (2020). Effect of zinc doses on yield and quality in coriander (*Coriandrum sativum* L.) plant. *ISPEC Journal of Agricultural Sciences*, 4(3), 550–564.

Pirbalouti, A. G., Oraie, M., Pouriamehr, M., & Babadi, E. S. (2013). Effects of drying methods on qualitative and quantitative of the essential oil of Bakhtiari savory (*Satureja bachtiarica* Bunge.). *Industrial Crops and Products*, *46*, 324–327.

Pirbalouti, A. G., Salehi, S., & Craker, L. (2017). Effect of drying methods on qualitative and quantitative properties of essential oil from the aerial parts of coriander. *Journal of Applied Research on Medicinal and Aromatic Plants*, *4*, 35–40.

Rahimmalek, M., & Goli, S. A. H. (2013). Evaluation of six drying treatments with respect to essential oil yield, composition and color characteristics of *Thymys daenensis* subsp. daenensis. Celak leaves. *Industrial Crops and Products*, *42*, 613–619.

Rajesh, K., Singh, M. K., Kumar, V., Verma, R. K., Kushwah, J. K., & Pal, M. (2015). Effect of nutrient supplementation through organic sources on growth, yield and quality of coriander (*Coriandrum sativum* L.). *Indian Journal of Agricultural Research*, *49*(3), 278–281.

Ramezani, S., Rasouli, F., & Solaimani, B. (2009). Changes in essential oil content of coriander (*Coriandrum sativum* L.) aerial parts during four phonological stages in Iran. *Journal of Essential Oil Bearing Plants*, *12*(6), 683–689.

Sangwan, N. S., Farooqi, A. H. A., Shabih, F., & Sangwan, R. S. (2001). Regulation of essential oil production in plants. *Plant Growth Regulation*, 34, 3–21.

Sellami, I. H., Wannes, W. A., Bettaieb, I., Berrima, S., Chahed, T., Marzouk, B., & Limam, F. (2011). Qualitative and quantitative changes in the essential oil of *Laurus nobilis* L. leaves as affected by different drying methods. *Food Chemistry*, *126*(2), 691–697.

Sifola, M. I., & Barbieri, G. (2006). Growth, yield and essential oil content of three cultivars of basil grown under different levels of nitrogen in the field. *Scientia Horticulturae*, *108*(4), 408–413.

Singh, P., Mor, V. S., Kumar, S., & Bhuker, A. (2017). Correlation and regression analysis of viability and vigour parameters in coriander (*Coriandrum sativumL.*). *International Journal of Plant and Soil Science*, *20*(2), 1–8.

Singh, M. (2012). Effect of vermicompost and chemical fertilizers on growth, yield and quality of coriander (*Coriandrum sativum* L.) in a semi-arid tropical climate. *Journal of Spices and Aromatic Crops*, *20*(1).

Song, X., Wang, J., Li, N., Yu, J., Meng, F., Wei, C., & Wang, X. (2020). Deciphering the high-quality genome sequence of coriander that causes controversial feelings. *Plant Biotechnology Journal*, *18*(6), 1444–1456.

Telci, I., Demirtas, I., Bayram, E., Arabaci, O., & Kacar, O. (2010). Environmental variation on aroma components of pulegone/piperitone rich spearmint (*Mentha spicata* L.). *Industrial Crops and Products*, *32*(3), 588–592.

Telci, İ., Toncer, Ö. G., & Sahbaz, N. (2006). Yield, essential oil content and composition of *Coriandrum sativum* varieties (var. vulgare Alef and var. microcarpum DC.) grown in two different locations. *Journal of Essential Oil Research*, *18*(2), 189–193.

Teneva, D., Denkova, Z., Goranov, B., Denkova, R., Kostov, G., Atanasova, T., & Merdzhanov, P. (2016). Chemical composition and antimicrobial activity of essential oils from black pepper, cumin, coriander and cardamom against some pathogenic microorganisms. *Acta Universitatis Cibiniensis. Series E: Food Technology*, *20*(2), 39–52.

Unlukara, A., Beyzi, E., Arif, I. P. E. K., & Gurbuz, B. (2016). Effects of different water applications on yield and oil contents of autumn sown coriander (*Coriandrum sativum* L.). *Turkish Journal of Field Crops*, *21*(2), 200–209.

Wahba, H. E., Abd Rabbu, H. S., & Ibrahim, M. E. (2020). Evaluation of essential oil isolated from dry coriander seeds and recycling of the plant waste under different storage conditions. *Bulletin of the National Research Centre*, *44*(1), 1–7.

Weisany, W., Tahir, N. A. R., & Schenk, P. M. (2021). Coriander/soybean intercropping and mycorrhizae application lead to overyielding and changes in essential oil profiles. *European Journal of Agronomy*, *126*, 126283.

Zheljazkov, V. D., Pickett, K. M., Caldwell, C. D., Pincock, J. A., Roberts, J. C., & Mapplebeck, L. (2008). Cultivar and sowing date effects on seed yield and oil composition of coriander in Atlantic Canada. *Industrial Crops and Products*, *28*(1), 88–94.

36 Antimicrobial Activity of Coriander and Its Application in the Food Industry

Miroslava Kačániová

CONTENTS

36.1 INTRODUCTION

Currently, the achievement of food security is based on food access, food stability, food utilization, and, most importantly, its preservation to avoid further contamination. These four food pillars constitute socioeconomic background and influence affordable food requirement. Regarding food insecurity, microbes and their associated toxins are prime driving factors for major food spoilage and biodeterioration due to their long-term impact along the food chain and food web. In several developing countries, 25–30% losses of food have been reported due to microbial contamination (Bondi et al., 2017). Microbial contamination of foods in different stages of production and processing causes different foodborne diseases. In addition to microbial contamination, some bacteria and fungi produce toxins. Many of these microbial toxins are thermostable and are not destroyed by high temperatures during cooking or food processing (Rajkovic, 2014). Bacteria produced two different types of toxins, *viz.* endotoxins and exotoxins. Exotoxins are proteinaceous substances secreted by *Clostridium botulinum*, *Staphylococcus aureus*, *Bacillus cereus*, and *Clostridium perfringens* (Josić et al., 2017; Rajkovic et al., 2020), whereas endotoxins are lipopolysaccharide (LPS) components and more powerful as well as specific to their target site (Rešetar et al., 2015). In addition to food pathogenic bacteria, different fungal species also play an active role in food deterioration by sporulation and the production of mycotoxins. Mycotoxins are low molecular weight secondary metabolites mainly synthesized by different fungal genera such as *Penicillium*, *Aspergillus*, and *Fusarium*, contaminating several stored food items with variable toxic effects, *viz.* carcinogenicity, teratogenicity, neurotoxicity, and hepatotoxicity. More than 400 different food-contaminating mycotoxins have been identified and characterized. Among them, aflatoxins, fumonisins, ochratoxins, zearalenone, and trichothecenes exhibit widespread occurrence based on substrate type, relative humidity, moisture content, and water activity of the substrate causing maximum health-related problems; these are a key factor in the prime destruction of the worldwide agricultural economy (Reddy et al., 2010).

The International Agency for Research on Cancer (2012) has classified aflatoxin B_1, aflatoxin B_2, aflatoxin G_1, aflatoxin G_2, and aflatoxin M_1 as class 1 human carcinogens (mycotoxins that are entirely carcinogenic for humans). In addition, ochratoxin A, fumonisin B_1, fumonisin B_2,

DOI: 10.1201/9781003204626-40

sterigmatocystin, and fusarin C have been categorized as class 2B human carcinogens (mycotoxins that are possibly carcinogenic to humans). In contrast, deoxynivalenol, zearalenone, patulin, and citrinin have been grouped as class 3 human carcinogens (mycotoxins are not classifiable based on their carcinogenicity).

36.2 ANTIMICROBIAL ACTIVITY

Antimicrobial susceptibility tests can be classified as diffusion, dilution, or bioautographic methods (Burt, 2004). The CLSI methods for antibacterial susceptibility testing have been adopted by researchers, with slight modifications, to test the antimicrobial and antifungal activity of essential oils (Hammer et al., 1999; Duarte et al., 2012; Soares et al., 2012). Due to the absence of a standardized protocol to test essential oil antimicrobial activity, the comparison between results from published works is complicated since the outcome of the test can be affected by numerous factors (Burt, 2004). When evaluating essential oil activity, the term minimum inhibitory concentration (MIC) is cited by most researchers as a measure of the antimicrobial performance of the oil (Delaquis et al., 2002; Duarte et al., 2012; Silva et al., 2011a; Silva et al., 2011b, Silva et al., 2017). Associated with MIC value determination, the authors sometimes refer to the minimum bactericidal or lethal concentrations (MBC or MLC) (Silva et al., 2011a; Silva et al., 2011b). So far, there is no consensus regarding the acceptable inhibition level for natural compounds when compared to standard antimicrobials. For instance, Duarte and co-authors proposed several degrees of bioactivity for natural compounds based on the MIC values: strong activity (up to 0.5 mg/mL), moderate activity (0.51–1.0 mg/mL), and weak activity (above 1.1 mg/mL) (Duarte et al., 2005). Coriander oil has shown broad antimicrobial activity against bacteria and fungi using diverse antimicrobial activity assays such as agar dilution, well or disc diffusion, and broth macro or microdilution that will be summarized in the following sections.

The antibacterial activity of *Coriandrum sativum* essential oil proved effective against a wide range of foodborne and clinically relevant Gram-positive and Gram-negative bacteria using several assays such as disk diffusion, agar, and broth dilution. Using disk diffusion assays, the amount of coriander oil loaded onto the paper disks able to cause bacterial growth inhibition ranged from 24 mg to 0.09 μg. Many bacterial species were inhibited in these studies, such as Shiga and non-Shiga toxin-producing *E. coli*, several *Pseudomonas* species, *B. gladioli*, *X. campestris*, *B. megaterium*, and *Y. enterocolitica*, *L. monocytogenes*, *S. typhimurium*, methicillin-sensitive *S. aureus*, and *P. mirabilis*. The results also showed that coriander oil seemed more active against Gram-positive than Gram-negative bacteria. For instance, in the work of Lo Cantore and collaborators, although coriander oil could inhibit the bacterial growth of Gram-positive and Gram-negative bacteria, higher activity was verified against Gram-positive bacteria and some species of the Gram-negative bacterium *Xanthomonas* and lower activities were obtained for *Pseudomonas* spp. (Lo Cantore et al., 2004). Contradicting these results, another study showed that coriander oil was effective against Gram-negative bacteria but had no effect on the Gram-positive bacterium *L. plantarum* (Elgayyar et al., 2001). Using both agar and broth dilution methods, it is possible to assess the level of bacterial susceptibility by determining MIC values and even MBC values using broth dilution methods. Due to the limited aqueous solubility of coriander oil, as a result of the high percentage of hydrophobic constituents, the oil is commonly solubilized in a culture medium containing a small percentage of DMSO, Tween 20, or Tween 80 (Hammer et al., 1999; Rattanachaikunsopon and Phumkhachorn, 2010; Silva et al., 2011b). Overall, the results obtained by these two methods are very similar, with MIC values ranging from 0.03 to 1.6% (v/v). The results obtained by these assays corroborated some of the results obtained with disk diffusion assays, showing coriander oil effectiveness against, for example, *C. jejuni*, methillin-sensitive and methicillin-resistant *S. aureus*, vancomycin-resistant *Enterococcus*, *K. pneumoniae*, non-Shiga and Shiga toxin-producing *E. coli*, and *A. baumannii*, among other pathogens. Some of these studies proved that Gram-positive bacteria were more susceptible to coriander oil. Delaquis and co-authors demonstrated that coriander oil was effective against *L. monocytogenes* and *S. aureus* but had no effect on *P. fragi* or *S. typhimurium* (Delaquis et al., 2002).

Other studies revealed a high activity of coriander oil against Gram-negative bacteria. For example, it was found that the Gram-negative bacterium *C. jejuni* was very susceptible to coriander oil (MIC values of 0.03–0.06%) (Rattanachaikunsopon and Phumkhachorn, 2010) and, also, that coriander oil was able to have a bactericidal effect on Gram-negative bacteria and not on Gram-positive bacteria, such as *B. cereus* and *E. faecalis* (Silva et al., 2011b). It is widely accepted that the antimicrobial activity of essential oils depends on major constituents and their concentrations. The inhibitory effects of essential oils are mainly due to their major components, but the minor components might also contribute to the antimicrobial activity. It has been shown that different compound classes have different antimicrobial activities: phenolic compounds possess the highest antimicrobial activities, followed by alcohols, aldehydes, ketones, ethers, and hydrocarbons (Ferdes and Ungureanu, 2012). The hydroxyl (–OH) group on phenolic compounds is thought to be responsible for these compounds' antimicrobial activity (Ferdes and Ungureanu, 2012). Some coriander oil constituents such as linalool, α-pinene, *p*-cymene, γ-terpinene, limonene, and linalyl acetate proved to be effective against several Gram-positive and Gram-negative bacteria (Cristani et al., 2007; Di Pasqua et al., 2006; Ozek et al., 2010; Sonboli et al., 2006; Trombetta et al., 2005). However, the susceptibility levels obtained for these compounds seemed to be slightly higher than those obtained for the essential oil, meaning that the oil's antimicrobial activity is more than a "sum of its parts" probably due to complex interactions between all individual components.

36.3 ANTIFUNGAL ACTIVITY

Some studies showed that coriander essential oil had antifungal activity against several types of fungi, such as yeasts, dermatophytes, and filamentous fungi, using agar and broth dilution as well as disk diffusion assays. Although some of the studies have focused on the oil's antifungal activity against *Candida* species (Furletti et al., 2011; Hammer et al., 1999; Silva et al., 2011a), probably due to its clinical relevance, the oil also inhibited the growth of other fungi such as *M. canis*, *A. niger*, *S. cerevisiae*, *G. candidum*, and *K. fragilis* (Delaquis et al., 2002; Elgayyar et al., 2001; Soares et al., 2012; Toroglu, 2011). Coriander oil was able to inhibit *Candida* spp. growth at concentrations ranging from 0.0008% to 0.4%, *Sacharomyces cerevisiae* at a concentration of 0.13%, *M. canis* at concentrations of 78–620 µg/mL (0.009–0.07%), and the other fungi were inhibited at 1.74 or 24 mg of coriander oil per disk. Some of these studies also evaluated the fungicidal activity of coriander oil, yielding MLC values ranging from 0.017 to 0.4% (Silva et al., 2011a; Soares et al., 2012).

36.4 OTHER ANTIMICROBIAL ACTIVITIES

Many years of increasing applications of antimicrobial agents have created a situation leading to the rise in the number of multiple-antibiotic-resistant bacteria and fungi (Teuber, 1999). To fight these new "superbugs", researchers have focused their attention on the possible combination of natural compounds with common antimicrobials, expecting that the natural compound could potentiate the antibiotic activity or even reverse the resistance to the antibiotic. Some researchers have evaluated coriander oil's potential synergism or additive effect against common antibiotics and antifungal drugs. Despite some results stating that coriander oil had an antagonistic effect with some antibiotics against both Gram-positive and Gram-negative bacteria (Toroglu, 2011), others revealed that coriander oil had synergetic effects with gentamicin and ceftriaxone for *S. aureus*, with ceftriaxone for *M. smegmatis* (Toroglu, 2011), and with chloramphenicol, ciprofloxacin, gentamicin, and tetracycline against *A. baumannii* (Duarte et al., 2012). The oil's combination with piperacillin and cefoperazone or gentamicin yielded only an additive effect against *A. baumannii* (Duarte et al., 2012) or *S. aureus* (Toroglu, 2011), respectively. It is known that the organization of microbial cells in biofilms increases their resistance to antimicrobials and disinfectants (Duarte et al., 2013), probably as a consequence of the delay in the antimicrobial's diffusion through the biofilms matrix, of the altered growth rate of biofilms cells, or other physiological modifications in cells while growing in a biofilm (Donlan and Costerton, 2002). Most studies on the antimicrobial activity of coriander oil have focused on planktonic cells; notwithstanding, the ability of the microorganism to form biofilms points to the need to evaluate the oil's anti-biofilm activity. A study evaluating the action of coriander oil on biofilm formation by *Candida albicans* showed an apparent effect of the oil on the formation of biofilms, characterized by an increased lag phase and a decrease in biofilm growth at a concentration of 0.125 mg/mL (Furletti et al., 2011). Another study evaluated coriander oil's ability to inhibit the formation or eradication of *Acinetobacter baumannii* biofilms (Duarte et al., 2013). The authors concluded that the oil could inhibit biofilm formation at concentrations ranging from two to four times the MIC value (0.2–1.6%), causing an 85% reduction in total biofilm biomass and metabolic activity. When applied to preformed biofilms, the same oil concentrations could eradicate preformed biofilms, leading to a 75–90% reduction in total biomass and metabolic activity after 24 h of incubation (Duarte et al., 2013). These results encourage the use of coriander oil as an anti-biofilm compound since, typically, antibiotics are about 1000 times less effective against cells in biofilms than planktonic cells (Melchior et al., 2006). Another essential feature of essential oils is their volatility, but there is a general lack of scientific information about the effectiveness of essential oils in the vapor phase. The evaluation of the antimicrobial activity of essential oils in the vapor phase has been gaining interest in recent years. Studies by Lopez and co-authors revealed that, in general, essential oils and some oils constituents are less effective in the vapor phase; for example, camphor, 1,8-cineole, *p*-cymene, and limonene were not effective against common food-borne pathogens, while linalool showed slight growth inhibition of *S. choleraesuis* and *C. albicans* (Lopez et al., 2005; Lopez et al., 2007b). This is because lipophilic molecules in the aqueous phase associate to form micelles, suppressing the attachment of essential oils' components to the microorganisms, whereas the vapor phase allows free attachment, thus increasing the oils' effectiveness (Martinez-Abad et al., 2013).

It is well known that the antimicrobial efficacy of essential oils is greatly reduced in food systems compared to *in vitro* work, as the presence of fats, carbohydrates, proteins, salts, and pH strongly influences the activity of the oils (Burt, 2004). Also, in a complex biological system such as food, various interactions may occur between the additives used, food constituents, and the food matrix (Michalczyk et al., 2012). Additionally, food is generally colonized by many different species of bacteria, and their interactions should be considered since they can be mutually antagonistic or even synergetic (Gram et al., 2002). Therefore, the disturbance of this equilibrium by adding a food preservative must be taken into account. As a result, larger amounts of essential oils are required to be added to food systems to attain the same antimicrobial properties, thus causing major shifts in the organoleptic properties of the food item (Busatta et al., 2008), exceeding the sensorially acceptable level (Michalczyk et al., 2012). The direct incorporation of essential oils in foods was one of the early approaches for food preservation using these compounds. Several essential oils, such as the ones obtained from sage, oregano, marjoram, and thyme, were incorporated in meat and fish products to improve these foods' shelf-life (Lucera et al., 2012). These studies proved that the addition of essential oils to foods, alone or in combination with modified atmosphere packaging (MAP) (Sellamuthu et al., 2013), was able to improve the sensory shelf-life of the products due to a reduction in microbial growth, thus exerting a bacteriostatic effect (Lucera et al., 2012). With the breakthroughs in the packaging industry, with the development of active and intelligent packaging systems, it was thought that the incorporation of the essential oil in plastic or biopolymer (polysaccharide-based, protein-based, lipid-based, or composites) film could overcome the addition of such high amounts of essential oils to foods, as a result of the gradual release of the oil from the film onto the food surface (Cran et al., 2010; Zivanovic et al., 2005). As a result of their properties such as their low cost, good processability, and mechanical and physical features (Kuorwel et al., 2011), plastic films have been described for the development of antimicrobial films with essential oils, proving to be effective in the improvement of the shelf-life of several foods such as meat and salads (Muriel-Galet et al., 2013). Nowadays, biopolymer-based films are preferred over plastic films since they can be edible and biodegradable, and some have intrinsic antimicrobial properties, as is verified for chitosan films (Cha and Chinnan, 2004; Zivanovic et al., 2005). Another relevant type of film is the type involved in paper packaging since many foods and drink packages are made of several papers and boards (Rodriguez et al., 2007). Studies have successfully incorporated clove, cinnamon, and oregano essential oils in paper packaging materials to inhibit fungal growth in fresh fruits (Rodriguez-Lafuente et al., 2010; Rodriguez et al., 2007). The data obtained proved that paper packages containing cinnamon oil were very effective against several fungi such as *C. albicans*, *A. flavus*, *Rhizopus stolonifer*, and *Alternaria alternata* (Rodriguez-Lafuente et al., 2010; Rodriguez et al., 2007; Rodriguez et al., 2008), showing bacteriostatic action as no complete growth inhibition was achieved. The use of films based on antimicrobial polymers can further reduce the amount of essential oil incorporated while still being effective as a preservation method (Wang et al., 2011). One of these strategies is based on the synergistic effect observed between some films, such as chitosan films and essential oils (Khanjari et al., 2013), and between two or more combined essential oils (Goni et al., 2009), allowing a reduction in the amount of essential oil incorporated (Wang et al., 2011). Additionally, the oil quantity incorporated in the packaging material can be reduced if we consider the vapor-phase activity of the oil, a parameter greatly influenced by the oil's polarity (Licciardello et al., 2013), since the MIC concentrations in the vapor phase can be 30–100 times lower than in the liquid phase (Martinez-Abad et al., 2013). Due to the vast array of film materials available for the development of antimicrobial packages, when designing new active packaging, one should investigate the kinetics of the oil release from the film, as the antimicrobial activity of the film might depend on its ability to release the oil (Gutierrez et al., 2010; Mercier et al., 2002). Overall, when analyzing the data available, the antimicrobial activity of active films containing essential oils seems to be similar to that obtained *in vitro*: antimicrobial films containing essential oils are, in general, very effective against yeasts and molds and more active against Gram-positive bacteria (Ghasemlou et al., 2013; Lopez et al., 2007a; Martinez-Abad et al., 2013) than Gram-negative bacteria, with *P. aeruginosa* being one of the most resistant bacteria to essential oils (Lopez et al., 2007a).

36.5 CORIANDER OIL IN FOOD PRESERVATION

Coriander oil is extensively used as a food additive or adjuvant in all sorts of foods, such as alcoholic beverages, tobacco, candy, pickles, dairy products, chewing gum, meat sausage, and pickles, with use levels ranging from 0.1 to 100 ppm. The *in vitro* effectiveness of coriander oil against numerous foodborne pathogens such as *S. aureus*, *C. jejuni*, Shiga and non-Shiga toxin-producing *E. coli*, *L. monocytogenes*, *Y. enterocolitica*, and *S. thypimurium*, conducted to the exploitation of the oil and major constituent, linalool. The efficacy of coriander oil and linalool in inhibiting microbial growth in foods was tested through direct addition and incorporation into films. The direct addition of coriander oil (0.02% v/w) to minced beef caused a reduction in Enterobacteriaceae counts and was able to inhibit undesirable sensory changes due to meat spoilage, although myoglobin oxidation was not prevented (Michalczyk et al., 2012). In another study, the direct addition of coriander oil to ground chicken meat and beef (0.5% v/w) resulted in complete *C. jejuni* cell death after 30 min of contact time. In comparison, 2- and 4-log reductions in bacterial loads were obtained for lower oil concentrations (0.1 and 0.25%) (Rattanachaikunsopon and Phumkhachorn, 2010). Although these two strategies were successful, when coriander oil was incorporated into a chitosan film, its microbial efficacy against *L. monocytogenes* and Shiga-toxin producing *E. coli* was limited and lower than that obtained with other essential oils, as the inhibitory effects of essential oil incorporated into the chitosan film were lower than the ones of the pure oil (Zivanovic et al., 2005). So far, the only strategy described for using linalool in food preservation was based on the development of low-density polystyrene (LPDE) films containing this compound to reduce microbial contamination in cheddar cheese (Suppakul et al., 2008). It was observed that linalool (0.34% w/w) LPDE films could cause a significant reduction in total aerobic bacteria, thus reducing natural microbial contamination in cheese samples (Suppakul et al., 2008). The films were also able to reduce microbial counts in cheese artificially contaminated with *L. innocua* or *E. coli*. Even after long-term storage (1 year), these films inhibited *E. coli* growth (Suppakul et al., 2011). Furthermore, sensory analysis of these films revealed that linalool, at the percentage used, may not present a problem in altering the sensory properties of the food item (Suppakul et al., 2008). Although food deterioration can be caused by microbiological contamination, a significant part of this process results from chemical alterations in the food product. One of the most relevant chemical processes is oxidative deterioration by the degradation of fats and pigments. Therefore, the antioxidant activity of essential oils, together with their antimicrobial activity, plays a key role in preventing food spoilage and improving the shelf-life of foods (Bentayeb et al., 2007; Lopez de Dicastillo et al., 2011; Nerin et al., 2006). Due to its described antioxidant activity, coriander oil was tested as an antioxidant in food products to increase their oxidative stability. Coriander oil was effectively used as an antioxidant for the preservation of Italian salami, being able to reduce lipid oxidation to a different extent than a synthetic antioxidant, hence improving the shelf-life of the product with no significant alterations in the sensory profile (Marangoni and de Moura, 2011a; Marangoni and de Moura, 2011b). In another study, coriander oil was used to prevent the deterioration of ghee: although its antioxidant activity was not so effective as one of the synthetic antioxidants during storage, during frying, the addition of coriander oil proved to result in the highest antioxidant activity (Patel et al., 2013). Conclusions and future trends: an attractive application of essential oils and their constituents is to prolong the shelf-life of food by controlling the growth and survival of microorganisms. The organoleptic impact of essential oils and their components in food products is still holding back their application to foods. Synergistic interactions of essential oils with other preservatives, their incorporation into antimicrobial packaging films, and their combined use with other preservation technologies such as irradiation or modified atmosphere have been proposed to reduce the sensory alterations caused by essential oils. Among all the essential oils with described antimicrobial properties, several coriander oil features make it an attractive and valuable choice for the development of natural-based food preservation techniques such as its antioxidant activity, broad antimicrobial effectiveness, and bactericidal activity within 30 min of contact against several foodborne pathogens and its safety as a food ingredient. Also, the numerous biological properties of coriander oil can be advantageous to consumers to promote their health, thus adding value to the food item. Although coriander oil is the second most used essential oil worldwide, there is still a need for more research and clinical studies to evaluate the safety of its consumption and prove its biological activities *in vivo*. Also, for the use as a food preservative, its efficacy against common foodborne fungi, such as *Fusarium* spp., *Aspergillus ochraceus*, *Penicillium verrucosum*, *Aspergillus flavus*, and *Aspergillus parasiticus*, and vapor-phase antimicrobial activities should be evaluated. Overall, the research on coriander oil application or linalool as food preservatives is still scarce, yielding very different outcomes in terms of success. So far, only the direct addition of coriander oil to meat, its incorporation in a chitosan film, and linalool's incorporation in LPDE film have been described. Due to the vast array of technologies for food packaging and preservation made available in recent years, the possibility of using coriander oil as a food preservative still needs extensive investigation to improve its efficacy in food media and reduce the undesirable alterations in the organoleptic properties of food.

Essential oils in food preservation: the use of essential oils as food preservatives is not controversial since they are plant-based materials and have long been used in culinary and folk medicine throughout the world. However, the possibility of applying the oils on an industrial scale for food preservation remains unclear. Their applicability is determined by two key factors: their effectiveness in terms of product and the acceptance level for the consumers of the modified product. In addition, consumers' approval is based on the sensory qualities of the end-product and the fact that no sensitization or allergic reaction is induced by its consumption (Michalczyk et al., 2012).

The quality and safety of prepared or processed foods are of prime importance in the food industry. The microorganisms present in food can lead to spoilage and deterioration of the quality of food products, and if ingested by humans, can cause infection and illness. Thus food manufacturers try to reduce or eliminate microorganisms from food products. It has been estimated that about one-third of the world's food production is lost annually due to microbial spoilage or contamination (Alboofetileh et al., 2014). The essential oil from coriander plants has exhibited excellent antimicrobial effects against bacteria, yeasts, fungi, and viruses. Coriander essential oil (CEO) and its various fractional distillates were effective antimicrobial agents, particularly against *Listeria monocytogenes* because of the presence of long-chain (C6-C10) alcohols and aldehydes. The mixing of different fractions showed additive, synergistic, or antagonistic effects against individual test microorganisms (Delaquis et al., 2002).

Similarly, Matasyoh et al. (2009) obtained essential oil from the leaves by hydrodistillation and evaluated it for *in vitro* antimicrobial activity. The oil was dominated by aldehydes and alcohols, which accounted for 56.1% and 46.3% of the oil. The extracted oil was screened for antimicrobial activity against Gram-positive (*Staphylococcus aureus, Bacillus* spp.) and Gram-negative (*Escherichia coli, Salmonella typhi, Klebsiella pneumonia, Proteus mirabilis, Pseudomonas aeruginosae*) bacteria and a pathogenic fungus, *Candida albicans*. Only *P. aeruginosae* showed resistance to CEO, while other tested bacteria were highly affected.

Rattanachaikunsopon and Phumkhachorn (2010) studied the 12 essential oils for antimicrobial activities against several *Campylobacter jejuni*, a pathogen causing foodborne diseases worldwide. The authors showed that CEO exhibited the strongest antimicrobial activity against all the tested strains. In addition, the antimicrobial potency of coriander oil against *C. jejuni* in beef and chicken meat at 4 and 32°C was also studied. It was found that the oil reduced the bacterial cell load in a dose-dependent manner; however, the type of meat and temperature did not influence the antimicrobial activity of the essential oil. This study indicates the potential of CEO to serve as a natural antimicrobial compound against *C. jejuni* in food.

In another study, the essential oils extracted from coriander and other plants were evaluated for their antimicrobial activity against 11 different bacterial and 3 fungal strains belonging to species reported to be involved in food poisoning and food decay. These include *S. aureus, E. coli, Salmonella enterica, L. monocytogenes, Bacillus cereus, C. albicans*, and *Aspergillus niger*. Coriander essential oil showed the best antibacterial activity in all, while thyme and spearmint oils better inhibited the fungal species (Lixandru et al., 2010). The antifungal activity of essential oils of some species of the family Apiaceae was tested against the fungus *Aspergillus flavus*. Coriander oil was the most effective against fungal growth and aflatoxin production at all concentrations studied (Abou El-Soud et al., 2012). The authors showed that an amount of 1000 ppm as a food additive protects the spices from bio-deteriorating fungi and aflatoxin contamination.

It is clear from the above-mentioned literature that coriander oil's antioxidant, antibacterial, and antifungal activities are linked to its uses in food flavoring and preservation and medicinal applications. Thus, using CEO for the above applications is recommended. The potential of the practical application of coriander oil was finally evaluated in bread over 14 days of storage. Fresh bread was baked following the fundamental formula. The moisture content obtained for this bread was 41.46%, and a_w was 0.945. The results of the performed *in-situ* antifungal analysis on bread indicate good antifungal activity, as the MID_{50} and MID_{90} of the coriander essential oil for *Penicillium expansum* on the bread after 14 days were 367 and 445 µL/L of air, respectively (Kačániová et al., 2020).

36.6 CONCLUSION

The coriander essential oils can be used in diverse applications in food and industries. However, coriander essential oil as antimicrobial and food preservative agents is of concern because of several reported side effects of synthetic oils. Coriander essential oils can be used as a food preservative for cereals, grains, pulses, fruits, and vegetables. In this chapter, we briefly describe the results in the relevant literature and summarize the uses of CEOs, emphasizing their antibacterial, bactericidal, antifungal, fungicidal, and food preservative properties. CEOs have pronounced antimicrobial and food preservative properties because they consist of various active constituents that have great significance in the food industry. Thus, the various properties of essential oils offer the possibility of using natural, safe, eco-friendly, cost-effective, renewable, and readily biodegradable antimicrobials for food commodity preservation in the near future.

REFERENCES

Abou El-Soud, N.H., Deabes, M.M., Abou El-Kassem, L.T. and Khalil, M.Y. 2012. Antifungal activity of family apiaceae essential oils. *J Appl Sci Res*. 8: 4964–4973.

Alboofetileh, M., Rezaei, M., Hosseini, H. and Abdollahi, M. 2014. Antimicrobial activity of alginate/clay nanocomposite films enriched with essential oils against three common foodborne pathogens. *Food Control*. 36: 1–7.

Bentayeb, K., Rubio, C., Sanchez, C., Batlle, R. and Nerin, C. 2007. Determination of the shelf life of a new antioxidant active packaging. *Ital J Food Sci*. 19: 110–115.

Bondi, M., Lauková, A., de Niederhausern, S., Messi, P. and Papadopoulou, C. 2017. Natural preservatives to improve food quality and safety. *J Food Qual*. 2017: 1090932. doi: 10.1155/2017/1090932.

Burt, S. 2004. Essential oils: Their antibacterial properties and potential applications in foods-a review. *Int J Food Microbiol*. 94: 223–253.

Busatta, C., Vidal, R.S., Popiolski, A.S., Mossi, A.J., Dariva, C., Rodrigues, M.R.A., Corazza, F.C., Corazza, M.L., Oliveira, J.V. and Cansian, R.L. 2008. Application of *Origanum majorana* L. essential oil as an antimicrobial agent in sausage. *Food Microbiol*. 25: 207–211.

Cha, D.S. and Chinnan, M.S. 2004. Biopolymer-based antimicrobial packaging: A review. *Crit Rev Food Sci Nutr*. 44: 223–237.

Cran, M.J., Rupika, L.A.S., Sonneveld, K., Miltz, J. and Bigger, S.W. 2010. Release of naturally derived antimicrobial agents from LDPE films. *J Food Sci*. 75: E126–E133.

Cristani, M., D'Arrigo, M., Mandalari, G., Castelli, F., Sarpietro, M.G., Micieli, D., Venuti, V., Bisignano, G., Saija, A. and Trombetta, D. 2007. Interaction of four monoterpenes contained in essential oils with model membranes: Implications for their antibacterial activity. *J Agric Food Chem*. 55: 6300–6308.

Delaquis, P.J., Stanich, K., Girard, B. and Mazza, G. 2002. Antimicrobial activity of individual and mixed fractions of dill, cilantro, coriander and eucalyptus essential oils. *Int J Food Microbiol*. 74: 101–109.

Di Pasqua, R., Hoskins, N., Betts, G. and Mauriello, G. 2006. Changes in membrane fatty acids composition of microbial cells induced by addiction of thymol, carvacrol, limonene, cinnamaldehyde, and eugenol in the growing media. *J Agric Food Chem*. 54: 2745–2749.

Donlan, R.M. and Costerton, J.W. 2002. Biofilms: Survival mechanisms of clinically relevant microorganisms. *Clin Microbiol Rev*. 15: 167–193.

Duarte, A., Ferreira, S., Oliveira, R. and Domingues, F.C. 2013. Effect of coriander oil (*Coriandrum sativum*) on planktonic and biofilm cells of *Acinetobacter baumannii*. *Nat Prod Commun*. 8: 673–678.

Duarte, A., Ferreira, S., Silva, F. and Domingues, F.C. 2012. Synergistic activity of coriander oil and conventional antibiotics against *Acinetobacter baumannii*. *Phytomedicine*. 19: 236–238.

Duarte, M.C., Figueira, G.M., Sartoratto, A., Rehder, V.L. and Delarmelina, C. 2005. AntiCandida activity of Brazilian medicinal plants. *J Ethnopharmacol*. 97: 305–311.

Elgayyar, M., Draughon, F.A., Golden, D.A. and Mount, J.R. 2001. Antimicrobial activity of essential oils from plants against selected pathogenic and saprophytic microorganisms. *J Food Prot*. 64: 1019–1024.

Ferdes, S. and Ungureanu, C. 2012. Antimicrobial activity of essential oils against four foodborne fungal strains. *U P B Sci Bull, series B*. 74: 87–98.

Furletti, V.F., Teixeira, I.P., Obando-Pereda, G., Mardegan, R.C., Sartoratto, A., Figueira, G.M., Duarte, R.M., Rehder, V.L., Duarte, M.C. and Hofling, J.F. 2011. Action of *Coriandrum sativum* L. essential oil upon oral *Candida albicans* biofilm formation. *Evid Based Complement Alternat Med*. 2011: 985832.

Ghasemlou, M., Aliheidari, N., Fahmi, R., Shojaee-Aliabadi, S., Keshavarz, B., Cran, M.J. and Khaksar, R. 2013. Physical, mechanical and barrier properties of corn starch films incorporated with plant essential oils. *Carbohydr Polym*. 98: 1117–1126.

Goni, P., Lopez, P., Sanchez, C., Gomez-Lus, R., Becerril, R. and Nerin, C. 2009. Antimicrobial activity in the vapour phase of a combination of cinnamon and clove essential oils. *Food Chem*. 116: 982–989.

Gram, L., Ravn, L., Rasch, M., Bruhn, J.B., Christensen, A.B. and Givskov, M. 2002. Food spoilage-interactions between food spoilage bacteria. *Int J Food Microbiol*. 78: 79–97.

Gutierrez, L., Batlle, R., Sanchez, C. and Nerin, C. 2010. New approach to study the mechanism of antimicrobial protection of an active packaging. *Foodborne Pathog Dis*. 7: 1063–1069.

Hammer, K.A., Carson, C.F. and Riley, T.V. 1999. Antimicrobial activity of essential oils and other plant extracts. *J Appl Microbiol.* 86: 985–990.

Josić, D., Rešetar, D., Peršurić, Ž., Martinović, T., and Pavelić, S.K. 2017. Detection of microbial toxins by-omics methods: A growing role of proteomics, in *Proteomics in Food Science*, ed Colgrave, M. (Brisbane, QLD: Academic Press), 485–506. doi: 10.1016/B978-0-12-804007-2.00029-1.

Kačániová, M., Galovičová, L., Ivanišová, E., Vukovic, N.L., Štefániková, J., Valková, V., Borotová, P., Žiarovská, J., Terentjeva, M., Felšöciová, S. and Tvrdá, E. 2020. Antioxidant, antimicrobial and anti-biofilm activity of coriander (*Coriandrum sativum* L.) essential oil for its application in foods. *Foods,* 9, 282.

Khanjari, A., Karabagias, I.K. and Kontominas, M.G. 2013. Combined effect of N,Ocarboxymethyl chitosan and oregano essential oil to extend shelf life and control *Listeria monocytogenes* in raw chicken meat fillets. *LWT-Food Sci Technol.* 53: 94–99.

Kuorwel, K.K., Cran, M.J., Sonneveld, K., Miltz, J. and Bigger, S.W. 2011. Essential oils and their principal constituents as antimicrobial agents for synthetic packaging films. *J Food Sci.* 76: R164–R177.

Licciardello, F., Muratore, G., Mercea, P., Tosa, V. and Nerin, C. 2013. Diffusional behaviour of essential oil components in active packaging polypropylene films by multiple headspace solid phase microextraction-gas chromatography. *Packag Technol Sci.* 26: 173–185.

Lixandru, B.E., Drăcea, N.O., Dragomirescu, C.C., Drăgulescu, E.C., Coldea, I.L., Anton, L., Dobre, E., Rovinaru, C. and Codiţă, I. 2010. Antimicrobial activity of plant essential oils against bacterial and fungal species involved in food poisoning and/or food decay. *Roum Arch Microbiol Immunol.* 69: 224–230.

Lo Cantore, P., Iacobellis, N.S., De Marco, A., Capasso, F. and Senatore, F. 2004. Antibacterial activity of *Coriandrum sativum* L. and *Foeniculum vulgare* Miller Var. vulgare (Miller) essential oils. *J Agric Food Chem.* 52: 7862–7866.

Lopez de Dicastillo, C., Nerin, C., Alfaro, P., Catala, R., Gavara, R. and Hernandez-Munoz, P. 2011. Development of new antioxidant active packaging films based on ethylene vinyl alcohol copolymer (EVOH) and green tea extract. *J Agric Food Chem.* 59: 7832–7840.

Lopez, P., Sanchez, C., Batlle, R. and Nerin, C. 2005. Solid- and vapor-phase antimicrobial activities of six essential oils: Susceptibility of selected foodborne bacterial and fungal strains. *J Agric Food Chem.* 53: 6939–6946.

Lopez, P., Sanchez, C., Batlle, R. and Nerin, C. 2007a. Development of flexible antimicrobial films using essential oils as active agents. *J Agric Food Chem.* 55: 8814–8824.

Lopez, P., Sanchez, C., Batlle, R. and Nerin, C. 2007b. Vapor-phase activities of cinnamon, thyme, and oregano essential oils and key constituents against foodborne microorganisms. *J Agric Food Chem.* 55: 4348–4356.

Lucera, A., Costa, C., Conte, A. and Del Nobile, M.A. 2012. Food applications of natural antimicrobial compounds. *Front Microbiol.* 3: 287.

Marangoni, C. and de Moura, N.F. 2011a. Antioxidant activity of essential oil from *Coriandrum sativum* L. Italian salami. *Cienia Tecnol Aliment.* 31: 124–128.

Marangoni, C. and de Moura, N.F. 2011b. Sensory profile of Italian salami with coriander (*Coriandrum sativum* L.) essential oil. *Ciencia Tecnol Aliment.* 31: 119–123.

Martinez-Abad, A., Sanchez, G., Fuster, V., Lagaron, J.M. and Ocio, M.J. 2013. Antibacterial performance of solvent cast polycaprolactone (PCL) films containing essential oils. *Food Control.* 34: 214–220.

Matasyoh, J.C., Maiyo, Z.C., Ngure, R.M., Chepkorir, R. 2009. Chemical composition and antimicrobial activity of the essential oil of *Coriandrum sativum*. *Food Chem.* 113, 526–529.

Melchior, M.B., Vaarkamp, H. and Fink-Gremmels, J. 2006. Biofilms: A role in recurrent mastitis infections? *Vet J.* 171: 398–407.

Mercier, R.C., Stumpo, C. and Rybak, M.J. 2002. Effect of growth phase and pH on the in vitro activity of a new glycopeptide, oritavancin (LY333328), against *Staphylococcus aureus* and *Enterococcus faecium*. *J Antimicrob Chemother.* 50: 19–24.

Michalczyk, M., Macura, R., Tesarowicz, I. and Banas, J. 2012. Effect of adding essential oils of coriander (*Coriandrum sativum* L.) and hyssop (*Hyssopus officinalis* L.) on the shelf life of ground beef. *Meat Sci.* 90: 842–850.

Muriel-Galet, V., Cerisuclo, J.R., Lopez-Carballo, G., Aucejo, S., Gavara, R. and HernandezMunoz, P. 2013. Evaluation of EVOH-coated PP films with oregano essential oil and citral to improve the shelf-life of packaged salad. *Food Control.* 30: 137–143.

Nerin, C., Tovar, L., Djenane, D., Camo, J., Salafranca, J., Beltran, J.A. and Roncales, P. 2006. Stabilization of beef meat by a new active packaging containing natural antioxidants. *J Agric Food Chem.* 54: 7840–7846.

Ozek, G., Demirci, F., Ozek, T., Tabanca, N., Wedge, D.E., Khan, S.I., Baser, K.H., Duran, A. and Hamzaoglu, E. 2010. Gas chromatographic-mass spectrometric analysis of volatiles obtained by four different techniques from *Salvia rosifolia* Sm., and evaluation for biological activity. *J Chromatogr A*. 1217: 741–748.

Patel, S., Shende, S., Arora, S. and Singh, A.K. 2013. An assessment of the antioxidant potential of coriander extracts in ghee when stored at high temperature and during deep fat frying. *Int J Dairy Technol*. 66: 207–213.

Rajkovic, A. 2014. Microbial toxins and low level of foodborne exposure. *Trends Food Sci Technol*. 38: 149–157. doi: 10.1016/j.tifs.2014.04.006.

Rajkovic, A., Jovanovic, J., Monteiro, S., Decleer, M., Andjelkovic, M., Foubert, A., et al. 2020. Detection of toxins involved in foodborne diseases caused by Gram-positive bacteria. *Comprehen. Rev. Food Sci. Food Saf*. 19: 1605–1657. doi: 10.1111/1541-4337.12571.

Rattanachaikunsopon, P. and Phumkhachorn, P. 2010. Potential of coriander (*Coriandrum sativum*) oil as a natural antimicrobial compound in controlling *Campylobacter jejuni* in raw meat. *Biosci Biotechnol Biochem*. 74: 31–35.

Reddy, K.R.N., Salleh, B., Saad, B., Abbas, H.K., Abel, C.A. and Shier, W.T. 2010. An overview of mycotoxin contamination in foods and its implications for human health. *Toxin Rev*. 29: 3–26. doi: 10.3109/15569541003598553.

Rešetar, D., Pavelić, S.K., and Josić, D. 2015. Foodomics for investigations of food toxins. *Curr. Opin. Food Sci*. 4: 86–91. doi: 10.1016/j.cofs.2015.05.004.

Rodriguez, A., Batlle, R. and Nerin, C. 2007. The use of natural essential oils as antimicrobial solutions in paper packaging. Part II. *Prog Org Coat*. 60: 33–38.

Rodriguez, A., Nerin, C., Battle, R. 2008. New Cinnamon-Based Active Paper Packaging against Rhizopus stolonifer Food Spoilage. *Journal of Agriculture and Food Chemistry*, 56: 6364–6369. doi: 10.1021/jf800699q.

Rodriguez-Lafuente, A., Nerin, C. and Batlle, R. 2010. Active paraffin-based paper packaging for extending the shelf life of cherry tomatoes. *J Agric Food Chem*. 58: 6780–6786.

Sellamuthu, P.S., Mafune, M., Sivakumar, D. and Soundy, P. 2013. Thyme oil vapour and modified atmosphere packaging reduce anthracnose incidence and maintain fruit quality in avocado. *J Sci Food Agric*. 93: 3024–3031.

Silva, F., Ferreira, S., Duarte, A., Mendonca, D.I. and Domingues, F.C. 2011a. Antifungal activity of *Coriandrum sativum* essential oil, its mode of action against Candida species and potential synergism with amphotericin B. *Phytomedicine*. 19: 42–47.

Silva, F., Ferreira, S., Queiroz, J.A. and Domingues, F.C. 2011b. Coriander (*Coriandrum sativum* L.) essential oil: Its antibacterial activity and mode of action investigated by flow cytometry. *J Med Microbiol*. 60: 1479–1486.

Silva, F. and Domingues, F.C. 2017. Antimicrobial activity of coriander oil and its effectiveness as food preservative. *Crit Rev Food Sci Nutr*. 57(1): 35–47.

Soares, B.V., Morais, S.M., dos Santos Fontenelle, R.O., Queiroz, V.A., Vila-Nova, N.S., Pereira, C.M., Brito, E.S., Neto, M.A., Brito, E.H., Cavalcante, C.S., Castelo-Branco, D.S. and Rocha, M.F. 2012. Antifungal activity, toxicity and chemical composition of the essential oil of *Coriandrum sativum* L. fruits. *Molecules*. 17: 8439–8448.

Sonboli, A., Babakhani, B. and Mehrabian, A.R. 2006. Antimicrobial activity of six constituents of essential oil from Salvia. *Zeitschrift Fur Naturforschung Section C-a Journal of Biosciences*. 61: 160–164.

Suppakul, P., Sonneveld, K., Bigger, S.W. and Miltz, J. 2008. Efficacy of polyethylene-based antimicrobial films containing principal constituents of basil. *LWT-Food Sci Technol*. 41: 779–788.

Suppakul, P., Sonneveld, K., Bigger, S.W., & Miltz, J. 2011. Loss of AM Additives from Antimicrobial Films during Storage. *Journal of Food Engineering*, 105(2): 270–276. 10.1016/j.jfoodeng.2011.02.031.

Teuber, M. 1999. Spread of antibiotic resistance with food-borne pathogens. *Cell Mol Life Sci*. 56: 755–763.

Toroglu, S. 2011. In-vitro antimicrobial activity and synergistic/antagonistic effect of interactions between antibiotics and some spice essential oils. *J Environ Biol*. 32: 23–29.

Trombetta, D., Castelli, F., Sarpietro, M.G., Venuti, V., Cristani, M., Daniele, C., Saija, A., Mazzanti, G. and Bisignano, G. 2005. Mechanisms of antibacterial action of three monoterpenes. *Antimicrob Agents Chemother*. 49: 2474–2478.

Wang, L.N., Liu, F., Jiang, Y.F., Chai, Z., Li, P.L., Cheng, Y.Q., Jing, H. and Leng, X.J. 2011. Synergistic antimicrobial activities of natural essential oils with chitosan films. *J Agric Food Chem*. 59: 12411–12419.

Zivanovic, S., Chi, S. and Draughon, A.F. 2005. Antimicrobial activity of chitosan films enriched with essential oils. *J Food Sci*. 70: M45–M51.

37 Encapsulation of Coriander Essential Oil

Muhammad Imran, Muhammad Kamran Khan,
Muhammad Haseeb Ahmad, Rabia Shabir Ahmad,
Haseeb Anwar, Muhammad Nadeem,
Muhammad Abdul Rahim, Ayesha Bibi, Khalid Abbas,
Muhammad Rashid, and Muhammad Noman

CONTENTS

37.1 INTRODUCTION

Plants have been recognized as a good source of bioactive or phytochemical compounds, and are currently recommended to be the best source of such ingredients. One of the most important sources of natural bioactive compounds and antioxidants is herbs and spices. Among them, coriander (*Coriandrum sativum* L.) is vital because of its versatile properties utilized as an herb and a spice in various processed food products. It is an annual dicotyledon herbaceous plant that belongs to the family Umbelliferae and is one of the earliest spice crops. "Kusthumbari" and "Dhanayaka" are common names for coriander in different states. It originates from Pakistan, India, Algeria, China, and Tunisia. Coriander leaves and seeds are mainly used as spices. In addition, coriander seeds and leaves are also known as medicinal herbs (Msaada et al., 2007; Ahmad et al., 2021).

Coriander seeds (CS) are fascinating as they contain flavoring compounds and essential oils. Coriander essential oil (CEO) is a triglyceride; it can be fractioned and isolated from the different parts of the coriander plant. The amount of essential oil in coriander seeds is up to 0.03% to 2.6%. CEO is extracted from CS using standard extraction methods such as hydrodistillation, supercritical fluid extraction, and microwave-assisted extraction techniques. These extraction methods give different essential oil yields in optimal processing conditions. Supercritical fluid extraction results in a higher essential oil yield with minimal loss of volatile compounds. CEO is slightly yellow or colorless (Eikani et al., 2007; Ghazanfari et al., 2020; Grosso et al., 2008). The volatile component linalool is present in a higher amount than other volatile compounds such as geraniol, lemanone, and monoterpenes. These compounds represent about 70% of the respective oil fractions. It is low in saturated fatty acids and high in unsaturated fatty acids. The fatty acids present in CEO are petroselinic acid, stearic acid (3.74%), palmitic acid (0.97%), arachidic acid (0.08%), and linoleic acid (14.7%). The presence of petroselinic acid gives rise to a wide range of opportunities with high potential for CEO (Orav et al., 2011). These fatty acids are considered adequate for curing various human physiological disorders. They have been shown to have a range of attractive properties like skin care benefits, anti-aging, reducing the risk of cardiovascular diseases, anti-inflammatory activity, and reducing metabolic disorders (Rajeshwari & Andallu, 2011).

The CEO is considered sensitive to the reaction of conversions and degradation. As a result, oxidative and polymerization processes may result in loss of quality, pharmaceutical, and pharmacological properties. Therefore, various encapsulation methods have been used to improve the oxidative stability of CEO. The most common methods used to prepare microcapsules are coacervation, spray drying, and fluidized-bed drying techniques. The stability of the CEO is enhanced after encapsulation, as the stability increases up to 79.7% (Dima et al., 2013).

37.2 PHYSICAL DESCRIPTION OF CORIANDER SEEDS (CS)

Coriander seeds (CS) have an almost elliptical round shape divided into several parts, such as the dry schizocarp and two mericarps, and the seed surface has many longitudinal ridges. CS are 3 to 5 mm long and usually brown when dry but can be a green straw in color. The CS are mostly sold after drying in the sun. CS have a sharply strong smell or taste like citrus with a hint of sage (Balasubramanian et al., 2012; Balasubramanian et al., 2021). They are used to prepare many traditional medicines to cure seasonal fever, insomnia, nausea, stomach disorders, cholesterol, and diabetes. They are also used as a drug for indigestion, reducing tension, anxiety, inducing calm, and repelling insects, and also used for anti-leukemic activity, atrophic arthritis, and pain in the joints (Emam & heydari, 2006; Dhanapakiam et al., 2007; Mechchate et al., 2021).

37.3 DIFFERENT EXTRACTION METHODS OF CORIANDER ESSENTIAL OIL (CEO)

The most important components of CS are CEO and fatty oil. The CEO content in dry CS ranges from 0.03% to 2.6%, and the fatty oil content ranges from 9.9% to 27%. The CEO is extracted from CS using the following extraction methods.

37.3.1 HYDRODISTILLATION

The principal method of extraction of oil from CS is hydrodistillation. In this process, hot water or steam is used as a heat source. The prepared CS samples are kept in an extractor and heated with steam to the optimum temperature. During heating, the volatile constituents first evaporate and then condense. Due to the low concentration of essential oil, it is difficult to measure the oil content in the extractor. Therefore, water is used to prepare the appropriate dilution ratio before the gas chromatography. Hydrodistillation also reduces decomposition risk, and approximately 0.15 mL of CEO is obtained. The oil produced in this way is a mixture of odorless, bioactive, and pigment compounds (Eikani et al., 2007).

37.3.2 MICROWAVE-ASSISTED HYDROTHIOLATION

Microwave-assisted hydrothiolation is another innovative method used to extract CEO from CS, as described in the research of Ghazanfari et al. (2020). This method uses microwave equipment to heat the prepared sample with adjusted ambient pressure at 100°C. Firstly, the CS is ground, then dissolved in distilled water. The amount of distilled water is adjusted according to the sample quantity. After that, the prepared solution is placed in the selected microwave oven under optimized conditions. A condenser is used on top of the solution to collect the obtained CEO in this process. This process is rapid, short in time, and requires less energy, and it is easy to extract the oil. Approximately 2.45 GHz microwaves are suitable for maximum yield or efficiency with 900 W power for the optimum time.

37.3.3 SUPERCRITICAL FLUID EXTRACTION

Nowadays, most researchers and different food processing industries use supercritical fluid extraction to extract the CEO from the prepared CS samples. The oil extracted has high integrity, good efficiency, higher yield, and good quality. Pressurized carbon dioxide gas is filled into a closed chamber in this method. The volume of carbon dioxide is estimated using a dry test meter. Furthermore, carbon dioxide gas is used as a solvent at optimal processing conditions. The major parts of the apparatus are a back pressure regulator to control the pressure during the extraction process, a diaphragm pump used to create the pressure, extraction vessels, and two separation vessels. The default temperature in the extraction vessel is achieved with the help of a water jacket (Grosso et al., 2008; Mhemdi et al., 2011; Zeković et al., 2015).

37.3.4 COMPARISON OF EXTRACTION METHODS

The content of the CEO from various sources of extraction indicates that hydrodistillation is not a suitable method for extracting oil due to high temperatures and acidic conditions. It is also time-consuming and requires a large amount of energy. Other methods such as supercritical fluid extraction and CO_2 supercritical extraction are also popular oil extraction methods as they extract essential oils with high-quality volatile compounds but are considered expensive. The microwave-assisted extraction is considered a time- and energy-saving method with increasing extraction yield. These methods often result in the extraction of undesired compounds such as nonaromatic active fats and waxes or pigments (Richter & Schellenberg, 2007).

37.4 COMPOSITION OF CEO

The composition of the CEO indicates that the alcohol group contains the maximum amount of lanolin and geraniol. Moreover, the hydrocarbon group contains γ-terpinene and r-cymene in the CEO. Terpinene-4-Ol, α-terpineol, limonene, a-pinene, camphene, myrcene, camphor, linalyl acetate, and geranyl acetate are also found in minor amounts in CEO. CEOs are extracted from the seed, flower, and leaves, leading to massive differences in their composition of volatile compounds. The oil from the entirely ripped and dried seeds is colorless or pale yellow with a mild odor and aromatic flavor. The CEO has an unpleasant odor from fresh herbs. Fresh herbs contain lanolin as an essential constituent in the CEO which is extracted from CS compared to aliphatic aldehydes. Linalool and monoterpenes hydrocarbons are also rich in oils extracted from coriander fruit (Msaada et al., 2007; Bhuiyan et al., 2009; Dos Santos et al., 2019). A significant correlation exists between the coriander fruit size and essential oil yield. The variety with tiny sized seeds is subjected to essential oil extraction, and the large sized seed variety is used as spices, which means small sized seeds have a high content of essential oil compared to large sized seeds (Beyzi & Gurbuz, 2014).

In a research work by Micić et al. (2019), CEO is a volatile liquid extracted from flowers, leaves, and seeds. The compounds that make up CEOs accumulate in various secretory glands and chemical components in plants. However, most common CEOs are derived from seeds and flowers. CEO can be derived from many materials of plant origin. Therefore, their mixture consists mainly of specific metabolites such as linalool, γ-terpinene, α-pinene, *p*-cymene, camphor, geranyl acetate, benzofuran, 2,3-dihydro hexadecanoic acid, methyl ester, 2,3,5,6-tetrafluroanisole, dodecanoic acid, 2-methoxy-4-vinylphenol, 2,6-dimethyl-3-amino benzoquinone, decanal, 2-decen-1-ol, *trans*-2-decenal, cyclodecane, dodecanal, and *cis*-2-dodecane, dodecan-1-ol. The chemical nature of these major components reveals that they are less soluble in polar solvents such as isopropanol, water, and methanol. Furthermore, they are more soluble in non-polar solvents, like chloroform, ethyl acetate, toluene, hexane, methylene chloride, pyridine, and hexane (Ramadan & Mörsel, 2002; Msaada et al., 2007).

37.5 FATTY ACID PROFILE AND OXIDATIVE STABILITY OF CEO

The fatty acid profile of the CEO was estimated using the gas chromatography method. This method of distribution and identification of fatty acid components in CEO indicates that petroselinic acid is the largest fatty acid compared to others, making up 70.1 to 73.2% of all fatty acids. The other major fatty acids in CEO are palmitic acid (4.43 to 5.18%), linoleic acid (13.12 to 14.1%), and oleic acid (5.13 to 6.32%). Myristic acid (0.06 to 0.09%), stearic acid (0.64 to 0.80%), α-linolenic acid (0.31 to 0.43%), and arachidic acid (0.09 to 0.16%) are also found in minor amounts in CEO (Neffati & Marzouk, 2008; Sriti et al., 2009; Nguyen et al., 2020).

The stability of the CEO is critical to predict the quality deterioration during various storage conditions. The oxidation of unsaturated fatty acids in the CEO not only produces off-flavors but can also reduce food safety by forming secondary reaction products. Temperature, light, and oxygen also significantly impact the integrity of essential oils. Different methods are used to determine the stability of essential oils during stabilization. These methods are based on physical and chemical properties. CEO is stored under oxidation conditions at 60°C for 21 days. The progress in oxidation is observed at 60°C by measuring the ultraviolet absorbability. The ability to absorb ultraviolet rays is calculated by the constant formation of oxidative products such as peroxide and *p*-anisidine values. Peroxide values and oxidative stability correlate as stability increases with decreased peroxide values. The production of conjugated dienes and polyenes gradually increases the absorption of ultraviolet radiation from 232 nm to 270 nm (Ramadan & Morsel, 2004). Another research work by Moser and Vaughn (2010) found that lowering the temperature and iodine value increased the oxidative stability of the CEO. Bag and Chattopadhyay (2018) reported that the peroxide value of CEO was increased during storage for 15 days from 1.26 to 12.1 mEq/kg oil.

37.6 FUNCTIONAL APPLICATIONS OF CEO

CEO is used as a flavoring agent in the food processing industries, but it also has a long history as the oldest medicine (Burdock & Carabin, 2009). Some essential biological characteristics that contribute to human nutrition are given below:

- CEO has an excellent antioxidant property; it can be utilized to replace synthetic antioxidants.
- CEO is used in spicy foods to prevent spoilage due to its antifungal and antibacterial activity. CEO also has fungicidal and bactericidal properties (Darughe et al., 2012).
- CEO has been shown to have antibacterial activity against various pathogenic and saprophytic bacteria, suggesting that it could be used as a disinfectant (Ramadan & Moersel, 2006).
- CEO has commonly been used as a flavoring agent in various food products like meat sauces, carbohydrates, alcoholic drinks, confectionary items, and tobacco. The average level of CEO usage in these food items is 0.1 to 100 parts per million (Elgayyar et al., 2001).
- CEO is utilized in various research work to reduce the risk of various disorders like diarrhea, acute gastritis, nausea, gas, headaches, and discomfort from various causes (Nadeem et al., 2013).

- It is used to prepare many domestic medicines to treat seasonal fever, nausea, and stomach disorders. It is also used as a drug for indigestion, against insects, atrophic arthritis, and pain in the joints.
- Its excellent phytonutrients are responsible for many of its healing qualities, and it is sometimes referred to as bioactive chemical storage (Rajeshwari & Andallu, 2011).

37.7 ENCAPSULATION OF CORIANDER ESSENTIAL OIL

Encapsulation is an innovative technique in the food processing industry, in which most of the oils are encapsulated using various wall materials, and microcapsules are manufactured. This technique can improve the functional properties of various oils, protect them from environmental factors, improve the thermos-stability, maintain the taste and flavor, improve oxidative stability, and enhance their bioactive properties. There are many encapsulation techniques based on their mechanical and physical processes (Figure 37.1). The methods used to make CEO microcapsules include complex coacervation, spray drying, and fluidized-bed drying techniques. In addition, some of these encapsulation methods are selected based on chemical changes in the material (Anandharamakrishnan, 2014; Alvim et al., 2016).

37.7.1 COMPLEX COACERVATION TECHNIQUE

The complex coacervation technique has been recognized as one of the oldest and most commonly used microencapsulation processes. In homogeneous compounds, complex coacervation consists of electrostatic attraction or repulsion between two biopolymers of the same charge or opposite charge (depending on the situation), forming or separating at a partial pH range. In this process, the liquid and coacervate phase has been used for separation. This process is only acceptable for food processing items. Furthermore, it has been divided into two categories, simple and complex coacervation (Timilsena et al., 2019). The complex coacervation method is used for encapsulation of the CEO. For the complex coacervation method, the emulsion is prepared using the polysaccharide and a protein as a biopolymer, while the temperature and pH have been adjusted to the optimum level. Then, the pH of the solution is reduced to the isoelectric point when the electrostatic attraction occurs between the anti-charged biopolymer; the phase of the soluble liquid separates from the insoluble biopolymer-rich phase. In the next step, the biopolymer-rich phase is assembled, producing a core around the hydrolytic droplets at the optimum temperature. Finally, the inclusion of a cross-linking agent is utilized to stiffen the cores of microcapsules in the last phase (Enascuta et al., 2018).

37.7.2 SPRAY DRYING TECHNIQUE

The spray drying microencapsulation technology has various advantages in the food sector, including low production costs, a broad spectrum of biopolymers that can be employed as encapsulating agents, effective volatile retention, and better final product stability. It has been commonly used for microencapsulation in the food processing industry. Recent research work is used to achieve the microencapsulation of CEO in three phases. After deciding the variety of core material used to preserve CEO from environmental factors, the first phase is to disperse the wall material in a suitable solvent (often with heating, depending on the material). After the distribution of the core material, the core material is added to produce a uniform emulsion by high-speed mixing through a homogenizer in the second phase of encapsulation by spray drying. When CEO is solubilized before being added to a suspension of wall material, its viscosity is lowered, allowing it to break down into tiny droplets and favoring coating by the core material. Emulsion formation is used to combine CEO into wall material suspension. The third stage is to pump the infeed dispersion to the spray dryer's drying chamber after being prepared. Co-current airflow is used to encapsulate CEO by spray drying, resulting in particles with a spherical shape. Depending on the feed material, spray drying generates

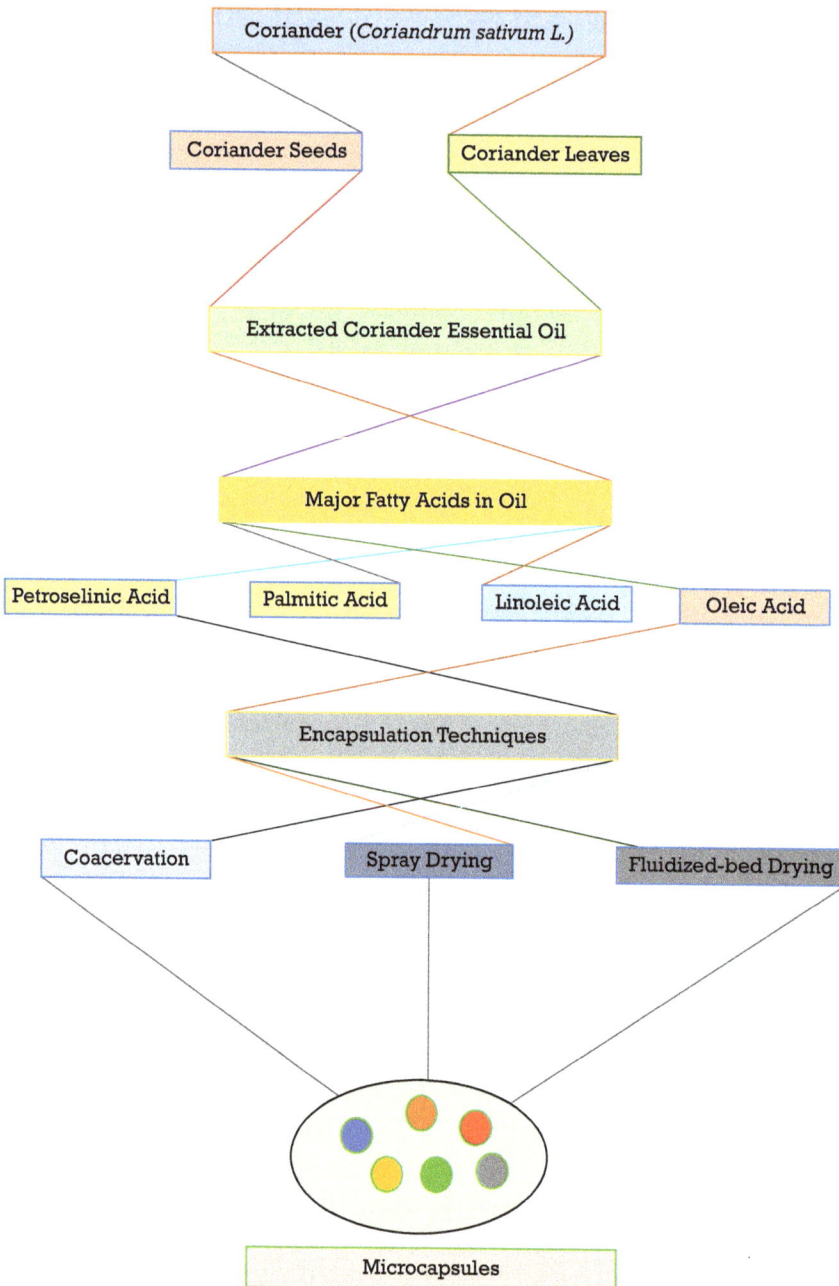

FIGURE 37.1   Characteristics of microcapsules from coriander oils.

very fine smaller particles or powder (10 to 50 mm) as well as bigger particles (2 to 3 mm) (Duman & Kaya, 2016; Dima et al., 2016; Rahim et al., 2022).

Another research work by Dima et al. (2013) developed CEO microcapsules using the spray drying technique. The emulsion was formulated with various wall or core materials for spray drying like gelatin, gum Arabic, and chitosan. The maximum efficiency was observed in the microcapsules prepared with alginate compared to others. Furthermore, the moisture content in the microcapsules was estimated at 1.12% and 2.48%. At the same time, microcapsules prepared with chitosan and inulin absorb water faster because they contain the maximum number of hydrophilic groups (–OH) of inulin, which can bind water molecules on the surface of the microcapsule.

37.7.3 FLUIDIZED-BED DRYING TECHNIQUE

The fluidized-bed drying technique is one of the most efficient drying methods, and it is increasingly used in the food, nutraceutical, pharmaceutical, and nutrition industries. For this method, raw material samples are dried and ground to form fine powder particles, which reduces material handling and processing times compared to other wet granular methods. Then, the powder is thoroughly mixed with the wall material in a volumetric flask or vessel. In this process, solvent-based coatings and hot melts are most commonly used for the microcapsulation of solids. The process of fluid bed encapsulation involves suspending solid particles in a temperature- and humidity-controlled upward-moving stream of air. Solid particles of prepared samples reach optimum temperature and move evenly. Then, the liquid coating material is sprayed on top with an atomizer, and the droplet size of microcapsules is smaller than the coated substrate. In the case of hot melts, the coating hardens with cold air until the walls have the desired thickness. In the case of solvent-based coatings, the coating hardens and evaporates the solvent with hot air (Hede et al., 2008; Nielsen, 2010). Popescu et al.'s (2016) fluidized bed granulator method was used to prepare capsules of CEO and other essential oils to improve the controlled release of bioactive principles. The results showed that the encapsulation efficiency of essential oils was ten times lower than that of a rotating drum due to the fluctuations of oil with the flow of air.

37.8 CONCLUSION

Coriander and its by-products contain a high amount of essential fatty acids and volatile oil contents. The stability of CEO declines with storage intervals. To improve CEO's stability, microencapsulation methods have been limited in use in the food and pharmaceutical industries. Therefore, there is a need to increase the large-scale production of microcapsules for the functionality of food products with maximum stability.

REFERENCES

Ahmad, R. S., Imran, M., Khan, M. K., Ahmad, M. H., Arshad, M. S., Ateeq, H., & Rahim, M. A. (2021). Introductory chapter: Herbs and spices-An overview. In *Herbs and Spices-New Processing Technologies.* IntechOpen, London, UK.

Alvim, I. D., Prata, A. S., & Grosso, C. R. (2016). Methods of encapsulation. In *Edible Films and Coatings* (pp. 317–334). CRC Press.

Anandharamakrishnan, C. (2014). *Techniques for Nanoencapsulation of Food Ingredients* (Vol. 8, pp. 65–67). New York: Springer.

Bag, A., & Chattopadhyay, R. R. (2018). Evaluation of antioxidant potential of essential oils of some commonly used Indian spices in in vitro models and in food supplements enriched with omega-6 and omega-3 fatty acids. *Environmental Science and Pollution Research, 25*(1), 388–398.

Balasubramanian, S., Awatar, R., & Singh, K. K. (2021). Grinding characteristics of coriander seed with reference to physical properties and energy consumption. *Food Science and Engineering,* 38–51.

Balasubramanian, S., Singh, K. K., & Kumar, R. (2012). Physical properties of coriander seeds at different moisture content. *International Agrophysics, 26*(4).

Beyzi, E., & Gurbuz, B. (2014). Effect of different fruit sizes on essential oil ratio and components of Coriander. *Journal of Essential Oil Bearing Plants, 17*(6), 1175–1180.

Bhuiyan, M. N. I., Begum, J., & Sultana, M. (2009). Chemical composition of leaf and seed essential oil of *Coriandrum sativum* L. from Bangladesh. *Bangladesh Journal of Pharmacology, 4*(2), 150–153.

Burdock, G. A., & Carabin, I. G. (2009). Safety assessment of coriander (*Coriandrum sativum* L.) essential oil as a food ingredient. *Food and Chemical Toxicology, 47*(1), 22–34.

Darughe, F., Barzegar, M., & Sahari, M. A. (2012). Antioxidant and antifungal activity of Coriander (*Coriandrum sativum* L.) essential oil in cake. *International Food Research Journal, 19*(3), 1253–1260.

Dhanapakiam, P., Joseph, J. M., Ramaswamy, V. K., Moorthi, M., & Kumar, A. S. (2007). The cholesterol lowering property of coriander seeds (*Coriandrum sativum*): Mechanism of action. *Journal of Environmental Biology, 29*(1), 53.

Dima, C., Cretu, R., Alexe, P., & Dima, S. (2013). Microencapsulation of coriander oil using complex coacervation method. *Scientific Study & Research. Chemistry & Chemical Engineering, Biotechnology, Food Industry, 14*(3), 155.

Dima, C., Pătrașcu, L., Cantaragiu, A., Alexe, P., & Dima, Ș. (2016). The kinetics of the swelling process and the release mechanisms of *Coriandrum sativum* L. essential oil from chitosan/alginate/inulin microcapsules. *Food Chemistry, 195*, 39–48.

Dos Santos, M. D. V., Neto, M. F. D. C., De Melo, A. C. G. R., Takahashi, J. A., Ferraz, V. P., Chagas, E. A., ... & De Melo Filho, A. A. (2019). Chemical composition of essential oil of Coriander seeds (*Coriandrum sativum*) cultivated in the Amazon Savannah, Brazil. *Chemical Engineering Transactions, 75*, 409–414.

Duman, F., & Kaya, M. (2016). Crayfish chitosan for microencapsulation of coriander (*Coriandrum sativum* L.) essential oil. *International Journal of Biological Macromolecules, 92*, 125–133.

Eikani, M. H., Golmohammad, F., & Rowshanzamir, S. (2007). Subcritical water extraction of essential oils from coriander seeds (*Coriandrum sativum* L.). *Journal of Food Engineering, 80*(2), 735–740.

Elgayyar, M., Draughon, F. A., Golden, D. A., & Mount, J. R. (2001). Antimicrobial activity of essential oils from plants against selected pathogenic and saprophytic microorganisms. *Journal of Food Protection, 64*(7), 1019–1024.

Emam, G. M., & heydari, H. G. (2006). *Sedative-hypnotic activity of extracts and essential oil of coriander seeds. Iranian Journal of Medical Sciences, 31*, 22–127.

Enascuta, C. E., Stepan, E., Oprescu, E. E., Radu, A., Alexandrescu, E., Stoica, R., ... & Niculescu, M. D. (2018). Microencapsulation of essential oils. *Revista de Chimie, 69*, 1612–1615.

Ghazanfari, N., Mortazavi, S. A., Yazdi, F. T., & Mohammadi, M. (2020). Microwave-assisted hydrodistillation extraction of essential oil from coriander seeds and evaluation of their composition, antioxidant and antimicrobial activity. *Heliyon, 6*(9), e04893.

Grosso, C., Ferraro, V., Figueiredo, A. C., Barroso, J. G., Coelho, J. A., & Palavra, A. M. (2008). Supercritical carbon dioxide extraction of volatile oil from Italian coriander seeds. *Food Chemistry, 111*(1), 197–203.

Hede, P. D., Bach, P., & Jensen, A. D. (2008). Two-fluid spray atomisation and pneumatic nozzles for fluid bed coating/agglomeration purposes: A review. *Chemical Engineering Science, 63*(14), 3821–3842.

Mechchate, H., Costa de Oliveira, R., Es-Safi, I., Vasconcelos Mourão, E. M., Bouhrim, M., Kyrylchuk, A., ... & Grafov, A. (2021). Antileukemic activity and molecular docking study of a polyphenolic extract from coriander seeds. *Pharmaceuticals, 14*(8), 770.

Mhemdi, H., Rodier, E., Kechaou, N., & Fages, J. (2011). A supercritical tuneable process for the selective extraction of fats and essential oil from coriander seeds. *Journal of Food Engineering, 105*(4), 609–616.

Micić, D., Ostojić, S., Pezo, L., Blagojević, S., Pavlić, B., Zeković, Z., & Đurović, S. (2019). Essential oils of coriander and sage: Investigation of chemical profile, thermal properties and QSRR analysis. *Industrial Crops and Products, 138*, 111438.

Moser, B. R., & Vaughn, S. F. (2010). Coriander seed oil methyl esters as biodiesel fuel: Unique fatty acid composition and excellent oxidative stability. *Biomass and Bioenergy, 34*(4), 550–558.

Msaada, K., Hosni, K., Taarit, M. B., Chahed, T., Kchouk, M. E., & Marzouk, B. (2007). Changes on essential oil composition of coriander (*Coriandrum sativum* L.) fruits during three stages of maturity. *Food Chemistry, 102*(4), 1131–1134.

Msaada, K., Hosni, K., Taarit, M. B., Chahed, T., & Marzouk, B. (2007). Variations in the essential oil composition from different parts of *Coriandrum sativum* L. cultivated in Tunisia. *The Italian Journal of Biochemistry, 56*(1), 47–52.

Nadeem, M., Anjum, F. M., Khan, M. I., Tehseen, S., El-Ghorab, A., & Sultan, J. I. (2013). Nutritional and medicinal aspects of coriander (*Coriandrum sativum* L.): A review. *British Food Journal, 115*, 743–755.

Neffati, M., & Marzouk, B. (2008). Changes in essential oil and fatty acid composition in coriander (*Coriandrum sativum* L.) leaves under saline conditions. *Industrial Crops and Products, 28*(2), 137–142.

Nguyen, Q. H., Talou, T., Evon, P., Cerny, M., & Merah, O. (2020). Fatty acid composition and oil content during coriander fruit development. *Food Chemistry, 326*, 127034.

Nielsen, A. F. (2010). *Microencapsulation by Spray Coating in Wurster Fluid Bed: Effects of Atomisation Related Variables*. Department of Pharmaceutics and Analytical Chemistry, Faculty of Pharmaceutical Sciences, University of Copenhagen.

Orav, A., Arak, E., & Raal, A. (2011). Essential oil composition of *Coriandrum sativum* L. fruits from different countries. *Journal of Essential Oil Bearing Plants, 14*(1), 118–123.

Popescu, M., Oancea, F., & Desliu-Avram, M. (2016). Vegetable oil conversion into core-shell bioproducts for stored grain protection. *Scientific Bulletin. Series F. Biotechnologies, 20*, 210–213.

Rahim, M. A., Imran, M., Khan, M. K., Ahmad, M. H., & Ahmad, R. S. (2022). Impact of spray drying operating conditions on encapsulation efficiency, oxidative quality, and sensorial evaluation of chia and fish oil blends. *Journal of Food Processing and Preservation, 46*(2), e16248.

Rajeshwari, U., & Andallu, B. (2011). Medicinal benefits of coriander (*Coriandrum sativum* L). *Spatula DD, 1*(1), 51–58.

Ramadan, M., & Mörsel, J. T. (2002). Oil composition of coriander (*Coriandrum sativum* L.) fruit-seeds. *European Food Research and Technology, 215*(3), 204–209.

Ramadan, M. F., & Mörsel, J. T. (2004). Oxidative stability of black cumin (*Nigella sativa* L.), coriander (*Coriandrum sativum* L.) and niger (*Guizotia abyssinica* Cass.) crude seed oils upon stripping. *European Journal of Lipid Science and Technology, 106*(1), 35–43.

Ramadan, M. F., & Moersel, J. T. (2006). Screening of the antiradical action of vegetable oils. *Journal of Food Composition and Analysis, 19*, 838–842.

Richter, J., & Schellenberg, I. (2007). Comparison of different extraction methods for the determination of essential oils and related compounds from aromatic plants and optimization of solid-phase microextraction/gas chromatography. *Analytical and Bioanalytical Chemistry, 387*(6), 2207–2217.

Sriti, J., Talou, T., Wannes, W. A., Cerny, M., & Marzouk, B. (2009). Essential oil, fatty acid and sterol composition of Tunisian coriander fruit different parts. *Journal of the Science of Food and Agriculture, 89*(10), 1659–1664.

Timilsena, Y. P., Akanbi, T. O., Khalid, N., Adhikari, B., & Barrow, C. J. (2019). Complex coacervation: Principles, mechanisms and applications in microencapsulation. *International Journal of Biological Macromolecules, 121*, 1276–1286.

Zeković, Z., Bušić, A., Komes, D., Vladić, J., Adamović, D., & Pavlić, B. (2015). Coriander seeds processing: Sequential extraction of non-polar and polar fractions using supercritical carbon dioxide extraction and ultrasound-assisted extraction. *Food and Bioproducts Processing, 95*, 218–227.

38 Non-Food Applications of *Coriandrum sativum* Essential Oil

Saba Ghufran, Muhammad Mohsin,
Qamar Ul Hassan, and Allah Nawaz Khan

CONTENTS

38.1 INTRODUCTION

Plants have been significant for man in providing beneficial compounds over the centuries. They were initially used for nutritional requirements, but with awareness of their medicinal characteristics, they became essential for disease control and health consequences (Vinatoru 2001). Therefore, they are a good source of a variety of essential oils. For thousands of years, essential oils have been used in different cultures for medical and health purposes. Uses of essential oils range from personal beauty care and household cleaning products to aromatherapy treatments and natural medicine. Essential oil benefits come from their antimicrobial, antioxidant, and anti-inflammatory traits. These therapeutic oils are gaining popularity because of their requirement in natural medicine and lack of side effects (Wahba, Abd Rabbu, et al. 2020).

The essential oils and extracts of aromatic plants and spices contain biologically active components with antibacterial, antifungal, and antioxidant activities that have allowed them to be used widely for pharmaceutical and food processing industry needs. Coriander (*Coriandrum sativum* L.) of the family Umbelliferae/Apiaceae is a glabrous, aromatic, herbaceous annual plant with a long history of applications in cooking and serving as a source of aroma compounds and essential oils. Coriander may be applied for food preparation as a flavoring agent and adjuvant and to prevent foodborne diseases and food spoilage because of the antibacterial properties of its essential oil (Ashraf, Ghufran, et al. 2020).

DOI: 10.1201/9781003204626-42

Coriander seed's essential oil is a good source of secondary plant metabolites such as poly-phenols, especially phenolic acids and flavonoids (Zeković, Vidović, et al. 2014). Therefore, it has industrial applications in pharmaceuticals and tobacco to eliminate unpleasant odors. In addition, it contains Δ6 palmitate desaturase enzyme, which assists coriander seeds oil in being accumulated with over 85% of the industrially worthwhile fatty acid and petroselinic acid (Jaworski and Cahoon 2003). Moreover, it has been regarded as a vital economic resource for the Egyptian national econ-omy (Wahba, Abd Rabbu, et al. 2020).

The essential oil extracted from coriander leaves contains significant compounds such as pheno-lic acids, polyphenols, and flavonoids, the most abundant bioactive compounds found in industrial, medical, and food applications. These compounds are harmless and have no side effects because their source is natural; these are more effective than the synthetic compounds prepared from chemi-cals in laboratories (Nambiar, Daniel, et al. 2010; Matasyoh, Maiyo, et al. 2009).

Essential oil of coriander possesses antibacterial (Erdogrul 2002), antioxidant (Rajeshwari, Siri, et al. 2012), antimicrobial (Delaquis, Stanich, et al. 2002), and antidiabetic (Aissaoui, Zizi, et al. 2011) activities. Coriander essential oil yields and components vary by age, origin, environmental factors, and ontogenesis. The main component of an essential oil varies in different plant parts. For example, the leading seed component is linalool, while in flowers, benzofuran,2,3-dihydro, and in leaves, decanal dodecanal is found as the main component (Turkmen, Bahadirli et al. 2016).

This chapter will discuss non-food applications of coriander essential oil on a commercial and domestic scale.

38.2 EXTRACTION AND PROCESSING OF SPECIFIC ESSENTIAL OIL

Essential oil from different parts of coriander can be extracted by simple processing. Freshly har-vested leaves and seeds are ground in a blender separately. Then, the ground leaves and seeds are subjected to hydro-distillation using a modified Clevenger-type glass apparatus for 4 h to isolate oils separately. The oil samples are stored at 0°C in air-tight containers after drying them over anhydrous sodium sulfate and filtered. Several techniques extract the essential oils from coriander (Richter and Schellenberg 2007). They are solvent extraction, hydro-distillation, Soxhlet dynamic headspace, steam distillation, supercritical water CO_2 extraction, micro-wave-assisted extraction, super-fluid extraction, and subcritical water extraction. Though these techniques provide a good yield of essential oils, supercritical water CO_2 extraction gives the greatest yield (Msaada, Taârit, et al. 2012).

38.3 CONCENTRATION AND CHEMISTRY OF CORIANDER ESSENTIAL OIL

Essential oil can be extracted from various parts of plants, including the leaves, flowers, stems, seeds, roots, and bark. The essential oil composition can vary among different parts of the same plant, for example, essential oil obtained from coriander seeds has a different composition from that of the flower and cilantro (immature leaves). The coriander oil from fully ripe and dried seeds is a colorless or pale yellow liquid with a characteristic odor and mild, sweet, warm, and aromatic fla-vor, and linalool is its principal constituent (Burdock and Carabin 2009). While aliphatic aldehydes (mainly C10-C16 aldehydes), with their unpleasant odor, are the main components of the volatile oil from the fresh herb (Ramezani, Rasouli, et al. 2009), linalool and other oxidized monoterpenes as well as monoterpene hydrocarbons predominate in the oil distilled from the fruit (Bhuiyan, Begum, et al. 2009). Coriander fruit contains about 0.2%–1.5% of volatile oil and 13%–20% of fatty acids (Olle, Bender, et al. 2010); however, it has been recorded that some cultivars contain up to 2.6% of volatile oil (Momin, Acharya, et al. 2012). As has been reported by Zawislak (2011), the essential oil content ranges from 1.87% to 2.33% in some of the coriander species. The fatty acid composi-tion of ripened fruits mainly includes petroselinic acid (68.8%), linoleic acid (16.6%), oleic acid (7.5%), and palmitic acid (3.8%). The hydro-distillation of coriander aerial parts gives essential oils in vegetative, full flowering, green fruit (immature), and brown fruit (mature) with a yield of 0.14%, 0.23%, 0.37%, and 0.31% (w/w), based on dry weight, respectively. Moreover, there is variation in seed yield and essential oil content of coriander cultivars grown at different locations. The essential oil content in coriander fruits is very different, 0.5%–2.5% (Mahendra and Bisht 2011, Zawiślak 2011), and it increases as the fruit ripens, while coriander leaves contain less essential oil than the fruit. In a report, the coriander leaves from Bangladesh have around 0.1% essential oil (Neffati and Marzouk 2008), and the plant harvested in Tunisia contains 0.12% essential oil in leaves dried in the air. The amount of essential oil in the coriander herb was, on average, 0.23 mL per 100 g, and it was higher in the generative phase (0.29 mL per 100 g) than in the vegetative phase (0.17 mL per 100 g) (Nurzynska-Wierdak 2013).

More specifically, coriander provides two types of herbal raw materials, fruits and leaves, the main biologically active substance of which is an essential oil, and these parts give a higher yield of essential oil than other parts. Coriander essential oil can be added to dishes as an aromatic spice, which at the same time acts as a digestive agent, accelerating the digestion process. The yield of coriander essential oil and its chemical composition changes ontogenesis (Bhuiyan, Begum, et al. 2009), which affects the plant's aroma, and thus the coriander fruit aroma is entirely different from the aroma of the herb (Neffati and Marzouk 2008). Immature fruits and leaves have an unpleasant odor called a "stink bug smell" due to *trans*-tridecen contained in the oil (Mandal and Mandal 2015).

Analysis of essential oil extracted from coriander seeds shows that seeds contain 0.8% more essential oil than other plant parts. This oil has a pleasant aroma due to the presence of oxygenated monoterpenes (80.5%) and monoterpene hydrocarbon (6.45%) (Pande, Pande, et al. 2010). This essential oil is colorless with a characteristic odor and soft, sweet, warm, and aromatic flavor, and linalool is its principal constituent (Świeca, Sęczyk, et al. 2014). More specifically, Shahwar, El-Ghorab et al. (2012) reported linalool, γ-terpinene, α-pinene, camphor, decanal geranyl acetate, limonene, geraniol, camphene, and d-limonene to be the main volatile compounds for seeds, while (E)-2-decenal, linalool, (E)-2-dodecenal, (E)-2-tetradecenal, 2-decen-1-ol, (E)-2-undecenal, dodecanal, (E)-2-tridecenal, (E)-2-hexadecenal, pentadecenal, and α-pinene were the main compounds of the coriander essential oils extracted from leaves. The growing seasons, growth stage, and climatic conditions are the main factors that affect the essential oil's concentration and chemical composition. Overall, the major volatile compounds of the coriander essential oil are β-linalool 66.0%, camphor 8.34%, geranyl acetate 6.91%, and cymene 6.35% (Kačániová, Galovičová et al. 2020).

In another report, the essential oils from leaves and fruits of *Coriandrum sativum* were analyzed by gas chromatography-mass spectroscopy (GC-MS); the results showed that the leaf oil contained 44 compounds mostly of aromatic acids including 2-decenoic acid (30.8%), E-11-tetradecenoic acid (13.4%), capric acid (12.7%), undecyl alcohol (6.4%), tridecanoic acid (5.5%), and undecanoic acid (7.1%) as major constituents. On the other hand, the seed oil contained 53 compounds, where the major compounds are linalool (37.7%), geranyl acetate (17.6%), and DŽ-terpinene (14.4%). The compositions of both oils varied qualitatively and quantitatively (Bhuiyan, Begum, et al. 2009).

38.4 ETHNO-MEDICINAL IMPORTANCE OF CORIANDER ESSENTIAL OIL

Plants' diversity is huge, but medicinal plants have had much more importance and attraction for humans throughout history; a plant is said to be medicinal if it helps cure diseases and maintain human health. It is a well-known truth that plants play a vital role in human therapy (Srivastava, 2000). Self-treatment is the most popular treatment method globally, and people like to take alternative medicinal plants for therapies. The essential oil of coriander is widely used as an ethnomedicinal plant in Indian, Middle Eastern, African, Mediterranean, and Southeast Asian cuisine, and in Latin American areas (Mahendra & Bisht, 2011). The coriander plant, scientifically named *Coriandrum sativum*, is an important medicinal herb that has been used from ancient times in traditional medicine, and its essential oil has valuable medicinal impacts (Asgarpanah & Kazemivash, 2012).

Ethnomedicinal sources of medication are less expensive and much more effective with minimum side effects; people throughout history have relied on potential plant sources for curing many diseases, and coriander is one such plant as a rich potential source of treatment for human diseases (Yasir et al. 2019). It is an herbaceous plant widely used in the Indian ethnomedicinal industry; people use it for urinary, digestive, and respiratory disorders (Momin et al. 2012). The whole coriander plant is medicinal, but essential oil from the seeds of this plant has many good impacts on the health of diabetic patients; seed powder, after clinical trials, proved to help reduce blood glucose levels (Deepa & Anuradha, 2011). Moreover, coriander essential oil has been proved to be best in inhibiting the growth of many microscopic organisms that are the leading cause of diseases nowadays. It is used to inhibit bacterial growth in many parts of the world (Silva et al. 2011b).

In ethnomedicine, dried coriander seeds are utilized to cure various ailments (Chithra and Leelamma, 1999; Momin et al. 2012). Coriander essential oil is also used for curing diabetic patients (Swanston-Flatt et al. 1990). In addition, the leaves and seeds have been used globally for a long time for medicinal purposes (Nadeem et al. 2013).

The essential oil of coriander is believed to be the best for joint pain and indigestion (Wichtl, 1994). While fresh leaves have antibacterial properties (Kubo et al. 2004), they are taken orally for abdominal pain, while seed paste is effective for treating headaches (Keshri & Dutta, 2010). The essential oil extracted from seeds has been extensively used to prepare folk medicine because it contains carminative and diuretic properties. In addition, coriander is rich with essential oil and iron that helps in anemia. Coriander plants also work to cure swelling, mouth ulcers, small pox, diarrhea, and many skin disorders (Rajeshwari, & Andallu, 2011).

The essential oil and dried powder of coriander leaves are used for patients with high cholesterol levels, and they are found to be very effective (Dhanapakiam et al. 2007). Furthermore, coriander is used in the Iranian medicinal industry for gastric and other urinary disorders, and it also helps in folk medicines; people use this plant for many medical applications (Pathak Nimish et al. 2011). As coriander essential oil's medicinal and pharmaceutical importance is recognized worldwide and clinically proven, it has been used in folk medicine for many years for curing many diseases in humans (Devi et al. 2020). In addition, the coriander plant is widely used for treating anxiety, for which the powder form is used in which the dry seeds are crushed and orally taken with water, and it is believed to be the best therapy (Aćimović et al. 2011).

The herbal industry is as old as humanity, and coriander is a major herb that is widely used for food and medicine; in both industries, it has great importance due to its incredible results, and this plant needs to be further explored to achieve its maximum potential and help in industrial purposes (Chawla & Thakur, 2013).

38.5　APPLICATIONS IN THE PHARMACEUTICAL INDUSTRY

Drugs derived from plant sources are healthy. Drug-resistant pathogens are increasing, and there are limitations in the availability of antibiotic drugs. This has increased the awareness of medicinal plants and their derived essential oils due to the presence of active chemicals used for manufacturing a wide variety of pharmaceutical drugs. When exposed to insect pests and herbivores, secondary metabolites are naturally synthesized in plants, especially coriander. Secondary metabolites and combinations of other chemicals are used for various chemotherapies. Medicine companies are globally working to explore cheap and potential plants for manufacturing healthy drugs. The essential oil of coriander has reached worldwide importance due to enormous numbers of bioactive chemical compounds (Raut & Karuppayil, 2014; Perumal Samy & Gopalakrishnakone, 2010).

The phytochemistry of the coriander plant includes essential oil, terpenoids, alkaloids, tannins, phenolics, glycosides, and flavonoids. Previous studies about the pharmaceutical importance of coriander show that its pharmacological effects are anxiolytic, sedative-hypnotic, memory enhancing, neuroprotective, antifungal, insecticidal, antibacterial, anticonvulsant, antidiabetic, anticancer, and diuretic (Al-Snafi, 2016). Major chemicals that are pharmaceutically very important found in the essential oil of coriander are α-pinene, geranyl acetate, linalool, camphor, ɣ-terpinene, geraniol, and terpinolene, acetic acid (fruit), α-phellandrene, β-sitosterol, borneol, bornyl-acetate, caffeic acid, carvone, caryophyllene, niacin, oleic-acid, *p*-cymene, *p*-hydroxybenzoic acid, psoralen, quercetin, palmitic acid, pectin, angelicin, apigenin, ascorbic acid, β-carotene, protocatechuic acid, rhamnetin, rutin, sabinene, tannin, *trans*-anethole, umbelliferone, vanillic acid, zinc, β-pinene, chlorogenic-acid, chromium, *cis*-ocimene, citronellol, copper, dipentene, elemol, fiber, fructose, γ-terpinene, geranial, isoquercitrin, limone, linoleic acid, magnesium, myrcene, myristic-acid, myristicin, nerol, and nerolidol. These chemical compounds vary in different coriander species, but the average essential oil present in almost all coriander species is linalool 67.5%, γ-terpinene 9.0%, camphor 3.0%, α-pinene 10.5%, geraaniol 1.9%, and geranyl acetate 4.0% (Kassahun, 2020; Rajeshwari & Andallu, 2011).

38.5.1 ANTIMICROBIAL APPLICATIONS

Antimicrobial infections are common, and they cause severe hazards to human health. Microbes have a high-speed growth rate, and they multiply very quickly. Experiments on strains of *C. jejuni* and *E. coli* bacteria for exploring chemicals for drugs from coriander essential oil show that coriander essential oil is a potential source of antimicrobial properties due to an important chemical (linalool). In the laboratory, linalool effectively inhibited bacterial strains' growth (Duarte et al. 2016). The medically important chemical linalool from coriander has a unique mechanism of action as it first destroys the membrane of different harmful bacteria, such as Gram-negative and positive, resulting in the cell death of bacteria (Silva et al. 2011).

38.5.2 ANTIOXIDANT APPLICATIONS

Antioxidants play a vital role in the human body because they help prevent the reaction of free radicals. Many plants have been explored to make antioxidants; coriander essential oil contains terpene, linalool, and terpenoid, and they have active chemicals that show antioxidant potential (Darughe et al. 2012). There is an increase in demand for synthetic antioxidants like butylated hydroxyl anisole and butylated hydroxyl toluene, but plant-based antioxidants are less expensive and effective (Reddy et al. 2005). Analysis of chemicals that are could be used as antioxidants in the drug industry, present in the essential oil of coriander plant, by GC-MS showed that chemicals present in essential oil have great potential to scavenge for free radicals like inhibiting the effect of reactive oxygen species (Samojlik et al. 2010).

38.5.3 ANTICONVULSANT APPLICATIONS

A seizure is a sudden electrical disturbance in the nervous system, especially in the brain, that is uncontrolled. Seizures can produce much pain, behavior changes, and negative feelings, and the body gets stiff due to an imbalance of ions in the body. When there is an imbalance of ions, the body cannot contract or relax its muscles properly, resulting in a lot of pain and severe headache, and if seizures are prolonged, they lead to epilepsy. Different chemicals in the essential oil of coriander were tested to explore anticonvulsant properties; two tests, pentylene-tetrazole and electroshock maximal tests, were designed to test antiseizure action. Both tests proved that camphor and *p*-cymene chemicals are very effective in tonic seizures (Hosseinzadeh Hossein and Madanifard, 2005).

38.5.4 ANTHELMENTIC APPLICATIONS

The anthelmintic potential of essential oil of coriander was examined *in vitro* on adult and egg nematode parasites; essential oil was also examined *in vivo* in different sheep that were infected by the same nematode species; it was proved clinically that active chemical activity inhibited the growth of and hatching of eggs; the hydroalcoholic extract was even better *in vitro* for adult parasites (Pathak Nimish et al. 2011).

38.5.5 Sedative-Hypnotic Applications

Sedative properties are well known in coriander essential oil; chemical compounds present in essential oil are widely used in drugs of a sedative nature. After clinical trials, many experiments have been conducted to observe the sedative properties of aqueous and hydroalcoholic compounds, and an essential oil rich with potential sedative compounds was administered to different rats. It was concluded that chemicals like linalool in essential oil induced sleep (Emamghoreishi and Heidari, 2006).

38.5.6 Hypoglycemic Applications

Due to busy and sedentary lifestyles, diabetic patients are increasing in number worldwide; different healthcare precautions are prescribed to minimize the number of diabetic patients. Drugs are the secondary solution for diabetic patients. Coriander essential oil plays an essential role in making drugs to relieve diabetic patients. The active compounds in coriander essential oil are α-phellandrene, β-pinene, and sabinene. After experiments, further analysis has proven that these chemicals help reduce blood sugar and regulate the hormone for releasing insulin (Rajeshwari & Andallu, B. 2011).

38.5.7 Applications in Menstrual Disorders

Menstruation is the monthly shedding of the uterus lining in a woman. The menstrual cycle is critical in female life; it ensures a healthy reproductive system. Different hormones regulate the menstrual cycle, and different internal and external factors can alter the timing or disturb the menstrual cycle. Chemical compounds of coriander essential oil have been effectively used and tested via large-scale trials. γ-Terpinene, isoquercitrin, and myrcene have great potential to induce a proper menstrual cycle, and they help kill bacteria in the female vagina (Rajeshwari & Andallu, 2011).

38.5.8 Applications in Heart Diseases

The essential oil of coriander has a wide range of non-food applications. Heart problems are common nowadays due to unhealthy food and sedentary lifestyles; there is a vast death rate worldwide due to heart problems. People use expensive medication for heart disorders with side effects; plant-based medication is impressive due to its low cost. The essential oil of coriander contains petroselenic acid which different pharmaceutical companies have investigated. It contains great potential to reduce arachidonic levels in cardiac muscle, resulting in a healthy blood circulatory system and low chances of heart failure (Weber et al. 1995).

38.5.9 Applications in Drugs to Reduce Heavy Metal Contamination

Heavy metal contamination is the biggest problem globally as water, or any food material, is rarely free from these harmful chemicals. Heavy metals, including mercury, chromium, arsenic, and cadmium, are the most commonly found poisons that harm human health and bring physiological and morphological impacts. Arsenic is a leading cause of heart, kidney, liver, and stomach disorders. Humans are exposed to these heavy metals in different ways. In recent years it has been proved that bioactive compounds present in the essential oil of coriander are a potential source of future drugs to reduce the harmful impacts of heavy metals on human beings, as they react with heavy metals and inhibit their workings (Kadiri et al. 2017; Winarti et al. 2017).

REFERENCES

Aissaoui, A., et al. (2011). "Hypoglycemic and hypolipidemic effects of *Coriandrum sativum* L. in Meriones shawi rats." *Journal of Ethnopharmacology* 137(1): 652–661.

Al-Snafi, A. E. (2016). "A review on chemical constituents and pharmacological activities of *Coriandrum sativum*." *IOSR Journal of Pharmacy* 6(7): 17–42.

Ashraf, R., et al. (2020). "Cold pressed coriander (*Coriandrum sativum* L.) seed oil." In *Cold Pressed Oils*. Elsevier, 345–356.

Bhuiyan, M. N. I., et al. (2009). "Chemical composition of leaf and seed essential oil of *Coriandrum sativum* L. from Bangladesh." *Bangladesh Journal of Pharmacology* 4(2): 150–153.

Burdock, G. A. and Carabin, I. G. (2009). "Safety assessment of coriander (*Coriandrum sativum* L.) essential oil as a food ingredient." *Food and Chemical Toxicology* 47(1): 22–34.

Darughe, F., Barzegar, M. and Sahari, M. A. (2012). "Antioxidant and antifungal activity of Coriander (*Coriandrum sativum* L.) essential oil in cake." *International Food Research Journal* 19(3): 1253–1260.

Duarte, A., Luís, Â., Oleastro, M. and Domingues, F. C. (2016). "Antioxidant properties of coriander essential oil and linalool and their potential to control *Campylobacter* spp." *Food Control* 61: 115–122.

Emamghoreishi, M. and Heidari-Hamedani, G. (2006). "Sedative-hypnotic activity of extracts and essential oil of coriander seeds." *Iranian Journal of Medical Sciences* 31(1): 22–27.

Erdogrul, Ö. T. (2002). "Antibacterial activities of some plant extracts used in folk medicine." *Pharmaceutical Biology* 40(4): 269–273.

Hosseinzadeh, H. and Madanifard, M. (2005). "Anticonvulsant effect of *Coriander sativum* L. seed extracts in mice." *Iranian Journal of Pharmacy* 3: 1–4.

Jaworski, J. and Cahoon, E. B. (2003). "Industrial oils from transgenic plants." *Current Opinion in Plant Biology* 6(2): 178–184.

Kačániová, M., et al. (2020). "Antioxidant, antimicrobial and antibiofilm activity of coriander (*Coriandrum sativum* L.) essential oil for its application in foods." *Foods* 9(3): 282.

Kadiri, L., Lebkiri, A., Rifi, E. H., Essaadaoui, Y., Ouass, A., Lebkiri, I. and Hamad, H. (2017). "Characterization of coriander seeds "*Coriandrum sativum*." *International Journal of Scientific and Engineering Research* 8(7): 2303–2308.

Kassahun, B. M. (2020). "Unleashing the exploitation of coriander (*Coriander sativum* L.) for biological, industrial and pharmaceutical applications." *Academic Research Journal of Agricultural Science and Research* 8(6): 552–564.

Mahendra, P. and Bisht, S. (2011). "*Coriandrum sativum*: A daily use spice with great medicinal effect." *Pharmacognosy Journal* 3(21): 84–88.

Mandal, S. and Mandal, M. (2015). "Coriander (*Coriandrum sativum* L.) essential oil: Chemistry and biological activity." *Asian Pacific Journal of Tropical Biomedicine* 5(6): 421–428.

Matasyoh, J., et al. (2009). "Chemical composition and antimicrobial activity of the essential oil of *Coriandrum sativum*." *Food Chemistry* 113(2): 526–529.

Momin, A. H., et al. (2012). "*Coriandrum sativum*-review of advances in phytopharmacology." *International Journal of Pharmaceutical Sciences and Research* 3(5): 1233.

Msaada, K., et al. (2012). "Comparison of different extraction methods for the determination of essential oils and related compounds from coriander (*Coriandrum sativum* L.)." *Acta Chimica Slovenica* 59(4).

Nambiar, V. S., et al. (2010). "Characterization of polyphenols from coriander leaves (*Coriandrum sativum*), red amaranthus (*A. paniculatus*) and green amaranthus (*A. frumentaceus*) using paper chromatography and their health implications." *Journal of Herbal Medicine and Toxicology* 4(1): 173–177.

Neffati, M. and Marzouk, B. (2008). "Changes in essential oil and fatty acid composition in coriander (*Coriandrum sativum* L.) leaves under saline conditions." *Industrial Crops and Products* 28(2): 137–142.

Nurzynska-Wierdak, R. (2013). "Essential oil composition of the coriander (*Coriandrum sativum* L.) herb depending on the development stage." *Acta Agrobotanica* 66(1).

Olle, M., et al. (2010). "The content of oils in umbelliferous crops and its formation." *Agronomy Research* 8(3): 687–696.

Pande, K. K., et al. (2010). "Gas chromatographic investigation of *Coriandrum sativum* L. from Indian Himalayas." *New York Science Journal* 3(6): 43–47.

Pathak Nimish, L., Kasture Sanjay, B., Bhatt Nayna, M. and Rathod Jaimik, D. (2011). "Phytopharmacological properties of *Coriander sativum* as a potential medicinal tree: An overview." *Journal of Applied Pharmaceutical Science* 1(4): 20–25.

Perumal Samy, R. and Gopalakrishnakone, P. (2010). "Therapeutic potential of plants as anti-microbials for drug discovery." *Evidence-Based Complementary and Alternative Medicine* 7(3): 283–294.

Rajeshwari, C., et al. (2012). "Antioxidant and antiarthritic potential of coriander (*Coriandrum sativum* L.) leaves." *e-SPEN Journal* 7(6): e223–e228.

Rajeshwari, U. and Andallu, B. (2011). "Medicinal benefits of coriander (*Coriandrum sativum* L)." *Spatula DD* 1(1): 51–58.

Ramezani, S., et al. (2009). "Changes in essential oil content of Coriander (*Coriandrum sativum* L.) aerial parts during four phonological stages in Iran." *Journal of Essential Oil Bearing Plants* 12(6): 683–689.

Raut, J. S. and Karuppayil, S. M. (2014). "A status review on the medicinal properties of essential oils." *Industrial Crops and Products* 62: 250–264.

Reddy, V., Urooj, A. and Kumar, A. (2005). "Evaluation of antioxidant activity of some plant extracts and their application in biscuits." *Food Chemistry* 90: 317–321.

Richter, J. and Schellenberg, I. (2007). "Comparison of different extraction methods for the determination of essential oils and related compounds from aromatic plants and optimization of solid-phase microextraction/gas chromatography." *Analytical and Bioanalytical Chemistry* 387(6): 2207–2217.

Samojlik, I., Lakic, N., Mimica-Dukic, N., Đaković-Švajcer, K. and Bozin, B. (2010). "Antioxidant and hepatoprotective potential of essential oils of coriander (*Coriandrum sativum* L.) and caraway (*Carum carvi* L.)(Apiaceae)." *Journal of Agricultural and Food Chemistry* 58(15): 8848–8853.

Shahwar, M. K., et al. (2012). "Characterization of coriander (*Coriandrum sativum* L.) seeds and leaves: Volatile and non volatile extracts." *International Journal of Food Properties* 15(4): 736–747.

Silva, F., Ferreira, S., Queiroz, J. A. and Domingues, F. C. (2011). "Coriander (*Coriandrum sativum* L.) essential oil: Its antibacterial activity and mode of action evaluated by flow cytometry." *Journal of Medical Microbiology* 60(10): 1479–1486.

Świeca, M., et al. (2014). "Bread enriched with quinoa leaves-The influence of protein-phenolics interactions on the nutritional and antioxidant quality." *Food Chemistry* 162: 54–62.

Turkmen, M., et al. (2016). "Essential oil components of fresh coriander (*Coriandrum sativum* L.) herbs from different locations in Turkey." In International Conference on Advanced Materials and Systems (ICAMS), The National Research & Development Institute for Textiles and Leather-INCDTP.

Vinatoru, M. (2001). "An overview of the ultrasonically assisted extraction of bioactive principles from herbs." *Ultrasonics Sonochemistry* 8(3): 303–313.

Wahba, H. E., et al. (2020). "Evaluation of essential oil isolated from dry coriander seeds and recycling of the plant waste under different storage conditions." *Bulletin of the National Research Centre* 44(1): 1–7.

Weber, N., Richter, K. D., Schulte, E. and Mukherjee, K. D. (1995). "Petroselinic acid from dietary triacylglycerols reduces the concentration of arachidonic acid in tissue lipids of rats." *Journal of Nutrition* 125(6): 1563–1568.

Winarti, S., Pertiwi, C. N., Hanani, A. Z., Mujamil, S. I., Putra, K. A. and Herlambang, K. C. (2017). "Beneficial of coriander leaves (*Coriandrum sativum* L.) to reduce heavy metals contamination in Rod Shellfish." *Journal of Physics: Conf. Series* 953: 1–6.

Zawiślak, G. (2011). "The chemical composition of essential oil from the fruit of coriander (*Coriandrum sativum* L.)." *Annal Univers Mariae Curie-Sklodowska Sect DDD: Pharm* 24: 169–175.

Zeković, Z., et al. (2014). "Optimization of subcritical water extraction of antioxidants from *Coriandrum sativum* seeds by response surface methodology." *The Journal of Supercritical Fluids* 95: 560–566.

Section V

Coriander Extracts

Chemistry, Technology, Functionality, and Applications

39 Food Applications of Coriander Seed Extracts

Muhammad Haseeb Ahmad, Muhammad Imran,
Muhammad Kamran Khan, Rabia Shabir Ahmad,
Muhammad Sajid Arshad, Muhammad Faizan Afzal,
Zubair Javed, and Muhammad Abdul Rahim

CONTENTS

39.1 INTRODUCTION

Coriander, known as *Coriandrum sativum* L., belongs to the family Umbelliferae. It is an annually cultivated herbaceous plant that is commonly used as a dry powder and fresh leaves because of its aromatic and flavoring properties. Coriander has different names in different languages, such as English (coriander), Arabic (Kuzbara), Hindi (Dhania), Greek (korion), and Urdu (Dhania). Coriander seeds were discovered in the Egyptian tomb of Rameses II. In addition, the Egyptians know the coriander herb as a "spice of happiness" because of its aphrodisiac properties. The Romans and Greeks used this herb as a flavoring agent in wine and medicine (Laribi et al., 2015). South Asia is the largest producer of coriander globally and exports to other countries, including the Middle East, the USA, Europe, and Southeast Asia (Diederichsen, 1996). The major producers of coriander are Morocco, Argentina, Russia, Ukraine, Romania, and India. It is mainly cultivated in the Mediterranean countries and coriander seed extract is commonly used in the pharmaceutical, nutraceutical, and food processing industries. Coriander is traditionally used as a medicine for joint pain, gastrointestinal problems, and rheumatism, but advanced studies have also confirmed the medicinal effect of coriander, like its effects on carbohydrate metabolism and hypoglycemic action (Wangesteen et al., 2004). In addition, coriander seed extract has an antimicrobial effect against certain microorganisms (Lo Cantore et al., 2004). In the food sector, coriander seeds and leaves are used as a spice and flavoring agent in various commercial food products such as tea, liqueur, meat products, and pickles (Illés et al., 2000). Coriander is mainly used to prepare salads, sauces, and soups. In addition, it is used as an essential ingredient in Vietnam and Thai cuisine due to the unique aroma of its fresh green leaves (Burdock & Carabin 2009).

The most important extraction techniques, such as organic solvent extraction, supercritical liquid extraction, and steam extraction, commonly extract oil from coriander seeds. The advantages and disadvantages of each technique vary depending on its capital cost, operating cost, quality of extract, and production. The most widespread and cheapest technique is the method of steam distillation and solvent extraction, but at the same time, there are some limitations regarding extraction as it stimulates the chemical transformation of extract through the oxidation of many compounds. The organic solvent extraction technique mainly extracts volatile oil from the seeds and is considered moderate in operation and capital cost. New regulations regarding the removal of solvent residues in oleoresins and the release of organic solvents into the air, and the cleaning of the sample after extraction prevent the use of the technology (Catchpole & Grey, 1996). Supercritical fluid extraction is an innovative technique compared to other extraction techniques; it is crucial for producing oil and other natural products (Brunner, 2005, Temelli, 2009). Supercritical fluid extraction is considered to have moderate operating costs and the highest capital cost. The extract obtained from this technique is of high quality as there is no need to retain solvent residues in the sample, and at the same time, there is no chemical change in the product (Machmudah et al., 2008; Boutin & Badens, 2009; Donelian et al., 2009).

Coriander seed oil is used as an appetizer, an aromatic stimulant, and a carminative, and it stimulates the intestine and stomach functions. In general, it is best for the nervous system. It is mainly used to prepare profane medicine like laxatives. It is used as an herb to treat inflammation, piles, and headaches, while the fruit is beneficial for colic, conjunctivitis, and piles and is used in the Asia region. It is used to prevent animal fat rancidity due to the presence of antioxidants. The coriander seed oil contains substances that are helpful to kill meat-spoiling bacteria and fungi, as well as to prevent infection in wounds. It is also very effective for stomach problems like flatulence, diarrhea, and indigestion. Its tea can be used for colic in children and is also helpful to improve discomfort and provide good health for infants. Coriander seed oil is beneficial in remedying arthritis due to its anti-inflammatory property. The extracts of spices and aromatics are widely used in various fields such as pharmaceuticals, food preservation, natural therapies, and alternative medicine (Wilkinson, 1993; Mandal & Mandal, 2015).

39.2   NUTRITIONAL COMPOSITION OF CORIANDER

All parts of the coriander plant are edible, while the dried seeds and fresh coriander leaves are used commonly. The shrubbery part of coriander is frequently used as salad and vegetable, and it is a rich source of vitamins, fibers, protein, carbohydrates, and minerals such as Fe, P, and Ca. On the other hand, fresh coriander leaves and seeds contain essential oils and components used in food products as preservatives and flavor enhancers (Kalemba & Kunicka, 2003).

Coriander seed extract contains a monounsaturated fatty acid known as petroselinic acid. Thus, coriander seed extract is a triglyceride oil that is a major source of essential oil and lipids derived from the seeds and upper parts of the plant. The mature coriander leaves contain protein, carbohydrates, total ash, and moisture in the following ratio: 3.3%, 6.5%, 1.7%, and 87.9%, respectively (Ganesan et al., 2013).

The coriander seed extract is commercially offered as a food supplement for healthy adults at a maximum of 600 mg/daily (EFSA, 2013). However, it is considered more important for the food industry as it is the main source of lipids, constituting 28.4% of the total seed weight (Yeung & Bowra, 2011). It has also been used as a traditional medicine for many years (Burdock & Carabin,

2009). Hippocrates (460–377 BC) used coriander as traditional Greek medicine. In combination with other herbs, coriander seed powder is beneficial for convulsion, insomnia, anxiety, dyspeptic problems, and loss of appetite (Grieve, 1971). The coriander seed extract is also noted to be effective for blood glucose regulation, so it is used as an anti-hyperglycemic agent (Gallagher, 2003).

39.3 CORIANDER CONTAMINATION

The literature shows that coriander seed extract and its products are mainly contaminated with pesticides, mycotoxins, and other substances. Mycotoxins in coriander seed extract are known to be a major contaminant. Six coriander samples showed the presence of mycotoxins such as aflatoxin, ochratoxin, zearalenone, and citrinin. In a screening of 126 spice samples, the presence of mycotoxins was confirmed in 20 out of 50 coriander seed extract samples. In addition, hexchlorohexane (0.4 ppm) and DDT (0.36 ppm) are reported in coriander seed extract samples (Thirumala-Devi et al., 2001).

39.4 CHEMISTRY OF CORIANDER SEED EXTRACT

Essential oils can be extracted from various parts of plants such as flowers, seeds, roots, leaves, and bark. Coriander oil obtained from completely dried and ripe seeds has important properties, including a mild odor, colorless or light yellow liquid, aromatic flavor, and lanolin (Burdock & Carabin, 2009). However, fresh coriander herbs contain aliphatic aldehydes (C_{10}-C_{16} aldehydes), which have a strong odor as an essential component of volatile oil (Potter & Fagerson, 1990). At the same time, other organic compounds such as linalool, oxidized monoterpenes, and monoterpene hydrocarbons are mainly found in the distilled oil from the plant's fruit (Bhuiyan et al., 2009).

The coriander seeds contain 0.2% to 20% volatile oil (Olle et al., 2010). The essential oil of ripened seeds contains a different composition of fatty acids such as palmitic acid (3.8%), oleic acid (7.5%), linoleic acid (16.6%), and petroselinic acid (68.8%) (Momin et al., 2012). Based on dry weight, the aerial parts of coriander provide essential oil including green fruit (0.37%), brown fruit (0.31%), full flowering (0.23%), and vegetative (0.14%) (Ramezani et al., 2009).

The coriander seed and coriander seed extract yield vary depending on its cultivation at various sites. The essential oil content increased as coriander fruit ripened in a ratio of 0.5%–2.5% (Mahendra & Bisht, 2011). However, the essential oil content in coriander leaves is less than in its fruit (Msaada et al., 2007). Coriander seed has a different concentration of essential oil in different countries like Bangladesh and Tunisia, containing 0.1% and 0.12%, respectively (Neffati & Marzouk, 2008; Bhuiyan et al., 2009). The fresh coriander plant is composed of various compounds including α-pinene (5.5%), limonene (2.3%), camphor (3.7%), geranyl acetate (1.9%), p-cymene (1.5%), and l-terpinene (8.8%) (Mageed et al., 2012). The coriander seed extract depends on the maturity of the seed. Hence, the concentration of essential oil content in coriander at the generative phase is 0.29 mL/100, while at the stage of harvest, the concentration of EO is 0.17 mL/100 g (Nurzynska-Wierdak, 2013).

The coriander seed extract varies qualitatively and quantitatively in different plant parts. The coriander leaves extract contains 44 compounds, including aromatic acids like E-11-tetradecenoic acid (13.4%), capric acid (12.7%), 2-decenoic acid (30.8%), undecyl alcohol (6.4%), undecanoic acid (7.1%), and tridecanoic acid (5.5%). In addition, coriander seed extract is composed of pinene, linalool, decanal geranyl acetate, camphor, geraniol, D-limonene, and camphene as major volatile compounds (Shahwar et al., 2012).

39.5 CORIANDER EXTRACT AS SOURCE OF ESSENTIAL OILS

Coriander leaves, stems, and seeds are acclaimed for their pleasant and sweet aroma because these components are rich in volatile essential oil. In ripe and dried coriander fruits, fatty oils and essential oil are present in different ratios: 2.6% to 22.5% and 0.01% to 6.0%, respectively (Saxena & Agarwal, 2019). Therefore, coriander seeds, stems, roots, and leaves are analyzed through many studies for total oil contents. The GC-MS technique examined the major constituents of coriander seed extract. When the coriander seed extract was treated with chloroform (2:1, v/v), the total amount of lipid was about 28.4% of seed weight. In addition, there was 65.7% of petroselinic acid after linoleic acid in coriander seed extract. Coriander contains about 52 components, of which 98.4% found in coriander seed extract. The main components of coriander seed extract were geranyl acetate (8.12%), pinene (4.09%) and linalool (75.3%), linoleic acid (33.2%), palmitic acid (11.0%), and oleic acid (36.5%) (Singh et al., 2006).

Coriander seed extract is rich in linalool, camphor, citronellol, coriandrin, coriandrons A-E, borneol, *p*-cymene, α-pinene, monoterpenes, and flavonoids (Pathak et al., 2011). Linalool is the main constituent of coriander seed extract and is considered safe (Gil et al., 2002). Coriander seed extract also contains fatty acids like petroselinic, oleic, and linolenic acids, and these fatty acids account for two-thirds of the coriander seed extract. The compositional determination of coriander seed extract shows the presence of various compounds like linalool (60–80%), α-terpineol (0.5%), r-cymene (trace–3.5%), a-pinene (0.2–8.5%), camphene (1.4%), ketones (7–9%), linalyl acetate (0–2.7%), geraniol (1.2–4.6%), limonene (0.5–4%), and camphor (0.9–4.9%) (Grosso et al., 2008) Linalool is present in coriander seed oil but its percentage varies in different places and is affected by storage conditions as well (Misharina, 2001). Coriander seed extract obtained from New Zealand cultivated coriander contains 65.8% linalool which is less than Russian coriander seed extract (linalool 68%) (Gil et al., 2002; Misharina, 2001). It is a fragrant compound in many plants such as sweet basil, cinnamon, spices, coriander, bay leaf, fruits, and thyme. It is used to minimize the effect of trauma and is beneficial for immune response. It is widely used as an aroma in essential oils and fragrance in soaps, toiletries, and shampoos (Nadeem et al., 2013).

39.6   COMMERCIAL USES OF CORIANDER

39.6.1   USES AS A FOOD CONSTITUENT

Coriander seed extract is generally used as a flavoring agent in many foods such as candy, pickles, tobacco, alcoholic beverages, and meat sauces. The coriander seed extract is blended with clary, nutmeg, sage, cardamom, bergamot, and anise in flavoring compositions. The average level of coriander seed extract used is from 0.1 to 100 ppm. Coriander seed extract is also a disinfectant due to its antimicrobial effects against specific saprophytic and pathogenic microorganisms (Meena, 1994; Elgayyar et al., 2001). Coriander seed is one of the most well-known spices of the world and is very significant for its global trade (Small, 1997). Coriander seed extract is considered a vital essential oil after orange oil due to its high annual production estimated at 50 million dollars (Lawrence, 1993). Coriander seed extract is also helpful in a typical spice blend like Indian curry, but this spice blend cannot be substituted with an essential oil liquid mixture. Coriander seeds are extensively used in food items like puddings, bread, sausages, gin essence, spicy sauces, liqueurs, and cakes (Facciola, 1990). Coriander seed extract is very beneficial for food preservation due to its antioxidant properties, free radical scavenging, and preventing oxidative deterioration. It was reported in the literature that the efficacy of coriander seed extract as a free radical scavenger is greater than other essential oils, as follows: niger < olive oil < linseed < hemp seed < sunflower < peanut < cottonseed < black cumin < coriander seed extract.

39.6.2   NON-FOOD USES

Coriander seed oil extract is mainly used in food but it is also very beneficial beyond food. It is used in many perfumes such as 'Oriental type' perfume and shows a stimulating effect with olibanum and Ceylon cinnamon. The coriander seed extract is also used in cosmetic products like creams, soaps, perfumes, and body moisturizers. Coriander fruit and coriander seed extract are beneficial to enhance the taste and properties of other drugs. Coriander seed extract is famous for its extract use and use for aromatherapy and decoration. A coriander seed extract has many uses in pharmaceutical preparations as a flavoring agent and is used as a carminative, spasmolytic, and stomachic due to its fungicidal and bactericidal properties. The coriander seed extract is used in traditional medicine for several diseases like effectiveness against intestinal parasites and treatment of rheumatism and joint pain (Opdyke, 1973; Cooksley, 2003; Platel & Srinivasan, 2004).

39.6.3 FOOD PRESERVATION AND ANTI-SPOILAGE

Oxidative stress caused by lipid peroxidation in food leads to a pungent smell and taste, change in color, rancidity, and minimized nutritional value (Bhanger et al., 2008). Butylated hydroxyl anisole (BHA) and BHT are the most common synthetic antioxidants used in foods, but their safety is compromised due to their toxic effect like carcinogenicity and liver damage (Nanditha et al., 2009). Meanwhile, coriander contains admirable antioxidant activity and is highly stable concerning temperature. However, coriander is used as a synthetic antioxidant. The coriander seed extract is considered excellent in food preservation and the prevention of food spoilage due to its antifungal, antibacterial, and antioxidative characteristics. It was noted that food products containing coriander as an antioxidant showed excellent results compared to BHA.

39.6.4 ANTIBACTERIAL ACTIVITY

Coriander seed and leaf oil have antibacterial activity against food spoilage: Gram-positive and Gram-negative bacteria like *Listeria monocytogenes*, *Staphylococcus aureus*, *Yersinia enterocolitica*, and *Bacillus cereus* and *Salmonella typhimurium*, etc. (Delaquis et al. 2002). It was reported that coriander seed showed an inhibitory effect on degenerative bacteria, including *Pseudomonas aeruginosa*, *Escherichia coli* (*E. coli*), *Escherichia cloaca*, *Klebsiella pneumoniae*, and *Bacillus megaterium* (Keskin & Toroglu., 2011). The coriander seed oil showed the antibacterial effect by limiting the diameter of some bacteria such as *P. aeruginosa*, *E. coli*, and *Salmonella typhi* in the following ratio: 10, 18, and 25 mm, respectively (Teshale et al., 2013). Coriander seed oil exhibits bactericidal and bacteriostatic actions against *S. typhi* and *E. coli*, respectively. The main component of coriander seed extract, linalool, has an extraordinary effect against many bacterial strains like food poisoning bacteria Nanasombat and Lohasupthawee, *Enterobacter aerogenes*, *E. coli*, *K. pneumoniae*, and *Citrobacter freundii*.

39.7 PHARMACOLOGICAL PROPERTIES OF CORIANDER

Coriander seed is one of the main health-supporting spices, and it has been used as medicine for many years (Mathias, 1994). Coriander was firstly used as medicine by the ancient Egyptians, but later on, it was evident in classical Greek and Latin medicine. Coriander was also known as an "anti-diabetic" plant in many parts of Europe. Coriander was also used in some regions of India for its anti-inflammatory effect. Recently, coriander has been studied for its cholesterol-lowering properties. Coriander is considered a functional food and has high potential for nutraceuticals (Rathore et al., 2013).

Coriander seed extract obtained from the coriander plant possesses antimicrobial, anticonvulsant, diuretic, anti-diabetic, hypnotic activity, anti-mutagenic, antioxidant, and anthelmintic properties (Rajeshwari & Andallu, 2011). The coriander seed extract is used to cover the taste and correct the disgusting properties of medicines and is also employed in aromatherapy (Cooksley, 2003). Coriander seed extract is effective against embrocations for joint pain, rheumatism, inflammation, and intestinal parasites. Coriander seed extract is used as a flavoring agent in pharmaceutical products and food items like liquor, chocolate industries, and cocoa. Coriander seed extract is very beneficial in gastrointestinal problems like diarrhea, gastritis, and dyspepsia due to its stomachic and antibilious properties (Platel & Srinivasan, 2004).

39.7.1 HEPATOPROTECTIVE

Coriander has better hepatoprotective properties because the ethanol extract of these seeds contains a maximum number of phenolic compounds, alkaloids, and flavonoids. Iso-quercetin and quercetin compounds are present in coriander examined through high-performance chromatography (HPLC). The term hepatoprotection refers to the activities of SGOT, SGPT, and ALP, reduction of liver weight, and direct bilirubin of CCl_4 intoxicated animals. Coriander seed extract was shown to have hepatoprotective activities due to the antioxidant properties of phenolic compounds like caffeic acid, quercetin, and linalool, the most active constituent of coriander seeds. Coriander seed extract is also beneficial for anxiolytic, sedative, anti-depressant, antimutagenic, diuretic, and anti-hypertensive activities. Coriander was used as a traditional diuretic as well, as it was also known as traditional medicine in Iran for its anti-depressant, anticonvulsant, sedative, and anxiolytic properties (Janbaz et al., 2004).

39.8   BIOLOGICAL PROPERTIES OF CORIANDER SEED EXTRACT

39.8.1 ANTI-INFLAMMATION

There is an increased demand for anti-inflammatory drugs obtained from plants than non-steroidal anti-inflammatory drugs due to toxic and side effects like gastrointestinal tract irritation. Coriander is an anti-inflammatory agent; it is a major component of the traditional Ayurvedic formulation Maharasnadhi Quather (MRQ). Several edible plants are used to treat inflammatory conditions in traditional medicine. Coriander seed contains bioactive components with better anti-inflammatory effects than synthetic anti-inflammatory drugs. Coriander seeds showed an inhibitory effect against inflammatory gastrointestinal bowel diseases. Coriander seed extract is more effective for skin tolerance with mild anti-inflammatory activity. The oral intake of coriander seed oil has an excellent anti-inflammatory effect compared to other synthetic anti-inflammatory drugs (Graham et al., 1988).

39.8.2 PHENOL AND FLAVONOID

The bioactive compounds found in coriander seed extracts like polyphenols and flavonoids with chelating and antioxidant properties are very effective in treating cancer and other chronic diseases. Coriander seed extract with a high-fat diet leads to a decrease in free FA, peroxides level, and glutathione, and increased antioxidant enzyme activity. Coriander leaf extract exhibits a higher antioxidant capacity against coriander seed extract. In addition, coriander seed extract showed higher free radical scavenging capacity than other essential extracts. The use of coriander seed extract curtails the oxidative stress in the kidney and is a biomarker for lowering blood sugar levels and increasing insulin production in diabetic patients from beta cells (Wong & Kitts, 2006).

Coriander seed extract contains phytochemicals like polyphenols and flavonoids. The phenolic compounds of coriander seed extract are calculated by using 1,1-diphenyl-2-picrylhydrazyl (DPPH·). It is observed that the extract of leaves shows more phenolic content than the seed extract of coriander. Coriander seed extract obtained from Saudi Arabia has increased phenolic content compared to others. Coriander leaf, seed, and stem extract have many phenols and flavonoids with methanol extract. Coriander's dried stem and leaf extract with ethyl acetate extract showed more phenolic content than green stem and leaf extract. The concentration of phenols and flavonoids in green and dried plant extract varies with the different solvent extracts. The effectiveness of coriander seed extract polyphenols was evaluated on hydrogen peroxide (H_2O_2)-induced oxidative stress in human lymphocytes. Thus, the administration of polyphenols of coriander seed extract reduced the oxidative stress induced by hydrogen peroxide (H_2O_2) and restrained the oxidative stress in normal cells (Hashim et al. 2005).

Coriander seed extract shows high antioxidant activity due to the high phenolic content. The prime phenolic antioxidants contain compounds like protocathenic acid, glycitin, and caffeic acid (Melo, 2002; de Almeida Melo et al., 2005). The administration of coriander seed extract against ulcers has excellent results due to its free radical scavenging ability, phenolic content, and hydrophobic properties with antioxidants (Al-Mofleh et al., 2006). It was observed that the stability, radical scavenging, and functional activity of coriander seed extract blended with corn oil enhanced DPPH· free radical scavenging capacity compared to non-blended extracts (Ramadan & Wahdan, 2012). Fatty acid and antioxidants like tocopherols in coriander seed extract may cause this valuable oxidative stability. The use of cryogenic technology on spices, especially coriander, shows excellent flavor retention and medicinal properties for these spices (Saxena et al., 2015).

39.8.3 ANTIOXIDANT ACTIVITY

Antioxidants are used in the food industry to prevent spoilage and enhance the shelf life of foods. Like other spices, coriander also contains antioxidants that are beneficial for improving the shelf life and minimizing food deterioration. Coriander seed oil has better radical scavenging activity than coriander leaf oil. The methanol extract of coriander has a better antioxidant effect than essential oil, while the DPPH· radical scavenging ability of methanol extract of coriander seed is higher than butylated hydroxytoluene, which is a synthetic antioxidant. The concentration of linalool is higher in microwave-heated coriander seed oil than conventionally-heated coriander seed oil. Coriander seed oil has excellent antioxidant ability against DPPH· and β-carotene assays (Mageed et al., 2012). The coriander seed extract is used as a natural antioxidant in lipid-rich foods due to its significant radical scavenging activity (Ramadan et al., 2003). The oxidative spoilage of food can be minimized by using coriander leaves and seed oils rich in antioxidants (Wangensteen et al., 2004). Coriander seed contains several antioxidants such as tocopherol (0.18), total polyphenol (18.7), total ascorbate (0.28), riboflavin (0.0046), gallic acid (0.173), caffeic acid (0.08), kaempferol (0.233), and quercetin (0.608). Coriander leaves are also rich in phytochemicals like carotenoids and EO like linalool, polyphenols that have potential for ferric reducing antioxidant ability and free radical scavenging activity. Coriander fresh leaf juice has significant antioxidant activities such as high reducing power, scavenging of hydroxyl and superoxide radicals, increased glutathione, and preventing biological macromolecular oxidative impairment (Panjwani et al., 2010).

39.8.4 ANTI-DIABETIC PROPERTIES

Coriander seed extract is proposed as an anti-diabetic remedy. In several countries, coriander seed extract is used to treat hyperglycemia (Tahraoui et al., 2007). Early studies show that coriander seed extract was ineffective on fasting blood sugar levels, but it explained the mitigation of adrenaline-induced hyperglycemia. The literature demonstrates that the administration of coriander seed extract for an extended period decreased hyperglycemia. Linalool is the major constituent of coriander seed extract that has a hypoglycemic effect. A high dose of coriander seed extract taken orally is more effective in hyperglycemia. It was observed that patients with a higher level of blood glucose showed a significant decrease in blood glucose after treatment with coriander seed extract. Coriander seed extract has anti-hyperglycemic and insulin-releasing activity. The use of coriander seed extract caused changes in carbohydrate metabolism; for example, the hepatic glycogen concentration and the activity of glycogen synthase enzyme increased. Hence, the gluconeogenesis and glycogenolysis process decreased, but the glucose-6 phosphate dehydrogenase enzyme activity increased with other glycolytic enzymes, indicating the antihyperglycemic activity of coriander seed extract. This antihyperglycemic activity of coriander seed extract supports the use of coriander seed extract in anti-diabetic functional foods and nutraceuticals (Abou El-Soud et al., 2007)

39.9 CONCLUSIONS

It is concluded that the extract of coriander parts, especially seeds and leaves, plays an essential role in developing functional food as a functional ingredient. The most important extraction techniques like organic solvent extraction, supercritical fluid extraction, and steam distillation are generally used for coriander seed extract. Previous studies reported that fresh leaves and seeds of coriander contain essential oils and different bioactive compounds used as preservatives and flavor enhancers in food products. In addition, different coriander parts are vital to human health, with hepatoprotective, anti-inflammation, and antioxidant activity, and anti-diabetic properties.

REFERENCES

Abou El-Soud, N.H., Khalil, M.Y., Hussein, J.S., Oraby, F.S.H. & Farrag, A.H. (2007). Antidiabetic effects of fenugreek alkaloid extract in streptozotocin induced hyperglycemic rats. *Journal of Applied Sciences Research*, *3*(10), pp.1073–1083.

Al-Mofleh, I.A., Alhaider, A.A., Mossa, J.S., Al-Sohaibani, M.O., Rafatullah, S. & Qureshi, S. (2006). Protection of gastric mucosal damage by *Coriandrum sativum* L. pretreatment in Wistar albino rats. *Environmental Toxicology and Pharmacology*, *22*(1), pp.64–69.

Bhanger, M.I., Iqbal, S., Anwar, F., Imran, M., Akhtar, M. & Zia-ul-Haq, M. (2008). Antioxidant potential of rice bran extracts and its effects on stabilisation of cookies under ambient storage. *International Journal of Food Science & Technology*, *43*(5), pp.779–786.

Bhuiyan, M.N.I., Begum, J. & Sultana, M. (2009). Chemical composition of leaf and seed essential oil of *Coriandrum sativum* L. from Bangladesh. *Bangladesh Journal of Pharmacology*, *4*(2), pp.150–153.

Boutin, O. & Badens, E. (2009). Extraction from oleaginous seeds using supercritical CO_2: Experimental design and products quality. *Journal of Food Engineering*, *92*(4), pp.396–402.

Brunner, G. (2005). Supercritical fluids: Technology and application to food processing. *Journal of Food Engineering*, *67*(1–2), pp.21–33.

Burdock, G.A. & Carabin, I.G. (2009). Safety assessment of coriander (*Coriandrum sativum* L.) essential oil as a food ingredient. *Food and Chemical Toxicology*, *47*(1), pp.22–34.

Catchpole, O.J., Grey, J.B. & Smallfield, B.M. (1996). Near-critical extraction of sage, celery, and coriander seed. *The Journal of Supercritical Fluids*, *9*(4), 273–279.

Cooksley, V. (2003). An integrative aromatherapy intervention for palliative care. *The International Journal of Aromatherapy*, *2*(13), 128–137.

de Almeida Melo, E., Mancini Filho, J. & Guerra, N.B. (2005). Characterization of antioxidant compounds in aqueous coriander extract (*Coriandrum sativum* L.). *LWT-Food Science and Technology*, *38*(1), pp.15–19.

Delaquis, P.J., Stanich, K., Girard, B. & Mazza, G. (2002). Antimicrobial activity of individual and mixed fractions of dill, cilantro, coriander and eucalyptus essential oils. *International Journal of Food Microbiology*, *74*(1–2), 101–109.

Diederichsen, A. (1996). *Coriander (Coriandrum sativum L.) promoting the conservation and use of under-utilized and neglected crops. 3*. Rome: Institute of Plant Genetics and Crop Plant Research, Gatersleben. International Plant Genetic Resources Institute, 83.

Donelian, A., Carlson, L.H.C., Lopes, T.J. & Machado, R.A.F. (2009). Comparison of extraction of patchouli (*Pogostemon cablin*) essential oil with supercritical CO_2 and by steam distillation. *The Journal of Supercritical Fluids*, *48*(1), 15–20.

Elgayyar, M., Draughon, F.A., Golden, D.A. & Mount, J.R. (2001). Antimicrobial activity of essential oils from plants against selected pathogenic and saprophytic microorganisms. *Journal of Food Protection*, *64*(7), 1019–1024.

European Food Safety Authority (EFSA). (2013). Scientific opinion on the safety of coriander seed oil as a novel food ingredient. *EFSA Journal*, *11*(10), 3422.

Facciola, S. (1990). *Cornucopia: A source book of edible plants*. Vista, CA: Kampong Publications.

Gallagher, A.M., Flatt, P.R., Duffy, G.A.W.Y. & Abdel-Wahab, Y.H.A. (2003). The effects of traditional anti-diabetic plants on in vitro glucose diffusion. *Nutrition Research*, *23*(3), 413–424.

Ganesan, P., Phaiphan, A., Murugan, Y. & Baharin, B.S. (2013). Comparative study of bioactive compounds in curry and coriander leaves: An update. *Journal of Chemical and Pharmaceutical Research*, *5*(11), 590–594.

Gil, A., De La Fuente, E.B., Lenardis, A.E., López Pereira, M., Suárez, S.A., Bandoni, A., Van Baren, C., Di Leo Lira, P. & Ghersa, C.M. (2002). Coriander essential oil composition from two genotypes grown in different environmental conditions. *Journal of Agricultural and Food Chemistry*, *50*(10), 2870–2877.

Graham, D., Agrawal, N. & Roth, S. (1988). Prevention of NSAID-induced gastric ulcer with misoprostol: Multicentre, double-blind, placebo-controlled trial. *The Lancet*, *332*(8623), 1277–1280.

Grieve, M. (1971). Mandrake. *A Modern Herbal'*, New York: Dover publication INC, 2.

Grosso, C., Ferraro, V., Figueiredo, A.C., Barroso, J.G., Coelho, J.A. & Palavra, A.M. (2008). Supercritical carbon dioxide extraction of volatile oil from Italian coriander seeds. *Food Chemistry*, *111*(1), 197–203.

Hashim, M.S., Lincy, S., Remya, V., Teena, M. & Anila, L. (2005). Effect of polyphenolic compounds from *Coriandrum sativum* on H_2O_2-induced oxidative stress in human lymphocytes. *Food Chemistry*, *92*(4), 653–660.

Illés, V., Daood, H.G., Perneczki, S., Szokonya, L. & Then, M. (2000). Extraction of coriander seed oil by CO_2 and propane at super-and subcritical conditions. *The Journal of Supercritical Fluids*, *17*(2), 177–186.

Janbaz, K.H., Saeed, S.A. & Gilani, A.H. (2004). Studies on the protective effects of caffeic acid and quercetin on chemical-induced hepatotoxicity in rodents. *Phytomedicine*, *11*(5), 424–430.

Kalemba, D.A.A.K. & Kunicka, A. (2003). Antibacterial and antifungal properties of essential oils. *Current Medicinal Chemistry*, *10*(10), 813–829.

Keskin, D. & Toroglu, S. (2011). Studies on antimicrobial activities of solvent extracts of different spices. *Journal of Environmental Biology*, *32*(2), 251–256.

Laribi, B., Kouki, K., M'Hamdi, M. & Bettaieb, T. (2015). Coriander (*Coriandrum sativum* L.) and its bioactive constituents. *Fitoterapia*, *103*, 9–26.

Lawrence, B.M. (1993). A planning scheme to evaluate new aromatic plants for the flavor and fragrance industries. *New Crops*, *1*, 620–627.

Lo Cantore, P., Iacobellis, N.S., De Marco, A., Capasso, F. & Senatore, F. (2004). Antibacterial activity of *Coriandrum sativum* L. and *Foeniculum vulgare* Miller var. vulgare (Miller) essential oils. *Journal of Agricultural and Food Chemistry*, *52*(26), 7862–7866.

Machmudah, S., Kondo, M., Sasaki, M., Goto, M., Munemasa, J. & Yamagata, M. (2008). Pressure effect in supercritical CO_2 extraction of plant seeds. *The Journal of Supercritical Fluids*, *44*(3), 301–307.

Mageed, M.A.A.E., Mansour, A.F., El Massry, K.F., Ramadan, M.M., Shaheen, M.S. & Shaaban, H. (2012). Effect of microwaves on essential oils of coriander and cumin seeds and on their antioxidant and antimicrobial activities. *Journal of Essential Oil Bearing Plants*, *15*(4), 614–627.

Mahendra, P. & Bisht, S. (2011). *Coriandrum sativum*: A daily use spice with great medicinal effect. *Pharmacognosy Journal*, *3*(21), 84–88.

Mandal, S. & Mandal, M. (2015). Coriander (*Coriandrum sativum* L.) essential oil: Chemistry and biological activity. *Asian Pacific Journal of Tropical Biomedicine*, *5*(6), 421–428.

Mathias, M.E. (1994). Magic, myth and medicine. *Economic Botany*, *48*(1), pp.3–7.

Meena, M.F. (1994). Antimicrobial activity of essential oils from spices. *Journal of Food Science and Technology*, *31*, 68–70.

Melo, E.A. (2002). *Caracterização dos principais compostos antioxidantes presentes no coentro (Coriandrum sativum L.)*. Pernabuco, Recife: Universidade Federal de Pernambuco.

Misharina, T.A. (2001). Effect of conditions and duration of storage on composition of essential oil from coriander seeds. *Prikladnaia biokhimiia i mikrobiologiia*, *37*(6), 726–732.

Momin, A.H., Acharya, S.S. & Gajjar, A.V. (2012). *Coriandrum sativum*-review of advances in phytopharmacology. *International Journal of Pharmaceutical Sciences and Research*, *3*(5), 1233.

Msaada, K., Hosni, K., Taarit, M.B., Chahed, T., Kchouk, M.E. & Marzouk, B. (2007). Changes on essential oil composition of coriander (*Coriandrum sativum* L.) fruits during three stages of maturity. *Food Chemistry*, *102*(4), 1131–1134.

Nadeem, M., Anjum, F.M., Khan, M.I., Tehseen, S., El-Ghorab, A. & Sultan, J.I. (2013). Nutritional and medicinal aspects of coriander (*Coriandrum sativum* L.): A review. *British Food Journal*, *115*(5), 743–755.

Nanditha, B.R., Jena, B.S. & Prabhasankar, P. (2009). Influence of natural antioxidants and their carry-through property in biscuit processing. *Journal of the Science of Food and Agriculture*, *89*(2), 288–298.

Neffati, M. & Marzouk, B. (2008). Changes in essential oil and fatty acid composition in coriander (*Coriandrum sativum* L.) leaves under saline conditions. *Industrial Crops and Products*, *28*(2), 137–142.

Nurzynska-Wierdak, R. (2013). Essential oil composition of the coriander (*Coriandrum sativum* L.) herb depending on the development stage. *Acta Agrobotanica*, *66*(1).

Olle, M., Bender, I. & Koppe, R. (2010). The content of oils in umbelliferous crops and its formation. *Agronomy Research*, *8*(3), 687–696.

Opdyke, D. L. J. (1973). Monographs on fragrance raw materials: Coriander oil. *Food and Cosmetics Toxicology*, *11*(6), 1077.

Panjwani, D., Mishra, B. & Banji, D. (2010). Time dependent antioxidant activity of fresh juice of leaves of *Coriandrum sativum*. *International Journal of Pharmaceutical Sciences and Drug Research*, *2*(1), 63–66.

Pathak Nimish, L., Kasture Sanjay, B., Bhatt Nayna, M. & Rathod Jaimik, D. (2011). Phytopharmacological properties of *Coriander sativum* as a potential medicinal tree: An overview. *Journal of Applied Pharmaceutical Science*, *1*(4), 20–25.

Platel, K. & Srinivasan, K. (2004). Digestive stimulant action of spices: A myth or reality? *Indian Journal of Medical Research*, *119*(5), 167.

Potter, T.L. & Fagerson, I.S. (1990). Composition of coriander leaf volatiles. *Journal of Agricultural and Food Chemistry*, *38*(11), 2054–2056.

Rajeshwari, U. & Andallu, B. (2011). Medicinal benefits of coriander (*Coriandrum sativum* L). *Spatula DD*, *1*(1), 51–58.

Ramadan, M.F., Kroh, L.W. & Mörsel, J.T. (2003). Radical scavenging activity of black cumin (*Nigella sativa* L.), coriander (*Coriandrum sativum* L.), and niger (*Guizotia abyssinica* Cass.) crude seed oils and oil fractions. *Journal of Agricultural and Food Chemistry*, *51*(24), 6961–6969.

Ramadan, M.F. & Wahdan, K.M.M. (2012). Blending of corn oil with black cumin (Nigella sativa) and coriander (*Coriandrum sativum*) seed oils: Impact on functionality, stability and radical scavenging activity. *Food Chemistry*, *132*(2), 873–879.

Ramezani, S., Rasouli, F. & Solaimani, B. (2009). Changes in essential oil content of coriander (*Coriandrum sativum* L.) aerial parts during four phonological stages in Iran. *Journal of Essential Oil Bearing Plants*, *12*(6), 683–689.

Rathore, S.S., Saxena, S.N. & Singh, B. (2013). Potential health benefits of major seed spices. *International Journal of Seed Spices*, *3*(2), 1–12.

Saxena, S.N. & Agarwal, D. (2019). Pharmacognosy and phytochemistry of coriander (*Coriandrum sativum* L.). *International Journal of Seed Spices*, *9*(1), 1–13.

Saxena, S.N., Sharma, Y.K., Rathore, S.S., Singh, K.K., Barnwal, P., Saxena, R., Upadhyaya, P. & Anwer, M.M. (2015). Effect of cryogenic grinding on volatile oil, oleoresin content and anti-oxidant properties of coriander (*Coriandrum sativum* L.) genotypes. *Journal of Food Science and Technology*, *52*(1), 568–573.

Shahwar, M.K., El-Ghorab, A.H., Anjum, F.M., Butt, M.S., Hussain, S. & Nadeem, M. (2012). Characterization of coriander (*Coriandrum sativum* L.) seeds and leaves: Volatile and nonvolatile extracts. *International Journal of Food Properties*, *15*(4), 736–747.

Singh, G., Maurya, S., De Lampasona, M.P. & Catalan, C.A. (2006). Studies on essential oils, Part 41. Chemical composition, antifungal, antioxidant and sprout suppressant activities of coriander (*Coriandrum sativum*) essential oil and its oleoresin. *Flavour and Fragrance Journal*, *21*(3), 472–479.

Small, E. (1997). *Coriander, culinary herbs*. Ottawa: NRC Research.

Tahraoui, A., El-Hilaly, J., Israili, Z.H. & Lyoussi, B. (2007). Ethnopharmacological survey of plants used in the traditional treatment of hypertension and diabetes in south-eastern Morocco (*Errachidia province*). *Journal of Ethnopharmacology*, *110*(1), 105–117.

Temelli, F. (2009). Perspectives on supercritical fluid processing of fats and oils. *The Journal of Supercritical Fluids*, *47*(3), 583–590.

Teshale, C., Hussien, J. & Jemal, A. (2013). Antimicrobial activity of the extracts of selected Ethiopian aromatic medicinal plants. *Spatula DD*, *3*(4), 175–180.

Thirumala-Devi, K., Mayo, M.A., Reddy, G., Emmanuel, K.E., Larondelle, Y. & Reddy, D.V.R. (2001). Occurrence of ochratoxin A in black pepper, coriander, ginger and turmeric in India. *Food Additives & Contaminants*, *18*(9), 830–835.

Wangensteen, H., Samuelsen, A.B. & Malterud, K.E. (2004). Antioxidant activity in extracts from coriander. *Food Chemistry*, *88*(2), 293–297.

Wilkinson, G.T. (1993). *Essential oil from coriander seed*. University of South Australia, School of Chemical Technology.

Wong, P.Y. & Kitts, D.D. (2006). Studies on the dual antioxidant and antibacterial properties of parsley (*Petroselinum crispum*) and cilantro (*Coriandrum sativum*) extracts. *Food Chemistry*, *97*(3), 505–515.

Yeung, E.C. & Bowra, S. (2011). Embryo and endosperm development in coriander (*Coriandrum sativum*). *Botany*, *89*(4), 263–273.

40 Health-Promoting Activities of Coriander Seed Extracts

Mohammad Rafiqul Islam, Anower Jahid, and Imam Hasan

CONTENTS

40.1 INTRODUCTION

Various types of environmental contaminants are hazardous to humans and animals at various stages of their life. Among them, lead is the most common occupational and environmental toxicant, and it has unrelenting potential health hazards to humans (especially children) and animals. Lead is considered the most common toxic metal in the surroundings (Patra et al., 2011). Storage batteries, metal alloys, pigments, stabilizers, and binders in many industries are the common

DOI: 10.1201/9781003204626-45

sources of lead (Henretig, 2002). Lead is also used in paints and plastic formation (Saryan and Zen, 1994).

Lead toxicity affects soft tissues like the liver, kidney, and testis seriously. Lead causes considerable alterations in the nuclei of the hepatocytes (Piasek et al., 1989), chronic nephropathy (Odigie et al., 2004), and progressive vascular, tubular, and interstitial testicular damage (Moniem et al., 2010).

Many antioxidants, including vitamin C (Hsu et al., 1998) and vitamin E (Patra et al., 2011), have been used to prevent the incidence of and minimize oxidative stress in tissues. Vitamin E functions as a free radical scavenger and scavenges superoxide, hydrogen peroxide, and hydroxyl radicals (Kartikeya et al., 2009). In addition, vitamin E improves fertility by protecting sperm DNA from the oxidative stress of free radicals (Tarin et al., 1998). Hong et al. (2009) stated that vitamin E in the diet enhances the activities of some antioxidant enzymes and decreases nitric oxide content and lipid peroxidation products in the testis of Boer goat.

South Asian kitchens regularly use coriander seeds as valuable spices. Coriander is also popular for its wide range of healing properties. It aids in dispelling toxic mineral residue such as mercury and lead from the body through the urine or feces. Coriander (*Coriandrum sativum*) decreases lipid peroxidation in lead-induced mice tissues (liver, kidney, and testis). Chithra and Leelamma (1999) show that the formation of lipid peroxides declines, and the activities of antioxidant enzymes (catalase, glutathione peroxidase) increase in rats treated with coriander extracts.

The present study is designed to investigate the effects of lead toxicity on soft tissues of mammals and health-promoting activities produced by vitamin E and extract of coriander seeds supplementation.

40.2 MATERIALS AND METHODS

40.2.1 LOCATION AND STUDY PERIOD

The study was conducted in the Department of Anatomy and Histology, Faculty of Veterinary Science, Bangladesh Agricultural University, Mymensing-2202.

40.2.2 PREPARATION OF AQUEOUS EXTRACT OF CORIANDER (*CORIANDRUM SATIVUM*)

Dried coriander seeds were collected from the local market of Mymensingh. Then coriander extract was prepared at the Department of Pharmacology, Faculty of Veterinary Science, Bangladesh Agricultural University, Mymensingh. The extract was filtered and stored at 4°C. It was dissolved in distilled water whenever needed for experiments (Figure 40.1).

FIGURE 40.1   Coriander seeds (A) and coriander extract (B).

40.2.3 EXPERIMENTAL ANIMALS

The experimental Swiss albino mice (male) were collected from the Department of Pharmacy, Jahangirnagar University, Dhaka. Collected mice were 6 weeks of age and about 26–28 grams at the time of collection. All the mice had good health and were devoid of any external deformities (Figure 40.2).

40.2.4 ETHICAL PERMISSION

All experimental protocols were approved by the Animal Welfare and Ethical Committee (order no. sha 1/444/edu) Faculty of Veterinary Science, Bangladesh Agricultural University.

40.2.5 EXPERIMENTAL GROUPS

After 7 days, the mice were divided into different groups according to the experimental design. At first, there were Group A: Control group (10 mice) and Group B: Lead intoxicated group (25 mice). The control group received only normal water and feed. The intoxicated lead group was treated with 60 mg lead acetate per kg body weight every day orally for 6 weeks. After 6 weeks, samples were collected from five mice of the control group and five mice of the intoxicated group. The remaining five mice of the control group were kept as a control for the next 42 days. Five mice of the intoxicated group were further intoxicated for the next 42 days. The other 15 mice of the intoxicated group were kept for treatment purposes. These mice were divided into three groups (C, D, and E) with five mice. Group C was treated with 150 mg vitamin E (diluted in soya oil) per kg body weight every day orally for 6 weeks. Group D was treated with 300 mg coriander extract (diluted in distilled water) per kg body weight every day orally for 6 weeks. Group E was treated with 150 mg vitamin E (diluted in soya oil) and 300 mg coriander extract (diluted in distilled water) per kg body weight every day orally for 6 weeks. After completing the experiment, samples were collected from all the mice of different groups (Figure 40.3A).

40.2.6 GROSS OBSERVATION

The gross study considered parameters such as color, length, and weight. All kinds of abnormalities were also observed. The color of exposed organs (liver, kidney, and testis) was compared with the organs of the control group by eye observation. A graded scale measured the length of the liver, kidney, and testis of different groups. The unit of length measurement was millimeters. Weight was measured in grams by electronic balance.

FIGURE 40.2 Experimental mice.

FIGURE 40.3   Sample collection (A) and fixation of samples (B).

40.2.7   HISTOLOGICAL STUDY

After gross observation, samples were preserved in 10% formalin and Bouin's fluid (Figure 40.3B). After proper fixation, samples were processed for histological study. H&E staining protocol was applied. The detailed histological study was done using a light microscope. Necessary photography was done with an Olympus BX 51 photographic light microscope and placed to illustrate the result better.

40.2.8   STATISTICAL ANALYSIS

The mice's initial and final body weight, and the weight and length of the liver, kidney, and testis were taken during the study period. The collected data were then analyzed using Statistical Package for the Social Sciences (SPSS) software and the results displayed in tabular form. The chi-squared test was used for the analytical assessment. The differences were considered statistically significant when the p values were less than 0.05.

40.3   RESULTS

The effect of lead toxicity on the liver, kidney, and testis of mice and the potency of vitamin E and coriander extract have been presented in this chapter. The studied results have been presented in different tables and figures for better illustration.

40.3.1   BODY WEIGHT

The mean body weight of mice at the beginning of the experiment in the control group, intoxicated group, vitamin E-treated group, coriander extract-treated group, and vitamin E and coriander extract (combined)-treated group was 26.5 ± 0.18, 26.8 ± 0.22, 34.0 ± 0.64, 33.7 ± 0.76, and 33.9 ± 0.45 g, respectively (Table 40.1).

The mean body weight of mice at the end of the experiment in the control group, intoxicated group, vitamin E-treated group, coriander extract-treated group, and vitamin E and coriander extract (combined)-treated group was 38.4, 35.9, 37.6, 37.4, and 39.6 g, respectively (Table 40.1).

The body weight of mice in the intoxicated lead group was significantly ($p < 0.05$) reduced compared to the control group. On the other hand, the body weight of mice in the vitamin E and coriander extract combined-treated group was significantly ($p < 0.05$) increased in comparison to the intoxicated group, vitamin E-treated group, and coriander extract-treated group (Table 40.1).

TABLE 40.1
Body Weight of Different Groups of Mice in Grams

Group	Initial Body Weight		Final Body Weight	
	Mean	SE	Mean	SE
Control	26.57	0.18	38.47	0.18
Intoxicated	26.81	0.22	35.97*	0.35
Vitamin E treated	34.07	0.64	37.63	0.48
Coriander extract treated	33.78	0.76	37.42	0.50
Vitamin E and coriander extract treated	33.90	0.45	39.63**	0.49

SE = standard error; significant at 5% ($p < 0.05$) level.

* Significant in comparison to the control group.

** Significant compared to the intoxicated, vitamin E-treated, and coriander extract-treated groups.

Initial body weights of vitamin E-treated group, coriander extract-treated group, and vitamin E and coriander extract-treated group were considered at the beginning of treatment after the intoxication of 42 days.

40.3.2 WEIGHT AND LENGTH OF DIFFERENT ORGANS OF MICE

40.3.2.1 Liver
The mean weight of liver in the control group, intoxicated group, vitamin E-treated group, coriander extract-treated group, and vitamin E and coriander extract (combined)-treated group was 2.69, 2.21, 2.32, 2.36, and 2.35 g, respectively (Table 40.2).

The mean length of the liver in the control group, intoxicated group, vitamin E-treated group, coriander extract-treated group, and vitamin E and coriander extract (combined)-treated group was 30.4, 27.1, 30.5, 27.2, and 29.2 mm, respectively (Table 40.3).

40.3.2.2 Kidney
The mean weight of the left kidney in the control group, intoxicated group, vitamin E-treated group, coriander extract-treated group, and vitamin E and coriander extract (combined)-treated group was

TABLE 40.2
Weight of Different Organs of Mice in Grams

Organ		Control	Intoxicated	Vitamin E Treated	Coriander Extract Treated	Vitamin E and Coriander Extract Treated
Liver	Mean	2.69	2.21	2.32	2.36	2.35
	SE	0.10	0.06	0.12	0.17	0.15
Left kidney	Mean	0.30	0.24*	0.27	0.27	0.26
	SE	0.01	0.01	0.01	0.01	0.02
Right kidney	Mean	0.30	0.27	0.28	0.28	0.28
	SE	0.01	0.01	0.01	0.01	0.01
Left testis	Mean	0.13	0.10*	0.12	0.11	0.12
	SE	0.01	0.01	0.01	0.01	0.01
Right testis	Mean	0.12	0.10*	0.12	0.11	0.11
	SE	0.01	0.02	0.01	0.01	0.01

SE = standard error; significant at 5% ($p < 0.05$) level.

* Significant in comparison to the control group.

TABLE 40.3

Length of Different Organs of Mice in Millimeters

Organ		Control	Intoxicated	Vitamin E Treated	Coriander Extract Treated	Vitamin E and Coriander Extract Treated
Liver	Mean	30.43	27.14	30.50	27.29	29.29
	SE	0.88	0.46	0.59	1.19	0.81
Left	Mean	10.86	10.36	11.07	10.50	10.79
kidney	SE	0.34	0.18	0.25	0.31	0.26
Right	Mean	10.75	10.29	10.93	10.43	10.86
kidney	SE	0.27	0.15	0.23	0.20	0.26
Left testis	Mean	7.59	6.57*	7.14	6.58	7.07
	SE	0.23	0.17	0.24	0.14	0.20
Right testis	Mean	7.43	6.50*	6.93	6.79	7.07
	SE	0.17	0.15	0.20	0.10	0.20

SE = standard error; significant at 5% ($p < 0.05$) level.

* Significant in comparison with the control group.

0.30 ± 0.01, 0.24 ± 0.01, 0.27 ± 0.01, 0.27 ± 0.01, and 0.26 ± 0.02 g, respectively. The weight of the left kidney in the intoxicated lead group was significantly ($p < 0.05$) reduced in comparison to the control group (Table 40.2).

The mean weight of the right kidney in the control group, intoxicated group, vitamin E-treated group, coriander extract-treated group, and vitamin E and coriander extract (combined)-treated group was 0.30 ± 0.01, 0.27 ± 0.01, 0.28 ± 0.01, 0.28 ± 0.01, and 0.28 ± 0.01 g, respectively (Table 40.2).

The mean length of the left kidney in the control group, intoxicated group, vitamin E-treated group, coriander extract-treated group, and vitamin E and coriander extract (combined)-treated group was 10.8 ± 0.34, 10.3 ± 0.18, 11.0 ± 0.25, 10.5 ± 0.31, and 10.7 ± 0.26 mm, respectively (Table 40.3).

The mean length of the right kidney in the control group, intoxicated group, vitamin E-treated group, coriander extract-treated group, and vitamin E and coriander extract (combined)-treated group was 10.7 ± 0.27, 10.2 ± 0.15, 10.9 ± 0.23, 10.4 ± 0.20, and 10.8 ± 0.26 mm, respectively (Table 40.3).

40.3.2.3 Testis

The mean weight of the left testis in the control group, intoxicated group, vitamin E-treated group, coriander extract-treated group, and vitamin E and coriander extract (combined)-treated group was 0.13 ± 0.01, 0.10 ± 0.01, 0.12 ± 0.01, 0.11 ± 0.01, and 0.12 ± 0.01 g, respectively. The weight of the left testis in the intoxicated lead group was significantly ($p < 0.05$) reduced in comparison to the control group (Table 40.2).

The mean weight of the right testis in the control group, intoxicated group, vitamin E-treated group, coriander extract-treated group, and vitamin E and coriander extract (combined)-treated group was 0.12 ± 0.01, 0.10 ± 0.02, 0.12 ± 0.01, 0.11 ± 0.01, and 0.11 ± 0.01 g, respectively. The weight of the right testis in the intoxicated lead group was significantly ($p < 0.05$) reduced in comparison to the control group (Table 40.2).

The mean length of the left testis in the control group, intoxicated group, vitamin E-treated group, coriander extract-treated group, and vitamin E and coriander extract (combined)-treated group was 7.59 ± 0.23, 6.57 ± 0.17, 7.14 ± 0.24, 6.58 ± 0.14, and 7.07 ± 0.20 mm, respectively. The length of the left testis in the intoxicated lead group was significantly ($p < 0.05$) reduced in comparison to the control group (Table 40.3).

The mean length of the right testis in the control group, intoxicated group, vitamin E-treated group, coriander extract-treated group, and vitamin E and coriander extract (combined)-treated group was 7.43 ± 0.17, 6.50 ± 0.15, 6.93 ± 0.20, 6.79 ± 0.10, and 7.07 ± 0.20 mm, respectively. The length of the right testis in the intoxicated lead group was significantly ($p < 0.05$) reduced in comparison to the control group (Table 40.3).

40.3.3 GROSS OBSERVATION

40.3.3.1 Liver

The liver of the control group was reddish (Figure 40.4A). The liver was also reddish in the intoxicated group and treatment groups. However, the nodular lesion was found in the intoxicated group (Figure 40.4B). The nodular lesion was also found in the vitamin E-treated group (Figure 40.4C). A nodular lesion was not found in the coriander extract-treated group (Figure 40.4D). On the other hand, the appearance of the liver was found normal in the vitamin E and coriander extract-treated group (Figure 40.4E). Therefore, the nodular lesion was not observed in this group.

40.3.3.2 Kidney

The kidney of the control group was reddish-brown with a smooth and shiny surface (Figure 40.5A). The kidneys of the other groups were also reddish-brown. However, the intoxicated lead group had nodular lesions and abnormal shapes (Figure 40.5B). The appearance of the kidney was found normal in the vitamin E-treated group (Figure 40.5C), coriander extract-treated group (Figure 40.5D), and vitamin E and coriander extract (combined)-treated group (Figure 40.5E). The nodular lesion was not observed in these groups.

40.3.3.3 Testis

The appearance of the testis was found normal in the control group (Figure 40.6A), vitamin E-treated group (Figure 40.6C), coriander extract-treated group (Figure 40.6D), and vitamin E and

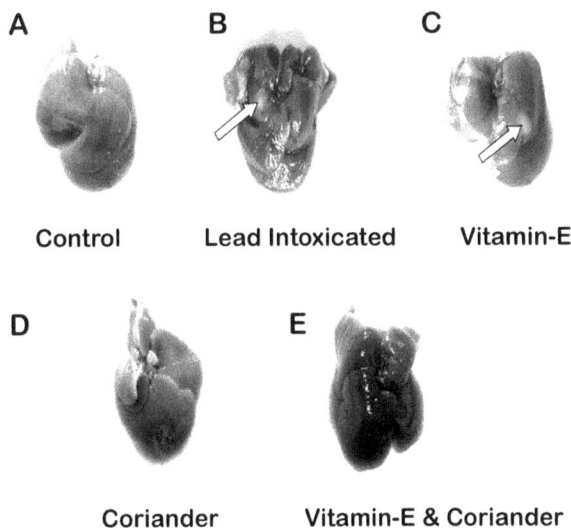

FIGURE 40.4 (A–E) Gross observation of liver in mice. Normal appearance of the liver was found in the control group (A). A nodular lesion (white arrow) was found in the intoxicated lead group (B). Nodular lesion (white arrow) was also found in the vitamin E-treated group (C). Normal appearance of the liver in coriander extract-treated group (D) and vitamin E and coriander extract (combined)-treated group (E). Scale bar = 10 mm.

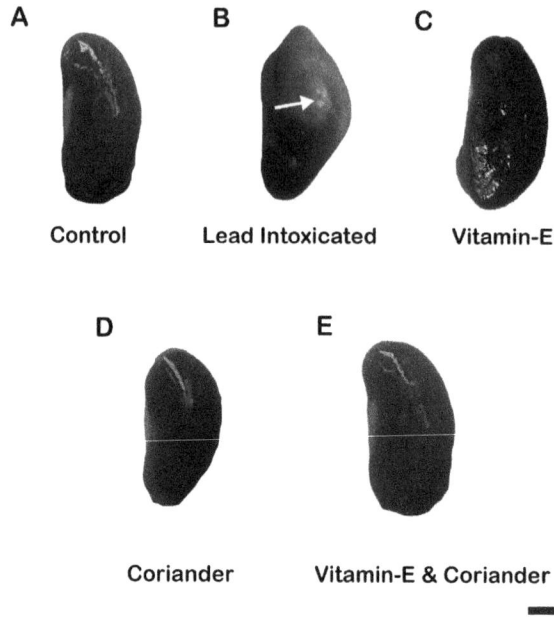

FIGURE 40.5 (A–E) Gross observation of the left kidney in mice.* Normal appearance of the kidney was found in the control group (A). Nodular lesions (white arrow) and abnormal shape were found in the intoxicated lead group (B). The appearance of the kidney was found normal in the vitamin E-treated group (C), coriander extract-treated group (D), and vitamin E and coriander extract-treated group (E). Scale bar = 10 mm. * Left kidneys of different groups are presented here because there was no gross difference between the left and right kidneys.

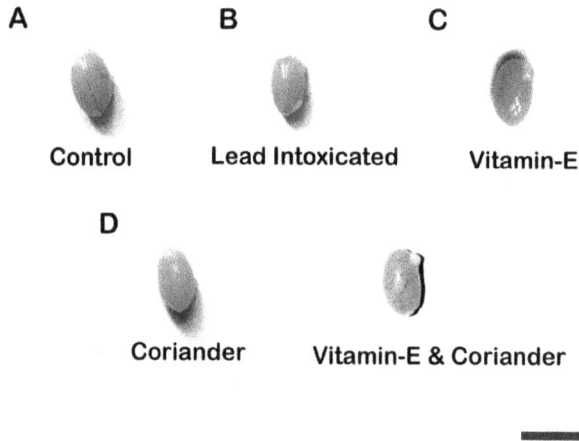

FIGURE 40.6 (A–E) Gross observation of left testis in mice.* Normal appearance of testis in the control group (A), vitamin E-treated group (C), coriander extract-treated group (D), and vitamin E and coriander extract-treated group (E). Testis was found in reduced size in the intoxicated lead group (B). Scale bar = 10 mm. * Only left testes of different groups are presented here because there was no gross difference between left and right testes.

coriander extract (combined)-treated group (Figure 40.6E). The testis was found in reduced size (Figure 40.6B).

40.3.4 Histological Observation

40.3.4.1 Liver

The present study found the liver with normal histological architecture in the control group (Figure 40.7A, 10X, and 40X). In the intoxicated lead group, congestion in the central vein (Figure 40.7B, 10X, and 40X) and nodules with fibrous covering (7C, 10X, and 40X) were found in some sections. The appearance of the liver was found normal in the vitamin E-treated group (Figure 40.7D, 10X, and 40X), coriander extract-treated group (Figure 40.7E, 10X, and 40X), and vitamin E and coriander extract (combined)-treated group (Figure 40.7F, 10X, and 40X). Congestion in the central vein and nodular lesions were not observed in these groups.

40.3.4.2 Kidney

The present study found the kidney with normal histological architecture in the control group (Figure 40.8A, 10X, and 40X). The epithelial cast was found in the urinary space (Figure 40.8B, 10X, and 40X). However, the appearance of the kidney was found normal in the vitamin E-treated group (Figure 40.8C, 10X, and 40X), coriander extract-treated group (Figure 40.8D, 10X, and 40X), and vitamin E and coriander extract (combined)-treated group (Figure 40.8E, 10X, and 40X). The epithelial cast was not observed in the urinary space in these groups.

40.3.4.3 Testis

In the present study, the appearance of seminiferous tubules was found normal in the control group. Spermatogenic cells were present in a normal pattern (Figure 40.9A, 10X). Primary spermatocytes

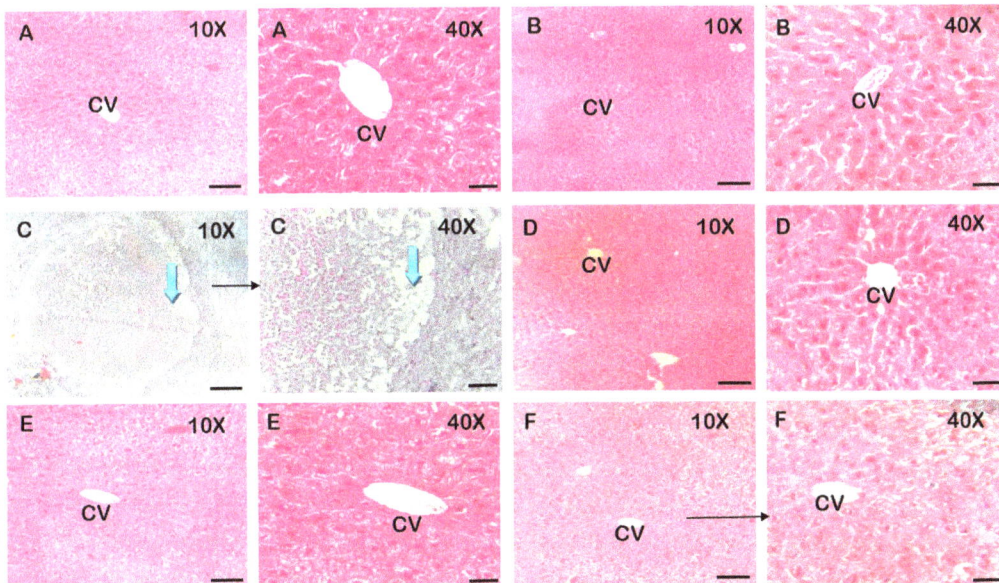

FIGURE 40.7 (A–F) Histological observation of liver in mice (H&E). Normal appearance of the liver was found in the control group (A, 10X, and 40X). In the intoxicated lead group, congestion in the central vein (white arrow) was found in some sections (B, 10X, and 40X). In addition, nodular lesion (blue arrow) was also found in some sections of this group (C, 10X, and 40X). Appearance of the liver was found normal in the vitamin E-treated group (D, 10X, and 40X), coriander extract-treated group (E, 10X, and 40X), and vitamin E and coriander extract (combined)-treated group (F, 10X, and 40X). CV = central vein. Scale bar: 5 μm (10X) and 1 μm (40X).

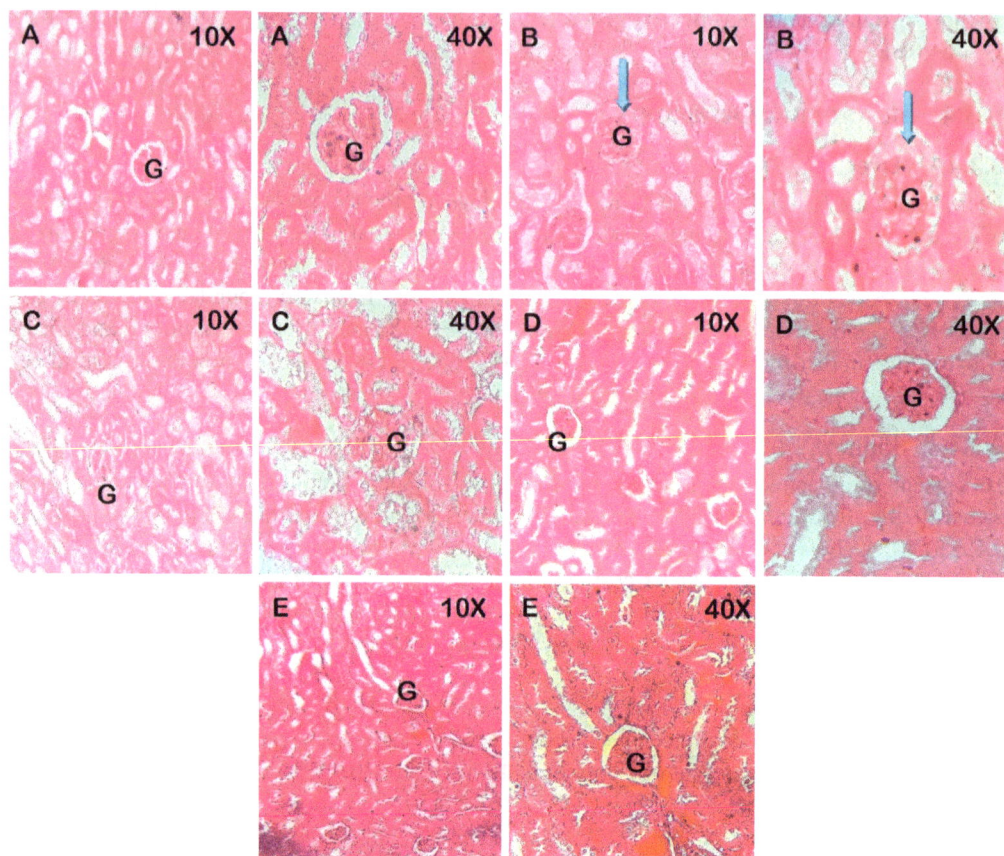

FIGURE 40.8 (A–E) Histological observation of kidney in mice (H&E).* Normal appearance of the kidney was found in the control group (A, 10X, and 40X). Epithelial cast (blue arrow) was found in the urinary space in the intoxicated lead group (B, 10X, and 40X). Appearance of the kidney was found normal in the vitamin E-treated group (C, 10X, and 40X), coriander extract-treated group (D, 10X, and 40X), and vitamin E and coriander extract-treated group (E, 10X, and 40X). * Histological observation of the left kidney is presented here because there was no histological differentiation between the left and right kidneys. G = glomerulus. Scale bar: 5 µm (10X) and 1 µm (40X).

in the intoxicated lead group were separated from spermatogonia (black arrow) (Figure 40.9B, 10X). In addition, morphological changes of seminiferous tubules (blue arrow) and irregular arrangement of spermatogenic cells (white arrow) were also observed in this group (Figure 40.9C, 10X). Seminiferous tubules were found normal in the vitamin E-treated group (Figure 40.9D, 10X), coriander extract-treated group (Figure 40.9E, 10X), and vitamin E and coriander extract (combined)-treated group (Figure 40.9F, 10X). Primary spermatocytes were not found separated from spermatogonia in these groups. Spermatogenic cells were present in a normal pattern. Morphological changes were not observed in the seminiferous tubules.

40.4   DISCUSSION

The present research aimed to study the efficacy of vitamin E and coriander extract on the liver, kidney, and testis of mice exposed to lead toxicity. Gross and histological studies were observed in this study and are discussed below.

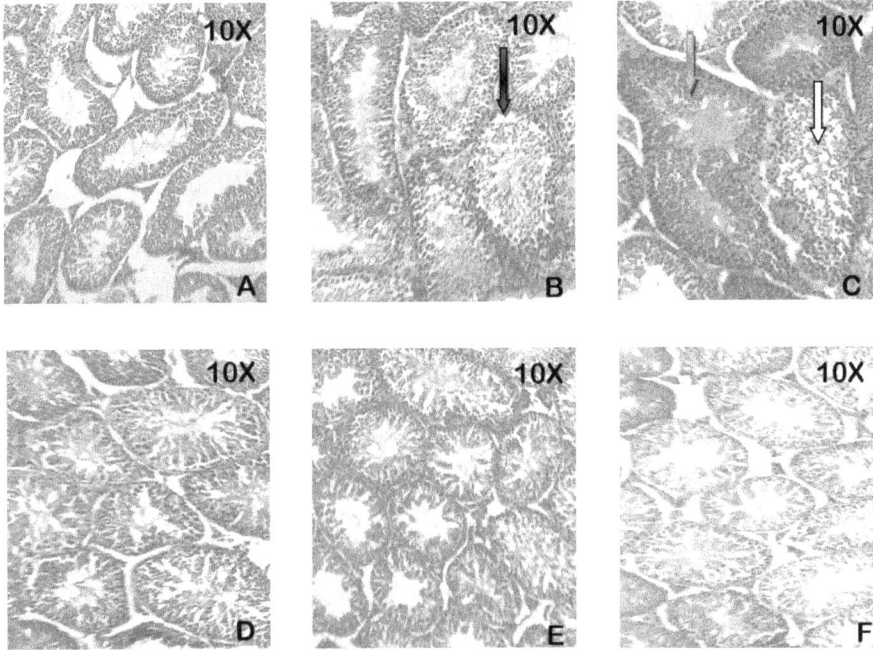

FIGURE 40.9 (A–F) Histological observation of testis in mice (H&E).* Normal appearance of seminiferous tubules was found in the control group. Spermatogenic cells were present in a normal pattern (A, 10X). Primary spermatocytes were separated from spermatogonia (black arrow) in the intoxicated lead group (Ba, 10X). In addition, morphological changes of seminiferous tubules (blue arrow) and irregular arrangement of spermatogenic cells (white arrow) were observed in this group (Bb, 10X). Appearance of seminiferous tubules was found normal in the vitamin E-treated group (C, 10X), coriander extract-treated group (D, 10X), and vitamin E and coriander extract-treated group (E, 10X). Spermatogenic cells were present in a normal pattern in these groups. Scale bar: 5 µm (10X). * Histological observation of left testis is presented here because there was no histological differentiation between left and right testes.

40.4.1 Body Weight

The mean body weight of mice in the intoxicated lead group was significantly ($p < 0.05$) reduced compared to the control group. This is similar to Nabil et al.'s (2012) report, which found that lead caused a decrease in the growth rate in rats when fed with lead. On the other hand, the body weight of mice in the vitamin E and coriander extract (combined)-treated group was significantly ($p < 0.05$) increased in comparison to the intoxicated group, vitamin E-treated group, and coriander extract-treated group. This result is due to the combined antioxidative properties of vitamin E and coriander extract. Antioxidant properties of vitamin E and coriander extract reduced the impacts of lead toxicity, and ultimately there was higher body weight gain.

40.4.2 Length and Weight of Liver, Kidney, and Testis

The mean length and weight of the liver, weight of the left kidney, and weight and length of both (left and right) testes in the intoxicated lead group were significantly ($p < 0.05$) reduced in comparison to the control group. This is partially supported by Nabil et al. (2012), who found a decreased growth rate in rats when fed with lead. More probably, the lower weight of the liver, kidney, and testis in the intoxicated lead group is due to the reduced body weight.

40.4.3 GROSS OBSERVATION

The nodular lesion was found in the liver and kidneys in the mice of the intoxicated lead group. Unfortunately, regarding nodular lesions, literature is not available. However, Zusman et al. (1991) reported carcinomatous lesions, which is partially consistent with my findings. In addition, they reported that nuclear polymorphism is seen in hepatic dysplasia and carcinomatous lesions due to lead toxicity. The actual mechanism of nodule formation is unknown. However, this may be due to increased cellular activity and nuclear interruption in the mechanism of lead detoxification.

The nodular lesion was found in the liver even after vitamin E treatment. However, there was a nodule in the kidneys after vitamin E treatment. So it indicates that the properties of vitamin E as an antioxidant are not always effective in reducing the nodular lesion.

In the present study, the nodular lesion was not observed in the coriander extract-treated group and the vitamin E and coriander extract combined-treated group. Therefore, this may be due to the combined actions of vitamin E and coriander extract. The gross architecture of the testis was found normal in the control group, vitamin E-treated group, coriander extract-treated group, and vitamin E and coriander extract combined-treated group in the present study. However, the testis was found in reduced size in the intoxicated lead group, similar to Hamadouche et al. (2013), who found that the testis, epididymis, and accessory sex glands were significantly affected in rats treated with lead compared to the control group.

40.4.4 HISTOLOGICAL OBSERVATION

In the intoxicated lead group, congestion was found in the central vein of the liver in some of the sections in the present study. In addition, nodules with fibrous covering were found in this group. Although the nodular lesion was not observed in the available literature, Zusman et al. (1991) reported the carcinomatous lesion, partially consistent with the present findings. The reason for congestion in the central vein is also not apparent in the present study. In the intoxicated lead group, the epithelial cast found in the urinary space of the kidney in the present study is similar to the report by Kansal et al. (2012), who found casts in the kidney of lead nitrated exposed mice. In the present study, the primary spermatocyte of testis was separated from spermatogonia in the intoxicated lead group. This group also found a morphological change in the seminiferous tubules and an irregular arrangement of spermatogenic cells. This result is similar or nearly similar to the findings of previous works. Nadia et al. (2013) reported that lead causes degeneration and necrosis of spermatogonia and interstitial cells and abnormal distribution of spermatozoa. Moniem et al. (2010) showed that lead exposure caused progressive vascular, tubular, and interstitial testicular damage.

Normal histology of the liver, kidney, and testis was found in the vitamin E-treated group, coriander extract-treated group, and vitamin E and coriander extract combined-treated group in the present study. There was no congestion in the central vein of the liver, no epithelial cast in the urinary space of the kidneys, and no abnormalities in the seminiferous tubules of the testis in these groups. This is due to the antioxidative properties of vitamin E and coriander extract. However, vitamin E has many other properties, preventing the incidence and reducing oxidative stress in tissues (Patra et al., 2011). Furthermore, Kansal et al. (2012) reported that the coriander-mediated suppression of the increased AST and ALT activities and cholesterol level suggests the possibility of the extract giving protection against hepatic, renal, and testicular injury upon lead induction.

40.5 CONCLUSION

The present findings revealed that lead has detrimental effects on mice's body weight, liver, kidney, and testis. The body weight of mice in the intoxicated lead group was significantly reduced compared to the control group. Lead was found to cause nodular lesions and congestion in the central vein in the liver of mice. In the kidney, lead caused nodular lesions and epithelial cast in the urinary space. Lead exerted its most harmful effects on the testis. Lead was found to cause the decreased weight of testis, separation of primary spermatocyte from spermatogonia, morphological changes of the seminiferous tubule, and irregular arrangement of spermatogenic cells in the seminiferous tubules. The antioxidative action of vitamin E and coriander extract effectively treated lead intoxication. The gross and histoarchitecture of the liver, kidney, and testis were found normal after treatment with vitamin E and coriander extract.

REFERENCES

Chithra, V., & Leelamma, S. (1999). *Coriandrum sativum* changes the levels of lipid peroxidase and activity of antioxidant enzymes in experimental animals. *Indian Journal of Biochemistry and Biophysics*, 36, 59–61.

Hamadouche, N. A., Sadi, N., Kharoubi, O., Slimani, M., & Aoues, A. (2013). The protective effect of vitamin E against genotoxicity of lead acetate intraperitoneal administration in male rat. *Archives of Biological Sciences*, 65, 1435–1445.

Henretig, F. M. (2002). Lead. In *Gold Frank's Toxicological Emergencies*. 7th ed. McGraw Hill Co., New York, pp. 1200–1237.

Hong, Z., Hailing, L., Hui, M., & Guijie, Z. (2009). Effect of vitamin E supplementation on development of reproductive organs in Boer goat. *Animal Reproduction Science*, 113, 93–101.

Hsu, P. C., Liu, M. Y., Hsu, C. C., Chen, L. Y., & Guo, Y. L. (1998). Effects of vitamin E and/or C on reactive oxygen species-related lead toxicity in the rat sperm. *Toxicology*, 128, 169–179.

Kansal, L., Sharma, A., & Lodi, S. (2012). Remedial effect of *Coriandrum sativum* (coriander extracts on lead induced oxidative damage in soft tissues of Swiss albino mice. *International Journal of Pharmacy and Pharmaceutical Sciences*, 4(3).

Kartikeya, M., Agarwal, A., & Sharma, R. (2009). Oxidative stress and male infertility. *Indian Journal of Medical Research*, 129, 357–367.

Moniem, A. E. A., Dkhil, M. A., & Al-Quraishy, S. (2010). Protective role of flaxseed oil against lead acetate induced oxidative stress in testes of adult rats. *African Journal of Biotechnology*, 9, 7216–7223.

Nabil, I. M., Eweis, E. A., El-Beltaqi, H. S., & Abdel-Mobdy, Y. E. (2012). Effect of lead acetate toxicity on experimental male albino rat. *Asian Pacific Journal of Tropical Biomedicine*, 41–46.

Nadia, A. H., Nesrine, S., Kharoubi, O., Slimani, M., & Aoues, A. (2013). The protective effect of vitamin E against genotoxicity of lead acetate intraperitoneal administration in male rat. *Archives of Biological Sciences*, 65(4), 1435–1445.

Odigie, I. P., Ladipo, C. O., Ettarh, R. R., & Izegbu, M. C. (2004). Effect of chronic exposure to low levels of lead on renal function and renal ultrastructure in SD rats. *Nigerian Journal of Physiological Sciences*, 19, 27–32.

Patra, R. C., Rautray, A. K., & Swarup, D. (2011). Oxidative stress in lead and cadmium toxicity and its amelioration. *Veterinary Medicine International*, 1–9.

Piasek, M., Kostial, K., & Bunarevic, A. (1989). The effect of lead exposure on pathohistological changes in the liver and kidney in relation to age in rats. *Arhiv za Higijenu Rada i Toksikologiju*, 40(1), 15–21.

Saryan, L. A., & Zen, Z. C. (1994). Lead and its compounds. In Occupational *Medicine*, Dickerson OB, Zenz C and Horvath EP, editors. 3rd ed. Mosby Inc., Orlando, FL, pp. 506–541.

Tarin, J. J., Brines, J., & Cano, A. (1998). Antioxidants may protect against infertility. *Human Reproduction*, 13, 1415–1416.

Zusman, I., Kozlenko, M., & Zimber, A. (1991). Nuclear polymorphism and nuclear size in precarcinomatous and carcinomatous lesions in rat colon and liver. *Cytometry*, 12(4), 302–307.

41 Coriander Extract in Biopolymer Applications

Vijayasarathy S. and Gayathri Mahalingam

CONTENTS

41.1 INTRODUCTION

Coriandrum sativum is an herb from the Apiaceae family that is grown annually. It contains a considerable number of promising compounds that have been proved antimicrobial. The family of herbs is known for their polyphenol composition, a polar secondary metabolite (Gantner *et al.*, 2017). At the outset, coriander extract is known for its antioxidant, antibacterial, antifungal, antidiabetic, hypocholesterolemic, and anticarcinogenic effects (Ganesan *et al.*, 2013). Seeds of *Coriandrum sativum* contain triglycerides, proteins, sugars, phenols, essential oils, etc., which are used in various applications (Beyzi *et al.*, 2017).

The use of coriander seeds and other plant parts is not new to humans. The European Commission [EC No 259/97] allowed the usage of oil from coriander seeds in food. Seeds of coriander are rich in uncommon isomers of oleic acid. Since fatty acids constitute 70% to 80% of the total extract and are unsaturated, these are safe and healthy (Nguyen *et al.*, 2020). Since early times the herb has been used in herbal medicine due to its antimicrobial activity. It also has shown effectiveness against cadmium and has immune system stimulating effects *in vivo* (Ahmed *et al.*, 2020). The herb contains multiple bioactive compounds that are of significant use in the pharma-based industries. Also, it is used in additives and packaging where it improves the stability of the food during storage. Thus, these are even infused with coatings to preserve food materials for longer durations (Omidi-Mirzaei *et al.*, 2020). The most abundant phenolic compound was *p*-coumaric acid, followed by chlorogenic acid. The analysis also revealed the presence of caffeic acid, gallic acid, luteolin, vanillic acid, rutin, and quercetin (Derouich *et al.*, 2020)

Biopolymers in recent years have come into the picture because society is becoming conscious of renewables, environmental protection, and biodegradability, reflecting concerns over environmental

DOI: 10.1201/9781003204626-46

problems. Hence, this has led to the fashion of using biopolymers; proteins, polysaccharides, lipids, and their derivatives in various applications are being explored (Rhim & Ng, 2007). These polymers offer various advantages by themselves in terms of safety and sustainability. These also have the advantage of being biodegradable. The properties conferred by petroleum-based polymers over biopolymers is that synthetic polymers have higher barrier properties and physical stability. Hence, biopolymers are mixed with unique compounds or treated with chemicals to obtain desirable characteristics (Othman, 2014). The major success seen in this is biopolymer application in the packaging and edible coating of food materials to prevent external interaction (Rhim & Ng, 2007). For instance, chitosan films demonstrate excellence in the properties used in food packaging. They are used for extending shelf life and are used as pure films, blends of chitosan and synthetic polymers, and derivatives of chitosan made into films (Wang *et al.*, 2018)

Using natural products in various applications progresses because synthetic chemicals cause multiple problems to the environment and humans and other living species. Therefore, identifying and developing natural compounds with the antimicrobial potential to be used in various applications is progressing quickly (Avci *et al.*, 2013).

41.2 EXTRACTS

The utilization of plants has become common due to their versatile applications and the chemicals they produce as part of their metabolism (Rasoanaivo *et al.*, 2011). Therefore, plant extracts are also referred to as botanically active substances. As per definition, a plant extract/botanically active substance may contain one or more chemical components/compounds that are found in plants, where these are extracted by putting through various parts of the plant through physical and/or chemical processes such as crushing, distillation, etc. (du Jardin, 2015). In addition, a plant extract also contains secondary metabolites, which can act as antibacterial, antifungal, and antioxidant chemicals. These can be used in various applications (Cardozo *et al.*, 2005). Figure 41.1 shows the steps involved in extracting plant extracts with the example of coriander seeds.

41.3 OBTAINING CORIANDER EXTRACT

Extracts from *Coriandrum sativum* are broadly classified into volatile and non-volatile compounds based on the methodologies implemented for the extraction. First, the leaves and seeds are subjected

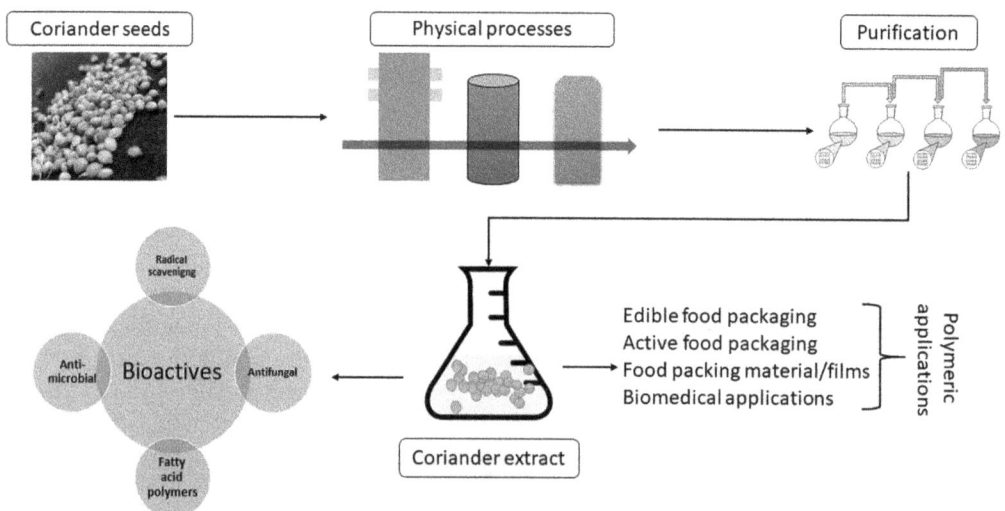

FIGURE 41.1   Schematic overview of extraction and the application of coriander extract.

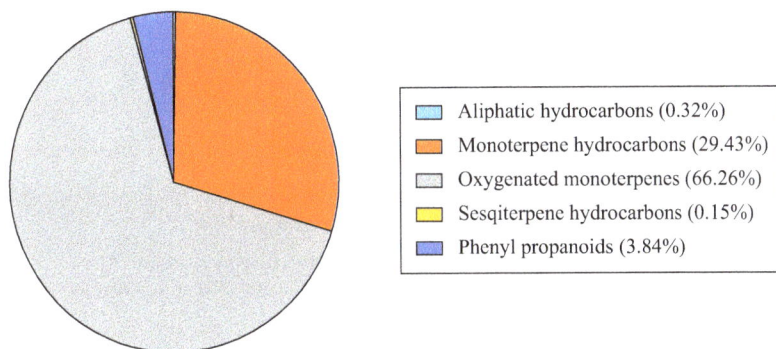

FIGURE 41.2 The overall components of essential oil from coriander extract.

to hydrodistillation to extract volatile compounds. Then, a Clevenger-type apparatus is used to heat the seeds or leaves in distilled water. Owing to the lower extraction rate of about only 0.1%, the distillate is put through a liquid-liquid extractor with dichloromethane as solvent. After this, the extract is evaporated, and the compounds are analyzed and used. In the case of non-volatile compound extraction, the solvent extraction method is used where the samples of leaves and seeds are taken and put in organic solvents like methanol or hexane. The formation of layers is observed while the sample is shaken at continuous intervals. Later the samples are filtered and evaporated under vacuum to remove the solvents from the sample. The solvent is then purified in an N_2 stream (Shahwar *et al.*, 2012).

A sequential extraction process is used to obtain the etheric extract of coriander. First, ethyl ether extracts the compounds by agitation, followed by centrifugation. Then, consecutively, ethanol and distilled water are added, and the process is repeated. Later the extract is evaporated under pressure and made into an emulsion with distilled water (De Almeida Melo *et al.*, 2003). Figure 41.2 gives an idea of the essential oils in the coriander extract. To obtain the aqueous coriander extract, the seeds are soaked in water and steam distillate. To get the volatile compounds, a separating funnel is used with dichloromethane as a solvent. The extract is then evaporated and distilled to achieve higher purity. For instance, to have an alcoholic, methanolic extract, a similar workflow outline is followed where the seeds are soaked in methanol for over 2 days and diluted with water. Then, it is extracted in a separating funnel using dichloromethane as solvent. Next, anhydrous salts are evaporated and distilled in a fractionation column (Mazza, 2002). Also, other solvents like *n*-hexane, chloroform, and ethyl acetate are used. Finally, the extraction is done by maceration. Finally, the solutions are vacuum dried in a rotary evaporator at low temperatures (Molina *et al.*, 2020).

Microwave-assisted hydrodistillation collects the essential oil from the coriander seeds, where the seeds are placed in a flask, and distilled water is added. Then, the flask is kept in a microwave oven. The extract is obtained and condensed outside the oven with appropriate pressure and temperature and then ventilated (Ghazanfari *et al.*, 2020). One of the newer technologies that comes into play in extracting essentials from plants is enzyme-assisted extraction. It has emerged as a cost-effective, green, and high bioactive yielding methodology by the use of an enzyme or combination of enzymes that break down undesired products and leave behind the desired bioactive. For example, Abbassi *et al.* used cellulase, hemicellulase, and their combination to extract essential oils from coriander seeds, where a 44% and 40% increase was seen in using cellulase and the combination, respectively (Abbassi *et al.*, 2018).

41.4 BIOACTIVES FROM CORIANDER USED IN POLYMER APPLICATIONS

The extract from coriander contains a wide variety of carbon compounds. These compounds possess various advantages and hence provide a multitude of applications. The two major biopolymers

TABLE 41.1

Biological Activities That Are Shown by Coriander Extract

Activity	Model Organism/ Assay Used	Extract Type and/or Method	Result	Reference
Antibacterial	*Salmonella choleraesuis*	Specific compound analysis	Acyclic aldehyde showed inhibition	(Kubo *et al.*, 2004)
	Staphylococcus aureus	Volatile compound extract-water distillation	Tested multiple species; strong inhibitory potential with *S. aureus*	(Hassanen *et al.*, 2015)
	Escherichia coli and *Pseudomonas aeruginosa*	Aqueous extract through hydrodistillation and microwave-assisted hydrodistillation	Both methods yielded extracts that inhibited *P. aeruginosa* with MIC 6.25 μL mL^{-1}	(Sourmaghi *et al.*, 2015)
Antifungal	*Penicillium lilacinum* and *Aspergillus niger*	Maceration	Inhibited growth of the fungal species at low concentrations	(Zardini *et al.*, 2012)
	Activity analyzed by using CEO in cakes	Aqueous extract using steam distillation	Showed properties of fungal growth inhibition	(Darughe *et al.*, 2012)
Antioxidant	DPPH, reducing power and β-carotene bleaching test	Methanolic extract (polyphenol extraction)	Antioxidant potential is high, and it is safe to be used as a natural antioxidant	(Msaada *et al.*, 2017)
	DPPH· and lipid peroxidation test	Supercritical fluid extract	Compared to other extraction methods, antioxidant activity using DPPH· recorded low value	(Zeković *et al.*, 2016)

in coriander extract are polyphenols and essential oils (Sivakanthan *et al.*, 2020). Table 41.1 outlines the application and the bioactivity in coriander extract. Assays like 2,2-diphenyl-1-picrylhydrazyl (DPPH·), ferric reducing antioxidant power assay (FRAP), and 2,2'-azino-bis(3-ethylbenzothiazoli ne-6-sulfonic acid) (ABTS) reveal that the coriander extract contains antioxidant compounds that scavenge oxygen and, thus, has the ability to act as an antioxidant solution (Hameed *et al.*, 2019). It was also observed that the ascorbic acid present in coriander extracts tends to possess abilities to cure serious ailments and also possesses medical advantages (Kaur *et al.*, 2018).

Many biopolymers have been in the development pipeline in recent years due to their advantages. Incorporating bioactive molecules into polymer is becoming an area of research. This provides additional functionalities to the polymer in applications. New properties such as colors, antimicrobial and antioxidant properties, and flavors are added to the polymers (Nogueira *et al.*, 2020). This proves most important in the food industry to preserve food items. Since natural compounds are added, it also does not compromise the organoleptic properties. Electrospraying is found to be an effective method in food coatings. Thus, intelligent packaging could be produced in one step (Niu *et al.*, 2020).

41.5 FOOD COATINGS

A notable preservation technique in food processing and production is the edible coating technique, where molecules such as proteins, lipids, and polysaccharides are applied to the product's surface, leading to the formation of a thin layer. This essentially preserves the organoleptic properties of the food particles by reducing respiration and controlling moisture transfer and oxidation processes (Sharif & Mustapha, 2017). Applying an edible coating of food materials aims to increase the shelf life and provide longevity to the food. Nowadays, these are incorporated with antimicrobials and antioxidants for additional functionalities (Kargozari *et al.*, 2018). Though essential oils have a potential bio-preservative effect, they cannot be applied directly onto the food matrix because they start degrading due to external factors and interactions. Hence, new developments encapsulate the bioactive in polymers, liposomes, or nanoparticles (Sharma *et al.*, 2021). Fernandez-Pan *et al.* proposed edible antimicrobial films made from whey protein isolate. The bioactive antimicrobial that was tested included coriander extract essential oil as well. The results showed that a higher concentration of extracts, when incorporated in the range of 5–9%, showed antimicrobial activity against *Staphylococcus aureus*, *Listeria innocua*, *Pseudomonas fragi*, and *Salmonella enteritidis* (Fernández-Pan *et al.*, 2012). Films made from alginate coriander seed oil extract prevented yeast and mold growth when applied over filleted chicken, making it an economical option to make coatings to prevent foods from spoiling (Kargozari *et al.*, 2018).

Similarly, chitosan films incorporating coriander extract showed inhibition against *L. monocytogenes* and *Escherichia coli* (Zivanovic, S., Chi, S., Draughon, 2005). Post-harvest extension of shelf life is also important. Lemons have a longer shelf life when harvested and stored when they are green. Edible coatings made up of guar gum-based polymers infused with ethanol, and methanol coriander seed extract delayed the ripening of fruits and had storage stability up to 15 weeks when stored at 10°C. The bacterial load on treated lemons was much less than those not coated with any polymeric coatings (Naeem *et al.*, 2019). A similar study also proved that guar gum-based coriander extract-incorporated coating of unripe green tomatoes preserved the post-harvest fruits for up to 60 days at 10°C. Changes in acidity, pH, and soluble solids in the fruits were not affected, and the microbial load decreased to a minimum. The study stated that the extract forms a semi-permeable polymeric membrane that reduces surface contact with the atmosphere, reducing spoilage chances. It also concluded that ethanolic extract of the coriander seeds slowed down the ripening of the fruits better than the methanolic extract (Ayeza Naeem *et al.*, 2018). Hydrocolloid gum Arabic modified with the antioxidants from coriander extract formed rigid thermal gelation and/or a cross-linking network when applied to potato strips before deep frying. This resulted in low-fat and acrylamide potato chips with enhanced sensory reception (Mousa, 2018).

41.6   FOOD PACKAGING

With developments demanding effective food packaging to reduce spoilage and contamination and improve shelf life, the incorporation of antimicrobials and antioxidants in packaging films is being considered. It is being done by incorporating extract that possesses favorable properties into the polymers used. The oil extract of coriander can be agonistic to a broad spectrum of bacteria and fungi contaminating food. Infusion of coriander oil extract with plastics and paper packaging has proven efficient. Nowadays, the direct use of antimicrobial bioactives is being investigated to make the packaging film since it can make the process economical and feasible (Silva & Domingues, 2017). Kostova *et al.* characterized active cellulose paper packaging, including bleached, unbleached, and recycled variants with coriander oil extract. This proved that all three types had the ability to inhibit microbial growth. The results showed that all the test organisms, including Gram-positive bacteria, Gram-negative bacteria, and fungi, were inhibited, probably through some antagonistic or synergistic interaction or solubility differences due to varied microfiber composition. Hence, the experimentation results showed that microbial growth prevention could be prevented by using coriander extracts (Kostova *et al.*, 2020). Besides just polysaccharides being used in packaging, protein-based active films have also come into the picture. Natural antioxidants are being looked into to prevent oxidation in food materials where incorporation of coriander essential oils into gelatin films slowed oxidation during storage of sliced cheddar cheese (Calva-Estrada *et al.*, 2019). As seen in Figure 41.3, foods packed with active compounds-infused polymer stay fresh longer. The major setback incorporating the extract is the poor stability and lack of efficiency. Silva *et al.* synthesized cyclodextrin nanosponges that can provide controlled release of antimicrobial agents incorporated into packaging

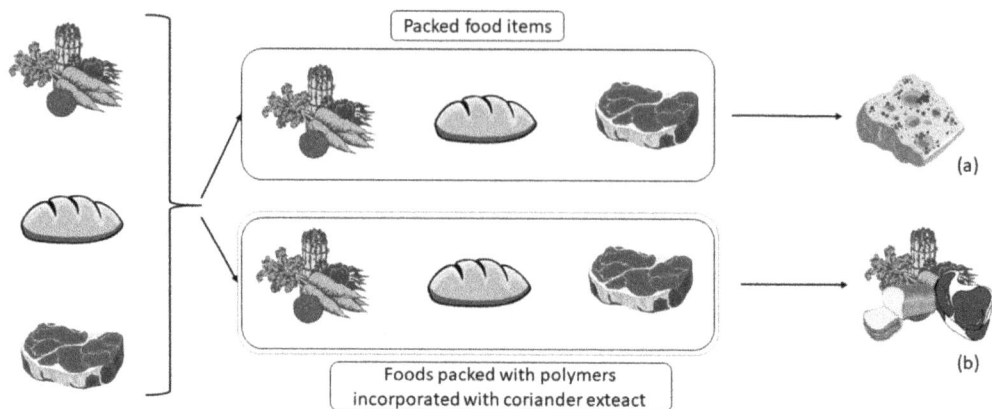

FIGURE 41.3 The effectiveness of extract-incorporated food packaging. (a) shows that the food has the chance of getting spoilt over the period and (b) shows that food packed with the extract-incorporated biopolymers remains fresh and edible for longer durations.

film polymers (Silva *et al.*, 2019). With the use of extracts, a noteworthy point is about the physical factors of the film. When coriander extract is added using high-speed homogenization, porcine meat and bone meat films slightly increased in tensile strength, and the elongation decreased. The changes are attributed to the interactions between the polymer and the extract (Lee *et al.*, 2015).

41.7   USE AS ANTIOXIDANTS

Polyphenols are groups of compounds that are synthesized in plants as secondary metabolites. These contain multiple aromatic groups that contain hydroxyl groups (de Araújo *et al.*, 2021; Sivakanthan *et al.*, 2020). The antioxidant activity of coriander extract is attributed to polyphenol biopolymers with multiple hydroxyl complexes. These hydroxyl groups scavenge oxygen and neutralize the free radicals (Mechchate *et al.*, 2021). They showed positive results when assays like DPPH·, ABTS, cupric reducing capacity, and ferric reducing capacity were performed with the phenolics from coriander extract. Since the compounds from plants are natural, they are employed in food science since they pose no harm even when consumed (Demir & Korukluoglu, 2020). Besides phenolic derivatives, flavonoids are the other compounds in higher amounts in coriander seeds. These compounds are proved to be potential antioxidants (Sriti *et al.*, 2012). Such flavonoids can be looked into for polymer preparation that could potentially be used in food preservation and medical applications.

41.8   BIOPOLYMERS FROM CORIANDER EXTRACT IN MEDICAL APPLICATIONS

Effective targeted drug delivery is becoming crucial in every disease. Improper drug availability and release result in loss of availability of the drug in the active site. To counter such drawbacks, drugs loaded on nanoparticles capped with polymers are used. For example, Madhusudhan *et al.* loaded doxorubicin on gold nanoparticles and capped it with a polymer called chitosan for better efficacy (Madhusudhan *et al.*, 2014). Figure 41.4 describes a few ways in which coriander extract can be put to use in medical applications. Decanoic acid extracted from coriander seed extract was polymerized with two groups to form new derivatives. The first derivative was obtained by reacting decanoic acid with thiourea, where later paracetamol was reacted with the mixture. Another derivative was synthesized by reacting the second amine in thiourea with maleic anhydride and mixing with various amino drugs. Thus, the conventional problem in the residence time can be overcome, and the drug's bioavailability can be prolonged (Hussein *et al.*, 2019). Due to tannins in coriander extract, oak and coriander extract were mixed and made into an herbal gel polymer to

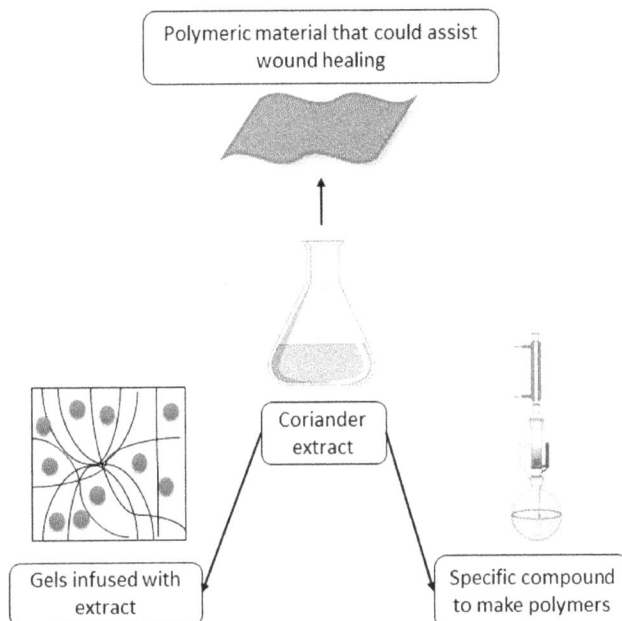

FIGURE 41.4 Use of coriander extract in medical applications.

treat periodontitis. This was designed as a local drug delivery polymer. The experiment's conclusion showed that the suggested system showed clinical improvements compared to placebo gel, but the effect was insignificant. However, effective treatment could be observed if the concentration increased (Naghsh *et al.*, 2014). In recent years, poly butylene succinate (PBSU), a biodegradable synthetic aliphatic polyester, is being studied for drug delivery and cartilage repair. The polymer is already widely used in packaging, agriculture, and catering. These polymers show promising results in wound dressing and skin substitutes in the medical sector. When essential oil from coriander extract was incorporated during electrospinning to develop antimicrobial PBSU, the results showed that a lower concentration did not show the desired result. However, the higher concentration indicated antimicrobial activity (Aliko *et al.*, 2021).

41.9 POLYMERIC APPLICATIONS OF SPECIFIC COMPOUNDS FROM CORIANDER EXTRACT

41.9.1 Petroselinic Acid

The picture of biopolymers starts to play when the human race understands the rate at which fossil fuels are being depleted and the increasing greenhouse gas emission rate. The implementation of green chemistry wherever possible makes efficient utilization of resources. Plant oils play an essential role in utilizing renewable resources in industries. Petroselinic acid is an unusual fatty acid that could potentially substitute petroleum-based polymers (Meier, 2009). Figure 41.5 clarifies that petroselinic acid is the principal constituent of the fatty acid fraction of the extract. The fatty acid is unsaturated at C6 and is an 18:1 acid, making it an option to explore novel reactivities in the proximity of the unsaturation (Metzger & Bornscheuer, 2006). Researchers from Queen's University, Belfast, in 2010, developed a composite membrane by embedding petroselinic acid in the cellulose acetate matrix. This was done to mimic the phenomenon of plant partitioning and *in vivo* polycyclic aromatic hydrocarbon uptake. The expected results were that the petroselinic acid-enforced cellulose acetate matrix (PECAM) had polyaromatic hydrocarbon accumulation and mimicked plant

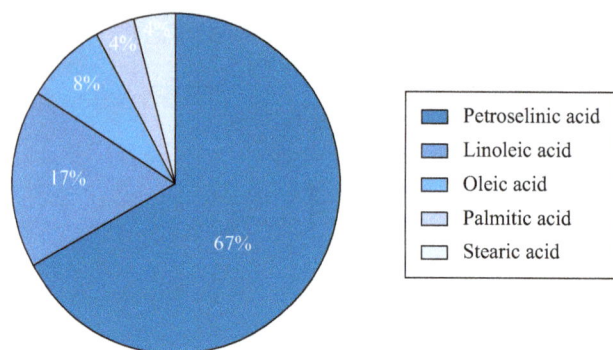

FIGURE 41.5 Composition of fatty acids in coriander extract.

partitioning (Li *et al.*, 2010). Petroselinic acid also has other applications where oxidation of the acid with ozone forms lauric acid and adipic acid. Here, adipic acid is mainly used in the textile industry, where it is the monomer for nylon6,6, and lauric acid is used in detergents. The fatty acid also finds application in the pharmaceutical industry because it has a higher melting temperature (Töpfer *et al.*, 1995).

41.9.2 Linalool

When coriander essential oils are separated into individual components, the major compound is identified in linalool. This compound is already proven to have antibacterial and antioxidant properties, and it is suggested to be a preservative and anti-biofilm agent (Duarte *et al.*, 2016). One of the methods to make linalool polymers or polymers fabricated with linalool is radio frequency (RF) plasma polymerization. The research groups also suggest polylinalool as an option in electronic applications after doping. When the polymer was tested, the physical properties of adhesion and Young's modulus had greater stability and proper adhesion, improving the interfacial bonding (Jacob *et al.*, 2013). A similar study in making polylinalool films observed that the film had many short polymer chains and presented strong methyl bands and weaker methyl bands in the Fourier transform infrared (FTIR) spectrum. Its properties, transparent with a refractive index of 1.5 and extinction coefficient of 0.001 at 500 nm, allow it to be used in encapsulating circuit boards as a protective covering. The film was also defect-free and smooth, revealed by atomic force microscopic imaging (Bazaka *et al.*, 2010). The application of linalool becomes restricted due to its volatility and lower stability. However, when emulsified droplets of linalool are precipitation polymerized by modifying the carbon double bonds with N-vinyl caprolactam (VCL), it is used in capsules with a high loading ratio. The capsule naturally had antibacterial activity. The linalool-HS-PVCL (poly VCL) capsuled had thermal and redox responsive properties. These also provided options for controlled release. Such work can open doors for green antibacterial capsules for various applications (Hu *et al.*, 2020). Linalool is also used to improve the properties of existing polymers. Chitosan oligosaccharide (COS) is a renewable polysaccharide polymer with excellent biocompatibility and is used in the pharmaceutical and food processing industries. Linalool grafted in COS by modifying its backbone in the presence of trimethylamine and dimethylformamide gives a thermally stable polymer compared to COS. The polymer also showed increased anti-inflammatory properties compared to COS (Mao *et al.*, 2021). Linalool has also found application in food packaging, where the diffusion of linalool over low-density polyethylene (LDPE) was studied. It showed that the film could be used as an antimicrobial food packaging material and the diffusion of the compound was not fully Fickian (Suppakul *et al.*, 2011). Linalool has also been incorporated into collagen to produce composite wound dressing. Oxidized amylose was used to introduce linalool into the matrix. The matrix had tunable pore size and could absorb exudates that came from the wound. The dressing also provided antibacterial activity due to the presence of linalool. The properties make it more notable in making biomaterials that can be used in wound healing (Lyu *et al.*, 2017).

41.9.3 RUTIN

Rutin is a bioactive flavonoid found in high content in coriander extract. It is reported to be an antioxidant and a compound with radioprotective potential (Han *et al.*, 2017). Applications using rutin become questionable due to its lower solubility in water. However, when the same is made into a polymer, the flavonoid in a polymeric form finds various applications. This polymer has proved to be efficient in scavenging biological radicals and influenced akin fibroblast positively *in vitro*. Therefore, this has proved to be potentially useful in wound healing and dressing. The polymer protects the cells from oxidative stress and helps wound healing (Elschner *et al.*, 2019).

Moreover, rutin also has been coupled with polymers to find biomedical applications. Rutin crosslinked with the epoxy group of poly(ethylene glycol) diglycidyl ether formed a rutin nanogel that was spherical and had an injectable size. This finds extensive applications such as providing iron homeostasis by reducing Fe^{3+} to Fe^{2+} and treating glycogen storage disease. Since the nanogel has been observed to interact less with fibrinogen, it can be more suitable for intravenous applications (Sahiner *et al.*, 2021). As already seen in Section 41.8, the specific compounds from extracts could be incorporated into polymers for multiple applications; rutin has been incorporated into chitosan to develop a biofilm. The film is bound to have application in food packaging by extending shelf life and improving the quality of the food. The formed biofilm also had enhanced physical properties like water resistance, UV barrier, and thermal stability (Narasagoudr *et al.*, 2020). Additionally, polymerized rutin also has an application in compound identification. A glassy carbon electrode polymerized with rutin has the potential ability to identify catechol and hydroquinone simultaneously. The proposed method is sensitive and used to test tap and wastewater. The limits at 31 nM and 53 nM for catechol and hydroquinone, respectively, could be a candidate for a sensor to detect the above-mentioned compounds. The polymeric electrode also exhibits higher electrochemical performance (Karabiberoğlu et al., 2019).

41.10 FUTURE DIRECTIONS

The number of studies that are being carried out on using plant extracts in polymer applications is overwhelming. However, the major challenge in using them in various applications is that the mechanism behind the bioactivity of coriander extract is not known or defined. When studies on mechanisms of action are done, it could reveal the pathway targeted by the drug in the microbe. This could open doors for a combination of plant extract used, improving functionalities and becoming effective. When the interactions of phytochemicals become apparent, it could help us understand if the interaction leads to some synergetic effect or antagonistic interaction. If a proper combination of compounds with a synergetic effect could be developed, it could lead to a mix-based polymer that provides maximum functionality. Characterization of the molecules present in the extract can unleash a possibility of application in pharmaceutical and other sectors. Proper evaluation and understanding of the structural and physical parameters of the polymer must be appropriately tested for enhanced functional properties (Sivakanthan *et al.*, 2020).

The applications of coriander seed extract are enormous. Its widespread implementation in food packing and coatings could prevent massive loss in the market. The losses could be in wastage, storage, spoilage, etc. Since these are natural extracts, they provide the best alternative for replacing synthetic chemicals and compounds (Sharma *et al.*, 2021). The potential to develop nanocomposites that are natural polymer-based is also foreseen, and this could diversify the application by providing higher functionalities and properties. When the natural polymer-based nanocomposite is infused with bioactive compounds, it can replace petroleum polymers (Rhim & Ng, 2007).

Though coriander extract has proven to be an antimicrobial, antifungal, and antioxidant compound, the concentration of the oil or extract incorporated while making the polymer is not standardized. It must be experimented with carefully due to the possibility that it could be toxic after a specific concentration. Also, when used in edible coatings, the extracts are consumed where *in vivo* studies validated. Note that the extract cannot be used because it satisfies the *in vitro* experimentation.

41.11   CONCLUSION

Research and development in polymer sciences and incorporating bioactive compounds to improve application have been accelerated. The drive to develop modified films with better functions for applications in the pharmaceutical, textile, biomedical, and food industries is increasing. The properties of maximum deliverables from the bioactive conjugation include a defect-free polymer surface with regular surface properties. The biopolymeric films do not involve the severe environmental hazards and health threats possible when petroleum-based polymers are used. Furthermore, biopolymeric materials are also less toxic and relatively biocompatible. The addition of coriander extract in biopolymers confers additional properties and does not affect the sensorial properties of the polymers. Apart from just being bioactive to inhibit microbial growth and scavenge oxygen species, the extract also adds to the physical properties of the polymer, like Young's modulus, tensile strength, and thermal stability. In addition to improving the quality in the case of food packaging and coating, it also enhances to the life of the polymer. The essential oil proactively maintains the micro-nutritional traits of the food material packed and increases shelf life.

REFERENCES

Abbassi, A., Mahmoudi, H., Zaouali, W., M'Rabet, Y., Casabianca, H., & Hosni, K. (2018). Enzyme-aided release of bioactive compounds from coriander (*Coriandrum sativum* L.) seeds and their residue by-products and evaluation of their antioxidant activity. *Journal of Food Science and Technology*, 55(8), 3065–3076. https://doi.org/10.1007/s13197-018-3229-4

Ahmed, S. A. A., Reda, R. M., & ElHady, M. (2020). Immunomodulation by *Coriandrum sativum* seeds (Coriander) and its ameliorative effect on lead-induced immunotoxicity in Nile tilapia (*Oreochromis niloticus* L.). *Aquaculture Research*, 51(3), 1077–1088. https://doi.org/10.1111/are.14454

Aliko, K., Aldakhlalla, M. B., Leslie, L. J., Worthington, T., Topham, P. D., & Theodosiou, E. (2021). Poly(butylene succinate) fibrous dressings containing natural antimicrobial agents. *Journal of Industrial Textiles*. https://doi.org/10.1177/1528083720987209

Avci, H., Monticello, R., & Kotek, R. (2013). Preparation of antibacterial PVA and PEO nanofibers containing *Lawsonia Inermis* (henna) leaf extracts. *Journal of Biomaterials Science, Polymer Edition*, 24(16), 1815–1830. https://doi.org/10.1080/09205063.2013.804758

Bazaka, K., Jacob, M. V., & Shanks, R. A. (2010). Fabrication and characterization of RF plasma polymerized thin films from 3,7-dimethyl-1,6-octadien-3-ol for electronic and biomaterial applications. *Advanced Materials Research*, 123–125, 323–326. https://doi.org/10.4028/www.scientific.net/AMR.123-125.323

Beyzi, E., Karaman, K., Gunes, A., & Buyukkilic Beyzi, S. (2017). Change in some biochemical and bioactive properties and essential oil composition of coriander seed (*Coriandrum sativum* L.) varieties from Turkey. *Industrial Crops and Products*, 109(April), 74–78. https://doi.org/10.1016/j.indcrop.2017.08.008

Calva-Estrada, S. J., Jiménez-Fernández, M., & Lugo-Cervantes, E. (2019). Protein-based films: Advances in the development of biomaterials applicable to food packaging. *Food Engineering Reviews*. https://doi.org/10.1007/s12393-019-09189-w

Cardozo, P. W., Calsamiglia, S., Ferret, A., & Kamel, C. (2005). Screening for the effects of natural plant extracts at different pH on in vitro rumen microbial fermentation of a high-concentrate diet for beef cattle. *Journal of Animal Science*, 83(11), 2572–2579. https://doi.org/10.2527/2005.83112572x

Darughe, F., Barzegar, M., & Sahari, M. A. (2012). Antioxidant and antifungal activity of coriander (*Coriandrum sativum* L.) essential oil in cake. *International Food Research Journal*, 19(3), 1253–1260.

De Almeida Melo, E., Martins Bion, F., Mancini Filho, J., & Barbosa Guerra, N. (2003). In vivo antioxidant effect of aqueous and etheric coriander (*Coriandrum sativum* L.) extracts. *European Journal of Lipid Science and Technology*, 105(9), 483–487. https://doi.org/10.1002/ejlt.200300811

de Araújo, F. F., de Paulo Farias, D., Neri-Numa, I. A., & Pastore, G. M. (2021). Polyphenols and their applications: An approach in food chemistry and innovation potential. *Food Chemistry*, 338, 127535. https://doi.org/10.1016/j.foodchem.2020.127535

Demir, S., & Korukluoglu, M. (2020). A comparative study about antioxidant activity and phenolic composition of cumin (*Cuminum cyminum* L.) and coriander (*Coriandrum sativum* L.). *Indian Journal of Traditional Knowledge*, 19(2), 383–393.

Derouich, M., Bouhlali, E. D. T., Bammou, M., Hmidani, A., Sellam, K., & Alem, C. (2020). Bioactive compounds and antioxidant, antiperoxidative, and antihemolytic properties investigation of three Apiacee species grown in the southeast of Morocco. *Scientifica*, 2020. https://doi.org/10.1155/2020/3971041

du Jardin, P. (2015). Plant biostimulants: Definition, concept, main categories and regulation. *Scientia Horticulturae*, 196, 3–14. https://doi.org/10.1016/j.scienta.2015.09.021

Duarte, A., Luís, Â., Oleastro, M., & Domingues, F. C. (2016). Antioxidant properties of coriander essential oil and linalool and their potential to control *Campylobacter* spp. *Food Control*, 61, 115–122. https://doi.org/10.1016/j.foodcont.2015.09.033

Elschner, T., Zagar, E., Pivec, T., Kargl, R., Maver, U., & Bra, M. (2019). Chemical structure - Antioxidant activity relationship of water-based enzymatic polymerized rutin and its wound healing potential. *Polymers*, 11(10), 1–21.

Fernández-Pan, I., Royo, M., & Ignacio Maté, J. (2012). Antimicrobial activity of whey protein isolate edible films with essential oils against food spoilers and foodborne pathogens. *Journal of Food Science*, 77(7). https://doi.org/10.1111/j.1750-3841.2012.02752.x

Ganesan, P., Phaiphan, A., Murugan, Y., & Baharin, B. S. (2013). Comparative study of bioactive compounds in curry and coriander leaves: An update. *Journal of Chemical and Pharmaceutical Research*, 5(11), 590–594.

Gantner, M., Guzek, D., Najda, A., Brodowska, M., Górska-Horczyczak, E., Wojtasik-Kalinowska, I., & Godziszewska, J. (2017). Oxidative and microbial stability of poultry meatballs added with coriander extracts and packed in cold modified atmosphere. *International Journal of Food Properties*, 20(11), 2527–2537. https://doi.org/10.1080/10942912.2016.1243125

Ghazanfari, N., Mortazavi, S. A., Yazdi, F. T., & Mohammadi, M. (2020). Microwave-assisted hydrodistillation extraction of essential oil from coriander seeds and evaluation of their composition, antioxidant and antimicrobial activity. *Heliyon*, 6(9), e04893. https://doi.org/10.1016/j.heliyon.2020.e04893

Hameed, S., Imran, A., Nisa, M. un, Arshad, M. S., Saeed, F., Arshad, M. U., & Asif Khan, M. (2019). Characterization of extracted phenolics from black cumin (*Nigella sativa* Linn), coriander seed (*Coriandrum sativum* L.), and fenugreek seed (*Trigonella foenum-graecum*). *International Journal of Food Properties*, 22(1), 714–726. https://doi.org/10.1080/10942912.2019.1599390

Han, X., Xue, X., Zhao, Y., Li, Y., Liu, W., Zhang, J., & Fan, S. (2017). Rutin-enriched extract from *Coriandrum sativum* L. Ameliorates ionizing radiation-induced hematopoietic injury. *International Journal of Molecular Sciences*, 18(5). https://doi.org/10.3390/ijms18050942

Hassanen, N. H. M., Eissa, A. M. F., & Hafez, S. A. M. (2015). Original research article antioxidant and antimicrobial activity of celery (*Apium graveolens*) and coriander (*Coriandrum sativum*) herb and seed essential oils. *International Journal of Current Microbiology and Applied Sciences*, 4(3), 284–296.

Hu, J., Liu, S., & Deng, W. (2020). Dual responsive linalool capsules with high loading ratio for excellent antioxidant and antibacterial efficiency. *Colloids and Surfaces B: Biointerfaces*, 190, 110978. https://doi.org/10.1016/j.colsurfb.2020.110978

Hussein, F. A., Awd, S. H., Abdulsahib, S., & Ali, L. A. M. (2019). Synthesis of polymerization of new derivative of 2-decanoic acid extracted from coriander seed. *Journal of Global Pharma Technology*, 11(2), 466–474.

Jacob, M. V., Olsen, N. S., Anderson, L. J., Bazaka, K., & Shanks, R. A. (2013). Plasma polymerised thin films for flexible electronic applications. *Thin Solid Films*, 546, 167–170. https://doi.org/10.1016/j.tsf.2013.05.023

Karabiberoğlu, Ş. U., Koçak, Ç. C., & Dursun, Z. (2019). An over-oxidized poly(Rutin) modified electrode for selective and sensitive determination of catechol and hydroquinone. *Journal of Electroanalytical Chemistry*, 850. https://doi.org/10.1016/j.jelechem.2019.113415

Kargozari, M., Hamedi, H., Amirnia, S. A., Montazeri, A., & Abbaszadeh, S. (2018). Effect of bioactive edible coating based on sodium alginate and coriander (*Coriandrum sativum* L.) essential oil on the quality of refrigerated chicken fillet. *Food & Health*, 1(6), 30–38.

Kaur, G., Kaur, P., & Kaur, A. (2018). Physico-chemical properties, bioactive compounds and color parameters of coriander puree: Effect of pretreatments and freezing. *Journal of Food Science and Technology*, 55(9), 3473–3484. https://doi.org/10.1007/s13197-018-3272-1

Kostova, I., Lasheva, V., Georgieva, D., Damyanova, S., Fidan, H., Stoyanova, A., & Gubenia, O. (2020). Characterization of active paper packaging materials with coriander essential oil (*Coriandrum sativum* L.). *Journal of Chemical Technology and Metallurgy*, 55(6), 2085–2093.

Kubo, I., Fujita, K. I., Kubo, A., Nihei, K. I., & Ogura, T. (2004). Antibacterial activity of coriander volatile compounds against *Salmonella choleraesuis*. *Journal of Agricultural and Food Chemistry*, 52(11), 3329–3332. https://doi.org/10.1021/jf0354186

Lee, J. H., Won, M., & Song, K. B. (2015). Physical properties and antimicrobial activities of porcine meat and bone meal protein films containing coriander oil. *LWT-Food Science and Technology*, 63(1), 700–705. https://doi.org/10.1016/j.lwt.2015.03.043

Li, X., Zhu, Y., Wu, T., Zhang, S., & Christie, P. (2010). Using a novel petroselinic acid embedded cellulose acetate membrane to mimic plant partitioning and in vivo uptake of polycyclic aromatic hydrocarbons. *Environmental Science and Technology*, 44(1), 297–301. https://doi.org/10.1021/es902283x

Lyu, Y., Ren, H., Yu, M., Li, X., Li, D., & Mu, C. (2017). Using oxidized amylose as carrier of linalool for the development of antibacterial wound dressing. *Carbohydrate Polymers*, 174, 1095–1105. https://doi.org/10.1016/j.carbpol.2017.07.033

Madhusudhan, A., Reddy, G. B., Venkatesham, M., Veerabhadram, G., Kumar, A. D., Natarajan, S., Yang, M. Y., Hu, A., & Singh, S. S. (2014). Efficient ph dependent drug delivery to target cancer cells by gold nanoparticles capped with carboxymethyl chitosan. *International Journal of Molecular Sciences*, 15(5), 8216–8234. https://doi.org/10.3390/ijms15058216

Mao, S., Liu, X., & Xia, W. (2021). Chitosan oligosaccharide-g-linalool polymer as inhibitor of hyaluronidase and collagenase activity. *International Journal of Biological Macromolecules*, 166, 1570–1577. https://doi.org/10.1016/j.ijbiomac.2020.11.036

Mazza, G. (2002). Minor volatile constituents of essential oil and extracts of coriander (*Coriandrum sativum* L.) fruits. *Sciences Des Aliments*, 22(5), 617–627. https://doi.org/10.3166/sda.22.617-627

Mechchate, H., Es-Safi, I., Amaghnouje, A., Boukhira, S., Alotaibi, A. A., Al-Zharani, M., Nasr, F. A., Noman, O. M., Conte, R., Amal, E. H. E. Y., Bekkari, H., & Bousta, D. (2021). Antioxidant, anti-inflammatory and antidiabetic proprieties of LC-MS/MS identified polyphenols from coriander seeds. *Molecules*, 26(2). https://doi.org/10.3390/molecules26020487

Meier, M. A. R. (2009). Metathesis with oleochemicals: New approaches for the utilization of plant oils as renewable resources in polymer science. *Macromolecular Chemistry and Physics*, 210(13–14), 1073–1079. https://doi.org/10.1002/macp.200900168

Metzger, J. O., & Bornscheuer, U. (2006). Lipids as renewable resources: Current state of chemical and biotechnological conversion and diversification. *Applied Microbiology and Biotechnology*, 71(1), 13–22. https://doi.org/10.1007/s00253-006-0335-4

Molina, R. D. I., Campos-Silva, R., Macedo, A. J., Blázquez, M. A., Alberto, M. R., & Arena, M. E. (2020). Antibiofilm activity of coriander (*Coriander sativum* L.) grown in Argentina against food contaminants and human pathogenic bacteria. *Industrial Crops and Products*, 151(March), 112380. https://doi.org/10.1016/j.indcrop.2020.112380

Mousa, R. M. A. (2018). Simultaneous inhibition of acrylamide and oil uptake in deep fat fried potato strips using gum Arabic-based coating incorporated with antioxidants extracted from spices. *Food Hydrocolloids*, 83, 265–274. https://doi.org/10.1016/j.foodhyd.2018.05.007

Msaada, K., Jemia, M. Ben, Salem, N., Bachrouch, O., Sriti, J., Tammar, S., Bettaieb, I., Jabri, I., Kefi, S., Limam, F., & Marzouk, B. (2017). Antioxidant activity of methanolic extracts from three coriander (*Coriandrum sativum* L.) fruit varieties. *Arabian Journal of Chemistry*, 10, S3176–S3183. https://doi.org/10.1016/j.arabjc.2013.12.011

Naeem, A., Abbas, T., Ali, T. M., & Hasnain, A. (2018). Effect of antioxidant and antibacterial properties of guar gum coating containing spice extracts and its application on tomatoes (*Solanum lycopersicum* L.). *Journal of Food Measurement and Characterization*, 12(4), 2725–2734. https://doi.org/10.1007/s11694-018-9890-5

Naeem, A., Abbas, T., Mohsin Ali, T., & Hasnain, A. (2019). Application of guar gum-based edible coatings supplemented with spice extracts to extend post-harvest shelf life of lemon (*Citrus limon*). *Quality Assurance and Safety of Crops and Foods*, 11(3), 241–250. https://doi.org/10.3920/QAS2018.1310

Naghsh, N., Shahabooei, M., Yaghini, J., Zadeh, M., Aslani, A., & Kiani, S. (2014). Efficacy of a local-drug delivery gel containing extracts of *Quercus brantii* and *Coriandrum sativum* as an adjunct to scaling and root planing in moderate chronic periodontitis patients. *Journal of Research in Pharmacy Practice*, 3(2), 67. https://doi.org/10.4103/2279-042x.137076

Narasagoudr, S. S., Hegde, V. G., Chougale, R. B., Masti, S. P., Vootla, S., & Malabadi, R. B. (2020). Physico-chemical and functional properties of rutin induced chitosan/poly (vinyl alcohol) bioactive films for food packaging applications. *Food Hydrocolloids*, 109, 106096. https://doi.org/10.1016/j.foodhyd.2020.106096

Nguyen, Q. H., Talou, T., Evon, P., Cerny, M., & Merah, O. (2020). Fatty acid composition and oil content during coriander fruit development. *Food Chemistry*, 326. https://doi.org/10.1016/j.foodchem.2020.127034

Niu, B., Shao, P., Luo, Y., & Sun, P. (2020). Recent advances of electrosprayed particles as encapsulation systems of bioactives for food application. *Food Hydrocolloids*, 99, 105376. https://doi.org/10.1016/j.foodhyd.2019.105376

Nogueira, G. F., de Oliveira, R. A., Velasco, J. I., & Fakhouri, F. M. (2020). Methods of incorporating plant-derived bioactive compounds into films made with agro-based polymers for application as food packaging: A brief review. *Polymers*, 12(11), 1–34. https://doi.org/10.3390/polym12112518

Omidi-Mirzaei, M., Hojjati, M., Alizadeh Behbahani, B., & Noshad, M. (2020). Modeling the growth rate of *Listeria innocua* influenced by coriander seed essential oil and storage temperature in meat using FTIR. *Quality Assurance and Safety of Crops & Foods*, 12(SP1), 1–8. https://doi.org/10.15586/qas.v12isp1.776

Othman, S. H. (2014). Bio-nanocomposite materials for food packaging applications: Types of biopolymer and nano-sized filler. *Agriculture and Agricultural Science Procedia*, 2, 296–303. https://doi.org/10.1016/j.aaspro.2014.11.042

Rasoanaivo, P., Wright, C. W., Willcox, M. L., & Gilbert, B. (2011). Whole plant extracts versus single compounds for the treatment of malaria: Synergy and positive interactions. *Malaria Journal*, 10(Suppl. 1), 1–12. https://doi.org/10.1186/1475-2875-10-S1-S4

Rhim, J. W., & Ng, P. K. W. (2007). Natural biopolymer-based nanocomposite films for packaging applications. *Critical Reviews in Food Science and Nutrition*, 47(4), 411–433. https://doi.org/10.1080/10408390600846366

Sahiner, M., Suner, S. S., & Enzyme, A. (2021). Poly(rutin) micro/nanogels for biomedical applications. *Hittite Journal of Science and Engineering*, 8(2), 179–187. https://doi.org/10.17350/HJSE19030000

Shahwar, M. K., El-Ghorab, A. H., Anjum, F. M., Butt, M. S., Hussain, S., & Nadeem, M. (2012). Characterization of coriander (*Coriandrum sativum* L.) seeds and leaves: Volatile and non volatile extracts. *International Journal of Food Properties*, 15(4), 736–747. https://doi.org/10.1080/10942912.2010.500068

Sharif, Z., & Mustapha, F. y J. Y. (2017). Revisión de métodos de preservación y conservantes naturales para extender la longevidad de los alimentos. *Ingeniería Química*, 19, 145–153. https://www.banglajol.info/index.php/CERB/article/view/33809

Sharma, S., Barkauskaite, S., Jaiswal, A. K., & Jaiswal, S. (2021). Essential oils as additives in active food packaging. *Food Chemistry*, 343, 128403. https://doi.org/10.1016/j.foodchem.2020.128403

Silva, F., Caldera, F., Trotta, F., Nerín, C., & Domingues, F. C. (2019). Encapsulation of coriander essential oil in cyclodextrin nanosponges: A new strategy to promote its use in controlled-release active packaging. *Innovative Food Science and Emerging Technologies*, 56(June), 102177. https://doi.org/10.1016/j.ifset.2019.102177

Silva, F., & Domingues, F. C. (2017). Antimicrobial activity of coriander oil and its effectiveness as food preservative. *Critical Reviews in Food Science and Nutrition*, 57(1), 35–47. https://doi.org/10.1080/10408398.2013.847818

Sivakanthan, S., Rajendran, S., Gamage, A., Madhujith, T., & Mani, S. (2020). Antioxidant and antimicrobial applications of biopolymers: A review. *Food Research International*, 136, 109327. https://doi.org/10.1016/j.foodres.2020.109327

Sourmaghi, M. H. S., Kiaee, G., Golfakhrabadi, F., Jamalifar, H., & Khanavi, M. (2015). Comparison of essential oil composition and antimicrobial activity of *Coriandrum sativum* L. extracted by hydro-distillation and microwave-assisted hydrodistillation. *Journal of Food Science and Technology*, 52(4), 2452–2457. https://doi.org/10.1007/s13197-014-1286-x

Sriti, J., Aidi Wannes, W., Talou, T., Ben Jemia, M., Elyes Kchouk, M., & Marzouk, B. (2012). Antioxidant properties and polyphenol contents of different parts of coriander (*Coriandrum sativum* L.) fruit. *Rivista Italiana Delle Sostanze Grasse*, 89(4), 253–262.

Suppakul, P., Sonneveld, K., Bigger, S. W., & Miltz, J. (2011). Diffusion of linalool and methylchavicol from polyethylene-based antimicrobial packaging films. *LWT-Food Science and Technology*, 44(9), 1888–1893. https://doi.org/10.1016/j.lwt.2011.03.024

Töpfer, R., Martini, N., & Schell, J. (1995). Modification of plant lipid synthesis. *Science*, 268(5211), 681–686. https://doi.org/10.1126/science.268.5211.681

Wang, H., Qian, J., & Ding, F. (2018). Emerging chitosan-based films for food packaging applications. *Journal of Agricultural and Food Chemistry*, 66(2), 395–413. https://doi.org/10.1021/acs.jafc.7b04528

Zardini, H. Z., Tolueinia, B., Momeni, Z., Hasani, Z., Hasani, M., & Branch, Y. (2012). Analysis of antibacterial and antifungal. *Gomal Journal of Medical Sciences*, 10(2), 167–171.

Zeković, Z., Pavlić, B., Cvetanović, A., & Đurović, S. (2016). Supercritical fluid extraction of coriander seeds: Process optimization, chemical profile and antioxidant activity of lipid extracts. *Industrial Crops and Products*, 94, 353–362. https://doi.org/10.1016/j.indcrop.2016.09.008

Zivanovic, S., Chi, S. and Draughon, A.F. (2005). Antimicrobial activity of chitosan films enriched with essential oils. *Journal of Food Science*, 70, M45–M51. https://doi.org/10.1111/j.1365-2621.2005.tb09045.x

42 Non-Food Applications of Coriander Seed Extracts

Mariam I. Gamal El-Din and Fadia S. Youssef

CONTENTS

42.1 INTRODUCTION

Since antiquity, nature has been a safe and perpetual source of medicines. Natural products from plants have been used for centuries as a rich resource for curing various ailments. More than 80% of the world population still implements natural products and herbs previously described in different traditions, including Chinese, Roman, Arabic, and Ayurveda (El-Din et al., 2021).

Coriander (*Coriandrum sativum* L.) is an annual herbaceous medicinal plant and spice belonging to Umbelliferae (Apiaceae). It is native to the Mediterranean and Middle Eastern regions and widely cultivated in Central and Eastern Europe, China, India, Bangladesh, Morocco, Malta, Egypt, and the Netherlands. The name "coriander" comes from the Greek word "Koris" for "bed-bug", as the smell of the green unripe fruits is comparable to that of a bug plague-ridden bed line. Coriander leaves closely resemble flat-leaf parsley. Hence, it is known as Chinese parsley. However, coriander foliage is characterized by its more distinct fragrance (Surya et al., 2018). Coriander was mentioned in the papyrus of Ebers (*ca.* 1550 BC) and the writings of Cato and Pliny. It was one of the ancient spice crops in the world (Evans, 2009). It was called the spice of happiness by the Egyptians,

FIGURE 42.1   Coriander fruits, extract, and essential oil.

probably for their belief in its aphrodisiac activity. Besides, they used it for stomach disorders and diarrhea. Greeks and ancient Romans also used coriander for medicinal uses and as a wine flavoring agent. South Asia and India, in particular, are considered the world's major producers of coriander, exporting it to the USA, South East Asia, the Middle East, and Europe (Nadeem et al., 2013).

The fruits of coriander, also known as coriander seeds, are sub-spherical-shaped cremocarps 3 to 4 mm in diameter. Each cremocarp consists of two hemispherical mericarps united by their margins. Their apices bear divergent styles, and externally they are characterized by primary and secondary ridges (Figure 42.1). According to the ripening stage and the degree of drying, the seeds are usually yellowish-brown in color but sometimes green, straw-colored, or off-white. Coriander seeds are characterized by their citrus-like flavor with a hint of sage and their spicy, aromatic taste (Coşkuner and Karababa, 2007).

42.2   ETHNOPHARMACOLOGY

Apart from the nutritional value of coriander leaves and seeds, being rich in vitamins, proteins, carbohydrates and fibers, fats, and oils, coriander was traditionally used in different cultures for numerous medicinal purposes. Different plant parts were used as carminative, stimulant, antispasmodic, diaphoretic, and flavoring agents. In the different traditional systems of medicine, formulations containing coriander seed extract were reported to be used as carminatives, antispasmodics, diuretics, and anti-rheumatic. Besides, the seeds were used as an appetizer, stomachic, diuretic, tonic, and aphrodisiac. A decoction of dried cardamom seeds was also used to cure flatulence, indigestion, vomiting, and other intestinal disorders. Moreover, coriander seed extract was applied topically for healing rheumatism and painful joints (Kumar and Jayaveera, 2014; Nair et al., 2012).

42.3   MEDICINAL ACTIVITIES OF CORIANDER SEED EXTRACTS

42.3.1   Antibacterial Activity

The antibacterial activity of coriander seed essential oil was evaluated by Silva et al. against a group of Gram-positive (*Enterococcus faecalis*, *Bacillus cereus*, and *Staphylococcus aureus*) and Gram-negative (*Klebsiella pneumonia*, *Salmonella typhimurium*, *Escherichia coli*, and *Pseudomonas aeruginosa A*) bacteria. Results demonstrated the efficient antibacterial potential of the coriander oil against all the tested strains except *Enterococcus faecalis* and *Bacillus cereus*. The antibacterial mode of action of the oil was suggested to disrupt the bacterial cell membrane causing cell death (Silva et al., 2011). Besides, Duarte et al. reported the effective synergistic antibacterial activity obtained by associating coriander essential oil with gentamicin, ciprofloxacin, tetracycline, and chloramphenicol. The antimicrobial activity was evaluated *in vitro* against *Acinetobacter baumannii* (Duarte et al., 2012). Moreover, Kačániová et al. reported the strong potent antibacterial activity of coriander seed oil against *B. subtilis*, *S. maltophilia*, and *Penicillium expansum*, demonstrating inhibition zones of 10.6, 9.22, and 8.99 mm, respectively, compared to gentamicin (26.3 mm) (Kačániová et al., 2020).

Meanwhile, Chaudhry and Tariq, in their comparative study of the antibacterial activities of aqueous decoctions of coriander, aniseed, black pepper, and bay leaf, reported the inactivity of the coriander decoction against the tested bacterial strains isolated from oral cavities (Chaudhry and Tariq, 2006). However, Zardini et al. reported the antibacterial activity of the aqueous extract of coriander seeds against the Gram-positive bacterium *Staphylococcus aureus* and the two Gram-negative bacteria *Pseudomonas aeruginosa* and *Klebsiella pneumonia*. The extract was most active against *Staphylococcus aureus*, followed by *Klebsiella pneumonia*, then *Pseudomonas aeruginosa*, demonstrating MIC values of 1.3, 2.65, and 3.2 mg/mL, respectively (Zardini et al., 2012).

42.3.2 ANTIFUNGAL ACTIVITY

Zardini et al. reported the antifungal activity of the aqueous extract of coriander seeds against *Aspergillus niger* (*A. niger*) and *Penicillium lilacinum* (*P. lilacinum*), demonstrating MIC of 2.3 and 2.5 mg/mL, respectively (Zardini et al., 2012). Moreover, the coriander oil and coriander oleoresin were reported by Singh et al. to possess potential antifungal activities when evaluated against eight different fungi. *Fusarium moniliforme*, *Fusarium oxysporum*, *Aspergillus terreus*, and *Curvularia palliscens* fungi were more sensitive to the essential oil of coriander seeds. However, *Aspergillus niger*, *Aspergillus terreus*, and *Fusarium oxysporum* were the most sensitive fungi to coriander seed oleoresin. It is worthy of mention that coriander oil demonstrated more potent fungi toxic activity than the oleoresin, which only displayed 100% inhibition against one fungus (Singh et al., 2006). In a comparative study performed by Dėnė et al., coriander essential oil demonstrated more prominent *in vitro* antifungal activity against *Fusarium culmorum* and *F. oxysporum* f. sp. *lycopersici* than did the coriander seed extract. The coriander extract, obtained by CO_2 extraction, was more potent against *F. oxysporum* than *F. culmorum*. For the essential oil, the most potent concentration was 1000 μL/L (Dėnė et al., 2021).

42.3.3 ANTIOXIDANT ACTIVITY

Numerous data reported the *in vitro* antioxidant potential of different coriander seed extracts (de Almeida Melo et al., 2005; Deepa and Anuradha, 2011; Kansal et al., 2011; Satyanarayana et al., 2004; Zeković et al., 2016a). Characterizing the antioxidant compounds in aqueous coriander extract, Melo et al. suggested phenolic acids to be the principal phytoconstituents responsible for the antioxidant potential of the aqueous extract of coriander seeds. Caffeic acid, glycitin, and protocatechinic acid were identified in high concentrations in different fractions of the coriander extract (de Almeida Melo et al., 2005). In a comparative study performed by Wangensteen et al., the antioxidant activities of different extracts of coriander seeds of different polarities were tested. The ethyl acetate extract demonstrated the most potent antioxidant activity as displayed by the three assays: inhibition of 15-lipoxygenase, diphenylpicrylhydrazyl (DPPH·) radical scavenging, and inhibition of Fe2þ-induced peroxidation assays. A correlation was demonstrated between the antioxidant activity and the total phenolic contents of the different extracts (Wangensteen et al., 2004). Moreover, Zeković et al. optimized the conditions for the microwave-assisted extraction (MAE) of polyphenols from coriander seeds to obtain maximum antioxidant activity. Extraction by 63% ethanol for 19 min with microwave irradiation power of 570 W, the optimized conditions, demonstrated maximum total flavonoid and total phenolic yields and antioxidant activities demonstrated by DPPH· and reducing power assays (Zeković et al., 2016b). Additionally, different *in vivo* experiments have supported the antioxidant potential of coriander seed extracts (de Almeida Melo et al., 2003; Moustafa et al., 2014). Oral administration of the aqueous and ethanol extracts of coriander seeds at the dose of 250 and 600 mg/kg, respectively, reduced the lead nitrate-induced oxidative damage in the kidneys and livers of treated animals. Besides, a noticeable increase in glutathione levels was observed in addition to enhanced activities of different antioxidant enzymes (Kansal et al., 2011). Meanwhile, coriander oil has also been reported for its antioxidant activity in a vast number of studies that explained its different promising activities as hepatoprotective, cardioprotective, neuroprotective, antidiabetic, anti-inflammatory, antiaging, and in cancer (Cioanca et al., 2013; Darughe et al., 2012; Duarte et al., 2016; Mechchate et al., 2021b).

42.3.4 ANXIOLYTIC ACTIVITY

The ethanol extract of *Coriandrum sativum* seeds was investigated for its antidepressant and anxiolytic activities at 100 and 200 mg/kg doses. Using the forced swim test, the ethanol extract demonstrated a significant reduction in the immobility time by 70.9% at the dose of 200 mg/kg compared to the standard diazepam treatment group, which produced a 58.7% reduction in the immobility time. Besides, the extract at the 200 mg/kg dose displayed a significant reduction in locomotion by 59.6% compared to the diazepam group, demonstrating 33.9% (Pathan et al., 2015). Moreover, Pathan et al. reported the anxiolytic activity of the aqueous extract of coriander seeds at 50, 100, and 200 mg/kg utilizing an elevated plus-maze animal model. The aqueous extract at 200 mg/kg displayed a remarkable anxiolytic effect increasing the percentage of open arm entries and increasing the time spent on open arms compared to the saline-treated control group (Pathan et al., 2011). Emamghoreishi et al. also evaluated the anxiolytic activity of the aqueous extract of coriander seeds at different doses: 10, 25, 50, 100 mg/kg in male albino mice using elevated plus-maze. Results demonstrated a significant reduction in mice's neuromuscular coordination and spontaneous activity with 50, 100, and 500 mg/kg doses compared to the control. Besides, the extract at the dose of 100 mg/kg displayed a remarkable anxiolytic effect demonstrated by the increased time spent on open arms and the percentage of open arm entries. The study indicated the central depressant activity of the aqueous coriander extract mediated via the inhibitory neurotransmitter system, the $GABA_A$ receptor complex (Emamghoreishi et al., 2005).

Meanwhile, the hydro-alcoholic extract of *Coriandrum sativum* was investigated by Mahendra and Bisht for its anxiolytic activity in mice at doses of 50, 100, and 200 mg/kg. Various animal models were utilized for assessment, including an open field test, light and dark test, elevated plus-maze, and social interaction test. The hydro-alcoholic extract demonstrated significant anxiolytic potential at 100 and 200 mg/kg doses comparable to the anxiolytic effect of standard diazepam drugs. However, the extract did not display any activity at 50 mg/kg in any utilized models (Mahendra and Bisht, 2011). Moreover, linalool, the principal component of coriander essential oil, has proved its anxiolytic activity *via* numerous studies (Cioanca et al., 2014; Gastón et al., 2016; Harada et al., 2018; Souto-Maior et al., 2011).

42.3.5 SEDATIVE AND HYPNOTIC ACTIVITIES

Emam and Heydari investigated the *in vivo* sedative effects of the aqueous, hydro-alcoholic extract in albino mice at 100, 200, 400, and 600 mg/kg. Results revealed prolonged pentobarbital-induced sleeping time demonstrated by all doses of the aqueous extract and the two doses (400 and 600 mg/kg) of the hydro-alcoholic extract compared to the control group (Emam and Heydari, 2006). Moreover, the coriander essential oil and its major component, linalool, have been reported for their potent sedative effects at 8.6 and 86 µg. The research was performed on the behavioral influence of the administration of the essential oil and linalool on emotionality and locomotor activity of neonatal chicks utilizing an open field test. The oil and linalool administration significantly reduced attempted escapes, squares crossed number, distress calls, and defecation number. Besides, both linalool and coriander essential oil demonstrated a significant increase in the sleeping posture compared with the saline group. The sedative effects of the coriander essential oil and its principal constituent were comparable to standard diazepam drugs (Gastón et al., 2016).

42.3.6 EFFECT ON LEARNING AND MEMORY

The ethanol extract of *Coriandrum sativum* seed was evaluated by Zargar-Nattaj et al. for its impact on learning and memory processes in second-generation mice. The extract demonstrated a negative effect at short periods (for the first hour), possessing benzodiazepine-like effects on memory. This effect was found to be antagonized by low doses of caffeine. However, potentiation of mice learning was evidenced in later investigations (24 h and 1 week post-training), indicating the ability of coriander metabolites to enhance long-term memory (Zargar-Nattaj et al., 2011). Besides, coriander essential oil inhalation has been reported to enhance special memory in the Aβ(1-42) rat model of Alzheimer's disease. The essential oil daily inhalation by rats at 1% and 3% for three weeks demonstrated significant sustainability of memory formation. Besides, the essential oil administration demonstrated a significant reduction of lactate dehydrogenase and superoxide dismutase activities, elevated glutathione peroxidase specific activity, and reduction of the elevated malondialdehyde level. The study suggested the high neuroprotective potency of coriander essential oil mediated *via* its antioxidant and antiapoptotic potentials (Cioanca et al., 2013).

42.3.7 ANTI-CANCER ACTIVITY

Mechchate et al. evaluated the *in vitro* cytotoxic activity of the phenolic extract of *Coriandrum sativum* seeds on chronic (K562), acute (HL60) myeloid leukemia cell lines, and normal Vero cell lines *via* methyl thiazole tetrazolium assay. The extract demonstrated potent cytotoxic activity against leukemia cell lines with IC_{50} of 16.86 μM and 11.7 μM against K562 and HL60, respectively. Moreover, the OECD 423 acute toxicity model was adopted to investigate the extract *in vivo* cytotoxic activity on Swiss albino mice, but no toxicity was demonstrated. *In silico* study followed the leukemia development receptors, including ABL kinase, ABL1, BCL2, and FLT3. The extract components, especially rutin, catechins, and flavonoids, demonstrated the least binding energy, hence the highest affinity with the targeted receptors (Mechchate et al., 2021a). Another study reported the protective role of coriander seeds against the injurious effects in lipid metabolism demonstrated in 1,2-dimethylhydrazine (DMH) induced colon cancer. Coriander seeds rat fed group demonstrated a significant reduction in cholesterol levels and cholesterol to phospholipid ratio at the end of the 30 weeks compared with the DMH control group. Results suggested the inhibiting effect of coriander on the proliferative activity of colonic epithelium and hence an inhibiting effect on tumor promotion. Besides, the coriander seeds fed group demonstrated a remarkable elevation in fecal bile acids and neutral sterols and increased dry fecal weight indicating the maintenance of the colon membrane integrity fluidity and function by coriander administration (Chithra and Leelamma, 2000). Moreover, linalool, the major component of coriander essential oil, has been proved to possess promising anticancer activity against prostate cancer. A study reported linalool's dose- and time-dependent inhibitory activity against DU145 prostate cancer cells. The induction of DNA fragmentation, apoptosis, and cell cycle arrest were the proposed mechanisms of linalool anticancer activity (Sun et al., 2015).

42.3.8 ANTIDIABETIC ACTIVITIES

The polyphenol-rich fraction of *Coriandrum sativum* seeds was reported by Mechchate et al. to possess significant antihyperglycemic activity at the 25 and 50 mg/kg doses. Using mice's oral glucose tolerance test, the coriander fraction significantly reduced fasting and postprandial blood glucose levels. Results were comparable to the antihyperglycemic control drug, glibenclamide (Mechchate et al., 2021b). Moreover, the aqueous extract of coriander at the dose of 1 mg/mL was studied for its insulin-like activity and its insulin-releasing potential by Gray and Flatt. The extract administration demonstrated significant antihyperglycemic activity in streptozotocin-diabetic mice, evidenced by the enhanced glucose oxidation, enhanced transport of 2-deoxyglucose, and increased glucose incorporation into glycogen by 1.4, 1.6, and 1.7 fold, respectively. Besides, the extract administration elicited a stepwise stimulation of insulin secretion by 1.3–5.7 fold from the colon B-cell line. Sequential extraction of coriander seeds with different solvents revealed the insulin-releasing activity in water and hexane fractions suggesting the possible cumulative effect of multiple coriander phytoconstituents of different polarities (Gray and Flatt, 1999). Besides, the ethanol extract of coriander seeds demonstrated significant hypoglycemic activity in streptozotocin-diabetic mice. The extract at 0.5 and 1% displayed potent antioxidant activity elevating the glutathione level, enhancing the activities of superoxide dismutase and kidney catalase. Besides, reduced malondialdehyde levels were demonstrated, reducing the risk of oxidative stress, liver damage, and other diabetic complications (Naveen and Khanum, 2012).

42.3.9 ANTI-INFLAMMATORY ACTIVITY

The hydro-alcohol extract and the essential oil of *Coriandrum sativum* seeds were compared for their anti-inflammatory activities in rodents. The hydro-alcohol extract significantly reduced pleural edema in carrageenan-induced animals, as evidenced by pleurisy tests. Besides, the topical application of coriander extract reduced the ear edema and cell migration in the croton oil-induced model. However, the essential oil was reported to lack the potent anti-inflammatory potential of the hydro-alcohol extract (Zanusso-Junior et al., 2011). Meanwhile, Mechchate et al. reported the potent anti-inflammatory activity of the polyphenol-rich fraction of *Coriandrum sativum* seeds in carrageenan-induced paw edema Wistar rats. The coriander fraction administration at the 25 and 50 mg/kg doses demonstrated a significant inhibition rate of paw edema volume, reaching 29% and 48%, respectively, within the first hour and increasing to 87% and 92%, respectively, at the end of the test. Results were comparable to standard diclofenac, with 34% and 95% inhibition rates at the start and end of the test, respectively (Mechchate et al., 2021b).

42.3.10 ANTI-HYPERLIPIDEMIC ACTIVITY

The polyphenol-rich fraction of *Coriandrum sativum* seeds was reported by Mechchate et al. to possess significant antihyperlipidemic activity in alloxan-induced diabetic mice at 25 and 50 mg/kg. Furthermore, the coriander fraction attenuated the altered serum lipid profile in diabetic mice assessed by estimating the biochemical parameters TC, TG, HDL, and LDL after four weeks of administration. Besides, marked improvement of lipid metabolism was evidenced, protecting against diabetic complications (Mechchate et al., 2021b).

42.3.11 ANALGESIC AND ANTINOCICEPTIVE ACTIVITY

Pathan et al. reported the potent analgesic activity of the aqueous extract of coriander seeds at 50, 100, and 200 mg/kg utilizing the hot plate method (Pathan et al., 2011). Another study evaluated the antinociceptive activity of the chloroform, ethanol, and the aqueous extracts of *C. sativum*. The three extracts at the different doses (20, 100, and 500 mg/kg) were evaluated using a hot plate test for their *in vivo* antinociceptive activities. The extracts demonstrated a remarkable analgesic effect comparable to the standard morphine drug. However, the chloroform and ethanol extract demonstrated more potent activity than the aqueous extract. Moreover, the activities of the extracts were found to be attenuated by pretreatment with naloxone. Hence, the mechanism of analgesia of the coriander extracts was suggested via interaction with the opioid system (Kazempor, 2015).

Meanwhile, Bhat et al. investigated the analgesic activity of the ethanol and aqueous extracts of *Coriandrum sativum* seeds by thermal pain stimulus. The extracts were investigated at the doses of 100 mg/kg, 250 mg/kg, and 500 mg/kg for the mean response time by Eddy's hot plate method. Statistical significance was evidenced in the analgesic effects of both extracts. Besides, both extracts displayed dose-dependent analgesic responses (Bhat et al., 2014). Moreover, linalool, the major constituent of coriander essential oil, has been reported for its potent analgesic activity. Peana et al. reported the antinociceptive activity of (−)-linalool demonstrated in two different pain models in mice: acid-induced writhing and heat-induced pain models. Linalool at doses ranging from 25 to 75 mg/kg demonstrated a significant reduction of the acid-induced writhing where that effect was reversed by the muscarinic receptor antagonist atropine and the opioid receptor antagonist naloxone. Meanwhile, (−)-linalool demonstrated significant suppression of the heat-induced pain only at the dose of 100 mg/kg. The activation of cholinergic and opioid systems was suggested to play a crucial role in (−)-linalool-induced analgesia (Peana et al., 2003). Other studies supported the efficacy of linalool as a potent analgesic in different models (Nascimento et al., 2014; Peana et al., 2004; Quintans-Júnior et al., 2013).

42.3.12 ANTIARTHRITIC ACTIVITY

Coriander seed oil was evaluated by Deepa et al. for its *in vivo* antiarthritic potential. The oil significantly reduced paw volume and joint diameter in complete Freund's adjuvant-induced arthritic animals (Deepa et al., 2020). Moreover, Nair et al. reported the *in vivo* antiarthritic activity of the hydro-alcoholic extract of coriander seeds using the two models: complete Freund's adjuvant (CFA) formaldehyde-induced arthritis. The coriander extract demonstrated dose-dependent inhibition of joint swelling in both models. The promising activity was related to the extract modulatory activity for the pro-inflammatory cytokines in the synovium (Nair et al., 2012).

42.3.13 ANTHELMINTIC ACTIVITY

The hot water extract of *Coriander sativum* seeds was investigated for its anthelmintic activity against the Indian adult earthworms, *Pheretima posthuma,* at 25, 50, and 75 mg/mL. The worm was selected for its physiological and anatomical resemblance to the human intestinal worm parasite. The extract demonstrated dose-dependent inhibition and promising activity, especially at the higher dose (75 mg/mL), comparable to the albendazole standard (Madhavan and Tharakan, 2017). Moreover, the alcohol extract of *C. sativum* seeds was evaluated for its *in vivo* and *in vitro* anthelmintic activity against *Hymenolepis nana* infection. The extract at the doses of 250, 500, and 750 mg/kg demonstrated efficacy of 60.9, 100, and 100%, respectively, after 21 days, compared with Niclosamid (50 mg/kg), whose efficacy reached 100% after 11 days indicating the potent activity of the two doses 500 and 750 mg/kg against the worm. Besides, the alcohol extract of *C. sativum* seeds demonstrated potent *in vitro* inhibition of *Hymenolepis nana* worms within 30 min compared to 1000 mg/mL of Niclosamid, completely killing the worms within 5 min (Hosseinzadeh et al., 2016). The hydro-alcohol and aqueous extracts of *Coriandrum sativum* seeds were evaluated for their anthelmintic activities against the adult nematode parasite *Haemonchus contortus* and its egg. Both extracts demonstrated promising inhibition against egg hatching. However, the hydro-alcoholic extract displayed better anthelmintic activity against the adult nematode than the aqueous extract (Eguale et al., 2007).

42.4 NON-MEDICINAL ACTIVITIES OF CORIANDER SEEDS

42.4.1 PRESERVATIVE

Darughe et al. evaluated coriander essential oil's antioxidant and antifungal activities in cake through 60 days' storage at room temperature. The oil demonstrated potent preservative activity inhibiting the formation rates of primary and secondary oxidation products at concentrations of

0.05, 0.10, and 0.15%, comparable to the preservative activity of butylated hydroxyanisole (BHA). Besides, the coriander oil at 0.05% prohibited fungal growth in the cake and preserved the cake's organoleptic characteristics comparable to the BHA control (Darughe et al., 2012). Another study by Patel et al. reported the promising capability of coriander seed extract and its oleoresin in enhancing the oxidative stability of ghee. The steam distilled extract was superior to the oleoresin in its antioxidant potential, as demonstrated by the DPPH· assay and β-carotene-linoleic acid model system. Evidenced by the thiobarbituric acid value, conjugated dienes, peroxide value, and the oxidative stability index after 21 days' storage, the coriander extract and its oleoresin both demonstrated significant effectiveness in retarding the deterioration of ghee compared to butylated hydroxyanisole control (Patel et al., 2013).

42.4.2 METAL CHELATION

Coriander seed powder was reported to possess a distinctive adsorption ability towards Cu(II), Zn(II), and Pb(II) ions from an aqueous solution. The influences of pH, the metal ion concentration, pH, adsorbent amount, and contact time were investigated by Rao and Kashifuddin. Optimum results were demonstrated when metal ion desorption occurred from a dilute solution affected by column methods (Rao and Kashifuddin, 2012). Besides, Karunasagar et al. (2005) reported that *Coriandrum sativum* acts as a sorbent that can efficiently clear aqueous solutions from inorganic and methyl mercury contaminants. Columns packed with silica-immobilized coriander proved efficient in removing considerable quantities of both forms of mercury from spiked groundwater without being affected by other ions. The sorption behavior was explained by the effect of carboxylic acid groups in binding the mercury. The study demonstrated that the sorbent could help remove inorganic and methyl mercury from contaminated water (Karunasagar et al., 2005).

42.5 CHEMISTRY OF *CORIANDRUM SATIVUM*

42.5.1 CORIANDER VOLATILE OIL

Coriander sativum seeds are reported to yield 0.3–1.2% essential oil. The essential oil distilled from the ripe and dried seeds of *C. sativum* is characterized by its pale yellow color, characteristic aromatic odor, and mild, warm, and aromatic flavor. Different studies reported higher oil content usually extracted from small coriander fruits than larger ones. Like all essential oils, the chemical composition of coriander essential is affected by different factors, including the cultivation region, the season of seed harvest (degree of maturity), and the method of oil extraction, in addition to different environmental factors, including the climate, the soil quality, plant diseases, and many other factors. Besides, the monoterpene alcohol linalool has been the principal constituent of coriander essential oil in most studies ranging in concentration between 0.2 and 1.3%. Other major constituents of coriander essential oil include α-pinene, γ-terpinene, geranyl acetate, and camphor. Ravi et al. investigated the volatile constituents of coriander seeds from eight different regions of India. GC-MS analysis identified 30 volatile constituents where linalool was the major constituent of all oil samples, constituting 56–75% of the oils. Geranyl acetate and α-pinene were other major constituents of the Indian coriander samples representing 9–24% and 2–23%, respectively (Ravi et al., 2007). The effect of regional variation on the oil chemical composition is demonstrated in Table 42.1.

The influence of the oil extraction method on the essential oil chemical composition has been extensively studied to determine the adequate conditions for producing the highest yield of coriander essential oil. Eikani et al. performed a comparative study of the efficacy of the subcritical water extraction (SCWE), hydrodistillation, and Soxhlet methods in extracting *Coriandrum sativum* essential oil. The conventional extraction methods, hydrodistillation, and Soxhlet extraction demonstrated better extraction efficacy than subcritical water extraction. However, SCWE produced essential oils richer in oxygenated volatile constituents. Moreover, the extraction efficiency of SCWE has been

TABLE 42.1

Essential Oil Yield of Hydrodistilled *Coriander sativum* Seeds from Different Countries, the Content of Linalool and Other Major Constituents

Region	Oil Yield (%)	Linalool (%)	Other Major Constituents	Reference
India	0.18–0.39	56.7–75.1	Geranyl acetate (8.95–24.5%) and α-pinene (2.36–23.2%)	(Ravi et al., 2007)
Turkey	0.2–0.5	69.4	*Cis*-ocimene (6.05%), neryl acetate (5.71%), γ-terpinene (4.34)	(İzgı et al., 2017; Kosar et al., 2005)
Egypt	0.7	70.9	α-Pinene (4.17%), *p*-cymene (3.63%), linalool acetate (4.78%)	(Khalil et al., 2018)
Pakistan	0.15	69.6	Geranyl acetate (4.99%), γ-terpinene (4.17%), α-pinene (1.63%), anethol (1.15%), and *p*-cymene (1.12%)	(Anwar et al., 2011)
Poland	1.20-1.35%	78.4	α-pinene (5.03%), camphor (3.90%), γ-terpinene (3.80%), D-limonene (2.58%), and geranyl acetate (2.13%)	(Huzar et al., 2018)
Iran	0.31%	49.0	α-pinene (1.5%), geranyl acetate (3.00%)	(Ghazanfari et al., 2020)
Bangladesh	0.4	37.7	Geranyl acetate (17.6%) and γ-terpinene (14.4%)	(Bhuiyan et al., 2009)

evaluated at different temperatures (100, 125, 150, and 175°C), water flow rates (1, 2, and 4 mL/min), and mean particle sizes (0.25, 0.50, and 1 mm). Two extraction yields were compared; the total essential oil yield and the linalool extraction yield were the major oil components. Results demonstrated that the optimum temperature, flow rate, and mean particle size were 125°C, 2 mL/min, and 0.5 mm, respectively (Eikani et al., 2007). In another study, a comparison was performed between the essential oils obtained by supercritical CO_2 extraction and steam distillation methods. Besides, the chemical composition of the obtained oils was compared with that of a commercial oil sample obtained by hydrodistillation of coriander fruits. The study demonstrated the remarkable similarity in the chemical composition of the oils obtained by supercritical fluid extraction and steam distillation. However, both essential oils demonstrated great variability in their chemical composition from the commercial oil sample extracted by hydrodistillation. The percentage of oxygenated volatile constituents was significantly higher in oil samples obtained by supercritical CO_2 extraction. Hence, the oil demonstrated a more intense fragrant aroma than commercial oil (Anitescu et al., 1997). Moreover, microwave-assisted hydrodistillation has been studied for extracting coriander sativum essential oil and compared with the hydrodistillation method. The study also compared the oils obtained from the whole and ground fruits. Results demonstrated a marked reduction in the content of the monoterpene alcohol, linalool representing 75% coriander essential oil obtained by microwave-assisted hydrodistillation of ground coriander fruits compared to the oil obtained by conventional hydrodistillation that contained linalool constituting 80% of the oil. However, microwave-assisted distillation demonstrated a marked elevation in fatty acid content like tetradecanoic acid (from 2.8% to 8.8%) and hexadecanoic acid (from 1.9% to 6.0%) in coriander oil compared to conventional hydrodistillation. Moreover, microwave-assisted hydrodistillation was characterized by its shorter extraction time and the short time taken to reach the boiling stage required for extraction (Kosar et al., 2005).

Different studies have also manipulated the effect of the seed's harvest time and the degree of maturity of coriander seeds on the chemical composition of coriander essential oil. Msaada et al. studied the variation of the essential oils composition of coriander seeds during different stages of maturity: immature, intermediate, and mature stages. GC-FID and GC-MS analysis of the hydrodistilled oils revealed a remarkable increase in the oil yield during the maturation process. Immature seeds produced essential oil containing mainly monoterpene esters (76.3%), alcohols (10.9%), and aldehydes (1.42%) represented by geranyl acetate, linalool, nerol, and neral. The essential oils produced from the hydrodistillation of intermediate and mature seeds demonstrated very similar chemical profiles but were markedly different from the oil of the immature seeds. Monoterpene alcohols (76.7%), ketones (3.43%), esters (2.85%), and ethers (1.87%) were the major classes dominating the essential oil of intermediate mature seeds. However, mature seeds produced essential oils dominated by monoterpene alcohols (88.5%) and ketones (2.61%) (Msaada et al., 2007). Another comparative study confirmed the qualitative and quantitative variation of the essential oil profile of Tunisian coriander fruits during different stages of maturity. The highest yield of essential oil was obtained at the final stage of maturity with linalool (80.6%), geranyl acetate (2.59%), geraniol (2.25%), *p*-cymene (1.33%), and thymol (1.40%) representing the major volatile constituents. The study also demonstrated the effect of regional variation on the chemical profile of essential oils produced by comparing the essential oils obtained from fruits grown in two different regions in Tunisia (Msaada et al., 2009c). Other studies investigated the variation of the essential oil composition of coriander fruits during the different developmental stages (Msaada et al., 2009b; Nurzynska-Wierdak, 2013).

42.5.2 Coriander Fixed Oil

Coriander vegetable oil obtained through solvent extraction of crushed seeds was fully characterized and reported by different studies. The oil usually contains about 96% of triglycerides (TAG), 1% of diglycerides (DAG), 0.1% of monoglycerides (MAG), and 1% of free fatty acids (FFA). Ramadan and Mörsel reported that the fatty acid (FA) content and vegetable oil extracted using chloroform/methanol mixture represented 28.4% of fresh seed weight. A total of 12 fatty acid methyl esters were identified. Petroselinic acid (9Z-octadecenoic) acid, the uncommon isomer of oleic acid found at high levels in a restricted range of seed oils mostly from the Apiaceae family, found as the major FA accounting for 65.7% the total FA methyl esters followed by linoleic acid (16.7%). Other acids identified included palmitic, stearic, and oleic (Ramadan and Mörsel, 2002). Moser and Vaughn identified petroselinic acid as the major prevailing fatty acid accounting for 68.5 wt%, followed by linoleic and oleic acids representing the majority of the remaining FA content accounting for 13.0 and 7.6 wt%, respectively. Other identified FA were vaccenic, stearic, palmitic, and palmitoleic acids (Moser and Vaughn, 2010).

The FA profile usually varies during maturation, as reported by Msaada et al., who studied the FA variation of the coriander fruits cultivated in Tunisia. Fruits were randomly harvested at different ripening stages from the cultivated coriander plants. The collection period ranged from 5 to 55 days after flowering (DAF), the essential time for complete maturity of coriander fruit. The oil content and the fatty acid profile were evaluated. Results demonstrated a steady rate of oil synthesis, starting with an oil yield of 9.6% in young fruits and reaching 26.4% at full maturity. At the early stages of maturity, petroselinic acid was the major component of the oil, accounting for 84.8%, and other acids like palmitoleic, erucic, gadoleic, and docosahexaenoic acids were not detected. A reduction in palmitic acid concentrations was observed, accompanied by an elevation in petroselinic acid concentration throughout fruit development. Besides, polyunsaturated and saturated fatty acids (PUFA) decreased markedly, and fruit maturation affected monounsaturated fatty acids (MUFA), which were observed to increase with maturation (Msaada et al., 2009a).

REFERENCES

Anitescu, G., Doneanu, C., Radulescu, V., 1997. Isolation of coriander oil: Comparison between steam distillation and supercritical CO_2 extraction. *Flavour and Fragrance Journal* 12, 173–176.

Anwar, F., Sulman, M., Hussain, A.I., Saari, N., Iqbal, S., Rashid, U., 2011. Physicochemical composition of hydro-distilled essential oil from coriander (*Coriandrum sativum* L.) seeds cultivated in Pakistan. *Journal of Medicinal Plants Research* 5, 3537–3544.

Bhat, S.P., Rizvi, W., Kumar, A., August, J., 2014. Dose-dependent effect of *Coriandrum sativum* Linn. seeds on thermal pain stimulus. *The Journal of Phytopharmacology* 3, 254–258.

Bhuiyan, M.N.I., Begum, J., Sultana, M., 2009. Chemical composition of leaf and seed essential oil of *Coriandrum sativum* L. from Bangladesh. *Bangladesh Journal of Pharmacology* 4, 150–153.

Chaudhry, N., Tariq, P., 2006. Bactericidal activity of black pepper, bay leaf, aniseed and coriander against oral isolates. *Pakistan Journal of Pharmaceutical Sciences* 19, 214–218.

Chithra, V., Leelamma, S., 2000. *Coriandrum sativum* effect on lipid metabolism in 1, 2-dimethyl hydrazine induced colon cancer. *Journal of Ethnopharmacology* 71, 457–463.

Cioanca, O., Hritcu, L., Mihasan, M., Hancianu, M., 2013. Cognitive-enhancing and antioxidant activities of inhaled coriander volatile oil in amyloid β (1–42) rat model of Alzheimer's disease. *Physiology & Behavior* 120, 193–202.

Cioanca, O., Hritcu, L., Mihasan, M., Trifan, A., Hancianu, M., 2014. Inhalation of coriander volatile oil increased anxiolytic-antidepressant-like behaviors and decreased oxidative status in beta-amyloid (1–42) rat model of Alzheimer's disease. *Physiology & Behavior* 131, 68–74.

Coşkuner, Y., Karababa, E., 2007. Physical properties of coriander seeds (*Coriandrum sativum* L.). *Journal of Food Engineering* 80, 408–416.

Darughe, F., Barzegar, M., Sahari, M., 2012. Antioxidant and antifungal activity of Coriander (*Coriandrum sativum* L.) essential oil in cake. *International Food Research Journal* 19, 1253–1260.

de Almeida Melo, E., Bion, F.M., Filho, J.M., Guerra, N.B., 2003. In vivo antioxidant effect of aqueous and etheric coriander (*Coriandrum sativum* L.) extracts. *European Journal of Lipid Science and Technology* 105, 483–487.

de Almeida Melo, E., Mancini Filho, J., Guerra, N.B., 2005. Characterization of antioxidant compounds in aqueous coriander extract (*Coriandrum sativum* L.). *LWT-Food Science and Technology* 38, 15–19.

Deepa, B., Acharya, S., Holla, R., 2020. Evaluation of antiarthritic activity of Coriander seed essential oil in Wistar albino rats. *Research Journal of Pharmacy and Technology* 13, 761–766.

Deepa, B., Anuradha, C., 2011. Antioxidant potential of *Coriandrum sativum* L. seed extract. *Indian Journal of Experimental Biology* 49, 30–38.

Dėnė, L., Steinkellner, S., Valiuškaitė, A., 2021. Antifungal properties of *Coriandrum sativum* extracts on *Fusarium* spp. in vitro. In *Proceedings of the International Scientific Conference "Rural Development"*, pp. 19–22.

Duarte, A., Ferreira, S., Silva, F., Domingues, F., 2012. Synergistic activity of coriander oil and conventional antibiotics against *Acinetobacter baumannii*. *Phytomedicine* 19, 236–238.

Duarte, A., Luís, Â., Oleastro, M., Domingues, F.C., 2016. Antioxidant properties of coriander essential oil and linalool and their potential to control *Campylobacter* spp. *Food Control* 61, 115–122.

Eguale, T., Tilahun, G., Debella, A., Feleke, A., Makonnen, E., 2007. In vitro and in vivo anthelmintic activity of crude extracts of *Coriandrum sativum* against *Haemonchus contortus*. *Journal of Ethnopharmacology* 110, 428–433.

Eikani, M.H., Golmohammad, F., Rowshanzamir, S., 2007. Subcritical water extraction of essential oils from coriander seeds (*Coriandrum sativum* L.). *Journal of Food Engineering* 80, 735–740.

El-Din, M.I.G., Youssef, F.S., Said, R.S., Ashour, M.L., Eldahshan, O.A., Singab, A.N.B., 2021. Chemical constituents and gastro-protective potential of *Pachira glabra* leaves against ethanol-induced gastric ulcer in experimental rat model. *Inflammopharmacology* 29, 317–332.

Emam, G.M., Heydari, H.G., 2006. Sedative-hypnotic activity of extracts and essential oil of coriander seeds. *Iranian Journal of Medical Sciences* 31(1), 22–37.

Emamghoreishi, M., Khasaki, M., Aazam, M.F., 2005. *Coriandrum sativum*: Evaluation of its anxiolytic effect in the elevated plus-maze. *Journal of Ethnopharmacology* 96, 365–370.

Evans, W.C., 2009. *Trease and Evans' Pharmacognosy*. Elsevier Health Sciences.

Gastón, M.S., Cid, M.P., Vázquez, A.M., Decarlini, M.F., Demmel, G.I., Rossi, L.I., Aimar, M.L., Salvatierra, N.A., 2016. Sedative effect of central administration of *Coriandrum sativum* essential oil and its major component linalool in neonatal chicks. *Pharmaceutical Biology* 54, 1954–1961.

Ghazanfari, N., Mortazavi, S.A., Yazdi, F.T., Mohammadi, M., 2020. Microwave-assisted hydrodistillation extraction of essential oil from coriander seeds and evaluation of their composition, antioxidant and antimicrobial activity. *Heliyon* 6, e04893.

Gray, A.M., Flatt, P.R., 1999. Insulin-releasing and insulin-like activity of the traditional anti-diabetic plant *Coriandrum sativum* (coriander). *British Journal of Nutrition* 81, 203–209.

Harada, H., Kashiwadani, H., Kanmura, Y., Kuwaki, T., 2018. Linalool odor-induced anxiolytic effects in mice. *Frontiers in Behavioral Neuroscience*, 241.

Hosseinzadeh, S., Ghalesefidi, M.J., Azami, M., Mohaghegh, M.A., Hejazi, S.H., Ghomashlooyan, M., 2016. In vitro and in vivo anthelmintic activity of seed extract of *Coriandrum sativum* compared to Niclosamid against *Hymenolepis nana* infection. *Journal of Parasitic Diseases* 40, 1307–1310.

Huzar, E., Dzieciol, M., Wodnicka, A., Orun, H., Icoz, A., Çiçek, E., 2018. Influence of hydrodistillation conditions on yield and composition of coriander (*Coriandrum sativum* L.) essential oil. *Polish Journal of Food and Nutrition Sciences* 68.

İzgı, M.N., Telci, İ., Elmastaş, M., 2017. Variation in essential oil composition of coriander (*Coriandrum sativum* L.) varieties cultivated in two different ecologies. *Journal of Essential oil Research* 29, 494–498.

Kačániová, M., Galovičová, L., Ivanišová, E., Vukovic, N.L., Štefániková, J., Valková, V., Borotová, P., Žiarovská, J., Terentjeva, M., Felšöciová, S., 2020. Antioxidant, antimicrobial and antibiofilm activity of coriander (*Coriandrum sativum* L.) essential oil for its application in foods. *Foods* 9, 282.

Kansal, L., Sharma, V., Sharma, A., Lodi, S., Sharma, S., 2011. Protective role of *Coriandrum sativum* (coriander) extracts against lead nitrate induced oxidative stress and tissue damage in the liver and kidney in male mice. *International Journal of Applied Biology and Pharmaceutical Technology* 2(3), 65–83.

Karunasagar, D., Krishna, M.B., Rao, S., Arunachalam, J., 2005. Removal and preconcentration of inorganic and methyl mercury from aqueous media using a sorbent prepared from the plant *Coriandrum sativum*. *Journal of Hazardous Materials* 118, 133–139.

Kazempor, S.F., 2015. The analgesic effects of different extracts of aerial parts of *Coriandrum sativum* in mice. *International Journal of Biomedical Science: IJBS* 11, 23.

Khalil, N., Ashour, M., Fikry, S., Singab, A.N., Salama, O., 2018. Chemical composition and antimicrobial activity of the essential oils of selected Apiaceous fruits. *Future Journal of Pharmaceutical Sciences* 4, 88–92.

Kosar, M., Özek, T., Göger, F., Kürkcüoglu, M., Hüsnü Can Baser, K., 2005. Comparison of microwave-assisted hydrodistillation and hydrodistillation methods for the analysis of volatile secondary metabolites. *Pharmaceutical Biology* 43, 491–495.

Kumar, G., Jayaveera, K., 2014. *A Textbook of Pharmacognosy and Phytochemistry*. Chand Publishing.

Madhavan, M., Tharakan, S.T., 2017. Study on phytochemicals, total phenols, antioxidant, anthelmintic activity of hot water extracts of *Coriandrum sativum* seeds. *World Journal of Pharmacy and Pharmaceutical Sciences*. https://doi.org/10.20959/wjpps20178-9931

Mahendra, P., Bisht, S., 2011. Anti-anxiety activity of *Coriandrum sativum* assessed using different experimental anxiety models. *Indian Journal of Pharmacology* 43, 574.

Mechchate, H., Costa de Oliveira, R., Es-Safi, I., Vasconcelos Mourão, E.M., Bouhrim, M., Kyrylchuk, A., Soares Pontes, G., Bousta, D., Grafov, A., 2021a. Antileukemic Activity and molecular docking study of a polyphenolic extract from coriander seeds. *Pharmaceuticals* 14, 770.

Mechchate, H., Es-Safi, I., Amaghnouje, A., Boukhira, S., Alotaibi, A. A., Al-Zharani, M., Nasr, F. A., Noman, O. M., Conte, R., Amal, E.H.E.Y., 2021b. Antioxidant, anti-inflammatory and antidiabetic proprieties of LC-MS/MS identified polyphenols from coriander seeds. *Molecules* 26, 487.

Moser, B.R., Vaughn, S.F., 2010. Coriander seed oil methyl esters as biodiesel fuel: Unique fatty acid composition and excellent oxidative stability. *Biomass and Bioenergy* 34, 550–558.

Moustafa, A.H.A., Ali, E.M.M., Moselhey, S.S., Tousson, E., El-Said, K.S., 2014. Effect of coriander on thioacetamide-induced hepatotoxicity in rats. *Toxicology and Industrial Health* 30, 621–629.

Msaada, K., Hosni, K., Taarit, M.B., Chahed, T., Hammami, M., Marzouk, B., 2009a. Changes in fatty acid composition of coriander (*Coriandrum sativum* L.) fruit during maturation. *Industrial Crops and Products* 29, 269–274.

Msaada, K., Hosni, K., Taarit, M.B., Chahed, T., Kchouk, M.E., Marzouk, B., 2007. Changes on essential oil composition of coriander (*Coriandrum sativum* L.) fruits during three stages of maturity. *Food Chemistry* 102, 1131–1134.

Msaada, K., Hosni, K., Taarit, M.B., Ouchikh, O., Marzouk, B., 2009b. Variations in essential oil composition during maturation of coriander (*Coriandrum sativum* L.) fruits. *Journal of Food Biochemistry* 33, 603–612.

Msaada, K., Taarit, M.B., Hosni, K., Hammami, M., Marzouk, B., 2009c. Regional and maturational effects on essential oils yields and composition of coriander (*Coriandrum sativum* L.) fruits. *Scientia Horticulturae* 122, 116–124.

Nadeem, M., Anjum, F.M., Khan, M.I., Tehseen, S., El-Ghorab, A., Sultan, J.I., 2013. Nutritional and medicinal aspects of coriander (*Coriandrum sativum* L.): A review. *British Food Journal.* 115, 755–743.

Nair, V., Singh, S., Gupta, Y., 2012. Evaluation of disease modifying activity of Coriandrum sativum in experimental models. *The Indian Journal of Medical Research* 135, 240.

Nascimento, S.S., Camargo, E.A., DeSantana, J.M., Araújo, A.A., Menezes, P.P., Lucca-Júnior, W., Albuquerque-Júnior, R.L., Bonjardim, L.R., Quintans-Júnior, L.J., 2014. Linalool and linalool complexed in β-cyclodextrin produce anti-hyperalgesic activity and increase Fos protein expression in animal model for fibromyalgia. *Naunyn-Schmiedeberg's Archives of Pharmacology* 387, 935–942.

Naveen, S., Khanum, F., 2012. Anti-diabetic, anti-oxidant, anti-dyslipidemic and hepatoprotective properties of coriander seed extract in streptozotocin induced diabetic rats. *Journal of Herbal Medicine and Toxicology* 6, 61–67.

Nurzynska-Wierdak, R., 2013. Essential oil composition of the coriander (*Coriandrum sativum* L.) herb depending on the development stage. *Acta Agrobotanica* 66.

Patel, S., Shende, S., Arora, S., Singh, A.K., 2013. An assessment of the antioxidant potential of coriander extracts in ghee when stored at high temperature and during deep fat frying. *International Journal of Dairy Technology* 66, 207–213.

Pathan, A., Alshahrani, A., Al-Marshad, F., 2015. Neurological assessment of seeds of *Coriandrum sativum* by using antidepressant and anxiolytic like activity on albino mice. *Inventi Impact: Ethnopharmacology* 3, 102–105.

Pathan, A., Kothawade, K., Logade, M.N., 2011. Anxiolytic and analgesic effect of seeds of *Coriandrum sativum* Linn. *International Journal of Research in Pharmacy and Chemistry* 1, 1087–1099.

Peana, A.T., De Montis, M.G., Nieddu, E., Spano, M.T., Paolo, S.D., Pippia, P., 2004. Profile of spinal and supra-spinal antinociception of (–)-linalool. *European Journal of Pharmacology* 485, 165–174.

Peana, A.T., Paolo, S.D., Chessa, M.L., Moretti, M.D., Serra, G., Pippia, P., 2003. (–)-Linalool produces antinociception in two experimental models of pain. *European Journal of Pharmacology* 460, 37–41.

Quintans-Júnior, L.J., Barreto, R.S., Menezes, P.P., Almeida, J.R., Viana, A.F.S., Oliveira, R.C., Oliveira, A.P., Gelain, D.P., de Lucca Júnior, W., Araújo, A.A., 2013. β -Cyclodextrin-complexed (–)-linalool produces antinociceptive effect superior to that of (–)-linalool in experimental pain protocols. *Basic & Clinical Pharmacology & Toxicology* 113, 167–172.

Ramadan, M., Mörsel, J.-T., 2002. Oil composition of coriander (*Coriandrum sativum* L.) fruit-seeds. *European Food Research and Technology* 215, 204–209.

Rao, R.A.K., Kashifuddin, M., 2012. Adsorption properties of coriander seed powder (*Coriandrum sativum*): Extraction and pre-concentration of Pb (II), Cu (II) and Zn (II) ions from aqueous solution. *Adsorption Science & Technology* 30, 127–146.

Ravi, R., Prakash, M., Bhat, K.K., 2007. Aroma characterization of coriander (*Coriandrum sativum* L.) oil samples. *European Food Research and Technology* 225, 367–374.

Satyanarayana, S., Sushruta, K., Sarma, G., Srinivas, N., Raju, G.S., 2004. Antioxidant activity of the aqueous extracts of spicy food additives-evaluation and comparison with ascorbic acid in vitro systems. *Journal of Herbal Pharmacotherapy* 4, 1–10.

Silva, F., Ferreira, S., Queiroz, J.A., Domingues, F.C., 2011. Coriander (*Coriandrum sativum* L.) essential oil: Its antibacterial activity and mode of action evaluated by flow cytometry. *Journal of Medical Microbiology* 60, 1479–1486.

Singh, G., Maurya, S., De Lampasona, M., Catalan, C.A., 2006. Studies on essential oils, Part 41. Chemical composition, antifungal, antioxidant and sprout suppressant activities of coriander (*Coriandrum sativum*) essential oil and its oleoresin. *Flavour and Fragrance Journal* 21, 472–479.

Souto-Maior, F.N., de Carvalho, F.L., de Morais, L.C.S.L., Netto, S.M., de Sousa, D.P., de Almeida, R.N., 2011. Anxiolytic-like effects of inhaled linalool oxide in experimental mouse anxiety models. *Pharmacology Biochemistry and Behavior* 100, 259–263.

Sun, X.-B., Wang, S.-M., Li, T., Yang, Y., 2015. Anticancer activity of linalool terpenoid: Apoptosis induction and cell cycle arrest in prostate cancer cells. *Tropical Journal of Pharmaceutical Research* 14, 619–625.

Surya, R., Geethumol, T., Anitha, P., 2018. Quality in coriander leaves as influenced by growing conditions. *Journal of Horticultural Sciences* 13, 188–191.

Wangensteen, H., Samuelsen, A.B., Malterud, K.E., 2004. Antioxidant activity in extracts from coriander. *Food Chemistry* 88, 293–297.

Zanusso-Junior, G., Melo, J., Romero, A., Dantas, J., Caparroz-Assef, S., Bersani-Amado, C., Cuman, R., 2011. Evaluation of the anti-inflammatory activity of coriander (*Coriandrum sativum* L.) in rodents. *Revista Brasileira de Plantas Medicinais* 13, 17–23.

Zardini, H.Z., Tolueinia, B., Momeni, Z., Hasani, Z., Hasani, M., 2012. Analysis of antibacterial and antifungal activity of crude extracts from seeds of *Coriandrum sativum*. *Gomal Journal of Medical Sciences* 10.

Zargar-Nattaj, S.S., Tayyebi, P., Zangoori, V., Moghadamnia, Y., Roodgari, H., Jorsaraei, S.G., Moghadamnia, A.A., 2011. The effect of *Coriandrum sativum* seed extract on the learning of newborn mice by electric shock: Interaction with caffeine and diazepam. *Psychology Research and Behavior Management* 4, 13.

Zeković, Z., Pavlić, B., Cvetanović, A., Đurović, S., 2016a. Supercritical fluid extraction of coriander seeds: Process optimization, chemical profile and antioxidant activity of lipid extracts. *Industrial Crops and Products* 94, 353–362.

Zeković, Z., Vladić, J., Vidović, S., Adamović, D., Pavlić, B., 2016b. Optimization of microwave-assisted extraction (MAE) of coriander phenolic antioxidants-response surface methodology approach. *Journal of the Science of Food and Agriculture* 96, 4613–4622.

Index

For Product Safety Concerns and Information please contact our EU
representative GPSR@taylorandfrancis.com
Taylor & Francis Verlag GmbH, Kaufingerstraße 24, 80331 München, Germany

9 781032 069333